Medicinal Natural Products

Medicinal Natural Products
A Biosynthetic Approach
3rd Edition

Paul M Dewick

formerly University of Nottingham, UK

A John Wiley and Sons, Ltd., Publication

Library of Congress Cataloging-in-Publication Data

Dewick, Paul M.
  Medicinal natural products : a biosynthetic approach / Paul M. Dewick. – 3rd ed.
     p. cm.
  Includes bibliographical references and index.
  ISBN 978-0-470-74168-9 (cloth) – ISBN 978-0-470-74167-2 (pbk. : alk. paper)
  1. Natural products. 2. Biosynthesis. 3. Pharmacognosy. I. Title.
  RS160.D48 2008
  615′.3 – dc22

                                                              2008044502

A catalogue record for this book is available from the British Library.

ISBN: 978-0-470-74168-9 (HB)   978-0-470-74167-2 (PB)

Typeset in 9pt/11pt Times Roman by Laserwords Private Limited, Chennai, India

Cover photograph
English yew (*Taxus baccata*). Photograph: Paul Dewick

for Jane

CONTENTS

1

ABOUT THIS BOOK, AND HOW TO USE IT

THE SUBJECT

This book has been written primarily for pharmacy students to provide a modern text to complement lecture courses dealing with pharmacognosy and the use of natural products in medicine. Nevertheless, it should be of value in other courses where the study of natural products is included, although the examples chosen are predominantly those possessing pharmacological activity.

For centuries, drugs were entirely of natural origin and composed of herbs, animal products, and inorganic materials. Early remedies may have combined these ingredients with witchcraft, mysticism, astrology, or religion, but it is certain that those treatments that were effective were subsequently recorded and documented, thus leading to the early Herbals. The science of pharmacognosy – the knowledge of drugs – grew from these records to provide a disciplined, scientific description of natural materials used in medicine. Herbs formed the bulk of these remedies. As chemical techniques improved, the active constituents were isolated from plants, were structurally characterized, and, in due course, many were synthesized in the laboratory. Sometimes, more active, better-tolerated drugs were produced by chemical modifications (semi-synthesis), or by total synthesis of analogues of the active principles.

Gradually, synthetic compounds superseded many of the old plant drugs, though certain plant-derived agents were never surpassed and remain as valued medicines to this day. Natural drugs derived from microorganisms have a much shorter history, and their major impact on medicine goes back only about 60 years to the introduction of the antibiotic penicillin. Microbially produced antibiotics now account for a very high proportion of the drugs commonly prescribed. There is currently a renewed interest in pharmacologically active natural products, be they from plants, microorganisms, or animals, terrestrial or marine, in the continued search for new drugs, particularly for disease states where our present range of drugs is less effective than we would wish. This is being reflected in a growing number of natural products or natural-product-inspired drugs entering medicine. Herbal remedies are also enjoying a revival as many sufferers turn away from modern drugs and embrace 'complementary medicine'.

THE AIM

Many university pharmacy courses include a pharmacognosy component covering a study of plant-derived drugs; traditionally, this area of natural products has been taught separately from the microbially derived antibiotics, or the animal related steroidal and prostanoid drugs. Such topics have usually formed part of a pharmaceutical chemistry course. The traditional boundaries may still remain, despite a general change in pharmacognosy teaching from a descriptive study to a phytochemical-based approach, a trend towards integrating pharmacognosy within pharmaceutical chemistry, and the general adoption of modular course structures. A chemistry-based teaching programme encompassing all types of natural products of medicinal importance, semi-synthetic derivatives, and synthetic analogues based on natural product templates is a logical development. This book provides a suitable text to complement such a programme, and attempts to break down the artificial divisions.

Medicinal Natural Products: A Biosynthetic Approach. 3rd Edition   Paul Dewick
© 2009 John Wiley & Sons, Ltd

THE APPROACH

This book provides a groundwork in natural product chemistry/phytochemistry by considering biosynthesis – the metabolic sequences leading to various selected classes of natural products. This allows application of fundamental chemical principles and displays the relationships between the diverse structures encountered in nature, thus providing a rationale for natural products and replacing a descriptive approach with one based more on deductive reasoning. It also helps to transform complicated structures into a comprehensible combination of simpler fragments; natural product structures can be quite complex. Subdivision of the topics is predominantly via biosynthesis, not by class or activity, and this provides a logical sequence of structural types and avoids a catalogue effect. There is extensive use of chemical schemes and mechanism, with detailed mechanistic explanations being annotated to the schemes, as well as outline discussions in the text. Lots of cross-referencing is included to emphasize links and similarities; it is not necessary to follow these to understand the current material, but they are used to stress that the concept has been met before, or that other uses will be met in due course. As important classes of compounds or drugs are reached, more detailed information is then provided in the form of short separate monographs in boxes, which can be studied or omitted as required, in the latter case allowing the main theme to continue. The monograph information covers sources, production methods, principal components, drug use, mode of action, semi-synthetic derivatives, synthetic analogues, etc., as appropriate. Those materials currently employed as drugs, or being tested clinically, are emphasized in the monographs by the use of bold type.

THE TOPICS

A preliminary chapter is used to outline the main building blocks, the basic construction mechanisms employed in the biosynthesis of natural products, and how metabolic pathways are deduced. Most of the fundamental principles should be familiar and will have been met previously in courses dealing with the basics of organic chemistry and biochemistry. These principles are then seen in action as representative natural product structures are described in the following chapters. The topics selected are subdivided initially into areas of metabolism fed by the acetate, shikimate, mevalonate, and methylerythritol phosphate pathways. The remaining chapters then cover alkaloids, peptides and proteins, and carbohydrates. Not all classes of natural products can be covered, and the book is intended as an introductory text, not a comprehensive reference work.

The book tries to include a high proportion of those natural products currently used in medicine, the major drugs that are derived from natural materials by semi-synthesis, and those drugs which are structural analogues. Some of the compounds mentioned may have a significant biological activity which is of interest, but not medicinally useful. The book is also designed to be forward looking and gives information on possible leads to new drugs and materials in clinical trials.

THE FIGURES

A cursory glance through the book will show that a considerable portion of the content is in the form of chemical structures and schemes. The schemes and figures are used to provide maximum information as concisely as possible. The following guidelines should be appreciated:

- A figure may present a composite scheme derived from studies in more than one organism.
- Comments in italics provide an explanation in chemical terms for the biochemical reaction; detailed enzymic mechanisms are not usually considered.
- Schemes in separate frames show a mechanism for part of the sequence, the derivation of a substrate, or perhaps structurally related systems.
- Although enzymic reactions may be reversible, single rather than reversible arrows are used, unless the transformation is one that may be implicated in both directions, e.g. amino acid/keto acid transaminations.
- E1, E2, etc., refer to enzymes catalysing the transformation, when known. Where no enzyme is indicated, the transformation may well have been determined by other methodology, e.g. isotope tracer studies. Speculative conversions may be included, but are clearly indicated.
- Enzyme names shown are the commonly accepted names; in general, only one name is given, even though alternative names may also be in current use.
- Proteins identified via the corresponding gene are often assigned a code name/number by researchers, and no systematic name has been proposed. This means that proteins carrying out the same transformation in different organisms may be assigned different codes.
- Double-headed curly arrows are used to represent an addition–elimination mechanism as follows:

FURTHER READING

A selection of articles suitable for supplementary reading is provided at the end of each chapter. In general, these are not chosen from the primary literature, but are recent review articles covering broader aspects of the topic. They are also located in easily accessible journals rather than books, and have been chosen as the most student friendly. In certain cases, the most recent reviews available may be somewhat less up to date than the information covered in this book. All of the selected articles contain information considered appropriate to this book, e.g. reviews on 'synthesis' may contain sections on structural aspects, biosynthesis, or pharmacology.

WHAT TO STUDY

Coverage is fairly extensive to allow maximum flexibility for courses in different institutions, and not all of the material will be required for any one course. However, because of the many subdivisions and the highlighted keywords, it should be relatively easy to find and select the material appropriate for a particular course. On the other hand, the detail given in monographs is purposely limited to ensure students are provided with enough factual information, but are not faced with the need to assess whether or not the material is relevant. Even so, these monographs will undoubtedly contain data which exceed the scope of any individual course. It is thus necessary to apply selectivity, and portions of the book will be surplus to immediate requirements. The book is designed to be user friendly, suitable for modular courses and student-centred learning exercises, and a starting point for later project and dissertation work. The information presented is as up to date as possible; undoubtedly, new research will be published that modifies or even contradicts some of the statements made. The reader is asked always to be critical and to maintain a degree of flexibility when reading the scientific literature, and to appreciate that science is always changing.

WHAT TO LEARN

The primary aim of the book is not to rely just on factual information, but to impart an understanding of natural product structures and the way they are put together by living organisms. Rationalization based on mechanistic reasoning is paramount. The sequences themselves are not important, whilst the names of chemicals and the enzymes involved in the pathways are even less relevant and included only for information; it is the mechanistic explanations that are the essence. Students should concentrate on understanding the broad features of the

sequences and absorb sufficient information to be able to predict how and why intermediates might be elaborated and transformed. The mechanistic explanations appended to the schemes should reinforce this approach. Anyone who commits to memory a sequence of reactions for examination purposes has missed the point. There is no alternative to memory for some of the material covered in the monographs, if it is required; wherever possible, information should be reduced to a concept that can be deduced, rather than remembered. The approach used here should help students to develop such deductive skills.

NOMENCLATURE

Natural product structures are usually quite complex, some exceedingly so, and fully systematic nomenclature becomes impracticable. Names are thus typically based on so-called trivial nomenclature, in which the discoverer of the natural product exerts their right to name the compound. The organism in which the compound is found is frequently chosen to supply the root name, e.g. hyoscyamine from *Hyoscyamus*, atropine from *Atropa*, or penicillin from *Penicillium*. Name suffixes might be -in to indicate 'a constituent of', -oside to show the compound is a sugar derivative, -genin for the aglycone released by hydrolysis of the sugar derivative, -toxin for a poisonous constituent, or may reflect chemical functionality, such as -one or -ol. Traditionally, -ine is always used for alkaloids (am*ines*).

Structurally related compounds are then named as derivatives of the original, using standard prefixes, such as hydroxy-, methoxy-, methyl-, dihydro-, homo-, etc. for added substituents, or deoxy-, demethyl-, demethoxy-, dehydro-, nor-, etc. for removed substituents. Homo- is used to indicate one carbon more, whereas nor- means one carbon less. The position of this change is then indicated by systematic numbering of the carbon chains or rings. Some groups of compounds, such as steroids, fatty acids, and prostaglandins, are named semi-systematically from an accepted root name for the complex hydrocarbon skeleton. In this book, almost all structures depicted in the figures carry a name; this is primarily to help identification, and, for the student, structural features should be regarded as more pertinent than the names used.

It will soon become apparent that drug names chosen by pharmaceutical manufacturers are quite random, and in most cases have no particular relationship to the chemical structure. However, some common stems are employed to indicate relationship to a group of therapeutically active drugs. Examples are -cillin for antibiotics of the penicillin group, cef- for antibiotics of the cephalosporin group, -mycin for antibiotics produced by *Streptomyces*, -caine for local anaesthetics, -stat for enzyme inhibitors, -vastatin for HMGCoA reductase inhibitors, prost for prostaglandins, and gest for progestogens. We are also currently still in a transitional period during which many established drug names are being changed to recommended international non-proprietary names (rINNs); both names are included here, with the rINN preceding the older name.

CONVENTIONS REGARDING ACIDS, BASES, AND IONS

In many structures, the abbreviation **OP** is used to represent the phosphate group and **OPP** the diphosphate (or pyrophosphate) group:

| a phosphate | a diphosphate (pyrophosphate) |
| **ROP** | **ROPP** |

At physiological pH values, these groups will be ionized as shown, but in schemes where structures are given in full, the non-ionized acids are usually depicted. This is done primarily to simplify structures, to eliminate the need for counter-ions, and to avoid mechanistic confusion. Likewise, amino acids are shown in non-ionized form, although they will typically exist as zwitterions:

| L-amino acid | zwitterion |

Ionized and non-ionized forms of many compounds are regarded as synonymous in the text; thus, acetate/acetic acid, shikimate/shikimic acid, and mevalonate/mevalonic acid may be used according to the author's whim and context and have no especial relevance.

SOME COMMON ABBREVIATIONS

5-HT	5-hydroxytryptamine
ACP	acyl carrier protein
ADP	adenosine diphosphate
Ara	arabinose
ATP	adenosine triphosphate
B:	general base
CoA	coenzyme A as part of a thioester, e.g. acetyl-CoA ($CH_3COSCoA$)
Dig	digitoxose
DMAPP	dimethylallyl diphosphate (dimethylallyl pyrophosphate)
DXP	1-deoxyxylulose 5-phosphate
Enz	enzyme (usually shown as thiol: EnzSH)
FAD	flavin adenine dinucleotide
$FADH_2$	flavin adenine dinucleotide (reduced)
FAS	fatty acid synthase
FH_4	tetrahydrofolic acid
FMN	flavin mononucleotide
$FMNH_2$	flavin mononucleotide (reduced)
FPP	farnesyl diphosphate (farnesyl pyrophosphate)
Fru	fructose
GABA	γ-aminobutyric acid
Gal	galactose
GFPP	geranylfarnesyl diphosphate (geranylfarnesyl pyrophosphate)
GGPP	geranylgeranyl diphosphate (geranylgeranyl pyrophosphate)
Glc	glucose
GPP	geranyl diphosphate (geranyl pyrophosphate)
HA	general acid
HSCoA	coenzyme A
IPP	isopentenyl diphosphate (isopentenyl pyrophosphate)
LT	leukotriene
Mann	mannose
MEP	methylerythritol phosphate
MVA	mevalonic acid
NAD^+	nicotinamide adenine dinucleotide
NADH	nicotinamide adenine dinucleotide (reduced)
$NADP^+$	nicotinamide adenine dinucleotide phosphate
NADPH	nicotinamide adenine dinucleotide phosphate (reduced)
NRPS	non-ribosomal peptide synthase
O	oxidation – in schemes
P	phosphate – in text
P	phosphate – in structures
PCP	peptidyl carrier protein
PEP	phosphoenolpyruvate
PG	prostaglandin
PKS	polyketide synthase
PLP	pyridoxal 5′-phosphate
PP	diphosphate (pyrophosphate) – in text
PP	diphosphate (pyrophosphate) – in structures
Rha	rhamnose
Rib	ribose

SAM	*S*-adenosyl methionine
TPP	thiamine diphosphate (thiamine pyrophosphate)
TX	thromboxane
UDP	uridine diphosphate
UDPGlc	uridine diphosphoglucose
UTP	uridine triphosphate
W–M	Wagner–Meerwein (rearrangement)
Xyl	xylose
Δ	heat
$h\nu$	electromagnetic radiation; usually UV or visible

FURTHER READING

Pharmacognosy, Phytochemistry, Natural Drugs

Books

Bruneton J (1999) *Pharmacognosy, Phytochemistry and Medicinal Plants*. Lavoisier, Paris.

Evans WC (2001) *Trease & Evans' Pharmacognosy*. Saunders, London.

Heinrich M, Barnes J, Gibbons S and Williamson EM (2004) *Fundamentals of Pharmacognosy and Phytotherapy*. Churchill Livingstone, London.

Samuelsson G (2004) *Drugs of Natural Origin. A Textbook of Pharmacognosy*. Swedish Pharmaceutical Press, Stockholm.

Reviews

Baker DD, Chu M, Oza U and Rajgarhia V (2007) The value of natural products to future pharmaceutical discovery. *Nat Prod Rep* **24**, 1225–1244.

Bode HB and Müller R (2005) The impact of bacterial genomics on natural product research. *Angew Chem Int Ed* **44**, 6828–6846.

Butler MS (2004) The role of natural product chemistry in drug discovery. *J Nat Prod* **67**, 2141–2153; errata (2006) **69**, 172.

Butler MS (2008) Natural products to drugs: natural product-derived compounds in clinical trials. *Nat Prod Rep* **25**, 475–516.

Chin Y-W, Balunas MJ, Chai HB and Kinghorn AD (2006) Drug discovery from natural sources. *AAPS J* **8**, 239–253.

Espín JC, García-Conesa MT and Tomás-Barberán FA (2007) Nutraceuticals: facts and fiction. *Phytochemistry* **68**, 2986–3008.

Kennedy J (2008) Mutasynthesis, chemobiosynthesis, and back to semi-synthesis: combining synthetic chemistry and biosynthetic engineering for diversifying natural products. *Nat Prod Rep* **25**, 25–34.

McChesney JD, Venkataraman SK and Henri JT (2007) Plant natural products: back to the future or into extinction? *Phytochemistry* **68**, 2015–2022.

Misiek M and Hoffmeister D (2007) Fungal genetics, genomics, and secondary metabolites in pharmaceutical sciences. *Planta Med* **73**, 103–115.

Newman DJ (2008) Natural products as leads to potential drugs: an old process or the new hope for drug discovery? *J Med Chem* **51**, 2589–2599.

Newman DJ and Cragg GM (2004) Marine natural products and related compounds in clinical and advanced preclinical trials. *J Nat Prod* **67**, 1216–1238.

Newman DJ and Cragg GM (2007) Natural products as sources of new drugs over the last 25 years. *J Nat Prod* **70**, 461–477.

Phillipson JD (2007) Phytochemistry and pharmacognosy. *Phytochemistry* **68**, 2960–2972.

Tietze LF, Bell HP and Chandrasekhar S (2003) Natural product hybrids as new leads for drug discovery. *Angew Chem Int Ed* **42**, 3996–4028.

2

SECONDARY METABOLISM: THE BUILDING BLOCKS AND CONSTRUCTION MECHANISMS

PRIMARY AND SECONDARY METABOLISM

All organisms need to transform and interconvert a vast number of organic compounds to enable them to live, grow, and reproduce. They need to provide themselves with energy in the form of ATP, and a supply of building blocks to construct their own tissues. An integrated network of enzyme-mediated and carefully regulated chemical reactions is used for this purpose, collectively referred to as **intermediary metabolism**, and the pathways involved are termed **metabolic pathways**. Some of the crucially important molecules of life are carbohydrates, proteins, fats, and nucleic acids. Apart from fats, these tend to be polymeric materials. Carbohydrates are composed of sugar units, whilst proteins are made up from amino acids, and nucleic acids are based on nucleotides. Organisms vary widely in their capacity to synthesize and transform chemicals. For instance, plants are very efficient at synthesizing organic compounds via photosynthesis from inorganic materials found in the environment, whilst other organisms, such as animals and microorganisms, rely on obtaining their raw materials in their diet, e.g. by consuming plants. Thus, many of the metabolic pathways are concerned with degrading materials taken in as food, whilst others are then required to synthesize specialized molecules from the basic compounds so obtained.

Despite the extremely varied characteristics of living organisms, the pathways for generally modifying and synthesizing carbohydrates, proteins, fats, and nucleic acids are found to be essentially the same in all organisms, apart from minor variations. These processes demonstrate the fundamental unity of all living matter, and are collectively described as **primary metabolism**, with the compounds involved in the pathways being termed **primary metabolites**. Thus, degradation of carbohydrates and sugars generally proceeds via the well-characterized pathways known as glycolysis and the Krebs/citric acid/tricarboxylic acid cycle, which release energy from the organic compounds by oxidative reactions. Oxidation of fatty acids from fats by the sequence called β-oxidation also provides energy. Aerobic organisms are able to optimize these processes by adding on a further process, namely oxidative phosphorylation. This improves the efficiency of oxidation by incorporating a more general process applicable to the oxidation of a wide variety of substrates rather than having to provide specific processes for each individual substrate. Proteins taken in via the diet provide amino acids, but the proportions of each will almost certainly vary from the organism's requirements. Metabolic pathways are thus available to interconvert amino acids, or degrade those not required and thus provide a further source of energy. Most organisms can synthesize only a proportion of the amino acids they actually require for protein synthesis. Those structures not synthesized, so-called essential amino acids, must be obtained from external sources.

In contrast to these primary metabolic pathways, which synthesize, degrade, and generally interconvert compounds commonly encountered in all organisms, there also exists an area of metabolism concerned with compounds which have a much more limited distribution in nature. Such compounds, called **secondary metabolites**,

Medicinal Natural Products: A Biosynthetic Approach. 3rd Edition   Paul Dewick
© 2009 John Wiley & Sons, Ltd

are found in only specific organisms, or groups of organisms, and are an expression of the individuality of species. Secondary metabolites are not necessarily produced under all conditions, and in the vast majority of cases the function of these compounds and their benefit to the organism are not yet known. Some are undoubtedly produced for easily appreciated reasons, e.g. as toxic materials providing defence against predators, as volatile attractants towards the same or other species, or as colouring agents to attract or warn other species, but it is logical to assume that all do play some vital role for the well-being of the producer. It is this area of **secondary metabolism** which provides most of the pharmacologically active natural products. It is thus fairly obvious that the human diet could be both unpalatable and remarkably dangerous if all plants, animals, and fungi produced the same range of compounds.

The above generalizations distinguishing primary and secondary metabolites unfortunately leave a 'grey area' at the boundary, so that some groups of natural products could be assigned to either division. Fatty acids and sugars provide good examples, in that most are best described as primary metabolites, whilst some representatives are extremely rare and found only in a handful of species. Likewise, steroid biosynthesis produces a range of widely distributed fundamental structures, yet some steroids, many of them with pronounced pharmacological activity, are restricted to certain organisms. Hopefully, the blurring of the boundaries will not cause confusion; the subdivision into primary metabolism ($\equiv$ biochemistry) or secondary metabolism ($\equiv$ natural products chemistry) is merely a convenience and there is considerable overlap.

THE BUILDING BLOCKS

The building blocks for secondary metabolites are derived from primary metabolism as indicated in Figure 2.1. This scheme outlines how metabolites from the fundamental processes of photosynthesis, glycolysis, and the Krebs cycle are tapped off from energy-generating processes to provide biosynthetic intermediates. The number of building blocks needed is surprisingly few, and as with any child's construction set, a vast array of objects can be built up from a limited number of basic building blocks. By far the most important building blocks employed in the biosynthesis of secondary metabolites are derived from the intermediates acetyl coenzyme A (acetyl-CoA), shikimic acid, mevalonic acid, and methylerythritol phosphate. These are utilized respectively in the **acetate, shikimate, mevalonate**, and **methylerythritol phosphate** pathways, which form the basis of succeeding chapters. **Acetyl-CoA** is formed by oxidative decarboxylation of

the glycolytic pathway product pyruvic acid. It is also produced by the β-oxidation of fatty acids, effectively reversing the process by which fatty acids are themselves synthesized from acetyl-CoA. Important secondary metabolites formed from the acetate pathway include phenols, prostaglandins, and macrolide antibiotics, together with various fatty acids and derivatives at the primary–secondary metabolism interface. **Shikimic acid** is produced from a combination of phosphoenolpyruvate, a glycolytic pathway intermediate, and erythrose 4-phosphate from the pentose phosphate pathway. The reactions of the pentose phosphate cycle may be employed for the degradation of glucose, but they also feature in the synthesis of sugars by photosynthesis. The shikimate pathway leads to a variety of phenols, cinnamic acid derivatives, lignans, and alkaloids. **Mevalonic acid** is itself formed from three molecules of acetyl-CoA, but the mevalonate pathway channels acetate into a different series of compounds than does the acetate pathway. **Methylerythritol phosphate** arises from a combination of two glycolytic pathway intermediates, namely pyruvic acid and glyceraldehyde 3-phosphate by way of deoxyxylulose phosphate. The mevalonate and methylerythritol phosphate pathways are together responsible for the biosynthesis of a vast array of terpenoid and steroid metabolites.

In addition to acetyl-CoA, shikimic acid, mevalonic acid, and methylerythritol phosphate, other building blocks based on amino acids are frequently employed in natural product synthesis. Peptides, proteins, alkaloids, and many antibiotics are derived from amino acids, and the origins of some of the more important amino acid components of these are briefly indicated in Figure 2.1. Intermediates from the glycolytic pathway and the Krebs cycle are used in constructing many of them, but the aromatic amino acids **phenylalanine, tyrosine**, and **tryptophan** are themselves products from the shikimate pathway. **Ornithine**, an amino acid not found in proteins, and its homologue **lysine**, are important alkaloid precursors and have their origins in Krebs cycle intermediates.

Of special significance is the appreciation that secondary metabolites can be synthesized by combining several building blocks of the same type, or by using a mixture of different building blocks. This expands structural diversity and, consequently, makes subdivisions based entirely on biosynthetic pathways rather more difficult. A typical natural product might be produced by combining elements from the acetate, shikimate, and methylerythritol phosphate pathways, for example. Many secondary metabolites also contain one or more sugar units in their structure, either simple primary metabolites,

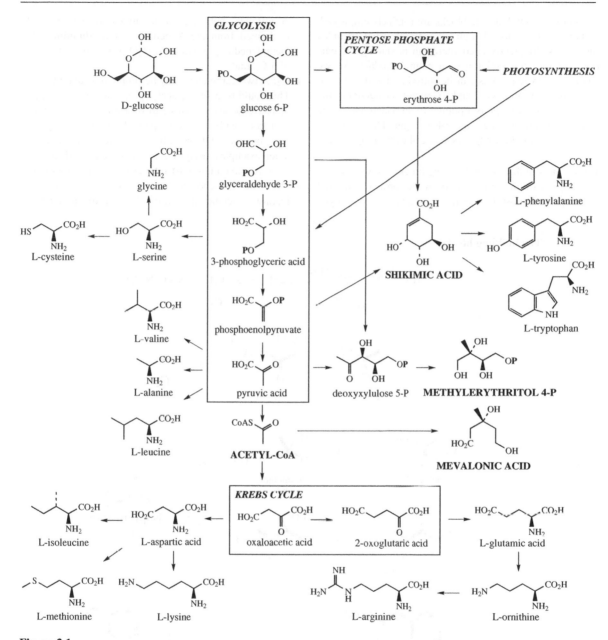

Figure 2.1

such as glucose or ribose, or alternatively substantially modified and unusual sugars. To appreciate how a natural product is elaborated, it is of value to be able to dissect its structure into the basic building blocks from which it is made up and to use fundamental chemical mechanisms to propose how these are joined together. With a little experience and practice, this becomes a relatively simple

process and it allows the molecule to be rationalized, thus exposing logical relationships between apparently quite different structures. In this way, similarities become much more meaningful than differences, and an understanding of biosynthetic pathways allows rational connecting links to be established. This forms the basic approach in this book.

Relatively few building blocks are routinely employed, and the following list, though not comprehensive, includes those most frequently encountered in producing the carbon and nitrogen skeleton of a natural product. As we shall see, oxygen atoms can be introduced and removed by a variety of processes, and so are not considered in the initial analysis, except as a pointer to an acetate (see page 101) or shikimate (see page 140) origin. The structural features of these building blocks are shown in Figure 2.2.

- **C_1:** The simplest of the building blocks is composed of a single carbon atom, usually in the form of a methyl group, and most frequently it is attached to oxygen

or nitrogen, but occasionally to carbon or sulfur. It is derived from the *S*-methyl of L-**methionine**. The methylenedioxy group ($-OCH_2O-$) is also an example of a C_1 unit.

- **C_2:** A two-carbon unit may be supplied by **acetyl-CoA**. This could be a simple acetyl group, as in an ester, but more frequently it forms part of a long alkyl chain (as in a fatty acid) or may be part of an aromatic system (e.g. phenols). Of particular relevance is that, in the latter examples, acetyl-CoA is first converted into the more reactive **malonyl-CoA** before its incorporation.
- **C_5:** The branched-chain C_5 'isoprene' unit is a feature of compounds formed from **mevalonate** or

The building blocks

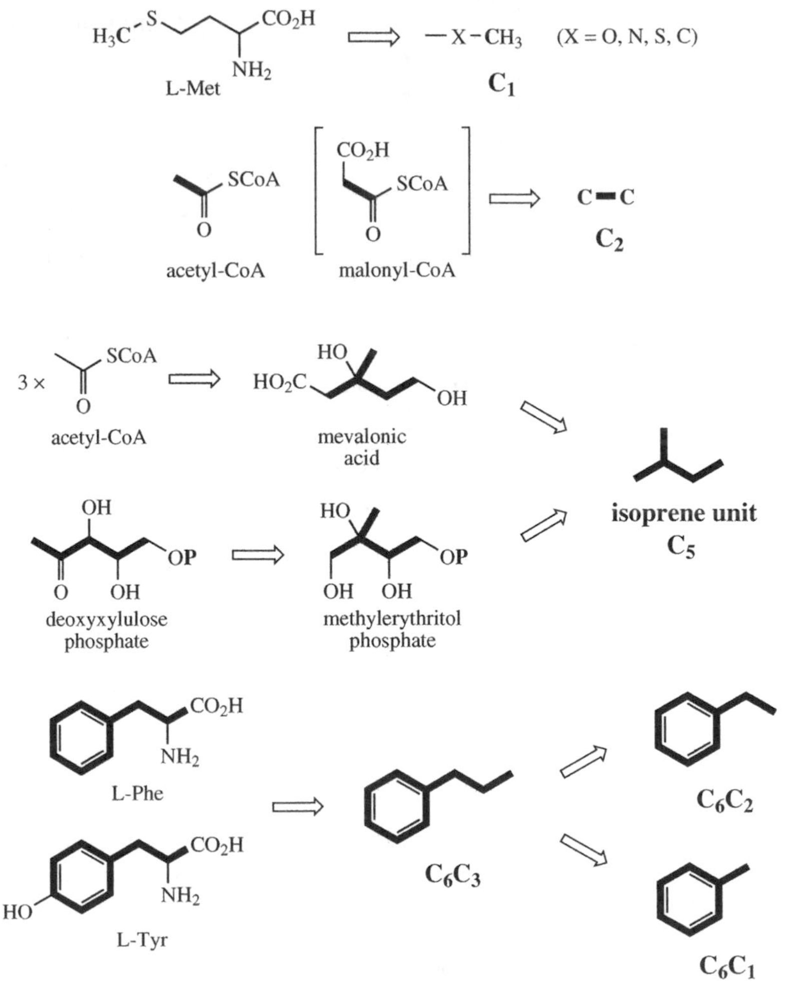

Figure 2.2

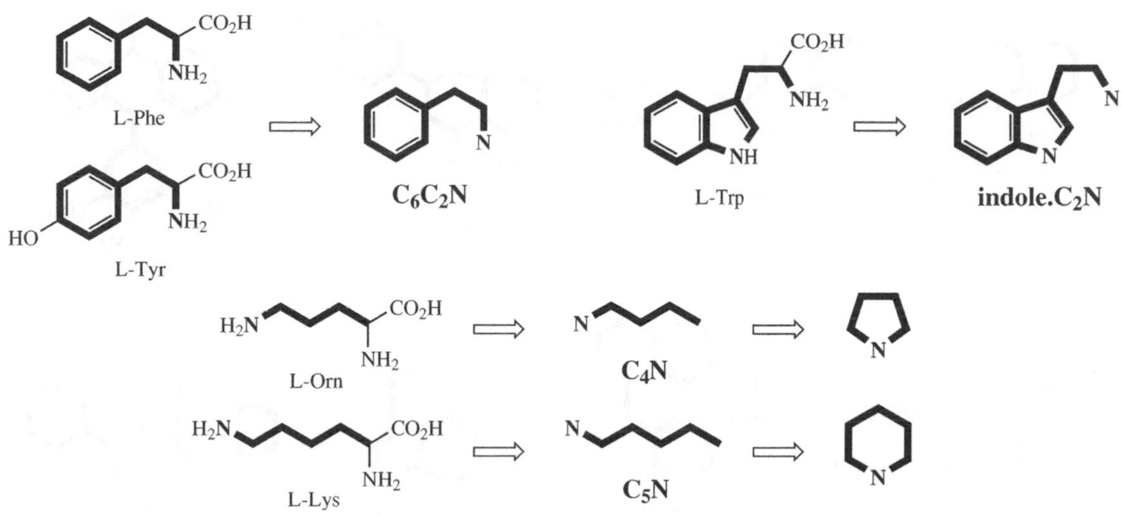

Figure 2.2 *(Continued)*

methylerythritol phosphate. Mevalonate itself is the product from three acetyl-CoA molecules, but only five of mevalonate's six carbon atoms are used, the carboxyl group being lost. The alternative precursor, methylerythritol phosphate, is formed from a straight-chain sugar derivative, deoxyxylulose phosphate, which undergoes a skeletal rearrangement to form the branched-chain isoprene unit.

- **C_6C_3:** This refers to a phenylpropyl unit and is obtained from the carbon skeleton of either L-**phenylalanine** or L-**tyrosine**, two of the shikimate-derived aromatic amino acids. This, of course, requires loss of the amino group. The C_3 side-chain may be saturated or unsaturated, and may be oxygenated. Sometimes the side-chain is cleaved, removing one or two carbon atoms. Thus, C_6C_2 and C_6C_1 units represent modified shortened forms of the C_6C_3 system.
- **C_6C_2N:** Again, this building block is formed from either L-**phenylalanine** or L-**tyrosine**, with L-tyrosine being by far the more common precursor. In the elaboration of this unit, the carboxyl carbon of the amino acid is removed.
- **indole.C_2N:** The third of the aromatic amino acids is L-**tryptophan**. This indole-containing system can undergo decarboxylation in a similar way to L-phenylalanine and L-tyrosine, so providing the remainder of the skeleton as an indole.C_2N unit.
- **C_4N:** The C_4N unit is usually found as a heterocyclic pyrrolidine system and is produced from the non-protein amino acid L-**ornithine**. In marked contrast to the C_6C_2N and indole.C_2N units described above,

ornithine supplies not its α-amino nitrogen, but the δ-amino nitrogen. The carboxylic acid function and the α-amino nitrogen are both lost.
- **C_5N:** This is produced in exactly the same way as the C_4N unit, but using L-**lysine** as precursor. The ε-amino nitrogen is retained, and the unit is commonly found as a piperidine ring system.

These eight building blocks will form the basis of many of the natural product structures discussed in the following chapters. Simple examples of how compounds can be visualized as a combination of building blocks are shown in Figure 2.3. At this stage, it is inappropriate to justify why a particular combination of units is used, but this aspect should become clear as the pathways are described. A word of warning is also necessary. Some natural products have been produced by processes in which a fundamental rearrangement of the carbon skeleton has occurred. This is especially common with structures derived from isoprene units, and it obviously disguises some of the original building blocks from immediate recognition. The same is true if one or more carbon atoms are removed by oxidation reactions.

THE CONSTRUCTION MECHANISMS

Natural product molecules are biosynthesized by a sequence of reactions which, with very few exceptions, are catalysed by enzymes. Enzymes are protein molecules which facilitate chemical modification of substrates by virtue of their specific binding properties conferred by

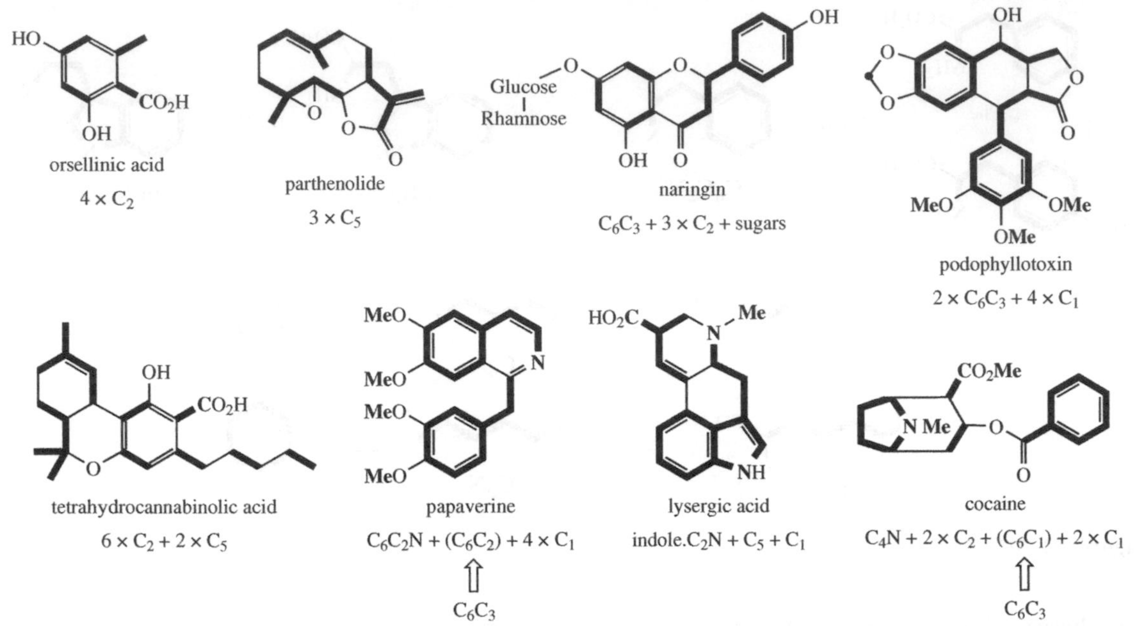

orsellinic acid

$4 \times C_2$

parthenolide

$3 \times C_5$

naringin

$C_6C_3 + 3 \times C_2$ + sugars

podophyllotoxin

$2 \times C_6C_3 + 4 \times C_1$

tetrahydrocannabinolic acid

$6 \times C_2 + 2 \times C_5$

papaverine

$C_6C_2N + (C_6C_2) + 4 \times C_1$

$\Uparrow$

C_6C_3

lysergic acid

indole.$C_2N + C_5 + C_1$

cocaine

$C_4N + 2 \times C_2 + (C_6C_1) + 2 \times C_1$

$\Uparrow$

C_6C_3

Figure 2.3

the particular combination of functional groups in the constituent amino acids. In many cases, a suitable cofactor, e.g. NAD^+, PLP, HSCoA (see below), as well as the substrate, may also be bound to participate in the transformation. Although enzymes catalyse some fairly elaborate and sometimes unexpected changes, it is generally possible to account for the reactions using established chemical principles and mechanisms. As we explore the pathways to a wide variety of natural products, the reactions will generally be discussed in terms of chemical analogies. Enzymes have the power to effect these transformations more efficiently and more rapidly than the chemical analogy, and also under very much milder conditions. Where relevant, they also carry out reactions in a stereospecific manner. Some of the important reactions frequently encountered are now described.

Alkylation Reactions: Nucleophilic Substitution

The C_1 methyl building unit is supplied from L-methionine and is introduced by a nucleophilic substitution reaction. In nature, the leaving group is enhanced by converting L-methionine into *S*-adenosylmethionine (**SAM**, AdoMet) [Figure 2.4(a)]. This gives a positively charged sulfur and facilitates the S_N2-type nucleophilic substitution mechanism [Figure 2.4(b)]. Thus, *O*-methyl and *N*-methyl linkages may be obtained using hydroxyl

and amino functions as nucleophiles. Methionine is subsequently regenerated by the methylation of homocysteine, using N^5-methyl-tetrahydrofolate (see page 144) as methyl donor. The generation of *C*-methyl linkages requires the participation of nucleophilic carbon. Positions *ortho* or *para* to a phenol group, or positions adjacent to one or more carbonyl groups, are thus candidates for *C*-methylation [Figure 2.4(c)].

A C_5 isoprene unit in the form of **dimethylallyl diphosphate** (**DMAPP**) may also act as an alkylating agent, since diphosphate is a good leaving group [Figure 2.4(d)]. Although a similar S_N2 displacement may be proposed, the available evidence points to an S_N1 process. DMAPP first ionizes to the resonance-stabilized allylic carbocation, and the nucleophile is able to attack either of the cationic centres. In the majority of cases, the nucleophile attacks the same carbon that loses the diphosphate. *C*-Alkylation at activated positions using DMAPP is analogous to the *C*-methylation process above, though by an S_N1 mechanism.

Alkylation Reactions: Electrophilic Addition

As indicated above, the C_5 isoprene unit in the form of **DMAPP** can be used to alkylate a nucleophile. In the elaboration of terpenoids and steroids, two or more C_5 units are joined together and the reactions are rationalized

Alkylation reactions: nucleophilic substitution

(a) formation of SAM

(b) *O*- and *N*-alkylation using SAM; regeneration of methionine

(c) *C*-alkylation using SAM

(d) *O*-alkylation using DMAPP

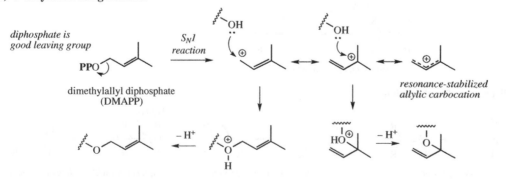

Figure 2.4

in terms of carbocation chemistry, including electrophilic addition of carbocations onto alkenes. DMAPP may ionize to generate a resonance-stabilized allylic carbocation as shown in Figure 2.4(d), and this can then react with an alkene, e.g. **isopentenyl diphosphate (IPP)**, as depicted in Figure 2.5(a). The resultant carbocation may then lose a proton to give the uncharged product geranyl diphosphate (GPP). Where the alkene and carbocation functions reside in the same molecule, this type of mechanism can also be responsible for cyclization reactions [Figure 2.5(a)].

The initial carbocation may be generated by a number of mechanisms, important examples being loss of a leaving group, especially diphosphate (i.e. S_N1-type ionization), protonation of an alkene, and protonation/ring opening of epoxides [Figure 2.5(b)]. **SAM** may also alkylate alkenes by an electrophilic addition mechanism, adding a C_1 unit, and generating an intermediate carbocation; this is simply an extension of the protonation reaction.

The final carbocation may be discharged by loss of a proton (giving an alkene or sometimes a cyclopropane ring) or by quenching with a suitable nucleophile, especially water [Figure 2.5(c)].

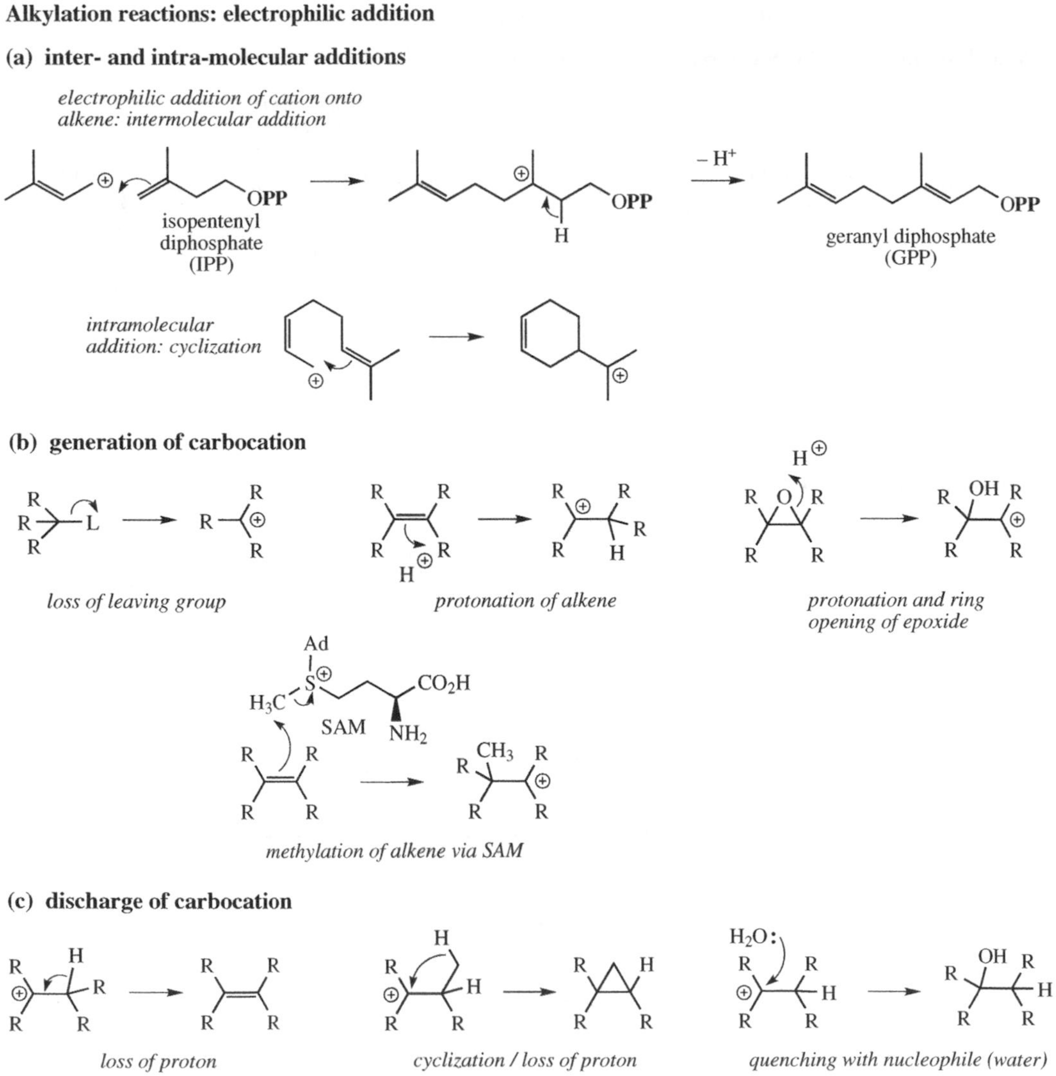

Alkylation reactions: electrophilic addition

(a) inter- and intra-molecular additions

electrophilic addition of cation onto alkene: intermolecular addition

(b) generation of carbocation

(c) discharge of carbocation

Figure 2.5

Wagner–Meerwein rearrangements

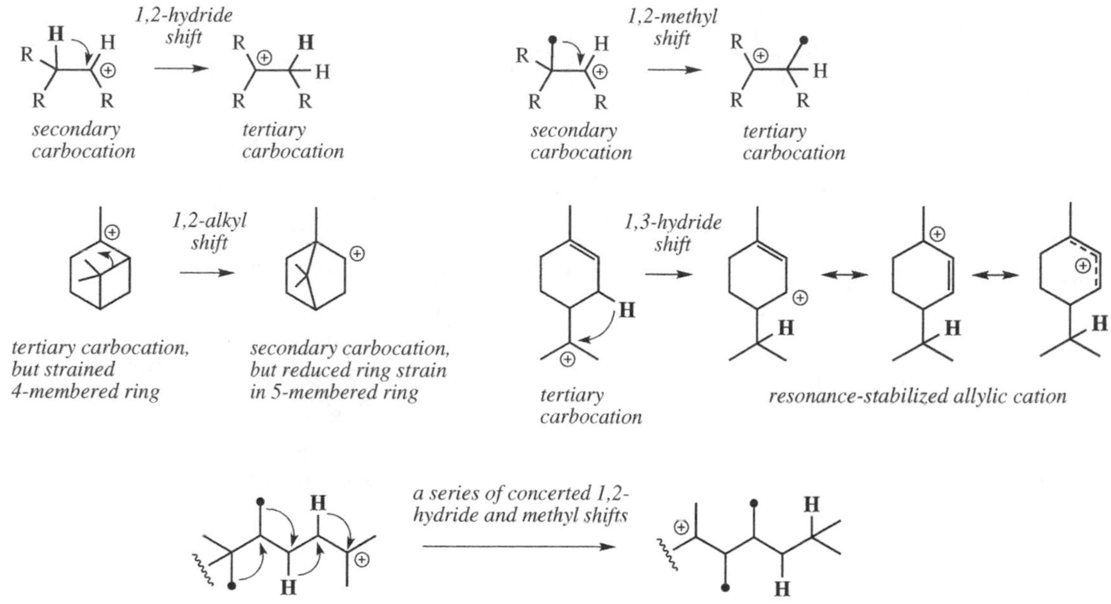

Figure 2.6

Wagner–Meerwein Rearrangements

A wide range of structures encountered in natural terpenoid and steroid derivatives can only be rationalized as originating from C_5 isoprene units if some fundamental rearrangement process has occurred during biosynthesis. These rearrangements have, in many cases, been confirmed experimentally, and are almost always consistent with the participation of carbocation intermediates. Rearrangements in chemical reactions involving carbocation intermediates, e.g. S_N1 and E1 reactions, are not uncommon and typically consist of 1,2-shifts of hydride, methyl, or alkyl groups. Occasionally, 1,3- or longer shifts are encountered. These shifts, termed **Wagner–Meerwein rearrangements**, are readily rationalized in terms of generating a more stable carbocation or relaxing ring strain (Figure 2.6). Thus, tertiary carbocations are favoured over secondary carbocations, and the usual objective in these rearrangements is to achieve tertiary status at the positive centre. However, a tertiary to secondary transition might be favoured if the rearrangement allows a significant release of ring strain. These general concepts are occasionally ignored by nature, but it must be remembered that the reactions are enzyme mediated and that carbocations may not exist as discrete species in the transformations. An interesting feature of some biosynthetic pathways, e.g.

that leading to steroids, is a remarkable series of concerted 1,2-migrations rationalized via carbocation chemistry, but entirely a consequence of the enzyme's participation (Figure 2.6).

Aldol and Claisen Reactions

The **aldol** and **Claisen** reactions both achieve carbon–carbon bond formation; in typical base-catalysed chemical reactions, this depends upon the generation of a resonance-stabilized enolate anion from a suitable carbonyl system (Figure 2.7). Whether an aldol-type or Claisen-type product is formed depends on the nature of X and its potential as a leaving group in the alkoxide anion intermediate. Thus, chemically, two molecules of acetaldehyde yield aldol, whilst two molecules of ethyl acetate can give ethyl acetoacetate. These processes are vitally important in biochemistry for the elaboration of both secondary and primary metabolites, but the enzyme catalysis obviates the need for strong bases, and probably means the enolate anion has little more than transitory existence. Nevertheless, the reactions do appear to parallel enolate anion chemistry and are frequently responsible for joining together of C_2 acetate groups.

In most cases, the biological reactions involve coenzyme A esters, e.g. **acetyl-CoA** (Figure 2.8). This is a

Aldol and Claisen reactions

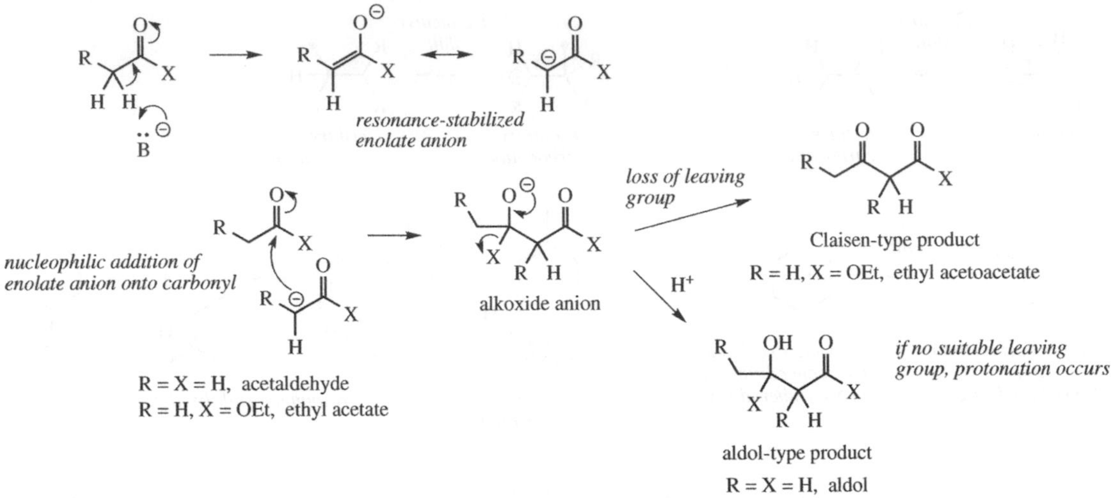

Figure 2.7

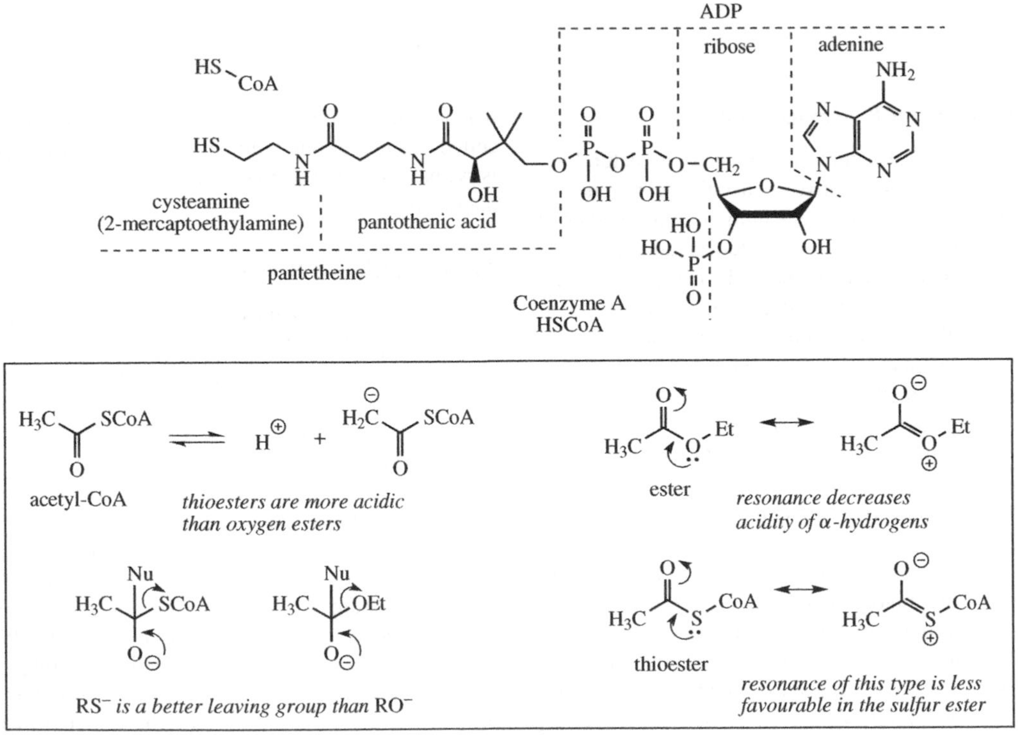

Figure 2.8

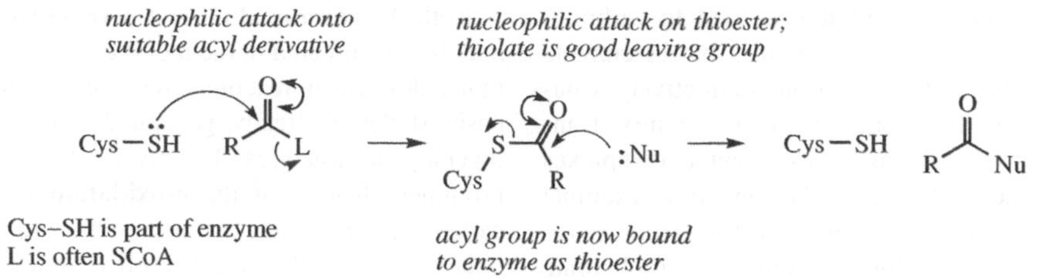

nucleophilic attack onto suitable acyl derivative

nucleophilic attack on thioester; thiolate is good leaving group

Cys–SH is part of enzyme
L is often SCoA

acyl group is now bound to enzyme as thioester

Figure 2.8 *(Continued)*

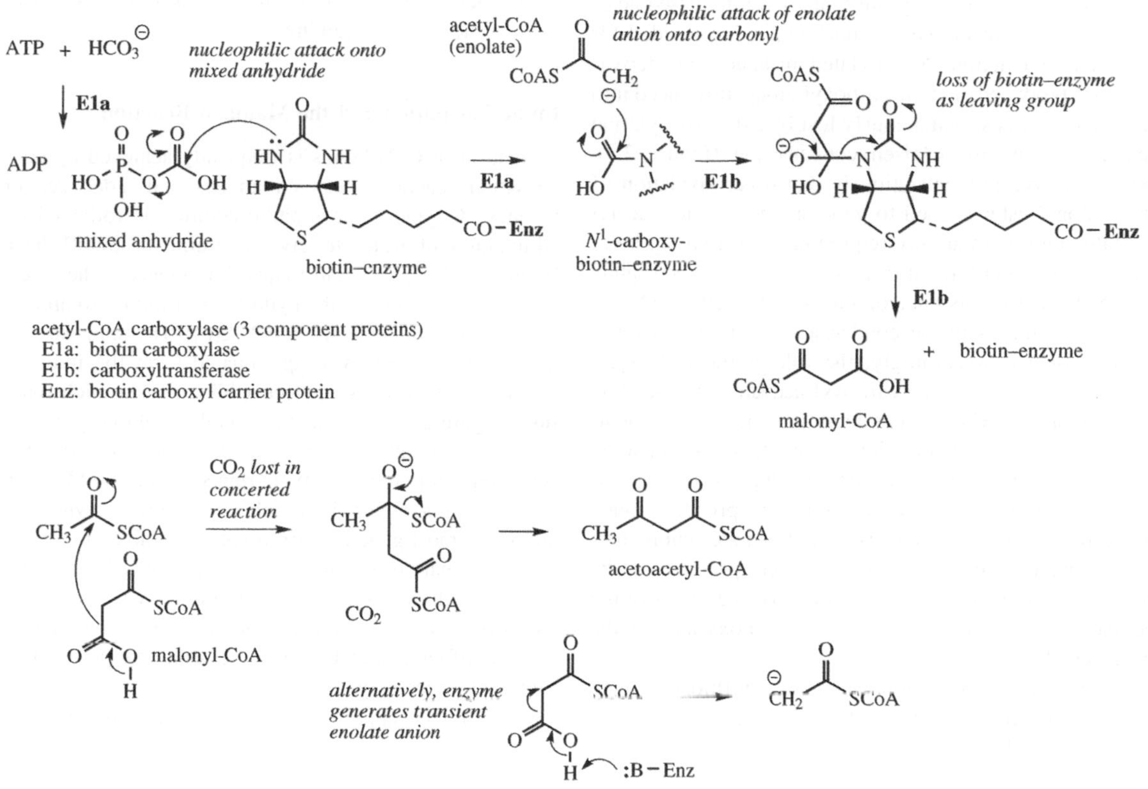

acetyl-CoA carboxylase (3 component proteins)
E1a: biotin carboxylase
E1b: carboxyltransferase
Enz: biotin carboxyl carrier protein

Figure 2.9

thioester of acetic acid, and it has significant advantages over oxygen esters, e.g. ethyl acetate, for two main reasons. First, the α-methylene hydrogen atoms are now more acidic, comparable in fact to those in the equivalent ketone, and this increases the likelihood of generating an enolate anion. This is explained in terms of electron delocalization in an ester function (Figure 2.8). This type of delocalization is more prominent in the oxygen ester than in the sulfur ester, due to oxygen's smaller size and closer proximity of the lone pair for overlap with carbon's orbitals. Second, the thioester has a much more favourable leaving group than the oxygen

ester. The combined effect is to increase the reactivity for both the aldol- and Claisen-type reactions. Thioester linkages also provide a means of covalently bonding suitable substrates to enzymes, prior to enzymic modification, followed by subsequent release (Figure 2.8).

Claisen reactions involving acetyl-CoA are made even more favourable by first converting acetyl-CoA into **malonyl-CoA** by a carboxylation reaction catalysed by the three-component enzyme acetyl-CoA carboxylase. The reaction requires CO_2, ATP and the coenzyme biotin (Figure 2.9). ATP and CO_2 (solubilized as bicarbonate, HCO_3^-) form the mixed anhydride, a reaction also catalysed by the enzyme, and this is used to carboxy-late the coenzyme which is bound in a biotin–enzyme complex. The carboxylation reaction is effectively a nu-cleophilic attack of the cyclic urea onto the mixed an-hydride. Fixation of CO_2 by biotin–enzyme complexes is not unique to acetyl-CoA; another important example occurs in the generation of oxaloacetate from pyru-vate in the synthesis of glucose from non-carbohydrate sources (gluconeogenesis). The conversion of acetyl-CoA into malonyl-CoA means the α-hydrogen atoms are now flanked by two carbonyl groups, and have increased acid-ity. Thus, a more favourable nucleophile is provided for the Claisen reaction. No acylated malonic acid deriva-tives are produced, and the carboxyl group introduced into malonyl-CoA is simultaneously lost by a decarboxylation reaction during the Claisen condensation (Figure 2.9). An alternative rationalization is that decarboxylation of the malonyl ester is used to generate the transient acetyl enolate anion without any requirement for a strong base. Thus, the product formed from acetyl-CoA as electrophile and malonyl-CoA as nucleophile is acetoacetyl-CoA, ex-actly the same as in the condensation of two molecules of acetyl-CoA. Accordingly, the role of the carboxyla-tion step is clear-cut: the carboxyl activates the α-carbon to facilitate the Claisen condensation and it is immedi-ately removed on completion of this task. By analogy, the chemical Claisen condensation using the enolate an-ion from diethyl malonate in Figure 2.10 proceeds much more favourably than that using the enolate anion from ethyl acetate. The same acetoacetic acid product can be formed in the malonate condensation by hydrolysis of the acylated malonate intermediate and decarboxylation of the *gem*-diacid.

Analogous carboxylations of some other thioesters may occur; for example, propionyl-CoA may be converted into methylmalonyl-CoA.

Both the **reverse aldol** and **reverse Claisen** reactions may be encountered in the modification of natural product molecules. Such reactions remove fragments from the basic skeleton already generated, but may extend the diversity of structures. The reverse Claisen reaction is a prominent feature of the **β-oxidation** sequence for the catabolic degradation of fatty acids (Figure 2.11): a C_2 unit as acetyl-CoA is cleaved off from a fatty acid chain, leaving it two carbon atoms shorter in length. Though the terminology β-oxidation strictly refers to the introduction of the new carbonyl group, it is usually understood to include the chain shortening.

Imine Formation and the Mannich Reaction

Formation of C–N bonds is frequently achieved by con-densation reactions between amines and aldehydes or ketones. A typical nucleophilic addition is followed by elimination of water to give an **imine** or **Schiff base** [Figure 2.12(a)]. Of almost equal importance is the rever-sal of this process, i.e. the hydrolysis of imines to amines and aldehydes/ketones [Figure 2.12(b)]. The imine so pro-duced, or more likely its protonated form the iminium cation, can then act as an electrophile in a **Mannich reac-tion** [Figure 2.12(c)]. The nucleophile might be provided by an enolate anion, or in many examples by a suitably activated centre in an aromatic ring system. The Mannich reaction is encountered throughout alkaloid biosynthesis, and in its most general form involves combination of an amine (primary or secondary), an aldehyde or ketone, and a nucleophilic carbon. Secondary amines will react with the carbonyl compound to give an iminium cation (quater-nary Schiff base) directly; thus, the additional protonation step is not necessary.

It should be appreciated that the Mannich-like addi-tion reaction in Figure 2.12(c) is little different from

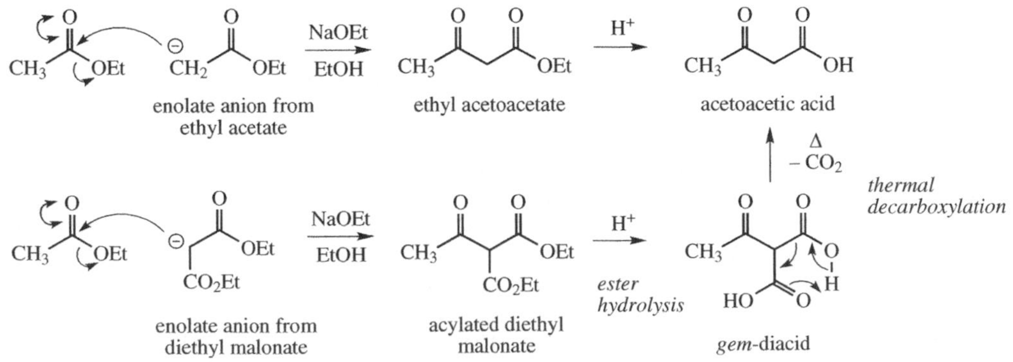

Figure 2.10

β-Oxidation of fatty acids

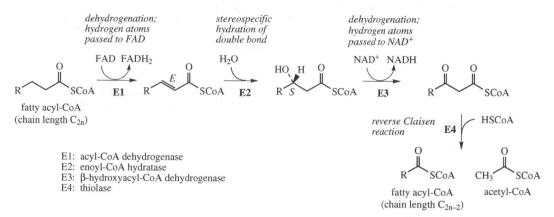

E1: acyl-CoA dehydrogenase
E2: enoyl-CoA hydratase
E3: β-hydroxyacyl-CoA dehydrogenase
E4: thiolase

Figure 2.11

(a) Imine formation

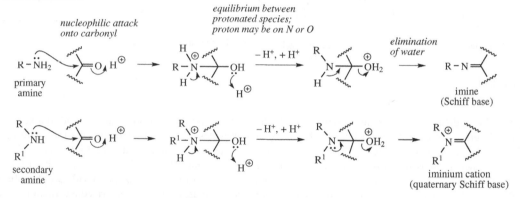

(b) Imine hydrolysis

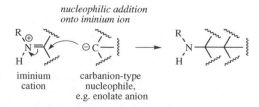

(c) Mannich reaction

Figure 2.12

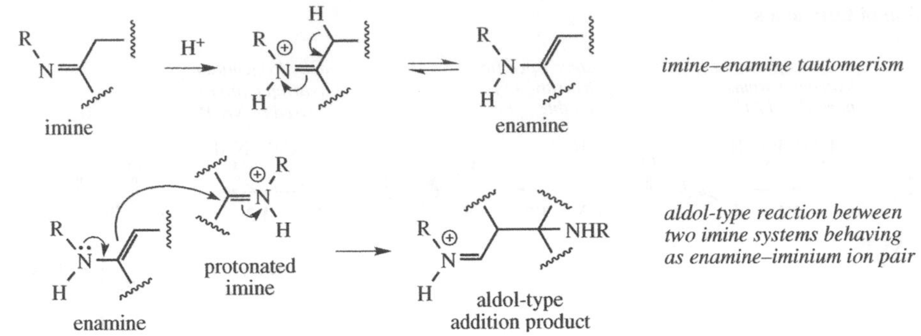

Figure 2.13

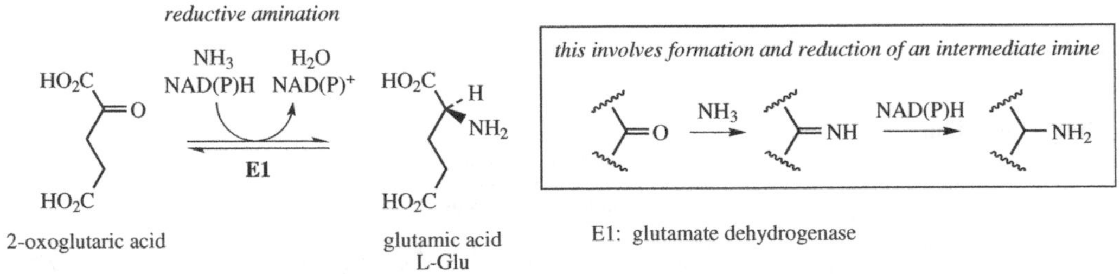

Figure 2.14

nucleophilic addition to a carbonyl group. Indeed, the imine/iminium cation is merely acting as the nitrogen analogue of a carbonyl/protonated carbonyl. To take this analogy further, protons on carbon adjacent to an imine group will be acidic, as are those α to a carbonyl group, and the isomerization of an imine to the enamine shown in Figure 2.13 is analogous to keto–enol tautomerism. Just as two carbonyl compounds can react via an aldol reaction, so can two imine systems; and this is indicated in Figure 2.13. Often, aldehyde/ketone substrates in enzymic reactions become covalently bonded to an amino group in the enzyme through imine linkages; in so doing they lose none of the carbonyl activation as a consequence of the new form of bonding.

Amino Acids and Transamination

The synthesis of amino acids depends upon the amination of the Krebs cycle intermediate **2-oxoglutaric acid** to **glutamic acid**, a process of reductive amination (Figure 2.14). The reaction involves imine formation and subsequent reduction. The reverse reaction is also important in amino acid catabolism.

Glutamic acid can then participate in the formation of other amino acids via **transamination**, the exchange of

the amino group from an amino acid to a keto acid. This provides the most common process for the introduction of nitrogen into amino acids, and for the removal of nitrogen from them. The reaction is catalysed by a transaminase enzyme using the coenzyme **pyridoxal phosphate (PLP)** and features an imine intermediate (aldimine) with the aldehyde group of PLP (Figure 2.15). Protonation of the pyridine nitrogen (as would occur at physiological pH) provides an electron sink and makes the α-hydrogen of the original amino acid now considerably more acidic. This can be removed, generating a dihydropyridine ring system. Reprotonation then produces a new imine (ketimine) and also restores aromaticity in the pyridine ring. However, because of the conjugation, it allows protonation at a position that is different from where the proton was originally lost. The net result is that the imine double bond has effectively moved to a position adjacent to its original position. Hydrolysis of the new imine function generates a keto acid and pyridoxamine phosphate. The remainder of the sequence is now a reversal of this process. This can now transfer the amine function from pyridoxamine phosphate to another keto acid.

The glutamic acid–2-oxoglutaric acid couple provides the usual donor–acceptor molecules for the amino group, and thus glutamate transaminase is the most important

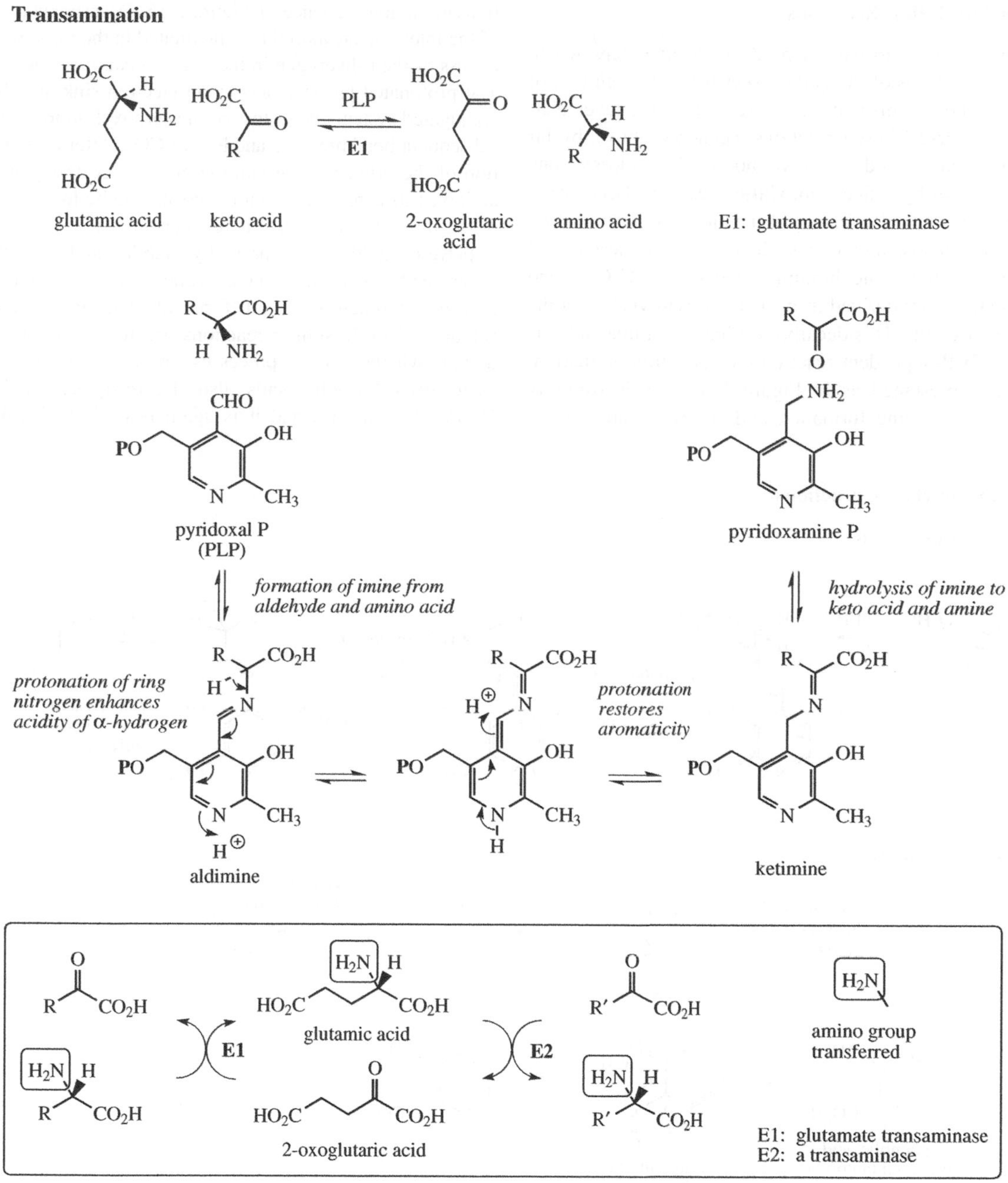

Transamination

Figure 2.15

of the transaminases. Transamination allows the amino group to be transferred from glutamic acid to a suitable keto acid, or from an amino acid to 2-oxoglutaric acid in the reverse mode. It is thus possible to transfer the amino group of one amino acid, which may be readily available from the diet, to provide another amino acid, which may be in short supply (Figure 2.15). For many organisms, this means it is no longer necessary to be able to carry out the ammonia-dependent reductive amination of 2-oxoglutaric acid.

Decarboxylation Reactions

Many pathways to natural products involve steps which remove portions of the carbon skeleton. Although two or more carbon atoms may be cleaved off via the reverse aldol or reverse Claisen reactions mentioned above, by far the most common degradative modification is loss of one carbon atom by a **decarboxylation** reaction. Decarboxylation is a particular feature of the biosynthetic utilization of amino acids, and it has already been indicated that several of the basic building blocks (e.g. C_6C_2N and indole.C_2N) are derived from an amino acid via loss of the carboxyl group. This decarboxylation of α-amino acids is also a **PLP**-dependent reaction (compare transamination) and is represented as in Figure 2.16(a). This similarly depends on imine formation and shares features of the

transamination sequence of Figure 2.15. Decarboxylation of the intermediate aldimine is facilitated in the same way as loss of the α-hydrogen in the transamination sequence. The protonated nitrogen acts as an electron sink and the conjugated system allows loss of the carboxyl proton with subsequent bond breaking and loss of CO_2. After protonation of the original α-carbon, the amine (decarboxylated amino acid) is released from the coenzyme by hydrolysis of the imine function; PLP is regenerated.

β-Keto acids are thermally labile and rapidly decarboxylated *in vitro* via a cyclic mechanism which proceeds through the enol form of the final ketone [Figure 2.16(b)]. Similar reactions are found in nature, though whether cyclic processes are necessary is not clear. *ortho*-Phenolic acids also decarboxylate readily *in vitro* and *in vivo*, and it is again possible to invoke

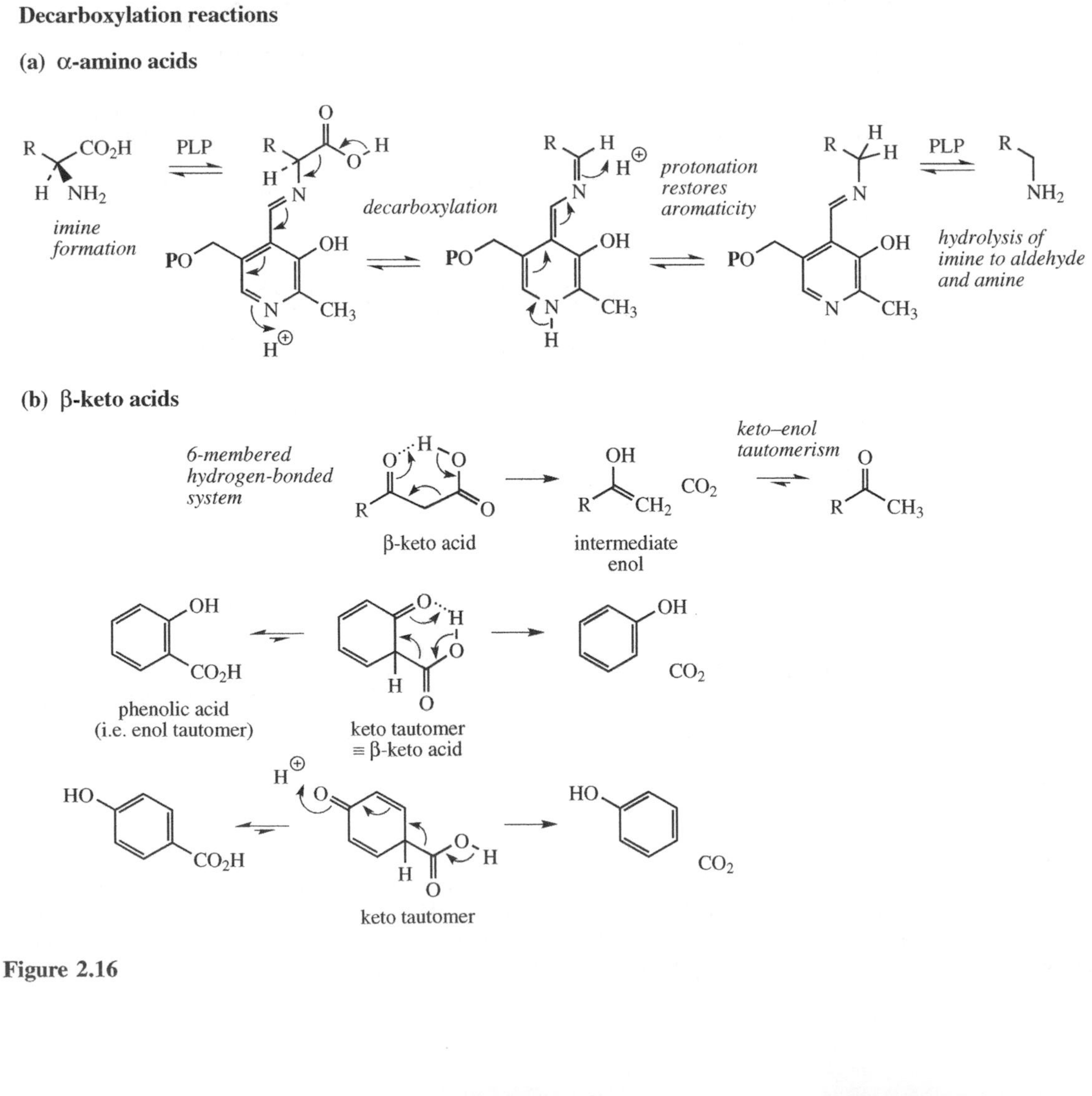

Decarboxylation reactions

(a) α-amino acids

(b) β-keto acids

Figure 2.16

(c) α-keto acids

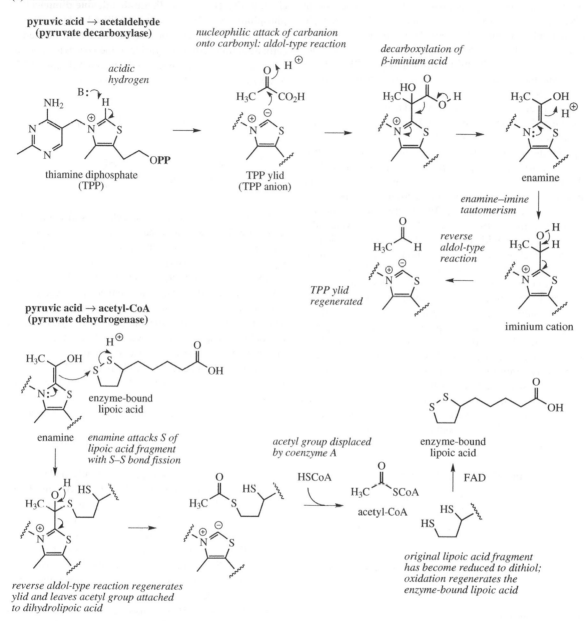

Figure 2.16 *(Continued)*

a cyclic β-keto acid tautomer of the substrate. The corresponding decarboxylation of *para*-phenolic acids cannot have a cyclic transition state, but the carbonyl group in the proposed keto tautomer activates the system for decarboxylation. The acetate pathway frequently yields structures containing phenol and carboxylic acid functions; thus, decarboxylation reactions may feature as further modifications. Although the carboxyl group may originate by hydrolysis of the thioester portion of the acetyl-CoA precursor, there are also occasions when a methyl group can be sequentially oxidized to a carboxyl, which then subsequently suffers decarboxylation.

Decarboxylation of α-keto acids is a feature of primary metabolism, e.g. pyruvic acid → acetaldehyde in glycolysis, and pyruvic acid → acetyl-CoA, an example of overall oxidative decarboxylation prior to entry of acetyl-CoA into the Krebs cycle. Both types of reaction depend upon **thiamine diphosphate (TPP)**. TPP is a coenzyme containing a thiazole ring in the form of a thiazolium salt. The proton in the thiazolium ring is relatively acidic and can be removed to generate a carbanion (strictly an ylid, a species with positive and negative charges on adjacent atoms). The ylid can act as a nucleophile and is also a reasonable leaving group. Addition to the carbonyl group of pyruvic acid is followed by decarboxylation, as depicted in Figure 2.16(c), with the positive nitrogen acting as an electron sink. The resulting molecule is an enamine. Tautomerism to the iminium cation is followed by a reverse aldol reaction, which also regenerates the ylid. In the oxidation step of oxidative decarboxylation, the enzyme-bound disulfide-containing coenzyme **lipoic acid** is also involved. The intermediate enamine in Figure 2.16(c), instead of accepting a proton, is used to attack a sulfur in the lipoic acid moiety with subsequent S–S bond fission, thereby effectively reducing the lipoic acid fragment. This allows regeneration of the TPP ylid by a reverse aldol reaction, leaving the acetyl group bound to the dihydrolipoic acid. This acetyl group is then released as acetyl-CoA by displacement with the thiol coenzyme A. The bound dihydrolipoic acid fragment is then reoxidized to restore its function. An exactly equivalent reaction is encountered in the Krebs cycle in the conversion of 2-oxoglutaric acid into succinyl-CoA.

Oxidation and Reduction Reactions

Changes to the oxidation state of a molecule are frequently carried out as a secondary metabolite is synthesized or modified. The processes are often complex and not always completely understood, but the following general features are recognized. The processes may be classified according to the type of enzyme involved and their mechanism of action.

Dehydrogenases

Dehydrogenases remove two hydrogen atoms from the substrate, passing them to a suitable coenzyme acceptor. The coenzyme system involved can generally be related to the functional group being oxidized in the substrate. Thus if the oxidation process is

$$CH-OH \longrightarrow C=O$$

then a pyridine nucleotide, **nicotinamide adenine dinucleotide (NAD$^+$)** or **nicotinamide adenine dinucleotide phosphate (NADP$^+$)**, tends to be utilized as hydrogen acceptor. One hydrogen from the substrate (that bonded to carbon) is transferred as hydride to the coenzyme and the other (as a proton) is passed to the medium (Figure 2.17). NAD(P)$^+$ may also be used in the oxidations:

$$C=O \longrightarrow -CO_2H$$
$$CH-NH_2 \longrightarrow C=NH$$

The reverse reaction, i.e. reduction, is also indicated in Figure 2.17, and may be compared with the chemical reduction process using complex metal hydrides, e.g. LiAlH$_4$ or NaBH$_4$, namely nucleophilic addition of hydride and subsequent protonation. The reduced forms NADH and NADPH are conveniently regarded as hydride-donating reducing agents. During the reduction sequence, there is stereospecific transfer of hydride from a prochiral centre on the dihydropyridine ring, and it is also delivered to the carbonyl compound in a stereospecific manner. In practice, NADPH is generally employed in reductive processes, whilst NAD$^+$ is used in oxidations.

Should the oxidative process be the conversion

$$-CH_2-CH_2- \longrightarrow -CH=CH-$$

the coenzyme used as acceptor is usually a flavin nucleotide, **flavin adenine dinucleotide (FAD)** or **flavin mononucleotide (FMN)**. These entities are bound to the enzyme in the form of a flavoprotein and take up two hydrogen atoms, represented in Figure 2.18 as being derived by addition of hydride from the substrate and a proton from the medium. Alternative mechanisms have also been proposed, however. Reductive sequences involving flavoproteins may be represented as the reverse reaction in Figure 2.18. NADPH may also be employed as a coenzyme in the reduction of a carbon–carbon double bond.

These oxidation reactions employing pyridine nucleotides and flavoproteins are especially important in primary metabolism in liberating energy from fuel molecules in the form of ATP. The reduced coenzymes formed in the process are normally reoxidized via the electron transport chain of oxidative phosphorylation, so that the hydrogen atoms eventually pass to oxygen, giving water.

Dehydrogenases: NAD⁺ and NADP⁺

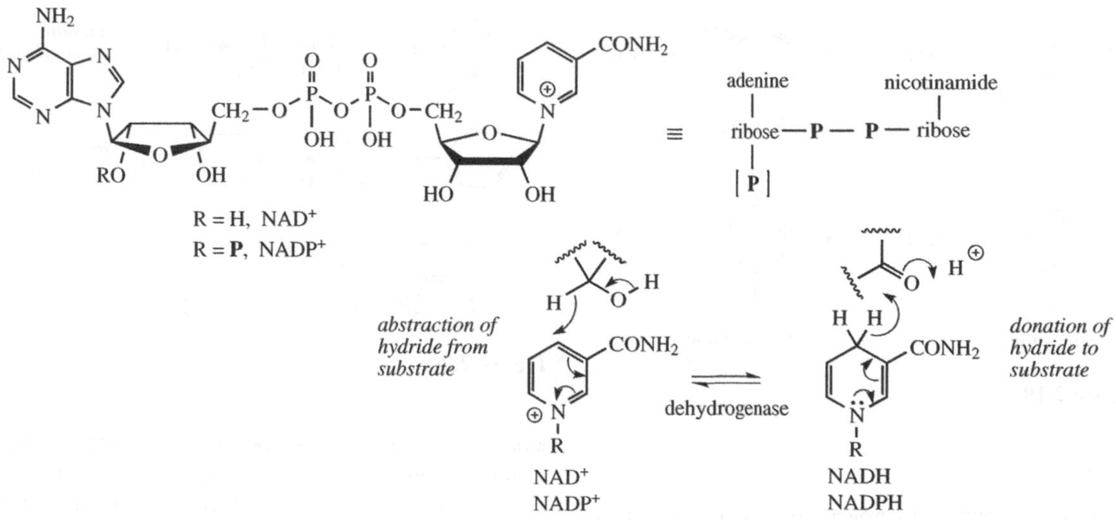

R = H, NAD⁺
R = P, NADP⁺

adenine   nicotinamide
| |
ribose — P — P — ribose
|
|P|

abstraction of hydride from substrate

donation of hydride to substrate

NAD⁺   NADH
NADP⁺   NADPH

dehydrogenase

Figure 2.17

Dehydrogenases: FAD and FMN

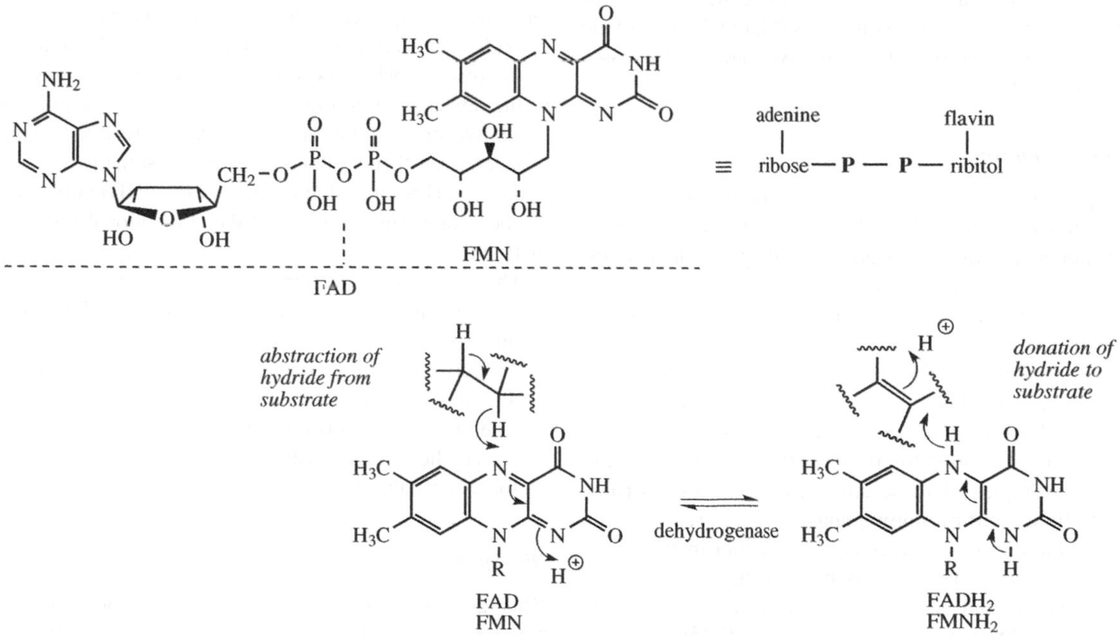

adenine   flavin
| |
ribose — P — P — ribitol

FMN

FAD

abstraction of hydride from substrate

donation of hydride to substrate

FAD   FADH₂
FMN   FMNH₂

dehydrogenase

Figure 2.18

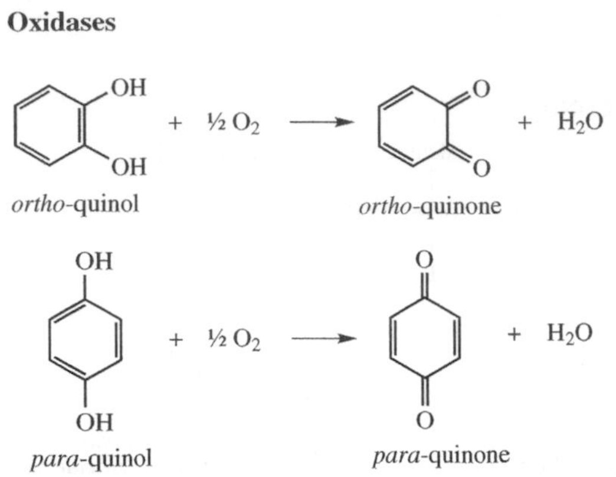

Oxidases

Figure 2.19

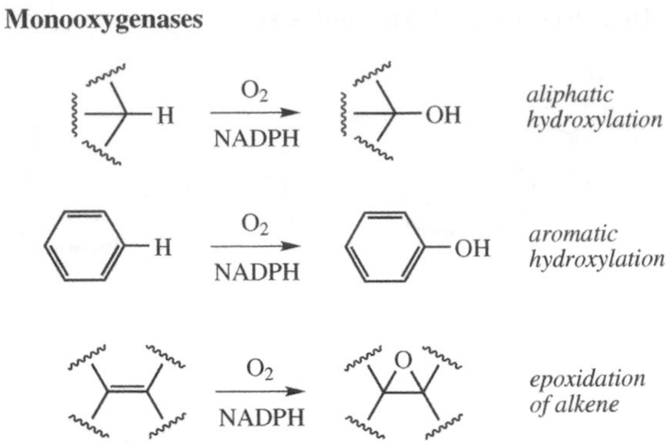

Monooxygenases

Figure 2.20

Oxidases

Oxidases also remove hydrogen from a substrate, but pass these atoms to molecular oxygen or to hydrogen peroxide, in both cases forming water. Oxidases using hydrogen peroxide are termed **peroxidases**. Mechanisms of action vary and need not be considered here. Important transformations in secondary metabolism include the oxidation of *ortho*- and *para*-quinols to quinones (Figure 2.19), and the peroxidase-induced phenolic oxidative coupling processes (see page 28).

Monooxygenases

Oxygenases catalyse the direct addition of oxygen from molecular oxygen to the substrate. They are subdivided into mono- and di-oxygenases according to whether just one or both of the oxygen atoms are introduced into the substrate. With **monooxygenases**, the second oxygen atom from O_2 is reduced to water by an appropriate hydrogen donor, e.g. NADH, NADPH, or **ascorbic acid** (vitamin C). In this respect they may also be considered to behave as oxidases, and the term 'mixed-function oxidase' is also used for these enzymes. Especially important examples of these enzymes are the **cytochrome P-450-dependent monooxygenases**. These are frequently involved in biological hydroxylations, either in biosynthesis or in the mammalian detoxification and metabolism of foreign compounds such as drugs, and such enzymes are thus termed **hydroxylases. Cytochrome P-450** is named after its intense absorption band at 450 nm when exposed to CO, which is a powerful inhibitor of these enzymes. It contains an iron–porphyrin complex (haem) which is bound to the enzyme. A redox change involving the Fe atom allows binding and the cleavage of molecular oxygen to oxygen atoms, with subsequent transfer of one

atom to the substrate. Many such systems have been identified, capable of hydroxylating aliphatic or aromatic systems, as well as producing epoxides from alkenes (Figure 2.20). In most cases, NADPH features as hydrogen donor.

Aromatic hydroxylation catalysed by monooxygenases (including cytochrome P-450 systems) probably involves arene oxide (epoxide) intermediates (Figure 2.21). An interesting consequence of this mechanism is that, when the epoxide opens up, the hydrogen atom originally attached to the position which becomes hydroxylated can migrate to the adjacent carbon on the ring. A high proportion of these hydrogen atoms are subsequently retained in the product, even though enolization allows some loss of this hydrogen. This migration is known as the **NIH shift**, having been originally observed at the US National Institutes of Health.

The oxidative cyclization of an *ortho*-hydroxymethoxy-substituted aromatic system giving a **methylenedioxy group** is also known to involve a cytochrome P-450-dependent monooxygenase. This enzyme hydroxylates the methoxy methyl to yield an intermediate that is a hemiacetal of formaldehyde; this can cyclize to the methylenedioxy bridge (the acetal of formaldehyde) by an ionic mechanism (Figure 2.22).

Dioxygenases

Dioxygenases introduce both atoms from molecular oxygen into the substrate, and are frequently involved in the cleavage of carbon–carbon bonds, including aromatic rings. Cyclic peroxides (dioxetanes) have been proposed as intermediates (Figure 2.23), and though evidence points against dioxetanes, they do provide a simple appreciation of the reaction outcome. Oxidative cleavage of aromatic rings typically employs catechol (1,2-dihydroxy) or

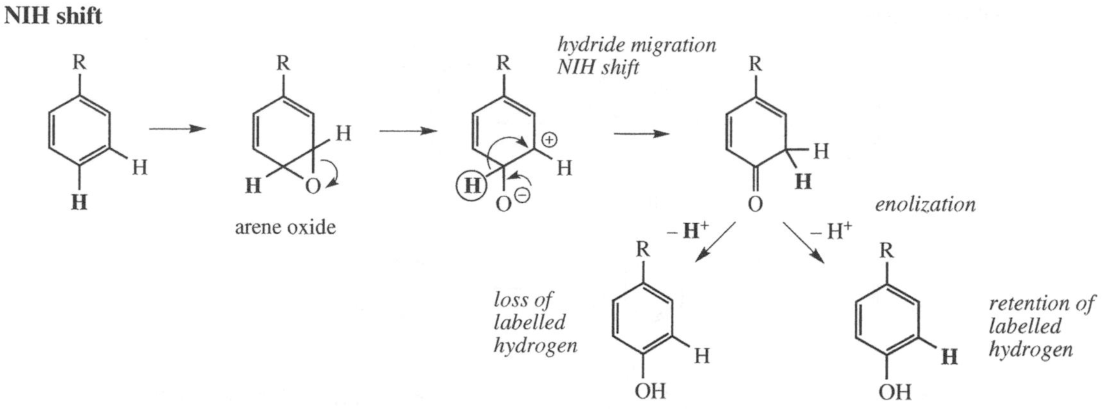

NIH shift

Figure 2.21

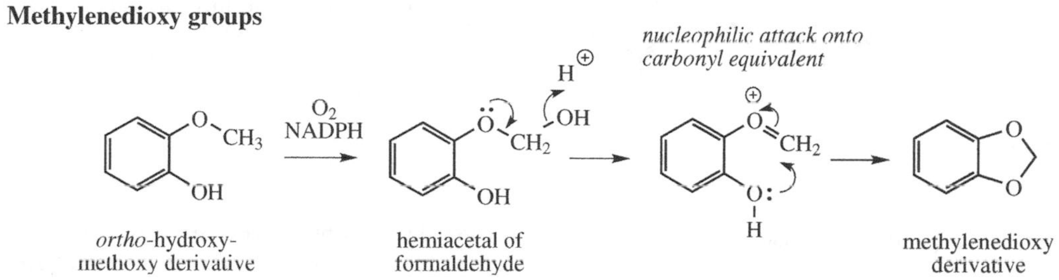

Methylenedioxy groups

Figure 2.22

quinol (1,4-dihydroxy) substrates, and in the case of cate-chols, cleavage may be between or adjacent to the two hydroxyls, giving products containing aldehyde and/or carboxylic acid functionalities (Figure 2.23).

Some dioxygenases utilize two acceptor substrates and incorporate one oxygen atom into each. Thus, **2-oxoglutarate-dependent dioxygenases** hydroxylate one substrate, whilst also transforming 2-oxoglutarate into succinate with the release of CO_2 (Figure 2.24). 2-Oxoglutarate-dependent dioxygenases also require as cofactors Fe^{2+} to generate an enzyme-bound iron–oxygen complex, and **ascorbic acid** (vitamin C) to reduce this complex subsequently.

Amine Oxidases

In addition to the oxidizing enzymes outlined above, those which transform an amine into an aldehyde, the **amine oxidases**, are frequently involved in metabolic pathways. These include **monoamine oxidases** and **diamine oxidases**. Monoamine oxidases utilize a flavin nucleotide, typically FAD, and molecular oxygen; the transformation involves initial dehydrogenation to an imine, followed by hydrolysis to the aldehyde and ammonia (Figure 2.25). Diamine oxidases require a diamine substrate and oxidize at one amino group using molecular oxygen to give the corresponding aldehyde. Hydrogen peroxide and ammonia are the other products formed. The aminoaldehyde so formed then has the potential to be transformed into a cyclic imine.

Baeyer–Villiger Monooxygenases

The chemical oxidation of ketones by peracids, the **Baeyer–Villiger oxidation**, yields an ester, and the process is known to involve migration of an alkyl group from the ketone (Figure 2.26). For comparable ketone → ester conversions known to occur in biochemistry, FAD-dependent monooxygenase enzymes requiring NADPH and O_2 appear to be involved. This leads to formation of a flavin–peroxide and a mechanism similar to that for the chemical Baeyer–Villiger oxidation may thus operate. The oxygen atom introduced thus originates from O_2.

Dioxygenases

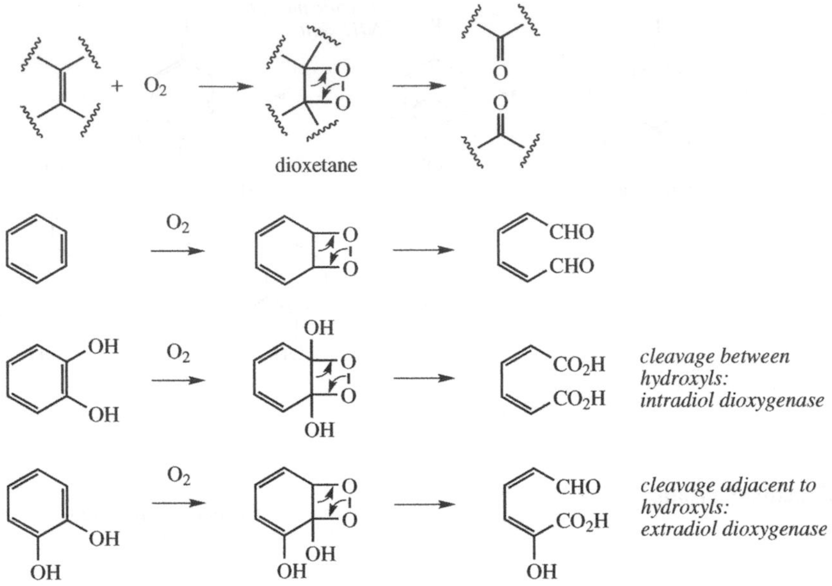

Figure 2.23

2-Oxoglutarate-dependent dioxygenases

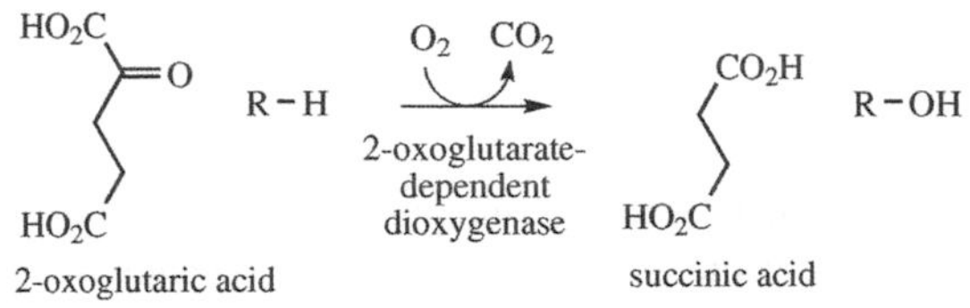

2-oxoglutaric acid

succinic acid

Figure 2.24

Amine oxidases

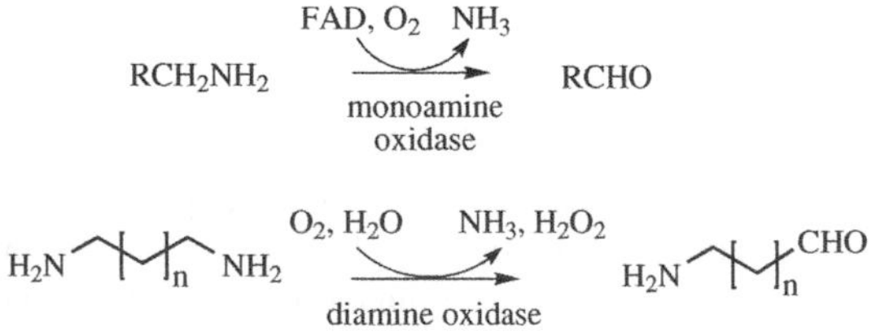

Figure 2.25

Phenolic Oxidative Coupling

Many natural products are produced by the coupling of two or more phenolic systems, in a process readily rationalized by means of radical reactions. The reactions can be brought about by oxidase enzymes, including peroxidase and laccase systems, known to be radical generators. Other enzymes catalysing **phenolic oxidative coupling** have been characterized as cytochrome P-450-dependent proteins, requiring NADPH and O_2 cofactors, but no oxygen is incorporated into the substrate. Hydrogen abstraction from a phenol (a one-electron oxidation) gives the radical, and the unpaired electron can then be delocalized via resonance forms in which the free electron is dispersed to positions *ortho* and *para* to the original oxygen function (Figure 2.27). These phenol-derived radicals do not propagate a radical chain reaction, but instead are quenched by coupling with other radicals. Coupling of two of these resonance structures in various combinations gives a range of dimeric systems, as exemplified in Figure 2.27. The final products indicated are then derived by enolization, which restores aromaticity to the rings. Thus, carbon—carbon bonds involving positions *ortho* or *para* to the original phenols, or ether linkages, may be formed. The reactive dienone systems formed as intermediates may in some cases be attacked by other nucleophilic groupings, extending the range of structures ultimately derived from this basic reaction sequence.

Halogenation Reactions

Halogenated natural products are widespread, though not particularly common; only a handful of examples are

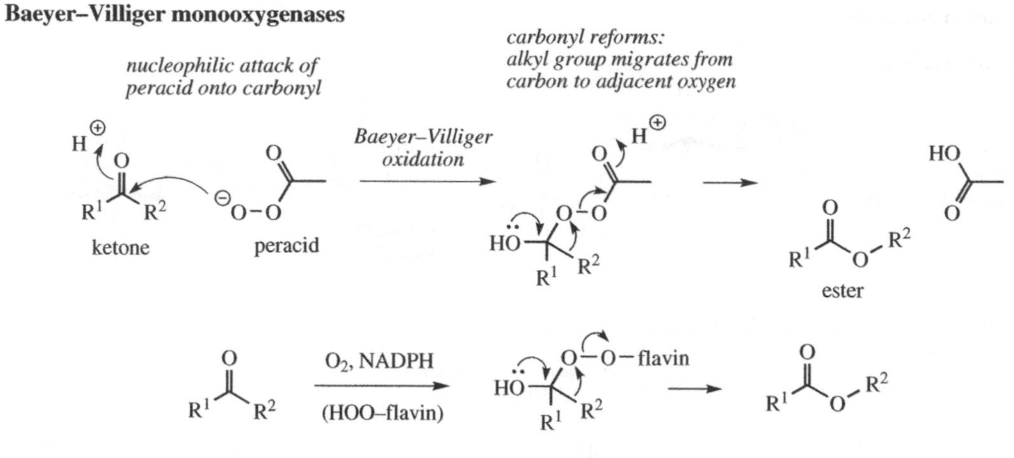

Baeyer–Villiger monooxygenases

Figure 2.26

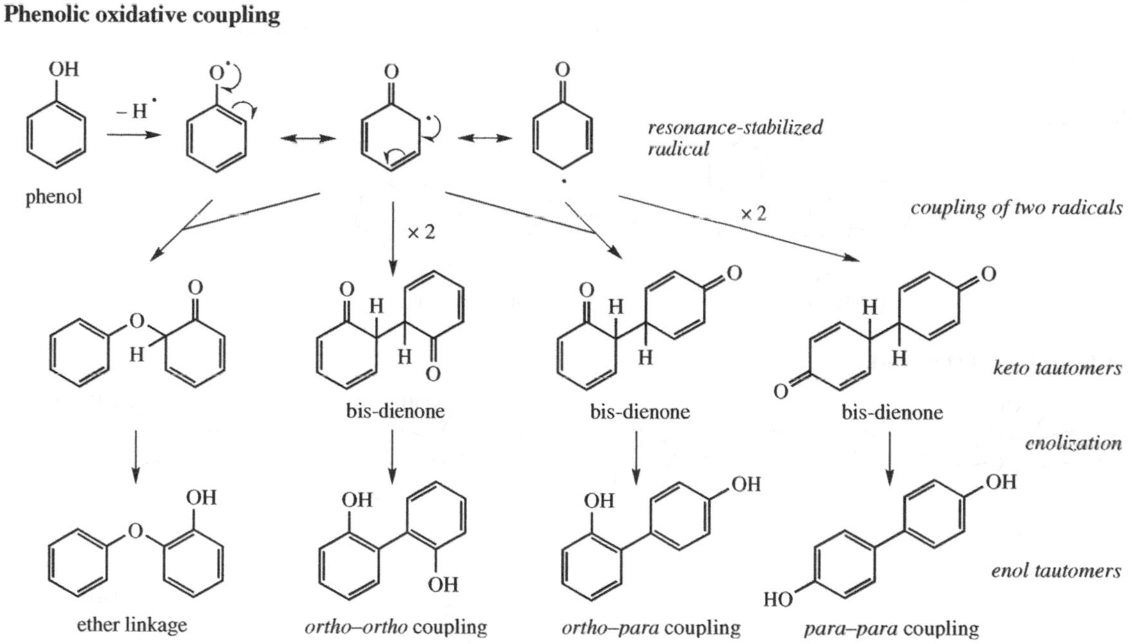

Phenolic oxidative coupling

Figure 2.27

found in this book. The nature of the halogen depends very much upon the relative amount of halide available in the environment, and marine organisms are more likely to synthesize brominated compounds than terrestrial organisms, which tend to produce chlorinated derivatives. Fluorine and iodine are also encountered. Despite halogenation being relatively uncommon, a number of different enzymic mechanisms have been identified for

the process. It is not considered appropriate to present detailed descriptions of these processes, but merely to indicate their general nature.

Hypohalite (HOX) is the halogenating species produced by **haloperoxidases** and flavin-dependent **halogenases**. This may be considered as a source of electron-deficient halogen (X^+) that can react with an electron-rich centre in the substrate. Free HOX is not

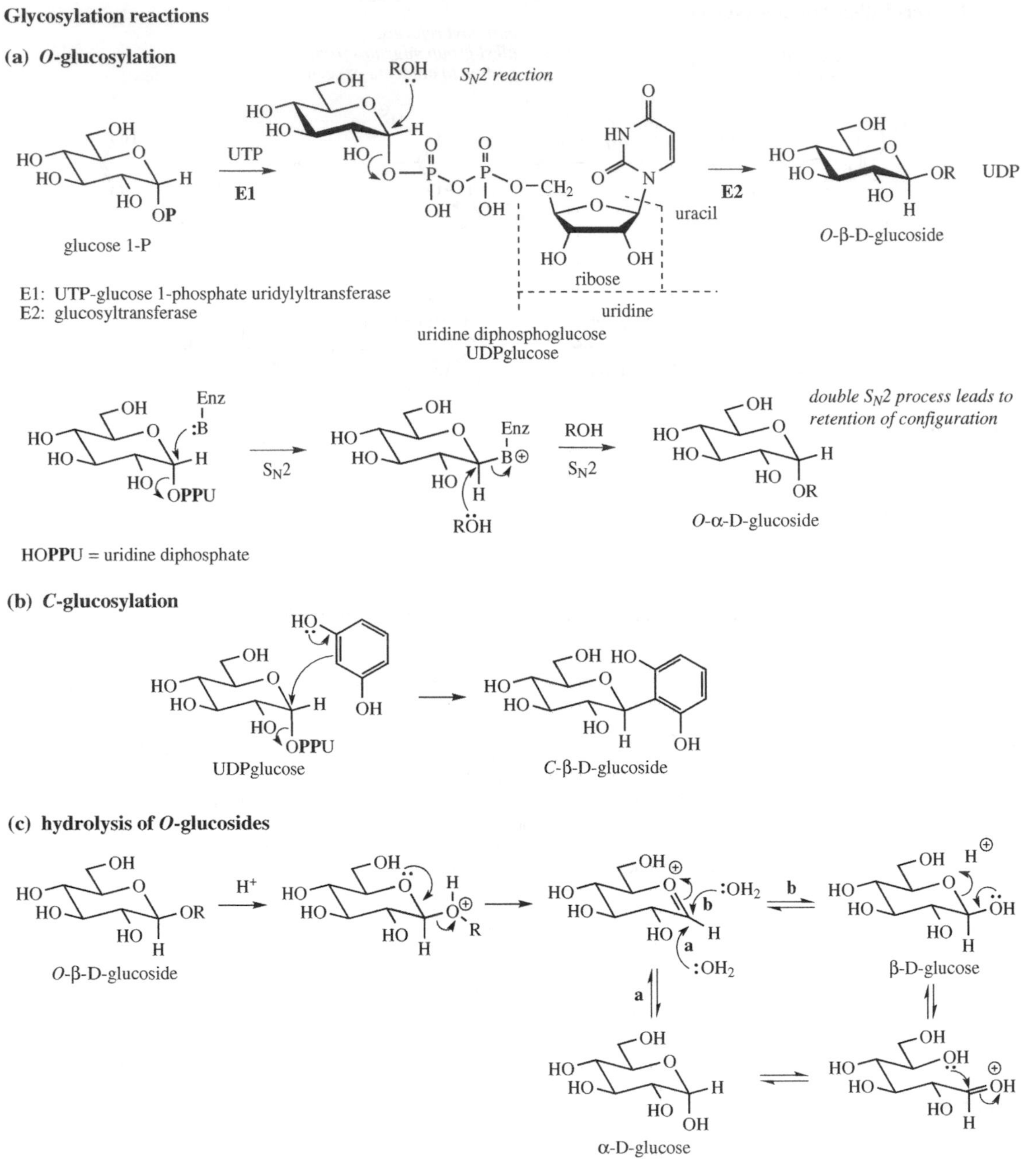
Glycosylation reactions
(a) O-glucosylation
glucose 1-P
UTP
E1
SN2 reaction
ROH
uracil
ribose
uridine
uridine diphosphoglucose
UDPglucose
E2
UDP
O-β-D-glucoside
E1: UTP-glucose 1-phosphate uridylyltransferase
E2: glucosyltransferase
Enz
SN2
ROH
double SN2 process leads to retention of configuration
O-α-D-glucoside
HOPPU = uridine diphosphate
(b) C-glucosylation
UDPglucose
C-β-D-glucoside
(c) hydrolysis of O-glucosides
O-β-D-glucoside
β-D-glucose
α-D-glucose

Figure 2.28

usually released; this would not allow regiospecific halogenation. The two types of enzyme require other cofactors, H_2O_2 in the case of haloperoxidases or reduced FAD and molecular oxygen for the flavin-dependent halogenases. Halogen radical ($X^{\cdot}$) appears to be the active species involved in halogenation catalysed by **2-oxoglutarate-dependent halogenases**. Sometimes, halide anion (X^-) can be employed as the halogenating species, acting as a nucleophile at a suitable electrophilic centre.

Glycosylation Reactions

The widespread occurrence of **glycosides** and **polysaccharides** requires processes for attaching sugar units to a suitable atom of an aglycone to give a glycoside, or to another sugar giving a polysaccharide. Linkages tend to be through oxygen, although they are not restricted to oxygen, since *S*-, *N*-, and *C*-glycosides are well known. The agent for glycosylation is a nucleoside diphosphosugar, e.g. **uridine diphosphoglucose** (UDPglucose). Uridine is the nucleoside most frequently employed, though others are utilized depending upon organism and metabolite. A wide variety of sugars, including uronic acids (see page 488), may be transferred via this general mechanism. UDPglucose is synthesized from glucose 1-phosphate and uridine triphosphate (UTP), and the glucosylation process can be envisaged as a simple nucleophilic displacement reaction of S_N2 type [Figure 2.28(a)]. Since UDPglucose has its leaving group in the α-configuration, the product has the β-configuration, as is most commonly found in natural glucosides. Note, however, that many important carbohydrates, e.g. sucrose and starch, possess α-linkages, and these appear to originate via a double S_N2 process [Figure 2.28(a)].

The *N*-, *S*-, and *C*-glycosides are produced in a similar manner with the appropriate nucleophile. In the case of *C*-glycosides, a suitable nucleophilic carbon is required, e.g. aromatic systems activated by phenol groups [Figure 2.28(b)], as noted in the *C*-alkylation process described earlier.

The hydrolysis of glycosides is achieved by specific hydrolytic enzymes, e.g. β-glucosidase for β-glucosides and β-galactosidase for β-galactosides. These enzymes mimic the readily achieved acid-catalysed processes [Figure 2.28(c)] and may retain or invert the configuration at the anomeric centre. Under acidic conditions, the product would be an equilibrium mixture of the α- and β-anomeric hemiacetal forms, which can also interconvert via the open-chain sugar. Of particular importance is to note that although *O*-, *N*-, and *S*-glycosides may be hydrolysed by acid, *C*-glycosides are stable to acid. *C*-Glycosylation introduces a new carbon–carbon linkage, and cleavage requires oxidation, not hydrolysis.

Box 2.1

Some Vitamins Associated with the Construction Mechanisms

Vitamin B₁

Vitamin B$_1$ (thiamine) (Figure 2.29) is a water-soluble vitamin with a pyrimidinylmethyl-thiazolium structure. It is widely available in the diet, with cereals, beans, nuts, eggs, yeast, and vegetables providing sources. Wheat germ and yeast have very high levels. Dietary deficiency leads to beriberi, characterized by neurological disorders, loss of appetite, fatigue, and muscular weakness. As thiamine diphosphate, vitamin B$_1$ is a coenzyme for pyruvate dehydrogenase, which catalyses the oxidative decarboxylation of pyruvate to acetyl-CoA (see page 23), and for 2-oxoglutarate dehydrogenase, which catalyses a similar reaction on 2-oxoglutarate in the Krebs cycle. It is also a cofactor for transketolase, which transfers a two-carbon fragment between carbohydrates in the pentose phosphate pathway (see page 486). Accordingly, this is a very important component in carbohydrate metabolism. **Thiamine** is produced synthetically, and foods such as cereals are often enriched. The vitamin is stable in acid solution, but decomposes above pH 5 and is also partially decomposed during normal cooking.

Vitamin B₂

Vitamin B$_2$ (riboflavin) (Figure 2.29) is a water-soluble vitamin having an isoalloxazine ring linked to D-ribitol. It is widely available in foods, including liver, kidney, dairy products, eggs, meat, and fresh vegetables. Yeast is a particularly rich source. It is stable in acid solution, not decomposed during cooking, but is sensitive to light. Riboflavin may be produced synthetically, or by fermentation using the yeast-like fungi *Eremothecium ashbyii* and *Ashbya gossypii*. Dietary deficiency is uncommon, but manifests itself by skin problems and eye disturbances. Riboflavin is a component of FMN and FAD, coenzymes which play a major role in oxidation–reduction reactions (see page 25). Many key enzymes containing riboflavin (flavoproteins) are involved in metabolic pathways. Since riboflavin contains ribitol and not ribose in its structure, FAD and FMN are not strictly nucleotides, though this nomenclature is commonly accepted and used.

Vitamin B₃

Vitamin B$_3$ (nicotinic acid, niacin; Figure 2.29) is a stable, water-soluble vitamin widely distributed in foodstuffs, especially meat, liver, fish, wheat germ, and yeast. However, in some foods, e.g. maize, it may be present in a bound form and is not readily

Box 2.1   (continued)

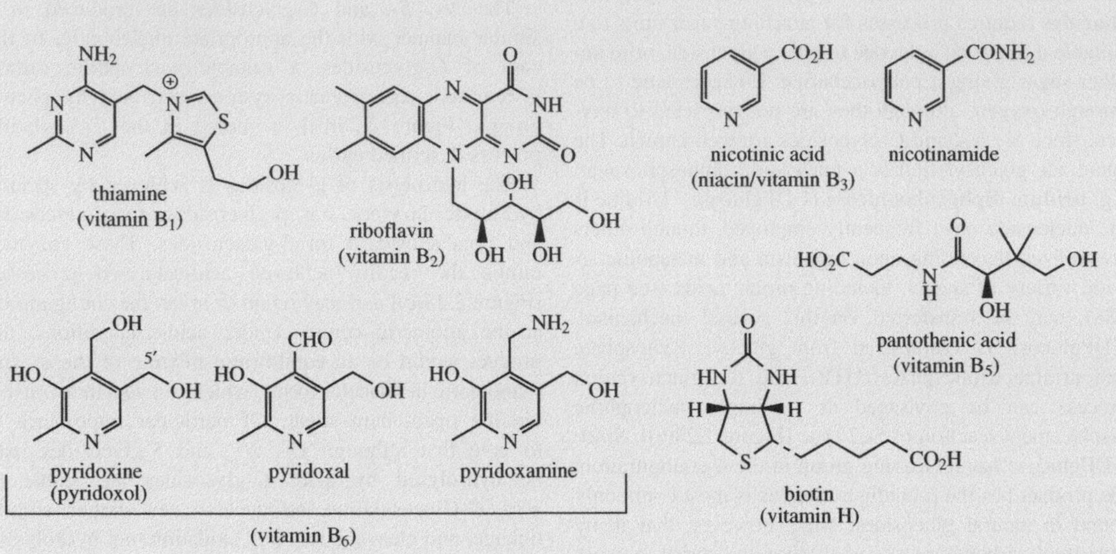

Figure 2.29

available. Accordingly, diets based principally on maize may lead to deficiencies. The amino acid tryptophan can be converted in the body into nicotinic acid (see Figure 6.32, page 332) and may provide a large proportion of the requirements. Nicotinic acid is converted into nicotinamide (Figure 2.29), though this compound also occurs naturally in many foods. The term vitamin B_3 is often used for the combined nicotinamide–nicotinic acid complement. In the form of the coenzymes NAD^+ and $NADP^+$, nicotinamide plays a vital role in oxidation–reduction reactions (see page 000) and is the most important electron carrier in primary metabolism. Deficiency in nicotinamide leads to pellagra, which manifests itself in diarrhoea, dermatitis, and dementia. Oral lesions and a red tongue may be more noticeable than the other symptoms. **Nicotinamide** is usually preferred over **nicotinic acid** for dietary supplements, since there is less risk of vasodilatation. Both are produced synthetically. It is common practice to enrich many foods, including bread, flour, corn, and rice products.

Nicotinic acid in large doses can lower both cholesterol and triglyceride concentrations by inhibiting their synthesis. It may be prescribed as an adjunct to statin treatment (see page 98).

Vitamin B_5

Vitamin B_5 (pantothenic acid; Figure 2.29) is a very widely distributed water-soluble vitamin, though yeast, liver, and cereals provide rich sources. Even though animals must obtain the vitamin through the diet, pantothenic acid deficiency is rare, since most foods provide adequate quantities. Its importance in metabolism is as part of the structure of coenzyme A (see page 16), the carrier molecule essential for carbohydrate, fat, and protein metabolism. Pantothenic acid is specifically implicated in enzymes responsible for the biosynthesis of fatty acids (see page 40), polyketides (see page 70) and some peptides (see page 438).

Vitamin B_6

Vitamin B_6 covers the three pyridine derivatives pyridoxine (pyridoxol), pyridoxal, and pyridoxamine, and also their 5′-phosphates (Figure 2.29). These are water-soluble vitamins, pyridoxine predominating in plant materials, whilst pyridoxal and pyridoxamine are the main forms in animal tissues. Meat, salmon, nuts, potatoes, bananas, and cereals are good sources. A high proportion of the vitamin activity can be lost during cooking, but a normal diet provides an adequate supply. Vitamin B_6 deficiency is usually the result of malabsorption, or may be induced by some drug treatments where the drug may act as an antagonist or increase renal excretion of the vitamin as a side-effect. Symptoms of deficiency are similar to those of niacin (vitamin B_3) and riboflavin (vitamin B_2) deficiencies, including eye, mouth, and nose lesions and neurological changes. Synthetic **pyridoxine** is used for

Box 2.1   (continued)

supplementation. Pyridoxal 5′-phosphate is a coenzyme for a large number of enzymes, particularly those involved in amino acid metabolism, e.g. in transamination (see page 20) and decarboxylation (see page 22). It thus plays a central role in the production of the neurotransmitters serotonin and noradrenaline (see page 335), and also of histamine (see page 398). The production of the neurotransmitter γ-aminobutyric acid (GABA) from glutamic acid is also an important pyridoxal-dependent reaction.

Vitamin B_{12}

Vitamin B_{12} (cobalamins; Figure 2.30) are extremely complex structures based on a corrin ring, which, although similar to the porphyrin ring found in haem, chlorophyll, and cytochromes, has two of the pyrrole rings directly bonded. The central metal atom is cobalt; haem and cytochromes have iron, whilst chlorophyll has magnesium. Four of the six coordinations are provided by the corrin ring nitrogen atoms, and a fifth by a dimethylbenzimidazole moiety. The sixth is variable, being cyano in **cyanocobalamin** (vitamin B_{12}), hydroxyl in **hydroxocobalamin** (vitamin B_{12a}), or other anions may feature. Cyanocobalamin is actually an artefact formed as a result of the use of cyanide in purification procedures. The physiologically active coenzyme form of the vitamin is 5′-deoxyadenosylcobalamin (coenzyme B_{12}). Vitamin B_{12} appears to be entirely of microbial origin, with intestinal flora contributing towards human dietary needs. The vitamin is then stored in the liver; animal liver extract has been a traditional source. Commercial supplies are currently obtained by semi-synthesis from the total cobalamin extract of *Streptomyces griseus, Propionibacterium* species, or other bacterial cultures. This material can be converted into cyanocobalamin or hydroxocobalamin. The cobalamins are stable when protected against light. Foods with a high vitamin B_{12} content include liver, kidney, meat, and seafood. Vegetables are a poor dietary source, and strict vegetarians may, therefore, risk deficiencies. Insufficient vitamin B_{12} leads to pernicious anaemia, a disease that results in nervous disturbances and low production of red blood cells, though this is mostly due to lack of the gastric glycoprotein (intrinsic factor) which complexes with the vitamin to facilitate its absorption. Traditionally, daily consumption of raw liver was used to counteract the problem. Cyanocobalamin (or preferably hydroxocobalamin, which has a longer lifetime in the body) may be administered orally or by injection to counteract deficiencies. Both agents are converted into coenzyme B_{12} in the body. Coenzyme B_{12} is a cofactor for a number of metabolic rearrangements, such as the conversion of methylmalonyl-CoA into succinyl-CoA in the oxidation of fatty acids with an odd number of carbon atoms, and for methylations, such as in the biosynthesis of methionine from homocysteine.

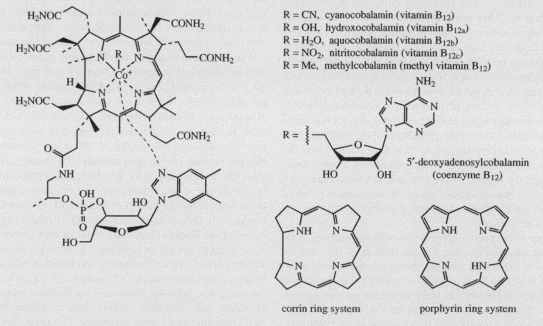

R = CN, cyanocobalamin (vitamin B_{12})
R = OH, hydroxocobalamin (vitamin B_{12a})
R = H_2O, aquocobalamin (vitamin B_{12b})
R = NO_2, nitritocobalamin (vitamin B_{12c})
R = Me, methylcobalamin (methyl vitamin B_{12})

5′-deoxyadenosylcobalamin
(coenzyme B_{12})

corrin ring system   porphyrin ring system

Figure 2.30

Box 2.1   (continued)

Vitamin H

Vitamin H (biotin; Figure 2.29) is a water-soluble vitamin found in eggs, liver, kidney, yeast, cereals, and milk; it is also produced by intestinal microflora, so that dietary deficiency is rare. Deficiency can be triggered by a diet rich in raw egg white, which contains the protein avidin that binds biotin so tightly that it is effectively unavailable for metabolic use. This affinity disappears on cooking which denatures the avidin. Biotin deficiency leads to dermatitis and hair loss. The vitamin functions as a carboxyl carrier, binding CO_2 via a carbamate link, then donating this in carboxylase reactions, e.g. carboxylation of acetyl-CoA to malonyl-CoA (see page 17), of propionyl-CoA to methylmalonyl-CoA (see page 55), and of pyruvate to oxaloacetate during gluconeogenesis.

ELUCIDATING BIOSYNTHETIC PATHWAYS

A biosynthetic pathway is a sequence of chemical transformations; with few exceptions, each of these reactions is catalysed by an enzyme. Chemical modification of a substrate depends upon the binding properties conferred by a particular combination of functional groups in the constituent amino acids of the protein. As a result, enzymes tend to demonstrate quite remarkable specificity towards their substrates, and usually catalyse only a single transformation. The main thrust of biosynthetic research has been in delineating the sequence of reactions and characterizing the enzymes involved. However, as the knowledge base increases, it is becoming possible through gene methodology to manipulate these processes to provide modified or new chemicals in the search for medicinal agents. This book summarizes the current knowledge about selected groups of natural products. How this knowledge has been acquired is the result of decades of meticulous research, often achieved without the benefits of modern technology. Not all pathways are known with the same level of detail; those shown in the figures present a composite picture of information accumulated by different methodologies at different times. A brief outline of the methods used for the elucidation of biosynthetic pathways helps to emphasize how these have changed over the years and should assist in interpretation of these figures. Though the methods have changed, it is most reassuring to realize that essentially all of the information gathered in the earlier studies has been confirmed correct by the newer techniques.

Any biosynthetic study begins with speculation, comparing structurally related compounds from the same or different organisms, and suggesting possible interrelationships based on known chemical reactions, i.e. application of 'paper chemistry'. In order to test the various suggestions, it is necessary to demonstrate that one compound is converted into another in a living organism, or, eventually, by an enzyme. This usually entails feeding an **isotopically labelled compound** to a plant, microorganism, animal tissue, etc, that produces the required metabolite. For success, it is an absolute requirement that the organism is actively synthesizing the compound during the feeding period. The metabolite is then isolated, purified, and analysed to detect presence of the isotope. The formation of a labelled product is cautiously interpreted as a demonstration of the precursor–product relationship, though it should also be confirmed that the position of labelling in the product is the same as that in the precursor; organisms can degrade an introduced chemical to smaller portions and utilize it in their general metabolism. Isotope detection methods need to be sensitive, since only small amounts of the added precursor may penetrate to the site of synthesis, and much may be metabolized on the way. Incorporations in microorganisms are typically very much higher than in plants. A realistic pathway can then be proposed by the combined results from a series of feeding experiments with different labelled substrates.

The biosynthetic origins of the animal sterol cholesterol are described in detail in Chapter 5 (see page 248), but for the present, can be used to illustrate some of the methodologies. In the early studies exploring the role of acetic acid in cholesterol biosynthesis in rats and rat liver, the isotope used was deuterium and mass spectroscopic detection was employed. When the radioactive isotopes **carbon-14** (^{14}C) and **tritium** (^{3}H) became available, the level of sensitivity was increased significantly. In due course, it was established that all carbon atoms in cholesterol are derived from acetic acid, as shown in Figure 2.31. This necessitated systematic chemical degradation of the labelled cholesterol to allow each carbon to be separately assessed for labelling, a painstaking and outstanding achievement. Cholesterol was found to originate from acetate, incorporating 15 methyl and 12 carboxyl carbon atoms. Actually, cholesterol biosynthesis requires 18 acetate units, and these numbers indicate loss of three methyl groups and six carboxyl groups. This is because an early precursor, mevalonic acid, is formed from three acetate groups (as acetyl-CoA), and six mevalonate groups

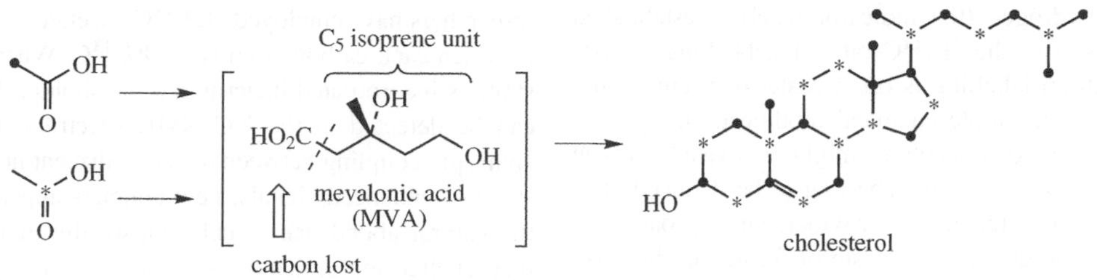

Figure 2.31

are used in producing squalene, a C_{30} precursor of cholesterol (C_{27}). It had initially been shown that feeding of squalene, found in shark liver, increased the cholesterol content of animal tissues. Mevalonate is the source of the biosynthetic C_5 isoprene unit that is used for steroid biosynthesis. The cyclization of squalene (as the oxide) provides the tetracyclic ring system of lanosterol; the process also involves some structural rearrangements involving methyl and hydride transfers, typical of carbocation chemistry, as briefly outlined in Figure 2.32. In due

course, lanosterol (C_{30}) is converted into cholesterol (C_{27}) with loss of three carbon atoms.

The characteristic β-radiations from ^{14}C and ^{3}H have different energies and can be measured separately in a mixed sample. This allowed **multiple-labelling studies**, tracking the retention or loss of hydrogen. To establish loss of hydrogen, it is pointless feeding a substrate to and then detecting no isotope in the product; this could mean no incorporation of the precursor. Instead, the carbon isotope is used as the reference label, and retention

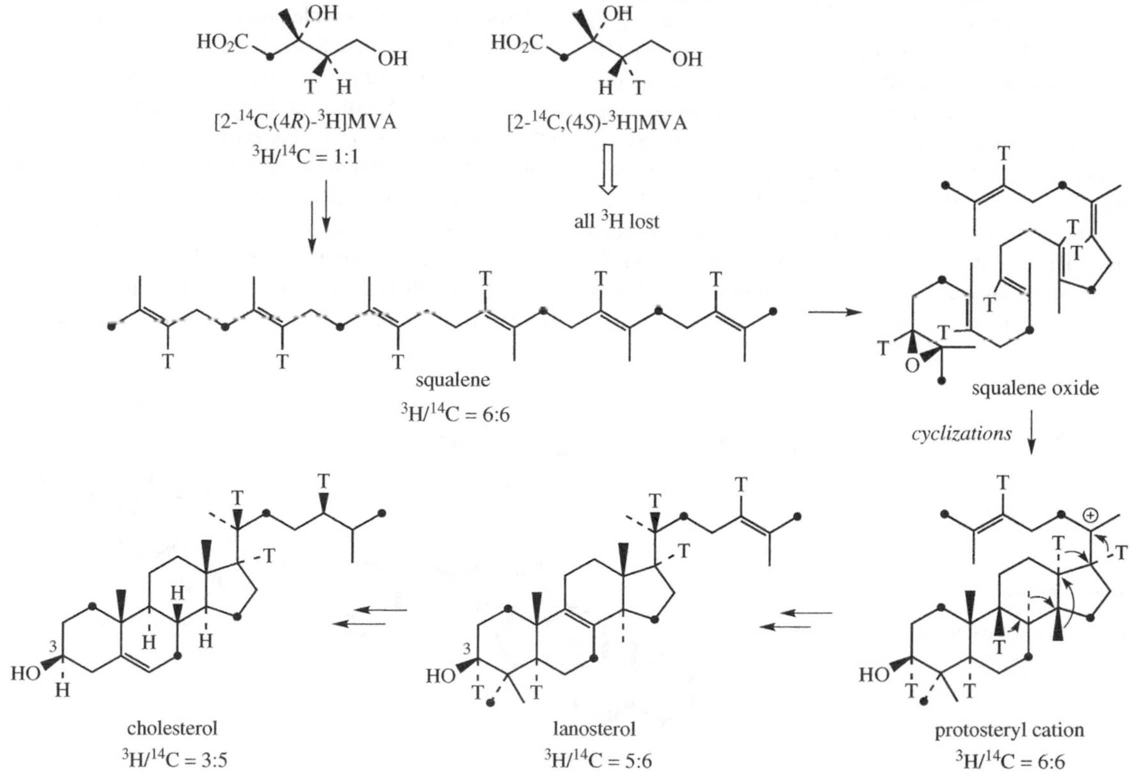

Figure 2.32

or loss of hydrogen (the indicator label) is established by comparison of the $^3H/^{14}C$ ratio. Establishing the position of tritium labelling is often easier than employing full degradation; simple chemical modification, e.g. oxidation of an alcohol to a ketone, might remove a hydrogen atom and allow the site of labelling to be deduced. The multiple-labelling technique provided an approach not only to the fine detail of cholesterol formation, but also to the mechanistic aspects for the mevalonate to squalene conversion, and the formation of the biosynthetic isoprene unit. In particular, the stereochemical aspects could be established by feeding experiments with a series of stereospecifically labelled mevalonates. Thus, in Figure 2.32, it can be seen that the 4-*pro-S* hydrogen is lost, whilst the 4-*pro-R* hydrogen is retained through subsequent reactions. The predicted rearrangements involving hydride migrations were also confirmed. A small detail in Figure 2.32 is the loss of tritium from C-3 in going from lanosterol to cholesterol; this was interpreted correctly as indicating the conversion probably involved a C-3 carbonyl group.

Labelling studies employing the non-radioactive isotope **carbon-13** (^{13}C) have made a huge impact on biosynthetic methodology. This has generally eliminated the need for degradations, and ^{13}C has effectively displaced the use of ^{14}C altogether. The natural abundance of ^{13}C is 1.1%, and modern NMR spectrometers can detect this in relatively small samples of a compound. Hence, any biosynthetic enrichment can be detected simply by an enhancement of the natural abundance signal. Furthermore, the position of labelling is immediately established from the spectrum via the assignment of the enhanced signal. The results of Figure 2.31 would be obtained in hours instead of years. The use of **double-labelled precursors** has made even greater impact. Most frequently,

researchers have employed $[1,2-^{13}C_2]$acetate, a compound in which each carbon atom is >99% ^{13}C. When this substrate is incorporated intact into a metabolite, the C_2 unit can be detected in the ^{13}C NMR spectrum because of spin–spin coupling between the two adjacent nuclides. Instead of enhanced signals, new doublets appear flanking the natural abundance signal. Statistically, it is unlikely that labelled molecules of precursor will be incorporated consecutively into a single molecule of product, so additional couplings that might complicate the spectrum will be negligible. Any acetate precursor that is cleaved during biosynthesis will end up enhancing a natural abundance signal, and not show any coupling. Cholesterol derived from $[^{13}C_2]$acetate would display couplings/enrichments as illustrated in Figure 2.33. There also exists the opportunity to employ $[^{13}CD_3]$acetate and to detect the deuterium by isotope-induced shifts in the ^{13}C spectrum. Note that $[^{14}C,^3H]$mevalonate in the previous paragraph refers to a mixture of ^{14}C-labelled and ^3H-labelled substrates; this is quite different from the double-labelled substrate considered here, and the NMR coupling observed is a consequence of the labelled atoms being adjacent in the same molecule.

Experiments with simple precursors such as acetate and mevalonate have their limitations, and further details of the pathway may require synthesis of more complex substrates in labelled form, and their observed transformation into final products. Thus, the biosynthesis of cholesterol from acetyl-CoA requires over 30 different enzymes; the transformation of lanosterol into cholesterol alone is a 19-step process requiring nine enzymes, some being used more than once. Often, the fine detail of the pathway can only be established via isolation and detailed study of the **enzymes** themselves. Approaches to enzymes have

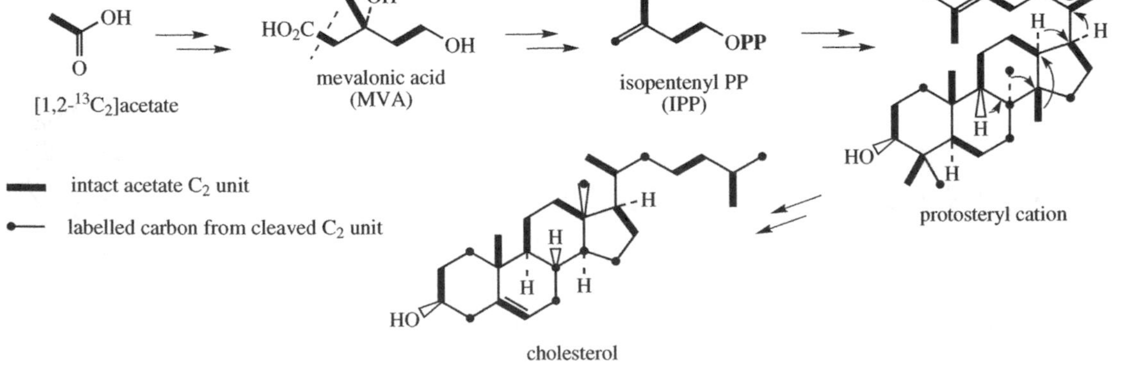

Figure 2.33

changed dramatically in recent years. Initially, it was necessary to obtain a crude protein extract from a tissue known to be capable of synthesizing the molecule under investigation. The protein extract was then subjected to extensive fractionation to provide a portion that contained the appropriate activity. Since enzymes only catalyse a limited chemical change, this necessitated supplying the extract with a substrate, usually in labelled form to facilitate assays, together with essential cofactors as required, and observing its conversion into the next intermediate on the pathway. Enzyme isolation and fractionation was tedious, and the amount of protein obtained was often very small indeed, limiting subsequent studies. Recent rapid progress in enzyme production is the result of significant advances in genetic techniques.

A **gene** is a segment of DNA that contains the information necessary for the synthesis of a particular protein/enzyme. Its identification initially required knowledge of the peptide sequence determined from the purified enzyme. Now, it is possible to search for likely genes in DNA sequences, produce them synthetically, and to express them in a suitable bacterium or yeast; to avoid complications with the normal biosynthetic machinery, this is usually different from the source organism. Recombinant proteins can then be tested for the expected enzyme activity. The search is facilitated by published gene sequences for similar enzymes, or by characteristic sequences that can be assigned to a particular class of enzyme, usually by the need to bind a specific cofactor, e.g. NADPH, SAM, cytochrome P-450, etc. In some organisms, especially bacteria and fungi, a group of genes involved in secondary metabolite biosynthesis may lie in close proximity as gene clusters. This again makes identification considerably easier, and can provide further information as the roles of all the genes are clarified. It should be appreciated that, although enzymes from different sources may catalyse the same reaction on the same substrate, the proteins may not have the same amino acid sequence, though they are likely to be identical or similar for most of sequence, especially the functional part. Hence, relevant gene sequences are also likely to vary, but, again, there is going to be a high proportion of identity or similarity.

These techniques allow accumulation of significant amounts of enzymes for mechanistic studies of the pathways. Various compounds can be tested as potential inhibitors or alternative substrates. For example, the stereochemistry of the side-chain at C-17 of the protosteryl cation in cholesterol biosynthesis was revised from 17α to 17β following studies with the enzyme oxidosqualene:lanosterol cyclase (Figure 2.34). When the oxygen analogue 20-oxa-2,3-oxidosqualene was used as substrate for the enzyme, the protosteryl cation analogue formed decomposed due to the presence of an oxygen function adjacent to the cationic centre. The stable ketone product obtained possessed a 17β-side-chain.

Modified genes can be synthesized to produce new proteins with specific changes to amino acid residues, thus shedding more light on the enzyme's mode of action. Specific genes can be damaged or deleted to prevent

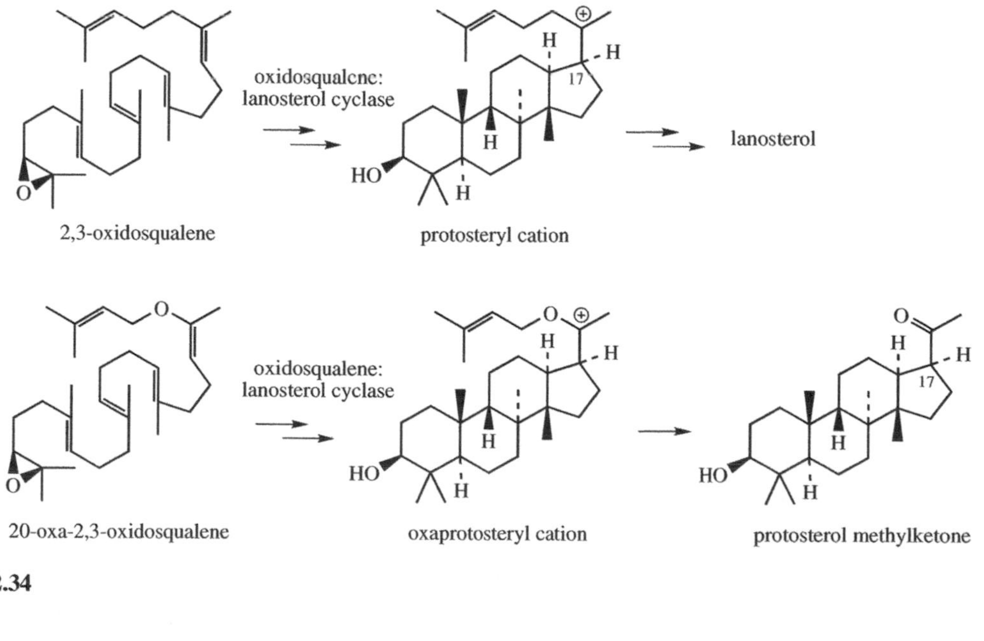

Figure 2.34

a particular enzyme being expressed in the organism. Genes from different organisms can be combined and expressed together so that an organism synthesizes abnormal combinations of enzyme activities, allowing production of modified products. However, an organism should not be viewed merely as a sackful of freely diffusible and always available enzymes; biosynthetic pathways are under sophisticated controls in which there may be restricted availability or localization of enzymes and/or substrates. Enzymes involved in the biosynthesis of many important secondary metabolites may be grouped together as enzyme complexes, or may form part of a multifunctional protein. Nevertheless, biosynthesis has indeed entered a new era; it has moved on from an information-gathering exercise to a technology in which the possibilities appear to be limitless.

FURTHER READING

Organic Chemistry, Mechanism

Dewick PM (2006) *Essentials of Organic Chemistry for Students of Pharmacy, Medicinal Chemistry and Biological Chemistry*. John Wiley & Sons, Ltd, Chichester.

McMurry J and Begley T (2005) *The Organic Chemistry of Biological Pathways*. Roberts, Englewood, CO.

Natural Products, Biosynthesis

Bugg TDH (2001) The development of mechanistic enzymology in the 20th century. *Nat Prod Rep* **18**, 465–493.

Stanforth SP (2006) *Natural Product Chemistry at a Glance*. Blackwell, Oxford.

Thomas R (2004) Biogenetic speculation and biosynthetic advances. *Nat Prod Rep* **21**, 224–248.

Vitamins

Amadasi A, Bertoldi M, Contestabile R, Bettati S, Cellini B, di Salvo ML, Voltattorni CB, Bossa F and Mozzarelli A (2007) Pyridoxal 5′-phosphate enzymes as targets for therapeutic agents. *Curr Med Chem* **14**, 1291–1324.

Bartlett MG (2003) Biochemistry of the water soluble vitamins: a lecture for first year pharmacy students. *Am J Pharm Educ* **67**, 1–7.

Begley TP (2006) Cofactor biosynthesis: an organic chemist's treasure trove. *Nat Prod Rep* **23**, 15–25.

Begley TP, Chatterjee A, Hanes JW, Hazra A and Ealick SE (2008) Cofactor biosynthesis – still yielding fascinating new biological chemistry. *Curr Opin Chem Biol* **12**, 118–125.

Kluger R and Tittmann K (2008) Thiamin diphosphate catalysis: enzymic and nonenzymic covalent intermediates. *Chem Rev* **108**, 1797–1833.

Rébeillé F, Ravanel S, Marquet A, Mendel RR, Smith AG and Warren MJ (2007) Roles of vitamins B_5, B_8, B_9, B_{12} and molybdenum cofactor at cellular and organismal levels. *Nat Prod Rep* **24**, 949–962.

Roje S (2007) Vitamin B biosynthesis in plants. *Phytochemistry* **68**, 1904–1921.

Smith AG, Croft MT, Moulin M and Webb ME (2007) Plants need their vitamins too. *Curr Opin Plant Biol* **10**, 266–275.

3

THE ACETATE PATHWAY: FATTY ACIDS AND POLYKETIDES

Polyketides constitute a large class of natural products grouped together on purely biosynthetic grounds. Their diverse structures can be explained as being derived from poly-β-keto chains, formed by coupling of acetic acid (C_2) units via condensation reactions, i.e.

$$n\text{CH}_3\text{CO}_2\text{H} \rightarrow -[\text{CH}_2\text{CO}]_n-$$

Included in such compounds are the fatty acids, polyacetylenes, prostaglandins, macrolide antibiotics and many aromatic compounds, e.g. anthraquinones and tetracyclines.

The formation of the poly-β-keto chain could be envisaged as a series of Claisen reactions, the reverse of which are involved in the β-oxidation sequence for the metabolism of fatty acids (see page 18). Thus, two molecules of acetyl-CoA could participate in a Claisen condensation giving acetoacetyl-CoA, and this reaction could be repeated to generate a poly-β-keto ester of appropriate chain length (Figure 3.1). However, a study of the enzymes involved in fatty acid biosynthesis showed this simple rationalization could not be correct and that a more complex series of reactions was operating. It is now known that fatty acid biosynthesis involves initial carboxylation of acetyl-CoA to malonyl-CoA, a reaction involving ATP, CO_2 (as bicarbonate, HCO_3^-), and the coenzyme biotin as the carrier of CO_2 (see page 17).

The conversion of acetyl-CoA into malonyl-CoA increases the acidity of the α-hydrogen atoms, thus providing a better nucleophile for the Claisen condensation. In the biosynthetic sequence, no acylated malonic acid derivatives are produced, and no label from [^{14}C]bicarbonate is incorporated, so the carboxyl group introduced into malonyl-CoA is simultaneously lost by a decarboxylation reaction during the Claisen condensation

(Figure 3.1). Accordingly, the carboxylation step helps to activate the α-carbon and facilitate Claisen condensation, and the carboxyl is immediately removed on completion of this task. An alternative rationalization is that decarboxylation of the malonyl ester is used to generate the acetyl enolate anion without any requirement for a strong base.

The pathways to fatty acids, macrolides, and aromatic polyketides branch early. The chain extension process of Figure 3.1 continues for aromatics, generating a highly reactive poly-β-keto chain that is stabilized by association with groups on the enzyme surface until chain assembly is complete and cyclization reactions occur. However, for fatty acids, the carbonyl groups are reduced before attachment of the next malonate group. Partial reduction processes, leading to a mixture of methylenes, hydroxyls, and carbonyls, are characteristic of macrolides (see page 68).

FATTY ACID SYNTHASE: SATURATED FATTY ACIDS

The processes of fatty acid biosynthesis are well studied and are known to be catalysed by the enzyme **fatty acid synthase** (FAS). FASs from various organisms show significant structural differences. In animals, FAS is a large multifunctional protein with seven discrete functional domains, providing all of the catalytic activities required. All domains are on a single polypeptide, encoded by a single gene, though the enzyme exists as a homodimer and requires both units for activity. Fungal FAS is also a multifunctional enzyme, but the seven component activities are distributed over two non-identical polypeptides α and β, and the enzyme is an $\alpha_6\beta_6$ dodecamer. The

Medicinal Natural Products: A Biosynthetic Approach. 3rd Edition   Paul Dewick
© 2009 John Wiley & Sons, Ltd

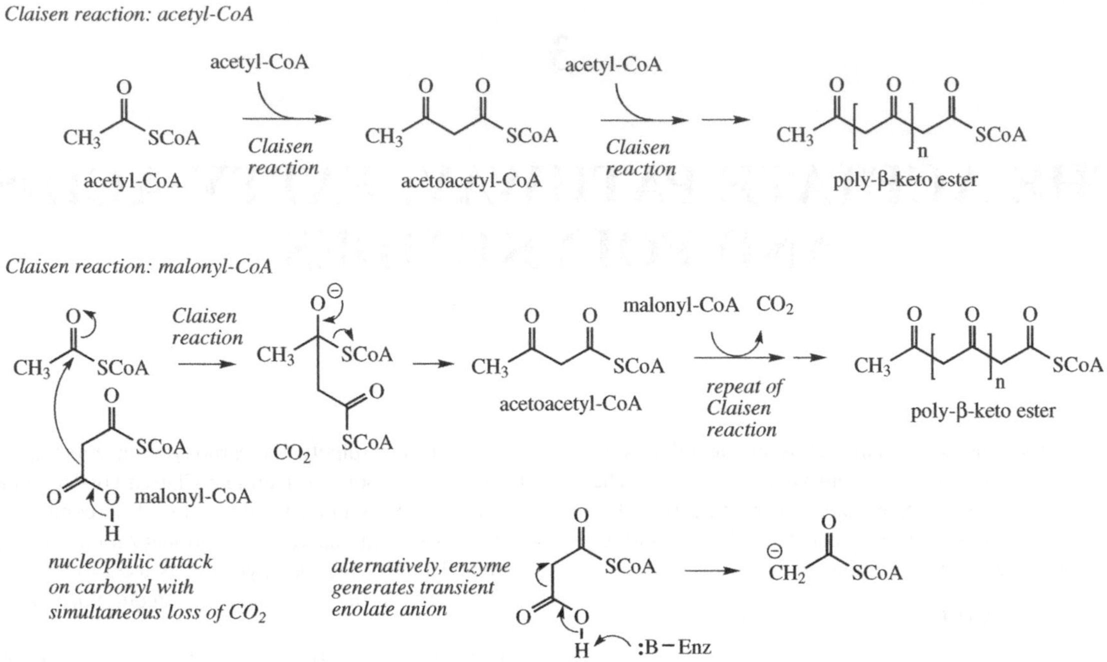

Claisen reaction: acetyl-CoA

Claisen reaction: malonyl-CoA

Figure 3.1

multifunctional protein systems in animals and fungi are termed **type I FASs**. Bacteria and plants possess a **type II FAS** that is an assembly of separable enzymes, encoded by seven different genes. Nevertheless, all of the different FAS systems perform effectively the same task employing the same mechanisms.

Acetyl-CoA and malonyl-CoA themselves are not involved in the condensation step: they are converted into enzyme-bound thioesters, the malonyl ester by means of an acyl carrier protein (ACP) (Figure 3.2). The Claisen reaction follows giving acetoacetyl-ACP (β-ketoacyl-ACP; R = H), which is reduced stereospecifically to the corresponding β-hydroxy ester, consuming NADPH in the reaction. Then follows elimination of water giving the *E* (*trans*) α,β-unsaturated ester. Reduction of the double bond again utilizes NADPH and generates a saturated acyl-ACP (fatty acyl-ACP; R = H) which is two carbon atoms longer than the starting material. This can feed back into the system, condensing again with malonyl-ACP and going through successive reduction, dehydration and reduction steps, gradually increasing the chain length by two carbon atoms for each cycle until the required chain length is obtained. At that point, the fatty acyl chain can be released as a fatty acyl-CoA or as the free acid. Fungi typically produce the CoA esters by a reversal of the loading reaction, whilst mammals release the free acid by the

action of a thioesterase (TE). The chain length actually elaborated is probably controlled by the specificity of the TE enzyme.

The FAS protein is known to contain an ACP binding site, and also an active-site cysteine residue in the ketosynthase (KS) domain. Acetyl and malonyl groups are successively transferred from coenzyme A esters and attached to the thiol groups of Cys and ACP respectively by the same malonyl/acetyl transferase (MAT; Figure 3.3). The Claisen condensation occurs (KS), and the processes of reduction (ketoreductase, KR), dehydration (dehydratase, DH), and reduction (enoyl reductase, ER) then occur whilst the growing chain is attached to ACP. The ACP carries a phosphopantetheine group exactly analogous to that in coenzyme A, and this provides a long flexible arm, enabling the growing fatty acid chain to reach the active site of each enzyme in the complex, allowing the different chemical reactions to be performed without releasing intermediates from the enzyme (compare polyketide synthesis on page 70 and peptide synthesis on page 438). The sequence of enzyme activities along the protein chain of the enzyme complex does not correspond with the order in which they are employed (see Figure 3.4). Then the chain is transferred to the thiol of Cys, and the process can continue. Making the process even more efficient, the dimeric animal FAS contains

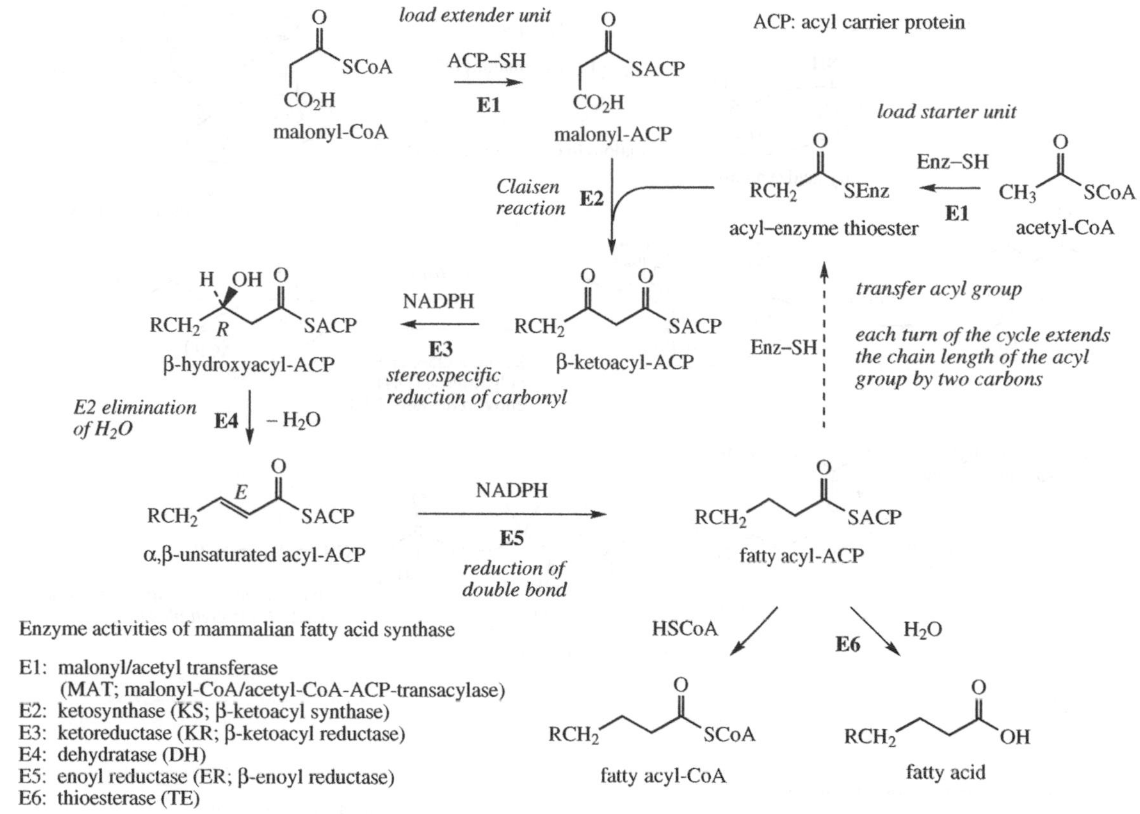

Figure 3.2

two catalytic centres and is able to generate two growing chains at the same time. The monomeric subunits are also arranged so that, although each monomer can catalyse the full set of reactions, there also exists the potential for utilizing domains on the other subunit for the initial reactions of substrate loading (MAT) and Claisen condensation (KS).

Although fungal FAS is also a type I multifunctional protein, there are some differences compared with the mammalian enzyme. Thus, the functional domains are arranged on two separate polypeptides (Figure 3.4), and the complex is a dodacamer rather than a dimer. Acetyl-CoA and malonyl-CoA are loaded onto the system by separate enzyme activities, namely acetyl transferase (AT) and malonyl/palmitoyl transferase (MPT), the latter also being involved in releasing the product palmitoyl-CoA from the ACP. A further enzyme activity is also included, phosphopantetheinyl transferase (PPT), which is required for activating the ACP by inserting the phosphopantetheine arm. The dissociated FAS II system found in most bacteria and plants consists of a series of discrete proteins, each of which catalyses an individual reaction of the fatty acid biosynthetic pathway. In some cases, two or more enzymes are able to perform the same chemical reaction, but have differing substrate specificities. For example, three distinct KS activities with characteristic chain-length specificities have been identified in plants. One condenses acetyl-CoA with malonyl-ACP, one is responsible for further elongation to C_{16}, and a third one elongates C_{16} to C_{18}. This allows both C_{16} and C_{18} metabolites to be tapped off for different functions. Unusually, the tuberculosis-causing bacillus *Mycobacterium tuberculosis* is known to possess both FAS I and FAS II systems. It uses FAS I to produce acyl chains in the range C_{14}–C_{16}, and FAS II to elongate these structures to mycolic acids, very long-chain fatty acids with 24–56 carbon atoms that are components of the mycobacterial cell wall. Because of the fundamental differences in mammalian FAS I and bacterial FAS II, there is considerable potential for exploiting these differences and developing selective inhibitors of fatty acid synthesis as antibacterial agents. Indeed, this approach has been

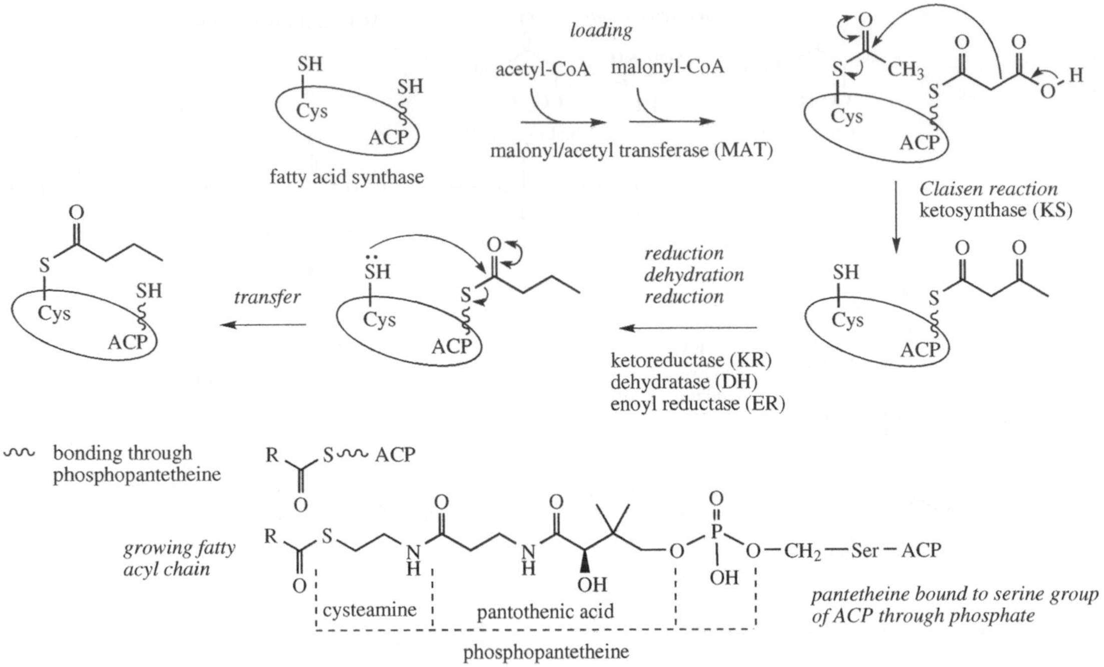

Figure 3.3

domain organization: mammalian FAS

— KS — MAT — DH — ER — KR — ACP — TE —

domain organization: fungal FAS

α   — ACP — KR — KS — PPT —

β   — AT — ER — DH — MPT —

ACP: acyl carrier protein
AT: acetyl transferase
DH: dehydratase
ER: enoyl reductase
KR: ketoreductase
KS: ketosynthase
MAT: malonyl/acetyl transferase
MPT: malonyl/palmitoyl transferase
PPT: phosphopantetheinyl transferase
TE: thioesterase

Figure 3.4

successful with agents such as isoniazid (antituberculosis) and triclosan (general biocide), both inhibitors of bacterial ER. Considerable effort is being invested in targeting FAS II in *Plasmodium falciparum*, the causative agent of malaria.

The combination of one acetate starter unit with seven malonate extender units would give the C_{16} fatty acid, palmitic acid, and with eight malonates the C_{18} fatty acid, stearic acid (Figure 3.5). Note that the two carbon atoms at the head of the chain (methyl end) are provided by acetate, not malonate, whilst the remainder are derived from malonate. However, malonate itself is produced by carboxylation of acetate. This means that all carbon atoms in the fatty acid originate from acetate, but malonate will only provide the C_2 chain extension units and not the C_2 starter group. The linear combination of acetate C_2 units as in Figure 3.2 explains why the common fatty acids are straight-chained and possess an even number of carbon atoms. Natural fatty acids may contain from 4 to 30, or even more, carbon atoms, the most abundant being those with 16 or 18 carbon atoms. Some naturally occurring fatty acids are shown in Figure 3.6. The rarer fatty acids containing an odd number of carbon atoms typically originate from incorporation of a different starter unit, e.g. propionic acid, or can arise by loss of one carbon from an even-numbered acid. Other structures (see page 53) can arise by utilizing different starter units and/or different extender units.

SCoA
acetyl-CoA
SCoA
CO_2H
malonyl-CoA
7 ×
CO_2H
starter unit
extender units
palmitic acid (16:0)
SCoA
acetyl-CoA
SCoA
CO_2H
malonyl-CoA
8 ×
CO_2H
starter unit
extender units
stearic acid (18:0)

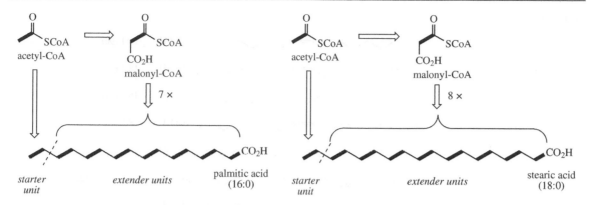

Figure 3.5

Fatty acids are mainly found in ester combination with glycerol in the form of triglycerides (Figure 3.7). These materials are called **fats** or **oils**, depending on whether they are solid or liquid at room temperature. If all three esterifying acids are the same, then the triglyceride is termed simple, whereas a mixed triglyceride is produced if two or more of the fatty acids are different. Most natural fats and oils are composed largely of mixed triglycerides. In this case, isomers can exist, including potential optical isomers, since the central carbon will become chiral if the primary alcohols of glycerol are esterified with different fatty acids. In practice, only one of each pair

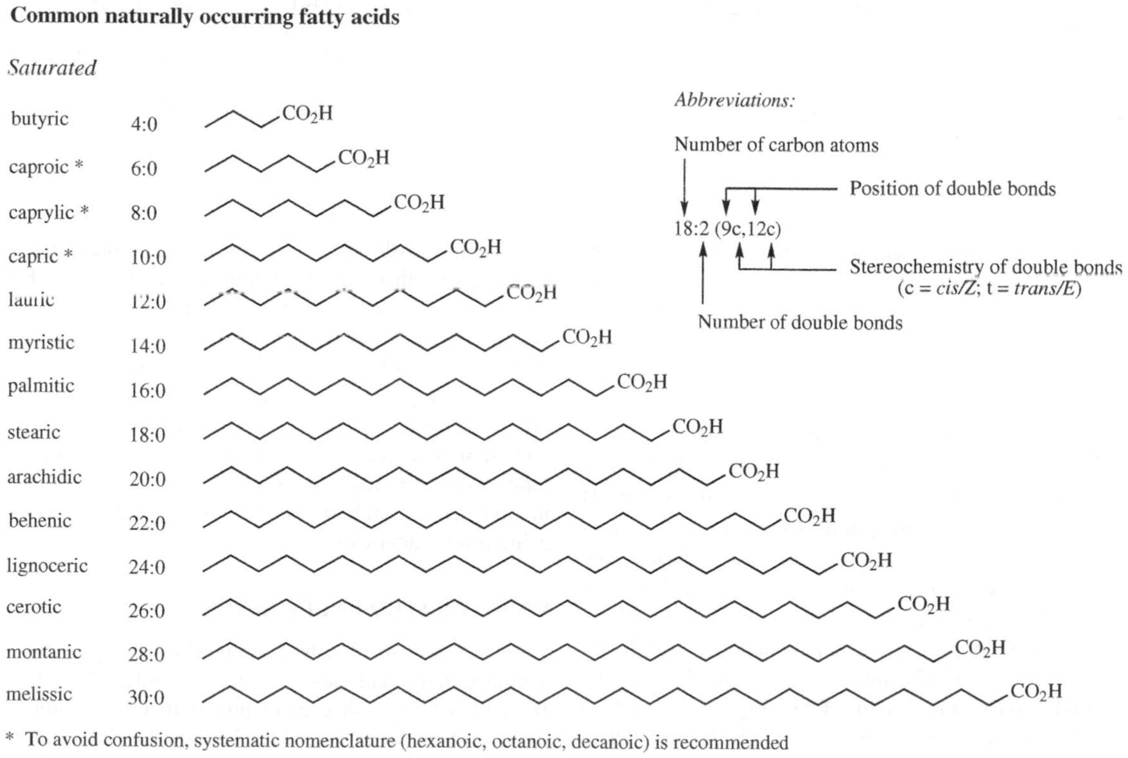

Common naturally occurring fatty acids
Saturated
butyric 4:0 CO_2H
caproic * 6:0 CO_2H
caprylic * 8:0 CO_2H
capric * 10:0 CO_2H
lauric 12:0 CO_2H
myristic 14:0 CO_2H
palmitic 16:0 CO_2H
stearic 18:0 CO_2H
arachidic 20:0 CO_2H
behenic 22:0 CO_2H
lignoceric 24:0 CO_2H
cerotic 26:0 CO_2H
montanic 28:0 CO_2H
melissic 30:0 CO_2H
Abbreviations:
Number of carbon atoms
Position of double bonds
18:2 (9c,12c)
Stereochemistry of double bonds
(c = cis/Z; t = trans/E)
Number of double bonds
* To avoid confusion, systematic nomenclature (hexanoic, octanoic, decanoic) is recommended

Figure 3.6 *(continued overleaf)*

Unsaturated

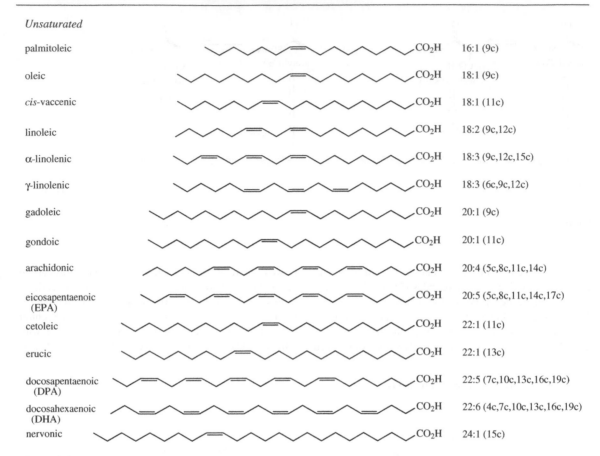

palmitoleic	16:1 (9c)
oleic	18:1 (9c)
cis-vaccenic	18:1 (11c)
linoleic	18:2 (9c,12c)
α-linolenic	18:3 (9c,12c,15c)
γ-linolenic	18:3 (6c,9c,12c)
gadoleic	20:1 (9c)
gondoic	20:1 (11c)
arachidonic	20:4 (5c,8c,11c,14c)
eicosapentaenoic (EPA)	20:5 (5c,8c,11c,14c,17c)
cetoleic	22:1 (11c)
erucic	22:1 (13c)
docosapentaenoic (DPA)	22:5 (7c,10c,13c,16c,19c)
docosahexaenoic (DHA)	22:6 (4c,7c,10c,13c,16c,19c)
nervonic	24:1 (15c)

Figure 3.6 (*continued*)

of enantiomers is formed in nature. Triglycerides are produced predominantly from glycerol 3-phosphate by esterification with fatty acyl-CoA residues, the phosphate being removed prior to the last esterification (Figure 3.7). The diacyl ester of glycerol 3-phosphate is also known as a **phosphatidic acid**, and is the basis of phospholipid structures. In these structures, the phosphate is also esterified with an alcohol, which is usually choline, ethanolamine, serine, or *myo*-inositol, e.g. **phosphatidyl choline** (Figure 3.8). **Phospholipids** are important structural components of cell membranes, and because of the polar and non-polar regions in their structure, they have detergent-like properties. They are also able to form liposomes, which have considerable potential as drug delivery systems. A particularly important natural phospholipid is **platelet-activating factor** (**PAF**; Figure 3.8), which resembles a phosphatidylcholine, though this compound possesses an ether linkage to a long-chain fatty alcohol, usually hexadecanol, rather than an ester linkage.

The central hydroxyl of glycerol is esterified, but to acetic acid rather than to a long-chain fatty acid. PAF functions at nanomolar concentrations, activates blood platelets, and contributes to diverse biological effects, including thrombosis, inflammatory reactions, allergies, and tissue rejection. Long-chain alcohols are reduction products from fatty acyl-CoA esters; they also feature in natural **waxes**. Waxes are complex mixtures of esters of long-chain fatty acids, usually C_{20}–C_{24}, with long-chain monohydric alcohols or sterols. The alcohol is esterified using a fatty acyl-CoA.

UNSATURATED FATTY ACIDS

Animal fats contain a high proportion of glycerides of saturated fatty acids and tend to be solids, whilst those from plants and fish contain predominantly unsaturated fatty acid esters and tend to be liquids. Some of the common naturally occurring unsaturated fatty acids are also included in Figure 3.6. A convenient shorthand

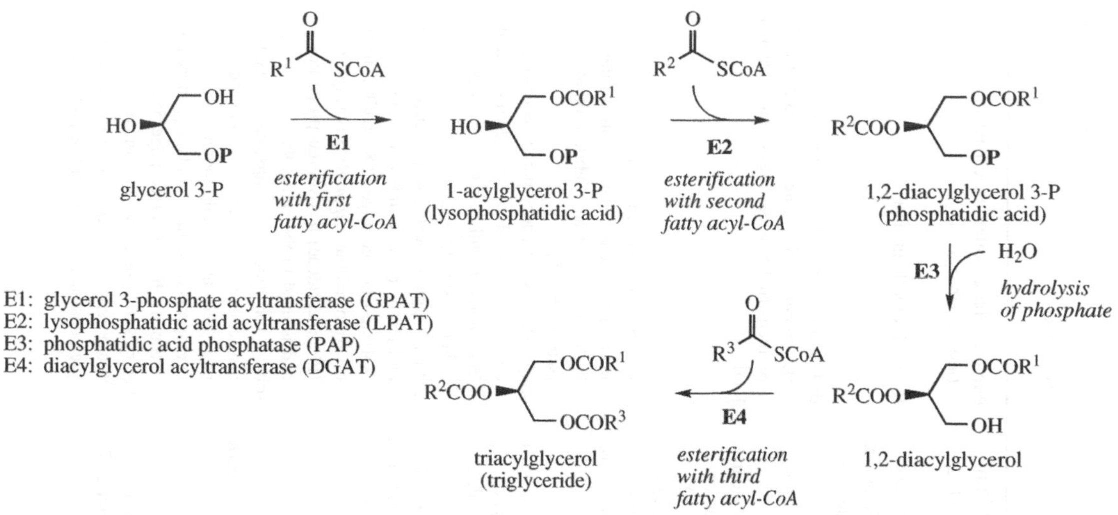
glycerol 3-P
esterification with first fatty acyl-CoA
1-acylglycerol 3-P (lysophosphatidic acid)
esterification with second fatty acyl-CoA
1,2-diacylglycerol 3-P (phosphatidic acid)
hydrolysis of phosphate
E1: glycerol 3-phosphate acyltransferase (GPAT)
E2: lysophosphatidic acid acyltransferase (LPAT)
E3: phosphatidic acid phosphatase (PAP)
E4: diacylglycerol acyltransferase (DGAT)
triacylglycerol (triglyceride)
esterification with third fatty acyl-CoA
1,2-diacylglycerol

Figure 3.7

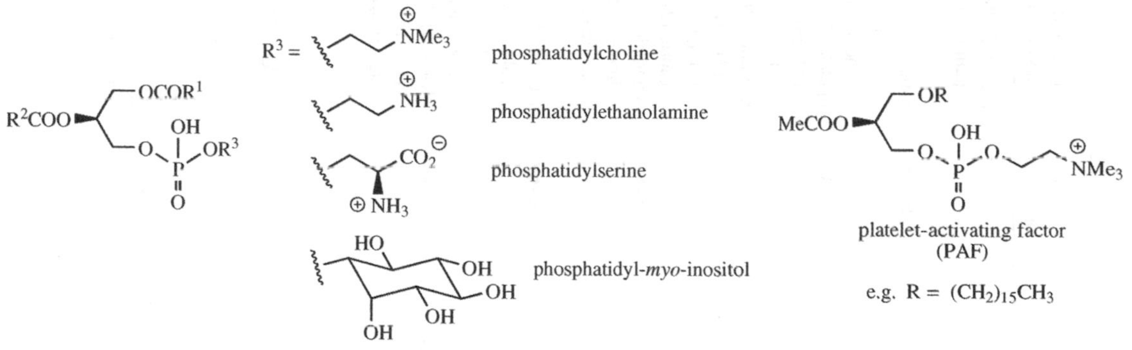
phosphatidylcholine
phosphatidylethanolamine
phosphatidylserine
phosphatidyl-*myo*-inositol
platelet-activating factor (PAF)
e.g. R = $(CH_2)_{15}CH_3$

Figure 3.8

representation for fatty acids indicating chain length with number, position and stereochemistry of double bonds is also presented in Figure 3.6. A less systematic numbering starting from the methyl (the ω end) may also be encountered. Major groups of fatty acids are designated ω–3 (omega-3), ω–6 (omega-6), ω–9 (omega-9), etc. (or sometimes n–3, n–6, n–9), if there is a double bond that number of carbon atoms from the methyl terminus. This has some value in relating structures when an unsaturated fatty acid is biosynthetically elongated from the carboxyl end as during prostaglandin biosynthesis (see page 50). Double bonds at position 9 are common, but unsaturation can occur at other positions in the chain. Polyunsaturated fatty acids tend to have their double bonds in a non-conjugated array as a repeating unit —$(CH{=}CHCH_2)_n$—. In virtually all cases, the stereochemistry of the double bond is *Z* (*cis*), thus introducing a 'bend' into the alkyl chain. This interferes with the close association and aggregation of molecules that is possible in saturated structures and helps to maintain the fluidity in oils and cellular membranes.

Fats and oils represent long-term stores of energy for most organisms, being subjected to oxidative metabolism as required. Major oils which are produced commercially for use as foods, toiletries, medicinals, or pharmaceutical formulation aids are listed in Table 3.1. Typical fatty acid analyses are shown, though it must be appreciated that these figures can vary quite widely. For instance, plant oils show significant variation according to the climatic conditions under which the plant was grown. In colder climates, a higher proportion of polyunsaturated fatty acids is produced, so that the plant can maintain the fluidity

Table 3.1 Fixed oils and fats.[a,b,c]

Oil	Source	Part used	Oil content (%)	Typical fatty acid composition (%)	Uses, notes
Almond	*Prunus amygdalus* var. *dulcis*, or var. *amara* (Rosaceae)	seed	40–55	oleic (62–86), linoleic (7–30), palmitic (4–9), stearic (1–2)	emollient base, toiletries, carrier oil (aromatherapy)
Arachis (groundnut, peanut)	*Arachis hypogaea* (Leguminosae/Fabaceae)	seed	45–55	oleic (35–72), linoleic (13–43), palmitic (7–16), stearic (1–7), behenic (1–5), arachidic (1–3)	food oil, emollient base
Borage (starflower)	*Borago officinalis* (Boraginaceae)	seed	28–35	linoleic (38), γ-linolenic (23–26), oleic (16), palmitic (11)	dietary supplement for γ-linolenic (gamolenic) acid content in treatment of premenstrual tension, breast pain, eczema
Butterfat	cow *Bos taurus* (Bovidae)	milk	2–5	palmitic (29), oleic (28), stearic (13), myristic (12), butyric (4), lauric (3), caproic (2), capric (2), palmitoleic (2)	food
Castor	*Ricinus communis* (Euphorbiaceae)	seed	35–55	ricinoleic (80–90), oleic (4–9), linoleic (2–7), palmitic (2–3), stearic (2–3)	emollient base, purgative, soap manufacture; castor seeds contain the highly toxic, but heat-labile protein ricin (see page 435)
Coconut	*Cocos nucifera* (Palmae/Arecaceae)	seed kernel	65–68	lauric (43–53), myristic (15–21), palmitic (7–11), caprylic (5–10), capric (5–10), oleic (6–8), stearic (2–4)	soaps, shampoos; fractionated coconut oil containing only short- to medium-length fatty acids (mainly caprylic and capric) is a dietary supplement
Cod-liver	cod *Gadus morrhua* (Gadidae)	fresh liver	50	oleic (24), DHA (14), palmitic (11), EPA (6), palmitoleic (7), stearic (4), myristic (3)	dietary supplement due to presence of EPA and DHA, plus vitamins A (see page 304) and D (see page 258); halibut-liver oil from halibut *Hippoglossus vulgaris* (Pleuronectideae) has similar properties and is used in the same way
Cottonseed	*Gossypium hirsutum* (Malvaceae)	seed	15–36	linoleic (33–58), palmitic (17–29), oleic (13–44), stearic (1–4)	solvent for injections, soaps; cotton seeds also contain 1.1–1.3% gossypol (see page 220) and small amounts of cyclopropenoid fatty acids, e.g. sterculic and malvalic acids (see page 55)
Evening primrose	*Oenothera biennis* (Onagraceae)	seed	24	linoleic (65–80), γ-linolenic (7–14), oleic (9), palmitic (7)	dietary supplement for γ-linolenic (gamolenic) acid content in treatment of premenstrual tension, breast pain, eczema

				Fatty acid composition (%)	Uses
Fish	various, including herring *Clupea harengus* (Clupeidae) (Europe) and menhaden *Brevoortia* spp. (Clupeidae) (N America)	whole fish – by-product of fishing and fish-meal industry	up to 20% (herring)	herring: cetoleic (7–30), oleic (9–25), gondoic (7–24), palmitic (10–19), EPA (4–15), palmitoleic (6–12), myristic (5–8), DHA (2–8), stearic (1–2)	dietary supplement for EPA and DHA (see page 49); refined and concentrated oils are produced containing higher amounts of EPA (15–30%) and DHA (10–20%)
Honesty	*Lunaria annua* (Cruciferae/Brassicaceae)	seed	30–40	erucic (43), nervonic (25), oleic (24)	nervonic acid is being investigated for the treatment of multiple sclerosis; the disease is characterized by a deficiency in nervonic acid
Lard	pig *Sus scrofa* (Suidae)	abdominal fat		oleic (45), palmitic (25), stearic (12), linoleic (10), palmitoleic (3)	food
Linseed (flaxseed)	*Linum usitatissimum* (Linaceae)	seed	35–44	α-linolenic (30–60), oleic (39), linoleic (15), palmitic (7), stearic (4)	liniments, dietary supplement for α-linolenic acid content; formerly the basis of paints, reacting with oxygen, polymerizing, and drying to a hard film
Maize (corn)	*Zea mays* (Graminae/Poaceae)	embryo	33–39	linoleic (34–62), oleic (19–50), palmitic (8–19), stearic (0–4)	food oil, dietary supplement, solvent for injections
Olive	*Olea europaea* (Oleaceae)	fruits	15–40	oleic (56–85), palmitic (8–20), linoleic (4–20), stearic (1–4)	food oil, emollient base
Palm kernel	*Elaeis guineensis* (Palmae/Arecaceae)	kernel	45–50	lauric (40–52), myristic (14–18), oleic (9–16), palmitoleic (6–10), caprylic (3–6), capric (3–5), stearic (1–4), linoleic (1–3)	soaps; fractionated palm oil is a solid obtained by fractionation and hydrogenation and is used as a suppository base
Rapeseed	*Brassica napus* (Cruciferae/Brassicaceae)	seed	40–50	erucic (30–60), oleic (9–25), linoleic (11–25), gadoleic (5–15), α-linolenic (5–12), palmitic (0–5)	food oil, using varieties producing lower levels of erucic acid where the main components are now oleic (48–60%), linoleic (18–30%), α-linolenic (6–14%), and palmitic (3–6%) acids; erucic acid is known to accumulate and cause lesions in heart muscle; erucic acid is used as a plasticizer in PVC clingfilm
Sesame	*Sesamum indicum* (Pedaliaceae)	seed	44–54	oleic (35–50), linoleic (35–50), palmitic (7–12), stearic (4–6)	food oil, soaps, solvent for injections, carrier oil (aromatherapy)

(continued overleaf)

Table 3.1 (*continued*)

Oil	Source	Part used	Oil content (%)	Typical fatty acid composition (%)	Uses, notes
Soya (soybean)	*Glycine max* (Leguminosae/ Fabaceae)	seed	18–20	linoleic (44–62), oleic (19–30), palmitic (7–14), α-linolenic (4–11), stearic (1–5)	food oil, dietary supplement, carrier oil (aromatherapy); soya oil contains substantial amounts of the sterols sitosterol and stigmasterol (see page 255)
Suet (mutton tallow)	sheep *Ovis aries* (Bovidae)	abdominal fat		stearic (32), oleic (31), palmitic (27), myristic (6)	foods
Suet (beef tallow)	cow *Bos taurus* (Bovidae)	abdominal fat		oleic (48), palmitic (27), palmitoleic (11), stearic (7), myristic (3)	foods
Sunflower	*Helianthus annuus* (Compositae/ Asteraceae)	seed	22–36	linoleic (50–70), oleic (20–40), palmitic (3–10), stearic (1–10)	food oil, carrier oil (aromatherapy)
Theobroma	*Theobroma cacao* (Sterculiaceae)	kernel	35–50	oleic (35), stearic (35), palmitic (26), linoleic (3)	suppository base, chocolate manufacture; theobroma oil (cocoa butter) is a solid

[a]The oil yields and fatty acid compositions given in the above table are typical values, and can vary widely. The quality of an oil is determined principally by its fatty acid analysis. Structures of the fatty acids are shown in Figure 3.6 (see page 43, 44)

[b]The term fat or oil has no precise significance, merely describing whether the material is a solid (fat) or liquid (oil) at room temperature. Most commercial oils are obtained from plant sources, particularly seeds and fruits, and the oil is extracted by cold or hot expression, or less commonly by solvent extraction with hexane. The crude oil is then refined by filtration, steaming, neutralization to remove free acids, washing and bleaching as appropriate. Many food oils are then partially hydrogenated to produce semi-solid fats. Animal fats and fish oils are usually extracted by steaming, the higher temperature deactivating enzymes which would otherwise begin to hydrolyse the glycerides.

[c]Oils and fats feature as important food components and cooking oils, some 80% of commercial production being used as human food, whilst animal feeds account for another 6%. Most of the remaining production is used as the basis of soaps, detergents, and pharmaceutical creams and ointments. A number of oils are used as diluents (carrier or base oils) for the volatile oils employed in aromatherapy.

of its storage fats and membranes. The melting points of these materials depend on the relative proportions of the various fatty acids, reflecting primarily the chain length and the amount of unsaturation in the chain. Saturation and increasing chain length in the fatty acids gives a more solid fat at room temperature. Thus, butterfat and cocoa butter (theobroma oil) contain a relatively high proportion of saturated fatty acids and are solids. Palm kernel and coconut oils are both semi-solids having a high concentration of the saturated C_{12} acid **lauric acid**. A characteristic feature of olive oil is its very high **oleic acid** (18:1) content, whilst rapeseed oil possesses high concentrations of long-chain C_{20} and C_{22} fatty acids, e.g. **erucic acid** (22:1). Typical fatty acids in fish oils have high unsaturation and also long chain lengths, e.g. **eicosapentaenoic acid (EPA)** (20:5) and **docosahexaenoic acid (DHA)** (22:6) in cod liver oil.

Unsaturated fatty acids can arise by more than one biosynthetic route, but in most organisms the common mechanism is by desaturation of the corresponding alkanoic acid, with further desaturation in subsequent steps. Most eukaryotic organisms possess a Δ^9-desaturase enzyme that introduces a *cis* double bond into a saturated fatty acid, requiring O_2 and NADPH or NADH cofactors. The mechanism of desaturation does not involve any intermediates hydroxylated at C-9 or C-10, and the requirement for O_2 is as an acceptor at the end of an electron transport chain. A stearoyl (C_{18}) thioester is the usual substrate giving an oleoyl derivative (Figure 3.9), coenzyme A esters being utilized by animal and fungal enzymes, and ACP esters by plant systems. In some systems, desaturation may take place on fatty acids bound as lipids or phospholipids.

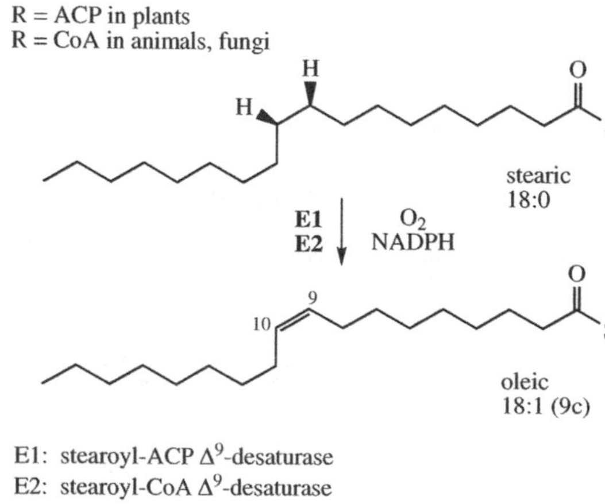

R = ACP in plants
R = CoA in animals, fungi

E1: stearoyl-ACP Δ^9-desaturase
E2: stearoyl-CoA Δ^9-desaturase

Figure 3.9

The position of further desaturation then depends very much on the organism. Non-mammalian enzymes tend to introduce additional double bonds between the existing double bond and the methyl terminus, e.g. oleic acid → linoleic acid → α-linolenic acid (Figure 3.10). Animals introduce new double bonds towards the carboxyl group. They also lack the Δ^{12} and Δ^{15} desaturase enzymes, so, although **linoleic acid** and **α-linolenic acid** are necessary for the synthesis of polyunsaturated acids that lead to prostaglandins (see page 59) and leukotrienes (see page 64), animals must obtain these materials in the diet, mainly from plants. Δ^6-Desaturation (towards the carboxyl) of linoleic acid leads to γ-linolenic acid, whilst analogous desaturation of α-linolenic acid gives stearidonic acid. Prostaglandins are derived from C_{20} polyunsaturated fatty acid precursors, so addition of two extra carbon atoms is required. This is achieved by an elongase enzyme acting upon γ-linolenic acid, giving **dihomo-γ-linolenic acid** ($\Delta^{8,11,14}$-eicosatrienoic acid). The additional two carbon atoms derive from malonate, by the usual chain extension mechanism. Dihomo-γ-linolenic acid is the precursor of prostaglandins in the 'one' series; the 'two' series derives from **arachidonic acid** ($\Delta^{5,8,11,14}$-eicosatetraenoic acid), formed by further Δ^5-desaturation, again towards the carboxyl terminus. The 'three' series of prostaglandins are derivatives of **$\Delta^{5,8,11,14,17}$-eicosapentaenoic acid (EPA)**. This is an analogue of arachidonic acid formed from α-linolenic acid via stearidonic acid with a similar chain extension process and Δ^5-desaturation.

Further chain extension of EPA gives **docosapentaenoic acid (DPA)**, and then Δ^4-desaturation gives **DHA**. DHA is vital for proper visual and neurological development in infants, and deficiency has been associated with cognitive decline and the onset of Alzheimer's disease in adults. Thus, a range of metabolites necessary for good health, including prostaglandins, leukotrienes, and these long-chain polyunsaturated fatty acids, are produced from the plant fatty acids linoleic acid and α-linolenic acid, which have to be obtained in the diet. Accordingly, these plant fatty acids are referred to as '**essential fatty acids**' (**EFAs**). Further, since beneficial fatty acids, including α-linolenic acid, EPA, DPA, and DHA, have a double bond three carbon atoms from the methyl end of the chain, they are grouped together under the term ω-3 fatty acids (**omega-3 fatty acids**). Marine fish represent a major source of the nutritionally relevant longer chain fatty acids, especially EPA and DHA, which are not found in seed oils of higher plants. Regular consumption of oily fish, e.g. herring, tuna, mackerel, or use of fish oil supplements, is claimed to reduce the risk of heart attacks and atherosclerosis. Fish obtain most of

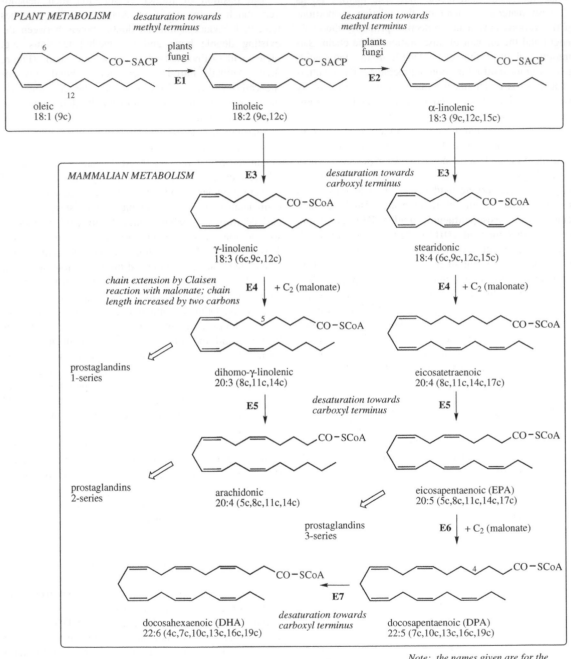

E1: Δ^{12}-desaturase E3: Δ^{6}-desaturase E5: Δ^{5}-desaturase E7: Δ^{4}-desaturase
E2: Δ^{15}-desaturase E4: C_{18} elongase E6: C_{20} elongase

Note: the names given are for the appropriate fatty acid; the structures shown are the thioesters involved in the conversions

Figure 3.10

these fatty acids by consumption of marine microalgae, which are considered to be the primary producers. It has also been found that these microalgae utilize another approach to synthesize unsaturated fatty acids, involving a multifunctional complex analogous to polyketide synthases (PKSs; see page 67) and a sequence that does not require desaturases and elongases. Some of these microalgae, e.g. *Crypthecodinium cohnii* and *Schizochytrium* spp., are currently exploited for the commercial production of DHA-enriched oils, which are also more palatable than fish-derived products. Limited fish stocks and expensive microbial culture processes encourage researchers to consider alternative plant-based production of these beneficial long-chain polyunsaturated fatty acids. There has already been considerable success in producing transgenic oilseed crops that synthesize fatty acids such as arachidonic acid, EPA, and DHA. Arachidonic acid itself has not been found in higher plants, but does occur in some algae, mosses, and ferns. All of the omega-3 fatty

acids are beneficial to health, but the plant and fish oils provide different compounds, and the distinction between the two groups should be recognized.

Although most plant-derived oils contain high amounts of unsaturated fatty acid glycerides, including those of linoleic and α-linolenic acids, the conversion of linoleic acid into **γ-linolenic acid** can be blocked or inhibited in certain conditions in humans. This restricts synthesis of prostaglandins. In such cases, the use of food supplements, e.g. **evening primrose oil** from *Oenothera biennis* (Onagraceae) or **borage oil** from *Borago officinalis* (Boraginaceae), which are rich in γ-linolenic esters (see Table 3.1), can be valuable and help in the disorder. These plants are somewhat unusual in their ability to desaturate linoleic esters towards the carboxyl terminus via a Δ^6-desaturase, rather than towards the methyl terminus as is more common in plants. Expression of Δ^6-desaturase genes, either from plants or suitable fungi, can be used

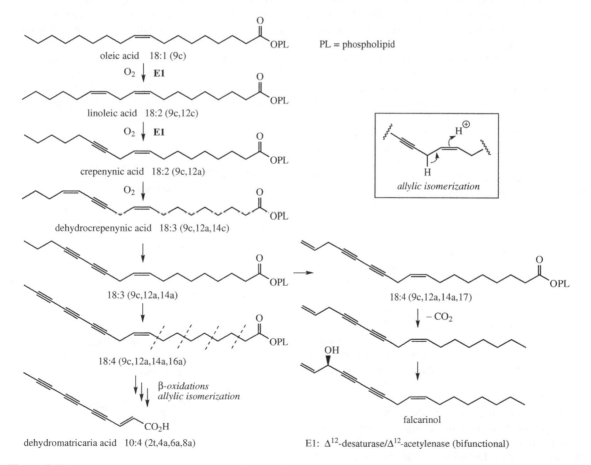

Figure 3.11

to increase the production of γ-linolenic and stearidonic acids in hosts such as soybeans.

In the vast majority of unsaturated fatty acids, the double bonds have the *Z*/*cis* configuration. Fatty acids with *trans* double bonds do occur naturally at relatively low levels in meat and dairy products as a by-product of fermentation in ruminant animals. However, *trans*-unsaturation may be introduced during the partial hydrogenation of polyunsaturated fats that is commonly practised during food processing to produce semi-solid fats from oils. There now seems to be good correlation between the consumption of *trans*-fats and the occurrence of coronary heart disease and atherosclerosis, though precise mechanisms are yet to be determined. Nevertheless, levels of *trans*-fats in foodstuffs are now closely monitored.

Many unsaturated compounds found in nature contain one or more acetylenic bonds, and these are predominantly produced by further desaturation of olefinic systems in fatty acid-derived molecules. They are surprisingly widespread in nature and are found in many organisms, but they are especially common in plants of the Compositae/Asteraceae, the Umbelliferae/Apiaceae, and fungi of the group Basidiomycetes. These compounds tend to be highly unstable and some are even explosive if sufficient amounts are accumulated. Since only very small amounts are present in plants, this does not present any widespread hazard. Whilst fatty acids containing several double bonds usually have these in a non-conjugated array, molecules containing triple bonds tend to possess conjugated unsaturation. This gives the compounds intense and highly characteristic UV spectra, which aids their detection and isolation.

The processes of desaturation are exemplified in Figure 3.11, in which oleic acid (probably bound as a phospholipid) features as a precursor of **crepenynic acid** and **dehydrocrepenynic acid**. A bifunctional desaturase/acetylenase system catalysing these two steps has been characterized in the plant *Crepis alpina* (Compositae/Asteraceae). The acetylenic bond is now indicated by 'a' in the semi-systematic shorthand nomenclature. Chain shortening by β-oxidation (see page 18) is often a feature of these pathways, and formation of the C_{10} acetylenic acid **dehydromatricaria acid** proceeds through C_{18} intermediates, losing eight carbon atoms, presumably via four β-oxidations. In the latter part of the pathway, the *Z* double bond from oleic acid moves into conjugation with the polyacetylene chain via an allylic isomerization, giving the more favoured *E*-configuration. Loss of the carboxyl function is also encountered, and **falcarinol** is an example of such structures, most likely derived via dehydrocrepenynic

acid. Some noteworthy acetylenic structures (though they are no longer acids and components of fats) are given in Figure 3.12. **Cicutoxin** from the water hemlock (*Cicuta virosa*; Umbelliferae/Apiaceae) and **oenanthotoxin** from the hemlock water dropwort (*Oenanthe crocata*; Umbelliferae/Apiaceae) are extremely toxic to mammals, causing persistent vomiting and convulsions, leading to respiratory paralysis. Ingestion of the roots of these plants may frequently lead to fatal poisoning. **Falcarinol** is a constituent of *Falcaria vulgaris* (Umbelliferae/Apiaceae), *Oenanthe crocata*, *Hedera helix* (Araliaceae), and several other plants, and is known to cause contact dermatitis in certain individuals when the plants are handled. Falcarinol (sometimes called panaxynol) and the structurally related **panaxytriol** are also characteristic polyacetylene components of ginseng (*Panax ginseng*; Araliaceae; see page 245). **Wyerone** from the broad bean (*Vicia faba*; Leguminosae/Fabaceae) has antifungal properties, and its structure exemplifies how the original straight chain may be cross-linked to produce a ring system. The furan ring is believed to originate from a conjugated diyne.

Primary amides of unsaturated fatty acids have been characterized in humans and other mammals, and although their biological role is not fully understood, they may represent a group of important signalling molecules.

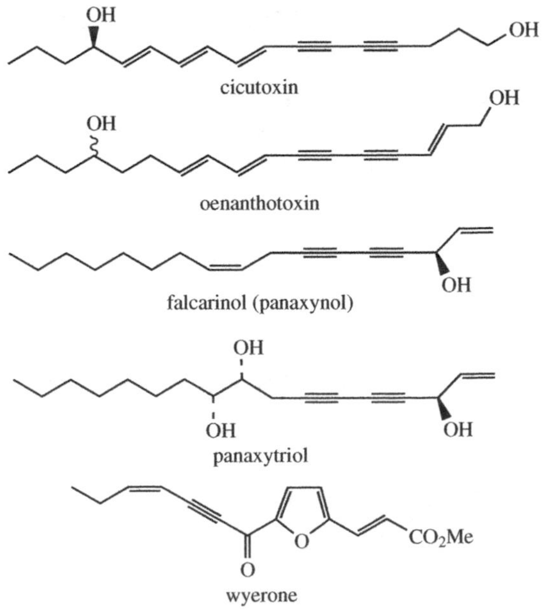

cicutoxin

oenanthotoxin

falcarinol (panaxynol)

panaxytriol

wyerone

Figure 3.12

Oleamide, the simple amide of oleic acid, has been shown to be a sleep-inducing lipid, and the amide of erucic acid, **erucamide**, stimulates the growth of blood vessels. The ethanolamide of arachidonic acid, **anandamide**, appears to be the natural ligand for receptors to which cannabinoids bind; this and related structures will be considered later (see page 122). The herbal preparation echinacea [Box 3.1] is derived from the roots

of *Echinacea purpurea* (Compositae/Asteraceae) and is used for its immunostimulant properties, particularly as a prophylactic and treatment for the common cold. At least some of its activity arises from a series of alkamides (also termed alkylamides), amides of polyunsaturated acids with isobutylamine. These acids are predominantly C_{11} and C_{12} diene-diynes (Figure 3.13).

Box 3.1

Echinacea

Echinacea consists of the dried roots of *Echinacea purpurea, Echinacea angustifolia,* or *Echinacea pallida* (Compositae/ Asteraceae), herbaceous perennial plants indigenous to North America, and widely cultivated for their large daisy-like flowers, which are usually purple or pink. Herbal preparations containing the dried root, or extracts derived from it, are hugely popular, being promoted as immunostimulants, particularly as prophylactics and treatments for bacterial and viral infections, e.g. the common cold. Tests have validated stimulation of the immune response, though the origins of this activity cannot be ascribed to any specific substance. Activity has variously been assigned to lipophilic alkamides, polar caffeic acid derivatives, high molecular weight polysaccharide material, or to a combination of these. Compounds in each group have been demonstrated to possess some pertinent activity, e.g. immunostimulatory, anti-inflammatory, antibacterial, or antiviral effects.

The alkamides comprise a complex mixture of unsaturated fatty acids as amides with 2-methylpropanamine (isobutylamine) or 2-methylbutanamine, amines which are probably decarboxylation products from valine and isoleucine respectively. The acid portions are predominantly C_{11} and C_{12} diene-diynes or tetraenes (Figure 3.13). These compounds are found throughout the plant, though relative proportions of individual components vary considerably. The root of *Echinacea purpurea* contains at least 12 alkamides (about 0.6%), of which C_{12} diene-diynes predominate; levels of these compounds fall significantly during drying and storage. Caffeic acid derivatives present include caffeic acid (see page 149), chlorogenic acid (5-*O*-caffeoylquinic acid, see page 150), caftaric acid (2-*O*-caffeoyltartaric acid), and cichoric acid (2,3-di-*O*-caffeoyltartaric acid) (Figure 3.13). Cichoric acid is a major component (0.6–2.1%) in *Echinacea purpurea*, but only minor in the other species.

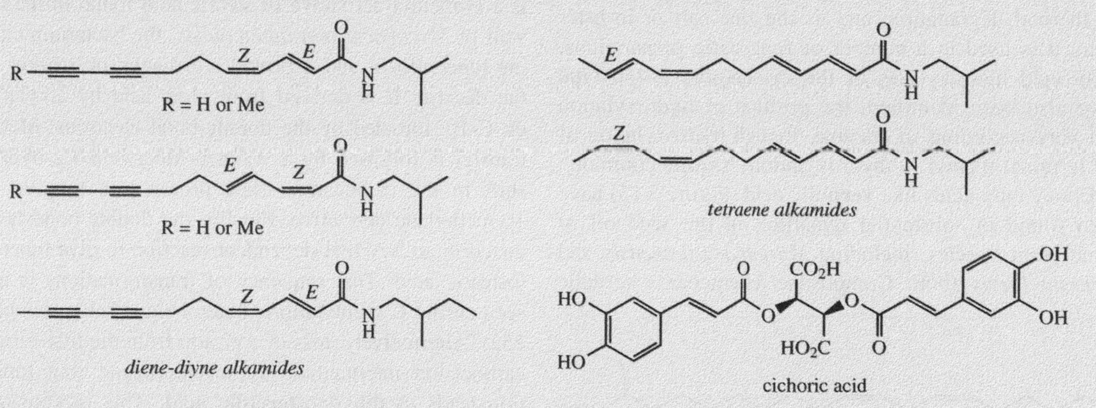

diene-diyne alkamides

tetraene alkamides

cichoric acid

Figure 3.13

UNCOMMON FATTY ACIDS

Most fatty acids are undoubtedly primary metabolites. However, a considerable amount of structural diversity is encountered in this group of compounds, with some structures having a rather limited natural distribution, so it is more appropriate to think of these as secondary metabolites. The distinction is actually unnecessary, since it is impossible to consider one group without the other; this is a typical 'grey' area. Some structures arise by further modification of the basic straight-chain systems

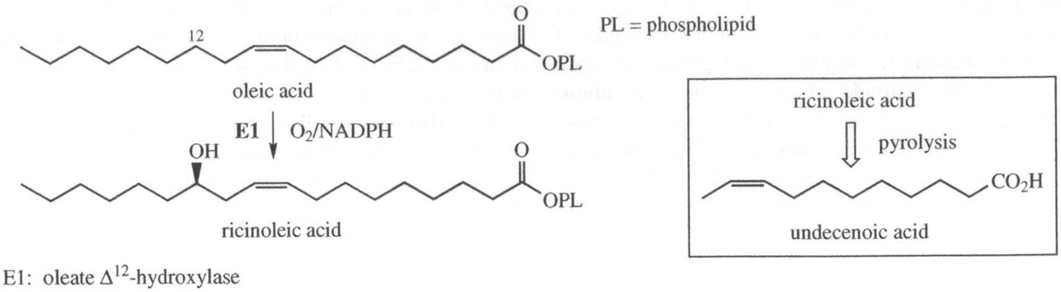

E1: oleate Δ^{12}-hydroxylase

Figure 3.14

already discussed, whilst others require a fundamental change in the nature of the starter and extender units employed in the biosynthesis.

Ricinoleic acid (Figure 3.14) is the 12-hydroxy derivative of oleic acid and is the major fatty acid found in castor oil, expressed from seeds of the castor oil plant (*Ricinus communis*; Euphorbiaceae). It is formed by direct hydroxylation of oleic acid, esterified as part of a phospholipid, by the action of an O_2- and NADPH-dependent mixed-function oxidase. This is not of the cytochrome P-450 type, but structurally and mechanistically resembles the fatty acid desaturase enzymes. Castor oil has a long history of use as a domestic purgative, but it is now mainly employed as a cream base. **Undecenoic acid** (Δ^9-undecenoic acid) can be obtained from ricinoleic acid by thermal degradation, and as the zinc salt or in ester form, it is used in a number of fungistatic preparations. Fatty acid hydroxylases of the cytochrome P-450 type have also been identified; the position of hydroxylation can vary according to enzyme, though hydroxylation of the terminal methyl (ω-hydroxylation) is quite common.

Epoxy fatty acids like **vernolic acid** (Figure 3.15) have been found in substantial quantities in the seed oil of some plant species, including *Vernonia galamensis* and *Stokesia laevis* (both Compositae/Asteraceae). Vernolic

acid is formed by direct epoxidation of linoleic acid (as a phospholipid ester), but, as with hydroxylation, different types of enzyme have been identified. These plants also exploit desaturase-like systems, whereas the same transformation in *Euphorbia lagascae* (Euphorbiaceae) is catalysed by a cytochrome P-450-dependent enzyme.

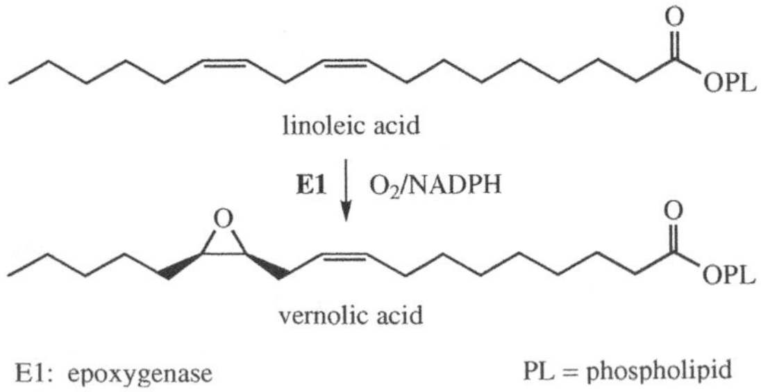

Whilst straight-chain fatty acids are the most common, branched-chain acids have been found to occur in mammalian systems, e.g. in wool fat and butter fat. They are also characteristic fatty acid constituents of the lipid part of cell walls in some pathogenic bacteria. Several mechanisms appear to operate in their formation. Methyl side-chains can be introduced by a *C*-alkylation mechanism using SAM. **Tuberculostearic acid** (Figure 3.16) is a *C*-methyl derivative of stearic acid found in the cell wall of *Mycobacterium tuberculosis*, the bacterium causing tuberculosis, and it provides a diagnostic marker for the disease. It is derived from oleic acid by alkylation on C-10, initiated by the double-bond electrons. Methyl transfer is followed by a Wagner–Meerwein 1,2-hydride shift in the carbocation, then proton loss to give the 10-methylene derivative. Finally, the double bond is reduced in an NADPH-dependent reaction to give tuberculostearic acid. This sequence of transformations is also seen during *C*-methylation of sterol side-chains (see page 252). Alternatively, loss of a proton from the first-formed carbocation intermediate via cyclopropane ring formation leads to **dihydrosterculic acid**. This is known to be dehydrogenated to **sterculic acid**, an unusual fatty acid containing a highly strained cyclopropene ring. Sterculic acid is present in the seed oil from *Sterculia foetida* (Sterculiaceae), and with similar cyclopropene acids, e.g. malvalic acid, is present in edible cottonseed oil from *Gossypium* species (Malvaceae). **Malvalic acid** is produced from sterculic acid by chain shortening from the carboxyl end (Figure 3.16). Sterculic acid is an inhibitor of the Δ^9-desaturase which converts stearic acid into oleic

linoleic acid

vernolic acid

E1: epoxygenase   PL = phospholipid

Figure 3.15

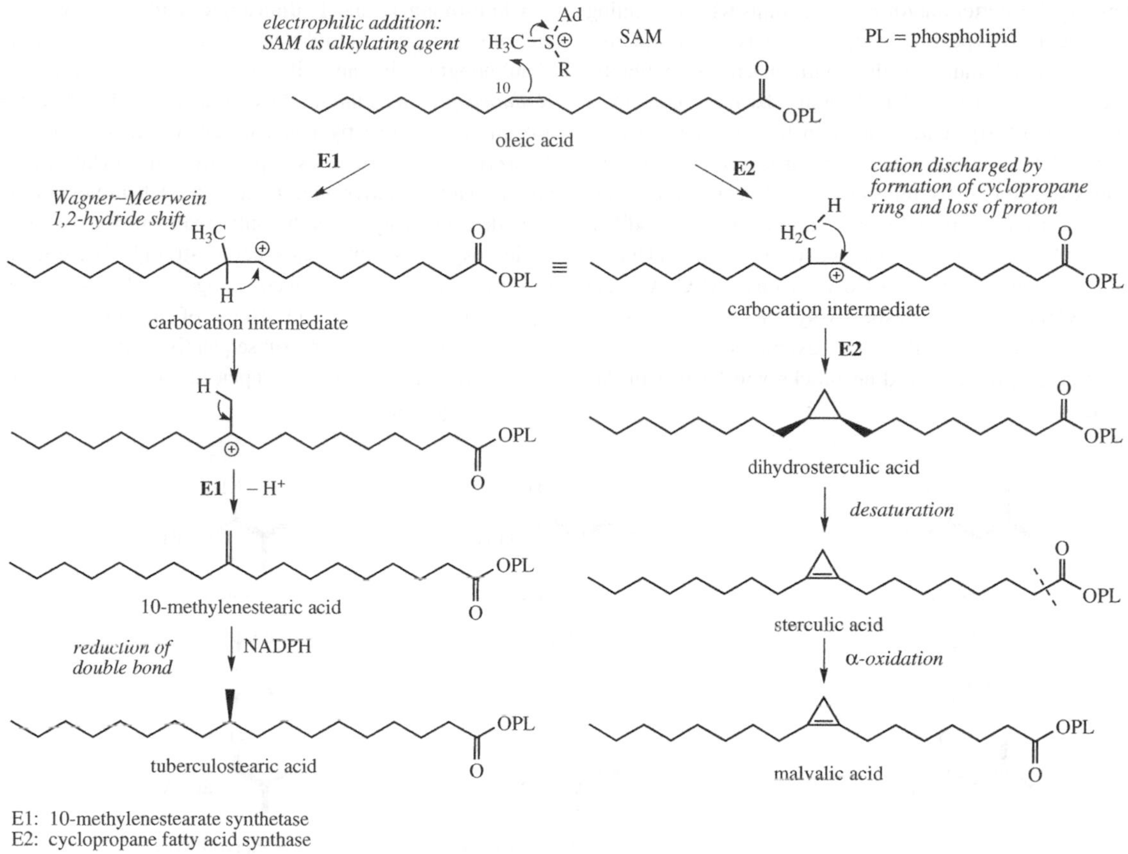

E1: 10-methylenestearate synthetase
E2: cyclopropane fatty acid synthase

Figure 3.16

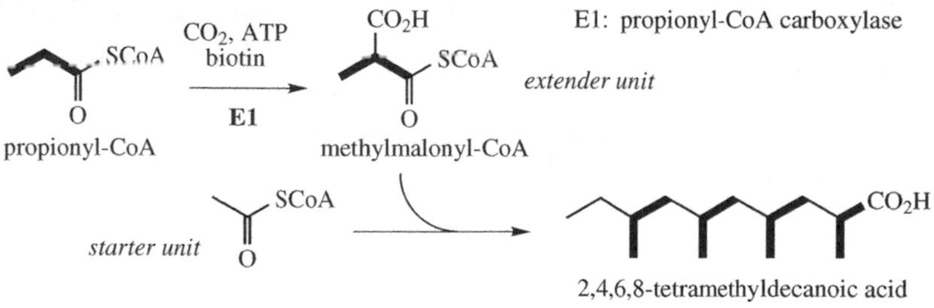

Figure 3.17

acid and is potentially harmful to humans in that it can alter membrane permeability and inhibit reproduction. Seed oils containing cyclopropene fatty acids destined for human consumption require suitable treatment to remove these undesirables.

Methyl side-chains can also be introduced by using methylmalonyl-CoA instead of malonyl-CoA as the chain

extending agent (Figure 3.17). Methylmalonyl-CoA arises by biotin-dependent carboxylation of propionyl-CoA in exactly the same way as malonyl-CoA was formed (see page 17). **2,4,6,8-Tetramethyldecanoic acid** found in the preen gland wax of the goose (*Anser anser*) is produced from an acetyl-CoA starter and four methylmalonyl-CoA chain extender units. This introduces the concept of changing the starter and/or extender units whilst retaining the same FAS type of mechanism. It is a concept that gets developed much further with macrolide antibiotic structures (see page 68). Various bacteria contain a range of **iso-fatty acids** that help to control membrane fluidity. These iso-fatty acids are produced by malonate chain extension of a number of different starter units derived from branched-chain amino acids, modified into CoA-esters. Thus, leucine, valine, and isoleucine may generate isovaleryl-CoA, isobutyryl-CoA, and 2-methylbutyryl-CoA respectively as starter groups (Figure 3.18). Several other examples where these amino acids provide building blocks will be met in due course.

Chaulmoogric and **hydnocarpic acids** (Figure 3.19) are uncommon cyclopentenyl fatty acids found in chaulmoogra oil, an oil expressed from seeds of *Hydnocarpus wightiana* (Flacourtiaceae). These acids are known to arise by malonate chain extension of the coenzyme A ester of 2-cyclopentenyl carboxylic acid as an alternative starter unit to acetate. Chaulmoogra oil provided for many years the only treatment for the relief of leprosy, these two acids being strongly bactericidal towards the leprosy infective agent *Mycobacterium leprae*. Purified salts and esters of hydnocarpic and chaulmoogric acids were subsequently employed, until they were then themselves replaced by more effective synthetic agents.

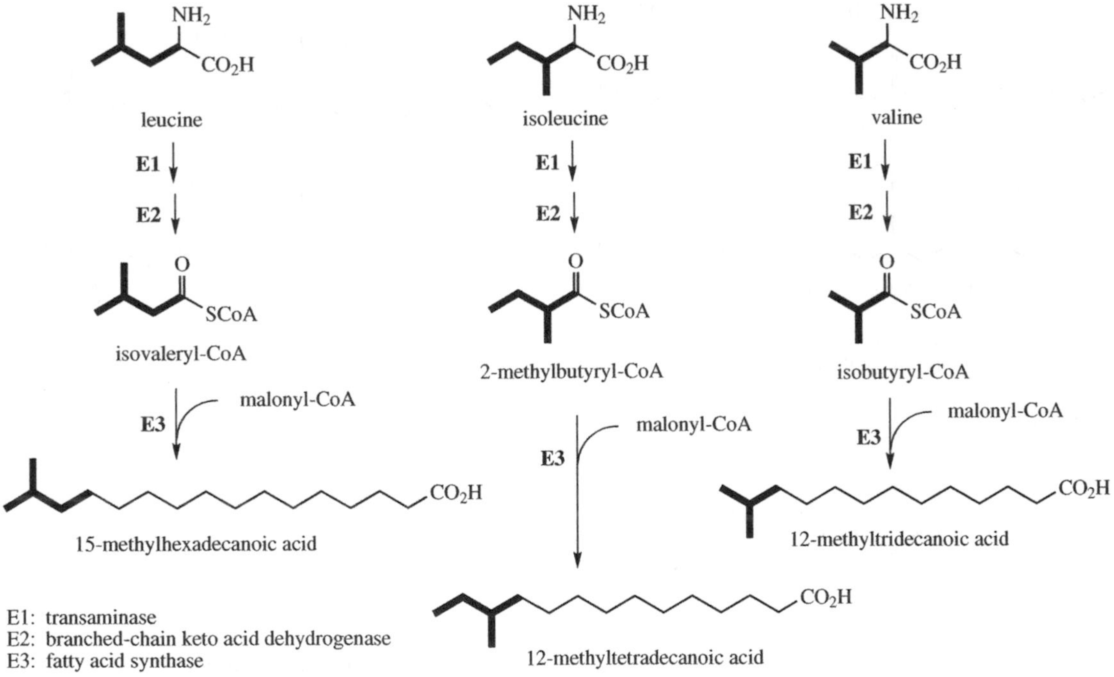

E1: transaminase
E2: branched-chain keto acid dehydrogenase
E3: fatty acid synthase

Figure 3.18

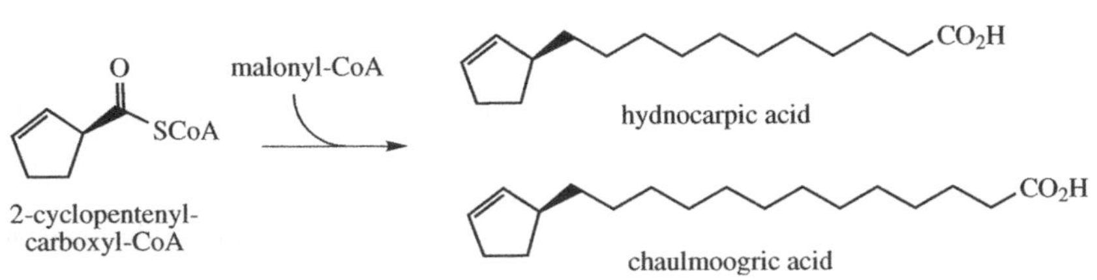

Figure 3.19

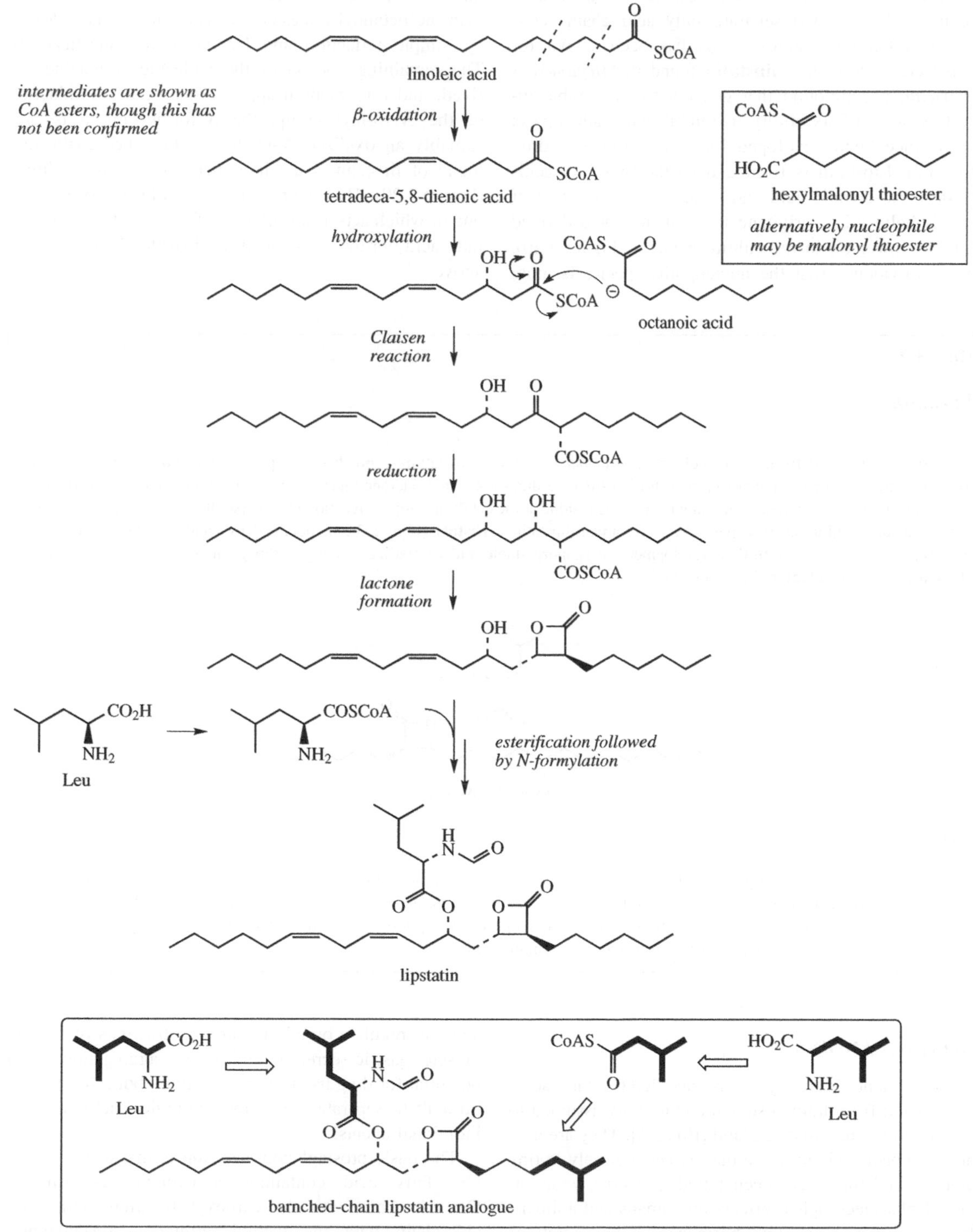

Figure 3.20

Some branched-chain systems can be rationalized as a combination of two separate fatty acid chains coupled in a Claisen reaction. This has been shown for the β-lactone derivative **lipstatin** found in *Streptomyces toxytricini*, a compound that generated considerable interest because of its ability to inhibit pancreatic lipase; it has since been developed into an anti-obesity drug [Box 3.2]. Lipstatin is formed from the two fatty acids tetradeca-5,8-dienoic acid and octanoic acid; the first of these is believed to originate from linoleic acid derived from sunflower oil in the culture medium (Figure 3.20). There is evidence that the nucleophilic species is more

likely to be the activated hexylmalonyl thioester rather than the octanoyl thioester. Claisen coupling is followed by simple reduction and lactonization (esterification). The remaining portion of the molecule is leucine derived, and esterification appears to precede introduction of the *N*-formyl group; the *N*-formyl group is presumably an oxidized *N*-methyl. A branched-chain analogue of lipstatin is a minor metabolite in the cultures (Figure 3.20). This incorporates two molecules of leucine, one of which acts as a starter unit for the methyloctanoate fatty acid portion, most likely via isovaleryl-CoA as seen above.

Box 3.2

Lipstatin

Lipstatin was isolated from the mycelium of *Streptomyces toxytricini* cultures and shown to possess marked inhibitory activity towards pancreatic lipase, the key enzyme for intestinal fat digestion. The β-lactone function in the lipophilic molecule irreversibly inactivates lipase by covalent reaction with a serine residue at the catalytic site; the reaction closely parallels that of serine residues with β-lactam antibiotics (see page 464). Tetrahydrolipstatin (**orlistat**) (Figure 3.21), obtained by catalytic hydrogenation of lipstatin, was selected for further development; it is more stable and crystallizes readily, although it is somewhat less active. Orlistat is now manufactured by total synthesis.

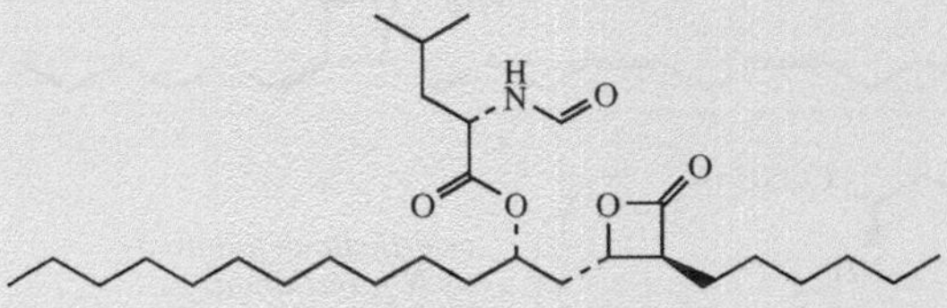

orlistat (tetrahydrolipstatin)

Figure 3.21

Orlistat reduces the absorption of dietary fat and is used in conjunction with a low-fat calorie-reduced diet to reduce body mass in obese patients. Only trace amounts of the drug are absorbed systemically, and the primary effect is local lipase inhibition within the digestive tract; dietary fat is thus excreted undigested. Absorption of fat-soluble vitamins, especially vitamin D, is also inhibited, and vitamin supplements are usually co-administered.

PROSTAGLANDINS

The prostaglandins are a group of modified C_{20} fatty acids first isolated from human semen and initially assumed to be secreted by the prostate gland [Box 3.3]. They are now known to occur widely in animal tissues, but only in tiny amounts, and they have been found to exert a wide variety of pharmacological effects on humans and animals. They are active at very low, hormone-like concentrations

and can regulate blood pressure, contractions of smooth muscle, gastric secretion, and platelet aggregation. Their potential for drug use is extremely high, but it has proved difficult to separate the various biological activities into individual agents.

The basic prostaglandin skeleton is that of a cyclized C_{20} fatty acid containing a cyclopentane ring, a C_7 side-chain with the carboxyl function, and a C_8 side-chain with the methyl terminus. Prostaglandins

dihomo-γ-linolenic ($\Delta^{8,11,14}$)   arachidonic ($\Delta^{5,8,11,14}$)   eicosapentaenoic ($\Delta^{5,8,11,14,17}$)

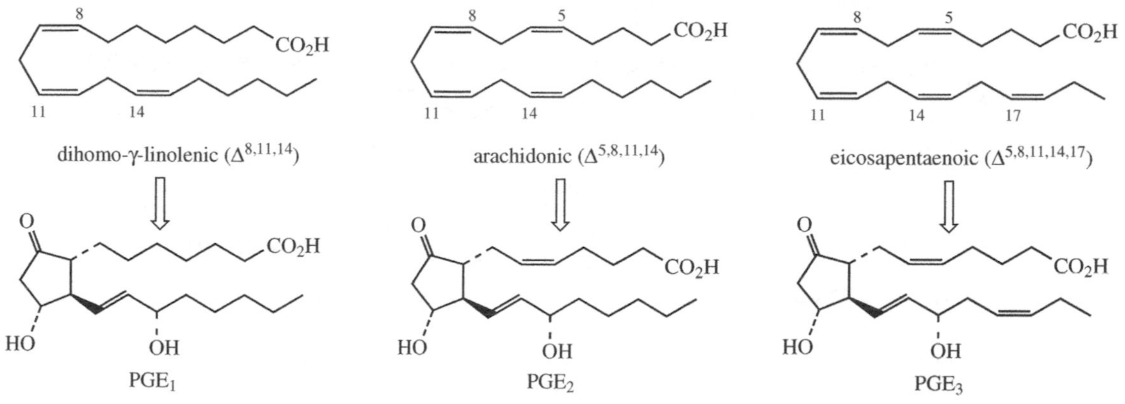

PGE$_1$ PGE$_2$ PGE$_3$

Figure 3.22

are biosynthesized from three EFAs, namely $\Delta^{8,11,14}$-**eicosatrienoic acid (dihomo-γ-linolenic acid)**, $\Delta^{5,8,11,14}$-**eicosatetraenoic acid (arachidonic acid)**, and $\Delta^{5,8,11,14,17}$-**eicosapentaenoic acid**, which yield prostaglandins of the 1-, 2-, and 3-series respectively (Figure 3.22; see Figure 3.24 below for principles of nomenclature). The three precursors lead to products of similar structure, but with varying levels of unsaturation in the two side-chains. Some of the structures elaborated from arachidonic acid are shown in Figure 3.23.

In the first reaction, arachidonic acid is converted into **prostaglandin H$_2$ (PGH$_2$)** by way of **prostaglandin G$_2$ (PGG$_2$)** by prostaglandin H synthase. This is a bifunctional enzyme comprised of an oxygenase (**cyclooxygenase, COX**) and a peroxidase, both requiring haem as cofactor. COX incorporates two molecules of oxygen, liberating a compound with both cyclic and acyclic peroxide functions, and also creating five chiral centres. In arachidonic acid, the methylene group flanked by two double bonds is susceptible to radical oxidation, initiated by a tyrosyl radical in the enzyme. This leads to incorporation of oxygen and formation of a cyclic peroxide function. Radical addition to an alkene function allows ring formation; then, a further incorporation of oxygen, in what is effectively a repeat of the earlier sequence, leads to an acyclic peroxide and produces PGG$_2$. The hydrogen atom terminating the synthesis is obtained from the same tyrosine function in the enzyme, thus propagating the radical reaction. The acyclic peroxide group in PGG$_2$ is then cleaved by the peroxidase component of prostaglandin H synthase to yield **PGH$_2$**. This unstable peroxide occupies a central role and can be modified in several different ways. Reductive cleavage gives **prostaglandin F$_{2\alpha}$ (PGF$_{2\alpha}$)**; other modifications can be rationally accommodated by initial cleavage of the cyclic peroxide to the diradical, though alternative ionic mechanisms may also be proposed. Quenching of the radicals by capture and loss of hydrogen atoms would provide either **prostaglandin E$_2$ (PGE$_2$)** or **prostaglandin D$_2$ (PGD$_2$)**. Reduction of PGE$_2$ is known to provide a minor pathway to PGF$_{2\alpha}$. The bicyclic system in **prostaglandin I$_2$ (PGI$_2$; prostacyclin)** is envisaged as arising by involvement of a side-chain double bond, then loss of a hydrogen atom, all catalysed by prostacyclin synthase, a cytochrome P-450-dependent enzyme. Prostaglandin structures representative of the 1-series, e.g. **PGE$_1$**, or of the 3-series, e.g. **PGE$_3$**, can be formed in a similar way from the appropriate fatty acid precursor (Figure 3.22).

The basic skeleton of the prostaglandins is termed **prostanoic acid**, and derivatives of this system are collectively known as prostanoids. The term **eicosanoids** is also used to encompass prostaglandins, thromboxanes, and leukotrienes, which are all derived from C$_{20}$ fatty acids (eicosanoic acids). Semi-systematic nomenclature of prostaglandins is based on the substitution pattern in the five-membered ring, denoted by a letter suffix (Figure 3.24), and the number of double bonds in the side-chains is given by a numerical subscript. Greek letters α and β are used to indicate the configuration at C-9, α indicating that the substituent is below the plane (as found in natural prostaglandins) and β indicating that the substituent is above the plane (as in some synthetic analogues). 'Prostaglandin' is usually abbreviated to PG. Prostaglandins A, B, and C are inactive degradation products from natural prostaglandins.

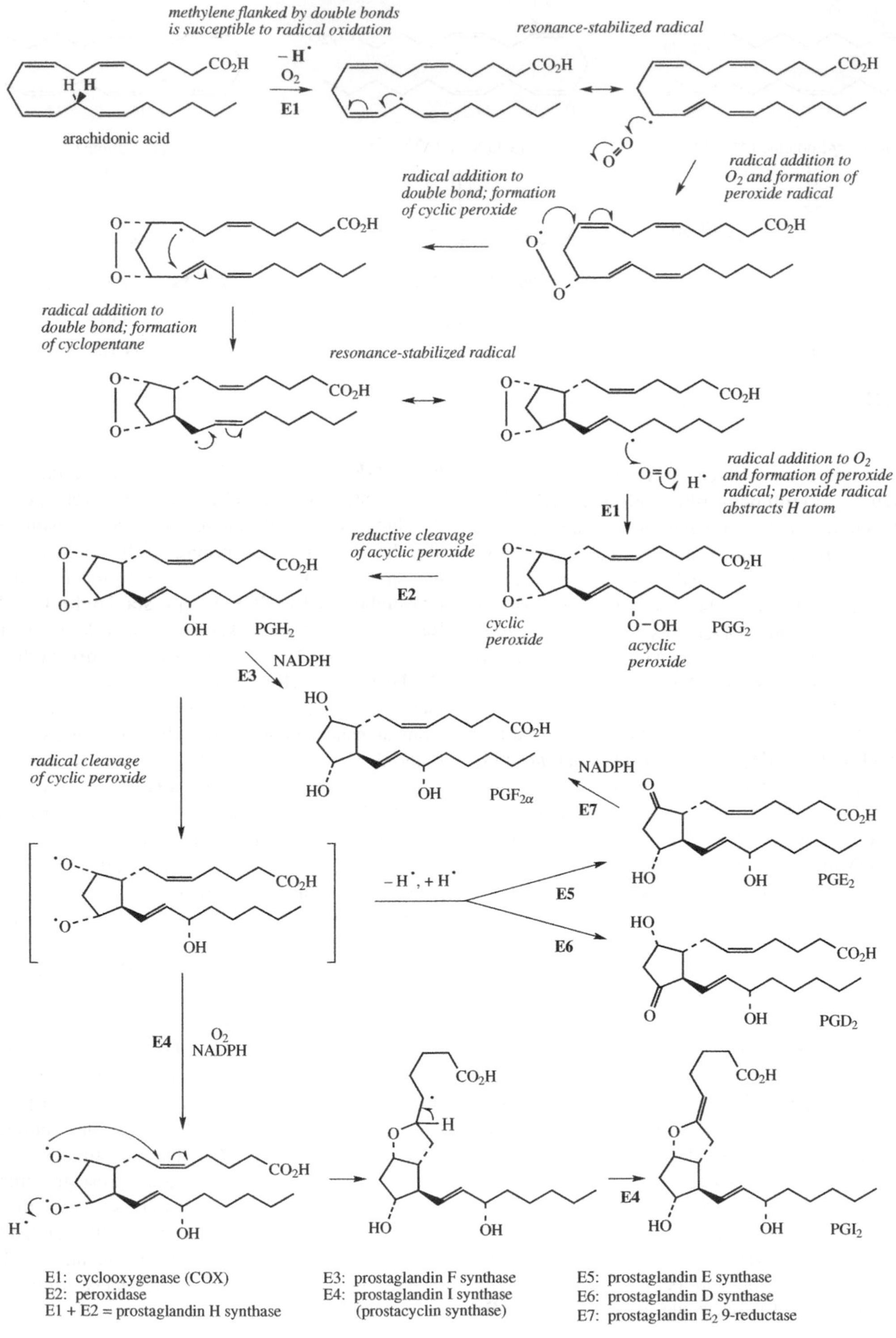
methylene flanked by double bonds is susceptible to radical oxidation
resonance-stabilized radical
arachidonic acid
E1
radical addition to O2 and formation of peroxide radical
radical addition to double bond; formation of cyclic peroxide
radical addition to double bond; formation of cyclopentane
resonance-stabilized radical
radical addition to O2 and formation of peroxide radical; peroxide radical abstracts H atom
reductive cleavage of acyclic peroxide
E2
PGH2
cyclic peroxide
acyclic peroxide
PGG2
NADPH
E3
PGF2α
radical cleavage of cyclic peroxide
E7
PGE2
E5
E6
PGD2
E4
NADPH
PGI2
E1: cyclooxygenase (COX)
E2: peroxidase
E1 + E2 = prostaglandin H synthase
E3: prostaglandin F synthase
E4: prostaglandin I synthase (prostacyclin synthase)
E5: prostaglandin E synthase
E6: prostaglandin D synthase
E7: prostaglandin E2 9-reductase

Figure 3.23

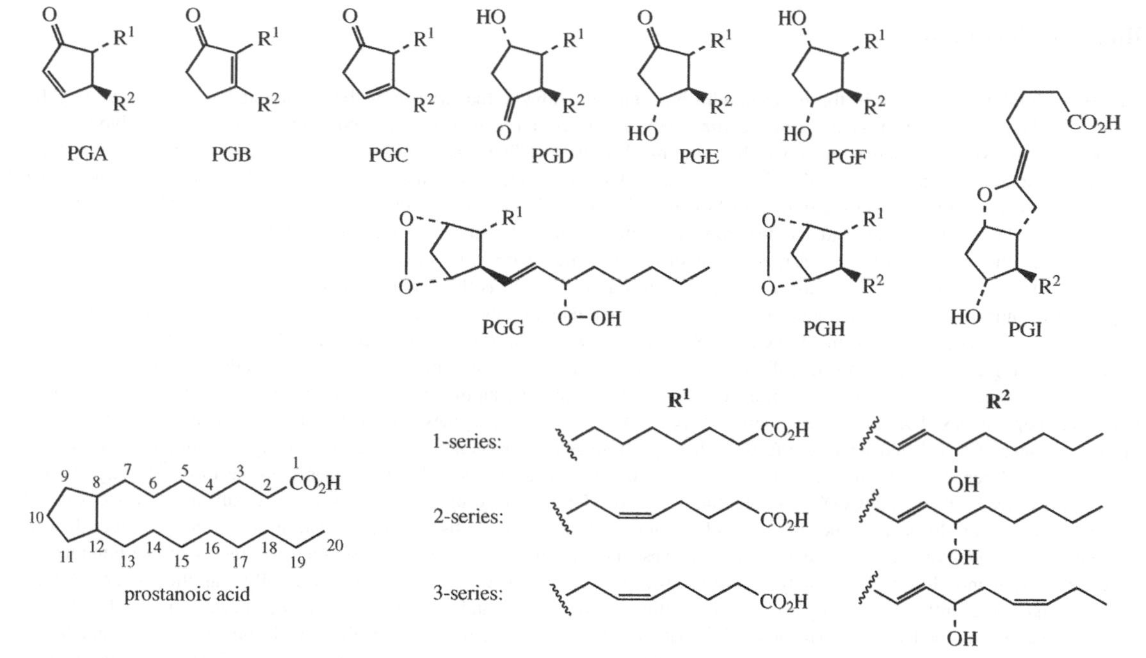

Figure 3.24

Box 3.3

Prostaglandins

Prostaglandins occur in nearly all mammalian tissues, but only at very low concentrations. PGE_1 and $PGF_{1\alpha}$ were initially isolated from sheep seminal plasma, but these compounds and PGD_2, PGE_2, and $PGF_{2\alpha}$ are widely distributed. Animal sources cannot supply sufficient amounts for drug usage. The soft coral *Plexaura homomalla* (sea whip) from the Caribbean has been identified as having very high (2–3%) levels of prostaglandin esters, predominantly the methyl ester of PGA_2 (1–2%) with related structures. Prostaglandins of the A-, E-, and F-types are widely distributed in soft corals, especially *Plexaura*, though many corals produce prostaglandin structures not found in animals. For several years, the sea whip coral was used as a source of research material, but this was neither a satisfactory nor renewable natural source. Considerable effort was exerted on the total synthesis of prostaglandins and their interconversions, and the high level of success achieved opened up the availability of compounds for pharmacological testing and subsequent drug use. Synthetic analogues have also been developed to modify or optimize biological activity. The studies have demonstrated that biological activity is effectively confined to the natural enantiomers; the unnatural enantiomer of PGE_1 had only 0.1% of the activity of the natural isomer.

The prostaglandins display a wide range of pharmacological activities, including contraction and relaxation of smooth muscle of the uterus, the cardiovascular system, the intestinal tract, and of bronchial tissue. They also inhibit gastric acid secretion, control blood pressure and suppress blood platelet aggregation, as well as acting as mediators of inflammation, fever, and allergy. Some of these effects are consistent with the prostaglandins acting as second messengers, modulating transmission of hormone stimulation and, thus, metabolic response. They are synthesized in response to various stimuli in a variety of cells, released immediately after synthesis, and act in the vicinity of their synthesis to maintain local homeostasis. Some prostaglandins in the A and J series have demonstrated potent antitumour properties.

Since the prostaglandins control many important physiological processes in animal tissues, their drug potential is high, but the chances of precipitating unwanted side-effects are also high, and this has so far limited their therapeutic use. There is, however, much additional scope for controlling the production of natural prostaglandins in body tissues by means of specific inhibitors. Indeed, it has been found that some established non-steroidal anti-inflammatory drugs (NSAIDs), e.g. aspirin, indometacin, and

Box 3.3   (continued)

ibuprofen, inhibit early steps in the prostaglandin biosynthetic pathway that transform the unsaturated fatty acids into cyclic peroxides. The local production of prostaglandins such as PGE_2 can sensitize pain nerve endings and increase blood flow, promoting pain, swelling, and redness. On the other hand, PGI_2 and PGE_2 are protective to the stomach, and inhibiting their formation explains the gastrointestinal toxicity associated with prolonged and high-dose use of NSAIDs. Aspirin is now known to inactivate irreversibly the COX activity (arachidonic acid → PGG_2), though not the peroxidase activity (PGG_2 → PGH_2), by selective acetylation of a serine residue of the enzyme; ibuprofen and indomecacin compete with arachidonic acid at the active site and are reversible inhibitors of COX. A more recent discovery is that two forms of the COX enzyme exist, designated COX-1 and COX-2. COX-1 is expressed constitutively in most tissues and cells and is thought to control synthesis of those prostaglandins important for normal cellular functions, such as gastrointestinal integrity and vascular homeostasis. In simplistic terms, COX-1 is considered a 'housekeeping' enzyme. COX-2 is not normally present, but is inducible in certain cells in response to inflammatory stimuli, resulting in enhanced prostaglandin release in the central nervous system and inflammatory cells with the characteristic inflammatory response. Its inhibition appears to produce the analgesic, antipyretic, and anti-inflammatory effects of NSAIDs. However, popular NSAIDs do not discriminate between the two COX enzymes, and so this leads to both therapeutic effects via inhibition of COX-2 and adverse effects such as gastrointestinal problems, ulcers, and bleeding via inhibition of COX-1. Because of differences in the nature of the active sites of the two enzymes, it has now been possible to develop agents that can inhibit COX-2 rather than COX-1 as potential new anti-inflammatory drugs. The first of these 'coxib' drugs, celecoxib and rofecoxib, were introduced for relief of pain and inflammation in osteoarthritis and rheumatoid arthritis, though rofecoxib was subsequently withdrawn because of significant cardiovascular side-effects. Second-generation agents etoricoxib and lumiracoxib have since been introduced. The widely used analgesic paracetamol (US: acetaminophen) is usually classified as an NSAID, though it has little anti-inflammatory activity. Its ability to inhibit pain and fever is suggested to emanate from inhibition of a COX-1 variant, termed COX-3. The anti-inflammatory activity of corticosteroids correlates with inhibition of phospholipase enzymes and preventing the release of arachidonic acid from storage phospholipids, but expression of COX-2 is also inhibited by glucocorticoids.

The role of of essential fatty acids (see page 49) such as linoleic and γ-linolenic acids, obtained from plant ingredients in the diet, can now readily be appreciated. Without a source of arachidonic acid, or compounds which can be converted into arachidonic acid, synthesis of prostaglandins would be compromised, and this would seriously affect many normal metabolic processes. A steady supply of prostaglandin precursors is required; prostaglandins are continuously being synthesized and then degraded. Prostaglandins are rapidly degraded by processes which include oxidation of the 15-hydroxyl to a ketone, reduction of the 13,14-double bond, and oxidative degradation of both side-chains.

A major area of application of prostaglandins as drugs is in obstetrics, where they are used to induce abortions during the early to middle stages of pregnancy, or to induce labour at term. Dinoprost ($PGF_{2\alpha}$) is not usually prescribed, since it is rapidly metabolized in body tissues (half-life less than 10 min), and the modified version **carboprost** (15-methyl $PGF_{2\alpha}$; Figure 3.25) has been developed to reduce deactivation by blocking oxidation at position 15. Carboprost is produced by oxidizing the 15-hydroxyl in a suitably protected $PGF_{2\alpha}$, then alkylating the 15-carbonyl with a Grignard reagent. Carboprost is effective at much reduced dosage than dinoprost and is of value in augmenting labour at term, especially in cases where ergometrine (see page 394) or oxytocin (see page 430) are ineffective. The natural prostaglandin structure **dinoprostone** (PGE_2) is also used primarily to induce labour; the 16-dimethyl prostaglandin **gemeprost** is currently the preferred prostaglandin for abortions. These agents are usually administered vaginally.

Alprostadil (PGE_1) differs from PGE_2 by having unsaturation in only one side-chain. Though having effects on uterine muscle, it also has vasodilator properties, and these are exploited for maintaining new-born infants with congenital heart defects, facilitating blood oxygenation prior to corrective surgery. The very rapid metabolism of PGE_1 means this drug must be delivered by continuous intravenous infusion. Alprostadil is also of value in male impotence, self-injectable preparations being used to achieve erection of the penis. An interesting modification to the structure of PGE_1 is found in the analogue **misoprostol**. This compound has had the oxygenation removed from position 15, transferred to position 16, plus alkylation at position 16 to reduce metabolism (compare 15-methyl $PGF_{2\alpha}$ above). These modifications result in an orally active drug which inhibits gastric secretion effectively and can be used to promote healing of gastric and duodenal ulcers. In combination with non-specific NSAIDs, it can significantly lower the incidence of gastrointestinal side-effects such as ulceration and bleeding. It also induces labour, and is occasionally used for this purpose.

Epoprostenol (PGI_2, prostacyclin) reduces blood pressure and also inhibits platelet aggregation by reducing calcium concentrations. It is employed to inhibit blood clotting during renal dialysis, but its very low half-life (about 3 min) again necessitates continuous intravenous administration. The tetrahydrofuran ring is part of an enol ether and is readily opened by hydration, leading to 6-ketoprostaglandin $F_{1\alpha}$ (Figure 3.26). **Iloprost** (Figure 3.25) is a stable carbocyclic analogue of potential use in the treatment of thrombotic diseases.

Box 3.3 (continued)

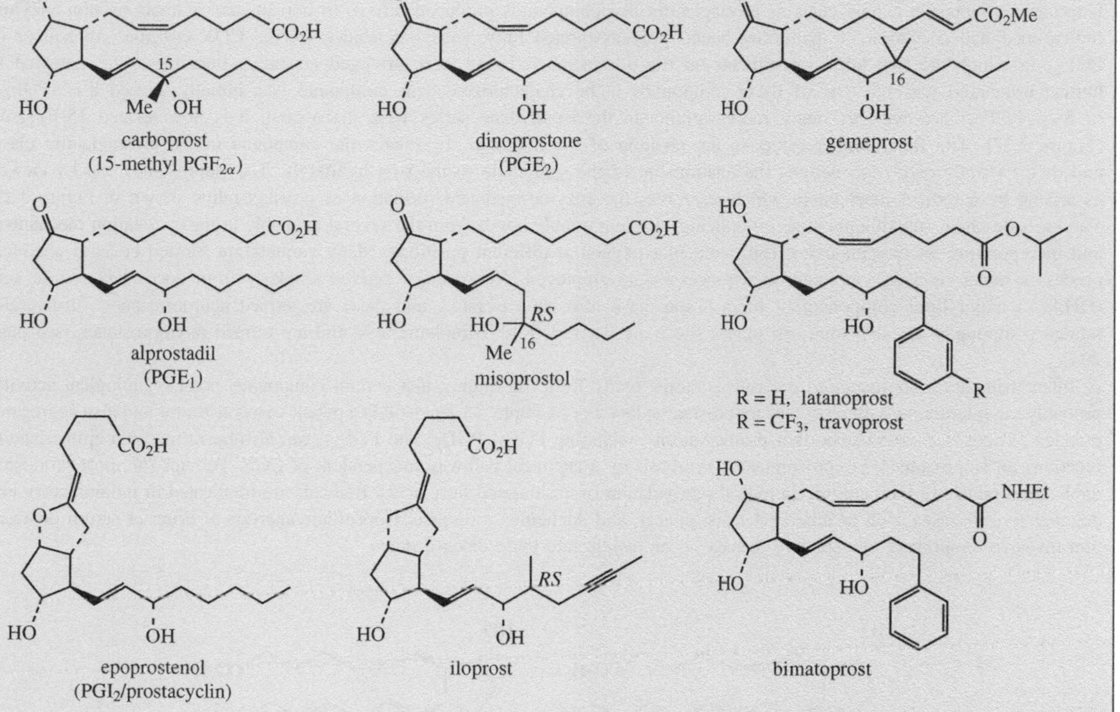

carboprost
(15-methyl PGF$_{2\alpha}$)

dinoprostone
(PGE$_2$)

gemeprost

alprostadil
(PGE$_1$)

misoprostol

R = H, latanoprost
R = CF$_3$, travoprost

epoprostenol
(PGI$_2$/prostacyclin)

iloprost

bimatoprost

Figure 3.25

PGI$_2$

*hydrolysis of
enol ether*

*hemiketal
formation*

*hydrolysis
of hemiketal*

6-keto PGF$_{1\alpha}$

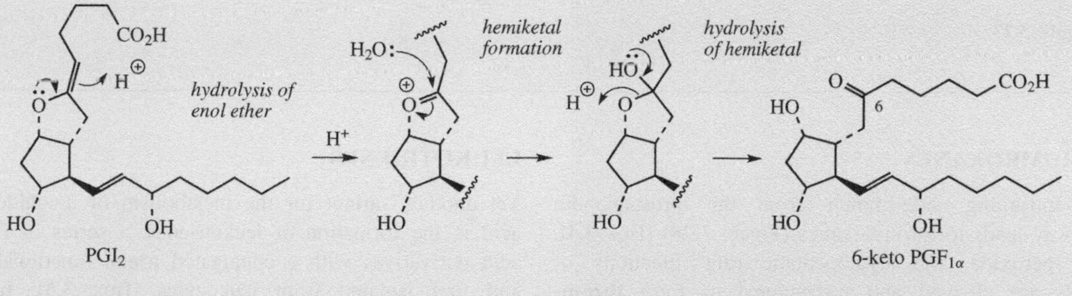

Figure 3.26

Latanoprost, travoprost, and **bimatoprost** (Figure 3.25) are recently introduced prostaglandin analogues which increase the outflow of aqueous humour from the eye. They are thus used to reduce intraocular pressure in the treatment of the eye disease glaucoma. Bimatoprost is an amide derivative related to prostaglandin amides (prostamides) obtained by the action of COX-2 on anandamide, the natural ligand for cannabinoid receptors (see page 122). The pharmacological properties of prostamides cannot readily be explained simply by interaction with prostaglandin receptors; bimatoprost also inhibits prostaglandin F synthase. An unusual side-effect of bimatoprost is also being exploited cosmetically: it makes eyelashes grow longer, thicker, and darker.

Box 3.3   (continued)

Isoprostanes

Isoprostanes represent a new class of prostaglandin-like compounds produced *in vivo* in humans and animals by non-enzymic radical-mediated oxidation of membrane-bound polyunsaturated fatty acids independent of the COX enzyme. An isomer of $PGF_{2\alpha}$ in which the two alkyl substituents on the five-membered ring were arranged *cis* rather than *trans* was detected in human urine and was the first of these compounds to be characterized. This compound was initially termed 8-*iso*-$PGF_{2\alpha}$, or 8-*epi*-$PGF_{2\alpha}$; however, as many more variants in the isoprostane series were discovered, it is now termed 15-F_{2t}-IsoP (Figure 3.27). The first number refers to the position of the hydroxyl, F_2 relates the compound to the prostaglandin class, and then t (*trans*) or c (*cis*) defines the relationship of the side-chain to the ring hydroxyls. The isoprostanes can be viewed as arising by a radical mechanism which resembles the enzyme-mediated formation of prostaglandins shown in Figure 3.23. Varying side-chain substituents arise by utilizing different double bonds from the several available in the cyclization mechanism and incorporating an oxygen atom from molecular oxygen at different positions. Many variants are formed because chemical processes rather than enzyme-controlled processes are employed. Isoprostanes derived similarly from docosahexaeneoic acid (DHA), a major lipid component of brain tissue, have also been isolated, and these are termed neuroprostanes. Structurally related compounds are also found in plants; these are derived from α-linolenic acid and are termed phytoprostanes (see page 204).

Interest in these isoprostanoid derivatives stems partly from the finding that certain compounds possess biological activity, probably via interaction with receptors for prostaglandins. For example, 15-F_{2t}-IsoP is a potent vasoconstrictor and also aggregates platelets. There is also evidence that prostaglandins, including PGE_2, PGD_2, and $PGF_{2\alpha}$, can also be formed via epimerization reactions on isoprostanes, i.e. prostaglandin synthesis by a chemical pathway independent of COX. Perhaps the most promising application relates to their origin via radical peroxidation of unsaturated fatty acids. Radicals are implicated in inflammatory and degenerative diseases such as atherosclerosis, cancer, and Alzheimer's disease. Isoprostane analysis of urine or serum provides non-invasive monitoring of oxidative damage as an insight into these disease states.

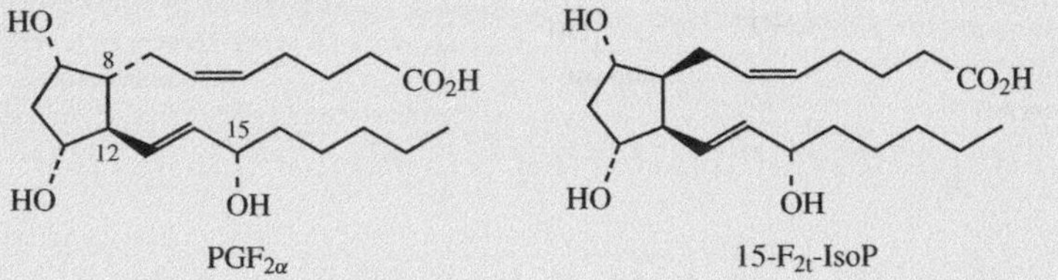

Figure 3.27

THROMBOXANES

An intriguing side-branch from the prostaglandin pathway leads to thromboxanes (Figure 3.28) [Box 3.4]. The peroxide and cyclopentane ring functions of PGH_2 are cleaved and restructured to form **thromboxane A_2 (TXA$_2$)**, which contains a highly strained four-membered oxetane ring. A hypothetical radical mechanism is shown; the enzyme also produces equimolar amounts of hydroxyheptadecatrienoic acid and malondialdehyde by a fragmentation reaction. TXA$_2$ is highly unstable and reacts readily with nucleophiles. In an aqueous environment, it reacts to yield the biologically inactive hemiacetal **thromboxane B_2 (TXB$_2$)**.

LEUKOTRIENES

Yet another variant for the metabolism of arachidonic acid is the formation of leukotrienes, a series of fatty acid derivatives with a conjugated triene functionality, and first isolated from leukocytes [Box 3.5]. In a representative pathway (others have been characterized) (Figure 3.29), **arachidonic acid** is converted into a hydroperoxide, the point of oxygenation being C-5, rather than C-11 as in the prostaglandin pathway (Figure 3.23). This compound loses water via formation of an epoxide ring, giving **leukotriene A_4 (LTA$_4$)**. This unstable allylic epoxide may hydrolyse by conjugate addition to give **leukotriene B_4 (LTB$_4$)**, or alternatively the epoxide may be

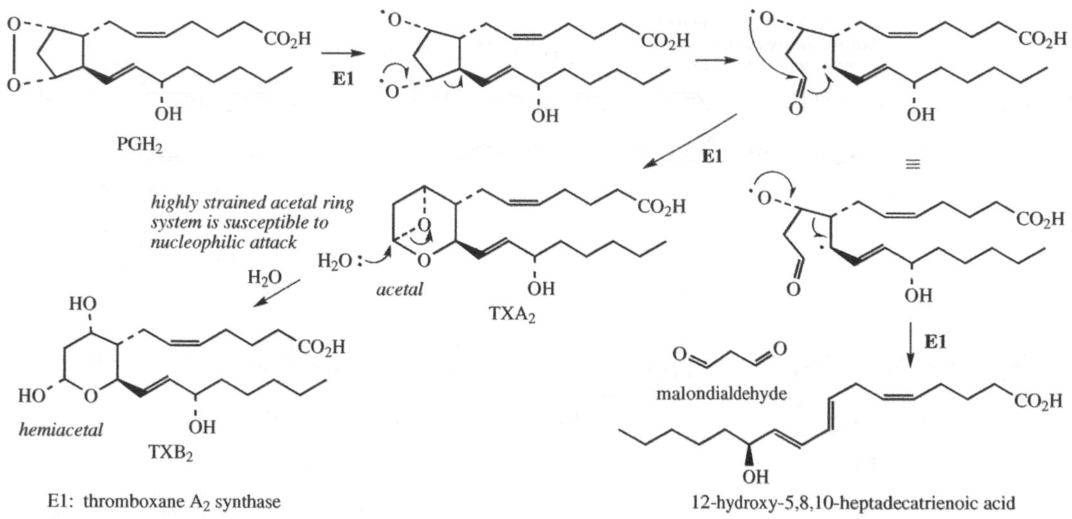

E1: thromboxane A$_2$ synthase

12-hydroxy-5,8,10-heptadecatrienoic acid

Figure 3.28

Box 3.4

Thromboxanes

The thromboxanes were isolated from blood platelets, and whilst TXA$_2$ showed high biological activity, TXB$_2$ was effectively inactive. TXA$_2$ is a potent stimulator of vasoconstriction and platelet aggregation. Aggregation of blood platelets to form a clot or thrombus is caused by increasing cytoplasmic calcium concentrations and thus deforming the platelets, which then fuse together. TXA$_2$ has the opposite effect to PGI$_2$, and presumably the development of thrombosis reflects an imbalance in the two activities. Both compounds are produced from the same precursor, PGH$_2$, which is converted in the blood platelets into TXA$_2$, and in the blood vessel wall into PGI$_2$. Thromboxanes A$_3$ and B$_3$ have also been isolated from blood platelets. These are derived from $\Delta^{5,8,11,14,17}$-eicosapentaenoic acid and relate structurally to prostaglandins in the 3-series. TXA$_3$ is not strongly aggregatory towards blood platelets. The highly unstable nature of the biologically active thromboxanes has made their synthesis difficult, and drug use of natural structures will probably be impracticable. It is likely that most efforts will be directed towards thromboxane antagonists to help reduce blood platelet aggregation in thrombosis patients. The value of aspirin in preventing cardiovascular disease is now known to be related to inhibition of thromboxane A$_2$ biosynthesis in platelets.

Box 3.5

Leukotrienes

The leukotrienes are involved in allergic responses and inflammatory processes. An antigen–antibody reaction can result in the release of compounds such as histamine (see page 398) or materials termed slow reacting substance of anaphylaxis (SRSA). These substances are then mediators of hypersensitive reactions such as hay fever and asthma. Structural studies have identified SRSA as a mixture of LTC$_4$, LTD$_4$, and LTE$_4$. These cysteine-containing leukotrienes are powerful bronchoconstrictors and vasoconstrictors, and induce mucus secretion, the typical symptoms of asthma. LTE$_4$ is some 10–100-fold less active than LTD$_4$, so that degradation of the peptide side-chain represents a means of eliminating leukotriene function. LTB$_4$ appears to facilitate migration of leukocytes in inflammation, and is implicated in the pathology of psoriasis, inflammatory bowel disease, and arthritis. The biological effects of leukotrienes are being actively researched to define the cellular processes involved. This may lead to the development of agents to control allergic and inflammatory reactions. Drugs inhibiting the formation of LTC$_4$ and LTB$_4$ are in clinical trials, whilst montelukast and zafirlukast have been introduced as orally active leukotriene (LTD$_4$) receptor antagonists for the prophylaxis of asthma.

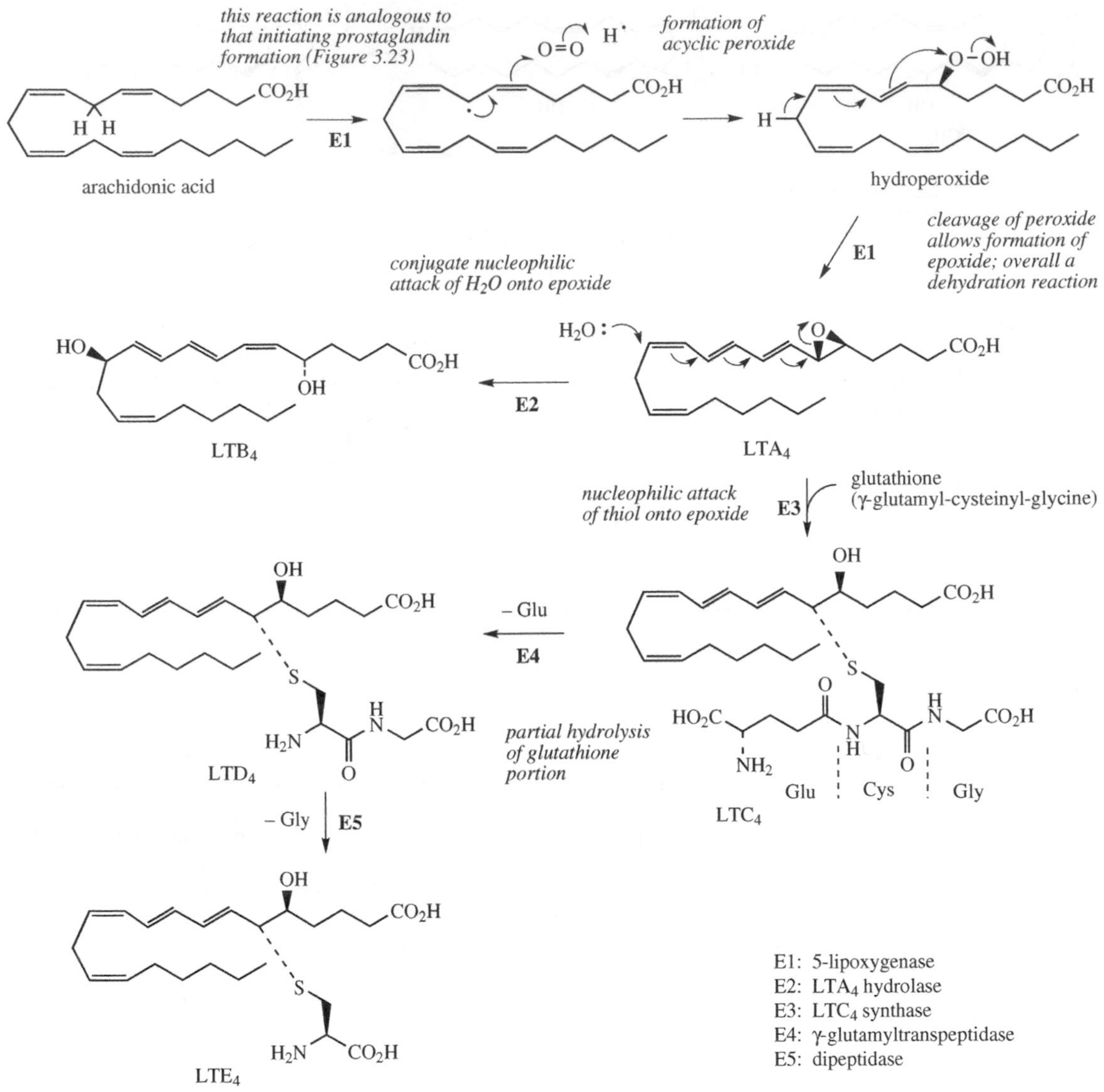

Figure 3.29

attacked directly by a nucleophile, in this case the sulfur atom of the tripeptide glutathione (γ-glutamyl-cysteinyl-glycine; Figure 3.29). The adduct produced in the latter reaction is termed **leukotriene C₄ (LTC₄)**. Partial hydrolysis in the tripeptide fragment then leads to **leukotriene D₄ (LTD₄)** and **leukotriene E₄ (LTE₄)**. Analogues, e.g. LTA₃ and LTA₅, are also known, and these are derived from $\Delta^{5,8,11}$-**eicosatrienoic acid** and $\Delta^{5,8,11,14,17}$-**eicosapentaenoic acid** respectively. The subscript numeral indicates the total number of double bonds in the leukotriene chain.

POLYKETIDE SYNTHASES: GENERALITIES

During fatty acid biosynthesis, the growing chain is constructed by a Claisen reaction, catalysed by the ketosynthase activity in the FAS complex. This reaction initially produces a β-ketoester, and the ketone group is reduced after each condensation step and before the next round of chain extension (Figure 3.2). The overall reduction is achieved by a three-stage process: reduction to an alcohol (KR), dehydration to the conjugated ester (DH), then reduction of the double bond (ER). These reactions

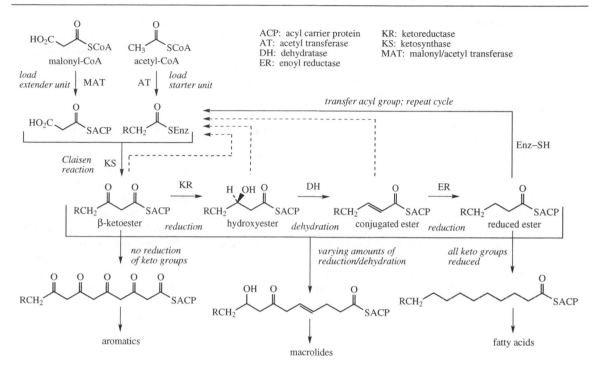

Figure 3.30

are shown again in Figure 3.30. This figure also shows how the same general processes may be modified to produce other polyketide systems, namely aromatics and macrolides; these processes are catalysed by **polyketide synthases (PKSs)**. For aromatics, the reduction sequence is omitted completely (or almost completely; see page 101) so that the β-ketoester re-enters the extension cycle, giving a poly-β-ketoester with alternate keto and methylene functions. This highly reactive poly-β-keto backbone undergoes further enzyme-catalysed intramolecular cyclization reactions which are responsible for generating a range of aromatic structures (see page 99). Macrolides are characterized by having carbon chains that have undergone partial, complete, or no reduction, according to the position in the chain. The chain may contain a mixture of hydroxyl groups, carbonyl groups, double bonds, and methylene groups. This means that the enzyme activities KR, DH, and ER are not all active during a particular extension cycle, and that a partially modified system, a β-ketoester, hydroxy ester, conjugated ester, or reduced ester, feeds back into the extension cycle. The order in which the modifications occur (or do not occur) is closely controlled by the enzyme.

A detailed study of the genes, protein amino acid sequences, and mechanistic similarities in various PKS

enzymes has led to three general types being distinguished. As with the type I and type II FASs (see page 40), **type I PKSs** are very large multifunctional proteins with individual functional domains, whilst **type II PKSs** are composed of a complex of individual monofunctional proteins. Type I systems can also be subdivided into 'iterative' (i.e. repeating) and 'non-iterative' categories. Iterative systems (like the FASs) use their functional domains repeatedly to produce a particular polyketide. Non-iterative systems possess a distinct active site for every enzyme-catalysed step. Type II systems are of the iterative type. Both types of system use ACP to activate acyl-CoA substrates and to channel the growing polyketide intermediates. Type I enzymes are responsible for macrolide biosynthesis, where the variability at each step in the biosynthetic pathway gives rise to much more structural diversity than encountered with fatty acids. The usual starter unit employed is either acetyl-CoA or propionyl-CoA, whilst malonyl-CoA or methylmalonyl-CoA is the main extender unit. Aromatic compounds are usually products from type II enzymes; though variation in starter units may be encountered, the chain extender unit is always malonyl-CoA. PKSs responsible for chain extension of cinnamoyl-CoA starter units leading to flavonoids and stilbenes (see page 116) do not fall into either of the above categories. These enzymes differ from the other types in that they are homodimeric proteins, they utilize coenzyme A esters rather than ACPs, and they employ a single active site to perform a series of decarboxylation, condensation, cyclization, and aromatization reactions. They have now been termed **type III PKSs** or chalcone synthase-like PKSs, though many examples outside of the flavonoid group are now recognized. Type I PKSs are found in bacteria and fungi, type II PKSs are restricted to bacteria, whilst type III PKSs are found in plants, bacteria, and fungi.

Despite the differences in molecular architecture, the chemical aspects of chain construction are effectively the same in all PKSs. In all cases, these natural products are produced by Claisen reactions involving thioesters, and a basic chain that is composed of a linear sequence of C_2 units is constructed. Even when the starter unit is modified, or methylmalonyl-CoA is used as an extender unit instead of malonyl-CoA, the fundamental chain is effectively the same and is produced in the same mechanistic manner. Therefore, notwithstanding the various building blocks employed, these compounds are all still considered as derived via the acetate pathway. It is also more fitting to discuss the compounds in groups according to structural aspects rather than the protein type involved in their construction.

POLYKETIDE SYNTHASES: MACROLIDES

The **macrolides** are a large family of compounds, many with antibiotic activity, characterized by a macrocyclic lactone (sometimes lactam) ring. Rings are commonly 12-, 14-, or 16-membered, though polyene macrolides (see page 81) are larger in the range 26–38 atoms. The largest natural macrolide structure discovered has a 66-membered ring. The macrolide antibiotics provide us with excellent examples of natural products conforming to the acetate pathway, but composed principally of propionate units, or mixtures of propionate and acetate units. There is now extensive genetic evidence from a variety of systems to show that macrolide assembly is

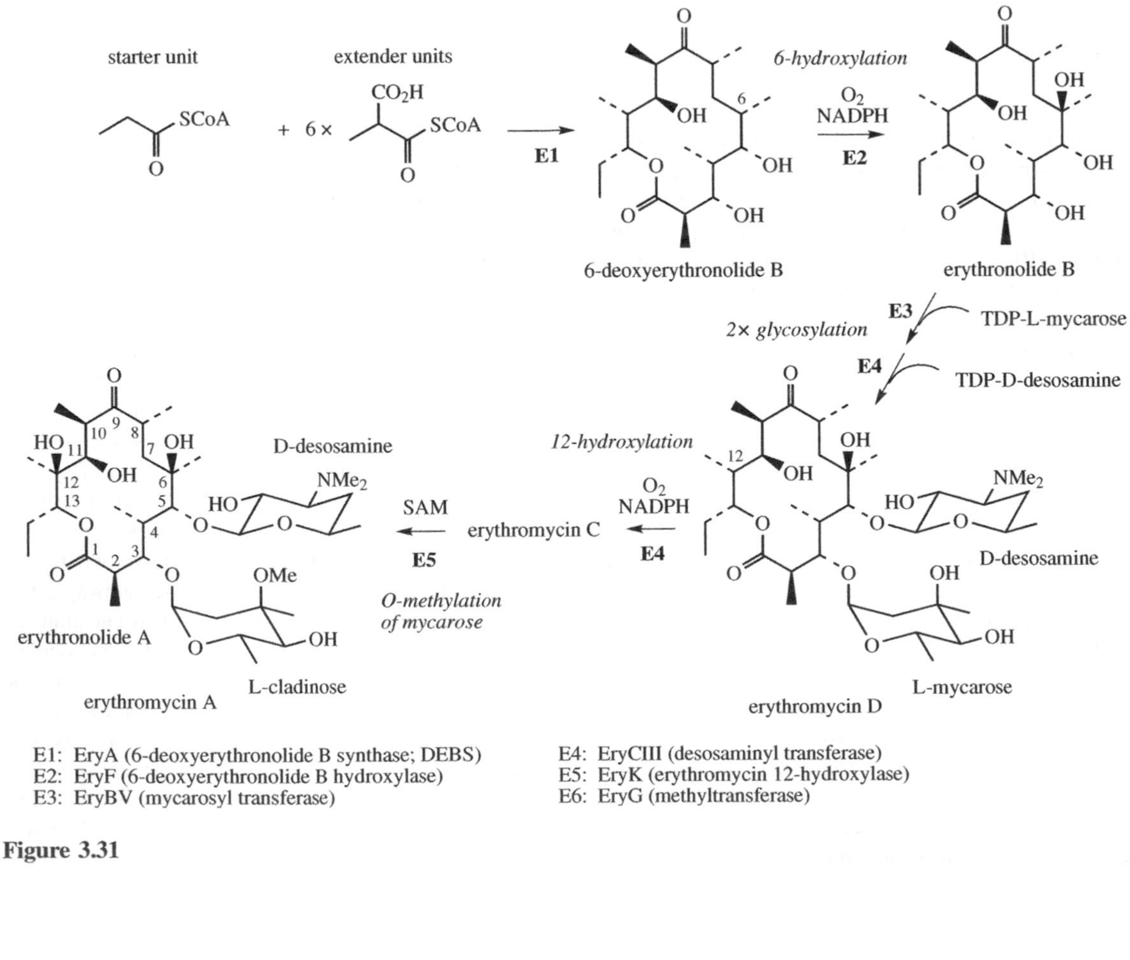
starter unit
extender units
SCoA
CO2H
+ 6×
SCoA
E1
6-hydroxylation
O2
NADPH
E2
6-deoxyerythronolide B
erythronolide B
2× glycosylation
E3
TDP-L-mycarose
E4
TDP-D-desosamine
12-hydroxylation
O2
NADPH
E4
erythromycin C
SAM
E5
O-methylation of mycarose
D-desosamine
NMe2
OMe
erythronolide A
L-cladinose
erythromycin A
L-mycarose
erythromycin D
E1: EryA (6-deoxyerythronolide B synthase; DEBS)
E2: EryF (6-deoxyerythronolide B hydroxylase)
E3: EryBV (mycarosyl transferase)
E4: EryCIII (desosaminyl transferase)
E5: EryK (erythromycin 12-hydroxylase)
E6: EryG (methyltransferase)

Figure 3.31

most often accomplished by non-iterative type I PKSs. Conceptually, polyketide synthesis by this group of enzymes is probably the easiest to envisage and understand, and provides the best grounding for appreciation of other systems. The PKS provides a biological production line of multifunctional proteins organized as discrete modules. Each module contains the enzyme activities for an extension cycle with its subsequent modifications. The developing polyketide chain attached to an ACP is thus modified according to the appropriate enzyme activities present, and is then passed on to another ACP prior to the next condensation and modification. The linear sequence of modules in the enzyme corresponds to the generated sequence of extender units in the polyketide product.

Erythromycin A (Figure 3.31) from *Saccharopolyspora erythraea* is a valuable antibacterial drug [Box 3.6] and contains a 14-membered macrocycle composed entirely of propionate units, both as starter and extension units, the latter via methylmalonyl-CoA. In common with many antibacterial macrolides, sugar units, including amino sugars, are attached through glycoside linkages. These unusual 6-deoxy sugars are frequently restricted to this group of natural products. In erythromycin A, the sugars are L-cladinose and D-desosamine. Chain extension and appropriate reduction processes lead to an enzyme-bound polyketide in which one carbonyl group has suffered total reduction, four have been reduced to alcohols, whilst one carbonyl is not reduced, and remains throughout the sequence. These processes ultimately lead to release of the modified polyketide as the macrolide ester **6-deoxyerythronolide B**, a demonstrated intermediate in the pathway to erythromycins (Figure 3.31). The stereochemistry in the chain is controlled by the condensation and reduction steps during chain extension, but a reassuring feature is that there appears to be a considerable degree of stereochemical uniformity throughout the known macrolide antibiotics. In the later stages of the biosynthesis of erythromycin, hydroxylations at carbon atoms 6 and 12, and addition of sugar units, are achieved. Methylation of a sugar hydroxyl is the last step in the sequence.

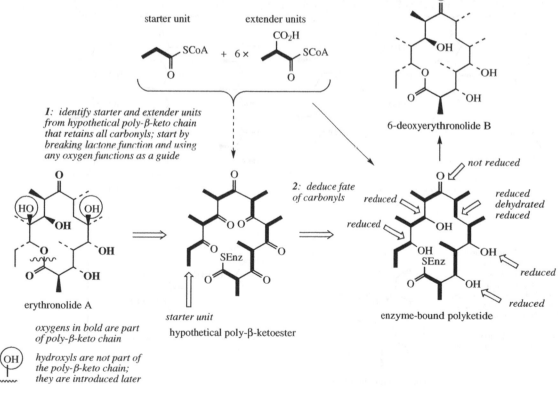

Figure 3.32

These latter modifications illustrate that features of the biosynthesis can be considered according to the timing of the steps in the sequence. The PKS controls a number of structural modifications on the basic polyketide, especially the reductive processes, whereas others are post-PKS changes. Here, hydroxylations, glycosylations, and methylation are post-PKS changes. It is usually possible to demonstrate these later conversions by appropriate interconversions and the isolation of enzymes catalysing the individual steps.

Since many macrolide structures are quite complex, and a variety of starter and extender units may be employed, it is appropriate here to demonstrate how simple analysis of the structure can help rationalize the biosynthetic origins of the basic macrolide. It also provides pointers to which modifications are likely to occur later in the pathway. Thus **erythronolide A**, the aglycone of erythromycin A, should be related to the hypothetical non-cyclized poly-β-ketoester, as shown in Figure 3.32. The positions of oxygen functions in erythronolide A are markers for the poly-β-keto system; the hydroxyls at C-6 and C-12 are not accommodated by alternate oxygenation, and can be deduced to be introduced later. Since a combination of acetyl starter and malonyl extender would give a straight-chain poly-β-ketoester, it is possible to identify the actual starter and extender

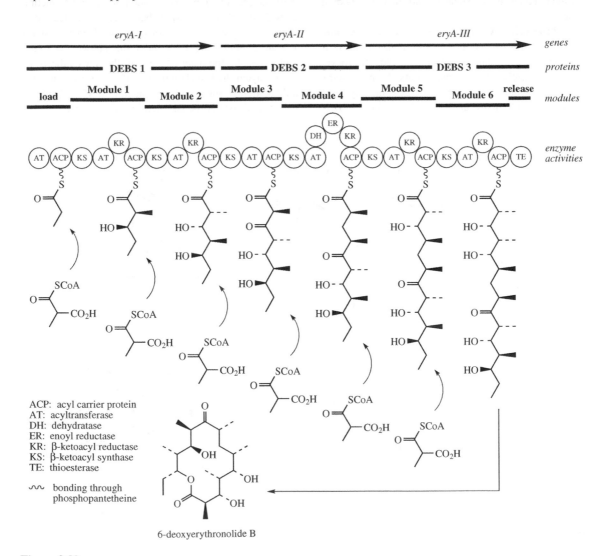

ACP: acyl carrier protein
AT: acyltransferase
DH: dehydratase
ER: enoyl reductase
KR: β-ketoacyl reductase
KS: β-ketoacyl synthase
TE: thioesterase

〰 bonding through phosphopantetheine

6-deoxyerythronolide B

Figure 3.33

units used from the extra substituents at the start and between the carbonyls. In this case, a combination of propionyl-CoA as starter and six methylmalonyl-CoA extenders can be deduced. Comparison with erythronolide A then allows the fate of the carbonyl groups during polyketide assembly to be appraised. Cyclization of this polyketide upon release from the enzyme should generate 6-deoxyerythronolide B.

6-Deoxyerythronolide B synthase (DEBS) is a modular type I PKS involved in erythromycin biosynthesis; its organization and function are illustrated in Figure 3.33. The enzyme contains three subunits (DEBS-1, DEBS-2, and DEBS-3), each encoded by a gene (*eryA*-I, *eryA*-II, and *eryA*-III). It has a linear sequence of six modules, each of which contains the activities needed for one cycle of chain extension, and a loading domain associated with the first module. A minimal module contains a β-ketoacyl synthase (KS), an acyltransferase (AT), and an ACP that

together would catalyse a two-carbon chain extension. The specificity of the AT for either malonyl-CoA or an alkyl-malonyl-CoA determines which two-carbon chain extender is used; in DEBS, the extender units are all methylmalonyl-CoA. The starter unit used is similarly determined by the specificity of the AT in the loading domain; for DEBS, this is propionyl-CoA. After each condensation reaction, the oxidation state of the β-carbon is determined by the presence of a β-ketoacyl reductase (KR), a KR plus a dehydratase (DH), or a KR plus DH plus an enoyl reductase (ER) in the appropriate module. Note that these modifications occur on the β-carbonyl, which is actually provided by the preceding module, i.e. module 1 modifies the oxidation state of the starter unit, etc. Thus, in DEBS, module 3 lacks any β-carbon-modifying domains, modules 1, 2, 5, and 6 contain KR domains and are responsible for production of hydroxy substituents, whereas module 4

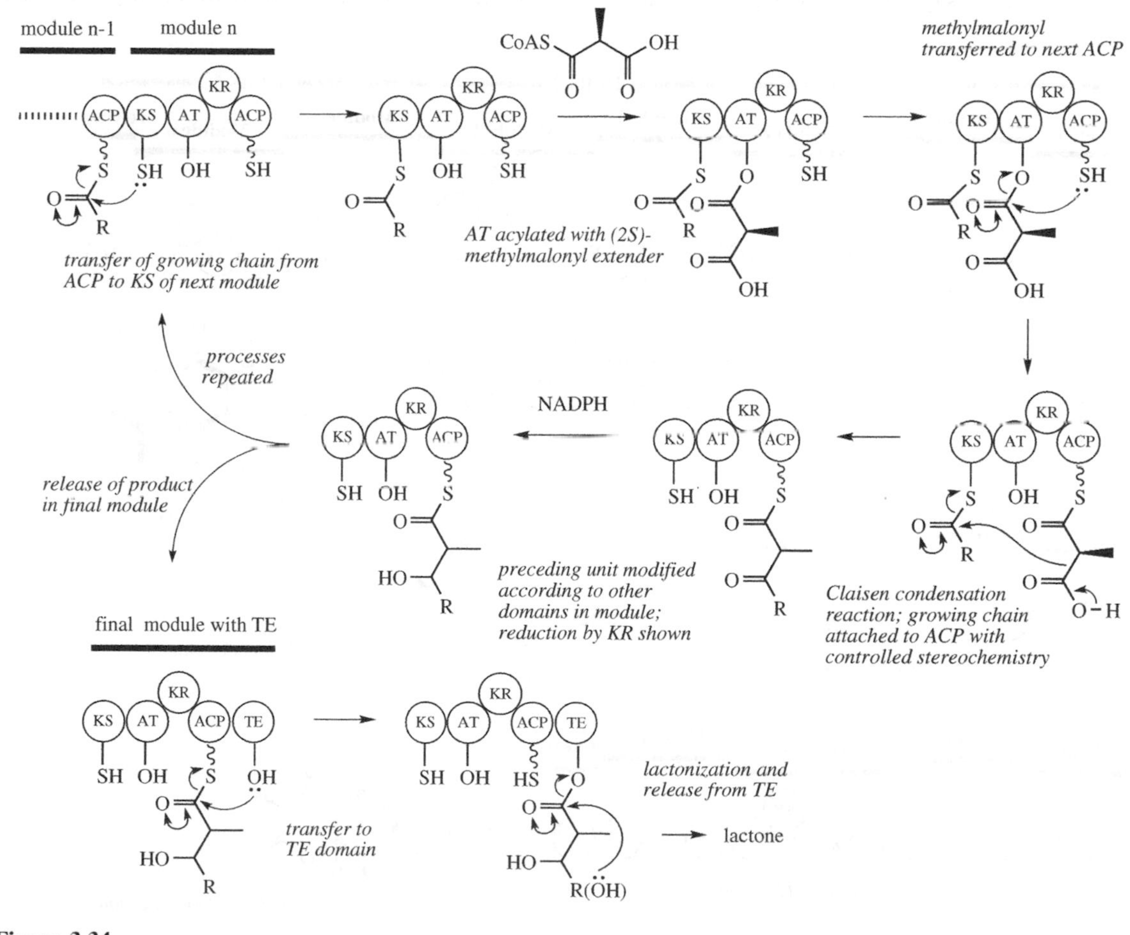

Figure 3.34

contains the complete KR, DH, and ER set, and results in complete reduction to a methylene. The chain is finally terminated by a thioesterase (TE) activity which releases the polyketide from the enzyme and allows cyclization.

The reactions involved in a typical module are outlined in Figure 3.34. The KS is responsible for the Claisen reaction and the growing chain is transferred to it from the ACP domain of the previous module. Initially, a starter unit, e.g. propionyl-CoA, would be bonded to the AT of the loading domain, then transferred to the loading domain's ACP – these steps are not shown in Figure 3.34, though analogous ones form part of the sequence given. The extender unit, here methylmalonyl-CoA, is attached to the AT domain, then transferred to the next downstream ACP. At this stage, the Claisen reaction is catalysed and the growing chain is thus attached to the ACP. The resultant stereochemistry of the methyl groups is also controlled by the enzyme; only (2S)-methylmalonyl-CoA is utilized, but epimerization occurs in some cycles, changing the stereochemistry. The reduction processes, if any,

are now catalysed according to the presence of KR, DH, and ER domains in this module. In the example shown, reduction of the ketone to a hydroxyl is catalysed by the KR domain. The ACP is linked to the chain through a phosphopantetheine group, allowing the various domains to be accessed, as seen in FASs. Transfer of the modified chain to KS in the next module allows the sequence to be repeated. Alternatively, at the end of the sequence, transfer to the TE allows lactonization (via one of the alcohol functions) and release from the enzyme.

Overall, the AT specificity and the catalytic domains on each module determine the structure and stereochemistry of each two-carbon extension unit, the order of the modules specifies the sequence of the units, and the number of modules determines the size of the polyketide chain. The vast structural diversity of natural polyketides arises from combinatorial possibilities of arranging modules containing the various catalytic domains, the sequence and number of modules, the stereochemistry of associated side-chains, and the post-PKS enzymes which

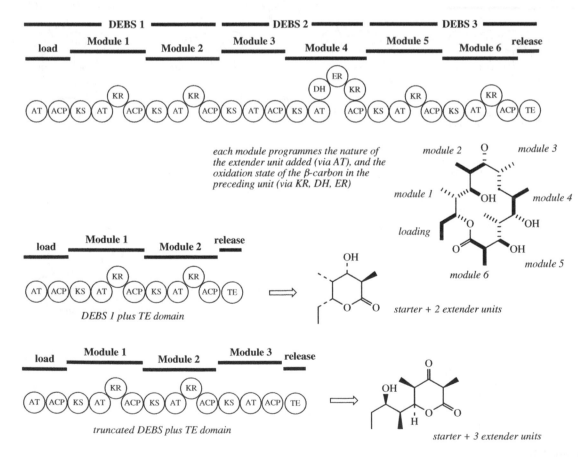

each module programmes the nature of the extender unit added (via AT), and the oxidation state of the β-carbon in the preceding unit (via KR, DH, ER)

DEBS 1 plus TE domain

starter + 2 extender units

truncated DEBS plus TE domain

starter + 3 extender units

Figure 3.35

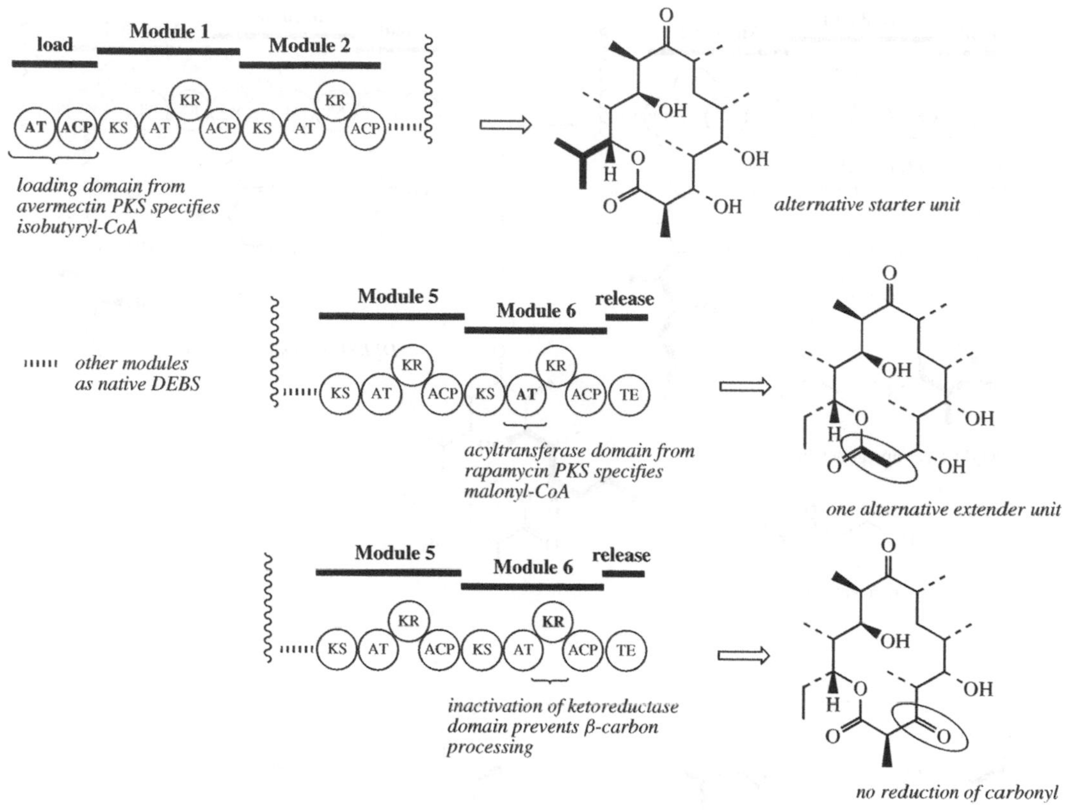

Figure 3.35 (*continued*)

subsequently modify the first-formed product by oxidation, glycosylation, etc., e.g. 6-deoxyerythronolide B → erythromycin A. Genetic engineering now offers vast opportunities for rational modification of the resultant polyketide structure, especially in PKS enzymes of the modular type I, which prove particularly adaptable. A few representative examples of more than a hundred successful experiments leading to engineered polyketides using the DEBS protein are shown in Figure 3.35.

Reducing the size of the gene sequence so that it encodes fewer modules results in the formation of smaller polyketides, characterized by the corresponding loss of extender units; in these examples, the gene encoding the chain-terminating TE also has to be attached to complete the biosynthetic sequence. Such products are naturally considerably smaller than erythromycin, but they have value as synthons containing defined functions and stereochemistry and can be used in the total synthesis of other natural product derivatives (see page 91). Replacing the loading domain of DEBS with that from another PKS, e.g. that producing avermectin (see page 80), alters the specificity of the enzyme for the starter unit. The loading module of the avermectin-producing PKS actually has a much broader specificity than that for DEBS; Figure 3.35 shows the utilization of isobutyryl-CoA, which features in the natural biosynthesis of avermectin B_{1b}. Other examples include the replacement of an AT domain (in DEBS specifying a methylmalonyl extender) with a malonyl-specific AT domain from the rapamycin-producing PKS (see page 84), and inactivation of a KR domain by modifying a catalytic amino acid, thus stopping any β-carbon processing for that module with consequent retention of a carbonyl group. Not all experiments in gene modification are successful, and even when they are, yields can be disappointingly lower than in the natural system. There is always a fundamental requirement that enzymes catalysing steps after the point of modification need to have sufficiently broad substrate specificities to accept and process the abnormal compounds being synthesized; this becomes more unlikely where two or more genetic changes have been made. Nevertheless,

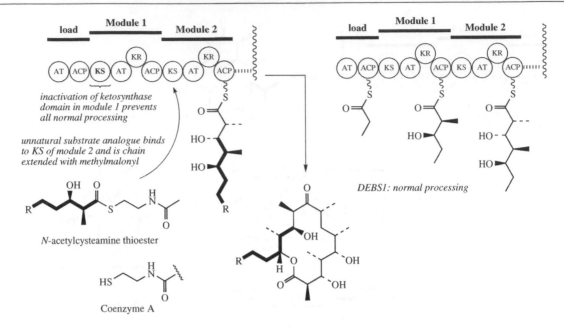

Figure 3.36

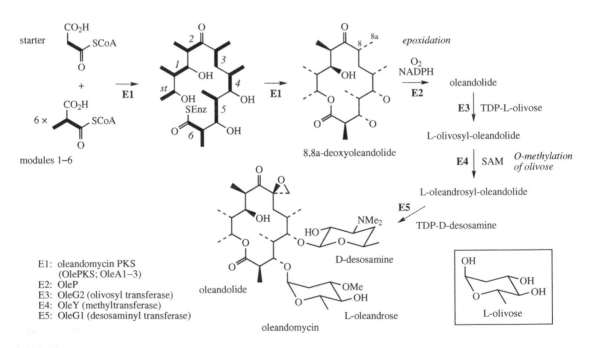

Figure 3.37

some multiple modifications have been successful, and it has also been possible to exploit changes in a combinatorial fashion using different expression vectors for the individual subunits, thus creating a library of polyketides, which may then be screened for potential biological activity.

Another approach is to disable a particular enzymic function, again by altering a specific amino acid, and to supply an alternative synthetic substrate analogue. Provided this substrate is not too dissimilar to the natural substrate, it may then be processed by the later enzymes. This is exemplified in Figure 3.36, where a number of artifical substrates may be utilized to produce a range of erythromycin derivatives inaccessible by semi-synthesis.

The KS domain in module 1 is disabled by replacement of the active site cysteine, and the alternative substrate is supplied in the form of an *N*-acetylcysteamine thioester. The *N*-acetylcysteamine group acts as a simple analogue of coenzyme A, which is acceptable to and processed by the PKS.

A combination of propionate and acetate units is used to produce the 14-membered macrocyclic ring of **oleandomycin** (Figure 3.37) from *Streptomyces antibioticus* [Box 3.6], but otherwise, many of the structural features and the stereochemistry of oleandomycin resemble those of erythromycin A. Structural analysis as described above for erthyromycin would predict that the starter unit should

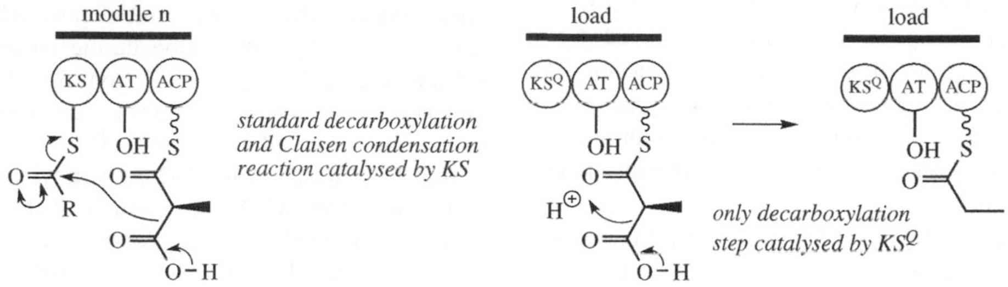

Figure 3.38

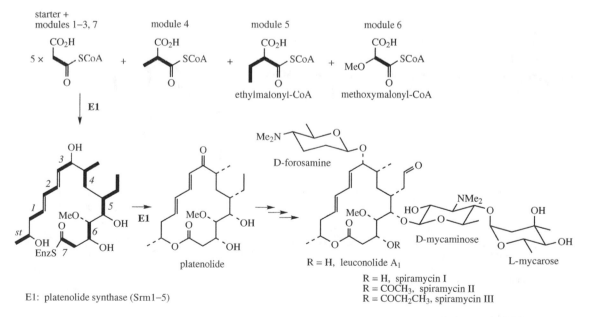

E1: platenolide synthase (Srm1–5)

ethylmalonyl-CoA

methoxymalonyl-CoA

D-forosamine

D-mycaminose

L-mycarose

platenolide

R = H, leuconolide A₁

R = H, spiramycin I
R = COCH₃, spiramycin II
R = COCH₂CH₃, spiramycin III

Figure 3.39

be acetyl-CoA, whilst six propionates, via methyl-malonyl-CoA, supply the extension units (Figure 3.37). The oleandomycin PKS (OlePKS) is organized essentially the same as DEBS, though with one significant difference. It has been discovered that the loading domain in the OlePKS contains an AT specific for malonyl-CoA (not acetyl-CoA) and contains an additional KS domain (KS^Q). In KS^Q, an active site Cys residue is replaced by Gln and cannot make the essential thioester linkage; Q is the single-letter abbreviation for glutamine (see page 423). Consequently, the KS^Q domain cannot participate in any Claisen reaction, and its role is to decarboxylate the ACP-bound substrate to acetyl-ACP for use as the starter (Figure 3.38); many other examples with this type of mechanism are known. Hence, oleandomycin is derived from malonyl-CoA and six methylmalonyl-CoAs. The 8-methyl group is subsequently modified by a P-450 monooxygenase to give an epoxide function. The sugar units in oleandomycin are D-desosamine (as in erythromycin) and L-oleandrose; as with erythromycin, *O*-methylation of one sugar unit occurs after attachment to the macrolide.

Spiramycin I (Figure 3.39) [Box 3.6] from *Streptomyces ambofaciens* has a 16-membered lactone ring and is built up from a combination of six acetate units (one as starter, but all derived from malonyl-CoA), one propionate extender (from methylmalonyl-CoA), together with two further variants. First, a butyrate chain extender is incorporated via ethylmalonyl-CoA and yields an extension unit having an ethyl side-chain. Second, a methoxymalonyl-CoA extender incorporates an extension unit with a methoxy side-chain; this function might plausibly be expected to arise from hydroxylation followed by methylation. The relationships are outlined in Figure 3.39; the first-formed product is platenolide. The platenolide PKS is composed of seven modules plus a loading domain, organized into five subunits: loading domain plus modules 1 and 2 on the first subunit, module 3 on the second, modules 4 and 5 on subunit 3, and modules 6 and 7 on subunits 4 and 5. Platenolide can be modified to a number of other macrolides. For the spiramycins, the ethyl side-chain is subsequently oxidized to generate an aldehyde. Spiramycin I also contains a conjugated diene, the result of carbonyl reductions being followed by dehydration during chain assembly. **Tylosin** (Figure 3.40) [Box 3.6] from *Streptomyces fradiae* has many structural resemblances to the spiramycins, but can be analysed as a propionate starter (actually derived from methylmalonyl-CoA) with chain extension from two malonyl-CoA units, four methylmalonyl-CoA units, and one ethylmalonyl-CoA unit. Oxidation at C-20 in the ethyl side-chain to give an aldehyde is known to occur after introduction of the first sugar, D-mycaminose, and this is followed by hydroxylation of the methyl C-23; both enzymes involved are cytochrome P-450 systems.

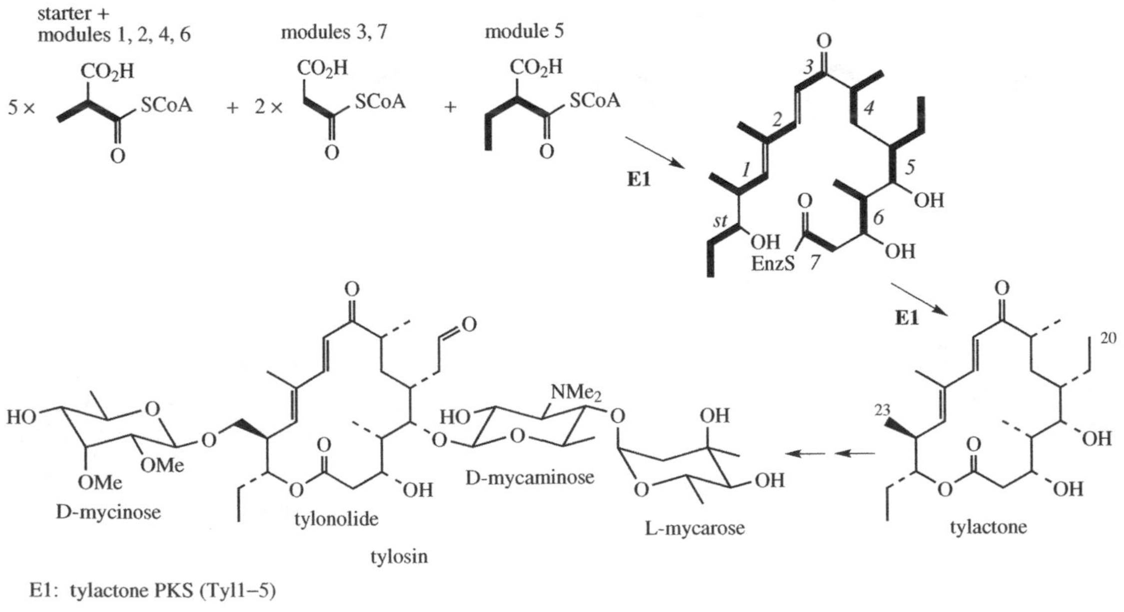

E1: tylactone PKS (Tyl1–5)

Figure 3.40

Box 3.6

Macrolide Antibiotics

The macrolide antibiotics are macrocyclic lactones with a ring size typically 12–16 atoms, and with extensive branching through methyl substituents. Two or more sugar units are attached through glycoside linkages; these sugars tend to be unusual 6-deoxy structures often restricted to this class of compounds. Examples include L-cladinose, L-mycarose, D-mycinose, and L-oleandrose. At least one sugar is an amino sugar, e.g. D-desosamine, D-forosamine, and D-mycaminose. These antibiotics have a narrow spectrum of antibacterial activity, principally against Gram-positive microorganisms. Their antibacterial spectrum resembles, but is not identical to, that of the penicillins, so they provide a valuable alternative for patients allergic to the penicillins. Erythromycin is the principal macrolide antibacterial currently used in medicine.

Erythromycins

The erythromycins (Figure 3.41) are macrolide antibiotics produced by cultures of *Saccharopolyspora erythraea* (formerly *Streptomyces erythreus*). The commercial product **erythromycin** is a mixture containing principally erythromycin A plus small amounts of erythromycins B and C (Figure 3.41). Erythromycin activity is predominantly against Gram-positive bacteria, and the antibiotic is prescribed for penicillin-allergic patients. It is also used against penicillin-resistant *Staphylococcus* strains, in the treatment of respiratory tract infections, and systemically for skin conditions such as acne. It is the antibiotic of choice for infections of *Legionella pneumophila*, the cause of legionnaire's disease. Erythromycin exerts its antibacterial action by inhibiting protein biosynthesis in sensitive organisms. It binds reversibly to the larger 50S subunit of bacterial ribosomes and blocks the translocation step in which the growing peptidyl-tRNA moves from the aminoacyl acceptor site to the peptidyl donor site on the ribosome (see page 422). The antibiotic is a relatively safe drug with few serious side-effects. Nausea and vomiting may occur, and if high doses are prescribed, a temporary loss of hearing might be experienced. Hepatotoxicity may also occur at high dosage. Unfortunately, the drug has a particularly vile taste.

Erythromycin is unstable under acidic conditions, undergoing degradation to inactive compounds by a process initiated by the 6-hydroxyl attacking the 9-carbonyl to form a hemiketal, and by a similar reaction involving the 12-hydroxyl. Dehydration to an enol ether may follow, though the principal product is the spiroketal anhydroerythromycin A (Figure 3.42). The keto form is the only material with significant antibacterial activity, and the enol ether shown is responsible for gastrointestinal side-effects. Thus, to protect oral preparations of erythromycin against gastric acid, they are formulated as enteric coated tablets, or as insoluble esters (e.g. ethyl succinate esters) which are then hydrolysed in the intestine. Esterification produces a taste-free product, and typically involves the hydroxyl of the amino sugar desosamine. To reduce acid instability, semi-synthetic analogues of erythromycin have also been developed. **Clarithromycin** (Figure 3.41) is a 6-*O*-methyl derivative of erythromycin A; this modification blocks 6,9-hemiketal formation as in Figure 3.42. **Azithromycin** (Figure 3.41) is a ring-expanded aza-macrolide in which the carbonyl function has been removed. In both analogues, the changes enhance activity compared with that of erythromycin, and cause fewer gastrointestinal side-effects. Clarithromycin is used as part of the drug regimen for eradication of *Helicobacter pylori* in the treatment of gastric and duodenal ulcers.

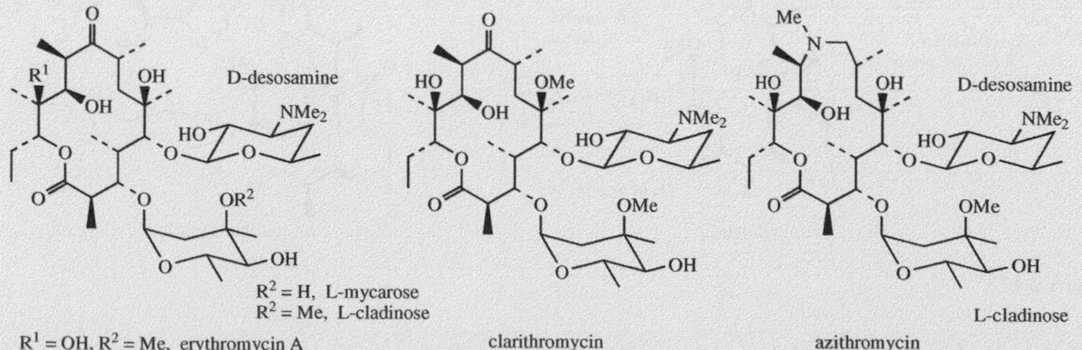

$R^1 = OH$, $R^2 = Me$, erythromycin A
$R^1 = H$, $R^2 = Me$, erythromycin B
$R^1 = OH$, $R^2 = H$, erythromycin C

$R^2 = H$, L-mycarose
$R^2 = Me$, L-cladinose

clarithromycin
(6-*O*-methyl erythromycin A)

azithromycin

Figure 3.41

Box 3.6   (continued)

Bacterial resistance to erythromycin has become significant and has limited its therapeutic use against many strains of *Staphylococcus*. Several mechanisms of resistance have been implicated, one of which is a change in permeability of the bacterial cell wall. Differences in permeability also appear to explain the relative insensitivity of Gram-negative bacteria to erythromycin when compared with Gram-positive bacteria. Resistant bacteria may also modify the chemical nature of the binding site on the ribosome, thus preventing antibiotic binding, and some organisms are now known to metabolize the macrolide ring to yield inactive products. New derivatives based upon 3-keto-6-*O*-methylerythromycin A, termed ketolides, are providing important advances in this respect. These agents retain potent antibacterial activity towards erythromycin-susceptible organisms, but improved activity towards MLS$_B$ (macrolide–lincosamide–streptogramin B)-resistant bacteria. In general, the ketolide structures differ from erythromycin by removal of the cladinose sugar and replacement with a 3-keto group, the presence of a 6-methoxy group, and attaching a cyclic carbamate at C-11/C-12. The first of these drugs in use is **telithromycin** (Figure 3.43).

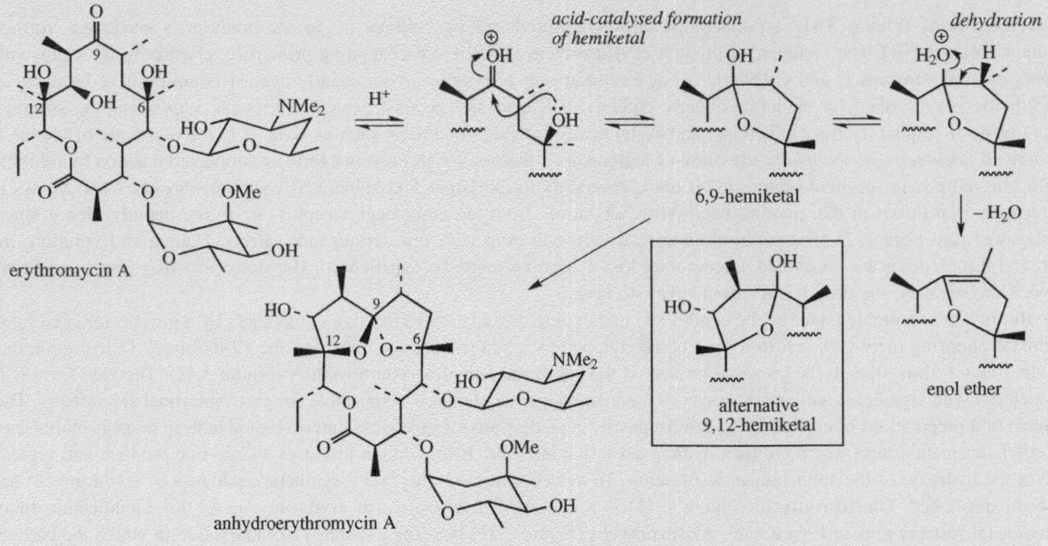

Figure 3.42

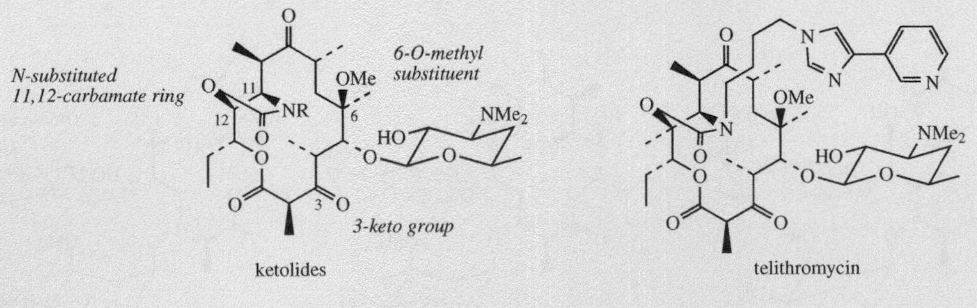

Figure 3.43

Oleandomycin and Spiramycin

Oleandomycin (Figure 3.37) is produced by fermentation cultures of *Streptomyces antibioticus* and has been used medicinally as its triacetyl ester **troleandomycin** against Gram-positive bacterial infections; it is rarely employed now. The **spiramycins**

Box 3.6 (continued)

(Figure 3.39) are macrolides produced by cultures of *Streptomyces ambofaciens*. The commercial antibiotic is a mixture containing principally spiramycin I, together with smaller amounts (10–15% each) of the acetyl ester spiramycin II and the propionyl ester spiramycin III. This antibiotic has recently been introduced into medicine for the treatment of toxoplasmosis, infections caused by the protozoan *Toxoplasma gondii*.

Tylosin

Tylosin (Figure 3.40) is an important veterinary antibiotic. It is produced by *Streptomyces fradiae*, and is used to control chronic respiratory diseases caused by *Mycoplasma galliseptum* in poultry and to treat Gram-positive infections in pigs.

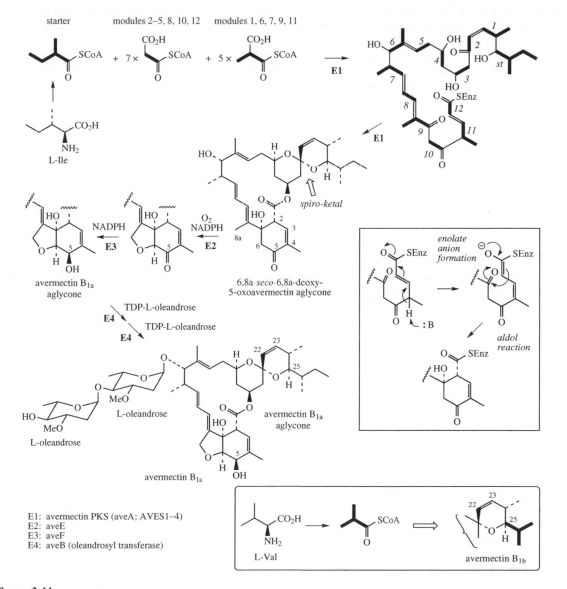

E1: avermectin PKS (aveA; AVES1–4)
E2: aveE
E3: aveF
E4: aveB (oleandrosyl transferase)

Figure 3.44

The **avermectins** (Figure 3.44) have no antibacterial activity, but possess anthelmintic, insecticidal, and acaricidal properties, and these are exploited in human and veterinary medicine [Box 3.7]. The avermectins are also 16-membered macrolides, but their structures are made up from a much longer polyketide chain (starter plus 12 extenders) which is also used to form oxygen heterocycles fused to the macrolide. **Avermectin B$_{1a}$** exemplifies a typical structure, and the basic carbon skeleton required to produce this can be postulated as in Figure 3.44. The starter unit in this case would be 2-methylbutyryl-CoA, which is derived from the amino acid L-isoleucine (compare iso-fatty acids, page 56). Both malonyl-CoA and methylmalonyl-CoA are then utilized as extender units. The simple structural analysis approach does not pick up two variants: extender 10 contains an unmodified keto function, and there is a double bond between extenders 11 and 12, not present in the final structure (C-2,3 in avermectin). Further, the product of the avermectin PKS is not a simple macrolide, but a structure in which additional cyclizations have also been catalysed. The spiroketal system is easily accounted for as a combination of ketone and two alcohol functions. The cyclohexenone ring arises by an aldol reaction as shown, and is dependent upon the 2,3-double bond and 5-keto function just mentioned. This reaction may precede lactone ring formation, based on the increased acidity of thioesters relative to oxygen lactones. The avermectin PKS is composed of four polypeptide subunits: load plus modules 1 and 2; modules 3–6; modules 7–9; modules 10–12. Intriguingly, module 7 contains a DH domain and module 10 a KR domain, but both of these are non-functional, so do not contribute to the oxygenation state of the extender unit added; in the other systems discussed so far, variation in extender modification is simply a result of the appropriate domain being absent. Post-PKS modifications include oxidative cyclization to produce the tetrahydrofuran ring, which appears to be catalysed by a single cytochrome P-450-dependent enzyme. Glycosylation is accomplished, unusually, by the repeated action of a single glycosyltransferase.

Box 3.7

Avermectins

The avermectins are a group of macrolides with strong anthelmintic, insecticidal, and acaricidal properties, but with low toxicity to animals and humans. They are produced by cultures of *Streptomyces avermectilis*. Some eight closely related structures have been identified, with avermectins B$_{1a}$ (Figure 3.44) and B$_{2a}$ (Figure 3.45) being the most active antiparasitic compounds. **Abamectin** (a mixture of about 85% avermectin B$_{1a}$ and about 15% avermectin B$_{1b}$) is used on agricultural crops to control mites and insects. **Ivermectin** (Figure 3.45) is a semi-synthetic 22,23-dihydro derivative of avermectin B$_{1a}$ (the commercial product also contains up to 20% dihydroavermectin B$_{1b}$) and was first used in veterinary practice against insects, ticks, mites, and roundworms. Although it is a broad spectrum nematocide against roundworms, it is inactive against tapeworms and flatworms, or against bacteria and fungi. It is an extremely potent agent, and is effective at very low dosages. It is now the drug of choice for use against filarial and several other worm parasites in humans, e.g. river blindness caused by the nematode *Onchocerca volvulus*. The avermectins and ivermectin target the glutamate-gated chloride channels unique to nematodes, insects, ticks, and arachnids, resulting in neuro-muscular paralysis and death.

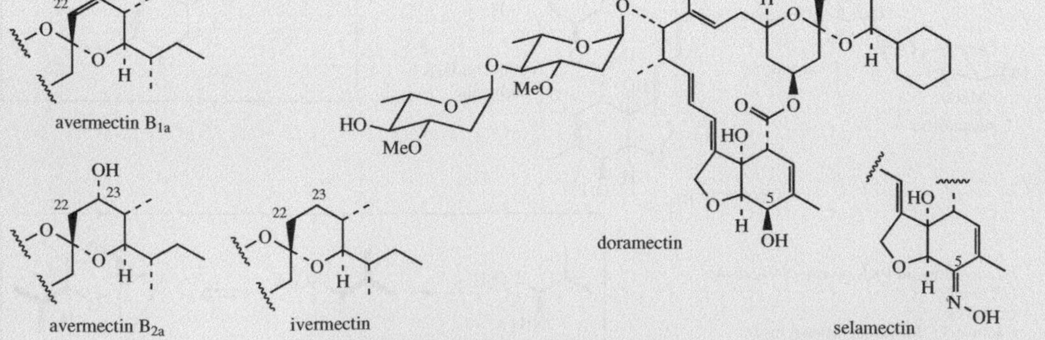

Figure 3.45

Box 3.7 (continued)

Doramectin is a newer analogue of avermectin B_{1a} (Figure 3.45) with extended biological activity, used to protect cattle against internal and external parasites. It is produced in a mutant of *Streptomyces avermectilis* that is unable to synthesize the branched-chain starter acids required for avermectin biosynthesis. The organism is able to utilize exogenously supplied acids as starter groups and incorporate them into modified avermectins; doramectin is produced by the use of cyclohexanecarboxylic acid in the culture medium (compare penicillins, page 461). It has also proved possible to express in this mutant biosynthetic genes for the production of cyclohexanecarbonyl-CoA, a natural compound involved in the biosynthesis of some other antibiotics. This engineered mutant synthesizes doramectin without the need for cyclohexanecarboxylic acid supplementation. **Selamectin** is the 5-oxime of doramectin, used to control fleas and internal parasites in dogs and cats.

Avermectins are usually isolated as a mixture in which the main *a* component has a 2-butyl group (derived from isoleucine) at C-25, whilst the minor *b* component has a 2-propyl group instead, e.g. **avermectin B_{1b}**. In this case, the starter group is 2-methylpropionyl-CoA, derived from the amino acid L-valine. The A-series of avermectins are the 5-methoxy analogues of the B-series.

Even larger macrolides are encountered in the **polyene macrolides**, most of which have antifungal properties, but not antibacterial activity [Box 3.8]. The macrolide ring size ranges from 26 to 38 atoms, and this also accommodates a conjugated polyene of up to seven *E* double bonds. Relatively few methyl groups are attached to the ring, and thus malonyl-CoA is utilized more frequently than methylmalonyl-CoA as chain extender. Typical examples are **amphotericin B** (Figure 3.46) from *Streptomyces nodosus* and **nystatin A_1** from *Streptomyces noursei*. These

have very similar structures and are derived from the same basic precursors (Figure 3.46). The ring size becomes contracted due to cross-linking by formation of a hemiketal; this modification occurs before or during release from the PKS. The two compounds have slightly different hydroxylation patterns, part of which is introduced by a post-PKS hydroxylation. The two areas of conjugation in nystatin A_1 are extended into a heptaene system in amphotericin B. Both compounds are carboxylic acids, a result of oxidation of a propionate-derived methyl group, and glycosylated with the amino sugar D-mycosamine. The PKSs involved in amphotericin and nystatin biosynthesis are also closely related. They are composed of six polypeptide subunits: load; modules 1 and 2; modules 3–8; modules 9–14; modules 15–17; module 18. Both proteins also contain some non-functional domains.

Box 3.8

Polyene Antifungals

The polyene antifungals are a group of macrocyclic lactones with very large 26–38-membered rings. They are characterized by the presence of a series of conjugated *E* double bonds and are classified according to the longest conjugated chain present. Medicinally important ones include the heptaene amphotericin B and the tetraene nystatin. There are relatively few methyl branches in the macrocyclic chain. The polyenes have no antibacterial activity but are useful antifungal agents. Their activity is a result of binding to sterols in the eukaryotic cell membrane; this action explains the lack of antibacterial activity, because bacterial cells do not contain sterol components (see page 241). Fungal cells are also attacked rather than mammalian cells, since the antibiotics bind about 10-fold more strongly to ergosterol, the major fungal sterol (see page 252), than to cholesterol, the main animal sterol component (see page 251). This binding modifies the cell wall permeability and leads to formation of transmembrane pores that allow K^+ ions, sugars, and proteins to be lost from the microorganism. Though binding to cholesterol is less than to ergosterol, it is responsible for the observed toxic side-effects of these agents on humans. The polyenes are relatively unstable, undergoing light-catalysed decomposition, and are effectively insoluble in water. This insolubility actually protects the antibiotic from gastric decomposition, allowing oral treatment of infections in the intestinal tract.

Amphotericin is an antifungal polyene produced by cultures of *Streptomyces nodosus* and contains principally the heptaene amphotericin B (Figure 3.46) together with structurally related compounds, e.g. the tetraene amphotericin A (about 10%) which is the 28,29-dihydro analogue of amphotericin B. Amphotericin A is much less active than amphotericin B. Amphotericin is active against most fungi and yeasts, but it is not absorbed from the gut, so oral administration is restricted to the treatment of intestinal candidiasis. It is administered intravenously for treating potentially life-threatening systemic fungal infections. However, it then becomes highly protein bound, resulting in poor penetration and slow elimination from the body. After parenteral administration, toxic side-effects, including nephrotoxicity, are relatively common. Close supervision and monitoring of the patient is thus

Box 3.8   (continued)

necessary, especially since the treatment may need to be prolonged. Lipid formulations of amphotericin are much less toxic and have proven a significant advance. *Candida* infections in the mouth or on the skin may be treated with appropriate formulations. More recently, amphotericin has been shown to be among the few agents that can slow the course of prion disease in animal models.

   Nystatin is a mixture of tetraene antifungals produced by cultures of *Streptomyces noursei*. The principal component is nystatin A_1 (Figure 3.46), but the commercial material also contains nystatin A_2 and A_3; these have additional glycoside residues. Nystatin is too toxic for intravenous use, but has value for oral treatment of intestinal candidiasis, as lozenges for oral infections, and as creams for topical control of *Candida* species.

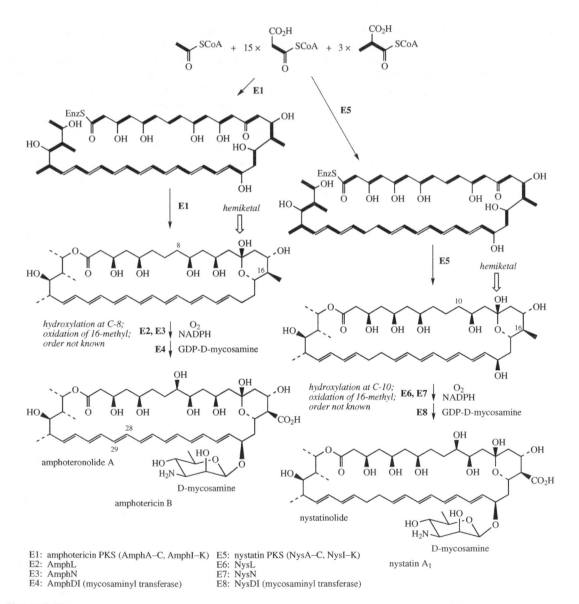

E1: amphotericin PKS (AmphA–C, AmphI–K)   E5: nystatin PKS (NysA–C, NysI–K)
E2: AmphL   E6: NysL
E3: AmphN   E7: NysN
E4: AmphDI (mycosaminyl transferase)   E8: NysDI (mycosaminyl transferase)

Figure 3.46

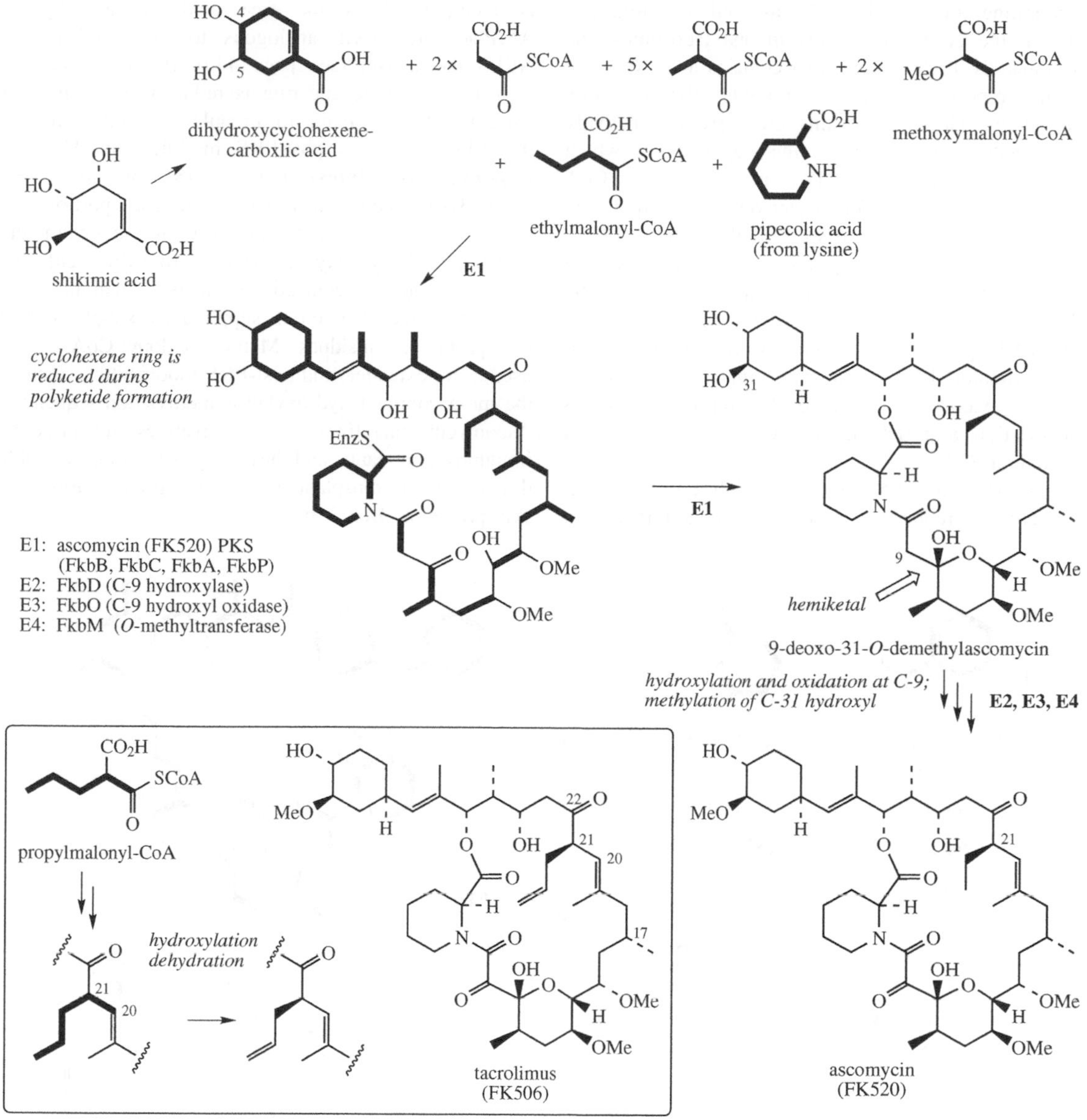

Figure 3.47

Two unusual and clinically significant macrolides are **ascomycin (FK520)**, from *Streptomyces hygroscopicus* var. *ascomyceticus*, and **tacrolimus (FK506)**, isolated from *Streptomyces tsukubaensis* (Figure 3.47). These compounds contain a 23-membered macrolactone that also incorporates an *N*-heterocyclic ring and are identical apart from the substituent at C-21. The piperidine ring and adjacent carbonyl are incorporated as pipecolic acid (see page 330) via an amide linkage onto the end of the growing chain. Chain assembly is catalysed by an enzyme combination that is partly a PKS and partly a non-ribosomal peptide synthetase (NRPS). Though PKSs and NRPSs utilize different substrates (carboxylic thioesters and amino acid thioesters respectively), they exhibit remarkable similarities in the modular arrangement of catalytic domains and product assembly mechanism (see page 438). PKS–NRPS hybrid systems appear to have evolved, and these produce metabolites incorporating amino acid units as well as carboxylic acids. In the cases of ascomycin and tacrolimus, the predominantly PKS system also contains an NRPS-like protein, responsible for incorporating the pipecolic acid moiety. The main chains are derived principally from acetate and propionate, the fragments of which can readily be identified. Two methoxymalonyl-CoA extenders are utilized for ascomycin production and most likely for tacrolimus as well. Methoxymalonyl-CoA is also an extender in spiramycin biosynthesis (see page 75), and its production from a C_3 glycolytic pathway intermediate such as glyceric acid is also part of the PKS system. Ethylmalonyl-CoA (for ascomycin) or propylmalonyl-CoA (for tacrolimus) provides the remaining extender. In the case of tacrolimus, post-PKS hydroxylation and dehydration accounts for the allyl side-chain at C-21.

The starter unit is 4,5-dihydroxycyclohex-1-enecarboxylic acid, a reduction product from shikimate; it is incorporated as the free acid, perhaps by an AMP-activated form analogous to that used in the NRPS mechanism (see page 440). A further surprise is that the cyclohexene ring is reduced during transfer from the load domain to module 1. Methylation of the 5-hydroxyl is a post-PKS modification. Although **rapamycin (sirolimus)** (Figure 3.48) contains a very large 31-membered macrocycle, several portions of the structure are identical to those in ascomycin and tacrolimus. Dihydroxycyclohexenecarboxylic acid and pipecolic acid are again utilized in its formation, whilst the rest of the skeleton is supplied by simple acetate and propionate residues. Methoxymalonyl-CoA is not used as an extender, and post-PKS modifications include the more common hydroxylation/methylation sequences. Ascomycin, tacrolimus, and derivatives are effective immunosuppressants, and they are proving to be valuable drugs in organ transplant surgery and in the treatment of skin problems [Box 3.9].

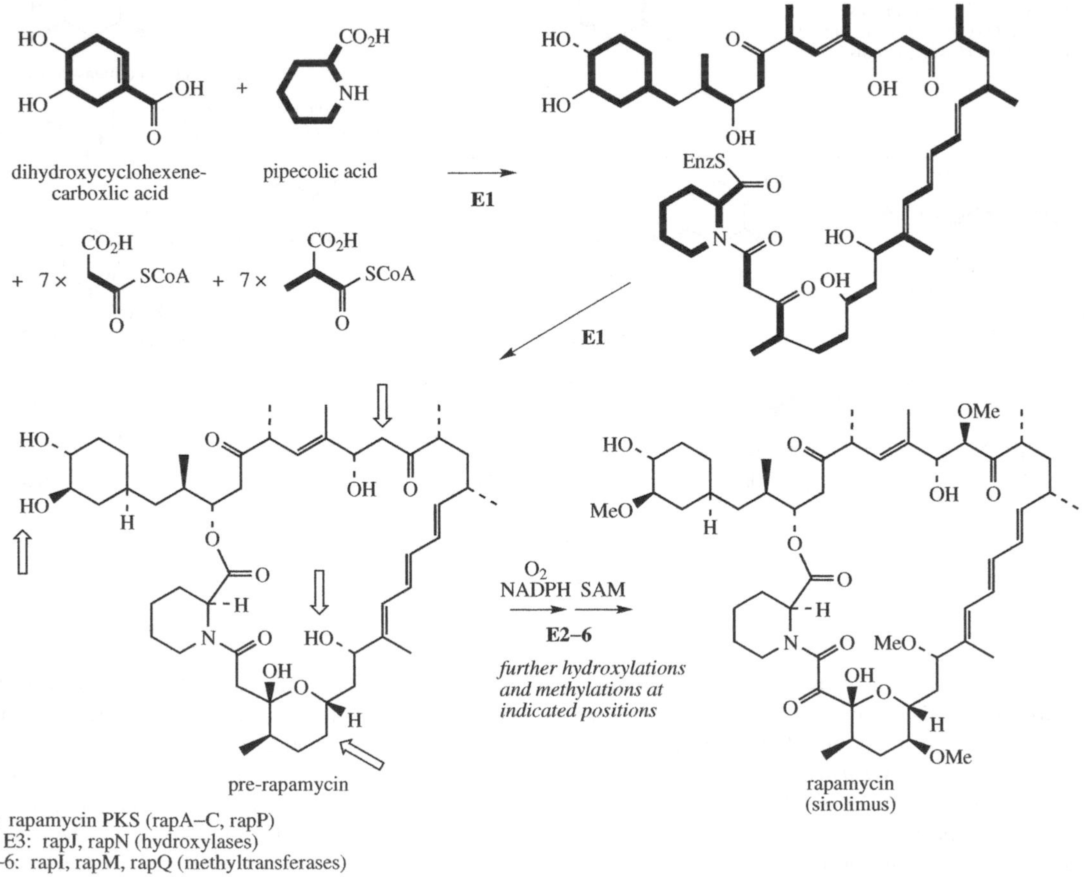

E1: rapamycin PKS (rapA–C, rapP)
E2, E3: rapJ, rapN (hydroxylases)
E4–6: rapI, rapM, rapQ (methyltransferases)

Figure 3.48

Box 3.9

Tacrolimus and Sirolimus

Tacrolimus (FK-506) (Figure 3.47) is a macrolide immunosuppressant isolated from cultures of *Streptomyces tsukubaensis*. It is used in liver and kidney transplant surgery. Despite the significant structural differences between tacrolimus and the cyclic peptide cyclosporin A (ciclosporin; see page 454), these two agents have a similar mode of action. They both inhibit T-cell activation in the immunosuppressive mechanism by binding first to a receptor protein giving a complex which then inhibits a phosphatase enzyme called calcineurin. The resultant aberrant phosphorylation reactions prevent appropriate gene transcription and subsequent T-cell activation. Structural similarities between the region C-17 to C-22 in tacrolimus and fragments of the cyclosporin A peptide chain have been postulated to account for this binding. Tacrolimus is up to 100 times more potent than cyclosporin A; it produces similar side-effects, including neurotoxicity and nephrotoxicity. Tacrolimus is also used topically to treat moderate to severe atopic eczema; the ascomycin derivative **pimecrolimus** (Figure 3.49) is also used for mild to moderate conditions.

Sirolimus (rapamycin) (Figure 3.48) is produced by cultures of *Streptomyces hygroscopicus* and is also used as an immunosuppressant drug. Although tacrolimus and sirolimus possess a common structural unit, and both inhibit T-cell activation, they appear to achieve this by somewhat different mechanisms. The first-formed sirolimus–receptor protein binds not to calcineurin, but to a different protein. It exerts its action later in the cell cycle by blocking growth-factor-driven cell proliferation. **Everolimus** (Figure 3.49) is a semi-synthetic derivative of sirolimus with better oral bioavailability now available for kidney and heart transplants. **Zotarolimus** (Figure 3.49) inhibits cell proliferation, preventing scar tissue formation following cardiovascular surgery; this agent is delivered via a coronary stent. Some semi-synthetic sirolimus derivatives have also been investigated as anticancer drugs. These also display modifications to the C-40 hydroxyl group in the side-chain cyclohexane ring. **Temsirolimus** is now approved for treatment of renal cell carcinoma, and **deforolimus** is in advanced clinical trials against a variety of cancers (Figure 3.49).

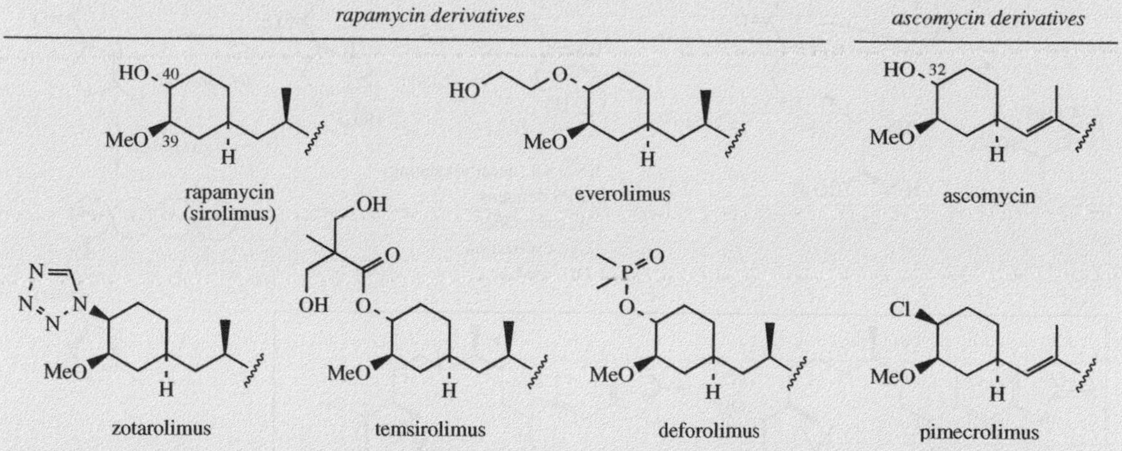

rapamycin derivatives   *ascomycin derivatives*

rapamycin (sirolimus)   everolimus   ascomycin

zotarolimus   temsirolimus   deforolimus   pimecrolimus

Figure 3.49

Attracting considerable interest at the present time are the **epothilones** (Figure 3.50), a group of macrolides produced by cultures of the myxobacterium *Sorangium cellulosum*. They are formed by a multienzyme system composed of nine PKS modules and one NRPS module distributed over six protein subunits. These compounds appear to employ an unusual starter unit containing a thiazole ring; this is constructed from the amino acid cysteine and an acetate unit by EpoB, the NRPS component of the enzyme (see also similar ring systems in bacitracin, page 443, and bleomycin, page 449). The loading module EpoA contains a decarboxylating KS domain, and the acetyl starter is thus derived from malonyl-CoA

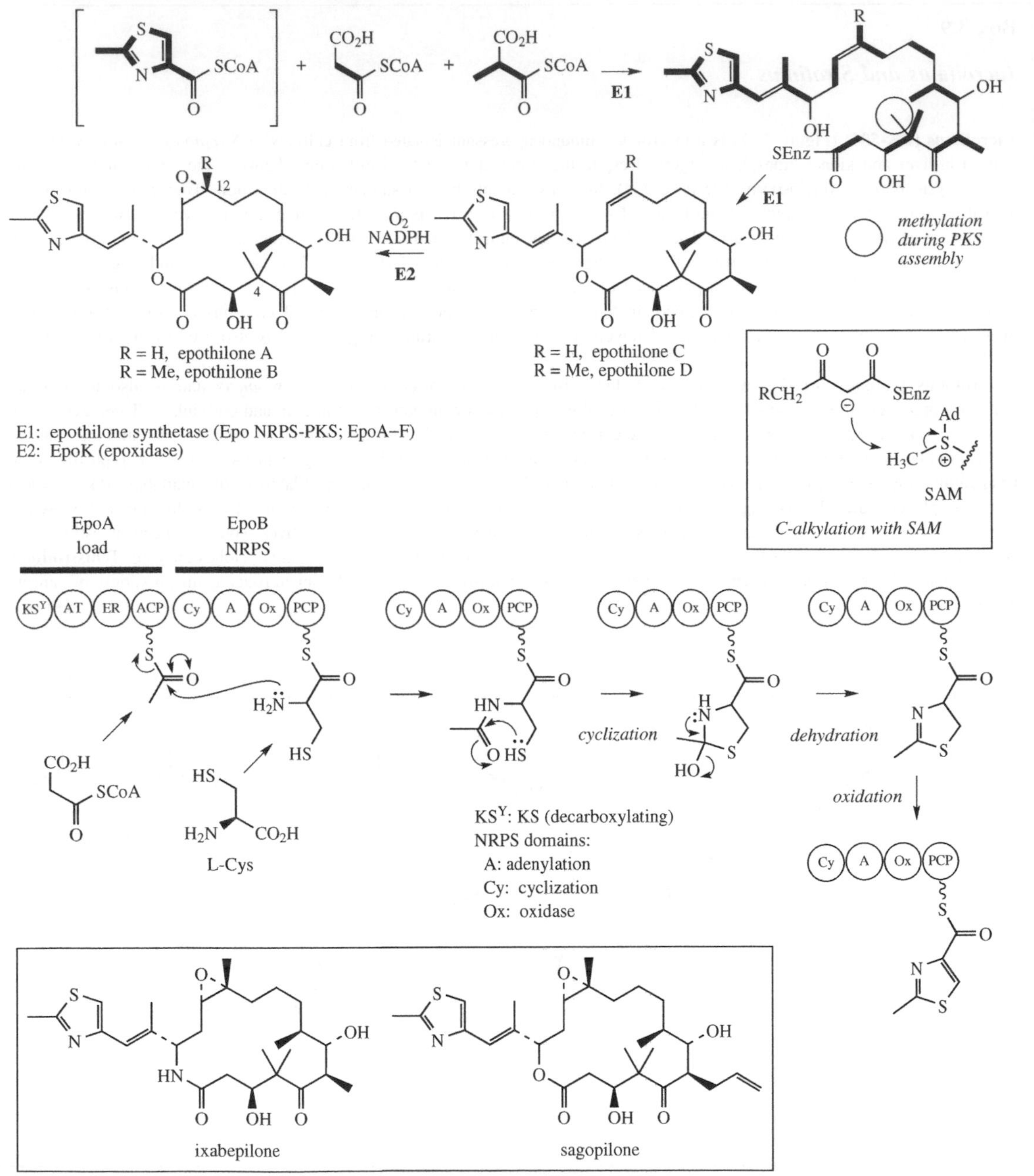
SCoA
CO_2H
SCoA
CO_2H
SCoA
E1
OH
SEnz
methylation during PKS assembly
E1
O_2
NADPH
E2
R = H, epothilone A
R = Me, epothilone B
R = H, epothilone C
R = Me, epothilone D
E1: epothilone synthetase (Epo NRPS-PKS; EpoA–F)
E2: EpoK (epoxidase)
RCH_2
SEnz
Ad
H_3C
SAM
C-alkylation with SAM
EpoA
load
EpoB
NRPS
KS^Y
AT
ER
ACP
Cy
A
Ox
PCP
L-Cys
cyclization
dehydration
oxidation
KS^Y: KS (decarboxylating)
NRPS domains:
A: adenylation
Cy: cyclization
Ox: oxidase
ixabepilone
sagopilone

Figure 3.50

(compare oleandomycin, page 73). The macrolide ring also contains an extra methyl group at C-4, the result of methylation during polyketide chain assembly by a methyltransferase domain in module 8, part of the EpoE protein; this undoubtedly occurs whilst a β-diketone function is present. The other interesting feature is that this bacterium produces epothilone A and epothilone B in a ratio of about 2:1. These compounds differ in the nature of the substituent at C-12, which is hydrogen in epothilone A but a methyl group in epothilone B.

Genetic evidence shows that the PKS enzyme can accept either malonyl-CoA or methylmalonyl-CoA extender units for this position. Thus, epothilone B is constructed from three malonate and five methylmalonate extender units, whilst epothiolone A requires four units of each type. The first-formed products are epothilones C and D; these alkenes are converted into epoxides by a post-PKS cytochrome P-450-dependent epoxidase activity. The epothilones display marked antitumour properties with a mode of action paralleling that of the highly successful anticancer drug taxol (see page 224). However, the epothilones have a much higher potency (2000–5000 times) and are active against cell lines which are resistant to taxol and other drugs. There appears to be considerable potential for developing the epothilones or analogues into valuable anticancer drugs, and several derivatives

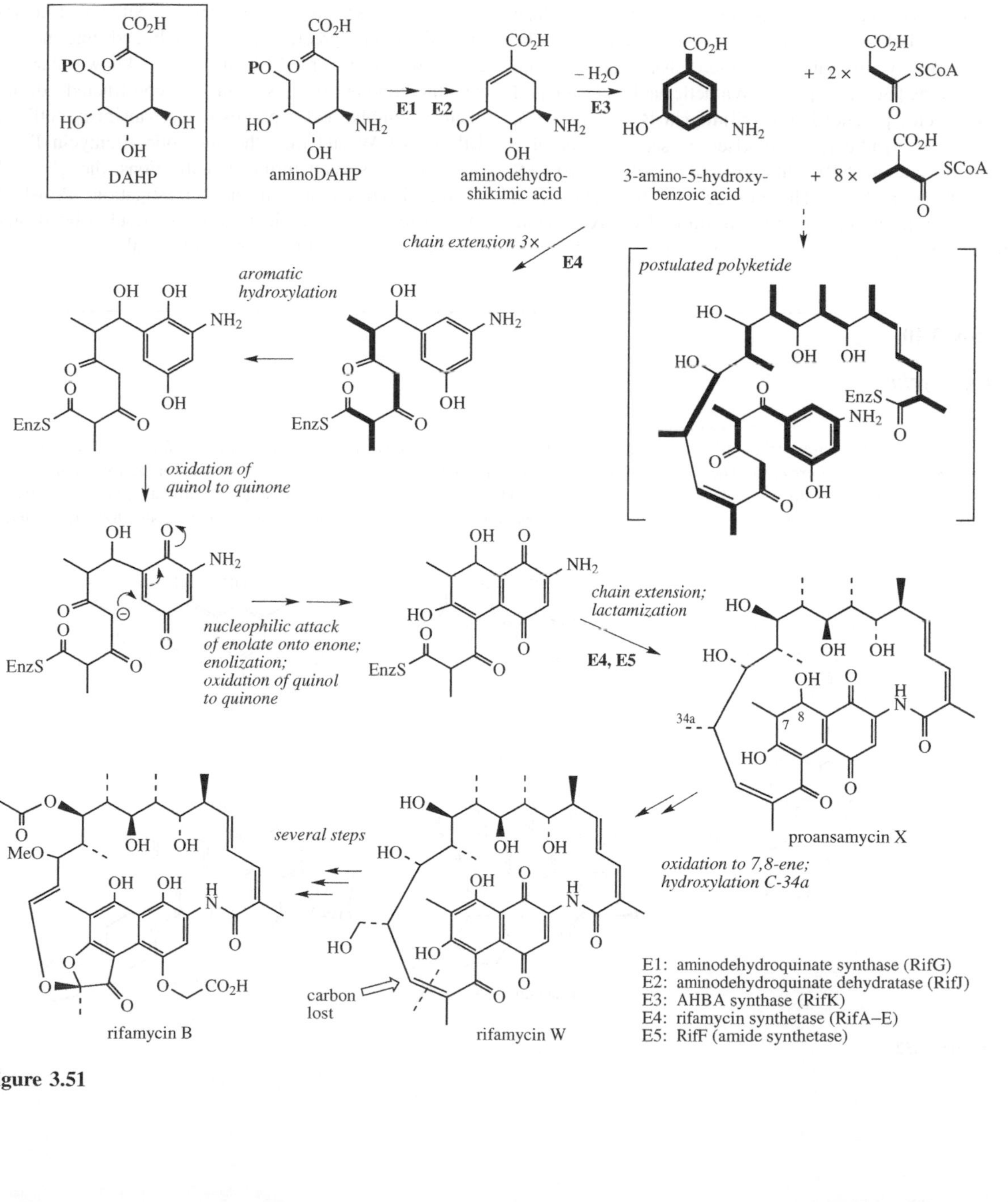
DAHP
aminoDAHP
E1 E2
aminodehydro-shikimic acid
− H2O
E3
3-amino-5-hydroxy-benzoic acid
+ 2 ×
+ 8 ×
SCoA
chain extension 3×
E4
postulated polyketide
aromatic hydroxylation
oxidation of quinol to quinone
nucleophilic attack of enolate onto enone; enolization; oxidation of quinol to quinone
chain extension; lactamization
E4, E5
34a
proansamycin X
oxidation to 7,8-ene; hydroxylation C-34a
several steps
carbon lost
rifamycin B
rifamycin W
E1: aminodehydroquinate synthase (RifG)
E2: aminodehydroquinate dehydratase (RifJ)
E3: AHBA synthase (RifK)
E4: rifamycin synthetase (RifA–E)
E5: RifF (amide synthetase)

Figure 3.51

are undergoing clinical trials. These include epothilone B (**patupilone**) and its allyl analogue **sagopilone**; the lactam **ixabepilone** has recently been approved for drug use in the treatment of breast cancer (Figure 3.50).

A further group of macrolides are termed **ansa macrolides**; these have non-adjacent positions on an aromatic ring bridged by a long aliphatic chain (Latin: *ansa* = handle). The aromatic portion may be a substituted naphthalene or naphthaquinone, or alternatively a substituted benzene ring. The macrocycle in the ansamycins is closed by an amide rather than an ester linkage, i.e. ansamycins are lactams. The only ansamycins currently used therapeutically are **rifamycins** [Box 3.10], semi-synthetic naphthalene-based macrocycles produced from **rifamycin B** (Figure 3.51). Rifamycin synthetase comprises a sequence of five proteins, with an NRPS-like loading module and 10 PKS elongation modules. The nitrogen of the lactam ring derives from the starter unit, **3-amino-5-hydroxybenzoic acid** (Figure 3.51), which is a simple phenolic acid

derivative produced by an unusual variant of the shikimate pathway (see Chapter 4). AminoDAHP is formed via aminosugar metabolism, and then the standard shikimate pathway continues with amino analogues. 3-Amino-5-hydroxybenzoic acid is produced from aminodehydroshikimic acid by dehydration. In the biosynthesis of **rifamycin B** (Figure 3.51) in *Amycolatopsis mediterranei*, this starter unit plus two malonyl-CoA and eight methylmalonyl-CoA extenders are employed to fabricate **proansamycin X** as the first product released from the enzyme. Simple structural analysis might predict an enzyme-bound intermediate as shown in Figure 3.51; it is now known that the naphthoquinone ring system is constructed during chain assembly, and a proposed mechansim is outlined. **Rifamycin W** and then the antibiotic rifamycin B are the result of several further modifications, though much of the detail requires further investigation. Post-PKS modifications include cleavage of the double bond, loss of one carbon, and formation of the ketal.

Box 3.10

Rifamycins

The rifamycins are ansamycin antibiotics produced by cultures of *Amycolatopsis mediterranei* (formerly *Nocardia mediterranei* or *Streptomyces mediterranei*). The crude antibiotic mixture was found to contain five closely related substances rifamycins A–E, but if the organism was cultured in the presence of sodium diethyl barbiturate (barbitone or barbital), the product was almost entirely rifamycin B (Figure 3.52). Rifamycin B has essentially no antibacterial activity, but may be converted chemically,

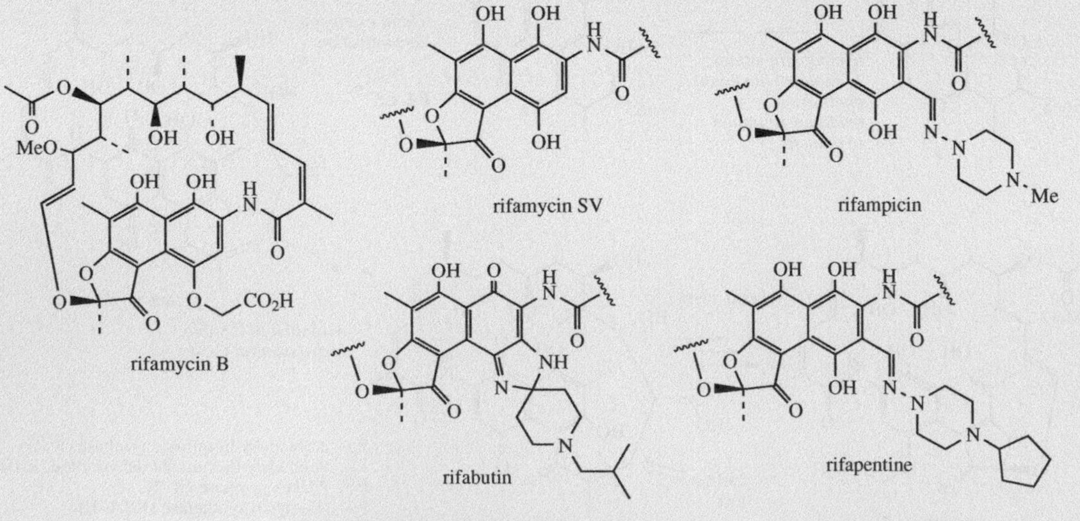

rifamycin B

rifamycin SV

rifampicin

rifabutin

rifapentine

Figure 3.52

Box 3.10 (continued)

enzymically, or by biotransformation into rifamycin SV (Figure 3.52), a highly active antibacterial agent, and the first rifamycin to be used clinically. Further chemical modifications of rifamycin SV have produced better clinically useful drugs. Rifamycin SV is actually the immediate biosynthetic precursor of rifamycin B, and this conversion can be genetically blocked, resulting in the accumulation of rifamycin SV. Strain optimization has not been fruitful, however, and most commercial fermentations still rely on producing rifamycin B.

The most widely used agent is the semi-synthetic derivative **rifampicin** (Figure 3.52). Rifampicin has a wide antibacterial spectrum, with high activity towards Gram-positive bacteria and a lower activity towards Gram-negative organisms. Its most valuable activity is towards *Mycobacterium tuberculosis*, and rifampicin is a key agent in the treatment of tuberculosis, usually in combination with at least one other drug to reduce the chances for development of resistant bacterial strains. It is also useful in control of meningococcal meningitis and leprosy. Rifampicin's antibacterial activity arises from inhibition of RNA synthesis by binding to DNA-dependent RNA polymerase. RNA polymerase from mammalian cells does not contain the peptide sequence to which rifampicin binds, so RNA synthesis is not affected. In contrast to the natural rifamycins, which tend to have poor absorption properties, rifampicin is absorbed satisfactorily after oral administration and is also relatively free of toxic side-effects. The most serious side-effect is disturbance of liver function. A trivial, but to the patient potentially worrying, side-effect is discoloration of body fluids, including urine, saliva, sweat, and tears, to a red–orange colour, a consequence of the naphthalene/naphthoquinone chromophore in the rifamycins. **Rifabutin** (Figure 3.52) is a newly introduced derivative, which also has good activity against the *Mycobacterium avium* complex frequently encountered in patients with AIDS. **Rifapentine** (Figure 3.52) has a longer duration of action than the other anti-tubercular rifamycins.

Zearalenone (Figure 3.53), a toxin produced by the fungus *Gibberella zeae* and several species of *Fusarium*, has a relatively simple structure which is derived entirely from acetate/malonate units. It can be envisaged as a cyclization product from a polyketide where consecutive retention of several carbonyl functions will then allow formation of an aromatic ring by aldol condensation and enolizations near the carboxyl terminus – the processes of aromatic ring formation will be considered in more detail shortly (see page 99). Other parts of the chain have suffered considerable reductive modification

(Figure 3.53). This simple structural analysis is essentially correct, though a novel feature here is that two proteins differing in their reducing nature are actually involved. Initial chain extension is carried out by a 'reducing' non-iterative type I PKS that constructs a C_{12} intermediate, which is then transferred to a second protein that lacks any KR, ER, and DH domains and carries out 'non-reducing' chain extension. The second protein possesses a starter unit-ACP transacylase (SAT) domain that loads the C_{12} chain for three further extensions.

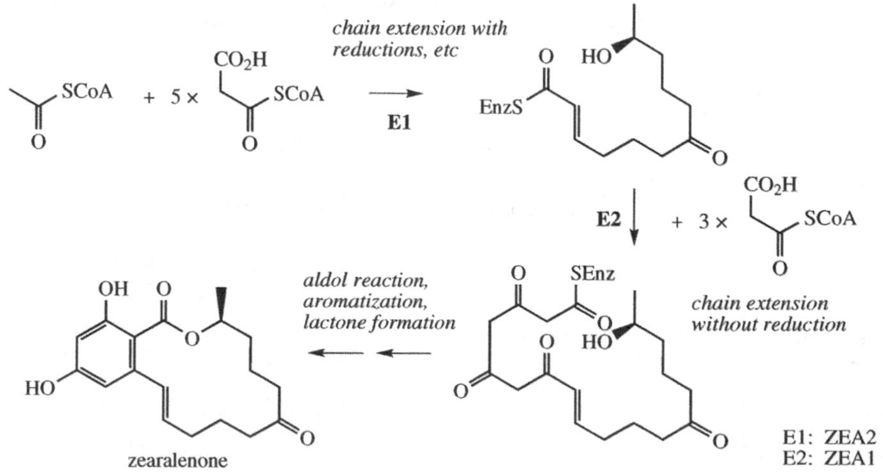

Figure 3.53

From a structural point of view, zearalenone is a remarkable example of an acetate-derived metabolite containing all types of oxidation level seen during the fatty acid extension cycle, i.e. carbonyl, secondary alcohol (eventually forming part of the lactone), alkene, and methylene, as well as having a portion which has cyclized to an aromatic ring because no reduction processes occur in that fragment of the chain.

POLYKETIDE SYNTHASES: LINEAR POLYKETIDES AND POLYETHERS

The macrolide systems described above are produced by formation of an intramolecular ester or amide linkage, utilizing appropriate functionalities in the growing polyketide chain. Macrolide formation does not always occur, and similar acetate/propionate precursors might also be expected to yield molecules which are essentially linear in nature. It is now possible to appreciate how some organisms, particularly marine microalgae, are able to synthesize polyunsaturated fatty acids such as EPA (Figure 3.54) directly by the use of PKS enzymes, instead of desaturating and elongating FAS-derived fatty acids (see page 49). The sequence proposed in the EPA-producing marine

bacterium *Shewanella pneumatophori* allows chain extension with malonyl-CoA, introducing unsaturation via the KR/DH components. However, this would normally lead to a conjugated system of *trans* double bonds, as seen in the polyene antifungals (see page 82); the polyunsaturated fatty acids usually have a non-conjugated array of *cis* double bonds. EPA formation in *Shewanella* requires two additional enzyme activities, a 2,3-isomerase and a 2,2-isomerase, probably acting in concert with the DH (Figure 3.54). The first of these is an allylic isomerization (see page 51), whilst the second is a *trans–cis* isomerization, usually considered to require energy absorption (compare coumarins, page 161, and retinol, page 304).

Discodermolide (Figure 3.55) does contain a lactone moiety, though this is in a six-membered ring. Nevertheless, the bulk of the molecule can be rationalized as a linear array of malonate- and methylmalonate-derived units. In common with the fatty acids, double bonds in the chain are also *cis*. This compound is synthesized by the deep-sea marine sponge *Discodermia dissoluta*, and is of considerable interest as the most potent natural promoter of tubulin assembly yet discovered. Taxol (see page 224) and the epothilones (see page 85) have a similar mode of action. It has also been

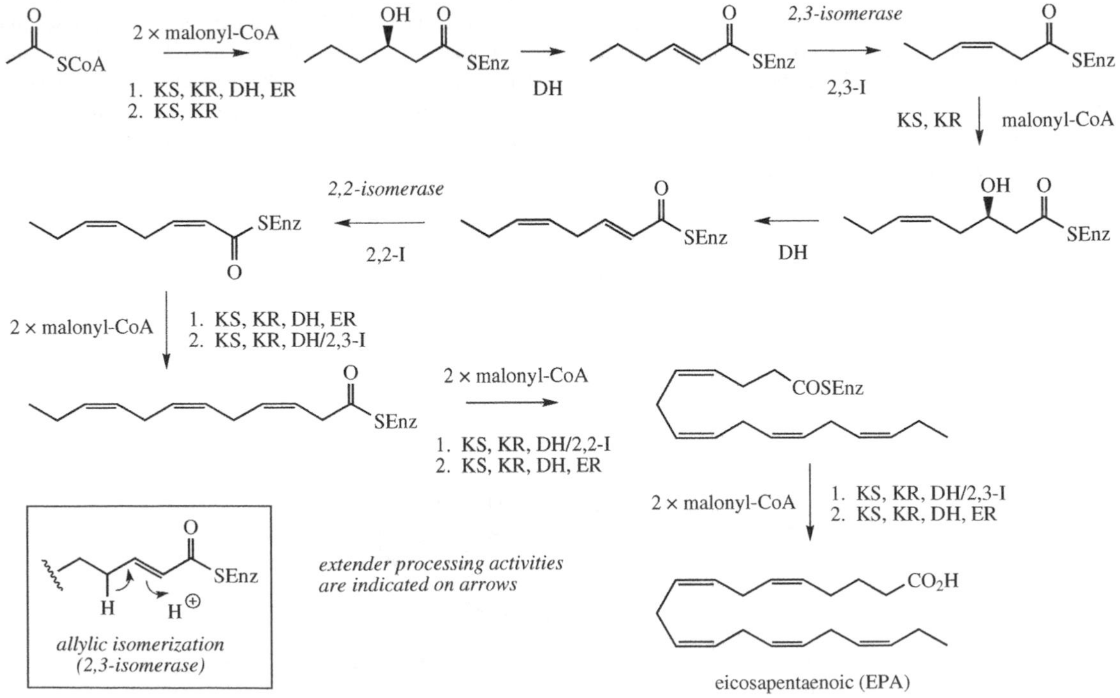

Figure 3.54

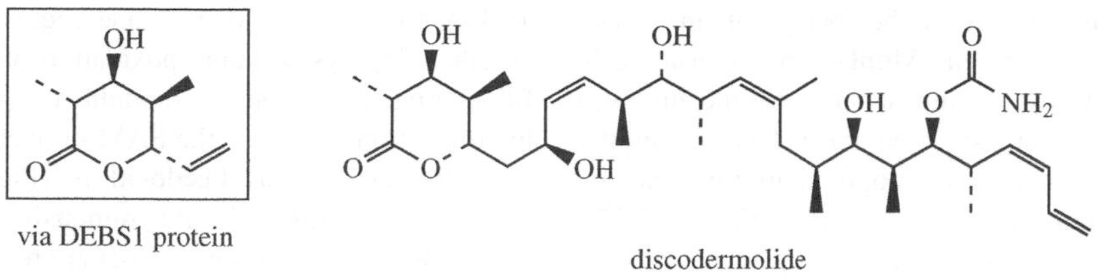

Figure 3.55

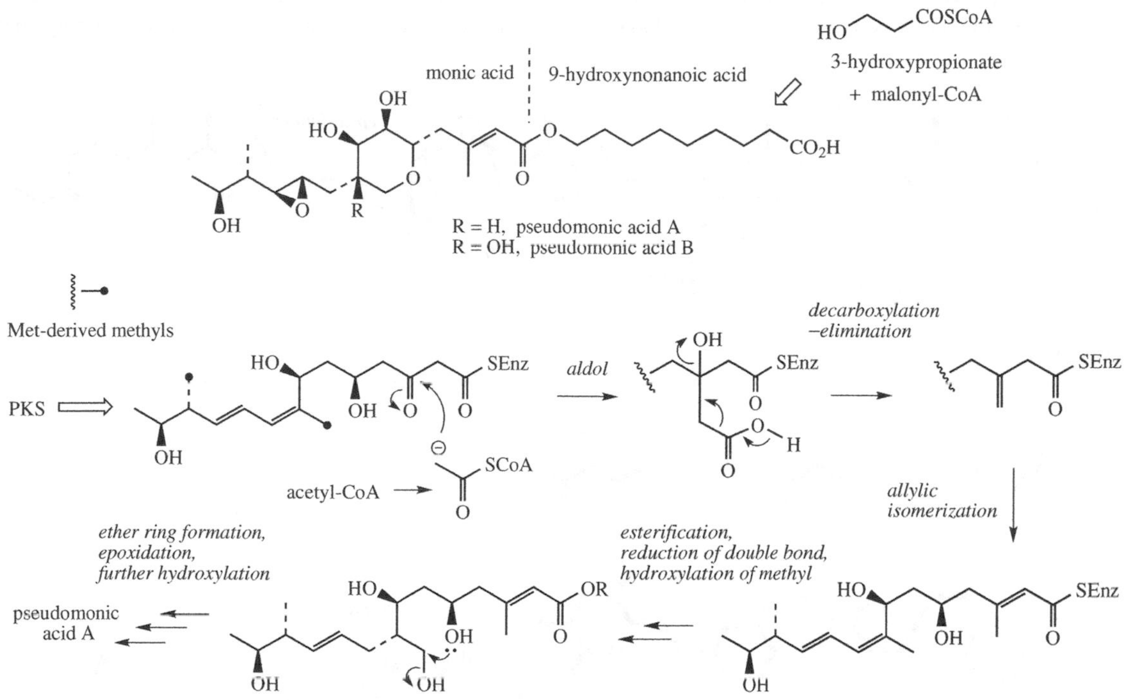

Figure 3.56

found that taxol and discodermolide display synergistic action in combination. Discodermolide itself has proved rather too toxic for drug use, and studies have been restricted by limited supplies. It has been possible to prepare key intermediates for total synthesis by using genetically modified DEBS1 proteins (see page 71) expressed in a suitable host.

Mupirocin is an antibiotic used clinically for the treatment of bacterial skin infections and for controlling *Staphylococcus aureus*, particularly methicillin-resistant *Staphylococcus aureus* (MRSA), when other antibiotics are ineffective. It has an uncommon mode of action, binding selectively to bacterial isoleucyl-tRNA synthase, which prevents incorporation of isoleucine into bacterial

proteins. Mupirocin is produced by *Pseudomonas fluorescens* and consists of a mixture of pseudomonic acids, the major components being **pseudomonic acid A** (about 90%) and pseudomonic acid B (about 8%) (Figure 3.56). These are esters of monic acids with a C_9 saturated fatty acid, 9-hydroxynonanoic acid. The mupirocin gene cluster contains several domains resembling bacterial type I PKS and type I FAS systems, and other genes probably responsible for production of precursors and for later modifications. Parts of the sequence are particularly noteworthy, however. During PKS-mediated elaboration of the monic acid fragment, methyltransferase domains incorporate methyl groups, as seen with epothilone (page 86). A further methyl group originates from the methyl of an acetate precursor. Since the protein MupH bears considerable similarity to HMG-CoA synthase, the enzyme involved in mevalonate biosynthesis (see page 190), a similar aldol addition mechanism is proposed in this reaction (see also streptogramin A formation, Figure 7.27). Decarboxylation–elimination (again compare mevalonate biosynthesis) catalysed by MupH is then followed by an allylic elimination. This 'methylation' looks much more complicated than an SAM-mediated reaction, but is a rather neat way of inserting a methyl group onto a carbonyl, which is not possible with SAM. Pseudomonic

acid A formation in due course requires formation of an ether ring system and epoxidation of a double bond. Ether ring formation is thought to proceed through hydroxylation of one of the SAM-derived methyls.

Molecules such as **lasalocid A** (Figure 3.57) from *Streptomyces lasaliensis* and **monensin A** (Figure 3.58) from *Streptomyces cinnamonensis* are representatives of a large group of compounds called **polyether antibiotics**. These and related compounds are of value in veterinary medicine, being effective in preventing and controlling coccidiae, and also having the ability to improve the efficiency of food conversion in ruminants. The polyether

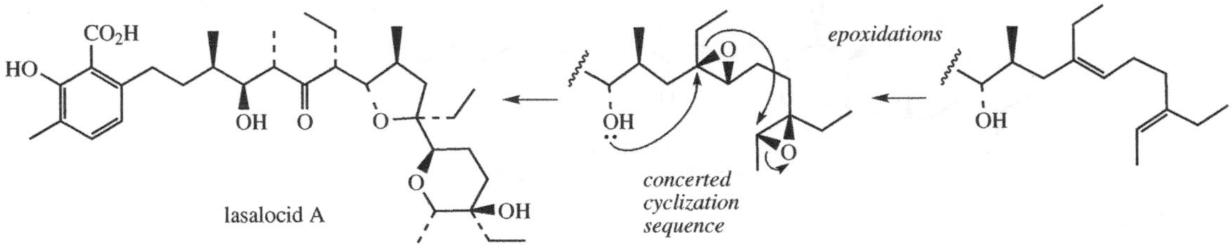

Figure 3.57

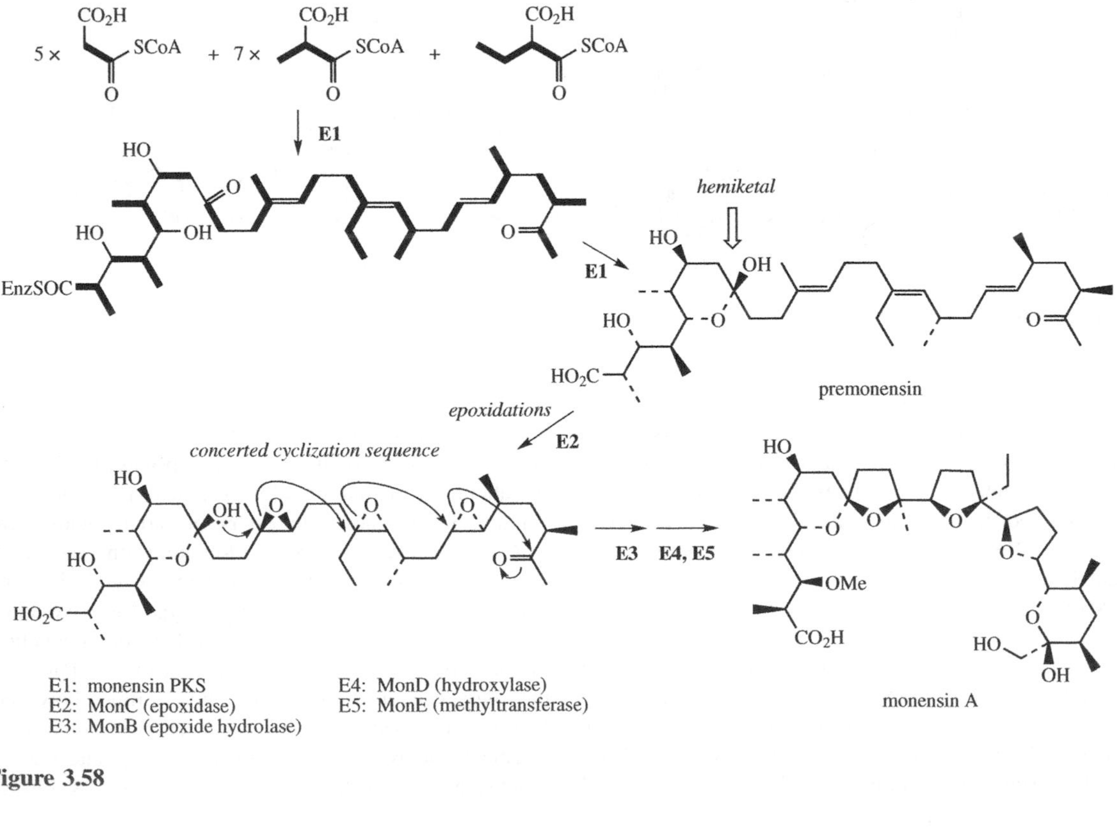

E1: monensin PKS
E2: MonC (epoxidase)
E3: MonB (epoxide hydrolase)
E4: MonD (hydroxylase)
E5: MonE (methyltransferase)

Figure 3.58

antibiotics are characterized by the presence of a number of tetrahydrofuran and/or tetrahydropyran rings along the basic chain. The polyether acts as an ionophore, increasing influx of sodium ions into the parasite, causing a resultant and fatal increase in osmotic pressure. Current thinking is that these ring systems arise via a cascade cyclization mechanism involving epoxide intermediates, as shown for lasalocid A (Figure 3.57). In the biosynthesis of **monensin A** (Figure 3.58), a modular PKS assembles a polyketide chain from malonate, methylmalonate, and ethylmalonate precursors to produce the triene shown, which is released as the hemiketal; the loading module contains a KS^Q domain and the starter is malonate (see page 73). Following triepoxide formation, a concerted stereospecific cyclization sequence initiated by the hemiketal hydroxyl could proceed as indicated. Hydroxylation of the starter methyl group and methylation of a hydroxyl are late steps.

Even more remarkable polyether structures are found in some toxins produced by marine dinoflagellates, which are in turn taken up by shellfish, thus passing on their toxicity to the shellfish. **Okadaic acid** (Figure 3.59) and related polyether structures from *Dinophysis* species are responsible for diarrhoeic shellfish poisoning in mussels, causing severe diarrhoea, nausea, and vomiting in consumers of contaminated shellfish in many parts of the world. Okadaic acid is now known to be a potent inhibitor of protein phosphatases, and causes dramatic increase in the phosphorylation of many proteins, leading to the observed cellular responses. It has become a useful pharmacological tool in the study of protein phosphorylation processes. **Brevetoxin A** and **brevetoxin B** (Figure 3.60) are examples of the toxins associated with 'red tide' blooms of dinoflagellates which affect fishing and also tourism, especially in Florida and the Gulf of Mexico. The red tide toxins are derived from *Karenia brevis* (formerly *Gymnodimium breve*) and are the causative agents of neurotoxic shellfish poisoning, leading to neurological disorders and gastrointestinal troubles. The toxins are known to bind to sodium channels, keeping them in an open state. Fatalities among marine life, e.g. fish, dolphins, and whales, and in humans are associated with these toxins synthesized by organisms at the base of the marine food chain. These compounds are postulated to be produced from a polyunsaturated fatty acid by epoxidation of double bonds, and then a concerted sequence of epoxide ring openings leading to the extended polyether structure of *trans*-fused rings (Figure 3.60). The carbon skeletons of okadaic acid and the brevetoxins do not seem to arise from acetate/propionate units in the systematic manner as seen with monensin A; biosynthetic studies have shown that fragments from the citric acid cycle and other precursors become involved. This may reflect some significant differences between terrestrial and marine microorganisms.

Ciguatoxin (Figure 3.60) is one of the more complex examples of a polyether structure found in nature, though it is certainly not the largest. This compound is found in the moray eel (*Gymnothorax javanicus*) and in a variety of coral reef fish, such as red snapper (*Lutjanus bohar*), though some 100 different species of fish may cause food poisoning (ciguatera) in tropical and subtropical regions. Ciguatoxin is remarkably toxic even at microgram levels; ciguatera is characterized by vomiting, diarrhoea, and neurological problems. Most sufferers slowly recover, though this may take a considerable time, and few cases are fatal due principally to the very low levels of toxin actually present in the fish. The dinoflagellate *Gambierdiscus toxicus* is ultimately responsible for polyether production, synthesizing a less toxic analogue which is passed through the food chain and eventually modified into the very toxic ciguatoxin by the fish. More than 20 ciguatoxin-like structures have been identified.

Halichondrins are both polyethers and macrolides, and were originally isolated from the Pacific marine sponge *Halichondria okadai*. They have received considerable attention because of their complex structures and their extraordinary antitumour activity. **Halichondrin B** (Figure 3.61) was shown to affect tubulin polymerization, though at a site distinct from the *Catharanthus* alkaloid site (see page 375). To

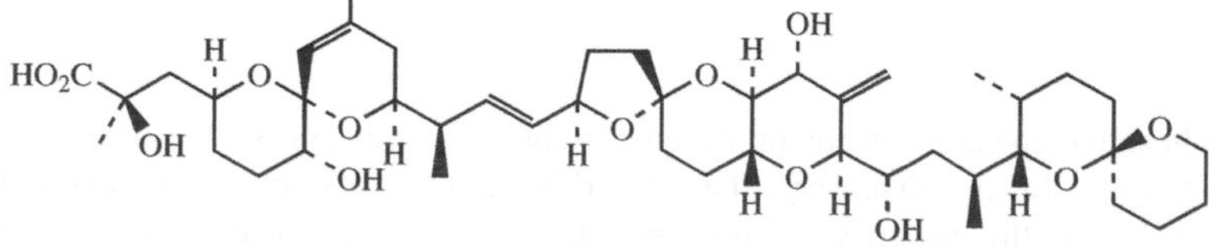

okadaic acid

Figure 3.59

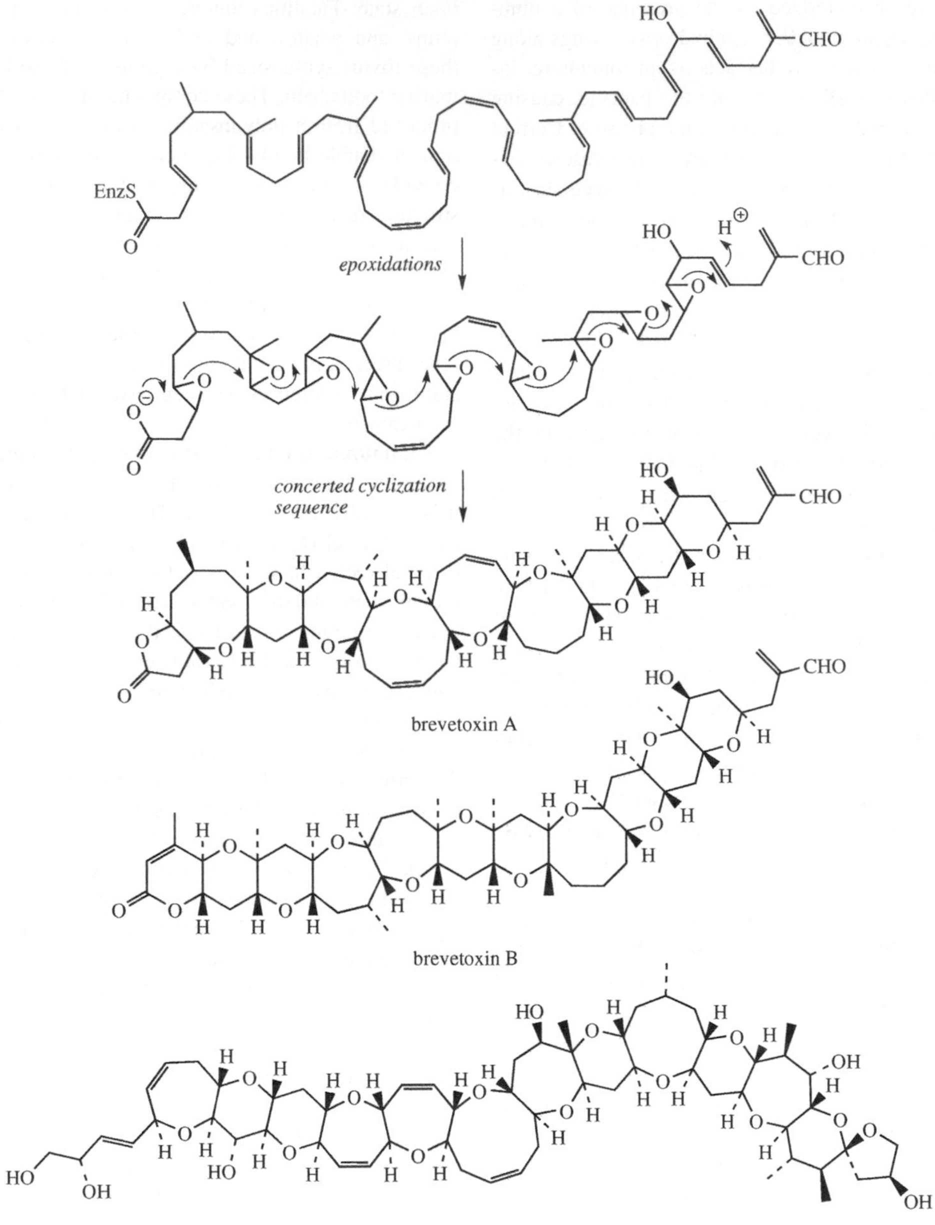

Figure 3.60

facilitate clinical studies, halichondrins were produced by total synthesis, leading to the discovery that it was only necessary to have the macrocyclic portion to maintain the biological activity; the macrocyle still contains several complex ether linkages, however. Substitution of a ketone for the lactone function increased

stability; in due course, the modified derivative **eribulin** (Figure 3.61) has proved successful in the treatment of advanced breast cancer and has been introduced as a drug.

The **zaragozic acids** (**squalestatins**) are structurally rather different from the polyethers discussed so far, but

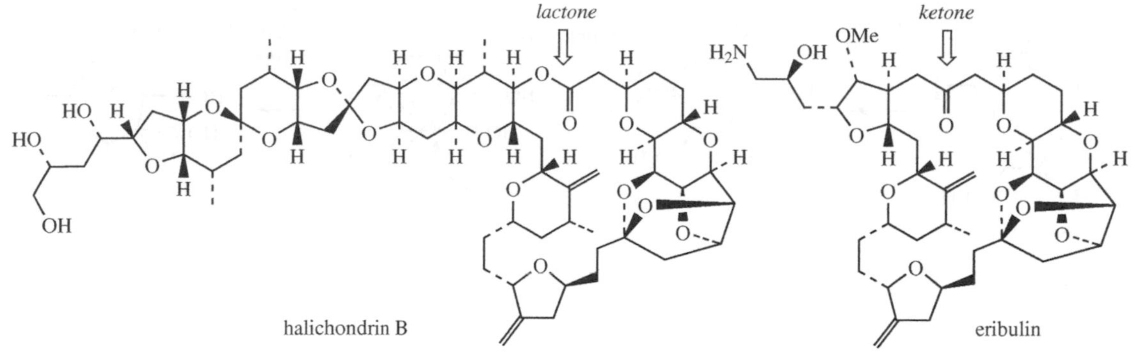

Figure 3.61

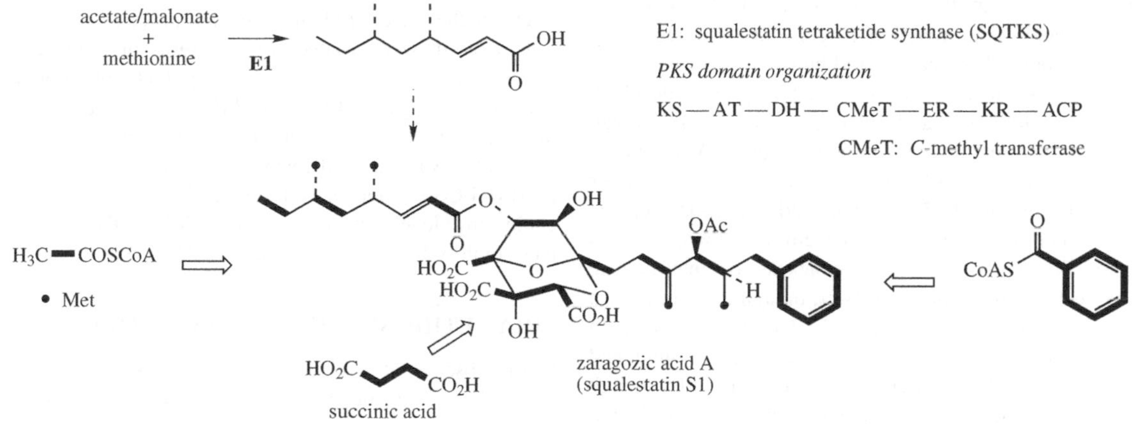

Figure 3.62

they are primarily acetate derived, and the central ring system is suggested to be formed by an epoxide-initiated process resembling the polyether derivatives just described. Thus, **zaragozic acid A** (squalestatin S1; Figure 3.62) is known to be constructed from two acetate-derived chains and a C_4 unit such as the Krebs cycle intermediate oxaloacetate (succinic acid can act as precursor). One chain has a benzoyl-CoA starter (from the shikimate pathway, see page 157), and both contain two methionine-derived side-chain substituents (Figure 3.62). An iterative (repeating) type I PKS catalysing synthesis of one of these has been characterized. In common with a number of fungal PKSs, a *C*-methyl transferase domain is included in the protein, and *C*-methylation is part of the PKS activity (see also epothilones, page 86, and mupirocin, page 91). Iterative PKSs do not have the modular system to control the order and composition of extender units added.

Instead, programming of the sequence is encoded in the PKS itself, in a manner that remains to be clarified.

The heterocyclic ring system can be envisaged as arising via nucleophilic attack onto oxaloacetic acid, formation of a diepoxide, then a concerted sequence of reactions as indicated (Figure 3.63). The zaragozic acids are produced by a number of fungi, including *Sporomiella intermedia* and *Leptodontium elatius*, and are of considerable interest since they are capable of reducing blood cholesterol levels in animals by acting as potent inhibitors of the enzyme squalene synthase (see page 235). This is achieved by mimicking the steroid precursor presqualene PP and irreversibly inactivating the enzyme; both compounds have lipophilic side-chains flanking a highly polar core. They have considerable medicinal potential for reducing the incidence of coronary-related deaths (compare the statins, below).

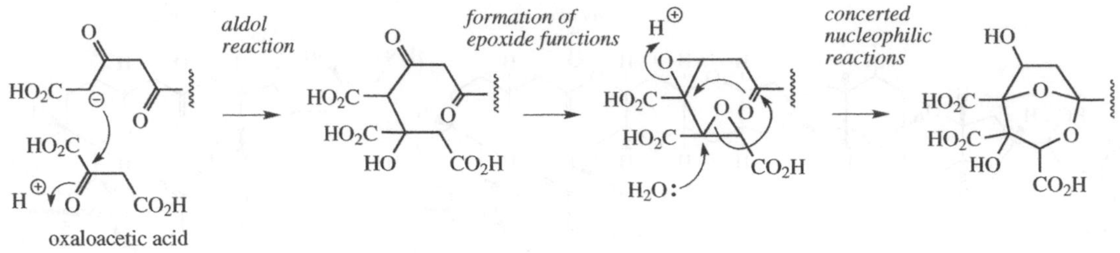

Figure 3.63

DIELS–ALDER CYCLIZATIONS

A number of cyclic structures, typically containing cyclohexane rings, are known to be formed via the acetate pathway and can be rationalized as involving an enzymic Diels–Alder reaction (Figure 3.64), though only one example will be considered here.

Thus, **lovastatin** arises from a C_{18} polyketide chain with *C*-methylation, and relatively few of the oxygen functions are retained in the final product. A secondary polyketide chain, with methylation, provides the ester side-chain (Figure 3.65). A Diels–Alder reaction accounts for formation of the decalin system as shown, and this is now known to occur at the hexaketide intermediate stage, catalysed by the PKS itself, though it is not known just how this is accomplished. The PKS (lovastatin nonaketide synthase) is an iterative type I system and carries a *C*-methyltransferase domain. Unusually, though the PKS carries an ER domain, this function is not activated unless an accessory protein LovC is also present. The product of lovastatin nonaketide synthase is the lactone dihydromonacolin L, and this is subjected to a sequence of post-PKS modifications, leading to monacolin J. The ester side-chain is produced by a non-iterative type I PKS, lovastatin diketide synthase, which also carries a *C*-methyltransferase domain. It carries no TE domain, and the fatty acyl chain appears to be delivered directly from the PKS protein in conjunction with the esterifying enzyme.

diene   dienophile

Diels–Alder reaction

Figure 3.64

The *C*-methylation processes, quite common with fungal PKS systems, are easily interpreted as in Figure 3.66. The ketoester intermediate may be methylated with SAM via an enolate anion mechanism; the product then re-enters the standard chain extension process.

Lovastatin was isolated from cultures of *Aspergillus terreus* and was found to be a potent inhibitor of hydroxymethylglutaryl-CoA (HMG-CoA) reductase, a rate-limiting enzyme in the mevalonate pathway (see page 189). Analogues of lovastatin (statins) find drug use as HMG-CoA reductase inhibitors, thus lowering blood cholesterol levels in patients [Box 3.11]. These drugs have rapidly become the top-selling drugs worldwide.

POLYKETIDE SYNTHASES: AROMATICS

In the absence of any reduction processes, the growing poly-β-keto chain needs to be stabilized on the enzyme surface until the chain length is appropriate, at which point cyclization or other reactions can occur. A poly-β-keto ester is very reactive, and there are various possibilities for undergoing intramolecular Claisen or aldol reactions, dictated of course by the nature of the enzyme and how the substrate is folded. Methylenes flanked by two carbonyls are activated, allowing formation of carbanions/enolates and subsequent reaction with ketone or ester carbonyl groups, with a natural tendency to form strain-free six-membered rings. Aromatic compounds are typical products from type II and type III PKSs, though there are a few examples where type I PKSs produce aromatic rings. Whilst the chemical aspects of polyketide synthesis are common to all systems, a very significant difference compared with most of the type I systems described under macrolides is that the processes are iterative rather than non-iterative. In a non-iterative system, the sequence of events is readily related to the sequence of enzymic domains as each is brought into play. It is less

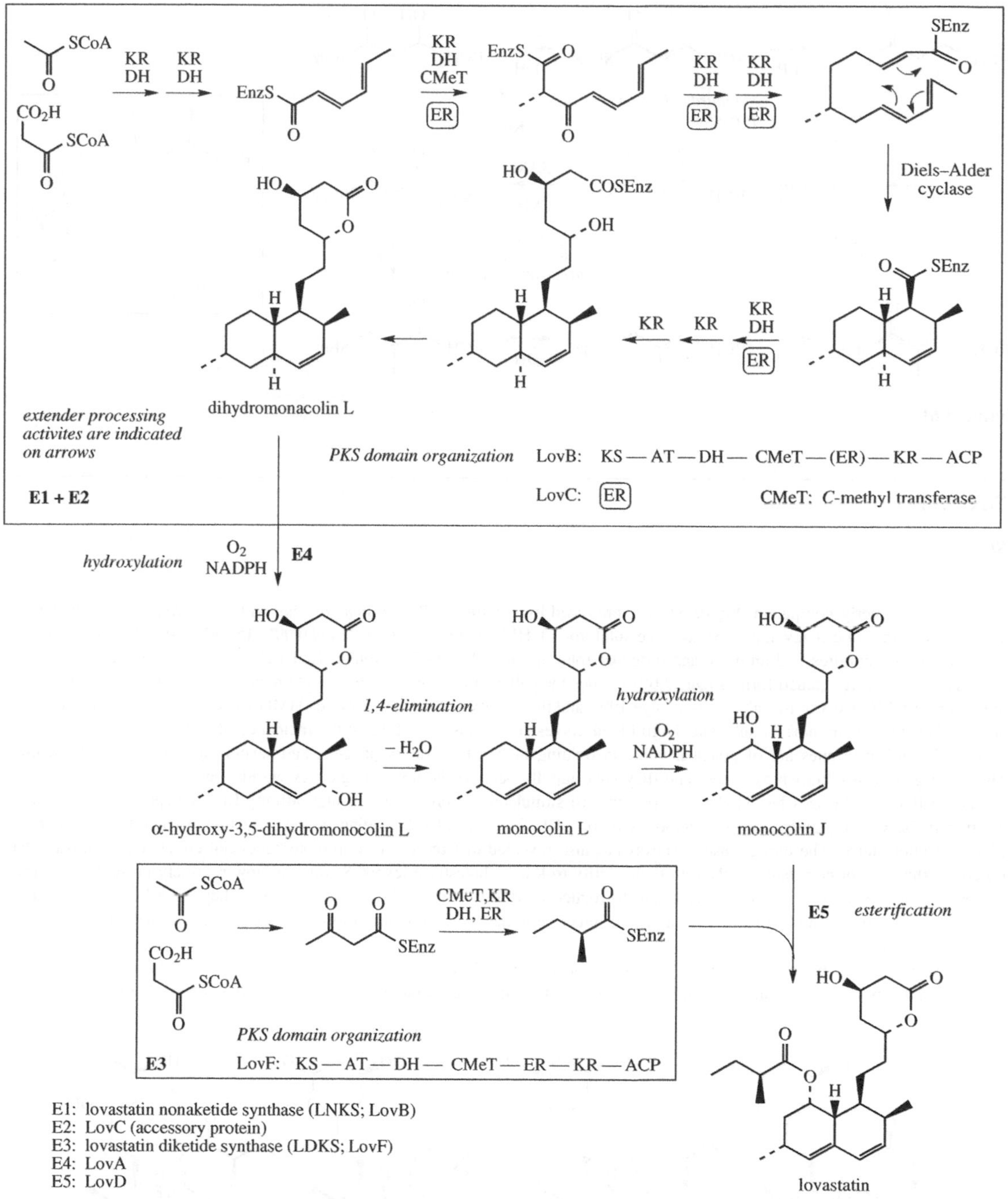
SCoA
KR DH
KR DH
EnzS
KR DH CMeT
ER
EnzS
KR DH
KR DH
ER
ER
SEnz
CO2H
SCoA
Diels–Alder cyclase
HO
COSEnz
OH
O
SEnz
H
KR
KR
KR DH
ER
H
dihydromonacolin L
extender processing activites are indicated on arrows
PKS domain organization
LovB: KS — AT — DH — CMeT — (ER) — KR — ACP
LovC: ER
CMeT: C-methyl transferase
E1 + E2
hydroxylation
O2 NADPH
E4
1,4-elimination
– H2O
hydroxylation
O2 NADPH
HO
OH
α-hydroxy-3,5-dihydromonocolin L
monocolin L
monocolin J
SCoA
CO2H
SCoA
CMeT,KR DH, ER
SEnz
SEnz
E5
esterification
PKS domain organization
E3
LovF: KS — AT — DH — CMeT — ER — KR — ACP
lovastatin
E1: lovastatin nonaketide synthase (LNKS; LovB)
E2: LovC (accessory protein)
E3: lovastatin diketide synthase (LDKS; LovF)
E4: LovA
E5: LovD

Figure 3.65

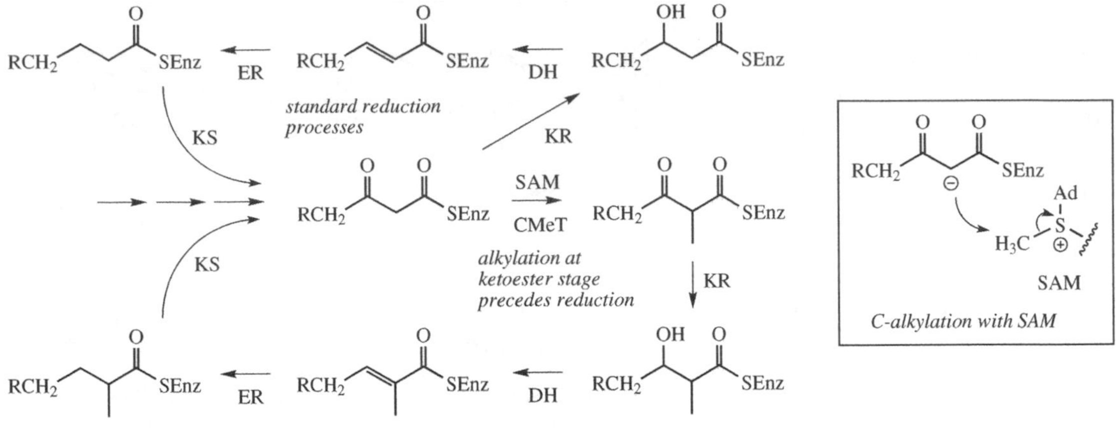

Figure 3.66

Box 3.11

Statins

Mevastatin (formerly compactin; Figure 3.67) is produced by cultures of *Penicillium citrinum* and *Penicillium brevicompactum*, and was shown to be a reversible competitive inhibitor of HMG-CoA reductase, dramatically lowering sterol biosynthesis in mammalian cell cultures and animals, and reducing total and low-density lipoprotein (LDL) cholesterol levels (see page 251). Mevastatin in its ring-opened form (Figure 3.68) mimics the half-reduced substrate mevaldate hemithioacetal during the two-stage reduction of HMG-CoA to mevalonate (see page 190), and the affinity of this agent towards HMG-CoA reductase is some 10^4-fold greater than that of the normal substrate. High blood cholesterol levels contribute to the incidence of coronary heart disease, so mevastatin and analogues are of potential value in treating high-risk coronary patients, and several agents are already in use. These are generically known as statins, and they have rapidly become the top-selling drugs worldwide.

As the liver synthesizes less cholesterol, this in turn stimulates the production of high-affinity LDL receptors on the surface of liver cells. Consequently, the liver removes more LDL from the blood, leading to the reduction of blood levels of both LDL and cholesterol. The clinical use of statins has also revealed that these drugs promote beneficial effects on cardiovascular functions that do not correlate simply with their ability to lower cholesterol levels. Statins are now routinely prescribed to slow the progression of coronary artery disease and to reduce mortality from cardiovascular disease. They have been suggested to have potentially useful anti-inflammatory and anticancer activities, and some are also being tested against Alzheimer's disease and osteoporosis.

Lovastatin (formerly called mevinolin or monacolin K; Figure 3.67) is produced by *Monascus ruber* and *Aspergillus terreus* and is slightly more active than mevastatin. It was the first statin to be marketed, but has since been superseded by more active

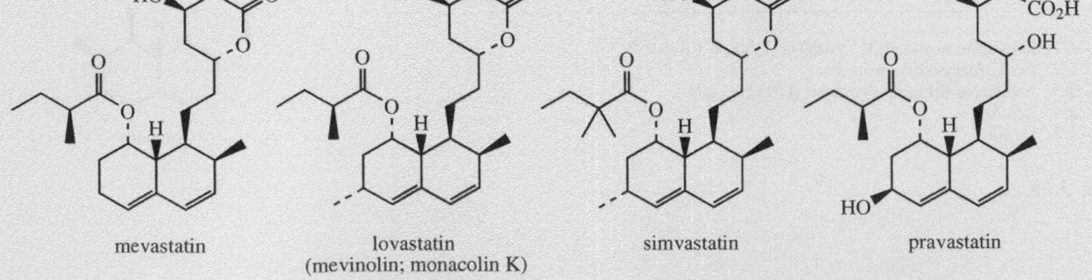

mevastatin	lovastatin (mevinolin; monacolin K)	simvastatin	pravastatin

Figure 3.67

Box 3.11 (continued)

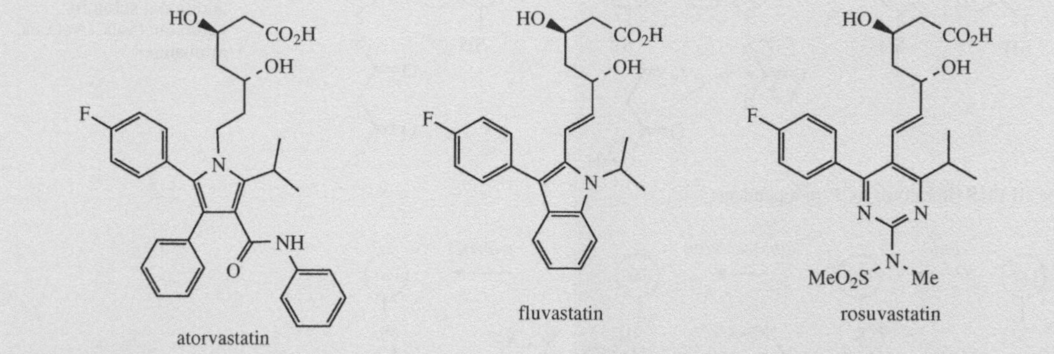

atorvastatin

fluvastatin

rosuvastatin

Figure 3.67 (*continued*)

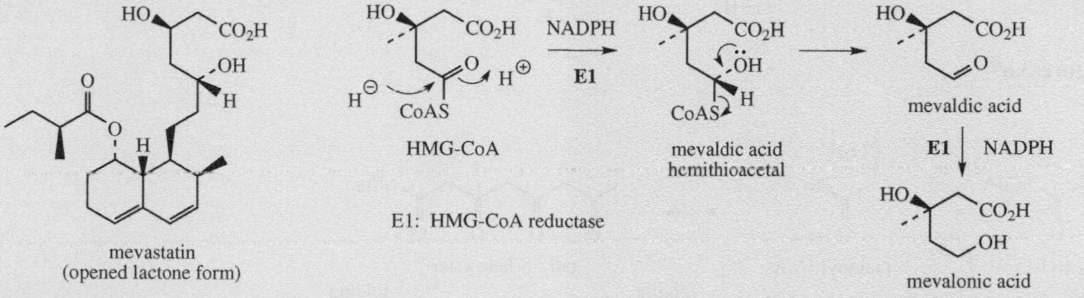

mevastatin
(opened lactone form)

HMG-CoA

mevaldic acid
hemithioacetal

mevaldic acid

mevalonic acid

E1: HMG-CoA reductase

Figure 3.68

agents. **Simvastatin** is obtained from lovastatin by ester hydrolysis and then re-esterification, and is two to three times as potent as lovastatin. **Pravastatin** is prepared from mevastatin by microbiological hydroxylation using *Streptomyces carbophilus* and is consequently more hydrophilic than the other drugs, with an activity similar to lovastatin. Lovastatin and simvastatin are both lactones, but at physiological pHs they exist in equilibrium with the open-ring hydroxy acids; only the open-ring form is biologically active. Other agents currently in use are synthetic, though they feature the same 3,5-dihydroxycarboxylic acid side-chain as in pravastatin. **Atorvastatin, fluvastatin**, and **rosuvastatin** have all been introduced recently, and others are in development.

clear how a repeating process can be closely programmed to produce the final product, and this is somehow encoded in the PKS itself. Thus, the mechanisms of Figure 3.69 will need to be repeated by reuse of the same domains. Type II systems do contain two KS domains (KS_α and KS_β), though only the KS_α domain is implicated in the Claisen condensation. KS_β lacks the active site cysteine and appears analogous to the KS^Q of some macrolide PKSs (see page 73), so it may be involved in loading and decarboxylation of malonyl-CoA. Additional domains for KR, cyclase, or aromatase activities may be present. Type III systems differ from the others in the use of coenzyme

A esters rather than ACPs (Figure 3.69). The PKS is a single protein, though it exists as a homodimer in its native form. It is not appropriate here to attempt subdivisions according to the PKS type, and compounds will be considered primarily according to structural features.

Cyclizations

The poly-β-keto ester (Figure 3.70), formed from four acetate units (one acetate starter group and three malonate chain extension units), is capable of being folded in at least two ways, A and B (Figure 3.70). For A, ionization

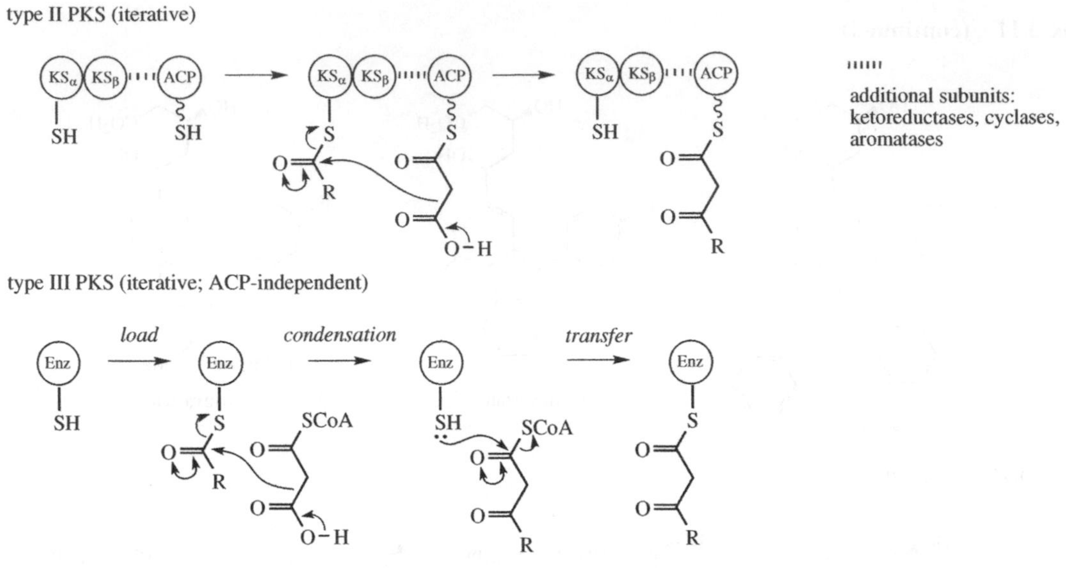

type II PKS (iterative)

additional subunits:
ketoreductases, cyclases,
aromatases

type III PKS (iterative; ACP-independent)

Figure 3.69

orsellinic acid

phloracetophenone

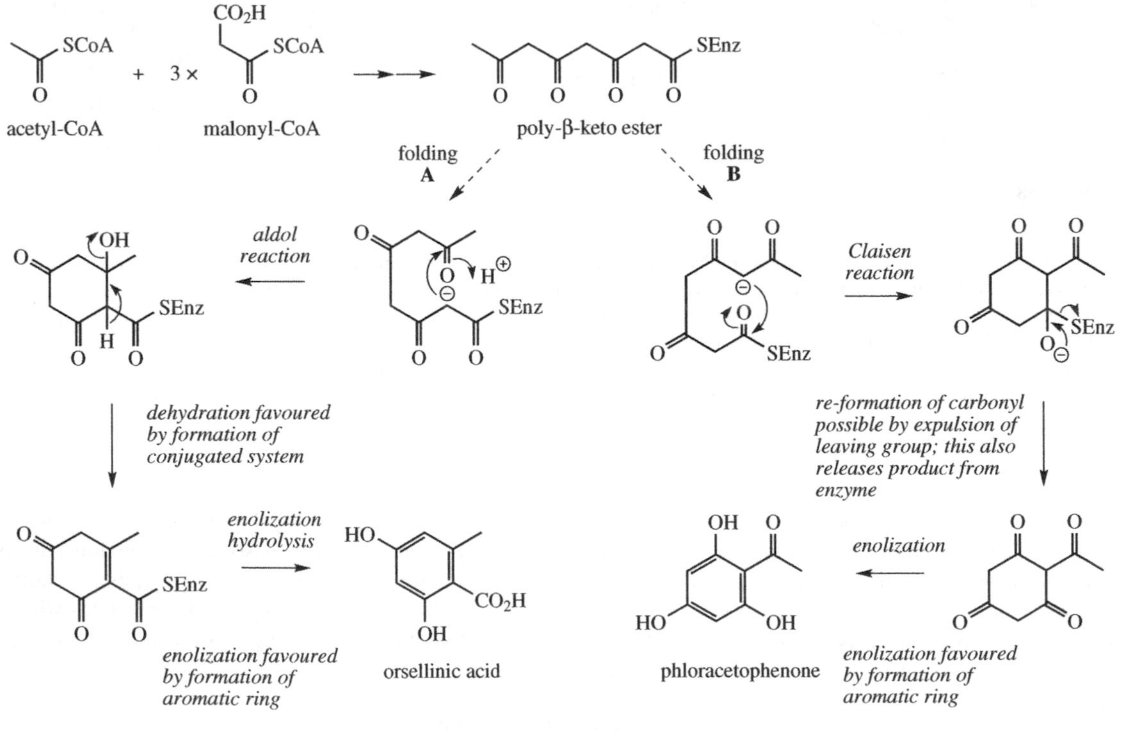

Figure 3.70

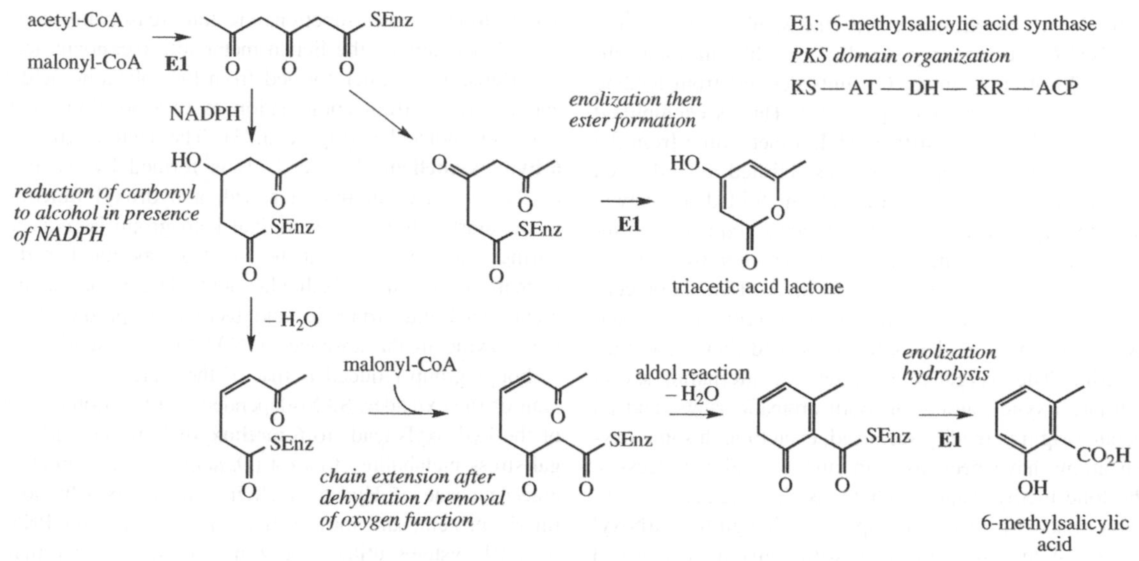

Figure 3.71

of the α-methylene allows aldol addition onto the carbonyl six carbon atoms distant along the chain, giving the tertiary alcohol. Dehydration occurs as in most chemical aldol reactions, giving the alkene, and enolization follows to attain the stability conferred by the aromatic ring. The enzyme–thioester linkage is then hydrolysed to produce **orsellinic acid**. Alternatively, folding of the polyketo ester as in B allows a Claisen reaction to occur, which, although mechanistically analogous to the aldol reaction, is terminated by expulsion of the thiol leaving group and direct release from the enzyme. Enolization of the cyclohexatrione produces **phloracetophenone**. As with other PKSs, the whole sequence of reactions is carried out by an enzyme complex which converts acetyl-CoA and malonyl-CoA into the final product without giving any detectable free intermediates, thus combining PKS and polyketide cyclase activities.

A distinctive feature of an aromatic ring system derived through the acetate pathway is that several of the carbonyl oxygens of the poly-β-keto system are retained in the final product. These end up on alternate carbon atoms around the ring system. Of course, one or more might be used in forming a carbon–carbon bond, as in orsellinic acid. Nevertheless, this oxygenation on alternate carbon atoms, a *meta* oxygenation pattern, is usually easily recognizable and points to the biosynthetic origin of the molecule. This *meta* oxygenation pattern contrasts with that seen on aromatic rings formed via the shikimate pathway (see page 140).

6-Methylsalicylic acid (Figure 3.71) is a metabolite of *Penicillium patulum*, and differs from orsellinic acid by the absence of a phenol group at position 4. It is also derived from acetyl-CoA and three molecules of malonyl-CoA, and the 'missing' oxygen function is removed during the biosynthesis. Orsellinic acid is not itself deoxygenated to 6-methylsalicylic acid. The enzyme 6-methylsalicylic acid synthase is one of the smallest PKSs known and is of the iterative type I group. It requires NADPH as cofactor and removes the oxygen function by employing its KR domain just once. Reduction of the ketone to an alcohol, followed by a dehydration step, occurs on a six-carbon intermediate after the second chain extension (Figure 3.71). Further chain extension is then followed by aldol condensation, enolization, and release from the enzyme to generate 6-methylsalicylic acid. Important evidence for reduction occurring at the C_6 stage comes from the formation of triacetic acid lactone (Figure 3.71) as a 'derailment product' if NADPH is omitted from the enzymic incubation. The PKS is iterative, the enzyme activities acting repeatedly, but there is some programming that allows KR to participate only after the second extension round. In the absence of NADPH, KR remains inactive. There is no TE domain; perhaps release from the enzyme is not simple hydrolysis.

The folding of a polyketide chain can be established by labelling studies, feeding carbon-labelled sodium acetate to the appropriate organism and establishing the position of labelling in the final product by chemical degradation and counting (for the radioactive isotope ^{14}C) or by NMR spectrometry (for the stable isotope ^{13}C). ^{13}C NMR spectrometry is also valuable in establishing the location of intact C_2 units derived from feeding $^{13}C_2$-labelled acetate (see page 36). This is exemplified in Figure 3.72, where **alternariol**, a metabolite from the mould *Alternaria tenuis*, can be established to be derived from a single C_{14} polyketide chain, folded as shown, and then cyclized. Whilst the precise sequence of reactions involved is not known, paper chemistry allows us to formulate the essential features. Two aldol condensations followed by enolization in both rings would give a biphenyl, and lactonization would then lead to alternariol. The oxygenation pattern in alternariol shows alternate oxygen atoms on both aromatic rings, and an acetate origin is readily surmised, even though some oxygen atoms have been used in ring formation processes. The lone methyl 'start-of-chain' is also usually very obvious in acetate-derived compounds, though the carboxyl 'end-of-chain' can often react with convenient hydroxyl functions, which may have arisen through enolization,

and lactone or ester functions are thus reasonably common. For example, the lichen metabolite **lecanoric acid** is a **depside** (an ester formed from two phenolic acids) and the result from combination of two orsellinic acid thioester molecules (Figure 3.73). The lactone ring in **6-hydroxymellein** (Figure 3.74) is formed by reaction of the end-of-chain thioester with a hydroxyl function in the starter group. The original keto group is reduced during chain extension, at the C_6 stage, as noted in the formation of 6-methylsalicylic acid. The same derailment metabolite, triacetic acid lactone, is produced by the enzyme in the absence of NADPH. Unusually, the carbonyl group reduced is that of the starter unit rather than of the extender. SAM-dependent methylation of one of the hydroxyls leads to **6-methoxymellein**, an antifungal stress metabolite of carrot (*Daucus carota*; Umbelliferae/Apiaceae). 6-Hydroxymellein synthase is a homodimeric protein, and may well be a type III plant PKS; type III systems utilize coenzyme A esters rather than ACPs.

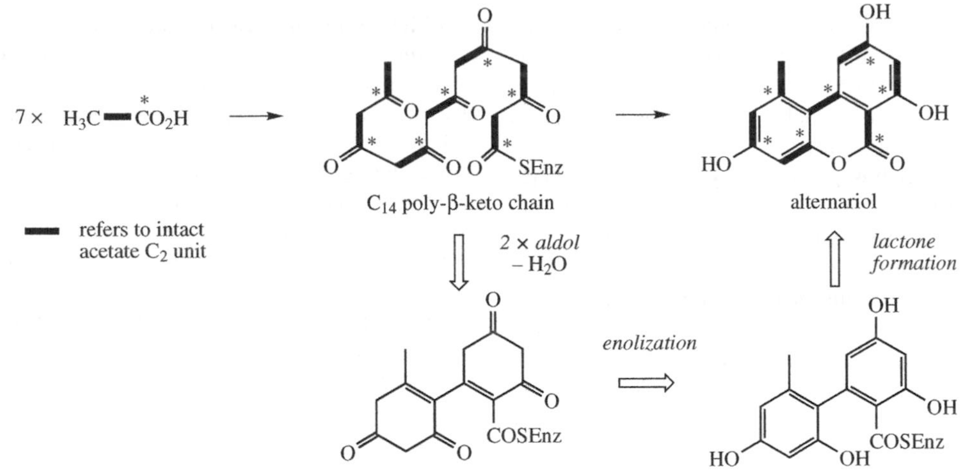

Figure 3.72

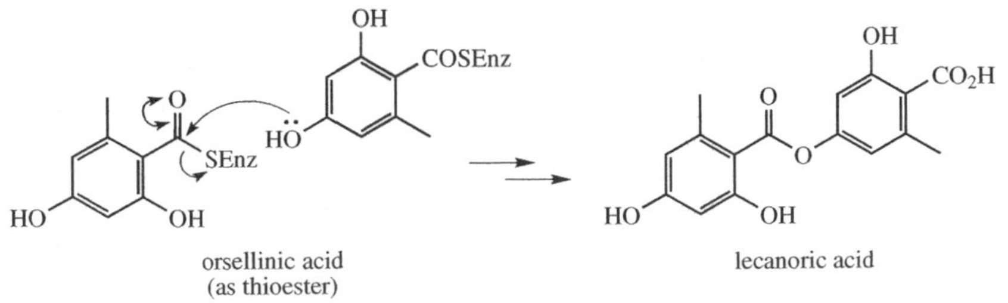

Figure 3.73

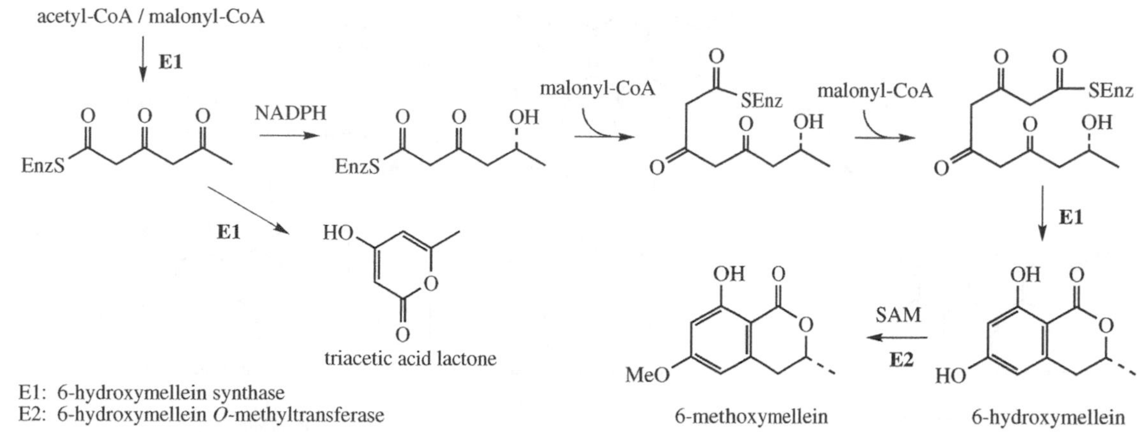

E1: 6-hydroxymellein synthase
E2: 6-hydroxymellein *O*-methyltransferase

Figure 3.74

Post-Polyketide Synthase Modifications

A number of natural anthraquinone derivatives are also excellent examples of acetate-derived structures. Structural analysis of **endocrocin** (Figure 3.75) found in species of *Penicillium* and *Aspergillus* fungi suggests folding of a polyketide containing eight C_2 units to form the periphery of the carbon skeleton. Three aldol-type condensations would give a hypothetical intermediate 1, and, except for a crucial carbonyl oxygen in the centre ring, endocrocin results by enolization reactions, one of which involves the vinylogous enolization $—CH_2—CH=CH—CO— \rightarrow —CH=CH—CH=C(OH)—$. The additional carbonyl oxygen must be introduced at some stage during the biosynthesis by an oxidative process, for which we have little information. **Emodin**, a metabolite of some *Penicillium* species, but also found in higher plants, e.g. *Rhamnus* and *Rumex* species, would appear to be formed from endocrocin by a simple decarboxylation reaction. This is facilitated by the adjacent phenol function (see page 22). *O*-Methylation of emodin would then lead to **physcion. Islandicin** is another anthraquinone pigment produced by *Penicillium islandicum*, and differs from emodin in two ways: one hydroxyl is missing and a new hydroxyl has been incorporated adjacent to the methyl. Without any evidence for the sequence of such reactions, the structure of hypotheical intermediate 2 shows the result of three aldol condensations with reduction of a carbonyl, though the latter would most likely occur during chain extension. A dehydration reaction, two oxidations, and a decarboxylation are necessary to attain the islandicin structure. In **chrysophanol, aloe-emodin** and **rhein**, the same oxygen function is lost by reduction as in islandicin, and decarboxylation also occurs. The three compounds are interrelated by a sequential oxidation of the methyl in chrysophanol to a hydroxymethyl in aloe-emodin and a carboxyl in rhein.

As noted in the case of macrolides, structural modifications undergone by the basic polyketide are also conveniently considered according to the timing of the steps in the synthetic sequence. Thus, 'missing' oxygen functions appear to be reduced out well before the folded and cyclized polyketide is detached from the enzyme, and are mediated by the KR component of the PKS enzyme complex. On the other hand, reactions like decarboxylation, *O*-methylation, and sequential oxidation of a methyl to a carboxyl are representative of post-PKS transformations occurring after the cyclization reaction. These later conversions can often be demonstrated by the isolation of separate enzymes catalysing the individual steps. Most of the secondary transformations are easily rationalized by careful consideration of the reactivity conferred on the molecule by the alternating and usually phenolic oxygenation pattern. These oxygen atoms activate adjacent sites, creating nucleophilic centres. Introduction of additional hydroxyl groups *ortho* or *para* to an existing phenol will be facilitated (see page 26), allowing the extra hydroxyl of islandicin to be inserted, for example. The *ortho*- or *para*-diphenols are susceptible to further oxidation in certain circumstances, and may give rise to *ortho*- and *para*-quinones (see page 26). The quinone system in anthraquinones is built up by an oxidation of the central cyclohexadienone ring, again at a nucleophilic centre activated by the enone system. Methyl groups on an aromatic ring are also activated towards oxidation, facilitating the chrysophanol → aloe-emodin oxidation, for example. Decarboxylation, e.g. endocrocin → emodin,

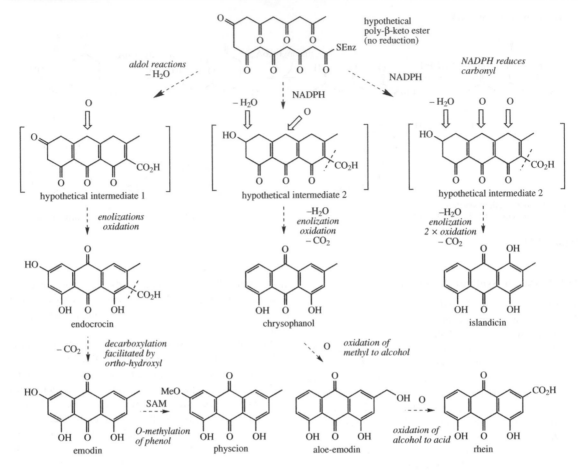

Figure 3.75

is readily achieved in the presence of an *ortho* phenol function, though a *para* phenol can also facilitate this (see page 22).

It is now appreciated that the assembly of the anthraquinone skeleton (and related polycyclic structures) is achieved in a stepwise sequence. After the polyketide chain is folded, the ring at the centre of the fold is formed first, followed in turn by the next two rings. The pathway outlined for the biosynthesis of endocrocin and emodin is shown in Figure 3.76. Mechanistically, there is little difference between this and the speculative pathway of Figure 3.75, but the sequence of reactions is altered. Decarboxylation appears to take place before aromatization of the last-formed ring system, and tetrahydroanthracene intermediates, such as atrochrysone carboxylic acid and atrochrysone, are involved. These dehydrate to the anthrones **endocrocin anthrone** and **emodin anthrone** respectively prior to introduction of the extra carbonyl

oxygen as a last transformation in the production of anthraquinones. This oxygen is derived from O_2.

The interrelationships of Figure 3.76 are derived mainly from feeding experiments in fungi. A type III octaketide synthase gene from the anthraquinone-producing plant *Aloe arborescens*, when expressed in *Escherichia coli*, did not synthesize anthraquinones, but instead the two products SEK 4 and SEK 4b, not normally found in *Aloe* (Figure 3.77). These obviously result from different folding arrangements of the polyketide precursor. Thus, the expressed protein seems to lack some agent to fold the chain correctly or perhaps a cyclase element; the observed products may result from non-enzymic cyclization reactions. This protein utilizes just malonyl-CoA, so the starter unit is generated by decarboxylation (see page 73).

Emodin, physcion, chrysophanol, aloe-emodin, and rhein form the basis of a range of purgative anthraquinone

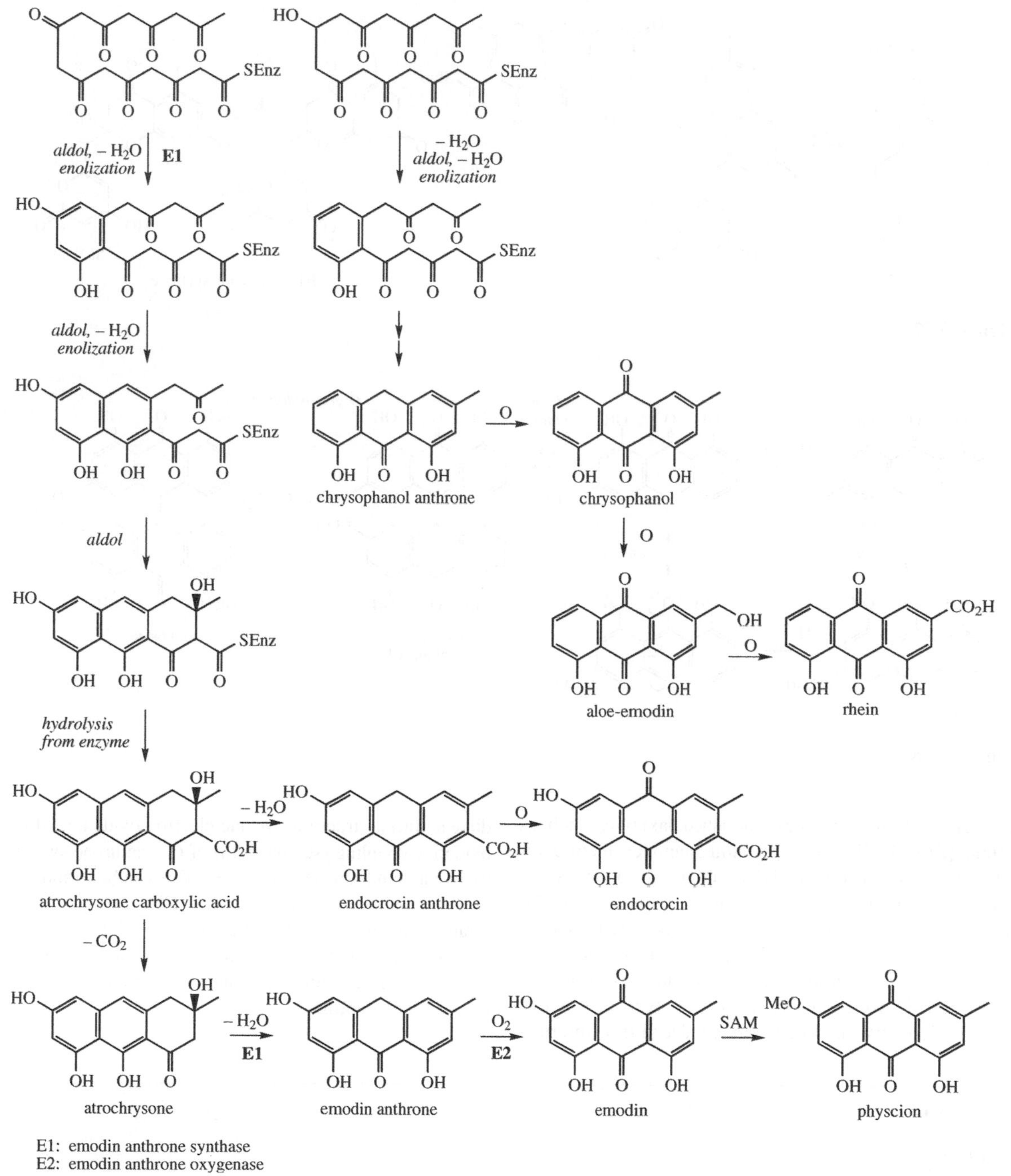
aldol, – H2O
enolization
E1
– H2O
aldol, – H2O
enolization
aldol, – H2O
enolization
aldol
hydrolysis from enzyme
chrysophanol anthrone
chrysophanol
aloe-emodin
rhein
atrochrysone carboxylic acid
endocrocin anthrone
endocrocin
– CO2
atrochrysone
emodin anthrone
emodin
physcion
SAM
E2
E1: emodin anthrone synthase
E2: emodin anthrone oxygenase

Figure 3.76

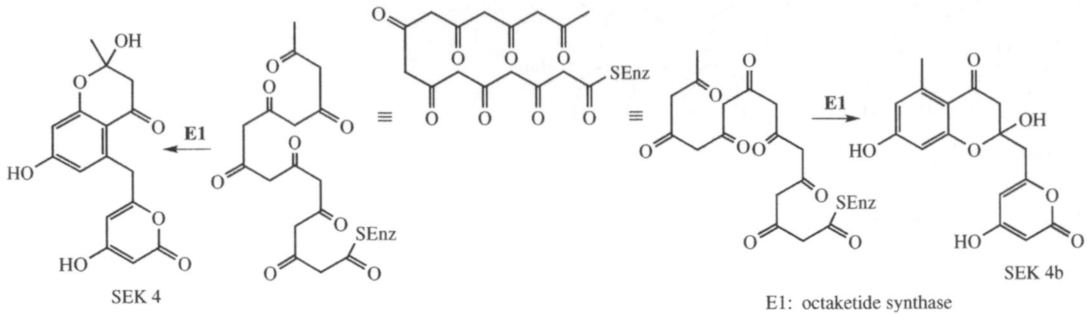

Figure 3.77

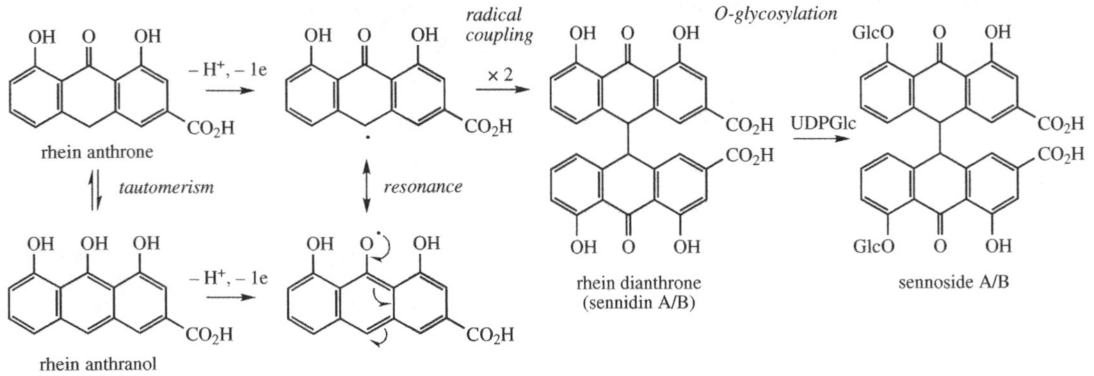

Figure 3.78

derivatives found in long-established laxatives such as senna [Box 3.12]. The free anthraquinones themselves have little therapeutic activity and need to be in the form of water-soluble glycosides to exert their action. Although simple anthraquinone *O*-glycosides are present in the drugs, the major purgative action arises from compounds such as **sennosides**, e.g. sennosides A and B (Figure 3.78), which are dianthrone *O*-glycosides. These types of derivative are likely to be produced from interme-

diate anthrone structures. A one-electron oxidation allows oxidative coupling (see page 28) of two anthrone systems to give a dianthrone (Figure 3.78). This can be formulated as direct oxidation at the benzylic $-CH_2-$, or via the anthranol, which is the phenolic tautomer of the anthrone. Glycosylation of the dianthrone system would then give a sennoside-like product; alternatively, glycosylation may precede the coupling.

Box 3.12

Senna

Senna leaf and fruit are obtained from *Cassia angustifolia* (Leguminosae/Fabaceae), known as Tinnevelly senna, or less commonly from *Cassia senna* (*syn Cassia acutifolia*), which is described as Alexandrian senna. The plants are low, branching shrubs, *Cassia angustifolia* being cultivated in India and Pakistan, and *Cassia senna* being produced in the Sudan, much of it from wild plants. Early harvests provide leaf material, whilst both leaf and fruit (senna pods) are obtained later on. There are no significant differences in the chemical constituents of the two sennas, or between leaf and fruit drug. However, amounts of the active constituents do vary, and this appears to be a consequence of cultivation conditions and the time of harvesting of the plant material.

Box 3.12 **(continued)**

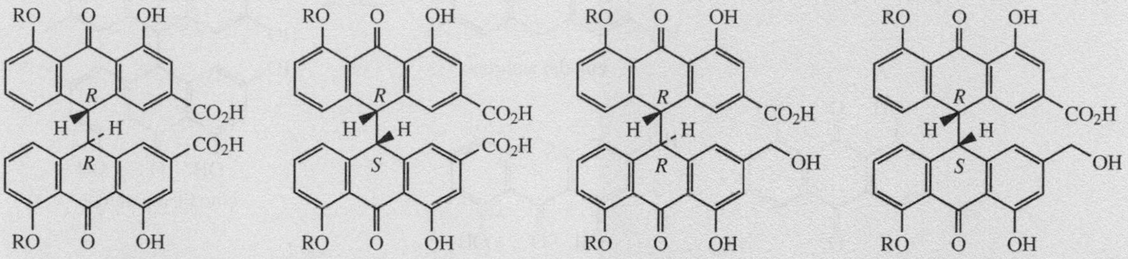

R = Glc, sennoside A
R = H, sennidin A

R = Glc, sennoside B
R = H, sennidin B

R = Glc, sennoside C
R = H, sennidin C

R = Glc, sennoside D
R = H, sennidin D

Figure 3.79

The active constituents in both senna leaf and fruit are dianthrone glycosides, principally sennosides A and B (Figure 3.79). These compounds are both di-*O*-glucosides of rhein dianthrone (sennidins A and B) and liberate these aglycones on acid hydrolysis; oxidative hydrolysis (e.g. aqueous HNO_3 or H_2O_2/HCl) produces the anthraquinone rhein (Figure 3.76). Sennidins A and B are optical isomers: sennidin A is dextrorotatory (+) whilst sennidin B is the optically inactive *meso* form. Minor constituents include sennosides C and D (Figure 3.79), which are also a pair of optical isomers, di-*O*-glucosides of heterodianthrones sennidins C and D. Sennidin C is dextrorotatory, whilst sennidin D is optically inactive, approximating to a *meso* form in that the modest change in substituent does not noticeably affect the optical rotation. Oxidative hydrolysis of sennosides C and D would produce the anthraquinones rhein and aloe-emodin (Figure 3.76). Traces of other anthraquinone glycoside derivatives are also present in the plant material. Much of the sennoside content of the dried leaf appears to be formed by enzymic oxidation of anthrone glycosides during the drying process. Fresh leaves and fruits also seem to contain primary glycosides which are more potent than sennosides A and B, and which appear to be partially hydrolysed to sennosides A and B (the secondary glycosides) by enzymic activity during collection and drying. The primary glycosides contain additional glucose residues.

Senna leaf suitable for medicinal use should contain not less than 2.5% dianthrone glycosides calculated in terms of sennoside B. The sennoside content of Tinnevelly fruits is between 1.2 and 2.5%, with that of Alexandrian fruits being 2.5–4.5%. Senna preparations, in the form of powdered leaf, powdered fruit, or extracts, are typically standardized to a given sennoside content. Non-standardized preparations have unpredictable action and should be avoided. Senna is a stimulant laxative and acts on the wall of the large intestine to increase peristaltic movement. After oral administration, the sennosides are transformed by intestinal flora into rhein anthrone (Figure 3.78), which appears to be the ultimate purgative principle. The glycoside residues in the active constituents are necessary for water solubility and subsequent transportation to the site of action. Although the aglycones, including anthraquinones, do provide purgative action, these materials are conjugated and excreted in the urine after oral administration rather than being transported to the colon. Senna is a purgative drug suitable for either habitual constipation or for occasional use, and is widely prescribed.

However, further oxidative steps on a dianthrone, e.g. **emodin dianthrone** (Figure 3.80), can create a dehydrodianthrone and then allow oxidative coupling of the aromatic rings through **protohypericin** to give a naphthodianthrone, in this case **hypericin**. The reactions of Figure 3.80 can also be achieved chemically by passing air into an alkaline solution of emodin anthrone, though yields are low. **Hypericin** is found in cultures of *Dermocybe* fungi and is also a constituent of St John's Wort, *Hypericum perforatum* (Guttiferae/Hypericaceae). A protein obtained by expression of a *Hypericum perforatum* gene in *Escherichia coli* has been observed to transform the anthraquinone emodin into hypericin, presumably by a sequence similar to that shown. However, since emodin rather than emodin anthrone was the substrate, some additional step must be involved. An aldol condensation between emodin and emodin anthrone could provide the dehydrodianthrone as an alternative to the oxidative coupling. *Hypericum perforatum* is a popular herbal medicine in the treatment of depression [Box 3.13]. The naphthodianthrones have no purgative action, but hypericin can act as a photosensitizing agent in a similar manner to furocoumarins (see page 165). Thus, ingestion of hypericin results in an increased absorption of UV light and can

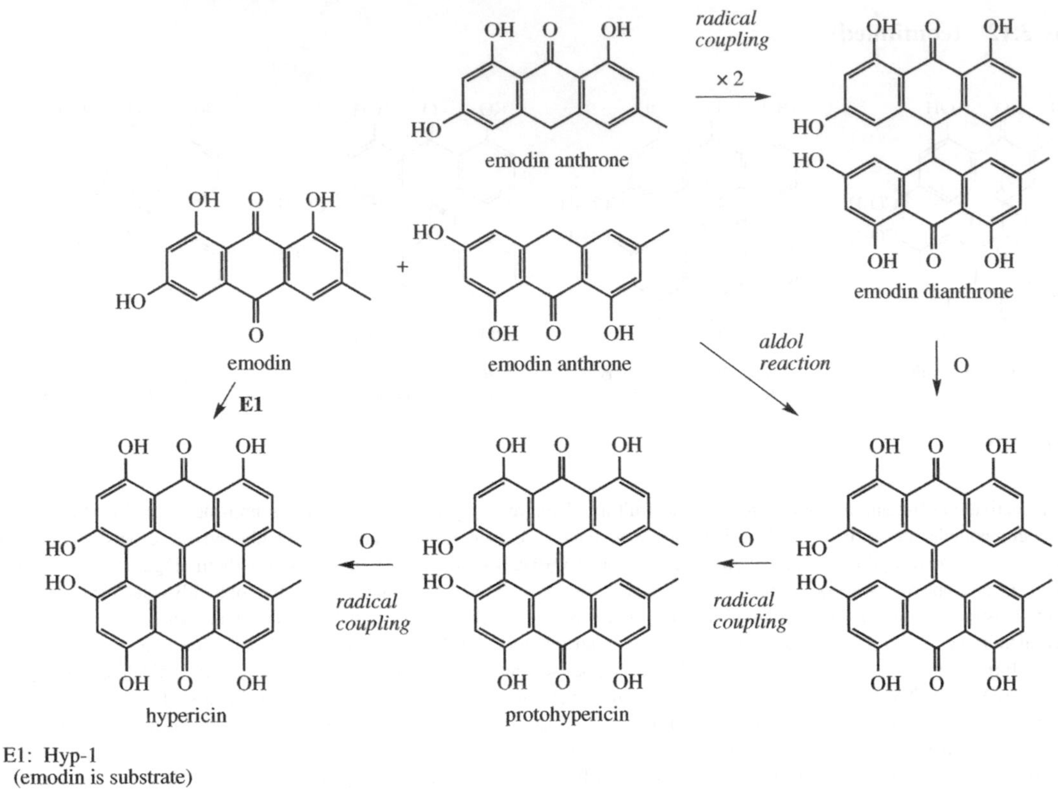

Figure 3.80

Box 3.13

Hypericum/St John's Wort

The dried flowering tops of St John's Wort (*Hypericum perforatum*; Guttiferae/Hypericaceae) have been used as a herbal remedy for many years, an extract in vegetable oil being employed for its antiseptic and wound healing properties. St John's Wort is now a major crop marketed as an antidepressant, which is claimed to be as effective in its action as the widely prescribed antidepressants of the selective serotonin re-uptake inhibitor (SSRI) class such as fluoxetine (Prozac®), and with fewer side-effects. There is considerable clinical evidence that extracts of St John's Wort are effective in treating mild to moderate depression and improving mood. However, to avoid potentially dangerous side-effects, St John's Wort should not be used at the same time as prescription antidepressants (see below). St John's Wort is a small to medium-height herbaceous perennial plant with numerous yellow flowers characteristic of this genus. It is widespread throughout Europe, where it is generally considered a weed, and has also become naturalized in North America. The tops, including flowers at varying stages of development which contain considerable amounts of the active principles, are harvested and dried in late summer.

The dried herb contains significant amounts of phenolic derivatives, including 4–5% of flavonoids (see page 170), though the antidepressant activity is considered to derive principally from naphthodianthrone structures such as hypericin (about 0.1%) and pseudohypericin (about 0.2%), and a prenylated acylphloroglucinol derivative hyperforin (Figure 3.81). The fresh plant also contains significant levels of protohypericin and protopseudohypericin (Figure 3.81), which are converted into hypericin and pseudohypericin during drying and processing, as a result of irradiation with visible light. Hyperforin is a major

Box 3.13 (continued)

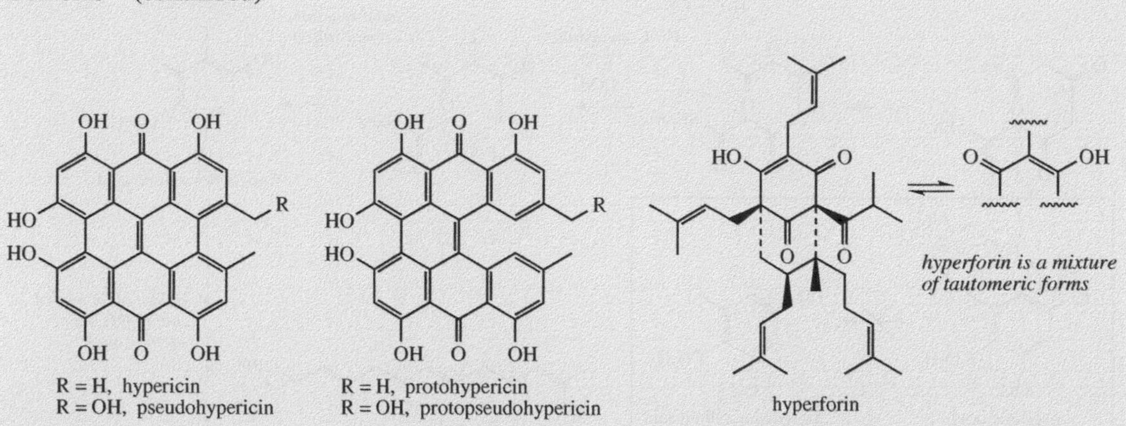

R = H, hypericin
R = OH, pseudohypericin

R = H, protohypericin
R = OH, protopseudohypericin

*hyperforin is a mixture
of tautomeric forms*

hyperforin

Figure 3.81

lipophilic constituent in the leaves and flowers (2–3%) and is now thought to be the major contributor to the antidepressant activity, as well as to the antibacterial properties of the oil extract. Studies show that the clinical effects of St John's Wort on depression correlate well with hyperforin content. Standardized aqueous ethanolic extracts containing 0.15% hypericin and 5% hyperforin are usually employed. The aqueous solubility of hypericin and pseudohypericin is markedly increased by the presence of flavonoid derivatives in the crude extract, particularly procyanidin B_2, a dimer of epicatechin (see page 171). Hyperforin has been demonstrated to be a powerful and non-selective inhibitor of amine reuptake, thus increasing levels of serotonin, noradrenaline (norepinephrine), and dopamine; the mechanism of its inhibition is different from conventional antidepressants. Through its enolized β-dicarbonyl system, hyperforin exists as a mixture of interconverting tautomers; this reactive system leads to hyperforin being relatively unstable in solution, though salts that are stable and storable may be produced. Smaller amounts of structurally related acylphloroglucinol derivatives are also present in the herb, e.g. adhyperforin (see page 113, Figure 3.86, which also includes the biosynthesis of hyperforin). *Hypericum perforatum* appears to be the only *Hypericum* species where hyperforin is present as a major component.

Hypericin also possesses extremely high toxicity towards certain viruses, a property that requires light and may arise via photo-excitation of the polycyclic quinone system. It is currently under investigation as an antiviral agent against HIV and hepatitis C. Antiviral activity appears to arise from an inhibition of various protein kinases, including those of the protein kinase C family. Hypericin and pseudohypericin are potent photosensitizers initiating photochemical reactions and are held responsible for hypericism, a photodermatosis seen in cattle that have consumed *Hypericum* plants present in pasture. Patients using St John's Wort as an antidepressant need to be warned to avoid overexposure to sunlight.

There is also considerable evidence that St John's Wort interacts with a number of prescription drugs, including oral contraceptives, warfarin (anticoagulant), digoxin (cardiac glycoside), theophylline (bronchodilator), indinavir (HIV protease inhibitor), and ciclosporin (immunosuppressant). In some cases, it is known to promote the cytochrome P-450-dependent metabolism of the co-administered drugs. The action of potentially life-saving drugs may thus be seriously jeopardized by co-administration of *Hypericum* extracts.

lead to dermatitis and burning. Hypericin is also being investigated for its antiviral activities, in particular for its potential activity against HIV.

A common feature of many natural products containing phenolic rings is the introduction of alkyl groups at nucleophilic sites. Obviously, the phenol groups themselves are nucleophilic, and with a suitable alkylating agent, *O*-alkyl derivatives may be formed (see page 13), e.g.

the *O*-methylation of emodin to physcion (Figure 3.76). However, a phenol group also activates the ring carbon atoms at the *ortho* and *para* positions, so that these positions similarly become susceptible to alkylation, leading to *C*-alkyl derivatives. The *meta* oxygenation pattern which is a characteristic feature of acetate-derived phenolics has the effect of increasing this nucleophilicity considerably, and the process of *C*-alkylation is very

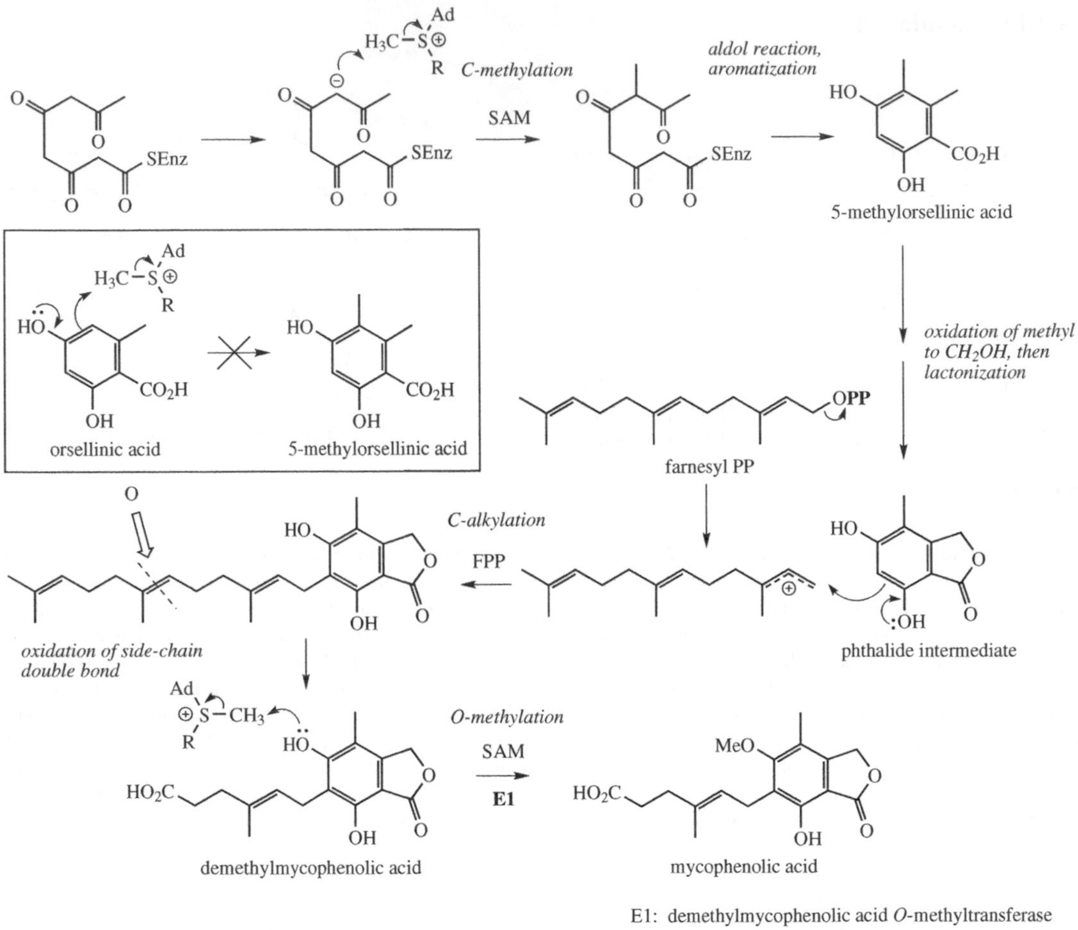

Figure 3.82

much facilitated (see page 13). Suitable natural alkylating agents are SAM and dimethylallyl diphosphate (DMAPP). Other polyprenyl diphosphate esters may also be encountered in biological alkylation reactions (e.g. see hyperforin, page 113, vitamin K, page 182). A minor inconsistency has been discovered, in that while *C*-alkylation with dimethylallyl and higher diphosphates is mediated *after* the initial polyketide cyclization product is liberated from the enzyme, there are examples where *C*-methylation occurs *before* cyclization and is catalysed by the PKS itself (compare macrolide PKSs, page 86). **5-Methylorsellinic acid** (Figure 3.82) is a simple *C*-methylated analogue of orsellinic acid found in *Aspergillus flaviceps*, and the extra methyl is derived from SAM. However, orsellinic acid (see page 101) is not a precursor of 5-methylorsellinic acid and, therefore, it is proposed that the poly-β-keto ester is methylated as part

of the reactions catalysed by the PKS (Figure 3.82). Similarly, 5-methylorsellinic acid, but not orsellinic acid, is a precursor of **mycophenolic acid** in *Penicillium brevicompactum* (Figure 3.82) [Box 3.14]. Further *C*-alkylation by farnesyl diphosphate (see page 210) proceeds *after* the aromatization step, and a phthalide intermediate is the substrate involved. The phthalide is a lactone derived from 5-methylorsellinic acid by hydroxylation of its starter methyl group and reaction with the end-of-chain carboxyl. The chain length of the farnesyl alkyl group is subsequently shortened by oxidation of a double bond, giving demethylmycophenolic acid, which is then *O*-methylated, again involving SAM, to produce mycophenolic acid. Note that the *O*-methylation step occurs only after the *C*-alkylations, so that the full activating benefit of two *meta*-positioned phenols can be utilized for the *C*-alkylation.

Box 3.14

Mycophenolic acid

Mycophenolic acid (Figure 3.82) is produced by fermentation cultures of the fungus *Penicillium brevicompactum*. It has been known for many years to have antibacterial, antifungal, antiviral, and antitumour properties. It has recently been introduced into medicine as an immunosuppressant drug, to reduce the incidence of rejection of transplanted organs, particularly in kidney and heart transplants. It is formulated as the *N*-morpholinoethyl ester **mycophenolate mofetil** (Figure 3.83), which is metabolized after ingestion to mycophenolic acid; it is usually administered in combination with another immunosuppressant, ciclosporin (see page 454). The drug is a specific inhibitor of mammalian inosine monophosphate (IMP) dehydrogenase and has an antiproliferative activity on cells due to inhibition of guanine nucleotide biosynthesis. IMP dehydrogenase catalyses the NAD^+-dependent oxidation of IMP to xanthosine monophosphate (XMP), a key transformation in the synthesis of guanosine triphosphate (GTP) (see also caffeine biosynthesis, page 413). Rapidly growing cells have increased levels of the enzyme, so this forms an attractive target for anticancer, antiviral, and immunosuppressive therapy. Mycophenolate mofetil may also be used to treat severe refractory eczema.

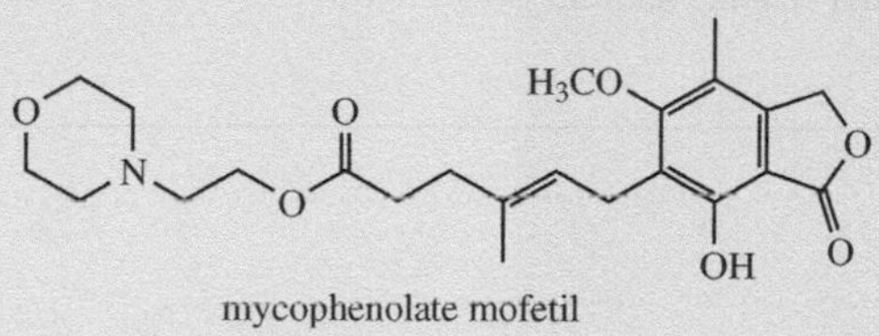

mycophenolate mofetil

Figure 3.83

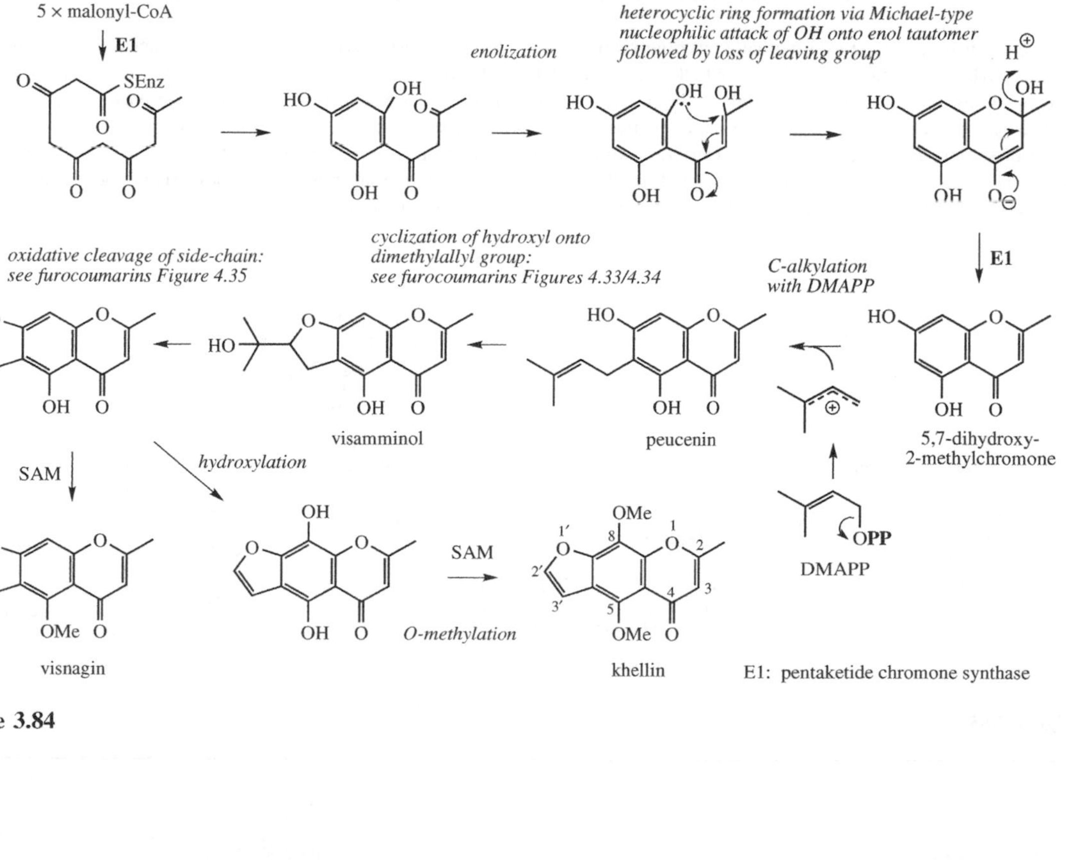

Figure 3.84

Khellin and **visnagin** (Figure 3.84) are furochromones found in the fruits of *Ammi visnaga* (Umbelliferae/Apiaceae), and the active principles of a crude plant drug which has a long history of use as an antiasthmatic agent. Figure 3.84 presents the sequence of steps utilized in the biosynthesis of these compounds, fully consistent with the biosynthetic rationale developed above. The two carbon atoms C-2′ and C-3′ forming part of the furan ring originate by metabolism of a five-carbon dimethylallyl substituent attached to C-6 (for a full discussion, see furocoumarins, page 163). The 8-methoxy group in khellin is absent in visnagin, so must be introduced late in the sequence. The key intermediate is thus **5,7-dihydroxy-2-methylchromone**. On inspection, this has the alternate acetate-derived oxygenation pattern and a methyl chain starter, so

is formed from a poly-β-keto chain through Claisen condensation. The heterocyclic ring is produced by an overall dehydration reaction, formulated as a Michael-like addition reaction (see also flavonoids, page 169). Indeed, a type III PKS derived from *Aloe arborescens* has been characterized and shown to catalyse formation of the chromone from five molecules of malonyl-CoA; the starter unit is derived by decarboxylation. After formation of the furan ring via the *C*-dimethylallyl derivative peucenin and then visamminol, **visnagin** can be obtained by *O*-methylation. Alternatively, further hydroxylation *para* to the free phenol followed by two methylations yields **khellin**. The antiasthmatic properties of khellin have been exploited by developing the more polar, water-soluble derivative **cromoglicate** [Box 3.15].

Box 3.15

Khellin and Cromoglicate

The dried ripe fruits of *Ammi visnaga* (Umbelliferae/Apiaceae) have a long history of use in the Middle East as an antispasmodic and for the treatment of angina pectoris. The drug contains small amounts of coumarin derivatives, e.g. visnadin (Figure 3.85) (compare *Ammi majus*, a rich source of furocoumarins, page 165), but the major constituents (2–4%) are furochromones, including khellin, visnagin, khellol, and khellol glucoside (Figure 3.85). Both khellin and visnadin are coronary vasodilators and spasmolytic agents, with visnadin actually being the more potent agent. Khellin has been used in the treatment of angina pectoris and bronchial asthma.

The synthetic analogue **cromoglicate** (**cromoglycate**; Figure 3.85) is a most effective and widely used agent for the treatment and prophylaxis of asthma, hay fever, and allergic rhinitis. Cromoglicate contains two chromone systems with polar carboxylic acid functions, joined by a glycerol linker. The mode of action is not fully established. It was believed to prevent the release of bronchospasm mediators by stabilizing mast cell membranes, but an alternative suggestion is that it may act by inhibiting the effect of sensory nerve activation, thus interfering with bronchoconstriction. It is poorly absorbed orally and is thus administered by inhalation or nasal spray. Eyedrops for relief of allergic conjunctivitis and nasal sprays for prophylaxis of allergic rhinitis are also available. The more potent pyridonochromone **nedocromil** (Figure 3.85) has also been introduced.

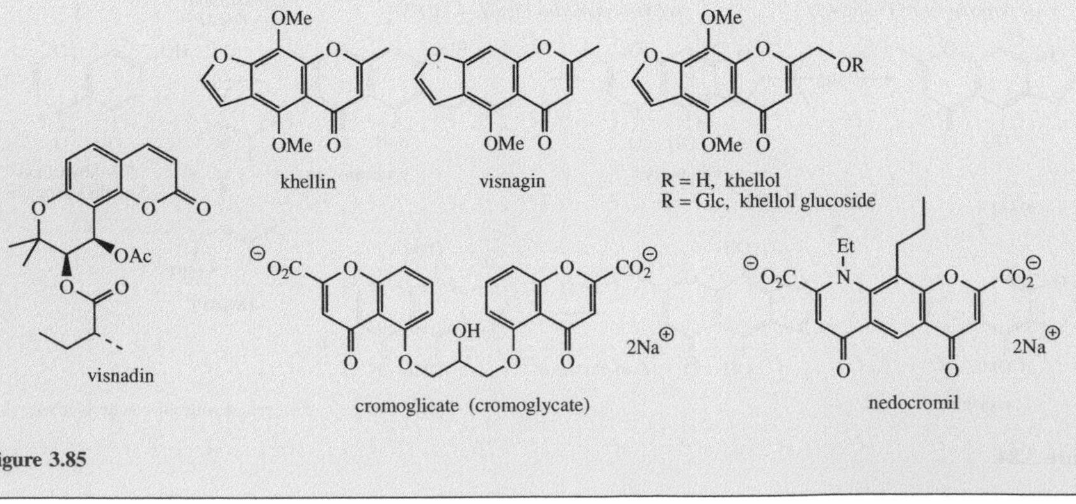

Figure 3.85

The polyketide nature of **hyperforin** (Figure 3.86) is almost entirely obscured by the added isoprenoid fragments. This compound is the predominant antidepressive agent in St John's Wort (*Hypericum perforatum*; Guttiferae/Hypericaceae) [Box 3.13], where it co-occurs with the naphthodianthrone hypericin (see page 108). The product of the type III PKS is **phlorisobutyrophenone**, an analogue of phloroacetophenone (see page 100) but using isobutyryl-CoA as starter instead of acetyl-CoA. This

starter is formed from the amino acid valine, as seen in the pathways to bacterial iso-fatty acids (see page 56).

Successive *C*-alkylations with DMAPP substitute both of the remaining aromatic sites. The next alkylating agent is geranyl diphosphate, and though the mechanism is exactly the same, this alkylation must destroy the aromaticity. One more alkylation occurs; this is best formulated as a carbocation addition to the geranyl double bond, generating a tertiary carbocation, then quenching

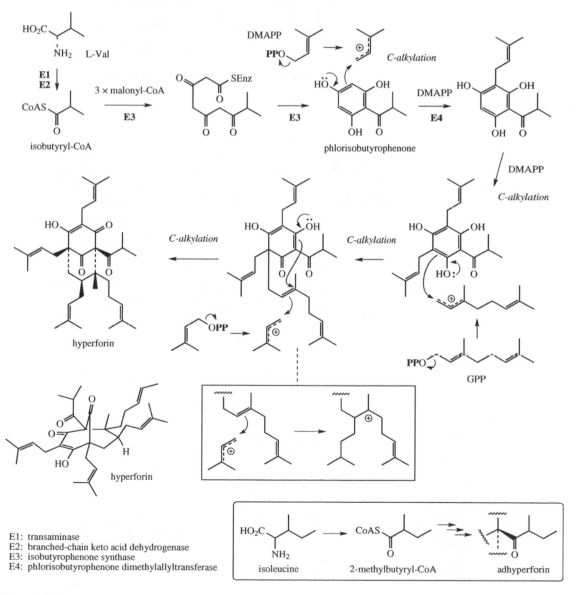

E1: transaminase
E2: branched-chain keto acid dehydrogenase
E3: isobutyrophenone synthase
E4: phlorisobutyrophenone dimethylallyltransferase

Figure 3.86

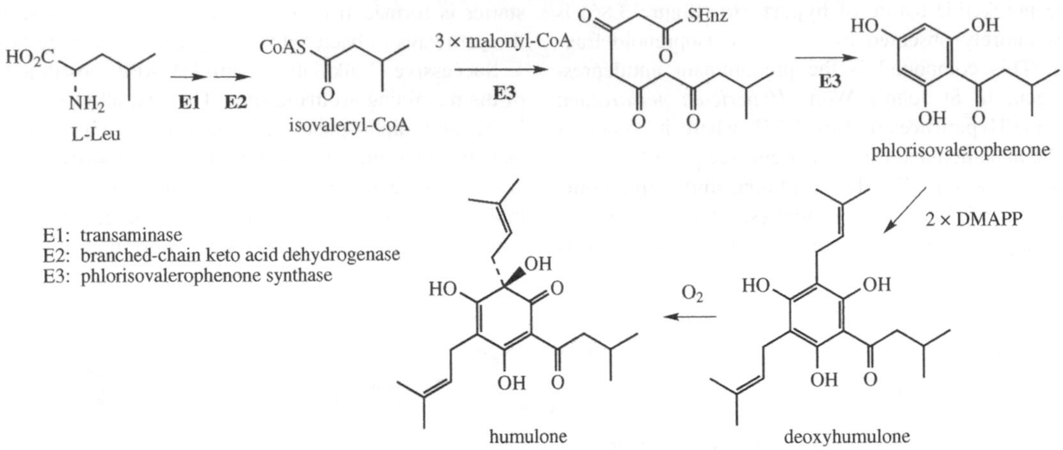

Figure 3.87

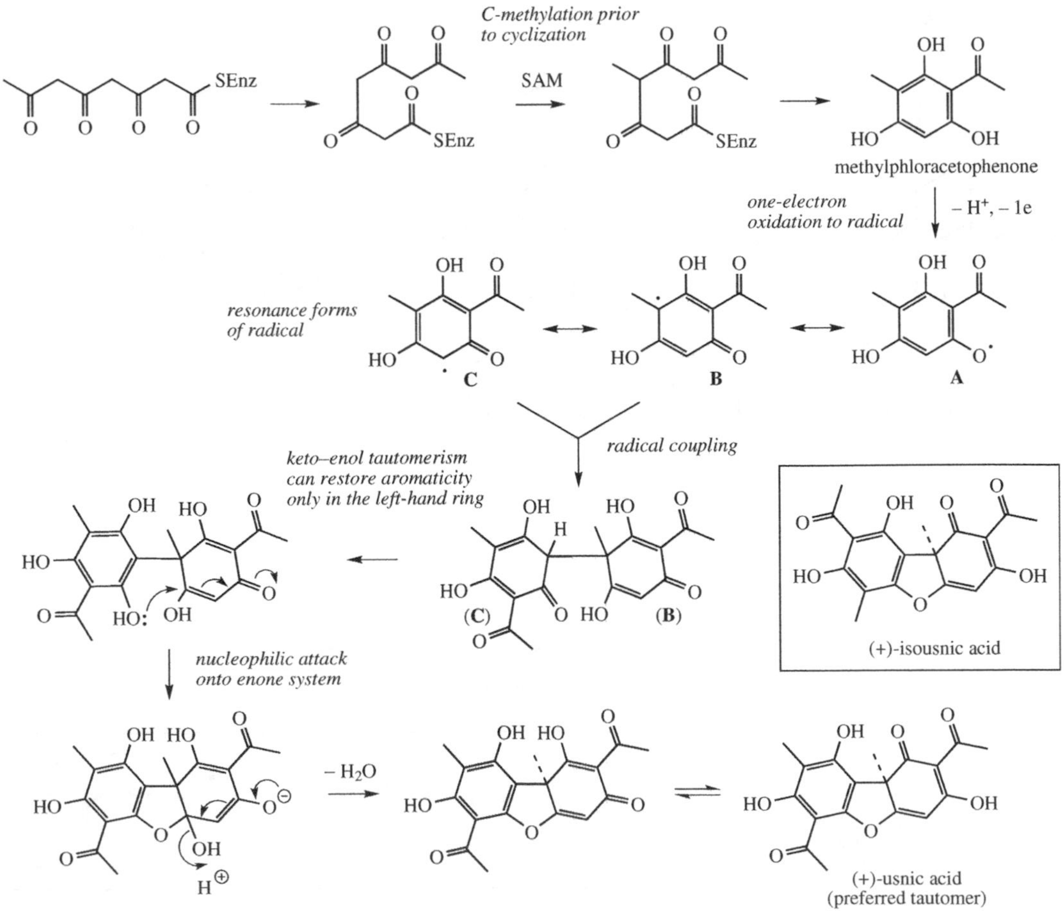

E1: transaminase
E2: branched-chain keto acid dehydrogenase
E3: phlorisovalerophenone synthase

Figure 3.88

this by yet another alkylation reaction on the ring system. The structurally related **adhyperforin** (Figure 3.86) is also found in *H. perforatum*. This compound is characterized by derivation from isoleucine rather than valine, and the starter unit is thus 2-methylbutyryl-CoA (compare iso-fatty acids, page 56). Somewhat less complex structures are found in hops (*Humulus lululus*; Cannabaceae), where they are representative of a group of compounds termed bitter acids. These are responsible for the typical bitter taste of beer, and also provide foam-stabilizing and antibacterial properties. **Humulone** (Figure 3.87) is typically the major bitter acid, and is formed by oxidative transformaton of deoxyhumulone. Here, the starter unit for the polyketide is leucine-derived isovaleryl-CoA.

C-Methylation also features in the biosynthesis of **usnic acid** (Figure 3.88), an antibacterial metabolite found in many lichens, e.g. *Usnea* and *Cladonia* species, which are symbiotic combinations of alga and fungus. However, the principal structural modification encountered involves phenolic oxidative coupling (see page 28). Two molecules of **methylphloracetophenone** are incorporated, and these are known to derive from a pre-aromatization methylation reaction, not by methylation of phloracetophenone

(Figure 3.88). The two molecules are joined together by an oxidative coupling mechanism which can be rationalized via the one-electron oxidation of a phenol group in methylphloracetophenone giving radical A, for which resonance forms B and C can be written. Coupling of B and C occurs. Only the left-hand ring can subsequently be restored to aromaticity by keto–enol tautomerism; this is not possible in the right-hand ring because coupling occurred on the position *para* to the original phenol and this position already contains a methyl. Instead, a heterocyclic ring is formed by attack of the phenol onto the enone system (see khellin, above). The outcome of this reaction is enzyme controlled; two equivalent phenol groups are present as potential nucleophiles, and two equivalent enone systems are also available. Potentially, four different products could be formed, but only one is typically produced. Loss of water then leads to usnic acid. Usnic acid is found in both (+)- and (−)-forms in nature, according to source. The product from using the alternative hydroxyl nucleophile in heterocyclic ring formation is **isousnic acid**, also known as a natural product in some species of lichens. Usnic acid is used as an antibacterial and preservative in toiletries.

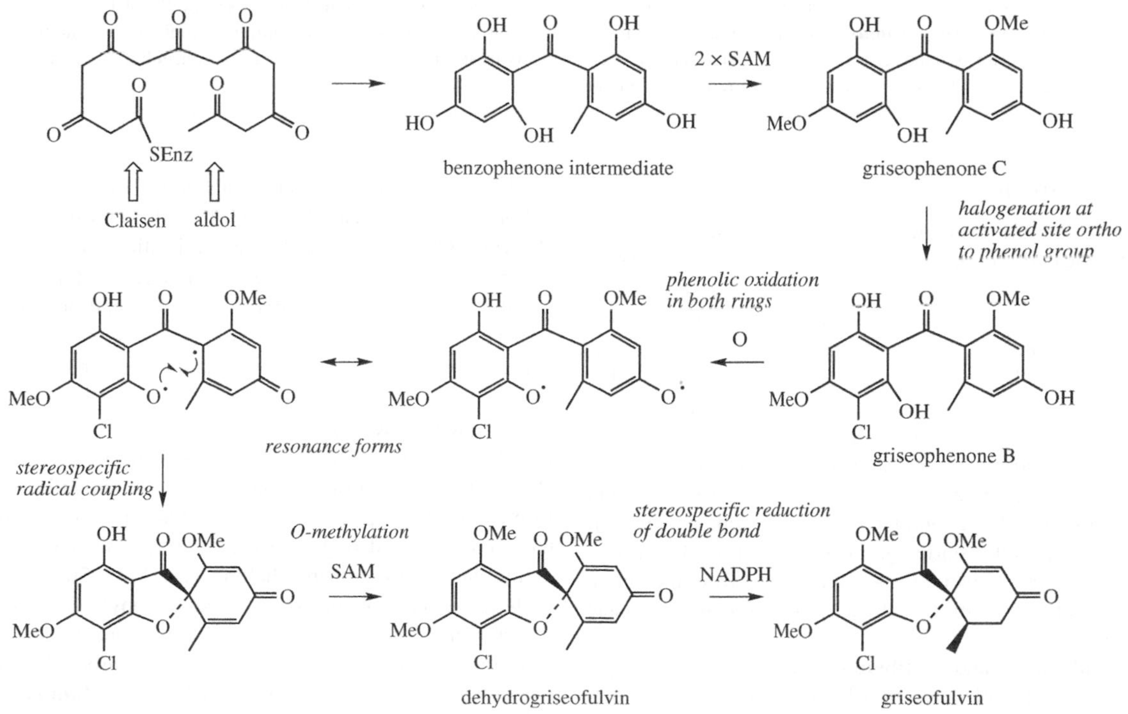

Figure 3.89

Phenolic oxidative coupling is widely encountered in natural product biosynthesis, and many other examples are described in subsequent sections. A further acetate-derived metabolite formed as a result of oxidative coupling is the antifungal agent **griseofulvin** (Figure 3.89) synthesized by cultures of *Penicillium griseofulvin* [Box 3.16]. The sequence of events leading to griseofulvin has been established in detail via feeding experiments; the pathway also includes *O*-methylation steps and the introduction of a halogen (chlorine) atom at one of the nucleophilic sites (Figure 3.89).

Initial inspection of the structure of griseofulvin shows the alternate oxygenation pattern, and also a methyl group

which identifies the start of the polyketide chain. Cyclization of the C_{14} poly-β-keto chain folded as shown allows both Claisen (left-hand ring) and aldol (right-hand ring) reactions to occur, giving a benzophenone intermediate. Two selective methylations lead to griseophenone C, which is the substrate for chlorination to griseophenone B; both these compounds appear as minor metabolites in *P. griseofulvin* cultures. One-electron oxidations on a phenolic group in each ring give a diradical; radical coupling in its resonance form generates the basic grisan skeleton with its spiro centre. **Griseofulvin** is then the result of methylation of the remaining phenol group and stereospecific reduction of the double bond in dehydrogriseofulvin.

Box 3.16

Griseofulvin

Griseofulvin is an antifungal agent produced by cultures of *P. griseofulvum* and a number of other *Penicillium* species, including *P. janczewski*, *P. nigrum* and *P. patulum*. Griseofulvin is used to treat widespread or intractable dermatophyte infections, though has to some extent been superseded by newer antifungals. It is less effective when applied topically, but it is well absorbed from the gut and is selectively concentrated into the keratin layers of the skin. It may thus be used orally to control dermatophytes such as *Epidermophyton*, *Microsporium* and *Trichophyton*. Treatment for some conditions, e.g. infections in fingernails, may have to be continued for several months, but the drug is generally free of side-effects. Griseofulvin is also used orally or topically to control ringworm (*Tinea capitis*). The antifungal action appears to be through disruption of the mitotic spindle, thus inhibiting mitosis. The drug is relatively safe for use, since the effects on mitosis in sensitive fungal cells occur at concentrations that are substantially below those required to inhibit mitosis in human cells.

Starter Groups

In most of the examples of aromatics so far discussed, the basic carbon skeleton has been derived from an acetate starter group (sometimes formed via PKS decarboxylation of malonate), with malonate acting as the chain extender. The molecule has then been made more elaborate, principally via alkylation reactions and the inclusion of other carbon atoms. However, the range of natural aromatic structures which are at least partly derived from acetate is increased enormously by altering the nature of the starter group from acetate to an alternative acyl-CoA system, with malonyl-CoA again providing the chain extender. For example, the starter group for hyperforin biosynthesis is a branched-chain CoA ester derived from the amino acid leucine, and this has already been discussed (see page 113).

Chalcones and **stilbenes** are relatively simple examples of molecules in which a suitable cinnamoyl-CoA C_6C_3 precursor from the shikimate pathway (see page 149) is used as a starter group for the PKS system. If **4-hydroxycinnamoyl-CoA** (Figure 3.90)

is chain extended with three malonyl-CoA units, the poly-β-keto chain can then be folded in two ways, allowing Claisen or aldol-type cyclizations to occur as appropriate. Chalcones, e.g. **naringenin-chalcone**, or stilbenes, e.g. **resveratrol**, are the end-products formed by the type III enzymes chalcone synthase and stilbene synthase respectively. Chalcone synthase and stilbene synthase enzymes share some 75–90% amino acid sequence identity, but cyclize the same substrate in different ways. The six-membered heterocyclic ring characteristic of most flavonoids, e.g. **naringenin**, is formed by nucleophilic attack of a phenol group from the acetate-derived ring onto the α,β-unsaturated ketone of the chalcone (compare khellin, page 111). Stilbenes, such as **resveratrol**, incorporate the carbonyl carbon of the cinnamoyl unit into the aromatic ring and typically lose the end-of-chain carboxyl by a decarboxylation reaction. Although some related structures, e.g. **lunularic acid** (Figure 3.90) from the liverwort *Lunularia cruciata*, still contain this carboxyl, these are rare in nature. Carboxylated stilbenes are not intermediates in the

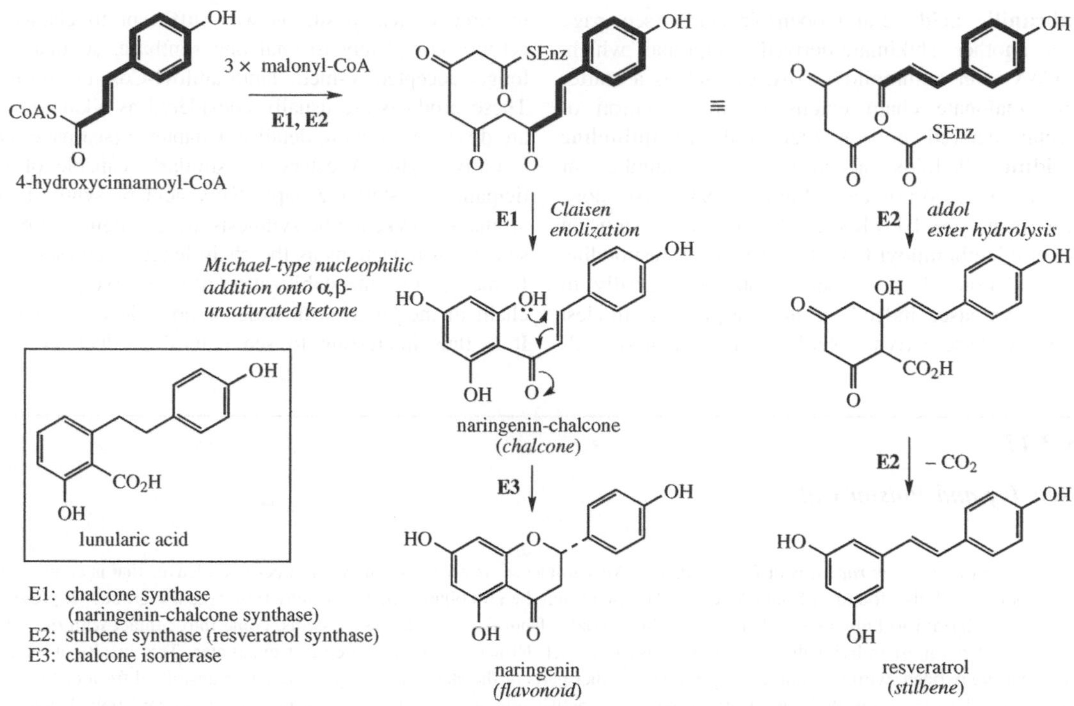

Figure 3.90

stilbene synthase reaction. Mutagenesis studies have confirmed that hydrolysis of the thioester bond and decarboxylation precede aromatization by the necessary dehydration/enolization reactions. Indeed, it appears that it is the TE activity component of the stilbene synthase enzyme that is primarily responsible for the stilbene-type cyclization (aldol reaction) rather than the chalcone-type cyclization (Claisen reaction). Through appropriate amino acid replacements it is possible to convert chalcone synthase into proteins with stilbene synthase activity. Flavonoids and stilbenes, and the further modifications that may occur, are traditionally treated as shikimate-derived compounds; they are discussed in more detail in Chapter 4 (see page 167).

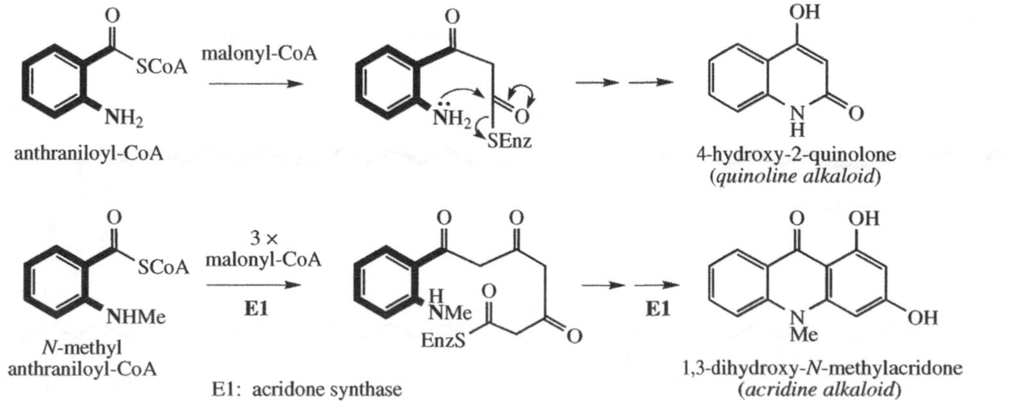

Figure 3.91

Anthranilic acid (2-aminobenzoic acid) (see page 141) is another shikimate-derived compound which, as its CoA ester anthraniloyl-CoA, can act as a starter unit for malonate chain extension. Aromatization of the acetate-derived portion then leads to **quinoline** or **acridine** alkaloids, according to the number of acetate units incorporated (Figure 3.91). Acridone synthase is a type III PKS catalysing chain extension of *N*-methylanthraniloyl-CoA to give the corresponding acridine alkaloid. It is closely related structurally to chalcone synthase; indeed, it also displays a modest chalcone synthase activity. Replacement of amino acids

in three critical positions was sufficient to change its activity completely to chalcone synthase, so that it no longer accepted *N*-methylanthraniloyl-CoA as substrate. These products are usually considered as alkaloids, and are discussed in more detail in Chapter 6 (see page 395).

Fatty acyl-CoA esters are similarly capable of participating as starter groups. Fatty acid biosynthesis and aromatic polyketide biosynthesis are distinguished by the sequential reductions as the chain length increases in the former, and by the stabilization of a reactive poly-β-keto chain in the latter, with little or no reduction involved. It is thus interesting to see natural product structures

---

Box 3.17

Poison Ivy and Poison Oak

Poison ivy (*Toxicodendron radicans* or *Rhus radicans*; Anacardiaceae) is a woody vine with three-lobed leaves that is common in North America and also found in South America. The plant may be climbing, shrubby, or may trail over the ground. It presents a considerable hazard to humans should the sap, which exudes from damaged leaves or stems, come into contact with the skin. The sap sensitizes most individuals, producing delayed contact dermatitis after a subsequent encounter. This results in watery blisters that break open, with the fluid quickly infecting other parts of the skin. The allergens may be transmitted from one person to another on the hands, on clothing, or by animals. The active principles are urushiols, a mixture of alkenyl polyphenols. In poison ivy, these are mainly catechols with C_{15} side-chains carrying varying degrees of unsaturation (Δ^8, $\Delta^{8,11}$, $\Delta^{8,11,14}$). Small amounts of C_{17} side-chain analogues are present. These catechols become oxidized to an *ortho*-quinone, which is then attacked by nucleophilic groups in proteins to yield an antigenic complex.

Poison oak (*Toxicodendron toxicaria* or *Rhus toxicodendron*; Anacardiaceae) is nearly always found as a low-growing shrub and has lobed leaflets similar to those of oak. It is also common throughout North America. There appears considerable confusion over nomenclature, and *Rhus radicans* may also be termed poison oak, and *R. toxicodendron* oakleaf poison ivy. Poison oak contains similar urushiol structures in its sap as poison ivy, though heptadecylcatechols (i.e. C_{17} side-chains) predominate over pentadecylcatechols (C_{15} side-chains).

Related species of *Toxicodendron*, e.g. *T. diversilobum* (Pacific poison oak) and *T. vernix* (poison sumach, poison alder, poison dogwood), are also allergenic with similar active constituents. These plants were all formerly classified under the genus *Rhus*, but the allergen-containing species have been reclassified as *Toxicodendron*; this nomenclature is not widely used. Dilute purified extracts containing urushiols may be employed to stimulate antibody production and, thus, build up some immunity to the allergens. These plants cause major problems to human health in North America, with many thousands of reported cases each year.

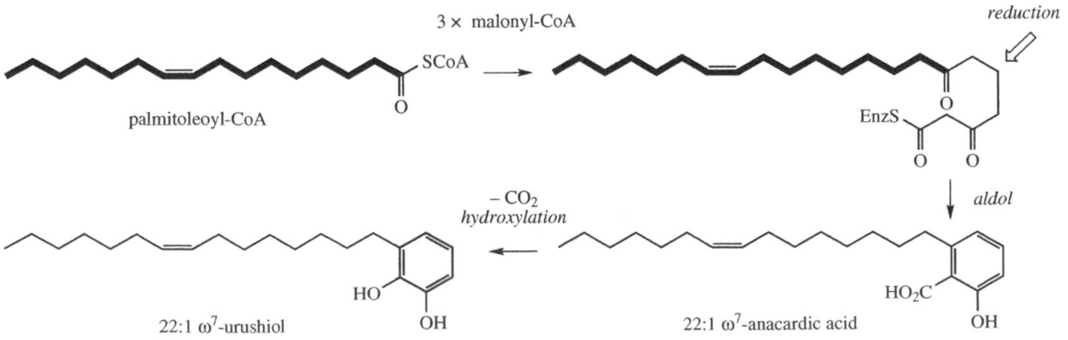

Figure 3.92

containing both types of acetate/malonate-derived chains (see also zearalenone, page 89). In plants of the Anacardiaceae, just such a pathway accounts for the formation of contact allergens called **urushiols** in poison ivy (*Toxicodendron radicans*) and poison oak (*Toxicodendron toxicaria*) [Box 3.17]. Thus, **palmitoleoyl-CoA** (Δ^9-hexadecenoyl-CoA) can act as starter group for extension by three malonyl-CoA units, with a reduction step during chain extension (Figure 3.92). Aldol cyclization then gives an **anacardic**

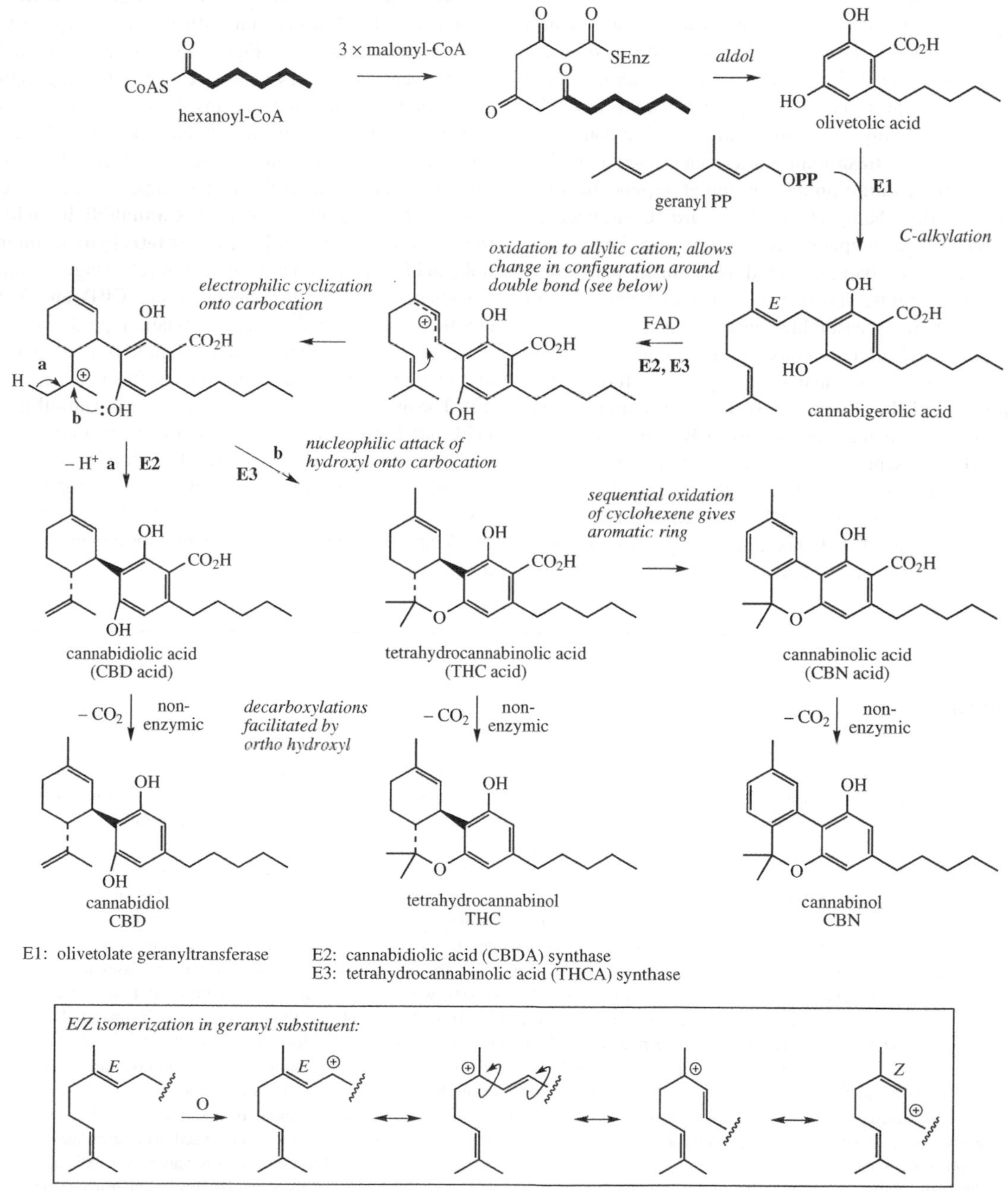

CoAS
hexanoyl-CoA
3 × malonyl-CoA
SEnz
aldol
olivetolic acid
geranyl PP
OPP
E1
C-alkylation
oxidation to allylic cation; allows change in configuration around double bond (see below)
FAD
E2, E3
cannabigerolic acid
electrophilic cyclization onto carbocation
− H+ a E2
E3
b
nucleophilic attack of hydroxyl onto carbocation
sequential oxidation of cyclohexene gives aromatic ring
cannabidiolic acid (CBD acid)
tetrahydrocannabinolic acid (THC acid)
cannabinolic acid (CBN acid)
− CO2 non-enzymic
decarboxylations facilitated by ortho hydroxyl
cannabidiol CBD
tetrahydrocannabinol THC
cannabinol CBN
E1: olivetolate geranyltransferase
E2: cannabidiolic acid (CBDA) synthase
E3: tetrahydrocannabinolic acid (THCA) synthase
E/Z isomerization in geranyl substituent:

Figure 3.93

acid, which is likely to be the precursor of a **urushiol** by decarboxylation/hydroxylation. However, enzymic evidence is not yet available. It is likely that different fatty acyl-CoAs can participate in this sequence, since urushiols from poison ivy can contain up to three double bonds in the C_{15} side-chain, whilst those from poison oak also have variable unsaturation in a C_{17} side-chain (i.e. a C_{18} acyl-CoA starter). The name anacardic acid relates particularly to the presence of large quantities of these compounds in the shells of cashew nuts (*Anacardium occidentale*; Anacardiaceae); these anacardic acids have C_{15} side-chains carrying one, two, or three double bonds.

A saturated C_6 **hexanoate** starter unit is used in the formation of the **cannabinoids**, a group of terpenophenolics found in Indian hemp (*Cannabis sativa*; Cannabaceae). This plant, and preparations from it, are known under a variety of names, including hashish, marihuana, pot, bhang, charas, and dagga, and have been used for centuries for the pleasurable sensations and mild euphoria experienced after its consumption, usually by smoking. The principal psychoactive component is **tetrahydrocannabinol** (**THC**; Figure 3.93), whilst structurally similar compounds such as **cannabinol** (**CBN**) and **cannabidiol** (**CBD**), present in similar or larger amounts, are effectively inactive. In recent years, the beneficial effects of cannabis, and especially THC, in alleviating nausea and vomiting in cancer patients undergoing chemotherapy, and in the treatment of glaucoma and multiple sclerosis,

has led to extensive study of cannabinoid analogues for potentially useful medicinal activity [Box 3.18].

All the cannabinoid structures contain a monoterpene C_{10} unit attached to a phenolic ring that carries a C_5 alkyl chain. The aromatic ring/C_5 chain originates from hexanoate and malonate, cyclization to a polyketide giving **olivetolic acid**, from which **cannabigerolic acid** can be obtained by *C*-alkylation with the monoterpene C_{10} unit geranyl diphosphate (Figure 3.93). Cyclization in the monoterpene unit necessitates a change in configuration of the double bond, and this involves an oxidative step in which FAD is a cofactor. It may be rationalized as involving the allylic cation, which will then allow electrophilic cyclization to proceed (compare terpenoid cyclization mechanisms, page 195). **Cannabidiolic acid** is then the result of proton loss, whilst **tetrahydrocannabinolic acid** is the product from heterocyclic ring formation via quenching with the phenol group. **CBD** and **THC** are then the respective decarboxylation products from these two compounds; decarboxylation is non-enzymic and facilitated by the adjacent phenol function (see page 22). The aromatic terpenoid-derived ring in **cannabinolic acid** and **CBN** can arise via a dehydrogenation process (compare thymol, page 204). The PKS for olivetolic acid formation has yet to be characterized; a related type III PKS system from *Cannabis* catalysed the formation of the decarboxylated derivative olivetol instead of olivetolic acid.

Box 3.18

Cannabis

Indian hemp, *Cannabis sativa* (Cannabaceae), is an annual herb indigenous to central and western Asia. It is cultivated widely in India and many tropical and temperate regions for its fibre (hemp) and seed (for seed oil). The plant is also grown for its narcotic and mild intoxicant properties, and in most countries of the world, its possession and consumption are illegal. Over many years, cannabis plants have been selected for either fibre production or drug use, the former resulting in tall plants with little pharmacological activity, whilst the latter tend to be short, bushy plants. Individual plants are almost always male or female, though the sex is not distinguishable until maturity and flowering. Seeds will produce plants of both sexes in roughly equal proportions. The active principles are secreted as a resin by glandular hairs, which are more numerous in the upper parts of female plants, and resin is produced from the time flowers first appear until the seeds reach maturity. However, all parts of the plant, both male and female, contain cannabinoids. In a typical plant, the concentration of cannabinoids increases in the order: large leaves, small leaves, flowers, bracts (which surround the ovaries), with stems containing very little. Material for drug use (ganja) is obtained by collecting the flowering tops (with little leaf) from female plants, though lower quality material (bhang) consisting of leaf from both female and male plants may be employed. By rubbing the flowering tops, the resin secreted by the glandular hairs can be released and subsequently scraped off to provide cannabis resin (charas) as an amorphous brown solid or semi-solid. A potent form of cannabis, called cannabis oil, is produced by alcoholic extraction of cannabis resin. A wide variety of names are used for cannabis products according to their nature and the geographical area. In addition to the Indian words above, the names hashish (Arabia), marihuana (Europe, USA), kief and dagga (Africa) are frequently used. The term 'assassin' is a corruption of 'hashishin', a group of 13th century murderous Persians who were said to have been rewarded for their activities with hashish. Names such as grass, dope, pot, hash, weed, and wacky backy are more likely to be in current usage. Cannabis for the illicit market is cultivated worldwide. Most of the resin originates from North Africa, Afghanistan, and Pakistan, whilst

Box 3.18 (continued)

cannabis herb comes mainly from North America and Africa. Canada, the United States, and the UK grow large quantities for legitimate use.

The quantity of resin produced by the flowering tops of high-quality Indian cannabis is about 15–20%. The amount produced by various plants is dependent on several features, however, and this will markedly alter biological properties. Thus, in general, plants grown in a tropical climate produce more resin than those grown in a temperate climate. The tall fibre-producing plants are typically low resin producers, even in tropical zones. However, the most important factor is the genetic strain of the plant, as the resin produced may contain high levels of psychoactive compounds or mainly inactive constituents. The quality of any cannabis drug is potentially highly variable.

The major constituents in cannabis are termed cannabinoids, a group of more than 60 structurally related terpenophenolics. The principal psychoactive agent is tetrahydrocannabinol (THC, Figure 3.93). This is variously referred to as Δ^1-THC or Δ^9-THC according to whether the numbering is based on the terpene portion or as a systematic dibenzopyran (Figure 3.94). Both systems are currently in use, though the systematic one is preferred. Also found, often in rather similar amounts, are cannabinol (CBN) and cannabidiol (CBD) (Figure 3.93), which have negligible psychoactive properties. These compounds predominate in the inactive resins. Many other cannabinoid structures have been characterized, including cannabigerol and cannabichromene (Figure 3.94). A range of cannabinoid acids, e.g. cannabidiolic acid, tetrahydrocannabinolic acid, and tetrahydrocannabinolic acid-B (Figure 3.94) are also present, as are some analogues of the other compounds mentioned, where a propyl side-chain replaces the pentyl group, e.g. tetrahydrocannabivarin (Figure 3.94). The latter compounds presumably arise from the use of butyrate rather than hexanoate as starter unit in the biosynthetic sequence.

The THC content of high-quality cannabis might be in the range 0.5–1% for large leaves, 1–3% for small leaves, 3–7% for flowering tops, 5–10% for bracts, 14–25% for resin, and up to 60% in cannabis oil. Higher amounts of THC are produced in selected strains known as skunk cannabis, so named because of its powerful smell; flowering tops from skunk varieties might contain 10–15% THC. The THC content in cannabis products tends to deteriorate on storage, an effect accelerated by heat and light. Cannabis leaf and resin stored under ordinary conditions rapidly lose their activity and can be essentially inactive after about 2 years. A major change which occurs is oxidation in the cyclohexene ring resulting in conversion of THC into CBN. THC is more potent when smoked than when taken orally, its volatility allowing rapid absorption and immediate effects, so smoking has become the normal means of using cannabis. Any cannabinoid acids will almost certainly be decarboxylated upon heating, and thus the smoking process will also effectively increase somewhat the levels of active cannabinoids available, e.g. THC acid → THC (Figure 3.93). The smoking of cannabis produces a mild euphoria similar to alcohol intoxication, inducing relaxation, contentment, and a sense of well-being, with some changes in perception of sound and colour. However, this is accompanied by a reduced ability to concentrate and do complicated tasks, and a loss of short-term memory. Users claim cannabis is much preferable to alcohol or tobacco, insisting it does not cause dependence, withdrawal symptoms, or lead to the use of other drugs,

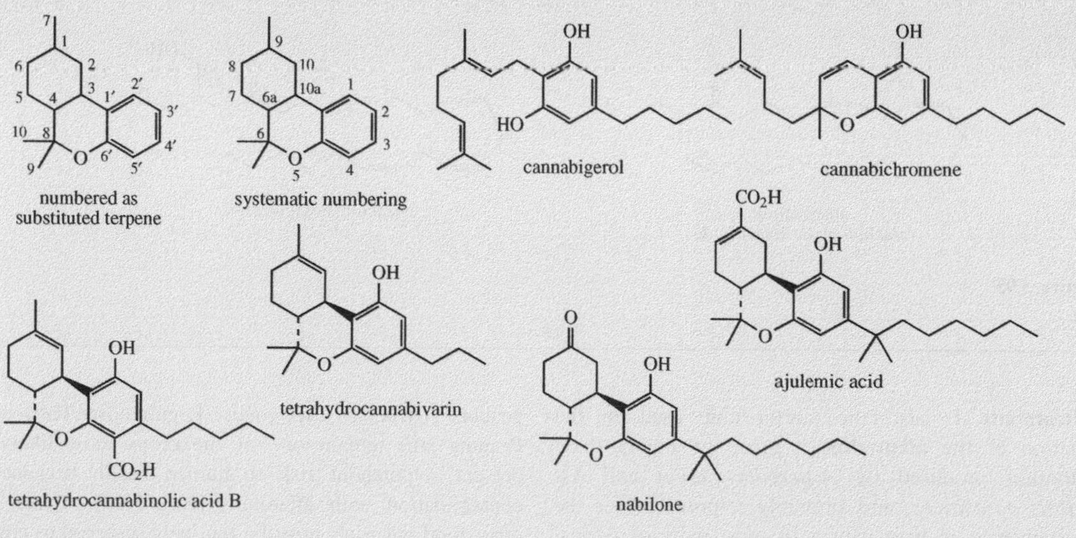

numbered as substituted terpene

systematic numbering

cannabigerol

cannabichromene

tetrahydrocannabinolic acid B

tetrahydrocannabivarin

nabilone

ajulemic acid

Figure 3.94

Box 3.18   (continued)

and they campaign vociferously for its legalization. However, psychological dependence does occur, and cannabis can lead to hallucinations, depression, anxiety, and panic, with the additonal risk of bronchitis and lung cancer when the product is smoked.

Cannabis has been used medicinally, especially as a mild analgesic and tranquillizer, but more effective and reliable agents replaced it, and even controlled prescribing was discontinued. In recent times, cannabis has been shown to have valuable anti-emetic properties which help to reduce the side-effects of nausea and vomiting caused by cancer chemotherapeutic agents. This activity stems from THC, and has resulted in some use of **dronabinol** (synthetic THC) and the prescribing of cannabis for a small number of patients. The synthetic THC analogue **nabilone** (Figure 3.94) has been developed as an anti-emetic drug for reducing cytotoxic-induced vomiting. Some of the psychoactive properties of THC, e.g. euphoria, mild hallucinations, and visual disturbances, may be experienced as side-effects of nabilone treatment. Cannabis has also been shown to possess properties which may be of value in other medical conditions. There is now ample evidence that cannabis can give relief to patients suffering from chronic pain, multiple sclerosis, glaucoma, asthma, migraine, epilepsy, and other conditions. Many sufferers who cannot seem to benefit from any of the current range of drugs are obtaining relief from their symptoms by using cannabis, but are breaking the law to obtain this medication. Current thinking is that cannabis offers a number of beneficial pharmacological responses and that there should be legal prescribing of cannabinoids or derivatives. Clinical trials have already confirmed the value of cannabis and/or THC taken orally for the relief of chronic pain and the painful spasms characteristic of multiple sclerosis, and in reducing intraocular pressure in glaucoma sufferers. An oral spray preparation of natural THC + CBD (**Sativex**®) for multiple sclerosis patients has been approved in some countries. In general, cannabis is only able to alleviate the symptoms of these diseases; it does not provide a cure. The non-psychoactive CBD has been shown to have anti-inflammatory properties potentially useful in arthritis treatment. The synthetic cannabinoid **ajulemic acid** (Figure 3.94) is in clinical trials for treatment of arthritic pain and inflammation; it does not produce any psychotropic actions at therapeutic dosage.

Two main cannabinoid receptors have been identified: CB_1, predominantly in the central nervous system, and CB_2, expressed mainly in the immune system. This was followed by discovery in animal brain tissue of a natural ligand for CB_1, namely anandamide (Figure 3.95), which is arachidonoylethanolamide; ananda is the Sanskrit word for bliss. Anandamide mimics several of the pharmacological properties of THC. The natural ligand of CB_2 is 2-arachidonoylglycerol (Figure 3.95); this also interacts with CB_1, and since levels of 2-arachidonoylglycerol in the brain are some 800 times higher than those of anandamide, it is now thought to be the physiological ligand for both receptors, rather than anandamide. These two compounds are the main ligands, but other related natural compounds from mammalian brain also function in the same way. These include polyunsaturated fatty acid ethanolamides, namely dihomo-γ-linolenoyl- (20:3) and adrenoyl- (22:4) ethanolamides, O-arachidonoylethanolamine (virodhamine) and 2-arachidonylglyceryl ether (noladin ether). The identification of these endogenous materials may open up other ways of exploiting some of the desirable pharmacological features of cannabis.

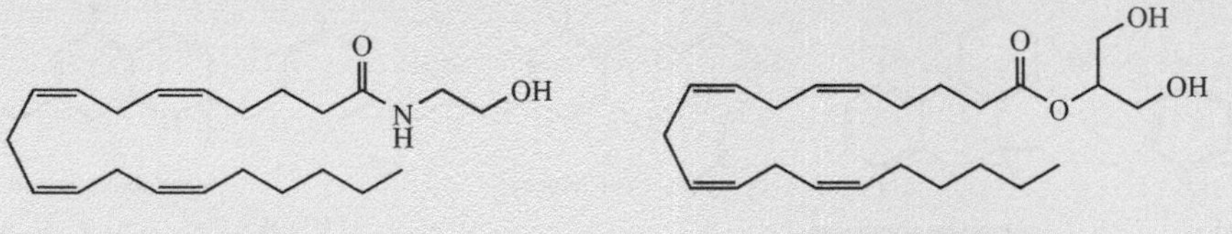

anandamide
(arachidonoylethanolamide)

2-arachidonoylglycerol

Figure 3.95

Hexanoate is also the starter unit used in the formation of the **aflatoxins**, a group of highly toxic metabolites produced by *Aspergillus flavus* and *Aspergillus parasiticus*, and probably responsible for the high incidence of liver cancer in some parts of Africa. These compounds were first detected following the deaths of young turkeys fed on mould-contaminated peanuts (*Arachis hypogaea*; Leguminosae/Fabaceae). Peanuts still remain one of the crops most likely to present a potential risk to human health because of contamination with aflatoxins [Box 3.19]. These and other food materials must be routinely screened to ensure levels of aflatoxins do not exceed certain set limits. The aflatoxin structures contain a bisfuran unit fused to an

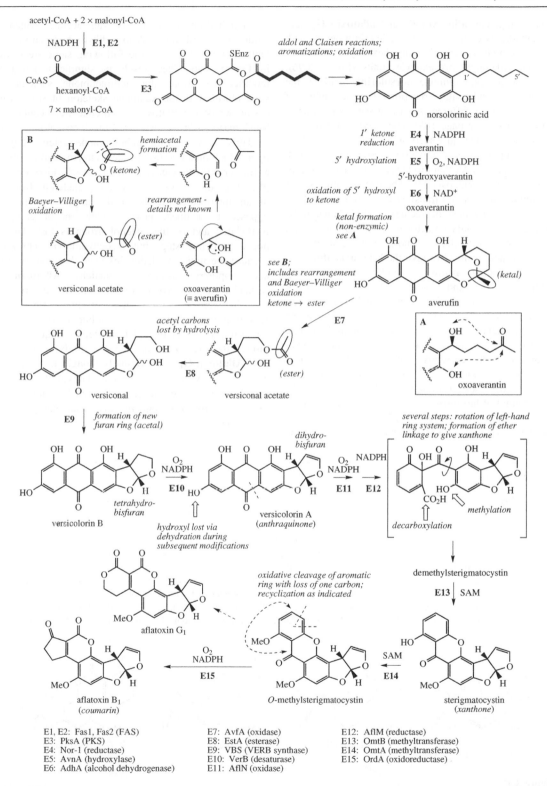

E1, E2: Fas1, Fas2 (FAS)
E3: PksA (PKS)
E4: Nor-1 (reductase)
E5: AvnA (hydroxylase)
E6: AdhA (alcohol dehydrogenase)

E7: AvfA (oxidase)
E8: EstA (esterase)
E9: VBS (VERB synthase)
E10: VerB (desaturase)
E11: AflN (oxidase)

E12: AflM (reductase)
E13: OmtB (methyltransferase)
E14: OmtA (methyltransferase)
E15: OrdA (oxidoreductase)

Figure 3.96

aromatic ring, e.g. **aflatoxin B$_1$** and **aflatoxin G$_1$**, and their remarkably complex biosynthetic origin begins with a poly-β-keto chain derived from a hexanoyl-CoA starter and seven malonyl-CoA extender units (Figure 3.96). This gives an anthraquinone **norsolorinic acid** by the now-familiar condensation reactions, but the folding of the chain is rather different from that seen with simpler anthraquinones (see page 105). The enzyme norsolorinic acid synthase utilizes just acetyl-CoA and malonyl-CoA, and is a complex of an iterative type I PKS, together with a yeast-like type I FAS that produces the hexanoate starter (compare zearalenone, page 89). The FAS is comprised of two subunit proteins and transfers the starter to the PKS without any involvement of free hexanoyl-CoA; naturally, NADPH is a necessary cofactor for the FAS component. The six-carbon side-chain of norsolorinic acid is then oxygenated and cyclized to give the ketal **averufin**.

The remaining pathway rapidly becomes very complicated indeed and is noteworthy for several major skeletal changes and the involvement of Baeyer–Villiger oxidations (see page 27), catalysed by cytochrome P-450-dependent monooxygenases. **Versiconal acetate** is another known intermediate; its formation involves a rearrangement, currently far from clear, and a Baeyer–Villiger oxidation. The latter oxidation achieves principally the transfer of a two-carbon fragment (the terminal ethyl of hexanoate) to become an ester function. These two carbon atoms can then be lost by hydrolysis, leading to formation of **versicolorin B**, now containing the tetrahydrobisfuran moiety (an acetal), which is oxidized in **versicolorin A** to a dihydrobisfuran system. Only minimal details of the following steps are shown in Figure 3.96. They result in significant structural modifications, from the anthraquinone skeleton of versicolorin A to the xanthone skeleton of sterigmatocystin, whilst the aflatoxin end-products are coumarins. **Sterigmatocystin** is derived from versicolorin A by oxidative cleavage of the anthraquinone system catalysed by a cytochrome P-450-dependent enzyme and involving a Baeyer–Villiger cleavage, then recyclization through phenol groups to give a xanthone skeleton. Rotation in an intermediate leads to the angular product as opposed to a linear product. One phenol group is methylated and another phenol group is lost. This is not a reductive step, but a dehydration, achievable because the initial oxidation involves the aromatic ring and temporarily destroys aromaticity. **Aflatoxin B$_1$** formation requires oxidative cleavage of an aromatic ring in sterigmatocystin, loss of one carbon, and recyclization exploiting the carbonyl functionality; interestingly, an *O*-methylation step is required first, though the methoxy group so formed is subsequently lost. **Aflatoxin G$_1$** was originally thought to be derived by a Baeyer–Villiger reaction on aflatoxin B$_1$, but it is now known to be produced via a branch pathway from *O*-methylsterigmatocystin; the fine detail has yet to be established.

Box 3.19

Aflatoxins

Aflatoxins are potent mycotoxins produced by the fungi *Aspergillus flavus* and *A. parasiticus*. Four main naturally occurring aflatoxins, aflatoxins B$_1$, B$_2$, G$_1$, and G$_2$ (Figure 3.97), are recognized, but these can be metabolized by microorganisms and animals to other aflatoxin structures, which are also toxic. Aflatoxin B$_1$ is the most commonly encountered member of the group, and is also the most acutely toxic and carcinogenic example. Aflatoxin B$_2$ is a dihydro derivative of aflatoxin B$_1$, whilst aflatoxins G$_1$ and G$_2$ are an analogous pair with a six-membered lactone rather than a five-membered cyclopentenone ring. These toxins are most commonly associated with peanuts (groundnuts), maize, rice, pistachio nuts, and Brazil nuts, though other crops can be affected, and although found worldwide, they are particularly prevalent in tropical and subtropical regions. Aflatoxin M$_1$ (Figure 3.97)

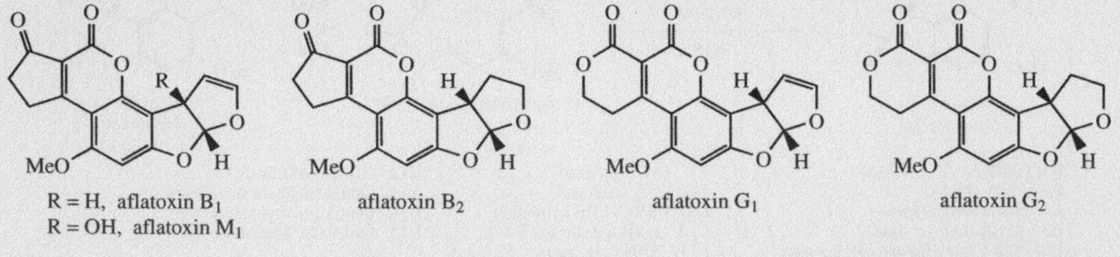

R = H, aflatoxin B$_1$
R = OH, aflatoxin M$_1$

aflatoxin B$_2$

aflatoxin G$_1$

aflatoxin G$_2$

Figure 3.97

Box 3.19 (continued)

is a hydroxy derivative of aflatoxin B_1 and equally toxic. It may occur in cow's milk as a result of mammalian metabolism of aflatoxin B_1 originally contaminating the animal's food. Because these compounds fluoresce strongly under UV light, they are relatively easily detected and monitored.

The aflatoxins primarily affect the liver, causing enlargement, fat deposition, and necrosis, at the same time causing cells of the bile duct to proliferate, with death resulting from irreversible loss of liver function. In the case of aflatoxin B_1, this appears to be initiated by cytochrome P-450-dependent metabolism in the body to the epoxide (Figure 3.98). The epoxide intercalates with DNA, and in so doing becomes orientated towards nucleophilic attack from guanine residues. This leads to inhibition of DNA replication and of RNA synthesis, and initiates mutagenic activity. Fortunately, endogenous glutathione is normally available to minimize damage by reacting with dangerous electrophiles. A glutathione–toxin adduct is formed in the same way (Figure 3.98), and the polar functionalities make this adduct water soluble and excretable. Aflatoxins are also known to cause hepatic carcinomas, this varying with the species of animal. The above normal incidence of liver cancer in Africa and Asia has been suggested to be linked to the increased amounts of aflatoxins found in foodstuffs, and a tolerance level of 30 ppb has been recommended. Acute hepatitis may result from food containing aflatoxin B_1 at levels of the order of 0.1 ppm; levels of more than 1 ppm are frequently encountered.

The biosynthesis of aflatoxins proceeds through intermediates sterigmatocystin and versicolorin (see Figure 3.96). Toxins related to these structures, but differing in aromatic substituents, are also produced by various fungi. The sterigmatocystins are synthesized by species of *Aspergillus* and *Bipolaris*, and contain a reduced bifuran fused to a xanthone, whilst the versicolorins from *Aspergillus versicolor* contain the same type of reduced bisfuran system but fused to an anthraquinone. Like the aflatoxins, the sterigmatocystins are acutely toxic and carcinogenic. The versicolorins are less toxic, though still carcinogenic.

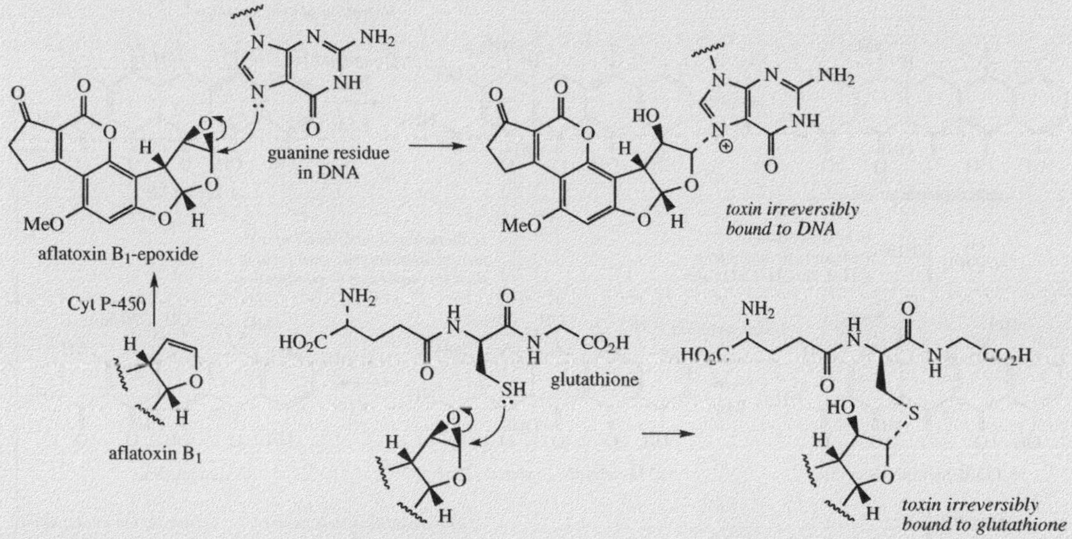

Figure 3.98

The **tetracyclines** are a group of broad-spectrum antibiotics produced by species of *Streptomyces*, and several natural and semi-synthetic members are used clinically [Box 3.20]. They contain a linear tetracyclic skeleton of polyketide origin in which the starter group is **malonamyl-CoA** (Figure 3.99), i.e. the coenzyme A ester of malonate semi-amide. Thus, all carbon atoms of the tetracycline skeleton are malonate derived, and the PKS also provides the amidotransferase activity that introduces the amino function into the starter group. The main features of the pathway (Figure 3.99) were deduced from extensive studies of mutant strains of *Streptomyces aureofaciens* with genetic blocks causing accumulation of mutant metabolites or the production of abnormal

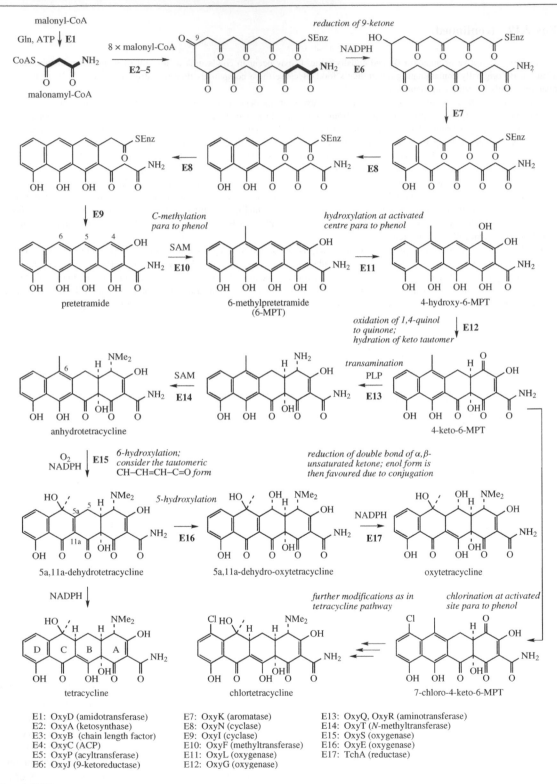

E1: OxyD (amidotransferase)
E2: OxyA (ketosynthase)
E3: OxyB (chain length factor)
E4: OxyC (ACP)
E5: OxyP (acyltransferase)
E6: OxyJ (9-ketoreductase)

E7: OxyK (aromatase)
E8: OxyN (cyclase)
E9: OxyI (cyclase)
E10: OxyF (methyltransferase)
E11: OxyL (oxygenase)
E12: OxyG (oxygenase)

E13: OxyQ, OxyR (aminotransferase)
E14: OxyT (*N*-methyltransferase)
E15: OxyS (oxygenase)
E16: OxyE (oxygenase)
E17: TchA (reductase)

Figure 3.99

tetracylines. This organism typically produces **chlortetra-cyline**, whilst the parent compound **tetracycline** is in fact an aberrant product synthesized in mutants blocked in the chlorination step. The use of mutants with genetic blocks also enabled the shikimate pathway (Chapter 4) to be delineated. In that case, since a primary metabolic pathway was affected, mutants tended to accumulate intermediates and could not grow unless later components of the pathway were supplied. With the tetracyclines, a secondary metabolic pathway is involved, and the relatively broad specificity of some of the enzymes concerned allows many of the later steps to proceed even if one step, e.g. the chlorination, is not achievable. This has also proved valuable for production of some of the clinical tetracycline antibiotics. More recently, the genetic details of the pathway have been clarified, with most attention being directed towards study of **oxytetracycline** production in *Streptomyces rimosus*. Nevertheless, the bulk of the pathway is effectively the same in both organisms. The PKS is an iterative type II system, i.e. the relevant enzyme activities are provided by individual separable proteins.

The appropriate poly-β-keto ester is constructed almost as expected (Figure 3.99). However, the 9-keto reductive step is catalysed after chain assembly is complete, rather than during chain extension. This is controlled by a specific 9-ketoreductase component; reduction at the position nine carbon atoms from the carboxy terminus is a remarkably consistent feature of a number of other type II PKS systems. The tetracene ring system is built up gradually, starting with the ring at the centre of the chain fold as with anthraquinones, and the first intermediate released from the enzyme is **pretetramide** (Figure 3.99). This is the substrate for introduction of a methyl group at C-6, giving 6-methylpretetramide (full tetracene numbering is shown in Figure 3.100). Hydroxylation in ring A followed by oxidation gives a quinone; to accommodate the quinone, this will necessitate tautomerism in ring B to the keto form. The quinone is then substrate for hydration at the A/B ring fusion. An amine group is subsequently introduced stereospecifically into ring A by a transamination reaction, followed by di-*N*-methylation using SAM to give **anhydrotetracycline**. In the last steps, C-6 and C-4 are hydroxylated, and NADPH reduction of the C-5a/11a double bond generates **oxytetracycline**. Omission of the C-4 hydroxylation would provide **tetracycline**. The pathway to **chlortetracycline** involves an addition chlorination step; this occurs on 4-keto-6-methylpretetramide at the nucleophilic site *para* to the phenol in ring D. The subsequent steps are then identical.

Box 3.20

Tetracyclines

The tetracyclines (Figure 3.100) are a group of broad-spectrum, orally active antibiotics produced by cultures of *Streptomyces* species. **Chlortetracycline** isolated from *Streptomyces aureofaciens* was the first of the group to be discovered, closely followed by **oxytetracycline** from cultures of *S. rimosus*. **Tetracycline** was found as a minor antibiotic in *S. aureofaciens*, but may be produced in quantity by utilizing a mutant strain blocked in the chlorination step (Figure 3.99). Similarly, the early C-6 methylation step can also be blocked, and such mutants accumulate 6-demethyltetracyclines, e.g. demeclocycline (demethylchlorotetracycline). These reactions can also be inhibited in the normal strain of *S. aureofaciens* by supplying cultures with either aminopterin (which inhibits C-6 methylation) or mercaptothiazole (which inhibits C-7 chlorination). Oxytetracycline from *S. rimosus* lacks the chlorine substituent, but has an additional 5α-hydroxyl group, introduced late in the pathway. Only minor alterations can be made to the basic tetracycline structure to modify the antibiotic activity, and these are at positions 5, 6, and 7. Other functionalities in the molecule are all essential to retain activity. Semi-synthetic tetracyclines used clinically include **methacycline**, obtained by a dehydration reaction from oxytetracycline, and **doxycycline**, via reduction of the 6-methylene in methacycline. **Minocycline** contains a 7-dimethylamino group and is produced by a sequence involving aromatic nitration. **Lymecycline** is an example of an antibiotic developed by chemical modification of the primary amide function at C-2.

Tetracyclines have both amino and phenolic functions, and are thus amphoteric compounds; they are more stable in acid than under alkaline conditions. They are thus suitable for oral administration and are absorbed satisfactorily. However, because of the sequence of phenol and carbonyl substituents in the structures, they act as chelators and complex with metal ions, especially calcium, aluminium, iron, and magnesium. Accordingly, they should not be administered with foods such as milk and dairy products (which have a high calcium content), aluminium- and magnesium-based antacid preparations, iron supplements, etc., otherwise erratic and unsatisfactory absorption will occur. A useful feature of doxycycline and minocycline is that their absorptions are much less affected by metal ions. Chelation of tetracyclines with calcium also precludes their use in children

Box 3.20   (continued)

Tetracyclines

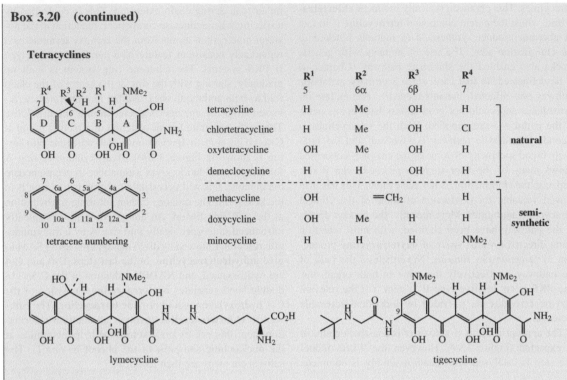

	R¹ 5	R² 6α	R³ 6β	R⁴ 7	
tetracycline	H	Me	OH	H	
chlortetracycline	H	Me	OH	Cl	natural
oxytetracycline	OH	Me	OH	H	
demeclocycline	H	H	OH	H	
methacycline	OH	=CH₂		H	
doxycycline	OH	Me	H	H	semi-synthetic
minocycline	H	H	H	NMe₂	

tetracene numbering

lymecycline

tigecycline

Figure 3.100

developing their adult teeth, and in pregnant women, since the tetracyclines become deposited in growing teeth and bone. In children, this would cause unsightly and permanent staining of teeth with the chelated yellow tetracycline.

Although the tetracycline antibiotics have a broad spectrum of activity spanning Gram-negative and Gram-positive bacteria, their value has decreased as bacterial resistance has developed in pathogens such as *Pneumococcus, Staphylococcus, Streptococcus*, and *E. coli*. These organisms appear to have evolved two main mechanisms of resistance: bacterial efflux and ribosome protection. In bacterial efflux, a membrane-embedded transport protein exports the tetracycline out of the cell before it can exert its effect. Ribosomal protection releases tetracyclines from the ribosome, the site of action (see below). Nevertheless, tetracyclines are the antibiotics of choice for infections caused by *Chlamydia, Mycoplasma, Brucella*, and *Rickettsia*, and are valuable in chronic bronchitis due to activity against *Haemophilus influenzae*. They are also used systemically to treat severe cases of acne, helping to reduce the frequency of lesions by their effect on skin flora. There is little significant difference in the antimicrobial properties of the various agents, except for **minocycline**, which has a broader spectrum of activity and, being active against *Neisseria meningitides*, is useful for prophylaxis of meningitis. The individual tetracyclines do have varying bioavailabilities, however, which may influence the choice of agent. **Tetracycline** and **oxytetracycline** are probably the most commonly prescribed agents, whilst chlortetracycline and methacycline have both been superseded. Tetracyclines are formulated for oral application or injection, as ear and eye drops, and for topical use on the skin. **Doxycycline** is the agent of choice for treating Lyme disease (caused by the spirochaete *Borellia burgdorferi*), and also finds use as a prophylactic against malaria in areas where there is widespread resistance to chloroquine and mefloquine (see page 382). A new generation of tetracyclines known as glycylcyclines has been developed to counter resistance and provide higher antibiotic activity. The first of these in general use is **tigecycline** (Figure 3.100), a glycylamido derivative of minocycline. It is synthesized via nitration of minocycline, itself produced from tetracycline. Tigecycline maintains antibiotic spectrum and potency, whilst overcoming resistance arising from both ribosomal protection and efflux mechanisms.

Their antimicrobial activity arises by inhibition of protein synthesis. This is achieved by interfering with the binding of aminoacyl-tRNA to acceptor sites on the ribosome by disrupting the codon–anticodon interaction (see page 422). Evidence points to a single strong binding site on the smaller 30S subunit of the ribosome. Although tetracyclines can also bind to mammalian ribosomes, there appears to be preferential penetration into bacterial cells, and there are few major side-effects from using these antibiotics.

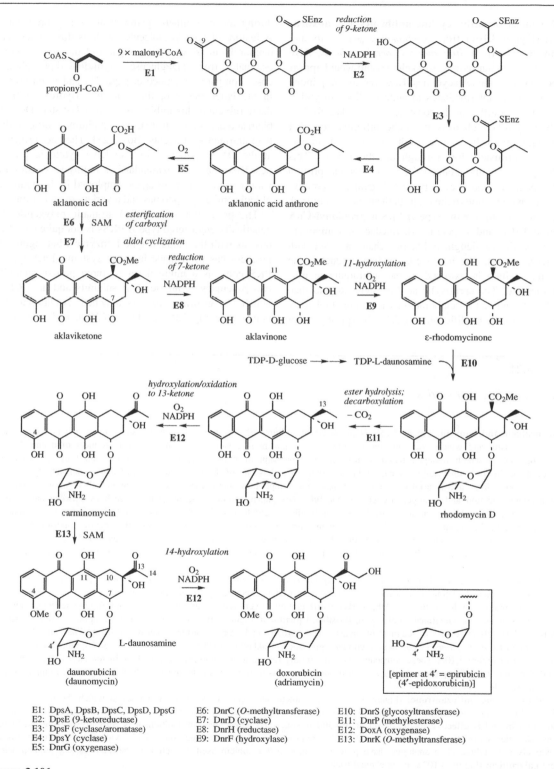

E1: DpsA, DpsB, DpsC, DpsD, DpsG
E2: DpsE (9-ketoreductase)
E3: DpsF (cyclase/aromatase)
E4: DpsY (cyclase)
E5: DnrG (oxygenase)

E6: DnrC (*O*-methyltransferase)
E7: DnrD (cyclase)
E8: DnrH (reductase)
E9: DnrF (hydroxylase)

E10: DnrS (glycosyltransferase)
E11: DnrP (methylesterase)
E12: DoxA (oxygenase)
E13: DnrK (*O*-methyltransferase)

Figure 3.101

A number of **anthracycline antibiotics**, e.g. **doxorubicin** (Figure 3.101) from *Streptomyces peuceticus* and **daunorubicin** from *Streptomyces coeruleorubicus*, have structurally similar tetracyclic skeletons and would appear to be related to the tetracyclines. However, anthraquinone derivatives are intermediates in anthracycline biosynthesis, and the fourth ring is constructed later. The folding of the poly-β-keto chain is also rather different from that seen with tetracyclines; as a result, the end-of-chain carboxyl is ultimately lost through decarboxylation. This carboxyl is actually retained for a considerable portion of the pathway, and is even protected against decarboxylation by methylation to the ester, until no longer required. The starter group for the type II PKS is **propionyl-CoA** (Figure 3.101) and a specific 9-ketoreductase component acts upon the full-length polyketide chain, as seen with the tetracyclines. The initial enzyme-free product is the anthrone, which is oxidized to the anthraquinone **alklanonic acid**. The carboxylic acid is esterified, and then the fourth ring can be elaborated by a simple aldol reaction. Most of the modifications which subsequently occur during the biosynthetic pathway are easily predictable. A feature of note in molecules such as doxorubicin and daunorubicin is the amino sugar L-daunosamine which originates from TDPglucose (thymidine diphosphoglucose; compare UDPglucose, page 31) and is introduced in the latter stages of the sequence. Hydroxylation of daunorubicin to doxorubicin is the very last step. Doxorubicin and daunorubicin are used as antitumour drugs rather than antimicrobial agents [Box 3.21]. They act primarily at the DNA level and so also have cytotoxic properties. Doxorubicin in particular is a highly successful and widely used antitumour agent, employed in the treatment of leukaemias, lymphomas and a variety of solid tumours.

The production of modified aromatic polyketides by genetically engineered type II PKSs is not quite so 'obvious' as with the modular type I enzymes, but significant progress has been made in many systems. Each type II PKS contains a minimal set of three protein subunits, two β-ketoacyl synthase (KS) subunits and an ACP to which the growing chain is attached. Additional subunits, including KRs, cyclases (CYC), and aromatases

---

Box 3.21

Anthracycline Antibiotics

Doxorubicin (adriamycin; Figure 3.101) is produced by cultures of *Streptomyces peucetius* var *caesius* and is one of the most successful and widely used antitumour drugs. The organism is a variant of *S. peucetius*, a producer of daunorubicin (see below), in which mutagen treatment resulted in expression of a latent hydroxylase enzyme and thus synthesis of doxorubicin by 14-hydroxylation of daunorubicin. Doxorubicin has one of the largest spectrum of antitumour activity shown by antitumour drugs and is used to treat acute leukaemias, lymphomas, and a variety of solid tumours. It is administered by intravenous injection and is largely excreted in the bile. The planar anthracycline molecule intercalates between base pairs on the DNA helix. The sugar unit provides further binding strength and also plays a major role in sequence recognition for the binding. Intercalation is central to doxorubicin's primary mode of action, inhibition of the enzyme topoisomerase II, which is responsible for cleaving and resealing of double-stranded DNA during replication (see page 155). Common toxic effects include nausea and vomiting, bone marrow suppression, hair loss, and local tissue necrosis, with cardiotoxicity at higher dosage.

Daunorubicin (Figure 3.101) is produced by *Streptomyces coeruleorubidus* and *S. peucetius*. Though similar to doxorubicin in its biological and chemical properties, it is no longer used therapeutically to any extent. It has a much less favourable therapeutic index than doxorubicin, and the markedly different effectiveness as an antitumour drug is not fully understood, though differences in metabolic degradation may be responsible. **Epirubicin** (Figure 3.101), the semi-synthetic 4′-epimer of doxorubicin, is particularly effective in the treatment of breast cancer, producing lower side-effects than doxorubicin. The antileukaemics **aclarubicin** from *Streptomyces galilaeus*, a complex glycoside of aklavinone (Figure 3.101), and the semi-synthetic **idarubicin** are shown in Figure 3.102. These compounds are structurally related to doxorubicin but can show increased activity with less cardiotoxicity. The principal disadvantage of all of these agents is their severe cardiotoxicity, which arises through inhibition of cardiac Na$^+$, K$^+$-ATPase.

Mitoxantrone (mitozantrone) (Figure 3.102) is a synthetic analogue of the anthracyclinones in which the non-aromatic ring and the aminosugar have both been replaced with aminoalkyl side-chains. This agent has reduced toxicity compared with doxorubicin and is effective in the treatment of solid tumours and leukaemias. In addition, it is currently proving useful in multiple sclerosis treatment, where it can reduce the frequency of relapses. Despite considerable synthetic research, however, relatively few anthracycline analogues have proved superior to doxorubicin itself, though some newer ones, e.g. **sabarubicin** and **galarubicin** (Figure 3.102), are in clinical trials.

Box 3.21   (continued)

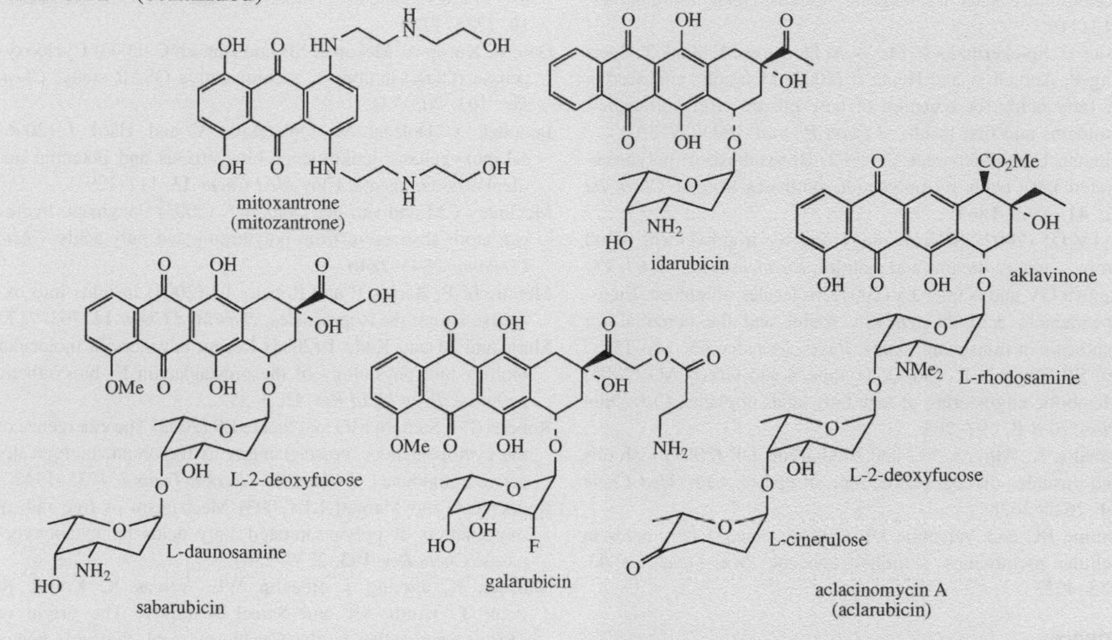

mitoxantrone
(mitozantrone)

idarubicin

aklavinone

L-rhodosamine

sabarubicin

L-2-deoxyfucose

L-daunosamine

galarubicin

L-2-deoxyfucose

L-cinerulose

aclacinomycin A
(aclarubicin)

Figure 3.102

(ARO) are responsible for modification of the nascent chain to form the final cyclized structure. Novel polyketides have been generated by manipulating type II PKSs, exchanging KS, CYC and ARO subunits among different systems. However, because of the highly reactive nature of poly-β-keto chains, the cyclizations that occur with the modified gene product frequently vary from those in the original compound. Compared with type I PKSs, the formation of new products with predictable molecular structures has proven less controllable.

FURTHER READING

Fatty Acid Synthase

Jenni S, Leibundgut M, Maier T and Ban N (2006) Architecture of a fungal fatty acid synthase at 5 Å resolution. *Science* **311**, 1263–1267.

Kolter T (2007) The fatty acid factory of yeasts. *Angew Chem Int Ed* **46**, 6772–6775.

Lu H and Tonge PJ (2008) Inhibitors of FabI, an enzyme drug target in the bacterial fatty acid biosynthesis pathway. *Acc Chem Res* **41**, 11–20.

Maier T, Jenni S and Ban N (2006) Architecture of mammalian fatty acid synthase at 4.5 Å resolution. *Science* **311**, 1258–1262.

Smith S (2006) Architectural options for a fatty acid synthase. *Science* **311**, 1251–1252.

Smith S and Tsai S-C (2007) The type I fatty acid and polyketide synthases: a tale of two megasynthases. *Nat Prod Rep* **24**, 1041–1072.

Surolia A, Ramya TNC, Ramya V and Surolia N (2004) 'FAS't inhibition of malaria. *Biochem J* **383**, 401–412; corrigenda **384**, 655.

Tasdemir D (2006) Type II fatty acid biosynthesis, a new approach in antimalarial natural product discovery. *Phytochem Rev* **5**, 99–108.

Wright HT and Reynolds KA (2007) Antibacterial targets in fatty acid biosynthesis. *Curr Opin Microbiol* **10**, 447–453.

Fatty Acids and Fats

Behrouzian B and Buist PH (2003) Bioorganic chemistry of plant lipid desaturation. *Phytochem Rev* **2**, 103–111.

Brouwer IA, Geelen A and Katan MB (2006) *n*–3 Fatty acids, cardiac arrhythmia and fatal coronary heart disease. *Prog Lipid Res* **45**, 357–367.

Buist PH (2004) Fatty acid desaturases: selecting the dehydrogenation channel. *Nat Prod Rep* **21**, 249–262.

Buist PH (2007) Exotic biomodification of fatty acids. *Nat Prod Rep* **24**, 1110–1127.

Chatgilialoglu C and Ferreri C (2005) Trans lipids: the free radical path. *Acc Chem Res* **38**, 411–448.

Cunnane SC (2003) Problems with essential fatty acids: time for a new paradigm? *Prog Lipid Res* **42**, 544–568.

Domergue F, Abbadi A and Heinz E (2005) Relief for fish stocks: oceanic fatty acids in transgenic oilseeds. *Trends Plant Sci* **10**, 112–116.

Drexler H, Spiekermann P, Meyer A, Domergue F, Zank T, Sperling P, Abbadi A and Heinz E (2003) Metabolic engineering of fatty acids for breeding of new oilseed crops: strategies, problems and first results. *J Plant Physiol* **160**, 779–802.

Kaulmann U and Hertweck C (2002) Biosynthesis of polyunsaturated fatty acids by polyketide synthases. *Angew Chem Int Ed* **41**, 1866–1869.

Murphy DJ (2002) Biotechnology and the improvement of oil crops – genes, dreams and realities. *Phytochem Rev* **1**, 67–77.

Sayanova OV and Napier JA (2004) Molecules of interest. Eicosapentaenoic acid: biosynthetic routes and the potential for synthesis in transgenic plants. *Phytochemistry* **65**, 147–158.

Singh SP, Zhou X-R, Liu Q, Stymne S and Green AG (2005) Metabolic engineering of new fatty acids in plants. *Curr Opin Plant Biol* **8**, 197–203.

Tziomalos K, Athyros VG and Mikhailidis DP (2007) Fish oils and vascular disease prevention: an update. *Curr Med Chem* **14**, 2622–2628.

Valentine RC and Valentine DL (2004) *Omega*-3 fatty acids in cellular membranes: a unified concept. *Prog Lipid Res* **43**, 383–402.

Echinacea

Woelkart K and Bauer R (2008) The role of alkamides as an active principle of *Echinacea*. *Planta Med* **73**, 615–623.

Lipstatin

Eisenreich W, Kupfer E, Stohler P, Weber W and Bacher A (2003) Biosynthetic origin of a branched chain analogue of the lipase inhibitor, lipstatin. *J Med Chem* **46**, 4209–4212.

Schuhr CA, Eisenreich W, Goese M, Stohler P, Weber W, Kupfer E and Bacher A (2002) Biosynthetic precursors of the lipase inhibitor lipstatin. *J Org Chem* **67**, 2257–2262.

Prostaglandins, Thromboxanes, Leukotrienes

Blobaum AL and Marnett LJ (2007) Structural and functional basis of cyclooxygenase inhibition. *J Med Chem* **50**, 1425–1441.

Das S, Chandrasekhar S, Yadav JS and Grée R (2007) Recent developments in the synthesis of prostaglandins and analogues. *Chem Rev* **107**, 3286–3337.

De Leval X, Hanson J, David J-L, Masereel B, Pirotte B and Dogne J-M (2004) New developments on thromboxane and prostacyclin modulators. Part II: prostacyclin modulators. *Curr Med Chem* **11**, 1243–1252.

Dogne J-M, de Leval X, Hanson J, Frederich M, Lambermont B, Ghuysen A, Casini A, Masereel B, Ruan K-H, Pirotte B and Kolh P (2004) New developments on thromboxane and prostacyclin modulators. Part I: thromboxane modulators. *Curr Med Chem* **11**, 1223–1241.

Dogné J-M, Supuran CT and Pratico D (2005) Adverse cardiovascular effects of the coxibs. *J Med Chem* **48**, 2251–2257.

Fam SS and Morrow JD (2003) The isoprostanes: unique products of arachidonic acid oxidation – a review. *Curr Med Chem* **10**, 1723–1740.

Garg R, Kurup A, Mekapati SB and Hansch C (2003) Cyclooxygenase (COX) inhibitors: a comparative QSAR study. *Chem Rev* **103**, 703–731.

Jampilek J, Dolezal M, Opletalova V and Hartl J (2006) 5-Lipoxygenase, leukotrienes biosynthesis and potential antileukotrienic agents. *Curr Med Chem* **13**, 117–129.

McGinley CM and van der Donk WA (2003) Enzymatic hydrogen atom abstraction from polyunsaturated fatty acids. *Chem Commun* 2843–2846.

Montuschi P, Barnes P and Roberts LJ (2007) Insights into oxidative stress: the isoprostanes. *Curr Med Chem* **14**, 703–717.

Murakami M and Kudo I (2004) Recent advances in molecular biology and physiology of the prostaglandin E_2-biosynthetic pathway. *Prog Lipid Res* **43**, 3–35.

Roberts SM, Santoro MG and Sickle ES (2002) The emergence of the cyclopentenone prostaglandins as important, biologically active compounds. *J Chem Soc, Perkin Trans 1* 1735–1742.

Rouzer CA and Marnett LJ (2003) Mechanism of free radical oxygenation of polyunsaturated fatty acids by cyclooxygenases. *Chem Rev* **103**, 2239–2304.

Valmsen K, Järving I, Boeglin WE, Varvas K, Koljak R, Pehk T, Brash AR and Samel N (2001) The origin of 15*R*-prostaglandins in the Caribbean coral *Plexaura homomalla*: molecular cloning and expression of a novel cyclooxygenase. *Proc Natl Acad Sci USA* **98**, 7700–7705.

Van der Donk WA, Tsai A-L and Kulmacz RJ (2002) The cyclooxygenase reaction mechanism. *Biochemistry* **41**, 15451–15458.

Warner TD and Mitchell JA (2002) Cyclooxygenase-3 (COX-3): filling in the gaps toward a COX continuum? *Proc Natl Acad Sci USA* **99**, 13371–13373.

Polyketide Synthases: General

Austin MB and Noel JP (2003) The chalcone synthase superfamily of type III polyketide synthases. *Nat Prod Rep* **20**, 79–110.

Cox RJ (2007) Polyketides, proteins and genes in fungi: programmed nano-machines begin to reveal their secrets. *Org Biomol Chem* **5**, 2010–2026.

Hill AM (2006) The biosynthesis, molecular genetics and enzymology of the polyketide-derived metabolites. *Nat Prod Rep* **23**, 256–320.

Liou GF and Khosla C (2003) Building-block selectivity of polyketide synthases. *Curr Opin Chem Biol* **7**, 279–284.

Moore BS and Hertweck C (2002) Biosynthesis and attachment of novel bacterial polyketide synthase starter units. *Nat Prod Rep* **19**, 70–99.

Shen B (2003) Polyketide biosynthesis beyond the type I, II and III polyketide synthase paradigms. *Curr Opin Chem Biol* **7**, 285–295.

Staunton J and Weissman KJ (2001) Polyketide biosynthesis: a millennium review. *Nat Prod Rep* **18**, 380–416.

Staunton J and Wilkinson B (2001) Combinatorial biosynthesis of polyketides and nonribosomal peptides. *Curr Opin Chem Biol* **5**, 159–164.

Walsh CT (2004) Polyketide and nonribosomal peptide antibiotics: modularity and versatility. *Science* **303**, 1805–1810.

Watanabe K and Oikawa H (2007) Robust platform for *de novo* production of heterologous polyketides and nonribosomal peptides in *Escherichia coli*. *Org Biomol Chem* **5**, 593–602.

Polyketide Synthases: Macrolides

Fischbach MA and Walsh CT (2006) Assembly-line enzymology for polyketide and nonribosomal peptide antibiotics: logic, machinery, and mechanisms. *Chem Rev* **106**, 3468–3496.

Katz L and Ashley GW (2005) Translation and protein synthesis: macrolides. *Chem Rev* **105**, 499–527.

Kopp F and Marahiel MA (2007) Macrocyclization strategies in polyketide and nonribosomal peptide biosynthesis. *Nat Prod Rep* **24**, 735–749.

McDaniel R, Welch M and Hutchinson CR (2005) Genetic approaches to polyketide antibiotics 1. *Chem Rev* **105**, 543–558.

Menzella HG and Reeves CD (2007) Combinatorial biosynthesis for drug development. *Curr Opin Microbiol* **10**, 238–245.

Pohl NL (2002) Nonnatural substrates for polyketide synthases and their associated modifying enzymes. *Curr Opin Chem Biol* **6**, 773–778.

Macrolide Antibiotics

Pal S (2006) A journey across the sequential development of macrolides and ketolides related to erythromycin. *Tetrahedron* **62**, 3171–3200.

Rodriguez E and McDaniel R (2001) Combinatorial biosynthesis of antimicrobials and other natural products. *Curr Opin Microbiol* **4**, 526–534.

Wu Y-J and Su W-G (2001) Recent developments on ketolides and macrolides. *Curr Med Chem* **8**, 1727–1758.

Avermectins

Ikeda H, Nonomiya T and Omura S (2001) Organization of biosynthetic gene cluster for avermectin in *Streptomyces avermitilis*: analysis of enzymatic domains in four polyketide synthases. *J Ind Microbiol Biotechnol* **27**, 170–176.

Polyene Antifungals

Aparicio JF, Mendes MV, Anton N, Recio E and Martin JF (2004) Polyene macrolide antibiotic biosynthesis. *Curr Med Chem* **11**, 1645–1656.

Caffrey P, Lynch S, Flood E, Finnan S and Oliynyk M (2001) Amphotericin biosynthesis in *Streptomyces nodosus*: deductions from analysis of polyketide synthase and late genes. *Chem Biol* **8**, 713–723; errata (2003) **10**, 93–94.

Zotchev SB (2003) Polyene macrolide antibiotics and their applications in human therapy. *Curr Med Chem* **10**, 211–223.

Ascomycin, Tacrolimus, Rapamycin

Gregory MA, Hong H, Lill RE, Gaisser S, Petkovic H, Low L, Sheehan LS, Carletti I, Ready SJ, Ward MJ, Kaja AL, Weston AJ, Challis IR, Leadlay PF, Martin CJ, Wilkinson B and Sheridan RM (2006) Rapamycin biosynthesis: elucidation of gene product function. *Org Biomol Chem* **4**, 3565–3568.

Lowden PAS, Wilkinson B, Böhm GA, Handa S, Floss HG, Leadlay PF and Staunton J (2001) Origin and true nature of the starter unit for the rapamycin polyketide synthase. *Angew Chem Int Ed* **40**, 777–779.

Mann J (2001) Natural products as immunosuppressive agents. *Nat Prod Rep* **18**, 417–430.

Wu K, Chung L, Revill WP, Katz L and Reeves CD (2000) The FK520 gene cluster of *Streptomyces hygroscopicus* var. *ascomyceticus* (ATCC 14891) contains genes for biosynthesis of unusual polyketide extender units. *Gene* **251**, 81–90.

Epothilones

Altmann K-H (2004) The merger of natural product synthesis and medicinal chemistry: on the chemistry and chemical biology of epothilones. *Org Biomol Chem* **2**, 2137–2152.

Altmann K-H and Gertsch J (2007) Anticancer drugs from nature – natural products as a unique source of new microtubule-stabilizing agents. *Nat Prod Rep* **24**, 327–357.

Feyen F, Cachoux F, Gertsch J, Wartmann M and Altmann K-H (2008) Epothilones as lead structures for the synthesis-based discovery of new chemotypes for microtubule stabilization. *Acc Chem Res* **41**, 21–31.

Nicolaou KC, Ritzén A and Namoto K (2001) Recent developments in the chemistry, biology and medicine of the epothilones. *Chem Commun* 1523–1535.

Rivkin A, Chou T-C and Danishefsky SJ (2005) On the remarkable antitumor properties of fludelone: how we got there. *Angew Chem Int Ed* **44**, 2838–2850.

Rifamycins

Floss HG and Yu T-W (2005) Rifamycins – mode of action, resistance, and biosynthesis. *Chem Rev* **105**, 621–632.

Zearalenone

Gaffoor I and Trail F (2006) Characterization of two polyketide synthase genes involved in zearalenone biosynthesis in *Gibberella zeae*. *Appl Environ Microbiol* **72**, 1793–1799.

Winssinger N and Barluenga S (2007) Chemistry and biology of resorcylic acid lactones. *Chem Commun* 22–36.

Discodermolide

Burlingame MA, Mendoza E, Ashley GW and Myles DC (2006) Synthesis of discodermolide intermediates from engineered polyketides. *Tetrahedron Lett* **47**, 1209–1211.

Smith AB and Freeze BS (2008) (+)-Discodermolide: total synthesis, construction of novel analogues, and biological evaluation. *Tetrahedron* **64**, 261–298.

Mupirocin

El-Sayed AK, Hothersall J, Cooper SM, Stephens E, Simpson TJ and Thomas CM (2003) Characterization of the mupirocin biosynthesis gene cluster from *Pseudomonas fluorescens* NCIMB 10586. *Chem Biol* **10**, 419–430.

Polyethers

Bhatt A, Stark CBW, Harvey BM, Gallimore AR, Demydchuk Y, Spencer JB, Staunton J and Leadlay PF (2005) Accumulation of an *E,E,E*-triene by the monensin-producing polyketide synthase when oxidative cyclization is blocked. *Angew Chem Int Ed* **44**, 7075–7078.

Dounay AB and Forsyth CJ (2002) Okadaic acid: the archetypal serine/threonine protein phosphatase inhibitor. *Curr Med Chem* **9**, 1939–1980.

Fernández JJ, Candenas ML, Souto ML, Trujillo MM and Norte M (2002) Okadaic acid, useful tool for studying cellular processes. *Curr Med Chem* **9**, 229–262.

Inoue M (2004) Convergent syntheses of polycyclic ethers. Illustrations of the utility of acetal-linked intermediates. *Org Biomol Chem* **2**, 1811–1817.

Nakata T (2005) Total synthesis of marine polycyclic ethers. *Chem Rev* **105**, 4314–4347.

Zaragozic Acids (Squalestatins)

Armstrong A and Blench TJ (2002) Recent synthetic studies on the zaragozic acids (squalestatins). *Tetrahedron* **58**, 9321–9349.

Cox RJ, Glod F, Hurley D, Lazarus CM, Nicholson TP, Rudd BAM, Simpson TJ, Wilkinson B and Zhang Y (2004) Rapid cloning and expression of a fungal polyketide synthase gene involved in squalestatin biosynthesis. *Chem Commun* 2260–2261.

Statins

Luo J-D and Chen AF (2003) Perspectives on the cardioprotective effects of statins. *Curr Med Chem* **10**, 1593–1601.

Müller M (2005) Chemoenzymatic synthesis of building blocks for statin side chains. *Angew Chem Int Ed* **44**, 362–365.

Oikawa H and Tokiwano T (2004) Enzymatic catalysis of the Diels–Alder reaction in the biosynthesis of natural products. *Nat Prod Rep* **21**, 321–352.

Rajanikant GK, Zemke D, Kassab M and Majid A (2007) The therapeutic potential of statins in neurological disorders. *Curr Med Chem* **14**, 103–112.

Roche VF (2005) Antihyperlipidemic statins: a self-contained, clinically relevant medicinal chemistry lesson. *Am J Pharm Educ* **69**, 546–560.

Sorensen JL and Vederas JC (2003) Monacolin N, a compound resulting from derailment of type I iterative polyketide synthase function en route to lovastatin. *Chem Commun* 1492–1493.

Stocking EM and Williams RM (2003) Chemistry and biology of biosynthetic Diels–Alder reactions. *Angew Chem Int Ed* **42**, 3078–3115.

Polyketide Synthases: Aromatics

Hertweck C, Luzhetskyy A, Rebets Y and Bechthold A (2007) Type II polyketide synthases: gaining a deeper insight into enzymatic teamwork. *Nat Prod Rep* **24**, 162–190.

Rix U, Fischer C, Remsing LL and Rohr J (2002) Modification of post-PKS tailoring steps through combinatorial biosynthesis. *Nat Prod Rep* **19**, 542–580.

Thomas R (2001) A biosynthetic classification of fungal and Streptomycete fused-ring aromatic polyketides. *ChemBioChem* **2**, 612–627.

Watanabe K, Praseuth AP and Wang CCC (2007) A comprehensive and engaging overview of the type III family of polyketide synthases. *Curr Opin Chem Biol* **11**, 279–286.

Anthraquinones

Gill M (2001) The biosynthesis of pigments in Basidiomycetes. *Aust J Chem* **54**, 721–734.

Hypericum/St John's Wort

Adam P, Arigoni D, Bacher A and Eisenreich W (2002) Biosynthesis of hyperforin in *Hypericum perforatum*. *J Med Chem* **45**, 4786–4793.

Barnes J, Anderson LA and Phillipson JD (2001) St John's wort (*Hypericum perforatum* L.): a review of its chemistry, pharmacology and clinical properties. *J Pharm Pharmac* **53**, 583–600.

Beerhues L (2006) Molecules of interest. Hyperforin. *Phytochemistry* **67**, 2201–2207.

Singh YN (2005) Potential for interaction of kava and St. John's wort with drugs. *J Ethnopharmacol* **100**, 108–113.

Usnic Acid

Ingólfsdóttir K (2002) Molecules of interest. Usnic acid. *Phytochemistry* **61**, 729–736.

Cannabis

Amar MB (2006) Cannabinoids in medicine: a review of their therapeutic potential. *J Ethnopharmacol* **105**, 1–25.

Ashton JC, Wright JL, McPartland JM and Tyndall Joel DA (2008) Cannabinoid CB1 and CB2 receptor ligand specificity and the development of CB2-selective agonists. *Curr Med Chem* **15**, 1428–1443.

Drysdale AJ and Platt B (2003) Cannabinoids: mechanisms and therapeutic applications in the CNS. *Curr Med Chem* **10**, 2719–2732.

Mechoulam R, Sumariwalla PF, Feldmann M and Gallily R (2005) Cannabinoids in models of chronic inflammatory conditions. *Phytochem Rev* **4**, 11–18.

Merzouki A and Mesa JM (2002) Concerning kif, a *Cannabis sativa* L. preparation smoked in the Rif mountains of northern Morocco. *J Ethnopharmacol* **81**, 403–406.

Raharjo TJ and Verpoorte R (2004) Methods for the analysis of cannabinoids in biological materials: a review. *Phytochem Anal* **15**, 79–94.

Taura F, Sirikantaramas S, Shoyama Y, Yoshikai K, Shoyama Y and Morimoto S (2007) Cannabidiolic-acid synthase, the chemotype-determining enzyme in the fiber-type *Cannabis sativa*. *FEBS Lett* **581**, 2929–2934.

Endocannabinoids

Ahn K, McKinney MK and Cravatt BF (2008) Enzymatic pathways that regulate endocannabinoid signaling in the nervous system. *Chem Rev* **108**, 1687–1707.

Di Marzo V, Bisogno T and De Petrocellis L (2007) Endocannabinoids and related compounds: walking back and forth between plant natural products and animal physiology. *Chem Biol* **14**, 741–756.

Lambert DM and Fowler CJ (2005) The endocannabinoid system: drug targets, lead compounds, and potential therapeutic applications. *J Med Chem* **48**, 5059–5087.

Lambert DM, Vandevoorde S, Jonsson K-O and. Fowler CJ (2002) The palmitoylethanolamide family: a new class of anti-inflammatory agents? *Curr Med Chem* **9**, 663–674.

Ueda N, Okamoto Y and Tsuboi K (2005) Endocannabinoid-related enzymes as drug targets with special reference to *N*-acylphosphatidylethanolamine-hydrolyzing phospholipase D. *Curr Med Chem* **12**, 1413–1422.

Wendeler M and Kolter T (2003) Inhibitors of endocannabinoid degradation: potential therapeutics for neurological disorders. *Angew Chem Int Ed* **42**, 2938–2941.

Aflatoxins

Bhatnagar D, Ehrlich KC and Cleveland TE (2003) Molecular genetic analysis and regulation of aflatoxin biosynthesis. *Appl Microbiol Biotechnol* **61**, 83–93.

Watanabe CMH and Townsend CA (2002) Initial characterization of a type I fatty acid synthase and polyketide synthase multienzyme complex NorS in the biosynthesis of aflatoxin B_1. *Chem Biol* **9**, 981–988.

Yabe K and Nakajima H (2004) Enzyme reactions and genes in aflatoxin biosynthesis. *Appl Microbiol Biotechnol* **64**, 745–755.

Tetracyclines

Petković H, Cullum J, Hranueli D, Hunter IS, Perić-Concha N, Pigac J, Thamchaipenet A, Vujaklija D and Long PF (2006) Genetics of *Streptomyces rimosus*, the oxytetracycline producer. *Microbiol Mol Biol Rev* **70**, 704–728.

Zhang W, Wenjun S, Wang CCC and Tang Y (2007) Investigation of early tailoring reactions in the oxytetracycline biosynthetic pathway. *J Biol Chem* **282**, 25717–25725.

Anthracyclines

Lee TS, Khosla C and Tang Y (2005) Engineered biosynthesis of aklanonic acid analogues. *J Am Chem Soc* **127**, 12254–12262.

4

THE SHIKIMATE PATHWAY: AROMATIC AMINO ACIDS AND PHENYLPROPANOIDS

The shikimate pathway provides an alternative route to aromatic compounds, particularly the aromatic amino acids L-**phenylalanine**, L-**tyrosine**, and L-**tryptophan**. This pathway is employed by microorganisms and plants, but not by animals; accordingly, the aromatic amino acids feature among the essential amino acids for man and have to be obtained in the diet. A central intermediate in the pathway is **shikimic acid** (Figure 4.1), a compound which had been isolated from plants of *Illicium* species (Japanese 'shikimi') many years before its role in metabolism had been discovered. Phenylalanine and tyrosine form the basis of C_6C_3 phenylpropane units found in many natural products, e.g. cinnamic acids, coumarins, lignans, and flavonoids, and along with tryptophan are precursors of a wide range of alkaloid structures. In addition, it is found that many simple benzoic acid derivatives, e.g. gallic acid and *p*-aminobenzoic acid (PABA; 4-aminobenzoic acid), are produced via branchpoints in the shikimate pathway. Shikimic acid is currently the raw material for synthesis of the antiviral drug oseltamivir (Tamiflu®), in demand as a defence against avian influenza (bird flu) (Figure 4.1); the main plant source is fruits of star anise (*Illicium verum*; Illiciaceae), though manufacturers are turning to cultures of genetically modified *Escherichia coli* as an alternative supply.

AROMATIC AMINO ACIDS AND SIMPLE BENZOIC ACIDS

The shikimate pathway begins with a coupling of phosphoenolpyruvate (PEP) from the glycolytic pathway and D-erythrose 4-phosphate from the pentose phosphate cycle to give the seven-carbon 3-deoxy-D-*arabino*-heptulosonic acid 7-phosphate (DAHP; Figure 4.2). This reaction, shown here as an aldol-type condensation, is known to be mechanistically more complex in the enzyme-catalysed version; several of the other transformations in the pathway have also been found to be surprisingly complex.

Elimination of phosphoric acid from DAHP followed by an intramolecular aldol reaction generates the first carbocyclic intermediate **3-dehydroquinic acid**. However, this is also an oversimplification. The elimination of phosphoric acid actually follows an NAD^+-dependent oxidation of the central hydroxyl, and this is then re-formed in an NADH-dependent reduction reaction on the intermediate carbonyl compound prior to the aldol reaction occurring. All these changes occur through the function of a single enzyme, 3-dehydroquinate synthase. Reduction of 3-dehydroquinic acid leads to **quinic acid**, a fairly common natural product found in the free form, as esters, or in combination with alkaloids such as quinine (see page 381). **Shikimic acid** itself is formed from 3-dehydroquinic acid via **3-dehydroshikimic acid** through dehydration (3-dehydroquinase) and reduction (shikimate dehydrogenase) steps. According to organism, two distinct types of dehydroquinase enzymes are recognized, and these employ different reaction mechanisms. A bifunctional enzyme catalyses the two steps in plants.

The simple phenolic acids **protocatechuic acid** (3,4-dihydroxybenzoic acid) and **gallic acid** (3,4,5-trihydroxybenzoic acid) can be formed by branchpoint reactions from 3-dehydroshikimic acid which involve dehydration and enolization, or, in the case of gallic acid, dehydrogenation and enolization. **Gallic acid** features as

Medicinal Natural Products: A Biosynthetic Approach. 3rd Edition   Paul Dewick
© 2009 John Wiley & Sons, Ltd

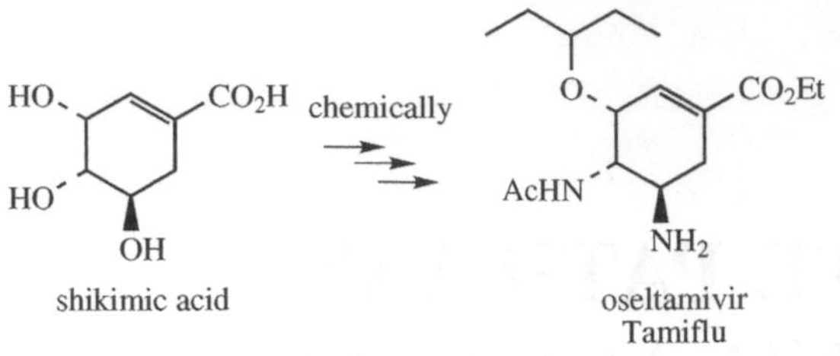

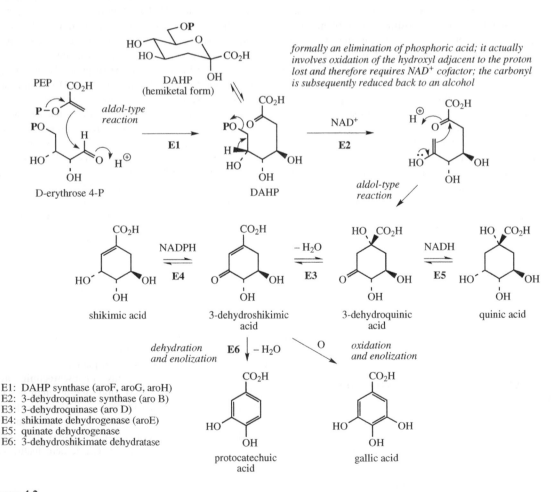

Figure 4.1

a component of many tannins, which are plant materials that have been used for thousands of years in the tanning of animal hides to make leather because of their ability to cross-link protein molecules. **Gallotannins** are esters of gallic acid with a polyalcohol, typically glucose, e.g. pentagalloylglucose, whilst **ellagitannins** contain one or more hexahydroxydiphenic acid functions, e.g. tellimagrandin II (Figure 4.3). Hexahydroxydiphenic acid esters are formed by phenolic oxidative coupling of galloyl functions catalysed by a laccase-type phenol oxidase enzyme (Figure 4.3). Chemical hydrolysis of an ellagitannin leads to formation of the dilactone ellagic acid. The glucose ester **β-glucogallin** is a crucial entity in the formation of gallotannins, acting as substrate for further esterification, and also as the donor of further galloyl groups (Figure 4.3). β-Glucogallin is an activated form of gallic acid and has similar reactivity to a CoA thioester. Tannins also contribute to the astringency of foods and beverages, especially tea, coffee, and wines, through their strong interaction with salivary proteins. In addition, they have beneficial antioxidant properties (see also condensed tannins, page 171).

E1: DAHP synthase (aroF, aroG, aroH)
E2: 3-dehydroquinate synthase (aro B)
E3: 3-dehydroquinase (aro D)
E4: shikimate dehydrogenase (aroE)
E5: quinate dehydrogenase
E6: 3-dehydroshikimate dehydratase

Figure 4.2

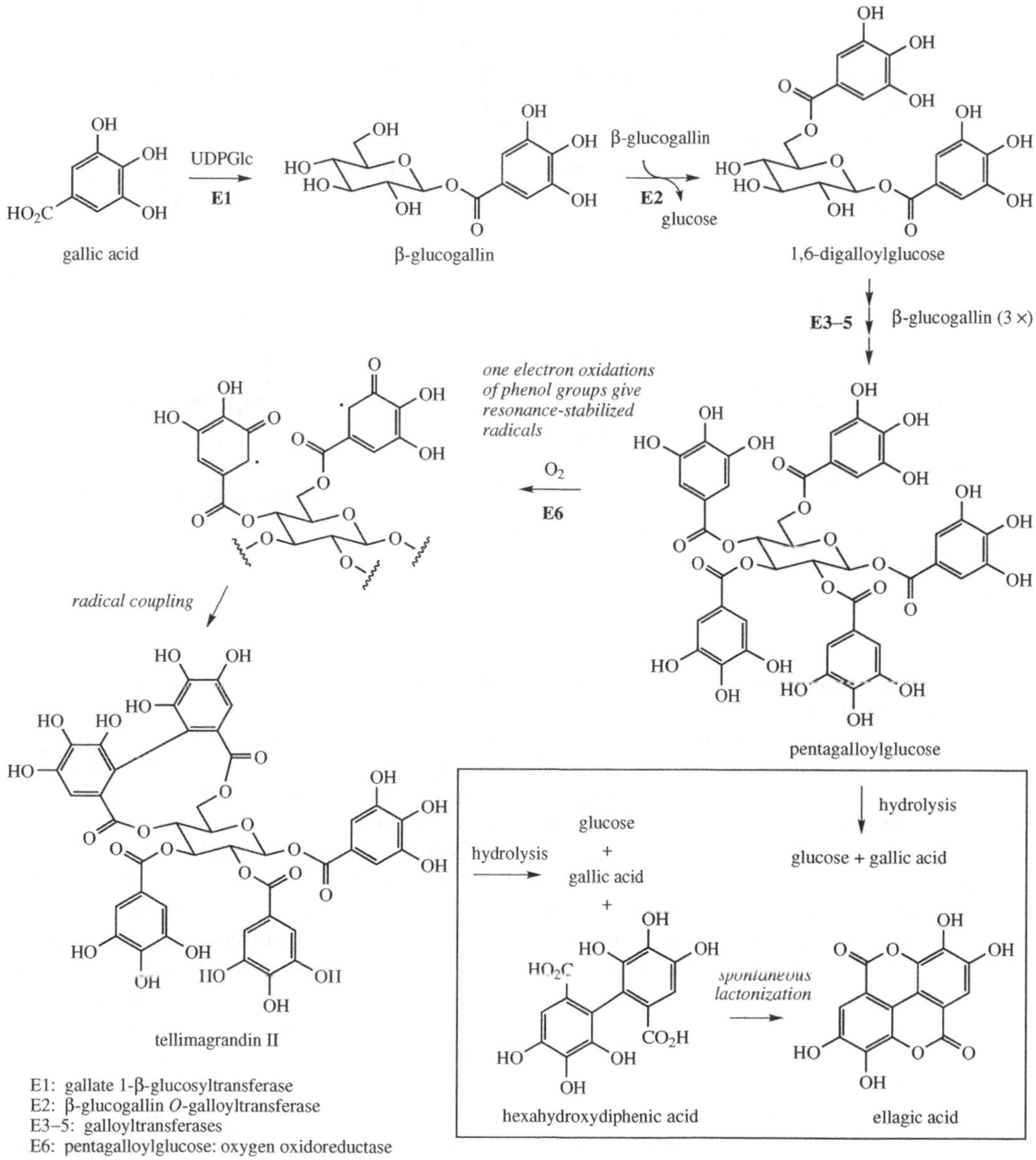

UDPGlc
E1
gallic acid
β-glucogallin
β-glucogallin
E2
glucose
1,6-digalloylglucose
E3–5
β-glucogallin (3 ×)
pentagalloylglucose
one electron oxidations of phenol groups give resonance-stabilized radicals
O_2
E6
radical coupling
tellimagrandin II
hydrolysis
glucose + gallic acid
hydrolysis
glucose
+
gallic acid
+
hexahydroxydiphenic acid
spontaneous lactonization
ellagic acid
E1: gallate 1-β-glucosyltransferase
E2: β-glucogallin *O*-galloyltransferase
E3–5: galloyltransferases
E6: pentagalloylglucose: oxygen oxidoreductase

Figure 4.3

A very important branchpoint compound in the shikimate pathway is **chorismic acid** (Figure 4.4), which has incorporated a further molecule of PEP as an enol ether side-chain. A simple ATP-dependent phosphorylation reaction produces shikimic acid 3-phosphate, and this is followed by an addition–elimination reaction with PEP to give **5-enolpyruvylshikimic acid 3-phosphate (EPSP)**. This reaction is catalysed by the enzyme EPSP synthase. The synthetic *N*-(phosphonomethyl)glycine derivative **glyphosate** (Figure 4.4) is a powerful inhibitor of this enzyme, and is believed to bind to the PEP binding site on the enzyme. Glyphosate finds considerable use as a broad-spectrum herbicide, the plant dying because of its inability to synthesize aromatic amino acids. The transformation of EPSP to **chorismic acid** (Figure 4.4) involves a 1,4-elimination of phosphoric acid, though

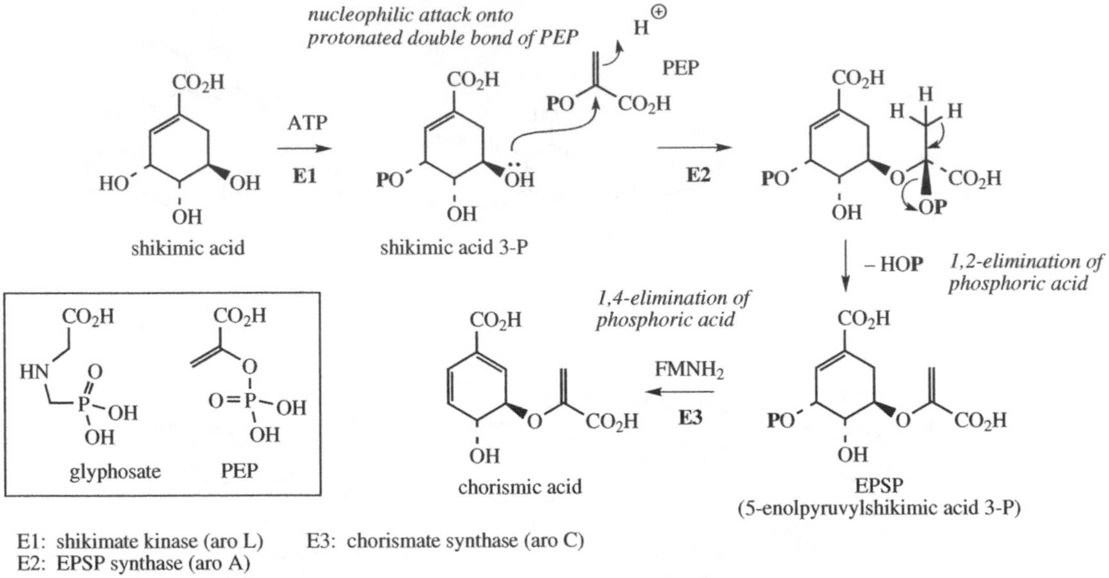

Figure 4.4

E1: shikimate kinase (aro L)   E3: chorismate synthase (aro C)
E2: EPSP synthase (aro A)

this is not a simple concerted elimination. Despite the reaction involving no overall change in oxidation state, the enzyme requires reduced FMN, which is not consumed in the reaction, and this coenzyme is believed to participate via reverse transfer of an electron to the substrate. The biosynthesis of chorismic acid from phosphoenolpyruvate and D-erythrose 4-phosphate involves seven enzyme activities. In most prokaryotes, these are monofunctional activities, whilst plants utilize the bifunctional dehydroquinase–shikimate dehydrogenase activity already mentioned. Fungi, however, possess a multifunctional enzyme complex (termed AROM) that catalyses five successive transformations (reactions 2–6).

4-Hydroxybenzoic acid (Figure 4.5) is produced in bacteria from chorismic acid by an elimination reaction, losing the recently introduced enolpyruvic acid side-chain. The stereochemistry of elimination is unusually *syn*, and so the mechanism is perhaps more complex than at first glance. In plants, 4-hydroxybenzoic acid is formed by a branch much further on in the pathway via side-chain degradation of cinnamic acids (see page 157). The three phenolic acids so far encountered, 4-hydroxybenzoic, protocatechuic, and gallic acids, demonstrate some of the hydroxylation patterns characteristic of shikimic acid-derived metabolites, i.e. a single hydroxy *para* to the side-chain function, dihydroxy groups arranged *ortho* to each other, typically 3,4- to the side-chain, and trihydroxy groups also *ortho* to each other and 3,4,5- to the side-chain. The single

para-hydroxylation and the *ortho*-polyhydroxylation patterns contrast with the typical *meta*-hydroxylation patterns characteristic of phenols derived via the acetate pathway (see page 101), and in most cases allow the biosynthetic origin (acetate or shikimate) of an aromatic ring to be deduced by inspection.

2,3-Dihydroxybenzoic acid and (in microorganisms, but not in plants; see page 161) **salicylic acid** (2-hydroxybenzoic acid) are derived from chorismic acid via its isomer **isochorismic acid** (Figure 4.5). The isomerization involves an S_N2'-type of reaction, an incoming water nucleophile attacking the diene system and displacing the hydroxyl. Salicyclic acid arises by an elimination reaction analogous to that producing 4-hydroxybenzoic acid from chorismic acid, again with *syn* stereochemistry. In the formation of 2,3-dihydroxybenzoic acid, the side-chain of isochorismic acid is lost first by hydrolysis, then dehydrogenation of the 3-hydroxy to a 3-keto allows enolization and formation of the aromatic ring. 2,3-Dihydroxybenzoic acid is a component of the powerful iron chelator (siderophore) **enterobactin** (Figure 4.6) found in *Escherichia coli* and many other Gram-negative bacteria. Such compounds play an important role in bacterial growth by making available sufficient concentrations of essential iron. Enterobactin comprises three molecules of 2,3-dihydroxybenzoic acid and three of the amino acid L-serine, in cyclic triester form. The enzymes involved in enterobactin biosynthesis are encoded by a six-gene

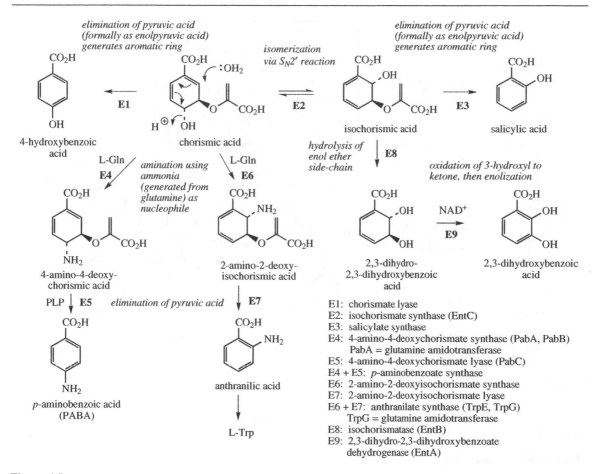

Figure 4.5

cluster; some enzymes have more than one function. Part of this cluster encodes for the isochorismic acid to 2,3-dihydroxybenzoic acid conversion and the remainder for a modular NRPS as seen in the non-ribosomal biosynthesis of peptides (see page 438), though at a minimal level with just two modules, since only a single peptide (amide) bond is created. 2,3-Dihydroxybenzoic acid and serine are activated through an ATP-mediated mechanism prior to attachment to the enzyme complex through a thioester linkage. A phosphopantetheine linker is also used (compare fatty acid biosynthesis, page 40). The lactone-forming steps are also catalysed, with a final intramolecular lactonization to give the trimer, which is released from the enzyme.

Simple amino analogues of the phenolic acids are produced from chorismic acid by related transformations in which ammonia, generated from glutamine, acts as a nucleophile (Figure 4.5). Chorismic acid can be aminated at C-4 to give 4-amino-4-deoxychorismic acid and then

p-**aminobenzoic acid (PABA)**, or at C-2 to give the isochorismic acid analogue which will yield **2-aminobenzoic (anthranilic) acid**. Amination at C-4 has been found to occur with retention of configuration, so perhaps a double inversion mechanism is involved. The *syn* elimination transforming 4-amino-4-deoxychorismic acid into PABA requires the cofactor pyridoxal phosphate and proceeds through an imine intermediate via the amino group. **PABA** forms part of the structure of **folic acid** (vitamin B_9; Figure 4.7) [Box 4.1]. The folic acid structure is built up (Figure 4.7) from a dihydropterin diphosphate which reacts with PABA to give dihydropteroic acid, an enzymic step for which the sulfonamide antibiotics are inhibitors. **Dihydrofolic acid** is produced from the dihydropteroic acid by amide formation incorporating glutamic acid, and reduction yields **tetrahydrofolic acid**. This reduction step is also necessary for the continual regeneration of tetrahydrofolic acid, and forms an important site of action for some antibacterial, antimalarial, and anticancer drugs.

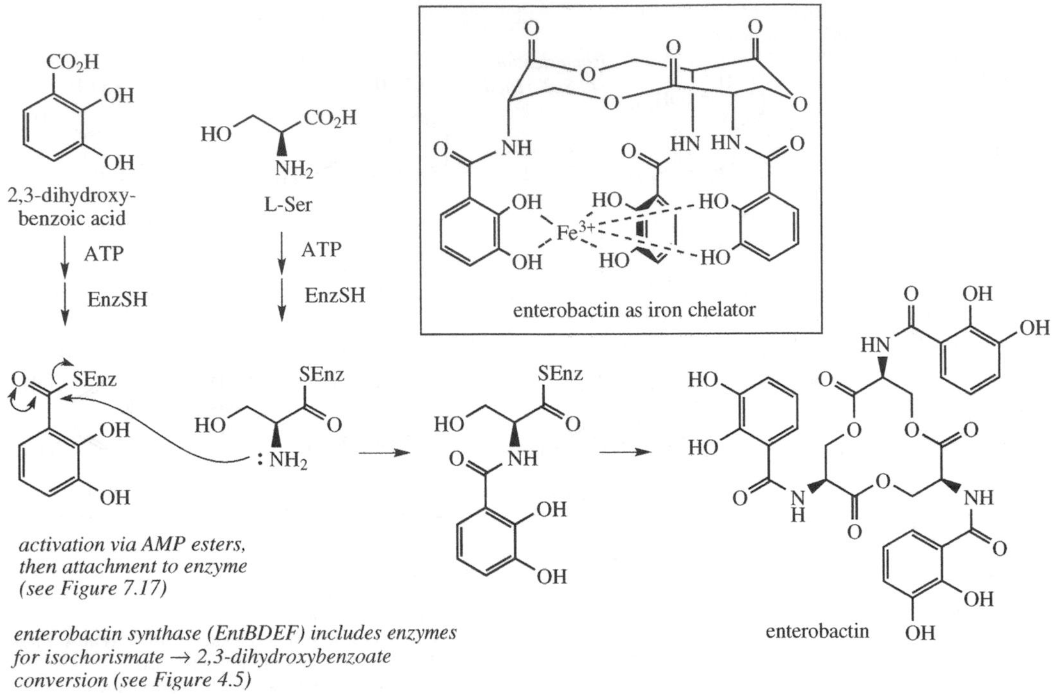

*activation via AMP esters,
then attachment to enzyme
(see Figure 7.17)*

*enterobactin synthase (EntBDEF) includes enzymes
for isochorismate → 2,3-dihydroxybenzoate
conversion (see Figure 4.5)*

enterobactin

Figure 4.6

Box 4.1

Folic acid (Vitamin B₉)

Folic acid (vitamin B_9; Figure 4.7) is a conjugate of a pteridine unit, *p*-aminobenzoic acid (PABA), and glutamic acid. It is found in yeast, liver, and green vegetables, though cooking may destroy up to 90% of the vitamin. Deficiency gives rise to anaemia, and it is also standard practice to provide supplementation during pregnancy to reduce the incidence of spina bifida. Otherwise, deficiency is not normally encountered unless there is malabsorption or chronic disease. Folic acid used for supplementation is usually synthetic. Through genetic engineering techniques it may be possible to enhance the folic acid content of common vegetables. Impressive increases (up to 25×) in folic acid content were found in fruits of tomato derived from a cross between two lines, one overexpressing the enzyme aminodeoxychorismate synthase, and the other GTP cyclohydrolase I, an enzyme from the pteridine pathway; these enzymes thus enhance the flow of metabolites through each contributing pathway.

In the body, folic acid becomes sequentially reduced in the pyrazine ring portion by the enzyme dihydrofolate reductase to give dihydrofolic acid (FH_2) and then tetrahydrofolic acid (FH_4) (Figure 4.7). Tetrahydrofolic acid then functions as a carrier of one-carbon groups for amino acid and nucleotide metabolism. The basic ring system is able to transfer methyl, methylene, methenyl, or formyl groups by the reactions outlined in Figure 4.8 Thus, a methyl group is transferred in the regeneration of methionine from homocysteine (see page 13), purine biosynthesis involves methenyl and formyl transfer, and pyrimidine biosynthesis utilizes methylene transfer. Tetrahydrofolate derivatives also serve as acceptors of one-carbon units in degradative pathways.

Mammals must obtain their tetrahydrofolate requirements from their diet, but microorganisms are able to synthesize this material. This offers scope for selective action and led to the use of sulfanilamide and other antibacterial sulfa drugs, compounds that competitively inhibit dihydropteroate synthase, the biosynthetic enzyme that incorporates PABA into the structure. Specific inhibitors of dihydrofolate reductase have also become especially useful as antibacterials, e.g. **trimethoprim**, and antimalarial

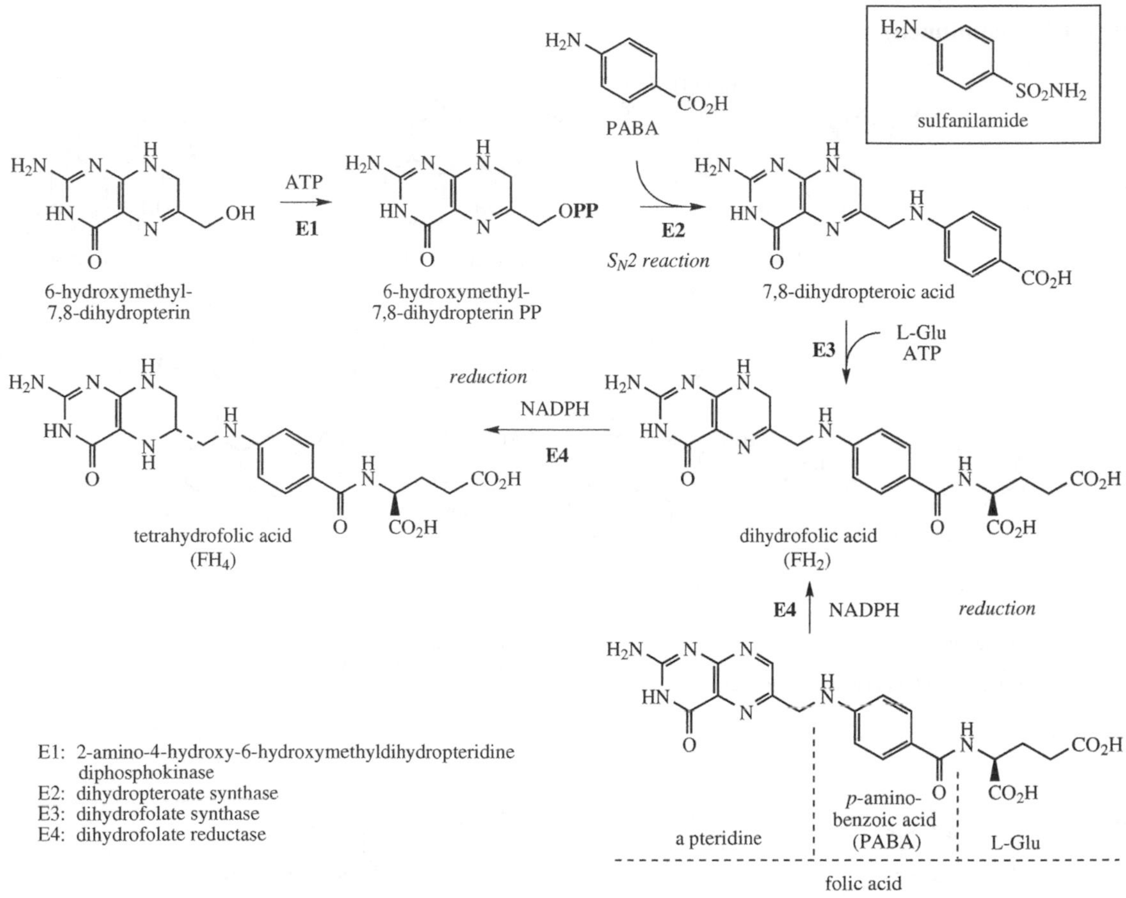

Figure 4.7

E1: 2-amino-4-hydroxy-6-hydroxymethyldihydropteridine diphosphokinase
E2: dihydropteroate synthase
E3: dihydrofolate synthase
E4: dihydrofolate reductase

Box 4.1 (continued)

drugs, e.g. **pyrimethamine** (Figure 4.9). These agents are effective because of differences in susceptibility between the enzymes in humans and in the infective organism.

Regeneration of tetrahydrofolate from dihydrofolate is vital for DNA synthesis in rapidly proliferating cells. The production of thymine from uracil requires methylation of the nucleotide deoxyuridylate (dUMP) to deoxythymidylate (dTMP) with N^5,N^{10}-methylene-FH$_4$ as the methyl donor; this is thereby transformed into dihydrofolate (Figure 4.10). To keep the reaction flowing, FH$_2$ is reduced to FH$_4$, and further N^5,N^{10}-methylene-FH$_4$ is produced using a one-carbon reagent. In this process, the one-carbon reagent comes from the amino acid serine, which is transformed into glycine. Anticancer agents based on folic acid, e.g. **methotrexate** (Figure 4.9), also inhibit dihydrofolate reductase, but they are less selective than the antimicrobial agents and rely on a stronger binding to the enzyme than the natural substrate has. They also block pyrimidine biosynthesis. Methotrexate treatment is potentially lethal to the patient, and is usually followed by 'rescue' with N^5-formyl-tetrahydrofolic acid (**folinic acid, leucovorin**) (Figure 4.8) to counteract the folate-antagonist action. The natural 6S isomer is termed **levofolinic acid (levoleucovorin)**; **folinic acid** in drug use is usually a mixture of the 6R and 6S isomers.

Box 4.1   (continued)

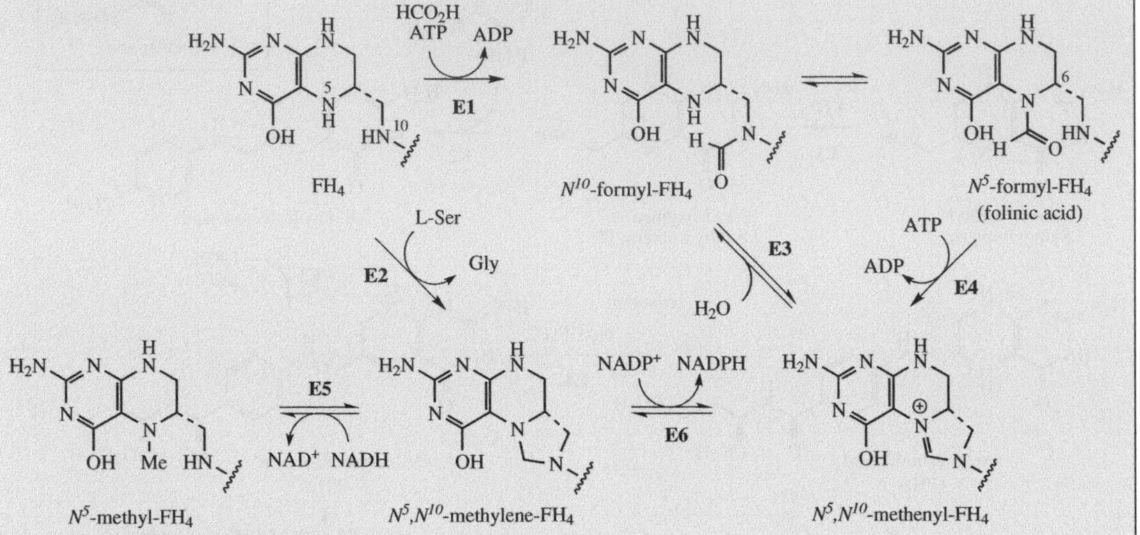

E1: formate:tetrahydrofolate ligase
E2: glycine hydroxymethyltransferase
E3: methenyltetrahydrofolate cyclohydrolase
E4: 5-formyltetrahydrofolate cyclo-ligase
E5: methylenetetrahydrofolate reductase
E6: methylenetetrahydrofolate dehydrogenase

Figure 4.8

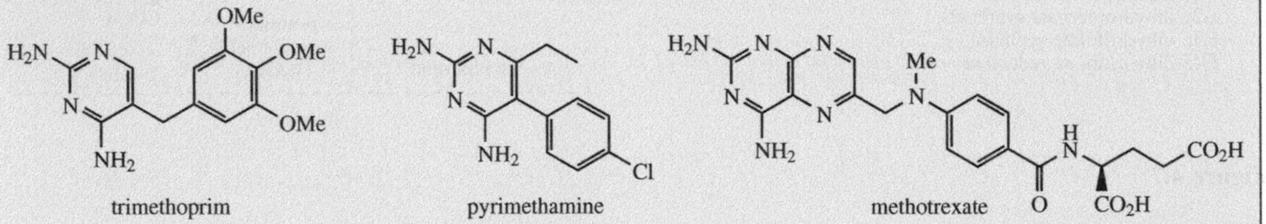

trimethoprim   pyrimethamine   methotrexate

Figure 4.9

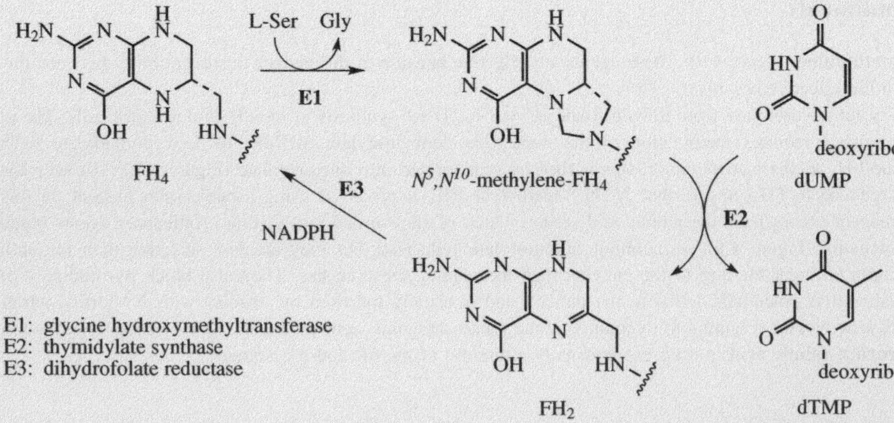

E1: glycine hydroxymethyltransferase
E2: thymidylate synthase
E3: dihydrofolate reductase

Figure 4.10

Anthranilic acid (Figure 4.5) is an intermediate in the biosynthetic pathway to the indole-containing aromatic amino acid L-**tryptophan** (Figure 4.11). In a sequence of complex reactions, which will not be considered in detail, the indole ring system is formed by incorporating two carbon atoms from phosphoribosyl diphosphate, with loss of the original anthranilate carboxyl. The remaining ribosyl carbon atoms are then removed by a reverse aldol reaction, to be replaced on a bound form of indole by those from L-serine, which then becomes the side-chain of L-tryptophan. The last two steps shown are catalysed

by the multi-enzyme complex tryptophan synthase. Although a precursor of L-tryptophan, anthranilic acid may also be produced by metabolism of tryptophan. Both compounds feature as building blocks for a variety of alkaloid structures (see Chapter 6).

Returning to the main course of the shikimate pathway, a singular rearrangement process occurs, transforming **chorismic acid** into **prephenic acid** (Figure 4.12). This reaction, a Claisen rearrangement, transfers the PEP-derived side-chain so that it becomes directly bonded to the carbocycle and so builds up the basic C_6C_3 carbon skeleton of phenylalanine and tyrosine. The reaction

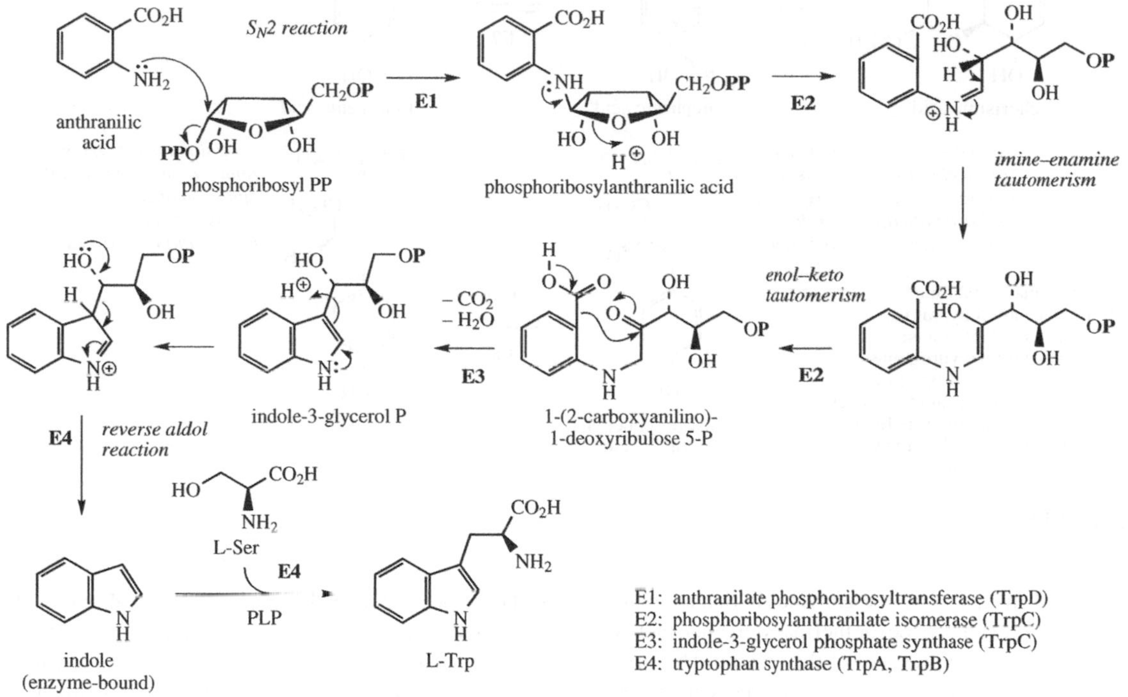

E1: anthranilate phosphoribosyltransferase (TrpD)
E2: phosphoribosylanthranilate isomerase (TrpC)
E3: indole-3-glycerol phosphate synthase (TrpC)
E4: tryptophan synthase (TrpA, TrpB)

Figure 4.11

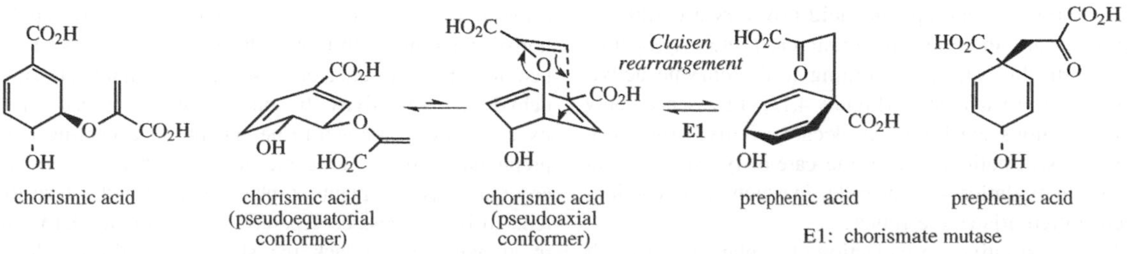

Figure 4.12

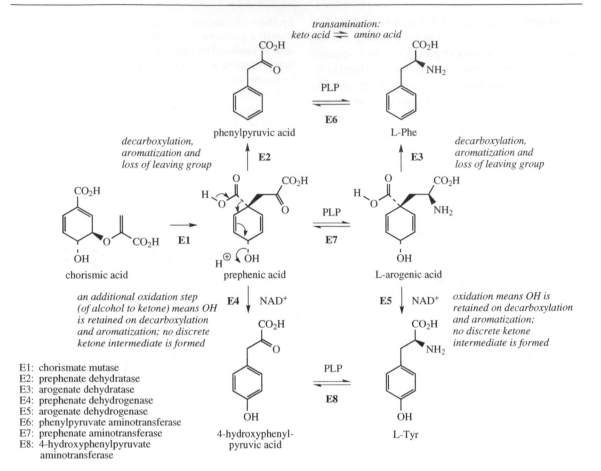

E1: chorismate mutase
E2: prephenate dehydratase
E3: arogenate dehydratase
E4: prephenate dehydrogenase
E5: arogenate dehydrogenase
E6: phenylpyruvate aminotransferase
E7: prephenate aminotransferase
E8: 4-hydroxyphenylpyruvate
     aminotransferase

Figure 4.13

is catalysed in nature by the enzyme chorismate mutase, and although it can also occur thermally, the rate increases some 10^6-fold in the presence of the enzyme. The enzyme achieves this by binding the pseudoaxial conformer of chorismic acid, allowing a transition state with chair-like geometry to develop.

Pathways to the aromatic amino acids L-**phenylalanine** and L-**tyrosine** via prephenic acid may vary according to the organism, and often more than one route may operate in a particular species according to the enzyme activities which are available (Figure 4.13). In essence, only three reactions are involved, decarboxylative aromatization, transamination, and, in the case of tyrosine biosynthesis, an oxidation, but the order in which these reactions occur differentiates the routes.

Decarboxylative aromatization of prephenic acid yields **phenylpyruvic acid**, and PLP-dependent transamination then leads to L-phenylalanine. In the presence of an

NAD^+-dependent dehydrogenase enzyme, decarboxylative aromatization occurs with retention of the hydroxyl function, though as yet there is no evidence that any intermediate carbonyl analogue of prephenic acid is involved. Transamination of the resultant **4-hydroxyphenylpyruvic acid** subsequently gives L-tyrosine. L-**Arogenic acid** is the result of transamination of prephenic acid occurring prior to the decarboxylative aromatization, and can be transformed into both L-phenylalanine and L-tyrosine depending on the absence or presence of a suitable enzymic dehydrogenase activity. In some organisms, broad activity enzymes are known to be capable of accepting both prephenic acid and arogenic acid as substrates. In microorganisms and plants, L-phenylalanine and L-tyrosine tend to be synthesized separately, as in Figure 4.13, but in animals, which lack the shikimate pathway, direct hydroxylation of L-phenylalanine to L-tyrosine, and of L-tyrosine to L-**dihydroxyphenylalanine** (L-**DOPA**) may

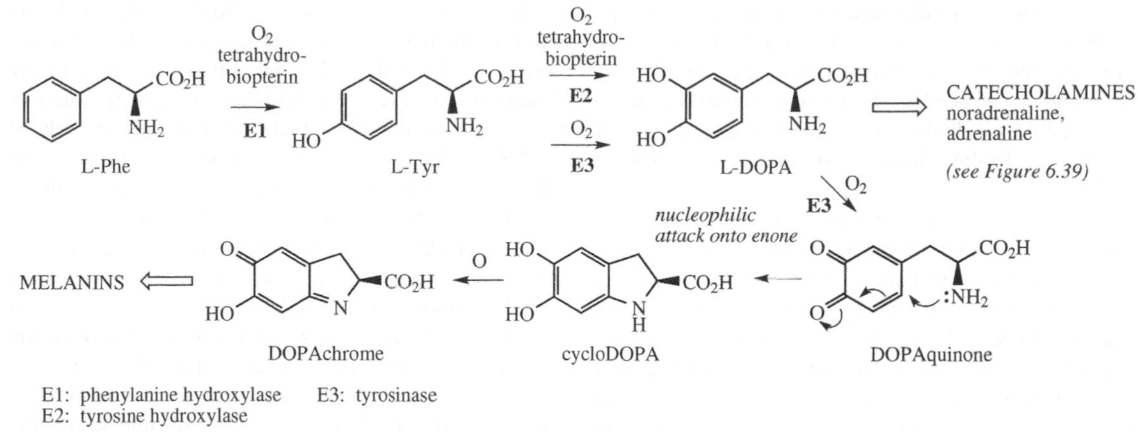

E1: phenylanine hydroxylase   E3: tyrosinase
E2: tyrosine hydroxylase

Figure 4.14

be achieved (Figure 4.14). These reactions are catal-ysed by tetrahydropterin-dependent hydroxylase enzymes, the hydroxyl oxygen being derived from molecular oxy-gen. L-DOPA is a precursor of the **catecholamines**, e.g. the neurotransmitter noradrenaline and the hormone

adrenaline (see page 335). Tyrosine and DOPA are also converted by oxidation reactions into a heterogeneous polymer **melanin**, the main pigment in mammalian skin, hair, and eyes. In this material, the indole system is not formed from tryptophan, but arises from DOPA

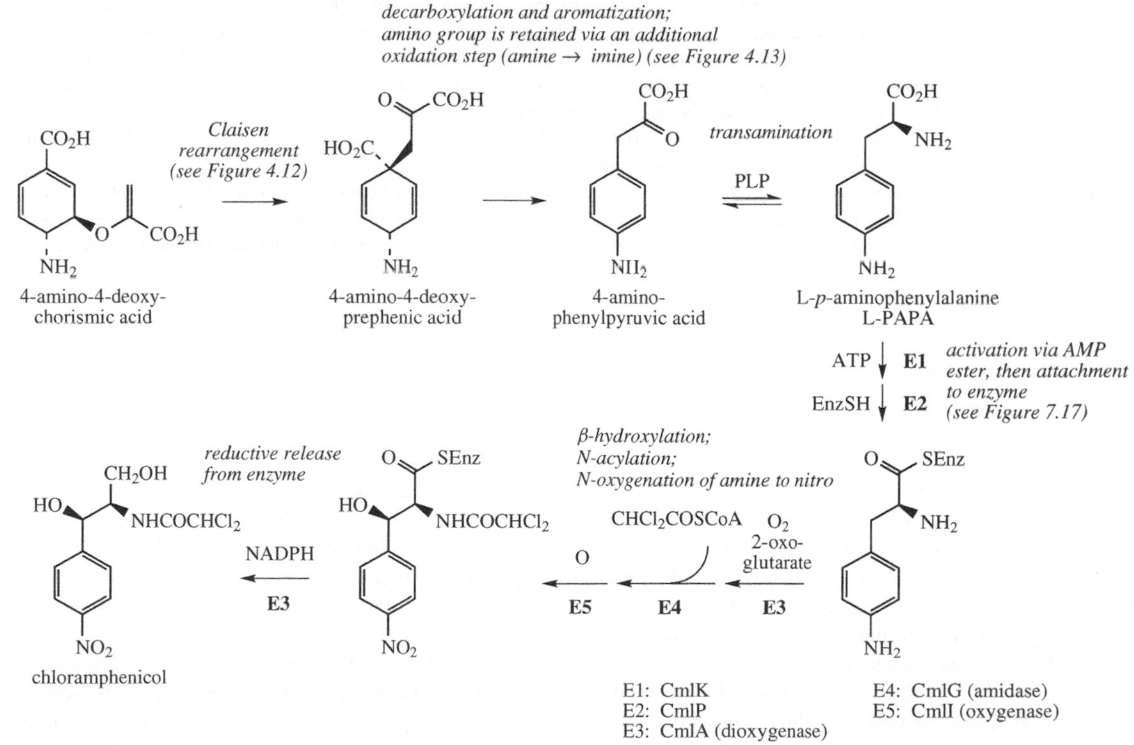

E1: CmlK
E2: CmlP
E3: CmlA (dioxygenase)

E4: CmlG (amidase)
E5: CmlI (oxygenase)

Figure 4.15

by cyclization of DOPAquinone, the nitrogen of the side-chain then attacking the *ortho*-quinone (Figure 4.14). In the melanin biosynthetic pathway, the enzyme tyrosinase, an O_2-dependent monophenol monooxygenase, is responsible for hydroxylation of tyrosine, oxidation of DOPA to DOPAquinone, and further steps beyond DOPAchrome.

Some organisms are capable of synthesizing an unusual variant of L-phenylalanine, the aminated derivative L-*p*-aminophenylalanine (L-PAPA; Figure 4.15). This is known to occur by a series of reactions paralleling those in Figure 4.13, but utilizing the PABA precursor 4-amino-4-deoxychorismic acid (Figure 4.5) instead of chorismic acid. Thus, amino derivatives of prephenic acid and pyruvic acid are elaborated. Although enzymes have not been characterized, appropriate genes have been identified. One important metabolite known to be formed from L-PAPA is the antibiotic **chloramphenicol**, produced by cultures of *Streptomyces venezuelae* [Box 4.2]. The late stages of the pathway (Figure 4.15) have been formulated by genetic analysis to involve a non-ribosomal synthase system related to NRPSs (see page 438). The enzyme contains an adenylation domain, a peptidyl carrier domain, and a domain that is homologous to an NAD^+-dependent dehydrogenase. It is proposed that activation and attachment to the enzyme is followed by β-hydroxylation and *N*-acylation in the side-chain, the latter reaction probably requiring a coenzyme A ester of dichloroacetic acid. This is followed by oxidation of the 4-amino group to a nitro, a fairly rare substituent in natural product structures; the final step is reductive release of the product from the enzyme.

Box 4.2

Chloramphenicol

Chloramphenicol (chloromycetin; Figure 4.15) was initially isolated from cultures of *Streptomyces venezuelae*, but is now obtained for drug use by chemical synthesis. It was one of the first broad-spectrum antibiotics to be developed, and exerts its antibacterial action by inhibiting protein biosynthesis. It binds reversibly to the 50S subunit of the bacterial ribosome, and in so doing it disrupts peptidyl transferase, the enzyme that catalyses peptide bond formation (see page 422). This reversible binding means that bacterial cells not destroyed may resume protein biosynthesis when no longer exposed to the antibiotic. Some microorganisms have developed resistance to chloramphenicol by an inactivation process involving enzymic acetylation of the primary alcohol group in the antibiotic. The acetate binds only very weakly to the ribosomes, so has little antibiotic activity. The value of chloramphenicol as an antibacterial agent has been severely limited by some serious side-effects. It can cause blood disorders, including irreversible aplastic anaemia in certain individuals, and these can lead to leukaemia and perhaps prove fatal. Nevertheless, it is still the drug of choice for some life-threatening infections, such as typhoid fever and bacterial meningitis. The blood constitution must be monitored regularly during treatment to detect any abnormalities or adverse changes. The drug is orally active, but may also be injected. Eye-drops are useful for the treatment of bacterial conjunctivitis.

PHENYLPROPANOIDS

Cinnamic Acids and Esters

L-Phenylalanine and L-tyrosine, as C_6C_3 building blocks, are precursors for a wide range of natural products. In plants, a frequent first step is the elimination of ammonia from the side-chain to generate the appropriate *trans* (*E*) cinnamic acid. In the case of phenylalanine, this would give **cinnamic acid**, whilst tyrosine could yield **4-coumaric acid** (*p*-coumaric acid; Figure 4.16). All plants appear to have the ability to deaminate phenylalanine via the enzyme phenylalanine ammonia lyase (PAL), but the corresponding transformation of tyrosine is more restricted, being mainly limited to members of the grass family (the Graminae/Poaceae). Whether a separate enzyme tyrosine ammonia lyase (TAL) exists in plants, or whether grasses merely have a broad specificity PAL also capable of deaminating tyrosine, is still debated. Tyrosine-specific TAL has been found in a number of bacteria. Those species that do not transform tyrosine to 4-coumaric acid synthesize this compound by direct hydroxylation of cinnamic acid, in a cytochrome P-450-dependent reaction, and tyrosine is often channelled instead into other secondary metabolites, e.g. alkaloids. Other cinnamic acids are obtained by further hydroxylation and methylation reactions, sequentially building up substitution patterns typical of shikimate pathway metabolites, i.e. an *ortho* oxygenation pattern

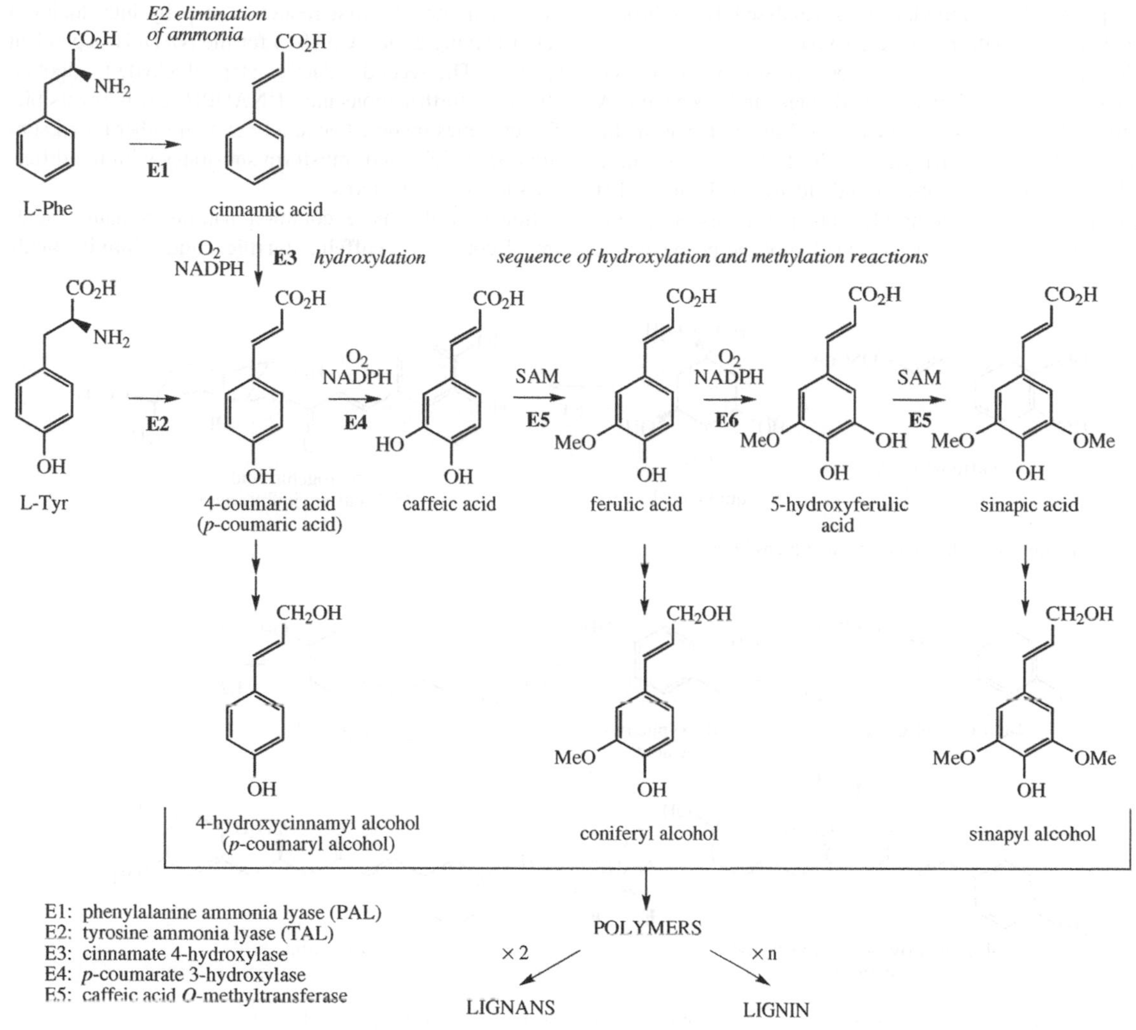

E2 elimination of ammonia
CO_2H
NH_2
L-Phe
E1
cinnamic acid
O_2
NADPH
E3 hydroxylation
sequence of hydroxylation and methylation reactions
L-Tyr
E2
4-coumaric acid (p-coumaric acid)
E4
caffeic acid
SAM
E5
ferulic acid
E6
5-hydroxyferulic acid
sinapic acid
HO
MeO
OMe
OH
CH_2OH
4-hydroxycinnamyl alcohol (p-coumaryl alcohol)
coniferyl alcohol
sinapyl alcohol
POLYMERS
× 2
× n
LIGNANS
LIGNIN
E1: phenylalanine ammonia lyase (PAL)
E2: tyrosine ammonia lyase (TAL)
E3: cinnamate 4-hydroxylase
E4: p-coumarate 3-hydroxylase
E5: caffeic acid O-methyltransferase

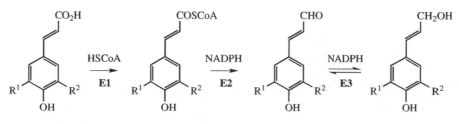

Figure 4.16

CO_2H
HSCoA
E1
COSCoA
NADPH
E2
CHO
E3
CH_2OH
R^1
R^2
OH
E1: 4-coumarate:CoA ligase
E2: cinnamoyl-CoA reductase
E3: cinnamyl alcohol dehydrogenase

Figure 4.17

(see page 140). Methylation is catalysed by a broad-specificity methyltransferase enzyme.

Substitution patterns are not necessarily elaborated completely at the cinnamic acid stage, and coenzyme A esters and aldehydes may also be substrates for aromatic hydroxylation and methylation. Reduction of cinnamic acids via coenzyme A esters and aldehydes (Figure 4.17) leads to the corresponding alcohols, precursors of lignans and lignin (see page 151). Formation of the coenzyme A ester facilitates the first reduction step by introducing a better leaving group (CoAS⁻) for the NADPH-dependent reaction. The second reduction step, aldehyde to alcohol, utilizes a further molecule of NADPH and is reversible. The enzymes involved generally exhibit rather broad substrate specificity and transform substrates with the different substitution patterns.

Some of the more common natural cinnamic acids are 4-coumaric, **caffeic, ferulic**, and **sinapic acids**

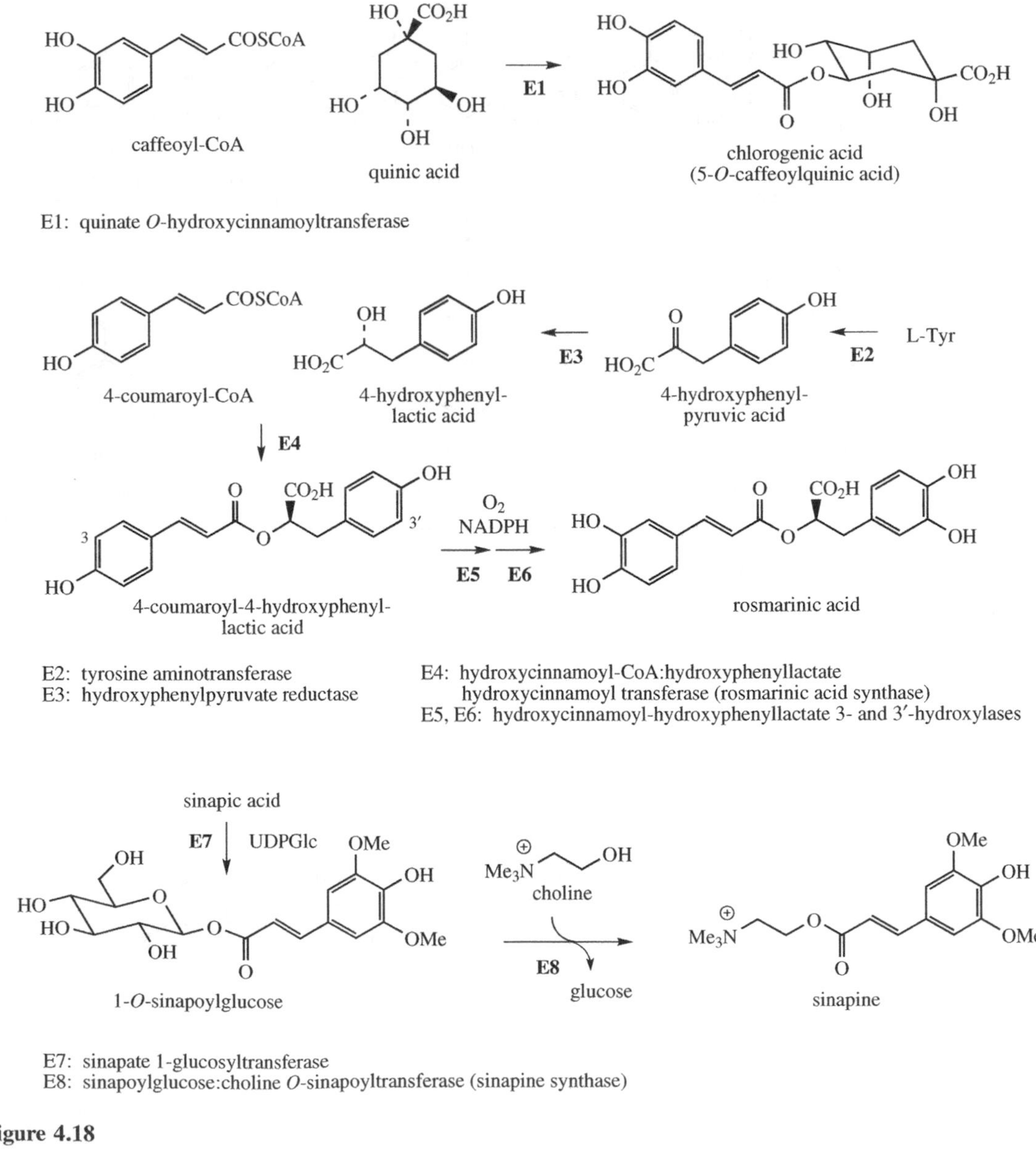

E1: quinate *O*-hydroxycinnamoyltransferase

E2: tyrosine aminotransferase
E3: hydroxyphenylpyruvate reductase

E4: hydroxycinnamoyl-CoA:hydroxyphenyllactate
      hydroxycinnamoyl transferase (rosmarinic acid synthase)
E5, E6: hydroxycinnamoyl-hydroxyphenyllactate 3- and 3′-hydroxylases

E7: sinapate 1-glucosyltransferase
E8: sinapoylglucose:choline *O*-sinapoyltransferase (sinapine synthase)

Figure 4.18

(Figure 4.16). These can be found in plants in free form and in a range of esterified forms, e.g. with quinic acid as in **chlorogenic acid** (5-*O*-caffeoylquinic acid; see coffee, page 414), with 3,4-dihydroxyphenyl-lactic acid as in **rosmarinic acid**, with glucose as in **1-*O*-sinapoylglucose**, and with choline as in **sinapine** (Figure 4.18). Esterification reactions require the cinnamic acid to be transformed initially to a suitably activated form, either a coenzyme A ester or a glucose ester. Coenzyme A esters are involved in the formation of chlorogenic acid in sweet potato (*Ipomoea batatas*; Convolvulaceae) and rosmarinic acid in *Coleus blumei* (Labiatae/Lamiaceae). Glucose esters are used in the biosynthesis of sinapine in rapeseed (*Brassica napus*; Cruciferae/Brassicaceae) (compare gallotannins, page 139). In the case of rosmarinic acid, further hydroxylation in the aromatic rings has been found to occur after the esterification reaction. This pathway also incorporates phenylalanine into one C_6C_3 unit, the 'acid' part of the ester, and tyrosine into the other. Rosmarinic acid, first isolated from rosemary (*Rosmarinus officinalis*; Labiatae/Lamiaceae), is found in a number of medicinal plants and has beneficial antioxidant, antibacterial, and anti-inflammatory properties.

Lignans and Lignin

The cinnamic acids also feature in the pathways to other metabolites based on C_6C_3 building blocks. Pre-eminent amongst these, certainly as far as nature is concerned, is the plant polymer **lignin**, a strengthening material for the plant cell wall which acts as a matrix for cellulose microfibrils (see page 494). Lignin represents a vast reservoir of aromatic materials, mainly untapped because of the difficulties associated with release of these metabolites. The action of wood-rotting fungi offers the most effective way of making these useful products more accessible. Lignin is formed by phenolic oxidative coupling of hydroxycinnamyl alcohol monomers, brought about by peroxidase enzymes (see page 28). The most important of these monomers are **4-hydroxycinnamyl alcohol** (*p*-coumaryl alcohol), **coniferyl alcohol**, and **sinapyl alcohol** (Figure 4.16), though the monomers used vary according to the plant type. Gymnosperms polymerize mainly coniferyl alcohol, dicotyledonous plants coniferyl alcohol and sinapyl alcohol, whilst monocotyledons use all three alcohols. The peroxidase enzyme then achieves one-electron oxidation of the phenol group. One-electron oxidation of a simple phenol allows delocalization of the unpaired electron, giving resonance forms in which the free electron resides at positions *ortho* and *para* to the oxygen function (see page 28). With cinnamic acid derivatives, conjugation allows the unpaired electron to

be delocalized also into the side-chain (Figure 4.19). Radical pairing of resonance structures can then provide a range of dimeric systems containing reactive quinonemethides which are susceptible to nucleophilic attack from hydroxyl groups in the same system, or by external water molecules. Thus, **coniferyl alcohol** monomers can couple, generating linkages as exemplified by **guaiacylglycerol β-coniferyl ether**(β-arylether linkage), **dehydrodiconiferyl alcohol** (phenylcoumaran linkage), and **pinoresinol** (resinol linkage). These dimers can react further by similar mechanisms to produce a lignin polymer containing a heterogeneous series of intermolecular bondings, as seen in the various dimers. In contrast to most other natural polymeric materials, lignin appears to be devoid of ordered repeating units, though some 50–70% of the linkages are of the β-arylether type. The dimeric materials are also found in nature and are called **lignans**. Some authorities like to restrict the term lignan specifically to molecules in which the two phenylpropane units are coupled at the central carbon of the side-chain, e.g. pinoresinol, whilst compounds containing other types of coupling, e.g. as in guaiacylglycerol β-coniferyl ether and dehydrodiconiferyl alcohol, are then referred to as **neolignans**. Lignan/neolignan formation and lignin biosynthesis are catalysed by different enzymes, and a consequence of this is that natural lignans/neolignans are normally enantiomerically pure because they arise from stereochemically controlled coupling. The control mechanisms for lignin biosynthesis are less well defined, but the enzymes appear to generate products lacking optical activity.

Further cyclization and other modifications can create a wide range of lignans of very different structural types. One of the most important of the natural lignans having useful biological activity is the aryltetralin lactone **podophyllotoxin** (Figure 4.20), which is derived from coniferyl alcohol via the dibenzylbutyrolactones **matairesinol** and **yatein**, cyclization probably occurring as shown in Figure 4.20. Matairesinol is known to arise by reductive opening of the furan rings of **pinoresinol**, followed by successive oxidations to yield first a lactol then a lactone. The substitution pattern in the two aromatic rings is built up further during the pathway, i.e. matairesinol → yatein, and does not arise by initial coupling of two different cinnamyl alcohol residues. The trisubstituted ring is elaborated before the methylenedioxy ring system. The latter grouping is found in many shikimate-derived natural products, and is formed by an oxidative reaction on an *ortho*-hydroxymethoxy pattern (see page 27). Podophyllotoxin and related lignans are found in the roots of *Podophyllum* species (Berberidaceae), and have clinically useful cytotoxic and anticancer activity [Box 4.3]. The

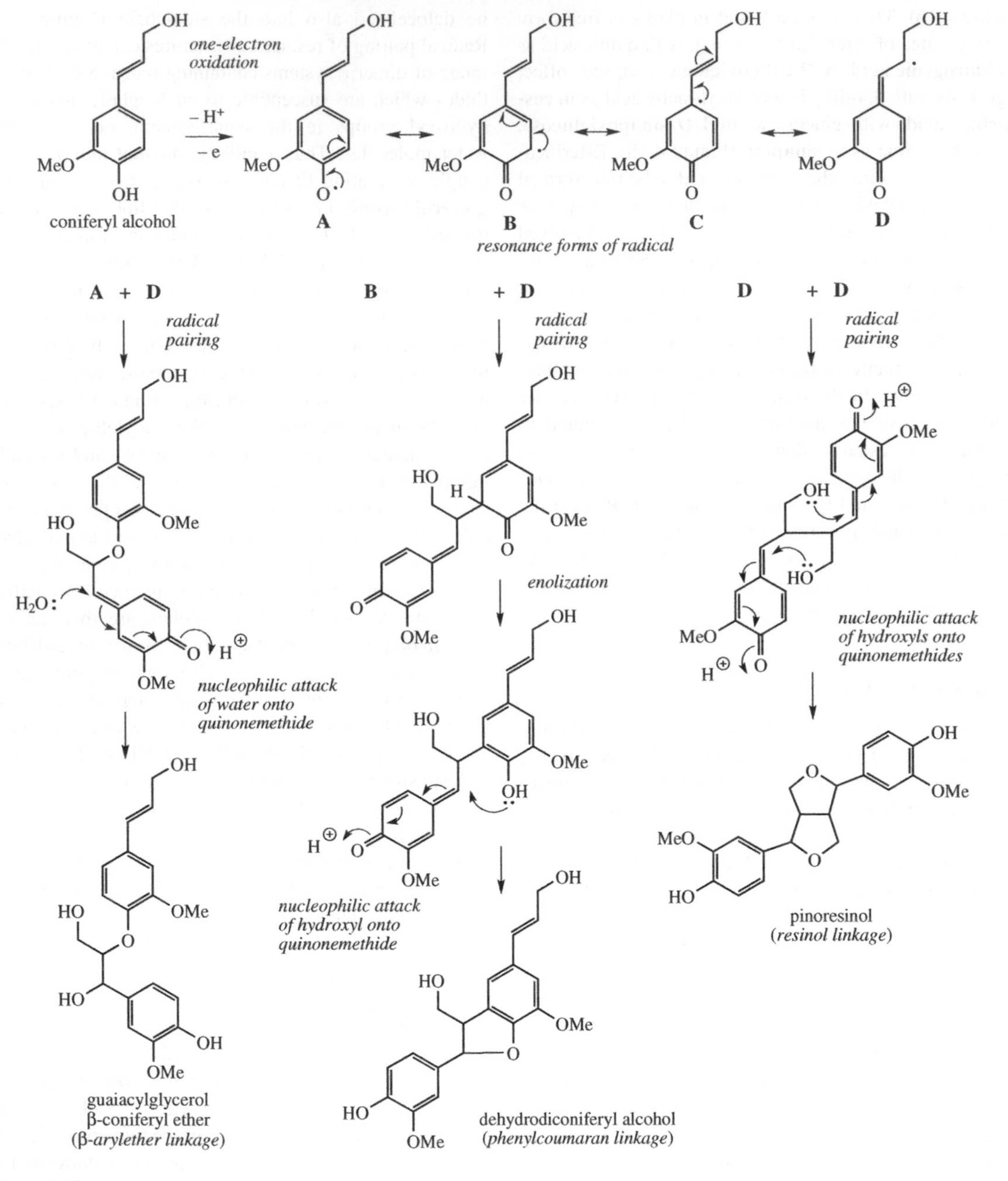

Figure 4.19

so-called 'mammalian' lignans **enterolactone** and **enterodiol** (Figure 4.21) were first discovered in human urine and were unusual in that each aromatic ring possessed only a *meta* hydroxyl group. They were subsequently shown to be derived from dietary plant lignans, especially matairesinol and secoisolariciresinol, by the action of intestinal microflora. Enterolactone and enterodiol have oestrogenic activity and have been implicated as contributing to lower levels of breast cancer amongst vegetarians (see phyto-oestrogens, page 177).

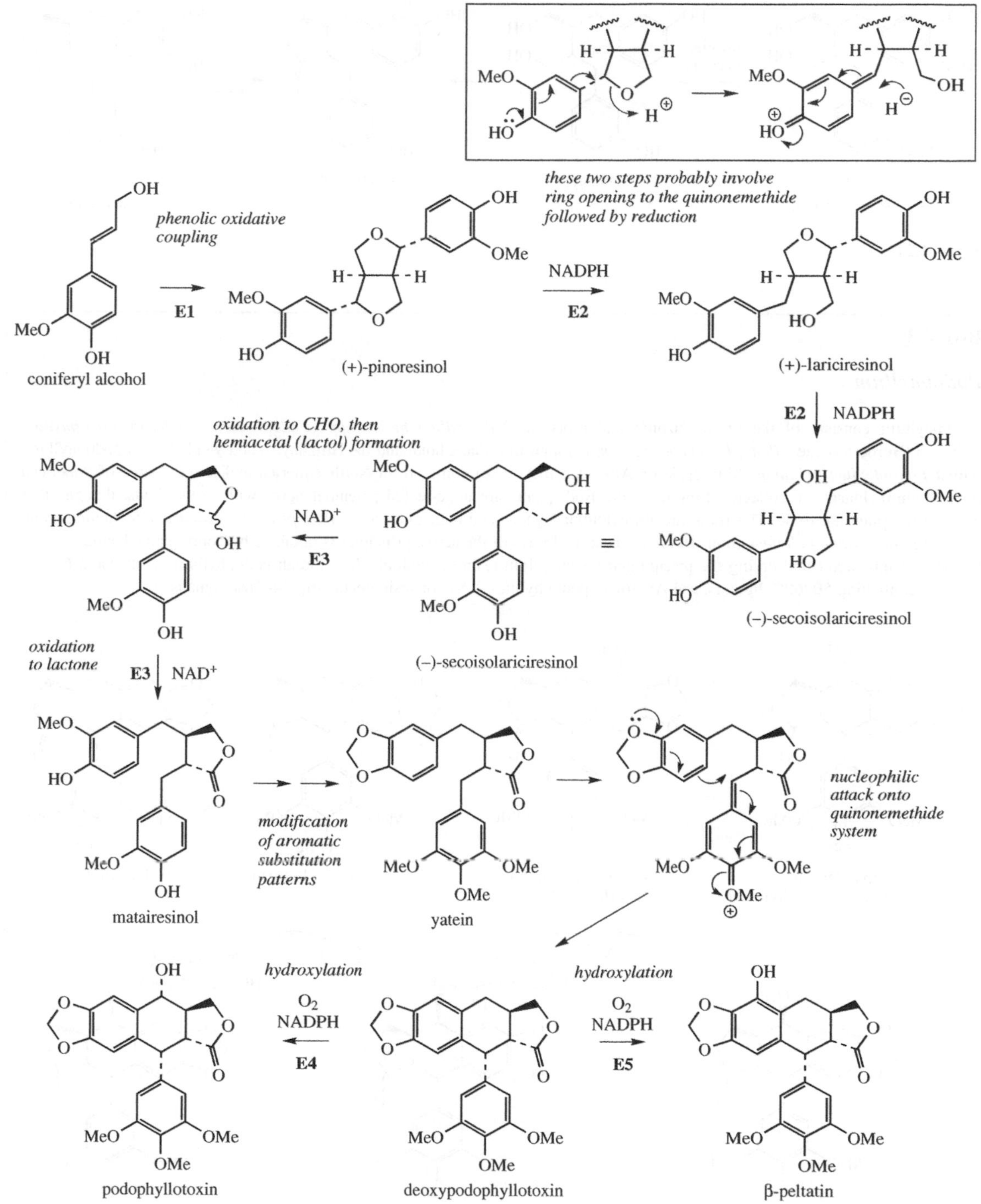

these two steps probably involve ring opening to the quinonemethide followed by reduction

E1: pinoresinol synthase
E2: pinoresinol-lariciresinol reductase
E3: secoisolariciresinol dehydrogenase
E4: deoxypodophyllotoxin 7-hydroxylase
E5: deoxypodophyllotoxin 6-hydroxylase

Figure 4.20

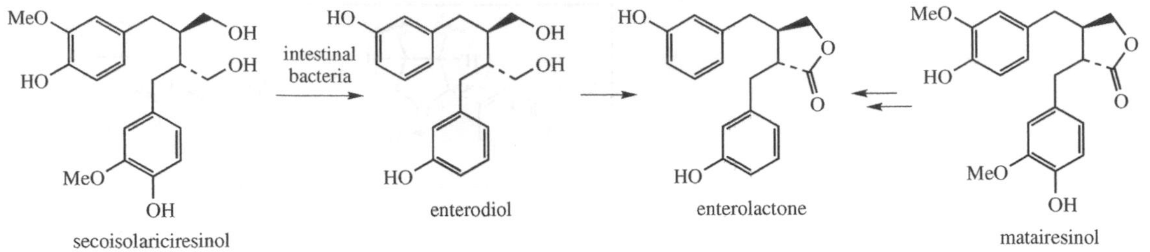

Figure 4.21

Box 4.3

Podophyllum

Podophyllum consists of the dried rhizome and roots of *Podophyllum hexandrum* (*Podophyllum emodi*) or *Podophyllum peltatum* (Berberidaceae). *Podophyllum hexandrum* is found in India, China, and the Himalayas, and yields Indian podophyllum, whilst *Podophyllum peltatum* (May apple or American mandrake) comes from North America and is the source of American podophyllum. Plants are collected from the wild. Both plants are large-leafed perennial herbs with edible fruits, though other parts of the plant are toxic. The roots contain cytotoxic lignans and their glucosides, *Podophyllum hexandrum* containing about 5% and *Podophyllum peltatum* about 1%. A concentrated form of the active principles is obtained by pouring an ethanolic extract of the root into water and drying the precipitated podophyllum resin or 'podophyllin'. Indian podophyllum yields about 6–12% of resin containing 50–60% lignans, and American podophyllum 2–8% of resin containing 14–18% lignans.

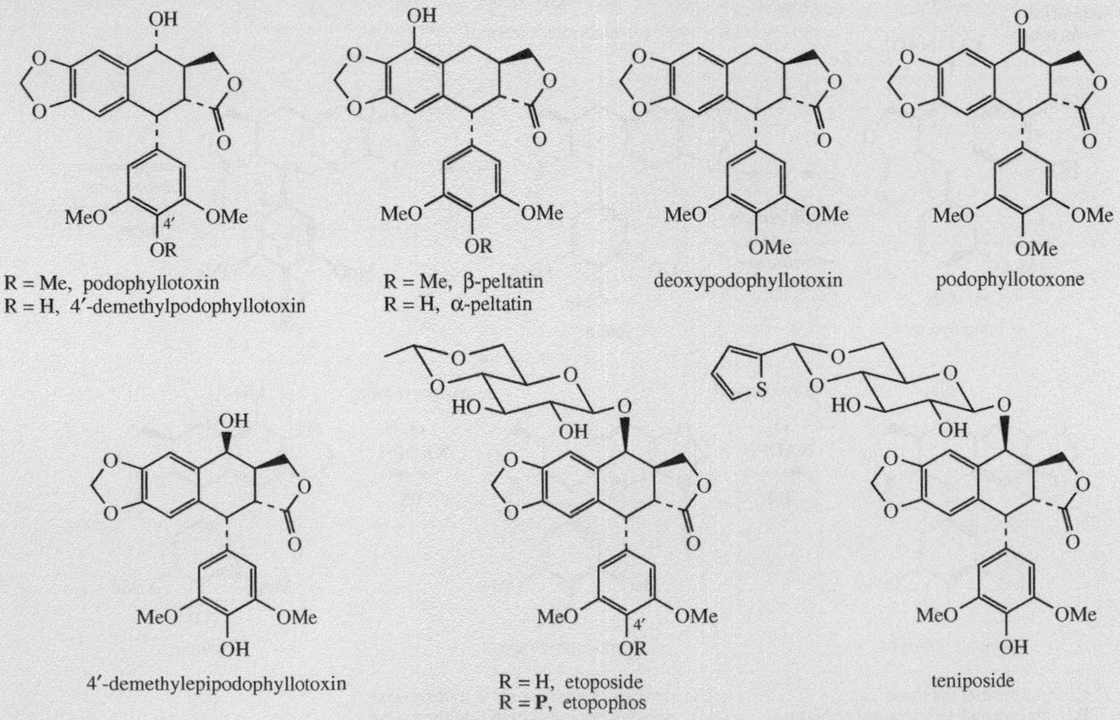

R = Me, podophyllotoxin
R = H, 4'-demethylpodophyllotoxin

R = Me, β-peltatin
R = H, α-peltatin

deoxypodophyllotoxin

podophyllotoxone

4'-demethylepipodophyllotoxin

R = H, etoposide
R = P, etopophos

teniposide

Figure 4.22

Box 4.3 (continued)

The lignan constituents of the two roots are the same, but the proportions are markedly different. The Indian root contains chiefly podophyllotoxin (Figure 4.22) (about 4%) and 4′-demethylpodophyllotoxin (about 0.45%). The main components in the American root are podophyllotoxin (about 0.25%), β-peltatin (about 0.33%) and α-peltatin (about 0.25%). Deoxypodophyllotoxin and podophyllotoxone are also present in both plants, as are the glucosides of podophyllotoxin, 4′-demethylpodophyllotoxin, and the peltatins, though preparation of the resin results in considerable losses of the water-soluble glucosides.

Podophyllum resin has long been used as a purgative, but the discovery of the cytotoxic properties of podophyllotoxin and related compounds has now made podophyllum a commercially important drug plant. Preparations of **podophyllum resin** (the Indian resin is preferred) are effective treatments for warts, and pure **podophyllotoxin** is available as a paint for venereal warts, a condition which can be sexually transmitted. The antimitotic effect of podophyllotoxin and the other lignans is by binding to the protein tubulin in the mitotic spindle, preventing polymerization and assembly into microtubules (compare vincristine, page 375, and colchicine, page 361). During mitosis, the chromosomes separate with the assistance of these microtubules, and after cell division, the microtubules are transformed back to tubulin. Podophyllotoxin and other *Podophyllum* lignans were found to be unsuitable for clinical use as anticancer agents due to toxic side-effects, but the semi-synthetic derivatives etoposide and teniposide (Figure 4.22), which are manufactured from natural podophyllotoxin, have proved excellent antitumour agents. They were developed as modified forms (acetals) of the natural 4′-demethylpodophyllotoxin glucoside. Attempted synthesis of the glucoside inverted the stereochemistry at the sugar–aglycone linkage, and these agents are thus derivatives of 4′-demethylepipodophyllotoxin (Figure 4.22). **Etoposide** is a very effective anticancer agent, and is used in the treatment of small-cell lung cancer, testicular cancer, and lymphomas, usually in combination therapies with other anticancer drugs. It may be given orally or intravenously. The water-soluble pro-drug **etopophos** (etoposide 4′-phosphate) is also available; this is efficiently converted into etoposide by phosphatase enzymes and is preferred for routine clinical use. **Teniposide** has similar anticancer properties and, though not as widely used as etoposide, has value in paediatric neuroblastoma.

Remarkably, the 4′-demethylepipodophyllotoxin series of lignans do not act via a tubulin-binding mechanism as does podophyllotoxin. Instead, these drugs inhibit the enzyme topoisomerase II, thus preventing DNA synthesis and replication. Topoisomerases are responsible for cleavage and resealing of the DNA strands during the replication process, and are broadly classified as type I or II according to their ability to cleave one or both strands, though sub-classes are now recognized. Camptothecin (see page 384) is an inhibitor of topoisomerase I. Etoposide is believed to inhibit strand-rejoining ability by stabilizing the topoisomerase II–DNA complex in a cleavage state, leading to double strand breaks and cell death. Rather than simply inhibiting the enzyme, etoposide converts topoisomerase II into a potent cellular toxin, and the term topoisomerase II poison is now often employed to distinguish the effect from an agent that merely alters catalytic activity. Development of other topoisomerase inhibitors based on podophyllotoxin-related lignans is an active area of research. Biological activity in this series of compounds is very dependent on the presence of the *trans*-fused five-membered lactone ring, this type of fusion producing a highly strained system. Ring strain is markedly reduced in the corresponding *cis*-fused system, and the natural compounds are easily and rapidly converted into these *cis*-fused lactones by treatment with very mild bases, via enol tautomers or enolate anions (Figure 4.23). Picropodophyllin is almost devoid of cytotoxic properties.

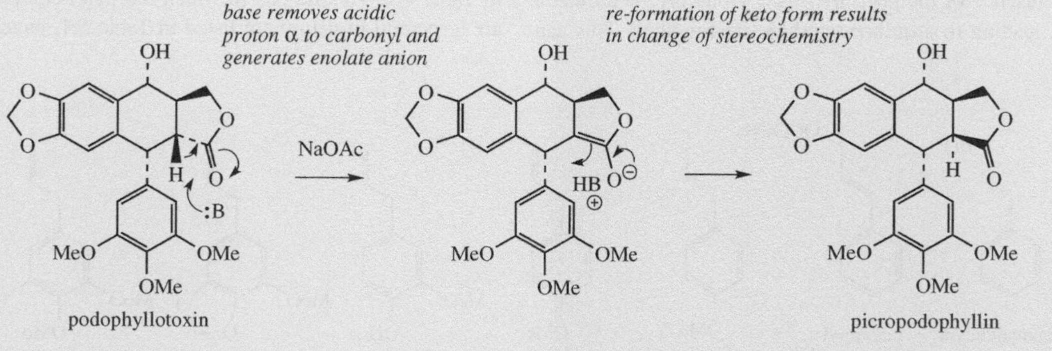

Figure 4.23

Box 4.3   (continued)

Podophyllotoxin is also found in significant amounts in the roots of other *Podophyllum* species, and in closely related genera such as *Diphylleia* (Berberidaceae). Fungi isolated from the roots of *Podophyllum* species have also been found to produce podophyllotoxin in culture, suggesting possible gene transfer between the plant and the endophytic fungus. These fungi include *Trametes hirsuta* from rhizomes of *Podophyllum hexandrum*, and *Phialocephala fortinii* from *Podophyllum peltatum*. Yields tend to be rather low, however, and would need considerable improvement to offer an alternative source to the plant material.

Phenylpropenes

The reductive sequence from an appropriate cinnamic acid to the corresponding cinnamyl alcohol is utilized for the production of various phenylpropene derivatives. Thus **cinnamaldehyde** (Figure 4.24) is the principal component in the oil from the bark of cinnamon (*Cinnamomum zeylanicum*; Lauraceae), widely used as a spice and flavouring. Fresh bark is known to contain high levels of the ester **cinnamyl acetate**, and cinnamaldehyde is released from this by fermentation processes which are part of commercial preparation of the bark, presumably by enzymic hydrolysis and participation of the reversible alcohol dehydrogenase. Cinnamon leaf, on the other hand, contains large amounts of **eugenol** (Figure 4.24) and much smaller amounts of cinnamaldehyde. Eugenol is also the principal constituent in oil from cloves (*Syzygium aromaticum*; Myrtaceae), used for many years as a dental anaesthetic as well as for flavouring.

Allylphenols (e.g. eugenol) or propenylphenols (e.g. **isoeugenol**) both originate from appropriate cinnamyl alcohols by way of acetate esters (Figure 4.25), though the two groups of compounds differ with respect to the position of the side-chain double bond. The acetate function provides a leaving group, loss of which is facilitated by the presence of the *para* hydroxyl group on the aromatic ring, leading to a quinonemethide intermediate. This can be reduced by addition of hydride (from NADPH) generating the different side-chains according to the position of attack; the genes and enzymes involved in the two processes have been characterized.

Myristicin (Figure 4.24) from nutmeg (*Myristica fragrans*; Myristicaceae) is a further example of an allylphenol found in flavouring materials. Myristicin also has a history of being employed as a mild hallucinogen via ingestion of ground nutmeg. It is probably metabolized in the body via an amination reaction to give an amfetamine-like derivative (see page 404). **Anethole** is the main component in oils from aniseed (*Pimpinella anisum*; Umbelliferae/Apiaceae), star anise (*Illicium verum*; Illiciaceae) and fennel (*Foeniculum vulgare*; Umbelliferae/Apiaceae). The presence of propenyl components in flavouring materials such as cinnamon, star anise, nutmeg, and sassafras (*Sassafras albidum*; Lauraceae) has reduced their commercial use somewhat, since these constituents have been shown to be weak carcinogens in laboratory tests on animals. In the case of **safrole** (Figure 4.26), the main component of sassafras oil, this toxicity has been shown to arise from hydroxylation in the side-chain followed by sulfation, giving an agent which binds to cellular macromolecules. Further data on volatile oils containing aromatic constituents isolated from these and other plant materials are given in Table 4.1. Volatile oils in which the main components are terpenoid in nature are listed in Table 5.1, page 158.

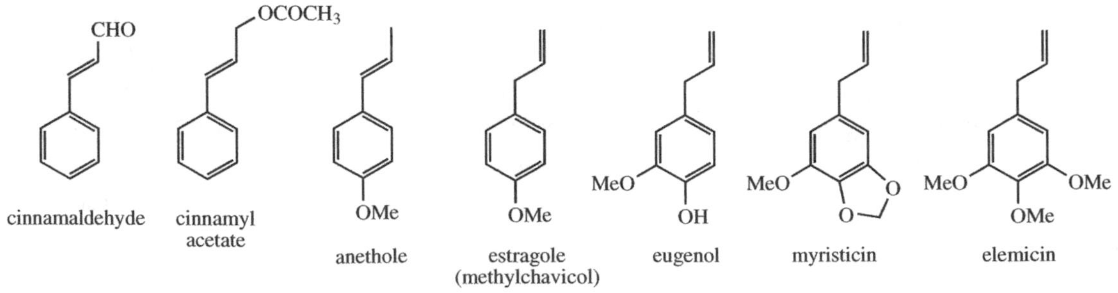

cinnamaldehyde   cinnamyl acetate   anethole   estragole (methylchavicol)   eugenol   myristicin   elemicin

Figure 4.24

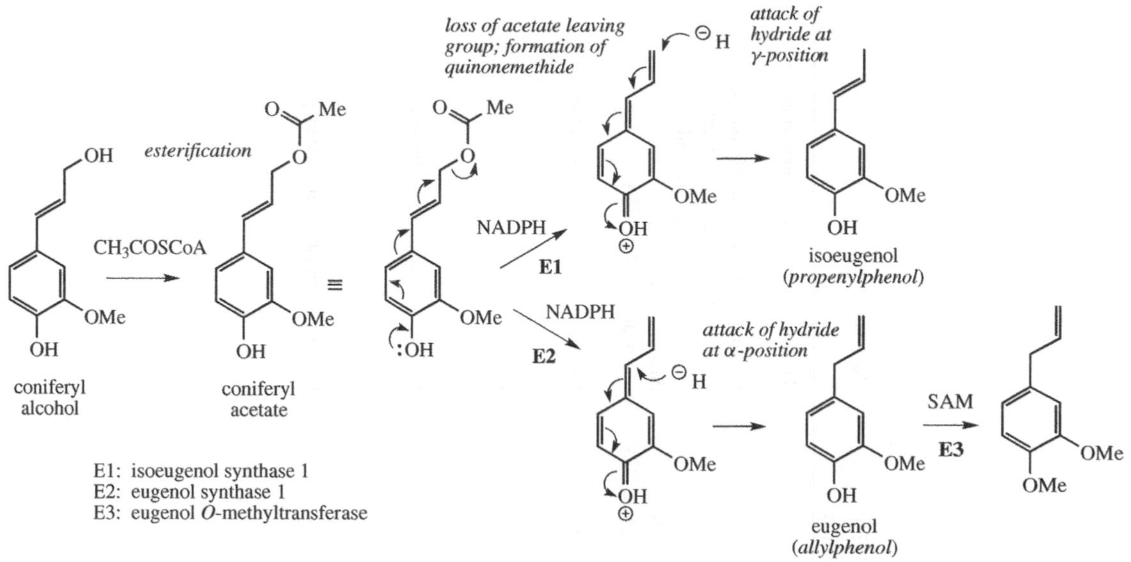

Figure 4.25

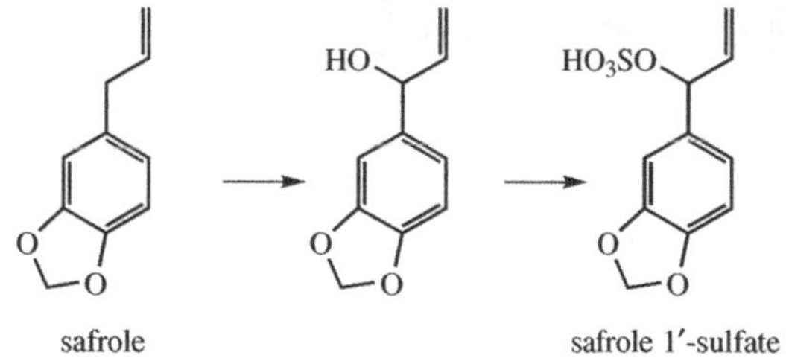

Figure 4.26

Benzoic Acids from C_6C_3 Compounds

Benzoic acids are structurally simple natural products, so it is quite surprising just how many different pathways may be employed for their biosynthesis. Pathways to the same compound may be quite different according to the organism, and sometimes more than one pathway may exist in a single organism. Some of the simple hydroxybenzoic acids (C_6C_1 compounds), such as gallic acid and 4-hydroxybenzoic acid, can be formed directly from intermediates early in the shikimate pathway, e.g. 3-dehydroshikimic acid (see page 138) or chorismic acid (see page 141), but alternative routes exist in which cinnamic acid derivatives (C_6C_3 compounds) are cleaved at the double bond and lose two carbon atoms from the side-chain. Thus, 4-coumaric acid may act as a precursor of **4-hydroxybenzoic acid**, and ferulic acid may

give **vanillic acid** (4-hydroxy-3-methoxybenzoic acid) (Figure 4.27). Alternatively, cinnamic acid itself may be converted into benzoic acid. A sequence analogous to that involved in the β-oxidation of fatty acids (see page 18) is possible, so that the double bond in the coenzyme A ester would be hydrated, the hydroxyl group oxidized to a ketone, and the β-ketoester would then lose acetyl-CoA by a reverse Claisen reaction, giving the coenzyme A ester of 4-hydroxybenzoic acid. Whilst this sequence has been generally accepted, there is also evidence to support another side-chain cleavage mechanism which is different from the fatty acid β-oxidation pathway (Figure 4.27). Coenzyme A esters are not involved, and though a similar hydration of the double bond occurs, chain shortening features a reverse aldol reaction, generating the appropriate aromatic aldehyde. The corresponding acid is then formed via an NAD^+-dependent oxidation step. Thus, aromatic aldehydes such as **vanillin**, the main flavour compound in vanilla (pods of the orchid *Vanilla planiflora*; Orchidaceae), would be formed from the correspondingly substituted cinnamic acid without proceeding through intermediate benzoic acids or esters. In some plants and microorganisms, yet another variant may be implicated. This resembles the latter route in employing a reverse aldol cleavage, but utilizes CoA esters and bifunctional hydratase/lyase enzymes (Figure 4.27).

Whilst the substitution pattern in these C_6C_1 derivatives is generally built up at the C_6C_3 cinnamic acid stage, there exists the possibility of further hydroxylations and/or methylations occurring at the C_6C_1 level prior to chain

Table 4.1 Volatile oils (i): containing principally aromatic compounds

Oil	Plant source	Plant part used	Oil content %	Major constituents with typical (%) composition	Uses, notes
Aniseed (Anise)	*Pimpinella anisum* (Umbelliferae/Apiaceae)	ripe fruit	2–3	anethole (80–90) estragole (1–6)	flavour, carminative, aromatherapy
Star anise	*Illicium verum* (Illiciaceae)	ripe fruit	5–8	anethole (80–90) estragole (1–6)	flavour, carminative; fruits contain substantial amounts of shikimic (about 3%) and quinic acids
Cassia	*Cinnamomum cassia* (Lauraceae)	dried bark, or leaves and twigs	1–2	cinnamaldehyde (70–90) 2-methoxycinnamaldehyde (12)	flavour, carminative; known as cinnamon oil in USA
Cinnamon bark	*Cinnamomum zeylanicum* (Lauraceae)	dried bark	1–2	cinnamaldehyde (70–80) eugenol (1–13) cinnamyl acetate (3–4)	flavour, carminative, aromatherapy
Cinnamon leaf	*Cinnamomum zeylanicum* (Lauraceae)	leaf	0.5–0.7	eugenol (70–95)	flavour
Clove	*Syzygium aromaticum* (*Eugenia caryophyllus*) (Myrtaceae)	dried flower buds	15–20	eugenol (75–90) eugenyl acetate (10–15) β-caryophyllene (3)	flavour, aromatherapy, antiseptic
Fennel	*Foeniculum vulgare* (Umbelliferae/Apiaceae)	ripe fruit	2–5	anethole (50–70) fenchone (10–20) estragole (3–20)	flavour, carminative, aromatherapy

Nutmeg	*Myristica fragrans* (Myristicaceae)	seed	5–16	sabinene (17–28), α-pinene (14–22), β-pinene (9–15), terpinen-4-ol (6–9), myristicin (4–8), elemicin (2)	flavour, carminative, aromatherapy; although the main constituents are terpenoids, most of the flavour comes from the minor aromatic constituents, myristicin, elemicin, etc.; myristicin is hallucinogenic (see page 156)
Wintergreen	*Gaultheria procumbens* (Ericacae) or *Betula lenta* (Betulaceae)	leaves bark	0.7–1.5 0.2–0.6	methyl salicylate (98)	flavour, antiseptic, antirheumatic; prior to distillation, plant material is macerated with water to allow enzymic hydrolysis of glycosides; methyl salicylate is now produced synthetically

Notes:

[a]Volatile or essential oils are usually obtained from the appropriate plant material by steam distillation. If certain components are unstable at these temperatures, however, then other, less harsh techniques, such as expression or solvent extraction, may be employed. These oils, which typically contain a complex mixture of low boiling-point components, are widely used in flavouring, perfumery, and aromatherapy. Only a small number of oils have useful therapeutic properties, e.g. clove and dill, though a wide range of oils is now exploited for aromatherapy. Most of those employed in medicines are simply added for flavouring purposes. Some of the materials are commercially important as sources of chemicals used industrially, e.g. turpentine.

[b]For convenience, the major oils listed are divided into two groups. Those oils which contain principally chemicals that are aromatic in nature and which are derived by the shikimate pathway are given in this table. Oils which are composed predominantly of terpenoid compounds are listed in Table 5.1 on page 200, since they are derived via the methylerythritol phosphate or mevalonate pathways. It must be appreciated that many oils may contain aromatic and terpenoid components, but usually one group predominates. The oil yields and the exact composition of any sample of oil will be variable, depending on the particular plant material used in its preparation. The quality of an oil and its commercial value is dependent on the proportion of the various components. Structures of aromatics are shown in Figure 4.23 (see page 156).

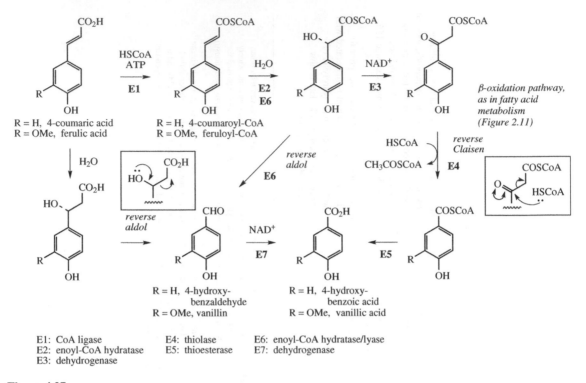

E1: CoA ligase   E4: thiolase   E6: enoyl-CoA hydratase/lyase
E2: enoyl-CoA hydratase E5: thioesterase   E7: dehydrogenase
E3: dehydrogenase

Figure 4.27

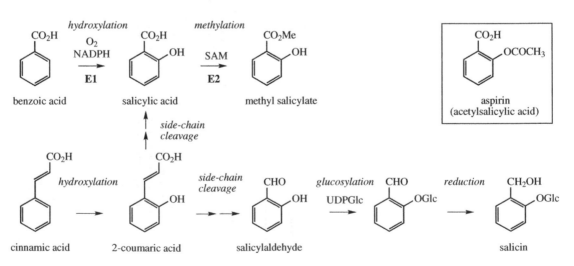

E1: benzoic acid 2-hydroxylase
E2: salicylic acid carboxyl methyltransferase

Figure 4.28

shortening, and this is known in certain examples. **Salicylic acid** (Figure 4.28) is synthesized in microorganisms directly from isochorismic acid (see page 141), but can arise in plants by two other mechanisms. It can be produced by hydroxylation of benzoic acid, or by side-chain cleavage of 2-coumaric acid, which itself is formed by an *ortho*-hydroxylation of cinnamic acid. **Methyl salicylate** is the principal component of oil of wintergreen from *Gaultheria procumbens* (Ericaceae), used for many years for pain relief. It is derived by SAM-dependent methylation of salicylic acid. The salicyl alcohol derivative **salicin**, found in many species of willow (*Salix* species; Salicaceae), is not derived from salicylic acid, but probably via glucosylation of salicylaldehyde and then reduction of the carbonyl (Figure 4.28). Salicin is responsible for the analgesic and antipyretic effects of willow barks, widely used for centuries. It provided the template for synthesis of acetylsalicylic acid (**aspirin**; Figure 4.28), a more effective and widely used pain-killer.

Coumarins

The hydroxylation of cinnamic acids *ortho* to the side-chain as seen in the biosynthesis of salicylic acid (Figure 4.28) is a crucial step in the formation of a group of cinnamic acid lactone derivatives, the **coumarins**. Whilst the direct hydroxylation of the aromatic ring of the cinnamic acids is common, hydroxylation generally involves initially the 4-position *para* to the side-chain, and

subsequent hydroxylations then proceed *ortho* to this substituent (see page 149). In contrast, for the coumarins, hydroxylation of cinnamic acid or 4-coumaric acid can occur *ortho* to the side-chain (Figure 4.29). In the latter case, the 2,4-dihydroxycinnamic acid produced confusingly seems to possess the *meta* hydroxylation pattern characteristic of phenols derived via the acetate pathway. Recognition of the C_6C_3 skeleton should help to avoid this confusion. The two 2-hydroxycinnamic acids then suffer a change in configuration in the side-chain, from the *trans* (*E*) to the less stable *cis* (*Z*) form. Whilst *trans–cis* isomerization would be unfavourable in the case of a single isolated double bond, the fully conjugated system in the cinnamic acids allows this process to occur quite readily, and UV irradiation, e.g. daylight, is sufficient to produce equilibrium mixtures which can be separated. The absorption of energy promotes an electron from the π-orbital to a higher energy state, i.e. the π*-orbital, thus temporarily destroying the double-bond character and allowing rotation. Loss of the absorbed energy then results in re-formation of the double bond, but in the *cis*-configuration. In conjugated systems, the π–π* energy difference is considerably less than with a non-conjugated double bond. Chemical lactonization can occur on treatment with acid. Both the *trans–cis* isomerization and the lactonization are enzyme mediated in nature, and light is not necessary for coumarin biosynthesis. However, the enzyme activities are poorly characterized. Cinnamic acid and 4-coumaric acid give rise to the coumarins **coumarin** and **umbelliferone** respectively (Figure 4.29). Other coumarins with

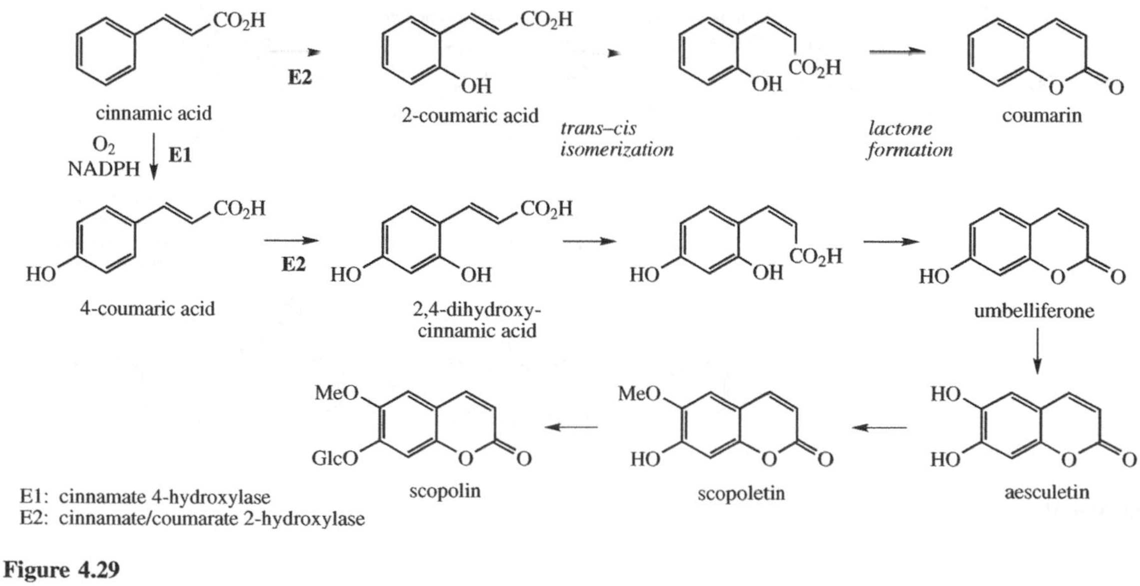

E1: cinnamate 4-hydroxylase
E2: cinnamate/coumarate 2-hydroxylase

Figure 4.29

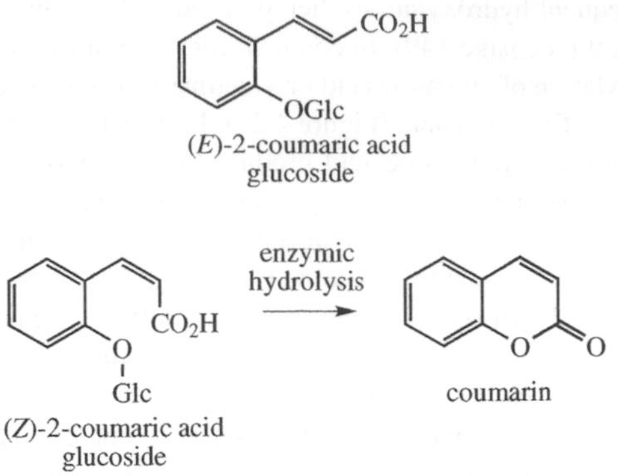

(*E*)-2-coumaric acid
glucoside

enzymic
hydrolysis

(*Z*)-2-coumaric acid
glucoside

coumarin

Figure 4.30

additional oxygen substituents on the aromatic ring, e.g. **aesculetin** (esculetin) and **scopoletin**, appear to be derived by modification of umbelliferone, rather than by a general cinnamic acid to coumarin pathway. This indicates that the hydroxylation *meta* to the existing hydroxyl, discussed above, is a rather uncommon occurrence and restricted to certain substrates.

Coumarins are widely distributed in plants, both in the free form and as glycosides, and are commonly found in families such as the Umbelliferae/Apiaceae and Rutaceae. **Coumarin** itself is found in sweet clover (*Melilotus alba*; Leguminosae/Fabaceae) and contributes to the smell of new-mown hay, though there is evidence that the plants actually contain the glucosides of (*E*)- and

(*Z*)-2-coumaric acid (Figure 4.30), and coumarin is only liberated as a result of enzymic hydrolysis and lactonization through damage to the plant tissues during harvesting and processing (Figure 4.30). If sweet clover is allowed to ferment, then 4-hydroxycoumarin is produced by the action of microorganisms on 2-coumaric acid (Figure 4.31). This can then react with formaldehyde, which is usually present from microbial degradative reactions, combining to give **dicoumarol**. Dicoumarol is a compound with pronounced blood anticoagulant properties which can cause internal bleeding and death of livestock, and is the forerunner of the warfarin group of rodenticides and medicinal anticoagulants [Box 4.4].

Many other natural coumarins have a more complex carbon framework and incorporate extra carbon atoms derived from an isoprene unit (Figure 4.33). For these compounds, there is considerably more evidence available from enzymic and genetic studies than for the simple coumarins. The aromatic ring in umbelliferone is activated at positions *ortho* to the hydroxyl group and can be alkylated by a suitable alkylating agent, in this case dimethylallyl diphosphate. The newly introduced dimethylallyl group in **demethylsuberosin** is then able to cyclize with the phenol group to give **marmesin**. This transformation is catalysed by a cytochrome P-450-dependent monooxygenase and requires cofactors NADPH and molecular oxygen. For many years, the cyclization had been postulated to involve an intermediate epoxide, so that nucleophilic attack of the phenol onto the epoxide group might lead to formation of either five-membered furan

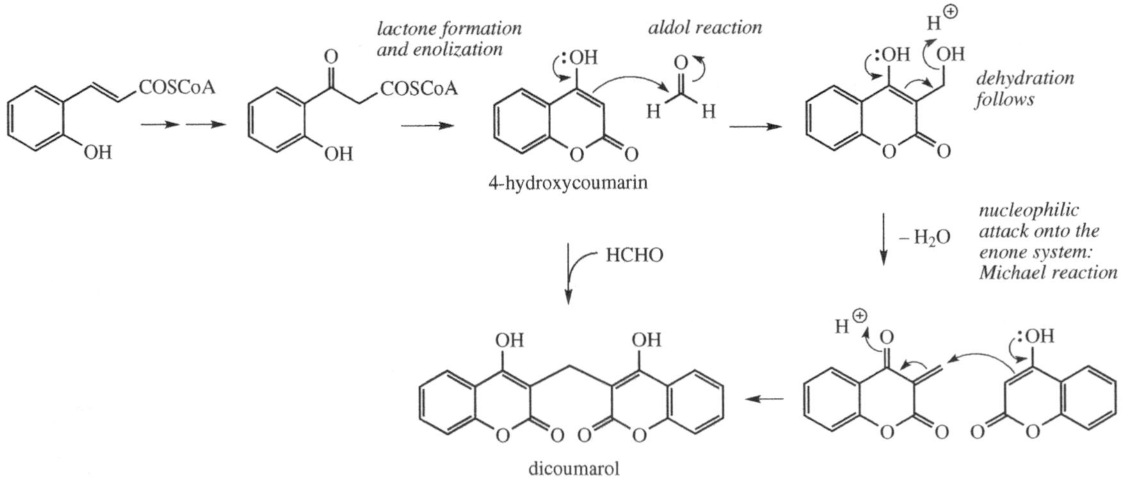

lactone formation and enolization

aldol reaction

dehydration follows

4-hydroxycoumarin

HCHO

nucleophilic attack onto the enone system: Michael reaction

− H₂O

dicoumarol

Figure 4.31

Box 4.4

Dicoumarol and Warfarin

The cause of fatal haemorrhages in animals fed spoiled sweet clover (*Melilotus officinalis*; Leguminosae/Fabaceae) was traced to dicoumarol (bishydroxycoumarin; Figure 4.31). This agent interferes with the effects of vitamin K in blood coagulation (see page 183), the blood loses its ability to clot, and thus minor injuries can lead to severe internal bleeding. Synthetic dicoumarol has been used as an oral blood anticoagulant in the treatment of thrombosis, where the risk of blood clots becomes life threatening. It has been superseded by salts of **warfarin** and in some cases the nitro analogue **acenocoumarol** (**nicoumalone**; Figure 4.32), which are synthetic developments from the natural product. An overdose of warfarin may be countered by injection of vitamin K_1.

Warfarin was initially developed as a rodenticide and has been widely employed for many years as the first-choice agent, particularly for destruction of rats. After consumption of warfarin-treated bait, rats die from internal haemorrhage. Other coumarin derivatives employed as rodenticides include **coumachlor** and **coumatetralyl** (Figure 4.32). In an increasing number of cases, rodents are developing resistance towards warfarin, an ability which has been traced to elevated production of vitamin K by their intestinal microflora. Modified structures **difenacoum, brodifacoum, bromadiolone**, and **flocoumafen** have been found to be more potent than warfarin, and are also effective against rodents that have become resistant to warfarin.

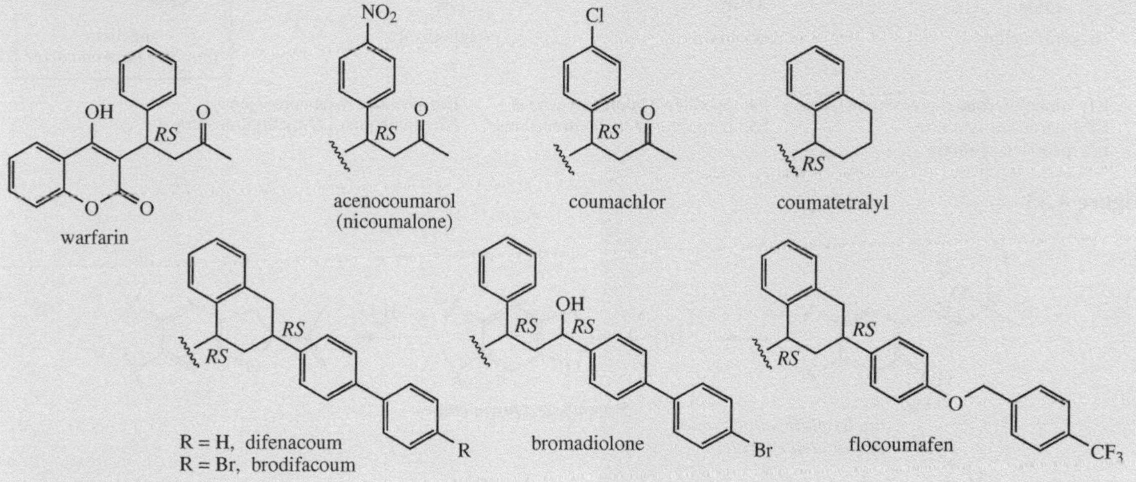

Figure 4.32

or six-membered pyran heterocycles, as commonly encountered in natural products (Figure 4.34). Although the reactions of Figure 4.34 offer a convenient rationalization for cyclization, epoxide intermediates have not been demonstrated in any of the enzymic systems so far investigated; therefore, some direct oxidative cyclization mechanism must operate.

A second cytochrome P-450-dependent monooxygenase enzyme then cleaves off the hydroxyisopropyl fragment (as acetone) from **marmesin** to give the furocoumarin **psoralen**. This does not involve any

hydroxylated intermediate, and cleavage is believed to be initiated by a radical abstraction process (Figure 4.35). Psoralen can act as a precursor for the further substituted furocoumarins **bergapten, xanthotoxin**, and **isopimpinellin** (Figure 4.33), such modifications occurring late in the biosynthetic sequence rather than at the cinnamic acid stage. These transformations are initiated by yet further cytochrome P-450-dependent enzymes, introducing hydroxyls into the aromatic ring. Psoralen, bergapten, and related compounds are termed 'linear' furocoumarins.

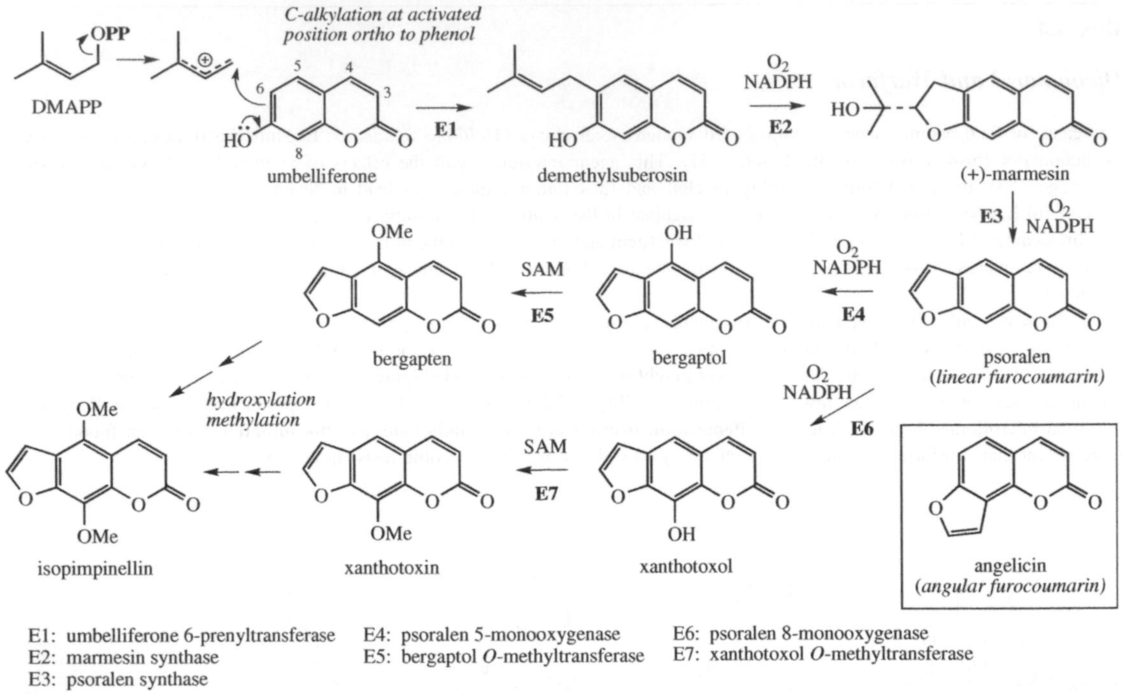
OPP
DMAPP
C-alkylation at activated position ortho to phenol
umbelliferone
E1
demethylsuberosin
O_2 NADPH
E2
(+)-marmesin
E3
psoralen (linear furocoumarin)
E4
bergaptol
SAM
E5
bergapten
E6
xanthotoxol
E7
xanthotoxin
hydroxylation methylation
isopimpinellin
angelicin (angular furocoumarin)
E1: umbelliferone 6-prenyltransferase
E2: marmesin synthase
E3: psoralen synthase
E4: psoralen 5-monooxygenase
E5: bergaptol O-methyltransferase
E6: psoralen 8-monooxygenase
E7: xanthotoxol O-methyltransferase

Figure 4.33

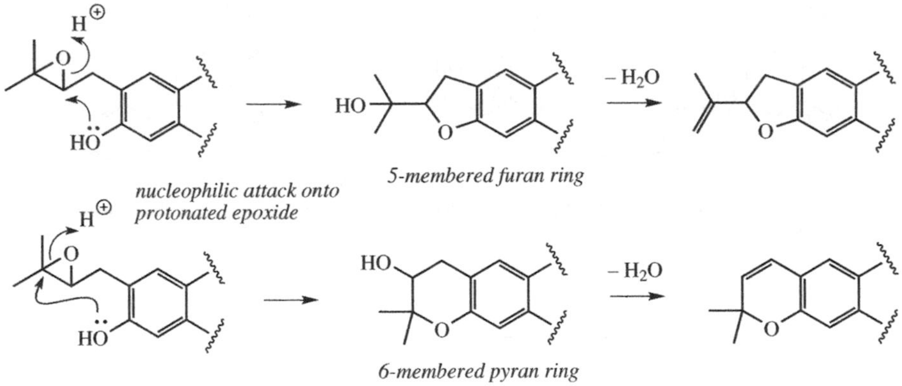
5-membered furan ring
nucleophilic attack onto protonated epoxide
6-membered pyran ring
$- H_2O$

Figure 4.34

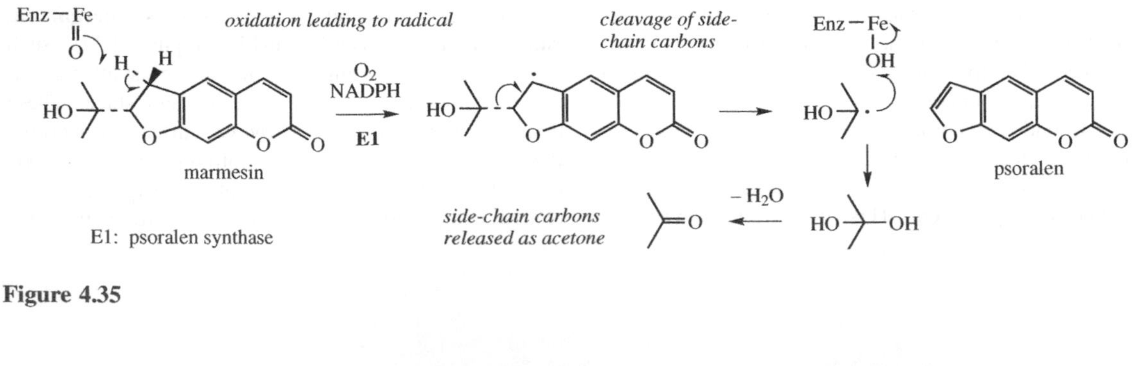
Enz — Fe
oxidation leading to radical
marmesin
O_2 NADPH
E1
cleavage of side-chain carbons
psoralen
side-chain carbons released as acetone
$- H_2O$
E1: psoralen synthase

Figure 4.35

'Angular' furocoumarins, e.g. **angelicin** (Figure 4.33), can arise by a similar sequence of reactions, but these involve initial dimethylallylation at the alternative position *ortho* to the phenol. An isoprene-derived furan ring system has already been noted in the formation of khellin (see page 112), though the aromatic ring to which it was fused was in that case a product of the acetate pathway. Linear furocoumarins (**psoralens**) can be troublesome to humans since they can cause photosensitization towards UV light, resulting in sunburn or serious blistering. Used medicinally, this effect may be valuable in promoting skin pigmentation and treating psoriasis [Box 4.5].

Box 4.5

Psoralens

Psoralens are linear furocoumarins which are widely distributed in plants, but are particularly abundant in the Umbelliferae/Apiaceae and Rutaceae. The most common examples are psoralen, bergapten, xanthotoxin, and isopimpinellin (Figure 4.33). Plants containing psoralens have been used internally and externally to promote skin pigmentation and suntanning. Bergamot oil obtained from the peel of *Citrus aurantium* ssp. *bergamia* (Rutaceae) (see page 165) can contain up to 5% bergapten and is frequently used in external suntan preparations. The psoralen, because of its extended chromophore, absorbs in the near UV and allows this radiation to stimulate formation of melanin pigments (see page 147).

Methoxsalen (xanthotoxin; 8-methoxypsoralen; Figure 4.36), a constituent of the fruits of *Ammi majus* (Umbelliferae/Apiaceae), is used medically to facilitate skin repigmentation where severe blemishes exist (vitiligo). An oral dose of methoxsalen is followed by long-wave UV irradiation, though such treatments must be very carefully regulated to minimize the risk of burning, cataract formation, and the possibility of causing skin cancer. The treatment is often referred to as PUVA (psoralen + UV-A). PUVA is also of value in the treatment of psoriasis, a widespread condition characterized by proliferation of skin cells. Similarly, methoxsalen is taken orally, prior to UV treatment. Reaction with psoralens inhibits DNA replication and reduces the rate of cell division. Because of their planar nature, psoralens intercalate into DNA, and this enables a UV-initiated cycloaddition reaction between pyrimidine bases (primarily thymine) in DNA and the furan ring of psoralens (Figure 4.36). A second cycloaddition can then occur, this time involving the pyrone ring, leading to interstrand cross-linking of the nucleic acid.

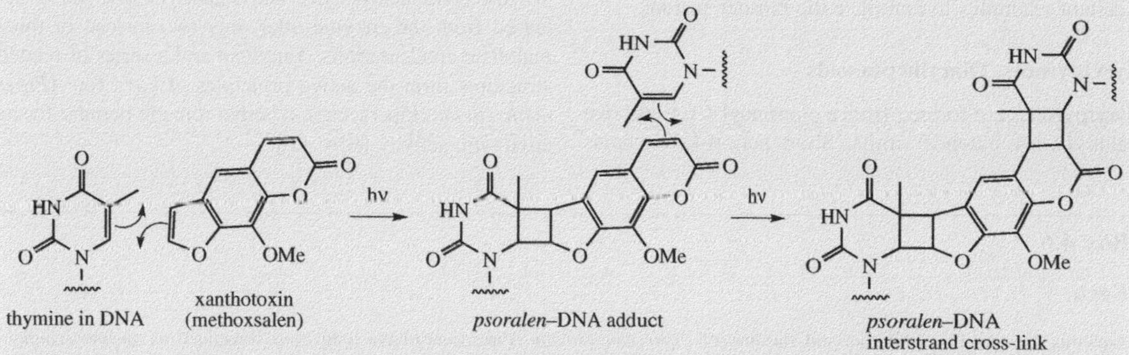

thymine in DNA   xanthotoxin (methoxsalen)   *psoralen–DNA adduct*   *psoralen–DNA interstrand cross-link*

Figure 4.36

A troublesome extension of these effects can arise from the handling of plants which contain significant levels of furocoumarins. Celery (*Apium graveolens*; Umbelliferae/Apiaceae) is normally free of such compounds, but fungal infection with the natural parasite *Sclerotinia sclerotiorum* induces the synthesis of furocoumarins (xanthotoxin and others) as a response to the infection. Field workers handling these infected plants may become very sensitive to UV light and suffer from a form of sunburn termed photophytodermatitis. Infected parsley (*Petroselinum crispum*) can give similar effects. Handling of rue (*Ruta graveolens*; Rutaceae) or giant hogweed (*Heracleum mantegazzianum*; Umbelliferae/Apiaceae), which naturally contain significant amounts of psoralen, bergapten, and xanthotoxin, can cause similar unpleasant reactions, or more commonly rapid blistering by direct contact with the sap. The giant hogweed can be particularly dangerous. Individuals vary in their sensitivity towards furocoumarins; some are unaffected, whilst others tend to become sensitized by an initial exposure and then develop the allergic response on subsequent exposures.

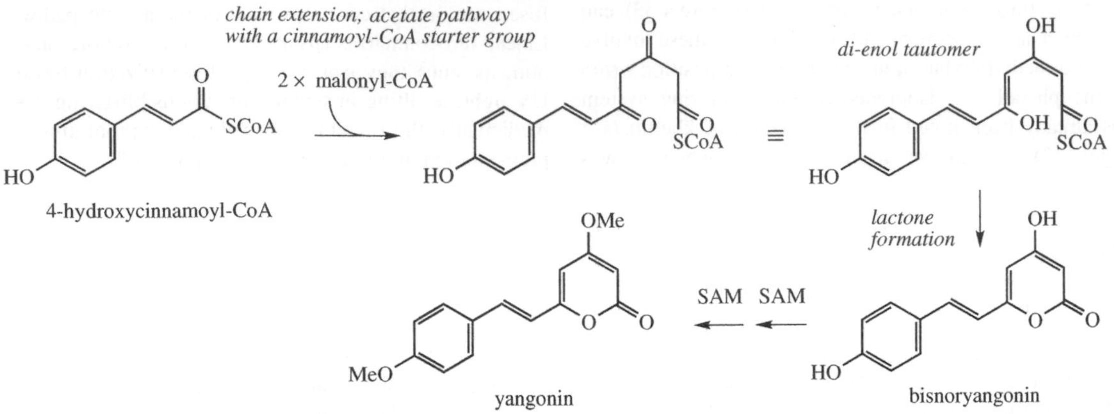

Figure 4.37

AROMATIC POLYKETIDES

Cinnamic acids, as their coenzyme A esters, may also function as starter units for chain extension with malonyl-CoA units via PKSs, thus combining elements of the shikimate and acetate pathways (see page 116). Most commonly, three C_2 units are added via malonate, giving rise to flavonoids and stilbenes (page 116). Products formed from a cinnamoyl-CoA starter plus one or two C_2 units from malonyl-CoA are rarer, but provide excellent examples to complete the broader picture.

Styrylpyrones, Diarylheptanoids

Styrylpyrones are formed from a cinnamoyl-CoA and two malonyl-CoA extender units. Short poly-β-keto chains frequently cyclize to form a lactone derivative (compare triacetic acid lactone, page 101). Thus, Figure 4.37 shows the proposed derivation of **yangonin** via cyclization of the di-enol tautomer of the polyketide formed from 4-hydroxycinnamoyl-CoA. Two methylation reactions complete the sequence. Indeed, the suggested intermediate **bisnoryangonin** is often produced in small amounts as a derailment product when recombinant chalcone synthase enzymes (see page 169) are incubated with 4-hydroxycinnamoyl-CoA and malonyl-CoA, being released from the enzyme after only two instead of three malonate condensations. Yangonin and a series of related structures form the active principles of kava root (*Piper methysticum*; Piperaceae), a herbal remedy popular for its anxiolytic activity [Box 4.6].

Box 4.6

Kava

Aqueous extracts from the root and rhizome of *Piper methysticum* (Piperaceae) have long been consumed as an intoxicating beverage by the peoples of Pacific islands comprising Polynesia, Melanesia, and Micronesia, and the name kava or kava-kava referred to this drink. In herbal medicine, the dried root and rhizome is now described as kava, and it is used for the treatment of anxiety, nervous tension, agitation, and insomnia. The pharmacological activity is associated with a group of styrylpyrone derivatives termed kavalactones, good-quality roots containing 5–8% kavalactones. At least 18 kavalactones have been characterized, the six major ones being four enolides kawain, methysticin and their corresponding dihydro derivatives, and the two dienolides yangonin and demethoxyyangonin (Figure 4.38). The enolides have a reduced pyrone ring and a chiral centre. Clinical trials have indicated kava extracts to be effective as an anxiolytic, the kavalactones also displaying anticonvulsive, analgesic, and central muscle-relaxing action. Several of these compounds have been shown to have an effect on neurotransmitter systems, including those involving glutamate, GABA, dopamine, and serotonin.

The safety of kava products has been questioned following hepatotoxic side-effects: severe liver damage in patients taking kava extracts has been reported. It is suggested that the traditional usage of aqueous extracts appears to produce little or no hepatotoxicity, and the problems stem from using commercial organic solvent extracts. Water extracts also contain glutathione (see page 125), which appears to provide protection against toxicity through reaction with the lactone ring. A number of human cytochrome P-450-dependent enzymes are also known to be inhibited by some of the kavalactones.

Box 4.6 (continued)

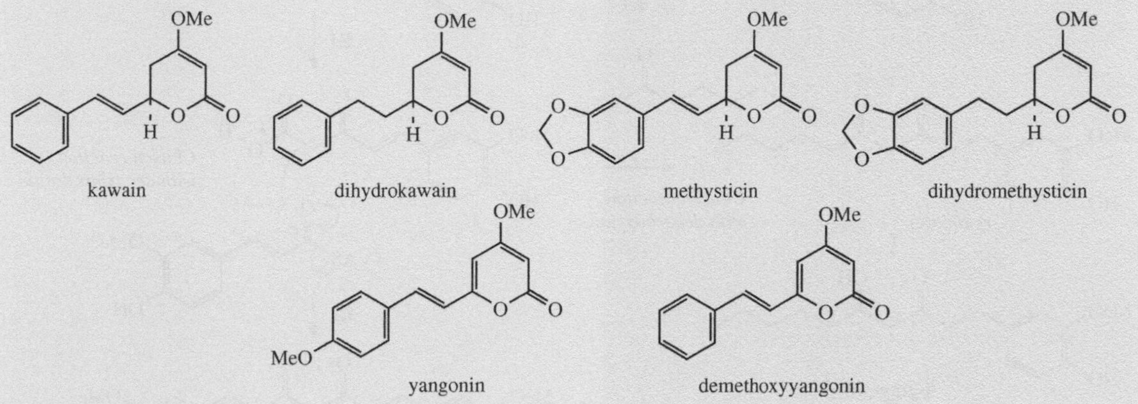

kawain

dihydrokawain

methysticin

dihydromethysticin

yangonin

demethoxyyangonin

Figure 4.38

Chain extension with a single malonyl-CoA unit is exemplified by **raspberry ketone** (Figure 4.39), the aroma constituent of ripe raspberry fruits (*Rubus idaeus*; Rosaceae). However, it is immediately apparent that the side-chain in this product is actually one carbon shorter than predicted; this PKS also catalyses hydrolysis and decarboxylation. A subsequent reduction generates raspberry ketone. Piperoyl-CoA is implicated in the biosynthesis of the alkaloid piperine (see page 328); this retains the two carbon atoms from the malonate extender.

Characteristic constituents of turmeric (*Curcuma longa*; Zingiberaceae), an essential spice ingredient in curries, are a group of curcuminoids, diarylheptanoids responsible for the yellow colour. The principle component, **curcumin** (Figure 4.40), is now attracting considerable interest because of its anti-inflammatory, antiulcer, antitumour, and cancer-preventative properties. Curcuminoids are produced by a type III PKS-like enzyme (see page 68); and as with the raspberry ketone example, a single extra carbon is incorporated from malonate. In this

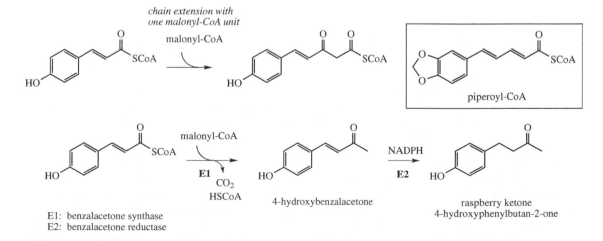

chain extension with
one malonyl-CoA unit

malonyl-CoA

piperoyl-CoA

malonyl-CoA

E1

CO_2
HSCoA

4-hydroxybenzalacetone

NADPH

E2

raspberry ketone
4-hydroxyphenylbutan-2-one

E1: benzalacetone synthase
E2: benzalacetone reductase

Figure 4.39

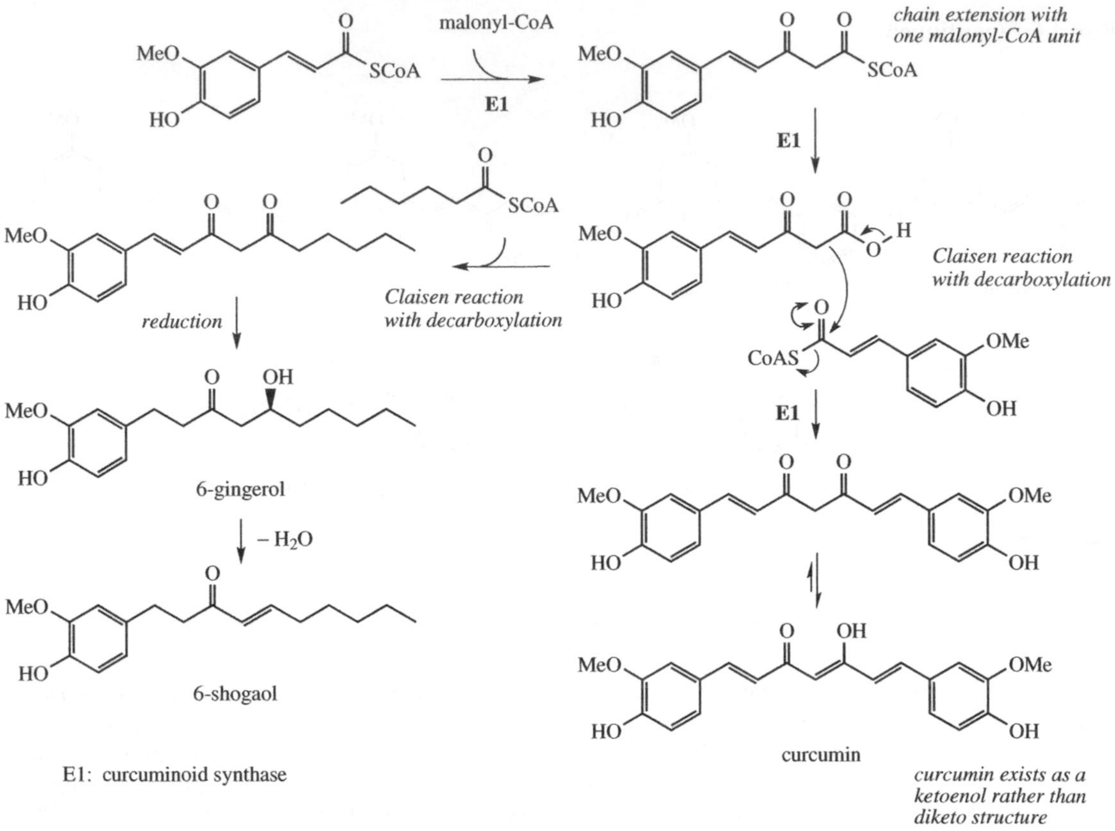

Figure 4.40

series of compounds, a further cinnamoyl group is also added. It seems likely that this unit is incorporated via a Claisen reaction with the β-ketoacid formed by hydrolysis of this β-keto ester. Decarboxylation is part of the Claisen process as with normal malonate chain extension. Related to the curcuminoids, and in some plants often co-occurring with them, are compounds such as the gingerols and shogaols. These are presumably derived by a variant of the curcuminoid pathway that utilizes a fatty acyl thioester rather than a cinnamoyl thioester in the further Claisen reaction (Figure 4.40). Chain lengths in this fatty acyl group vary; thus, **6-gingerol** has incorporated a six-carbon fragment, whilst other analogues, e.g. 8-gingerol and 10-gingerol, employ 8- or 10-carbon fatty acid esters. 6-Gingerol is a major pungent component in ginger root (*Zingiber officinale*; Zingiberaceae), widely employed as a flavouring and spice; the distinctive aroma of ginger arises from volatile sesquiterpenes (see page 211). Shogaols are dehydrated analogues of gingerols, and tend to be the predominant pungent agents in dried root,

so may be formed during the drying process. Ginger has useful anti-inflammatory activity due to the gingerols.

Flavonoids and Stilbenes

Flavonoids and stilbenes are products from a cinnamoyl-CoA starter unit, with chain extension using three molecules of malonyl-CoA. Reactions are catalysed by a type III PKS enzyme (see page 68). These enzymes do not utilize ACPs, but instead employ coenzyme A esters and have a single active site to perform the necessary series of reactions, e.g. chain extension, condensation, cyclization. Chain extension of 4-hydroxycinnamoyl-CoA with three molecules of malonyl-CoA gives initially a polyketide (Figure 4.41) which can be folded in two different ways. These allow aldol or Claisen-like reactions to occur, generating aromatic rings, as already seen in Chapter 3 (see page 100). Stilbenes (e.g. **resveratrol**) or chalcones (e.g. **naringenin-chalcone**) are the end-products formed

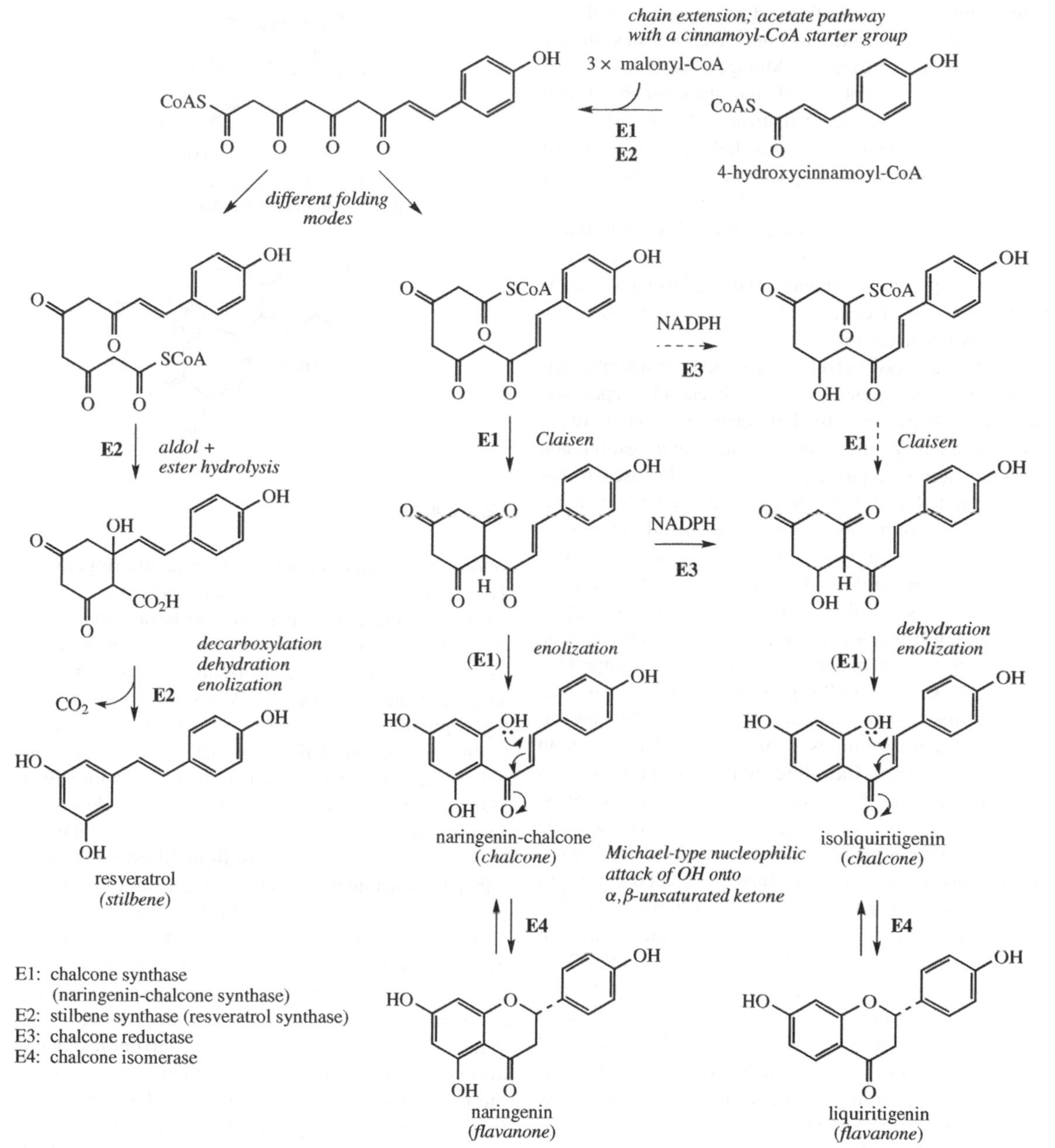

Figure 4.41

by enzymes stilbene synthase and chalcone synthase respectively. Both structures nicely illustrate the different characteristic oxygenation patterns in the two aromatic rings derived from the acetate or shikimate pathways. Though cyclizing the same substrate in different ways, chalcone synthase and stilbene synthase enzymes are found to share some 75–90% amino acid sequence identity. Whilst chalcone synthase enzymes appear to be found in all plants, stilbene synthase activity has a much more restricted distribution. With the **stilbenes**, the terminal ester function is no longer present; therefore, hydrolysis and decarboxylation have also taken place during this transformation. Carboxylated stilbenes are rare in nature and are not intermediates in the stilbene synthase reaction. Mutagenesis studies have confirmed that hydrolysis of the thioester bond and decarboxylation precede aromatization by the necessary dehydration/enolization reactions. Indeed, it appears that it is the thioesterase activity component of the stilbene synthase enzyme that is primarily responsible for the stilbene-type cyclization through an aldol reaction rather than the chalcone-type cyclization via a Claisen reaction. Through appropriate amino acid replacements it is possible to convert chalcone synthase into proteins with stilbene synthase activity.

The stilbene **resveratrol** has assumed considerable relevance in recent years as a constituent of grapes and wine, as well as other food products, with antioxidant, anti-inflammatory, inhibition of platelet aggregation, and protective activity against cancer and cardiovascular diseases. Coupled with the cardiovascular benefits of moderate amounts of alcohol, and the beneficial antioxidant effects of flavonoids (see page 171), red wine has now emerged as an unlikely but most acceptable medicinal agent. Resveratrol is available as a dietary supplement, though it is not currently regarded as a therapeutic agent. Commercial material is isolated from grapeskins or Japanese knotweed (*Polygonum cuspidatum*, syn. *Fallopia japonica*; Polygonaceae). Resveratrol also has antifungal properties and is formed as a phytoalexin in some plant species, including grapes and peanuts, as a result of infection. It is now feasible to increase resistance to fungi by expressing a stilbene synthase gene in a plant that normally does not produce resveratrol. The precursors normally destined for chalcone biosynthesis are thus channelled into stilbene synthesis. Another natural stilbene of medicinal interest is **combretastatin A-4** (Figure 4.42), isolated from bark of the tree *Combreta caffrum* (Solanaceae). Apart from the substitution pattern, this compound differs from resveratrol in that the double bond has the Z-configuration. Combretastatin A-4 is a potent cytotoxic agent which strongly inhibits polymerization of tubulin by binding to the colchicine site (see page 361); the structural similarity to colchicine is readily appreciated (Figure 4.42). The water-soluble phosphate pro-drug is currently showing considerable promise in clinical trials against various solid tumours; it appears to reduce blood flow in the tumour markedly.

Chalcones act as precursors for a vast range of **flavonoid** derivatives found throughout the plant kingdom. Enzymic and genetic studies for this group of compounds are now well advanced. Most flavonoids contain a six-membered heterocyclic ring, formed by Michael-type nucleophilic attack of a phenol group onto the unsaturated

R = H, combretastatin A-4
R = **P**, combretastatin A-4 phosphate

colchicine

Figure 4.42

ketone to give a **flavanone**, e.g. **naringenin** (Figure 4.41). This isomerization can occur chemically, with acid conditions favouring the flavanone and basic conditions the chalcone. In nature, however, the reaction is enzyme catalysed and stereospecific, resulting in formation of a single flavanone enantiomer. Many flavonoid structures, e.g. **liquiritigenin**, have lost one of the hydroxyl groups, so that the acetate-derived aromatic ring has a resorcinol oxygenation pattern rather than the phloroglucinol system. This modification has been tracked down to the action of a chalcone reductase enzyme concomitant with the chalcone synthase; thus, **isoliquiritigenin** is produced rather than naringenin-chalcone. Reduction of the linear poly-β-ketoester or of the cyclic trione might be formulated; evidence points to the cyclic trione as substrate. The last stage(s) in chalcone formation through enolization or dehydration/enolization are probably spontaneous and non-enzymic. Flavanones can then give rise to many variants on this basic skeleton, e.g. **flavones, flavonols, anthocyanidins**, and **catechins** (Figure 4.43). Most of the enzyme systems show rather broad substrate specificity and convert compounds with different oxygenation patterns in the aryl substituent. Modifications to the hydroxylation patterns in either of the two aromatic rings may occur, generally at the flavanone or dihydroflavonol stage, and methylation, glycosylation, and dimethylallylation are also possible, increasing the range of compounds enormously. A high proportion of flavonoids occur naturally as water-soluble glycosides.

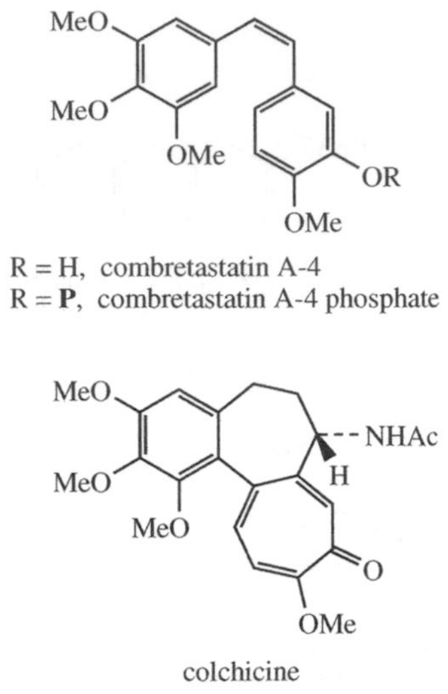

Considerable quantities of flavonoids are consumed daily in our vegetable diet, so adverse biological effects on man are not particularly intense. Indeed, there

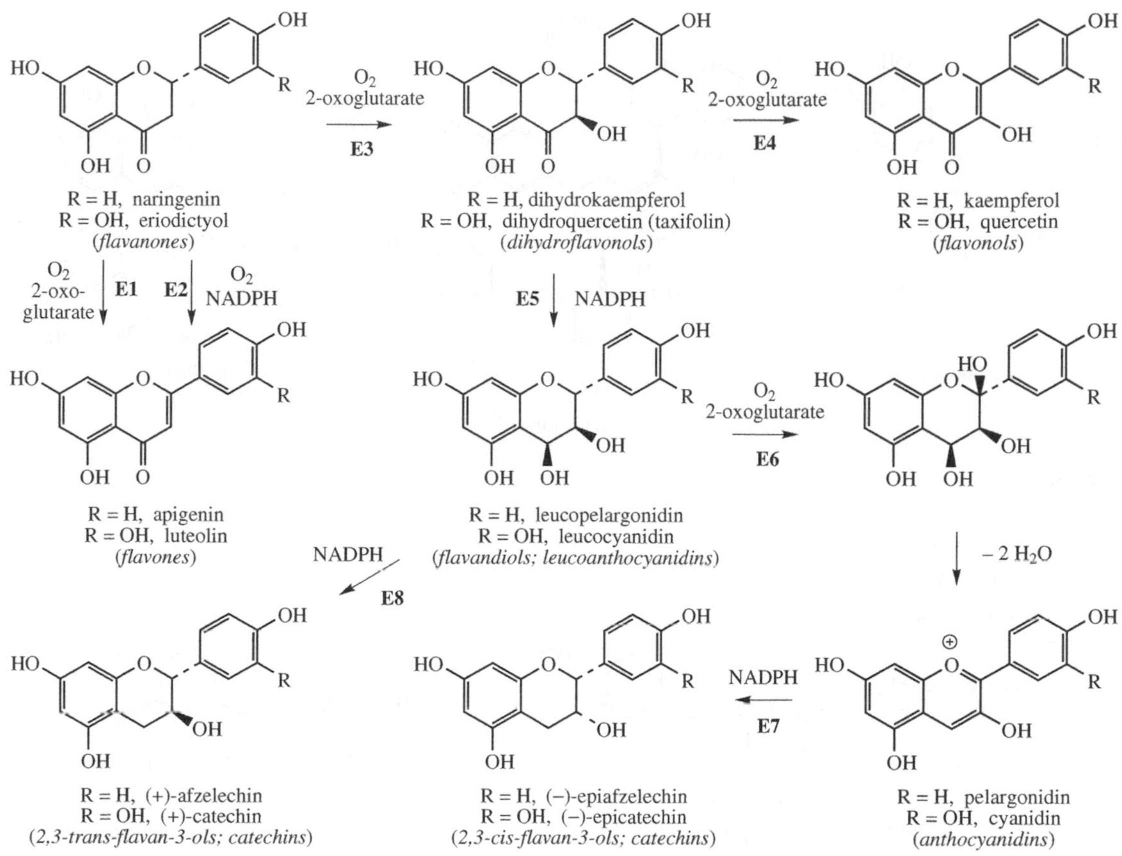

R = H, naringenin
R = OH, eriodictyol
(*flavanones*)

R = H, dihydrokaempferol
R = OH, dihydroquercetin (taxifolin)
(*dihydroflavonols*)

R = H, kaempferol
R = OH, quercetin
(*flavonols*)

R = H, apigenin
R = OH, luteolin
(*flavones*)

R = H, leucopelargonidin
R = OH, leucocyanidin
(*flavandiols; leucoanthocyanidins*)

R = H, (+)-afzelechin
R = OH, (+)-catechin
(*2,3-trans-flavan-3-ols; catechins*)

R = H, (−)-epiafzelechin
R = OH, (−)-epicatechin
(*2,3-cis-flavan-3-ols; catechins*)

R = H, pelargonidin
R = OH, cyanidin
(*anthocyanidins*)

E1: flavone synthase I
E2: flavone synthase II
E3: flavanone 3-hydroxylase
E4: flavonol synthase

E5: dihydroflavanol 4-reductase
E6: anthocyanidin synthase (leucoanthocyanidin dioxygenase)
E7: anthocyanidin reductase
E8: leucoanthocyanidin reductase

Figure 4.43

is growing belief that some flavonoids are particularly beneficial, acting as antioxidants and giving protection against cardiovascular disease, certain forms of cancer, and, it is claimed, age-related degeneration of cell components. Their polyphenolic nature enables them to scavenge injurious radicals such as superoxide and hydroxyl radicals. **Quercetin** in particular is almost always present in substantial amounts in plant tissues; it is a powerful antioxidant, chelating metals, scavenging radicals, and preventing oxidation of low density lipoprotein. Flavonoids in red wine (**quercetin, kaempferol**, and anthocyanidins) and in tea (**catechins** and catechin gallate esters) are also demonstrated to be effective antioxidants. Green tea contains **epigallocatechin gallate** (Figure 4.44), a particularly effective agent. Flavonoids contribute to plant

colours: yellows from chalcones and flavonols, and reds, blues, and violets from anthocyanidins. Even the colourless materials, e.g. flavones, absorb strongly in the UV and are detectable by insects, probably aiding flower pollination. Catechins form small polymers (oligomers), the **condensed tannins**, e.g. the epicatechin trimer (Figure 4.44), which contribute astringency to our foods and drinks, as do the simpler gallotannins (see page 138), and are commercially important for tanning leather. **Theaflavins**, antioxidants found in fermented tea (see page 415), are dimeric catechin structures in which oxidative processes have led to formation of a seven-membered tropolone ring.

The flavonol glycoside **rutin** (Figure 4.45) from buckwheat (*Fagopyrum esculentum*; Polygonaceae) and rue

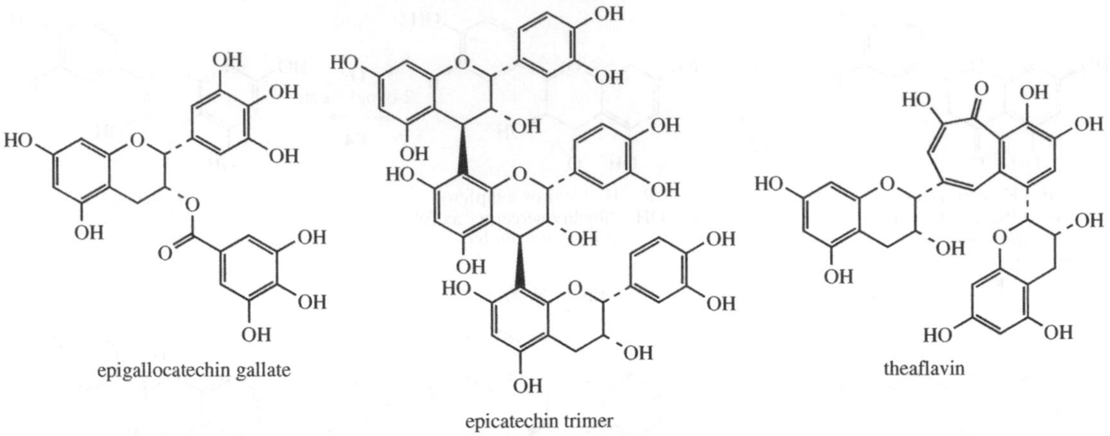

epigallocatechin gallate

epicatechin trimer

theaflavin

Figure 4.44

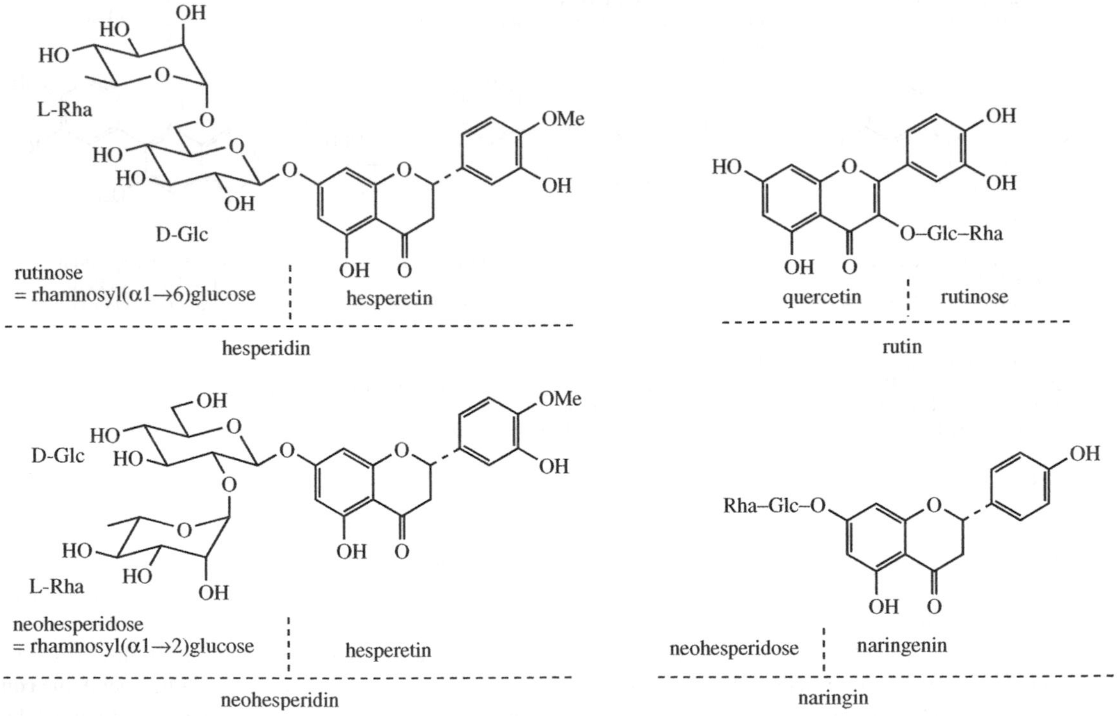

L-Rha

D-Glc

rutinose
= rhamnosyl(α1→6)glucose

hesperetin

hesperidin

quercetin

rutinose

rutin

D-Glc

L-Rha

neohesperidose
= rhamnosyl(α1→2)glucose

hesperetin

neohesperidin

neohesperidose

naringenin

naringin

Figure 4.45

(*Ruta graveolens*; Rutaceae), and the flavanone glycoside **hesperidin** from *Citrus* peels have been included in dietary supplements as vitamin P and claimed to be of benefit in treating conditions characterized by capillary bleeding; their therapeutic efficacy is far from conclusive. Useful anti-inflammatory properties have been

demonstrated, however. **Neohesperidin** (Figure 4.45) from bitter orange (*Citrus aurantium*; Rutaceae) and **naringin** from grapefruit peel (*Citrus paradisi*) are intensely bitter flavanone glycosides. It has been found that conversion of these compounds into **dihydrochalcones** by hydrogenation in alkaline solution (Figure 4.46) produces

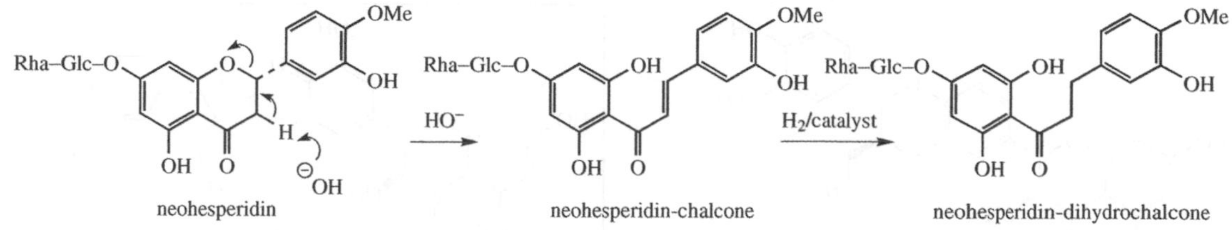

Figure 4.46

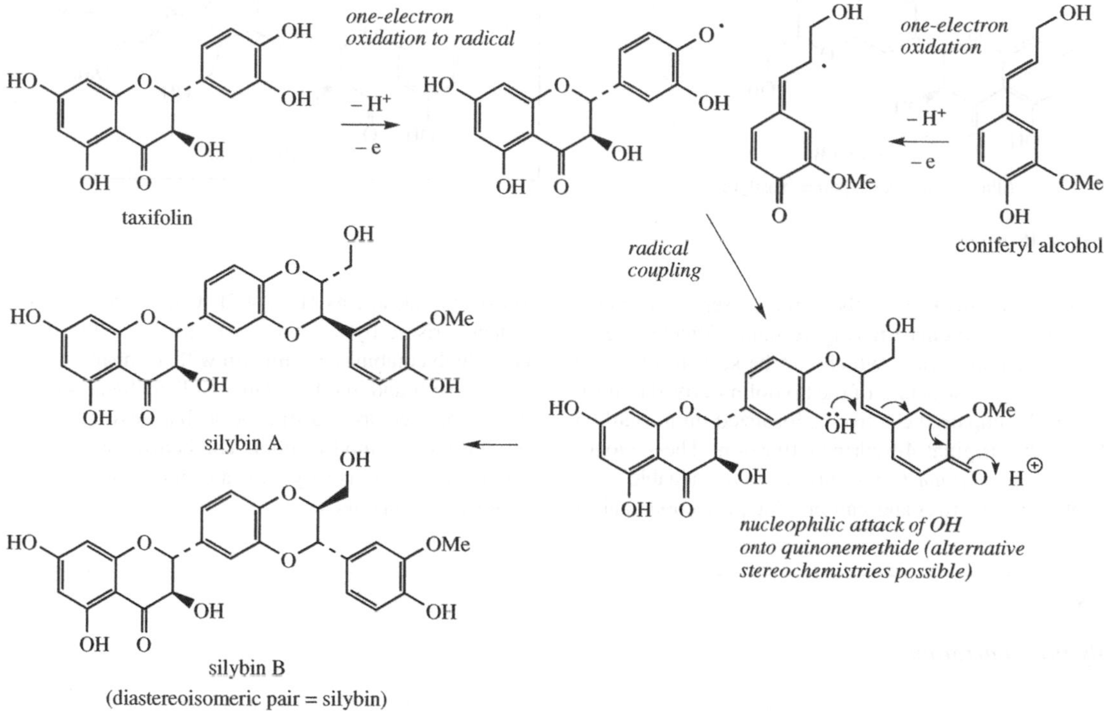

Figure 4.47

a remarkable change to their taste, and the products are now intensely sweet, being some 300–1000 times as sweet as sucrose. Neohesperidin-dihydrochalcone is used as a non-sugar sweetening agent.

Flavonolignans

An interesting combination of flavonoid and lignan-like structures is found in a group of compounds called **flavonolignans**. They arise by oxidative coupling processes between a flavonoid and a phenylpropanoid, the latter usually coniferyl alcohol. Thus, the dihydroflavonol **taxifolin** (dihydroquercetin) through one-electron oxidation may provide a radical, which may combine with the

radical generated from **coniferyl alcohol** (Figure 4.47). This would lead to an adduct which could cyclize by attack of the phenol nucleophile onto the quinone methide system provided by coniferyl alcohol. The product would be **silybin**, found in *Silybum marianum* (Compositae/Asteraceae) as a mixture of two *trans* diastereoisomers, silybins A and B, reflecting a lack of stereospecificity for the original radical coupling. In addition, the regioisomer **isosilybin** (Figure 4.48), again a mixture of *trans* diastereoisomers, is also found in *Silybum*. In keeping with the postulated biosynthetic origin, mixtures of silybin and isosilybin may be obtained by incubating taxifolin and coniferyl alcohol with a peroxidase preparation

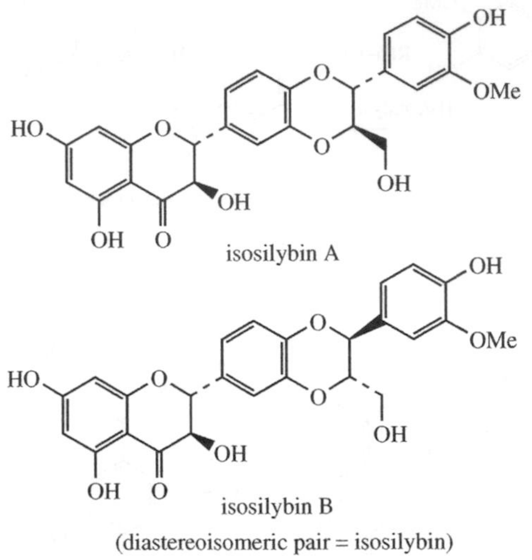

isosilybin A

isosilybin B

(diastereoisomeric pair = isosilybin)

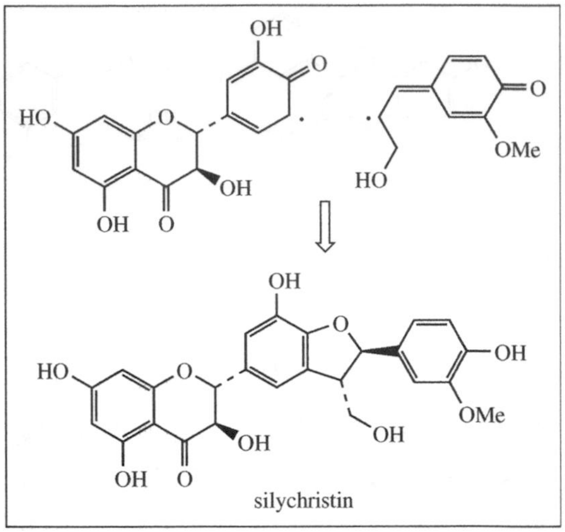

silychristin

Figure 4.48

from *Silybum marianum* cells in the presence of hydrogen peroxide. **Silychristin** (Figure 4.48) demonstrates a further structural variant which can be seen to originate from a resonance structure of the taxifolin-derived radical, in which the unpaired electron is localized on the carbon *ortho* to the original 4-hydroxyl function. The flavonolignans from *Silybum* (milk thistle) have valuable anti-hepatotoxic properties and can provide protection against liver-damaging agents [Box 4.7]. Coumarinolignans are products arising by a similar oxidative coupling mechanism which combines a coumarin with a cinnamyl alcohol and may be found in other plants. Stilbenolignans are also known; these combine a stilbene such as resveratrol (page 169) with a cinnamyl alcohol. The benzodioxane ring as seen in silybin and isosilybin is a characteristic feature of many such compounds.

Box 4.7

Silybum marianum

Silybum marianum (Compositae/Asteraceae) is a biennial thistle-like plant (milk thistle) common in the Mediterranean area of Europe. The seeds yield 1.5–3% of flavonolignans collectively termed silymarin. This mixture contains mainly silybin (Figure 4.47), together with silychristin (Figure 4.48), isosilybin (Figure 4.48), and related compounds. Both silybin and isosilybin are equimolar mixtures of two *trans* diastereoisomers. Commercial extracts typically contain about 80% flavonolignans. *Silybum marianum* is widely used in traditional European medicine, the fruits being used to treat a variety of hepatic and other disorders. Silymarin has been shown to protect animal livers against the damaging effects of carbon tetrachloride, thioacetamide, drugs such as paracetamol, and the toxins α-amanitin and phalloidin found in the death cap fungus (*Amanita phalloides*) (see page 434). Silymarin may be used in many cases of liver disease and injury, including hepatitis, cirrhosis, and jaundice, though it still remains peripheral to mainstream medicine. It can offer particular benefit in the treatment of poisoning by the death cap fungus. These agents appear to have two main modes of action. They act on the cellular membrane of hepatocytes, inhibiting absorption of toxins; second, because of their phenolic nature, they can act as antioxidants and scavengers for radicals. Such radicals originate from liver detoxification of foreign chemicals and can cause liver damage. Derivatives of silybin with improved water solubility and/or bioavailability have been developed, e.g. the bis-hemisuccinate and a phosphatidylcholine complex.

Isoflavonoids

The **isoflavonoids** form a quite distinct subclass of flavonoid compound, being structural variants in which the shikimate-derived aromatic ring has migrated to the adjacent carbon of the heterocycle. This rearrangement process is brought about by a cytochrome P-450-

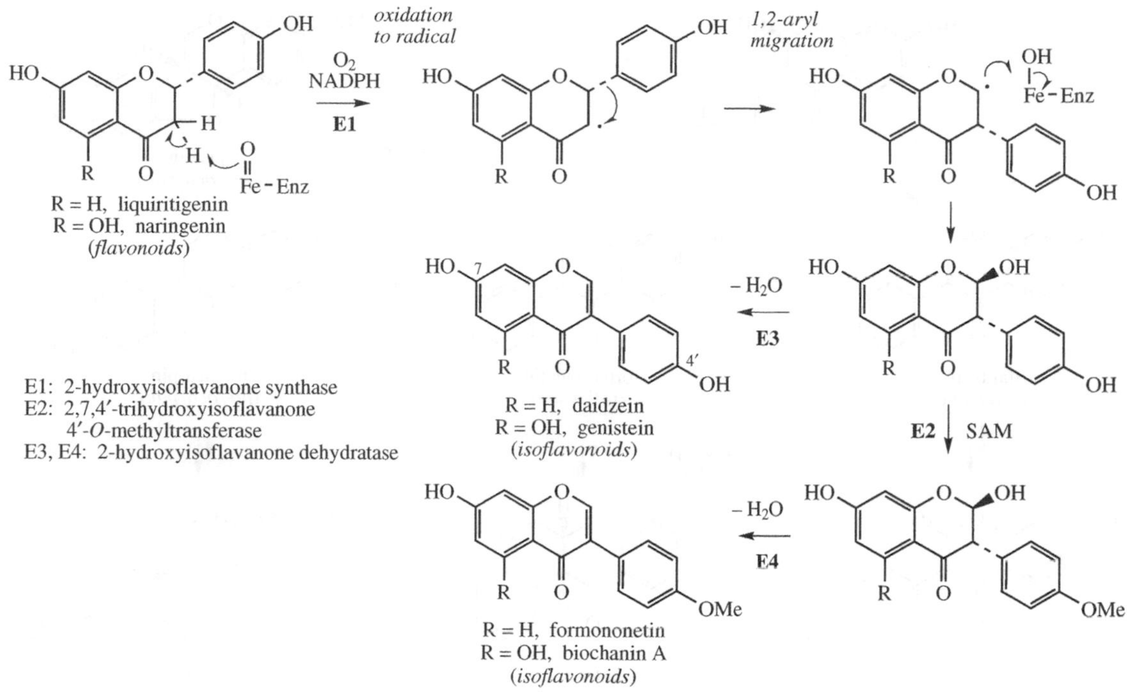

Figure 4.49

E1: 2-hydroxyisoflavanone synthase
E2: 2,7,4′-trihydroxyisoflavanone
     4′-O-methyltransferase
E3, E4: 2-hydroxyisoflavanone dehydratase

dependent enzyme requiring NADPH and O_2 cofactors, which transforms the flavanones **liquiritigenin** or **naringenin** into the isoflavones **daidzein** or **genistein** respectively via intermediate hydroxyisoflavanones (Figure 4.49). A radical mechanism has been proposed. Formation of isoflavones **formononetin** and **biochanin A** involves methylation at the hydroxyisoflavanone stage; the dehydratase enzymes appear to be specific for either hydroxy or methoxy substrates. The rearrangement step is quite rare in nature, and although flavonoids are found throughout the plant kingdom, isoflavonoids are almost entirely restricted to the Leguminosae/Fabaceae plant family. Nevertheless, many hundreds of different isoflavonoids have been identified, and structural complexity is brought about by hydroxylation and alkylation reactions, varying the oxidation level of the heterocyclic ring, or forming additional heterocyclic rings. Some of the many variants are shown in Figure 4.50; though no details are shown, the transformations, enzymes involved, and most of the genes have been characterized. **Pterocarpans**, e.g. **medicarpin** from lucerne (*Medicago sativa*) and **pisatin** from pea (*Pisum sativum*), and **isoflavans**, e.g. **vestitol** from lucerne, have antifungal activity and form part of the natural defence mechanism against fungal attack in these plants. Simple **isoflavones** (such as **daidzein**) and **coumestans** (such

as **coumestrol**) from lucerne and clovers (*Trifolium* species) have sufficient oestrogenic activity to affect the reproduction of grazing animals seriously and are termed **phyto-oestrogens** [Box 4.8]. These planar molecules undoubtedly mimic the shape and polarity of the steroid hormone estradiol (see page 291). The consumption of legume fodder crops by animals must, therefore, be restricted, or low isoflavonoid-producing strains have to be selected. Isoflavonoids in the human diet, e.g. from soya (*Glycine max*) products, are believed to give some protection against oestrogen-dependent cancers such as breast cancer, by restricting the availability of the natural hormone. In addition, they are employed as dietary oestrogen supplements for the reduction of menopausal symptoms, in a similar way to hormone replacement therapy (HRT; see page 294).

The **rotenoids** take their name from the first known example **rotenone**, and are formed by ring cyclization of a methoxyisoflavone (Figure 4.51). Rotenone itself contains a C_5 isoprene unit (as do virtually all the natural rotenoids) introduced via dimethylallylation of **demethylmunduserone**. The isopropenylfurano system of rotenone and the dimethylpyrano of **deguelin** are formed via rotenonic acid (Figure 4.51) without any detectable epoxide or hydroxy intermediates (compare furocoumarins,

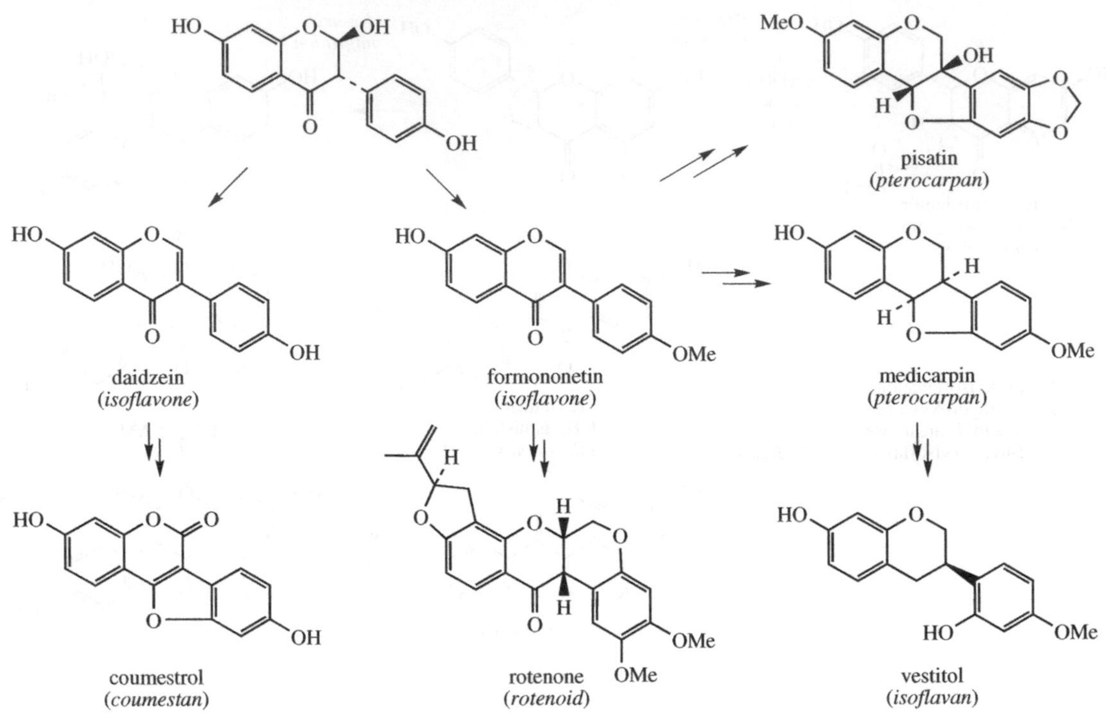

Figure 4.50

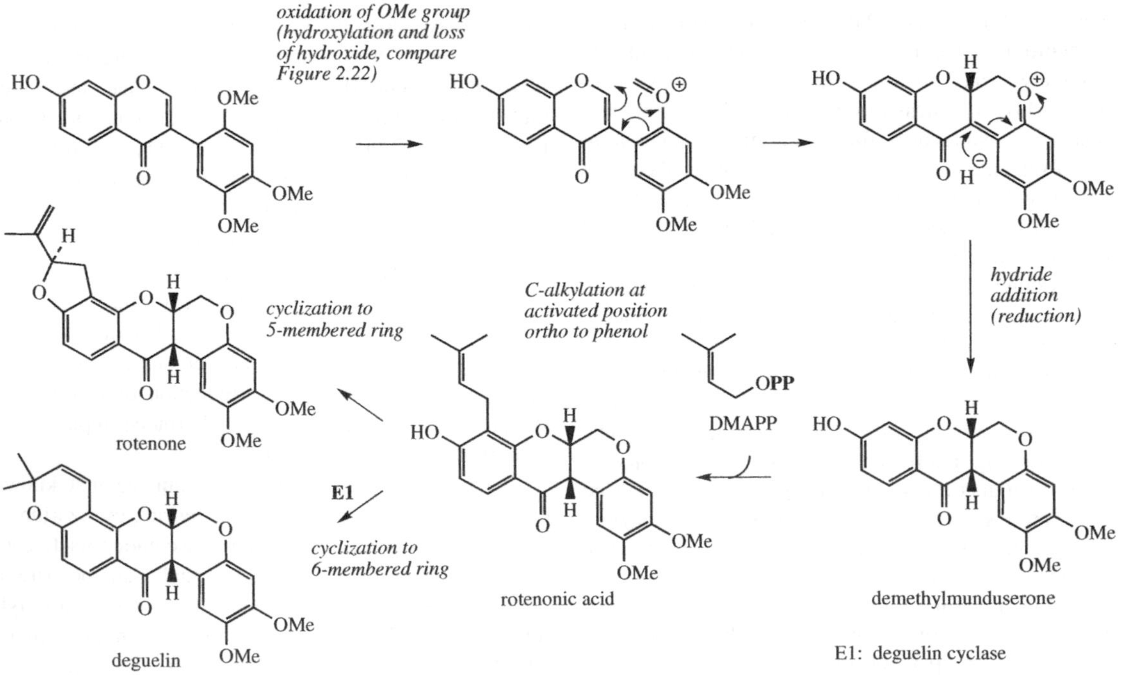

Figure 4.51

page 164). Rotenone and other rotenoids are powerful insecticidal and piscicidal (fish poison) agents, interfering with oxidative phosphorylation. They are relatively harmless to mammals unless they enter the bloodstream, being metabolized rapidly upon ingestion. Rotenone thus provides an excellent biodegradable insecticide and is used as such either in pure or powdered plant form. Roots of *Derris elliptica* or *Lonchocarpus* species are rich sources of rotenone [Box 4.8].

Box 4.8

Phyto-oestrogens

Phyto-oestrogen (phytoestrogen) is a term applied to non-steroidal plant materials displaying oestrogenic properties. Pre-eminent amongst these are isoflavonoids. These planar molecules mimic the shape and polarity of the steroid hormone estradiol (Figure 4.52; see page 290), and are able to bind to an oestrogen receptor, though their activity is much less than that of estradiol. In some tissues, they stimulate an oestrogenic response, whilst in others they can antagonize the effect of oestrogens. Such materials taken as part of the diet, therefore, influence overall oestrogenic activity in the body by adding their effects to normal levels of steroidal oestrogens (see page 293). Foods rich in isoflavonoids are valuable in countering some of the side-effects of the menopause in women, such as hot flushes, tiredness, and mood swings. In addition, there is mounting evidence that phyto-oestrogens also provide a range of other beneficial effects, helping to prevent heart attacks and other cardiovascular diseases, protecting against osteoporosis, lessening the risk of breast and uterine cancer, and in addition displaying significant antioxidant activity which may reduce the risk of Alzheimer's disease. Whilst some of these benefits may be obtained by the use of steroidal oestrogens, particularly via HRT (see page 294), phyto-oestrogens offer a dietary alternative.

The main food source of isoflavonoids is the soya bean (*Glycine max*; Leguminosae/Fabaceae) (see also page 255), which contains significant levels of the isoflavones daidzein and genistein (Figure 4.52), in free form and as their 7-*O*-glucosides. Total isoflavone levels fall in the range 0.1–0.4%, according to variety. Soya products such as soya milk, soya flour, tofu, and soya-based textured vegetable protein may all be used in the diet for their isoflavonoid content. Breads in which wheat flour is replaced by soya flour are also popular. Extracts from red clover (*Trifolium pratense*; Leguminosae/Fabaceae) are also used as a dietary supplement. Red clover isoflavones are predominantly formononetin (Figure 4.50) and daidzein, together with their 7-*O*-glucosides.

The lignans enterodiol and enterolactone (Figure 4.52) are also regarded as phyto-oestrogens. These compounds are produced by the action of intestinal microflora on lignans such as secoisolariciresinol or matairesinol ingested in the diet (see Figure 4.21, page 154). A particularly important precursor is secoisolariciresinol diglucoside from flaxseed (*Linum usitatissimum*; Linaceae), and flaxseed may be incorporated into foodstuffs along with soya products. Enterolactone and enterodiol were first detected in human urine, and their origins were traced back to dietary fibre-rich foods. Levels in the urine were much higher in vegetarians; a lower incidence of breast cancer in vegetarians has been related to these compounds. Enterolactone has been detected in sewage treatment plant water and even in tap water; some may arise from degradation of plant material and some from human excretions.

Several other biological activities are reported for isoflavones. One of significance is the modest anticancer activity of the isoflavone genistein. This has led to development of the simple isoflavene **phenoxodiol** (Figure 4.52), which has a promising profile in clinical trials. Phenoxodiol targets and inhibits NADH oxidase (tNOX) in cancer cells, which in turn leads to apoptosis; it is being investigated against late-stage chemoresistant ovarian cancers, as well as prostate and cervical cancers.

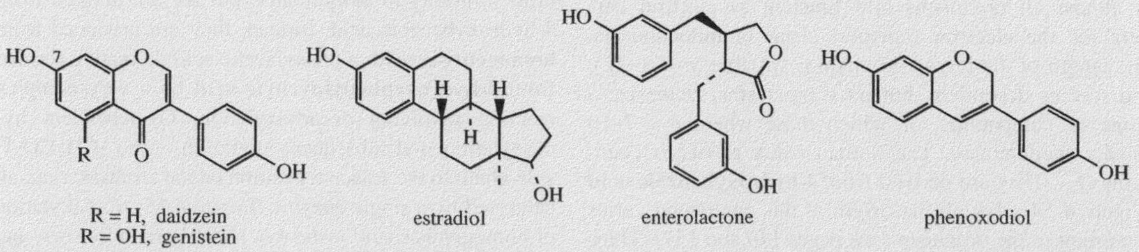

R = H, daidzein
R = OH, genistein

estradiol

enterolactone

phenoxodiol

Figure 4.52

Box 4.8   (continued)

Derris and Lonchocarpus

Species of *Derris* (e.g. *Derris elliptica, Derris malaccensis*) and *Lonchocarpus* (e.g. *Lonchocarpus utilis, Lonchocarpus urucu*) (Leguminosae/Fabaceae) have provided useful insecticides for many years. Roots of these plants have been employed as a dusting powder, or extracts have been formulated for sprays. *Derris* plants are small shrubs cultivated in Malaysia and Indonesia, whilst *Lonchocarpus* includes shrubs and trees, with commercial material coming from Peru and Brazil. The insecticidal principles are usually supplied as a black, resinous extract. Both *Derris* and *Lonchocarpus* roots contain 3–10% of rotenone (Figure 4.51) and smaller amounts of other rotenoids, e.g. deguelin (Figure 4.51). The resin may contain rotenone (about 45%) and deguelin (about 20%).

   Rotenone and other rotenoids interfere with oxidative phosphorylation, blocking transfer of electrons to ubiquinone (see page 178) by complexing with NADH:ubiquinone oxidoreductase of the respiratory electron transport chain. However, unless they enter the bloodstream they are relatively innocuous to mammals, being metabolized rapidly upon ingestion. Insects and also fish seem to lack this rapid detoxification. The fish poison effect has been exploited for centuries in a number of tropical countries, allowing lazy fishing by the scattering of powdered plant material on the water. The dead fish were collected and, when subsequently eaten, produced no ill effects on the consumers. More recently, rotenoids have been used in fish management programmes to eradicate undesirable fish species prior to restocking with other species. As insecticides, the rotenoids still find modest use, and are valuable for their selectivity and rapid biodegradability. However, they are perhaps inactivated too rapidly in the presence of light and air to compete effectively with other insecticides, such as the modern pyrethrin derivatives (see page 205).

TERPENOID QUINONES

Quinones are potentially derivable by oxidation of suitable phenolic compounds, catechols (1,2-dihydroxybenzenes) giving rise to *ortho*-quinones and quinols (1,4-dihydroxybenzenes) yielding *para*-quinones (see page 26). Accordingly, quinones can be formed from phenolic systems generated by either the acetate or shikimate pathways, provided a catechol or quinol system has been elaborated, and many examples are found in nature. A range of quinone derivatives and related structures containing a terpenoid fragment, as well as a shikimate-derived portion, are also widely distributed. Many of these have important biochemical functions in electron transport systems for respiration or photosynthesis, and some examples are shown in Figure 4.53.

   Ubiquinones (coenzyme Q; Figure 4.53) are found in almost all organisms and function as electron carriers for the electron transport chain in mitochondria. The length of the terpenoid chain is variable ($n = 1–12$), and species dependent, but most organisms synthesize a range of compounds, of which those where $n = 7–10$ usually predominate. The human redox carrier is coenzyme Q_{10}. They are derived from **4-hydroxybenzoic acid** (Figure 4.54), though the origin of this compound varies according to the organism (see pages 140 and 157). Thus, bacteria are known to transform chorismic acid by enzymic elimination of pyruvic acid, whereas plants and animals utilize a route from phenylalanine or tyrosine via 4-hydroxycinnamic acid. 4-Hydroxybenzoic acid is the substrate for *C*-alkylation *ortho* to the phenol group with a polyisoprenyl diphosphate of appropriate chain length (see page 306). The product then undergoes further elaboration, the exact sequence of modifications, i.e. hydroxylation, *O*-methylation, and decarboxylation, varying in eukaryotes and prokaryotes. Quinone formation follows in an O_2-dependent combined hydroxylation–oxidation process, and ubiquinone production then involves further hydroxylation, plus *O*- and *C*-methylation reactions. In yeast, *Saccharomyces cerevisiae*, nine genes have been identified as necessary for coenzyme Q biosynthesis, though not all of the encoded proteins have enzymatic functions. Coenzyme Q supplementation has been shown to be effective in treating patients with specific respiratory chain defects and to slow the progression of Parkinson's disease symptoms.

   Plastoquinones (Figure 4.53) bear considerable structural similarity to ubiquinones, but are not derived from 4-hydroxybenzoic acid. Instead, they are produced from **homogentisic acid**, a phenylacetic acid derivative formed from **4-hydroxyphenylpyruvic acid** by a very complex reaction involving decarboxylation, O_2-dependent hydroxylation, and subsequent migration of the $–CH_2CO_2H$ side-chain to the adjacent position on the aromatic ring, all catalysed by a single enzyme (Figure 4.55). *C*-Alkylation of homogentisic acid *ortho* to a phenol group follows, and involves a polyisoprenyl diphosphate with $n = 3–10$, but most commonly with $n = 9$, i.e. **solanesyl diphosphate**. However, during the alkylation reaction, the $–CH_2CO_2H$ side-chain of homogentisic acid suffers decarboxylation

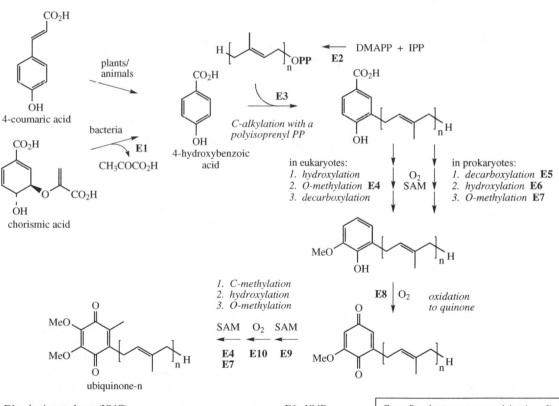

Figure 4.53

E1: chorismate lyase (UbiC)
E2: polyprenyl diphosphate synthase (Coq1, IspB)
E3: 4-hydroxybenzoate polyprenyltransferase (Coq2, UbiA)
E4: Coq3
E5: UbiD

E6: UbiB
E7: UbiG
E8: Coq6, UbiH
E9: Coq5, UbiE
E10: Coq7, UbiF

Coq: *Saccharomyces cerevisiae* (n = 6)
Ubi, Isp: *Escherichia coli* (n = 8)

Figure 4.54

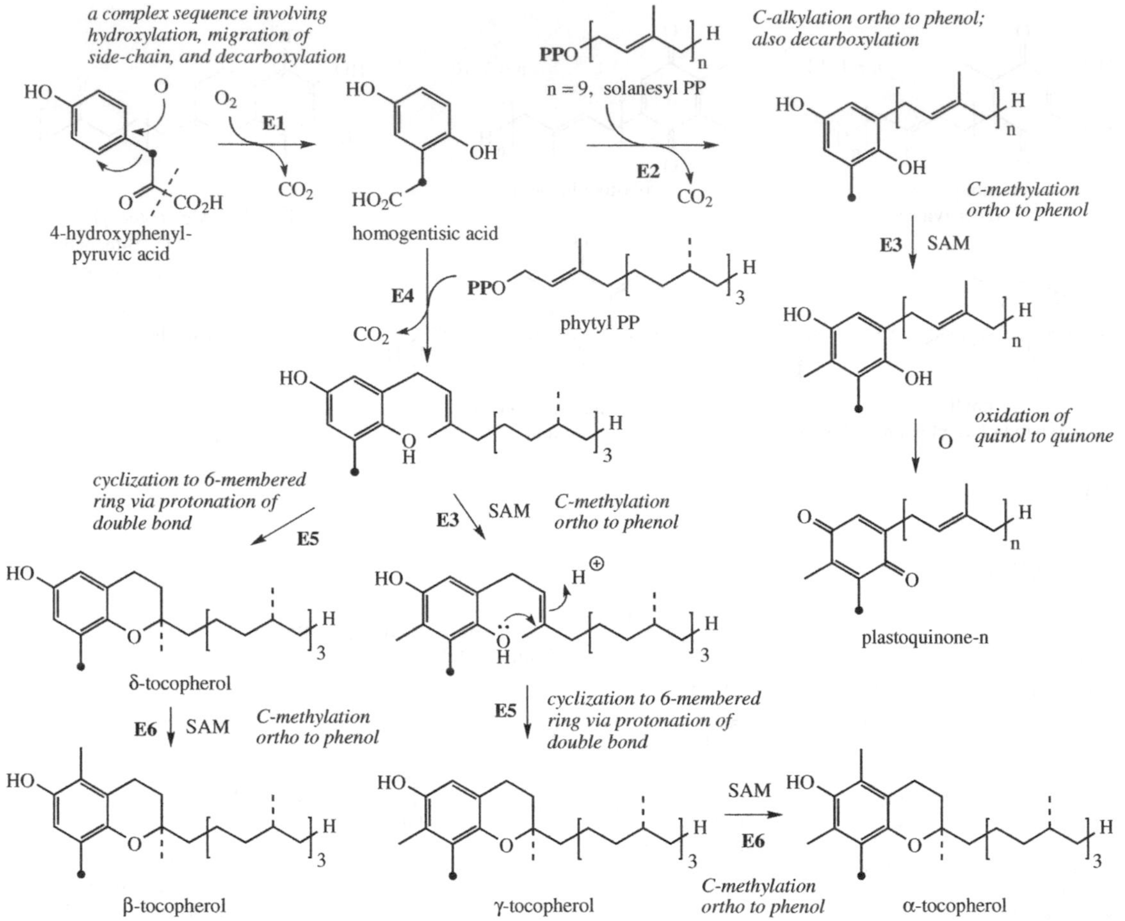

E1: 4-hydroxyphenylpyruvate dioxygenase
E2: homogentisate solanesyltransferase
E3: 2-methyl-6-phytyl-1,4-benzoquinol methyltransferase

E4: homogentisate phytyltransferase
E5: tocopherol cyclase
E6: γ-tocopherol methyltransferase

Figure 4.55

and the product is thus an alkyl methyl *p*-quinol derivative. Further aromatic methylation (via SAM) and oxidation of the *p*-quinol to a quinone follow to yield the plastoquinone. Thus, only one of the two methyl groups on the quinone ring of the plastoquinone is derived from SAM. Plastoquinones are involved in the photosynthetic electron transport chain in plants.

Tocopherols are also frequently found in the chloroplasts and constitute members of the vitamin E group [Box 4.9]. Their biosynthesis shares many of the features of plastoquinone biosynthesis, with an additional cyclization reaction involving the *p*-quinol and the terpenoid side-chain to give a chroman ring (Figure 4.55). Thus, the tocopherols, e.g. α-**tocopherol** and γ-**tocopherol**, are

not in fact quinones, but are indeed structurally related to plastoquinones. The isoprenoid side-chain added, from **phytyl diphosphate**, contains only four isoprene units, and three of the expected double bonds have suffered reduction. Again, decarboxylation of homogentisic acid co-occurs with the alkylation reaction. *C*-Methylation steps using SAM and the cyclization of the *p*-quinol to γ-tocopherol have been established as in Figure 4.55. Formation of δ-**tocopherol** and β-**tocopherol** occurs via a sequence in which cyclization precedes the methylation. Note again that one of the nuclear methyl groups is homogentisate-derived, whilst the others are supplied by SAM. A number of the enzymes characterized display broad substrate specificity.

Box 4.9

Vitamin E

Vitamin E refers to a group of fat-soluble vitamins, the tocopherols, e.g. α-, β-, γ-, δ-tocopherols (Figure 4.55), which are widely distributed in plants, with high levels in cereal seeds such as wheat, barley, and rye. Wheat germ oil is a particularly good source. The proportions of the individual tocopherols vary widely in different seed oils, e.g. principally β- in wheat oil, γ- in corn oil, α- in safflower oil, and γ- and δ- in soybean oil. Vitamin E deficiency is virtually unknown, with most of the dietary intake coming from food oils and margarine, though much can be lost during processing and cooking. Rats deprived of the vitamin display reproductive abnormalities. α-Tocopherol has the highest activity (100%), with the relative activities of β-, γ-, and δ-tocopherols being 50%, 10%, and 3% respectively. The acetate ester **α-tocopheryl acetate** is the main commercial form used for food supplementation and for medicinal purposes. The vitamin is known to provide valuable antioxidant properties, probably preventing the destruction by radical reactions of vitamin A and unsaturated fatty acids in biological membranes. It is used commercially to retard rancidity in fatty materials in food manufacturing, and there are also claims that it can reduce the effects of ageing and help to prevent heart disease. Its antioxidant effect is likely to arise by reaction with peroxyl radicals, generating by one-electron phenolic oxidation a resonance-stabilized radical that does not propagate the radical reaction. Instead, it mops up further peroxyl radicals (Figure 4.56). In due course, the tocopheryl peroxide is hydrolysed to the tocopherolquinone. A main function of vitamin C (see page 492) is to provide a regenerating system for tocopherol, allowing the tocopheryloxyl radical to abstract a hydrogen atom.

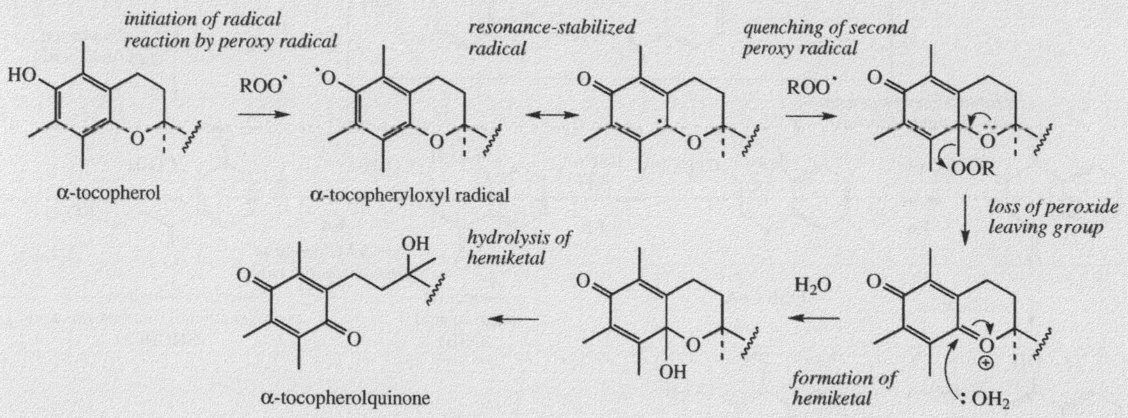

Figure 4.56

The **phylloquinones** (vitamin K_1) and **menaquinones** (vitamin K_2) are shikimate-derived naphthoquinone derivatives found in plants and algae (vitamin K_1) or bacteria and fungi (vitamin K_2) [Box 4.10]. The most common phylloquinone structure (Figure 4.53) has a diterpenoid (C_{20}) side-chain, whereas the range of menaquinone structures tends to be rather wider, with 1–13 isoprene units. These quinones are derived from chorismic acid via its isomer **isochorismic acid** (Figure 4.57). Additional carbon atoms for the naphthoquinone skeleton are provided by 2-oxoglutaric acid, which is incorporated by a mechanism involving the coenzyme thiamine diphosphate (TPP; see page 24).

2-Oxoglutaric acid is decarboxylated in the presence of TPP to give the TPP anion of succinic semialdehyde, which attacks isochorismic acid in a Michael-type reaction. Loss of the thiamine cofactor, elimination of pyruvic acid, and then dehydration yields the intermediate *o*-**succinylbenzoic acid** (OSB); the dehydration is an unusual *syn*-elimination, and is suggested to involve initial proton removal with carboxylate stabilization of the anion. OSB is activated by formation of a coenzyme A ester, and a Dieckmann-like condensation allows ring formation. The dihydroxynaphthoic acid coenzyme A ester is the more favoured aromatic tautomer, and hydrolysis of this leads to the free acid. This compound

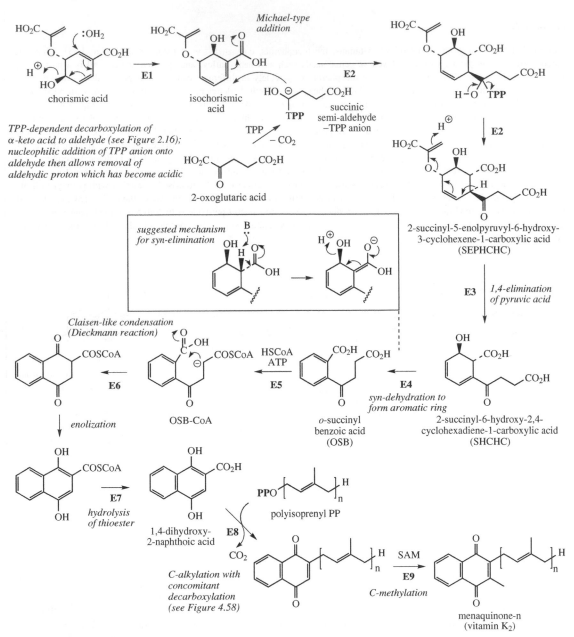

E1: isochorismate synthase (menF)
E2: 2-succinyl-5-enolpyruvyl-6-hydroxy-3-cyclohexene-
     1-carboxylate synthase (men D)
E3: 2-succinyl-6-hydroxy-2,4-cyclohexadiene-
     1-carboxylate synthase (menH)
E4: *o*-succinylbenzoate synthase (men C)

E5: *o*-succinylbenzoyl-CoA synthetase (men E)
E6: 1,4-dihydroxy-2-naphthoyl-CoA synthase (menB)
E7: thioesterase
E8: 1,4-dihydroxy-2-naphthoate phytyltransferase (menA)
E9: demethylmenaquinone methyltransferase (menG, ubiE)

Figure 4.57

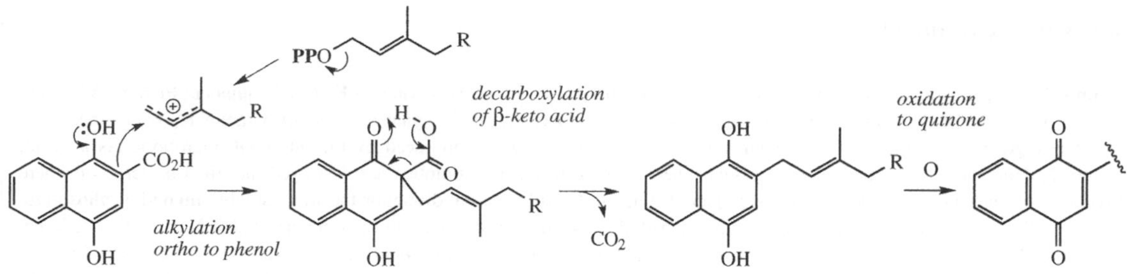

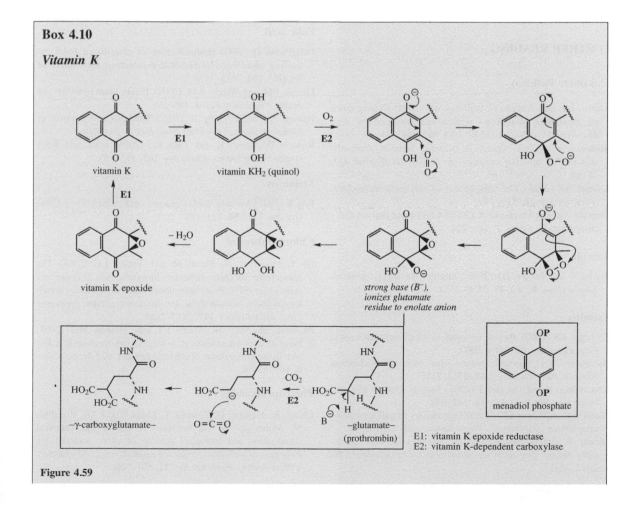

Figure 4.58

is now the substrate for alkylation and methylation, as seen with ubiquinones and plastoquinones. However, the terpenoid fragment is found to replace the carboxyl group, and the decarboxylated analogue is not involved. The transformation of **1,4-dihydroxynaphthoic acid** into the isoprenylated naphthoquinone appears to be catalysed by a single enzyme and can be rationalized

by the mechanism in Figure 4.58. This involves alkylation, decarboxylation of the resultant β-keto acid, and finally an oxidation to the *p*-quinone. OSB and 1,4-dihydroxynaphthoic acid have been implicated in the biosynthesis of a wide range of plant naphthoquinones and anthraquinones; other carbon atoms in the skeleton derive from a C_5 isoprene unit.

Box 4.10

Vitamin K

Figure 4.59

Box 4.10   (continued)

Vitamin K comprises a number of lipid-soluble naphthoquinone derivatives, with vitamin K_1 (phylloquinone; Figure 4.53) being of plant origin whilst the vitamins K_2 (menaquinones) are produced by microorganisms. Dietary vitamin K_1 is obtained from almost any green vegetable, whilst a significant amount of vitamin K_2 is produced by the intestinal microflora. As a result, vitamin K deficiency is rare and only necessary where prolonged use of antibiotics has destroyed intestinal bacteria, or where absorption of nutrients from the intestine is impaired. Vitamin K_1 (**phytomenadione**) or the water-soluble **menadiol phosphate** (Figure 4.59) may be employed as supplements. Menadiol is oxidized in the body to the quinone, which is then alkylated, e.g. with geranylgeranyl diphosphate, to yield a metabolically active product, and would be employed where malabsorption is a problem. Vitamin K supplements are given routinely to newborn infants, since they initially lack the intestinal bacteria capable of producing the vitamin.

Vitamin K is involved in normal blood clotting processes, and a deficiency would lead to haemorrhage. Blood clotting requires the carboxylation of glutamate residues in the protein prothrombin, generating bidentate ligands that allow the protein to bind to other factors. This carboxylation requires carbon dioxide, molecular oxygen, and the reduced quinol form of vitamin K. A mechanism for the process based on current information is shown in Figure 4.59. During the carboxylation, the reduced vitamin K is converted into an epoxide, and vitamin K is subsequently regenerated by reduction. The same protein is capable of reducing vitamin K to its quinol form. Anticoagulants such as dicoumarol and warfarin (see page 163) inhibit this reductase. In contrast, the polysaccharide anticoagulant heparin (see page 498) does not interfere with vitamin K metabolism, but acts by complexing with blood-clotting enzymes.

FURTHER READING

Shikimate Pathway

Alibhai MF and William C. Stallings WC (2001) Closing down on glyphosate inhibition – with a new structure for drug discovery. *Proc Natl Acad Sci USA* **98**, 2944–2946.

Anderson KS (2005) Detection of novel enzyme intermediates in PEP-utilizing enzymes. *Arch Biochem Biophys* **433**, 47–58.

Knaggs AR (2003) The biosynthesis of shikimate metabolites. *Nat Prod Rep* **20**, 119–136.

Mustafa NR and Verpoorte R (2005) Chorismate derived C_6C_1 compounds in plants. *Planta* **222**, 1–5.

Tamiflu

Farina V and Brown JD (2006) Tamiflu: the supply problem. *Angew Chem Int Ed* **45**, 7330–7334.

Tannins

Feldman KS (2005) Recent progress in ellagitannin chemistry. *Phytochemistry* **66**, 1984–2000.

Haslam E (2007) Vegetable tannins – lessons of a phytochemical lifetime. *Phytochemistry* **68**, 2713–2721.

Khanbabaee K and van Ree T (2001) Tannins: classification and definition. *Nat Prod Rep* **18**, 641–649.

Niemetz R and Gross GG (2005) Enzymology of gallotannin and ellagitannin biosynthesis. *Phytochemistry* **66**, 2001–2011.

Okuda T (2005) Systematics and health effects of chemically distinct tannins in medicinal plants. *Phytochemistry* **66**, 2012–2031.

Folic acid

DellaPenna D (2007) Biofortification of plant-based food: enhancing folate levels by metabolic engineering. *Proc Natl Acad Sci USA* **104**, 3675–3676.

Finglas PM and Wright AJA (2002) Folate bioavailability and health. *Phytochem Rev* **1**, 189–198.

Hanson AD and Gregory JF (2002) Synthesis and turnover of folates in plants. *Curr Opin Plant Biol* **5**, 244–249.

Kompis IM, Islam K and Then RL (2005) DNA and RNA synthesis: antifolates. *Chem Rev* **105**, 593–620.

Melanins

Roy S (2007) Melanin, melanogenesis, and vitiligo. *Prog Chem Org Nat Prod* **88**, 131–185.

Chloramphenicol

He J, Magarvey N, Piraee M and Vining LC (2001) The gene cluster for chloramphenicol biosynthesis in *Streptomyces venezuelae* ISP5230 includes novel shikimate pathway homologues and a monomodular non-ribosomal peptide synthetase gene. *Microbiology* **147**, 2817–2829.

Pacholec M, Sello JK, Walsh CT and Thomas MG (2007) Formation of an aminoacyl-*S*-enzyme intermediate is a key step in the biosynthesis of chloramphenicol. *Org Biomol Chem* **5**, 1692–1694.

Phenylpropanoids

Chaieb K, Hajlaoui H, Zmantar T, Kahla-Nakbi AB, Rouabhia M, Mahdouani K and Bakhrouf A (2007) The chemical composition and biological activity of clove essential oil, *Eugenia caryophyllata* (*Syzygium aromaticum* L. Myrtaceae): a short review. *Phytother Res* **21**, 501–506.

Dixon RA, Chen F, Guo D and Parvathi K (2001) The biosynthesis of monolignols: a 'metabolic grid', or independent pathways to guaiacyl and syringyl units? *Phytochemistry* **57**, 1069–1084.

Dixon RA and Reddy MSS (2003) Biosynthesis of monolignols. Genomic and reverse genetic approaches. *Phytochem Rev* **2**, 289–306.

Petersena M and Simmonds MSJ (2003) Molecules of interest. Rosmarinic acid. *Phytochemistry* **62**, 121–125.

Walton NJ, Mayer MJ and Narbad A (2003) Molecules of interest. Vanillin. *Phytochemistry* **63**, 505–515.

Lignin

Anterola AM and Lewis NR (2002) Trends in lignin modification: a comprehensive analysis of the effects of genetic manipulations/mutations on lignification and vascular integrity. *Phytochemistry* **61**, 221–294.

Boerjan W, Ralph J and Baucher M (2003) Lignin biosynthesis. *Annu Rev Plant Biol* **54**, 519–546.

Hatfield R and Vermerris W (2001) Lignin formation in plants. The dilemma of linkage specificity. *Plant Physiol* **126**, 1351–1357.

Ralph J, Lundquist K, Brunow G, Lu F, Kim H, Schatz PF, Marita JM, Hatfield RD, Ralph SA, Christensen JH and Boerjan W (2004) Lignins: natural polymers from oxidative coupling of 4-hydroxyphenylpropanoids. *Phytochem Rev* **3**, 29–60.

Lignans

Apers S, Vlietinck SA and Pieters L (2003) Lignans and neolignans as lead compounds. *Phytochem Rev* **2**, 201–217.

Botta B, Delle Monache G, Misiti D, Vitalia A and Zappia G (2001) Aryltetralin lignans: chemistry, pharmacology and biotransformations. *Curr Med Chem* **8**, 1363–1381.

Davin LB and Lewis NG (2003) An historical perspective on lignan biosynthesis: monolignol, allylphenol and hydroxycinnamic acid coupling and downstream metabolism. *Phytochem Rev* **2**, 257–288.

Fuss E (2003) Lignans in plant cell and organ cultures: an overview. *Phytochem Rev* **2**, 307–320.

Harmatha J and Dinan L (2003) Biological activities of lignans and stilbenoids associated with plant-insect chemical interactions. *Phytochem Rev* **2**, 321–330.

Saleem M, Kim HJ, Ali MS and Lee YS (2005) An update on bioactive plant lignans. *Nat Prod Rep* **22**, 696–716.

Smeds AI, Willför SM, Pietarinen SP, Peltonen-Sainio P and Reunanen MHT (2007) Occurrence of 'mammalian' lignans in plant and water sources. *Planta* **226**, 639–646.

Umezawa T (2003) Diversity in lignan biosynthesis. *Phytochem Rev* **2**, 371–390.

Westcott ND and Muir AD (2003) Flax seed lignan in disease prevention and health promotion. *Phytochem Rev* **2**, 401–417.

Podophyllum

Castro MA, del Corral JMM, Gordaliza M, Gómez-Zurita MA, García PA and San Feliciano A (2003) Chemoinduction of cytotoxic selectivity in podophyllotoxin-related lignans. *Phytochem Rev* **2**, 219–233.

Lee K-H and Xiao Z (2003) Lignans in treatment of cancer and other diseases. *Phytochem Rev* **2**, 341–362.

Meresse P, Dechaux E, Monneret C and Bertounesque E (2004) Etoposide: discovery and medicinal chemistry *Curr Med Chem* **11**, 2443–2466.

Coumarins

Bourgaud F, Hehn A, Larbat R, Doerper S, Gontier E, Kellner S and Matern U (2006) Biosynthesis of coumarins in plants: a major pathway still to be unravelled for cytochrome P450 enzymes. *Phytochem Rev* **5**, 293–308.

Kava

Abe I, Watanabe T and Noguchi H (2004) Enzymatic formation of long-chain polyketide pyrones by plant type III polyketide synthases. *Phytochemistry* **65**, 2447–2453.

Singh YN (2005) Potential for interaction of kava and St. John's wort with drugs. *J Ethnopharmacol* **100**, 108–113.

Whittona PA, Laua A, Salisbury A, Whitehouse J and Evans CS (2003) Kava lactones and the kava-kava controversy. *Phytochemistry* **64**, 673–679.

Flavonoids

Ayabe S-I and Akashi T (2006) Cytochrome P450s in flavonoid metabolism. *Phytochem Rev* **5**, 271–282.

De Pascual-Teresa S and Sanchez-Ballesta MT (2008) Anthocyanins: from plant to health. *Phytochem Rev* **7**, 281–299.

Garg, A, Garg S, Zaneveld LJD and Singla AK (2001) Chemistry and pharmacology of the citrus bioflavonoid hesperidin. *Phytother Res* **15**, 655–669.

Grotewold E (2006) The genetics and biochemistry of floral pigments. *Annu Rev Plant Biol* **57**, 761–780.

Hackman RM, Polagruto JA, Zhu QY, Sun B, Fujii H and Keen CL (2008) Flavanols: digestion, absorption and bioactivity. *Phytochem Rev* **7**, 195–208.

Kong J-M, Chia L-S, Goh N-K, Chia T-F and Brouillard R (2003) Analysis and biological activities of anthocyanins. *Phytochemistry* **64**, 923–933.

Martens S and Mithöfer A (2005) Flavones and flavone synthases. *Phytochemistry* **66**, 2399–2407; corrigendum **67**, 521.

Nagle DG, Ferreira D and Zhou Y-D (2006) Epigallocatechin-3-gallate (EGCG): chemical and biomedical perspectives. *Phytochemistry* **67**, 1849–1855.

Prior RL and Gu L (2005) Occurrence and biological significance of proanthocyanidins in the American diet. *Phytochemistry* **66**, 2264–2280.

Rice-Evans C (2001) Flavonoid antioxidants. *Curr Med Chem* **8**, 797–807.

Schijlen EGWM, de Vos CHR, van Tunen AJ and Bovy AG (2004) Modification of flavonoid biosynthesis in crop plants. *Phytochemistry* **65**, 2631–2648.

Springob K, Nakajima J-I, Yamazaki M and Saito K (2003) Recent advances in the biosynthesis and accumulation of anthocyanins. *Nat Prod Rep* **20**, 288–303.

Winkel-Shirley B (2001) Flavonoid biosynthesis. a colorful model for genetics, biochemistry, cell biology, and biotechnology. *Plant Physiol* **126**, 485–493.

Xie D-Y and Dixon RA (2005) Proanthocyanidin biosynthesis – still more questions than answers? *Phytochemistry* **66**, 2127–2144.

Stilbenes

Cirla A and Mann J (2003) Combretastatins: from natural products to drug discovery. *Nat Prod Rep* **20**, 558–564.

Nam N-H (2003) Combretastatin A-4 analogues as antimitotic antitumor agents. *Curr Med Chem* **10**, 1697–1722.

Roberti M, Pizzirani D, Simoni D, Rondanin R, Baruchello R, Bonora C, Buscemi F, Grimaudo S and Tolomeo M (2003) Synthesis and biological evaluation of resveratrol and analogues as apoptosis-inducing agents. *J Med Chem* **46**, 3546–3554.

Tron GC, Pirali T, Sorba G, Pagliai F, Busacca S and Genazzani AA (2006) Medicinal chemistry of combretastatin A4: present and future directions. *J Med Chem* **49**, 3033–3044.

Flavonolignans

Crocenzi FA and Roma MG (2006) Silymarin as a new hepatoprotective agent in experimental cholestasis: new possibilities for an ancient medication. *Curr Med Chem* **13**, 1055–1074.

Gažák R, Walterová D and Kren V (2007) Silybin and silymarin – new and emerging applications in medicine. *Curr Med Chem* **14**, 315–338.

Isoflavonoids, Phytoestrogens

Cornwell T, Cohick W and Raskin I (2004) Dietary phytoestrogens and health. *Phytochemistry* **65**, 995–1016.

Cos P, De Bruyne T, Apers S, Van den Berghe D, Pieters L and Vlietinck AJ (2003) Phytoestrogens: recent developments. *Planta Med* **69**, 589–599.

Dixon RA (2004) Phytoestrogens. *Annu Rev Plant Biol* **55**, 225–261.

Dixon RA and Ferreira D (2002) Molecules of interest. Genistein. *Phytochemistry* **60**, 205–211.

Heinonen S-M, Wähälä K and Adlercreutz H (2002) Metabolism of isoflavones in human subjects. *Phytochem Rev* **1**, 175–182.

Martin JHJ, Crotty S, Warren P and Nelson PN (2007) Does an apple a day keep the doctor away because a phytoestrogen a day keeps the virus at bay? A review of the anti-viral properties of phytoestrogens. *Phytochemistry* **68**, 266–274.

Ososki AL and Kennelly EJ (2003) Phytoestrogens: a review of the present state of research. *Phytother Res* **17**, 845–869.

Reinwald S and Weaver CM (2006) Soy isoflavones and bone health: a double-edged sword? *J Nat Prod* **69**, 450–459.

Vaya J and Tamir S (2004) The relation between the chemical structure of flavonoids and their estrogen-like activities. *Curr Med Chem* **11**, 1333–1343.

Veitch NC (2007) Isoflavonoids of the Leguminosae. *Nat Prod Rep* **24**, 417–464.

Terpenoid Quinones

Azzi A, Ricciarelli R and Zingg J-M (2002) Non-antioxidant molecular functions of α-tocopherol (vitamin E). *FEBS Lett* **519**, 8–10.

DellaPenna D (2005) A decade of progress in understanding vitamin E synthesis in plants. *J Plant Physiol* **162**, 729–737.

DellaPenna D and Pogson BJ (2006) Vitamin synthesis in plants: tocopherols and carotenoids. *Annu Rev Plant Biol* **57**, 711–738.

Dörmann P (2007) Functional diversity of tocochromanols in plants. *Planta* **225**, 269–276.

Herbers K (2003) Vitamin production in transgenic plants. *J Plant Physiol* **160**, 821–829.

Maeda H and DellaPenna D (2007) Tocopherol functions in photosynthetic organisms. *Curr Opin Plant Biol* **10**, 260–265.

Moran GR (2005) 4-Hydroxyphenylpyruvate dioxygenase. *Arch Biochem Biophys* **433**, 117–128.

Munné-Bosch S and Falk J (2004) New insights into the function of tocopherols in plants. *Planta* **218**, 323–326.

Niki E and Noguchi N (2004) Dynamics of antioxidant action of vitamin E. *Acc Chem Res* **37**, 45–51.

Nohl H, Kozlov AV, Staniek K, and Gille L (2001) The multiple functions of coenzyme Q. *Bioorg Chem* **29**, 1–13.

Traber MG (2004) Vitamin E, nuclear receptors and xenobiotic metabolism. *Arch Biochem Biophys* **423**, 6–11.

Upston JM, Kritharides L and Stocker R (2003) The role of vitamin E in atherosclerosis. *Prog Lipid Res* **42**, 405–422.

5

THE MEVALONATE AND METHYLERYTHRITOL PHOSPHATE PATHWAYS: TERPENOIDS AND STEROIDS

Terpenoids form a large and structurally diverse family of natural products derived from C_5 **isoprene units** (Figure 5.1) joined in a head-to-tail fashion. Typical structures contain carbon skeletons represented by $(C_5)_n$, and are classified as **hemiterpenes** (C_5), **monoterpenes** (C_{10}), **sesquiterpenes** (C_{15}), **diterpenes** (C_{20}), **sesterterpenes** (C_{25}), **triterpenes** (C_{30}), and **tetraterpenes** (C_{40}) (Figure 5.2). Higher polymers are encountered in materials such as rubber. Isoprene itself (Figure 5.1) was known as a decomposition product from various natural cyclic hydrocarbons, and had been suggested as the fundamental building block for these compounds, also referred to as 'isoprenoids'. Isoprene is produced naturally but is not involved in the formation of these compounds; the biochemically active isoprene units were subsequently identified as the diphosphate (pyrophosphate) esters **dimethylallyl diphosphate (DMAPP)** and **isopentenyl diphosphate (IPP)** (Figure 5.2). Relatively few of the natural terpenoids conform exactly to the simple concept of a linear head-to-tail combination of isoprene units as seen with **geraniol** (C_{10}), **farnesol** (C_{15}), and **geranylgeraniol** (C_{20}) (Figure 5.3). **Squalene** (C_{30}) and **phytoene** (C_{40}), although formed entirely of isoprene units, display a tail-to-tail linkage at the centre of the molecules. Most terpenoids are modified further by cyclization reactions, though the head-to-tail arrangement of the units can usually still be recognized, e.g. **menthol**, **bisabolene**, and **taxadiene**. The linear arrangement of isoprene units can be much more difficult to appreciate in many other structures when rearrangement reactions have taken place, e.g. steroids, where, in addition, several carbon atoms have been lost. Nevertheless, such compounds are formed by way of regular terpenoid precursors. Terpenoids comprise the largest group of natural products, with over 35 000 known members.

Many other natural products contain terpenoid elements in their molecules, in combination with carbon skeletons derived from other sources, such as the acetate and shikimate pathways. Many alkaloids, phenolics, and vitamins discussed in other chapters are examples of this. A particularly common terpenoid fragment in such cases is a single C_5 unit, usually a dimethylallyl substituent, and molecules containing these isolated isoprene units are sometimes referred to as '**meroterpenoids**'. Some examples include furocoumarins (see page 162), rotenoids (see page 175), and ergot alkaloids (see page 387). One should also note that the term '**prenyl**' is in general use to indicate the dimethylallyl substituent. Even macromolecules like proteins can be modified by attaching terpenoid chains. Cysteine residues in proteins are alkylated with farnesyl (C_{15}) or geranylgeranyl (C_{20}) groups, thereby increasing the lipophilicity of the protein and its ability to associate with membranes.

Medicinal Natural Products: A Biosynthetic Approach. 3rd Edition   Paul Dewick
© 2009 John Wiley & Sons, Ltd

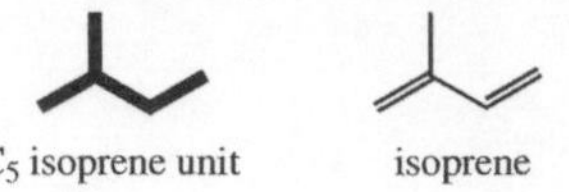

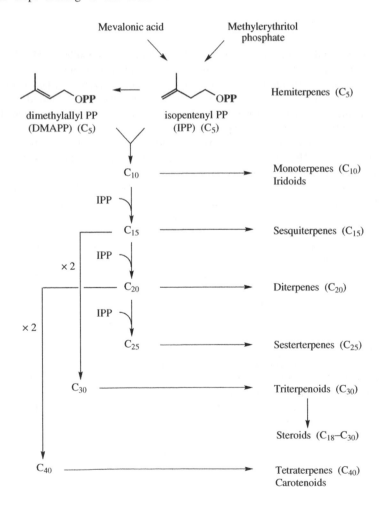

C₅ isoprene unit   isoprene

Figure 5.1

MEVALONIC ACID AND METHYLERYTHRITOL PHOSPHATE

The biochemical isoprene units may be derived by two pathways: by way of intermediates **mevalonic acid** (**MVA**) (Figure 5.4) or 2-*C*-methyl-D-erythritol 4-phosphate (**methylerythritol phosphate; MEP**; see Figure 5.6). MVA, itself a product of acetate metabolism, had been established as a precursor of the animal sterol cholesterol, and the steps leading to and from MVA

were gradually detailed in a series of painstakingly executed experiments. For many years, the early parts of the mevalonate pathway were believed to be common to the whole range of natural terpenoid derivatives in all organisms. However, after detailed investigation of inconsistencies in labelling patterns, it has since been proven that an alternative pathway to IPP and DMAPP exists, via MEP, and that this pathway is probably more widely utilized in nature than is the mevalonate pathway. This pathway is also referred to as the **mevalonate-independent pathway** or the **deoxyxylulose phosphate pathway**; the terminology **MEP pathway** is preferred, in that MEP is the first committed terpenoid precursor, whilst deoxyxylulose phosphate is also used for the biosynthesis of pyridoxal phosphate (vitamin B_6, page 32) and thiamine (vitamin B_1, page 31).

Figure 5.2

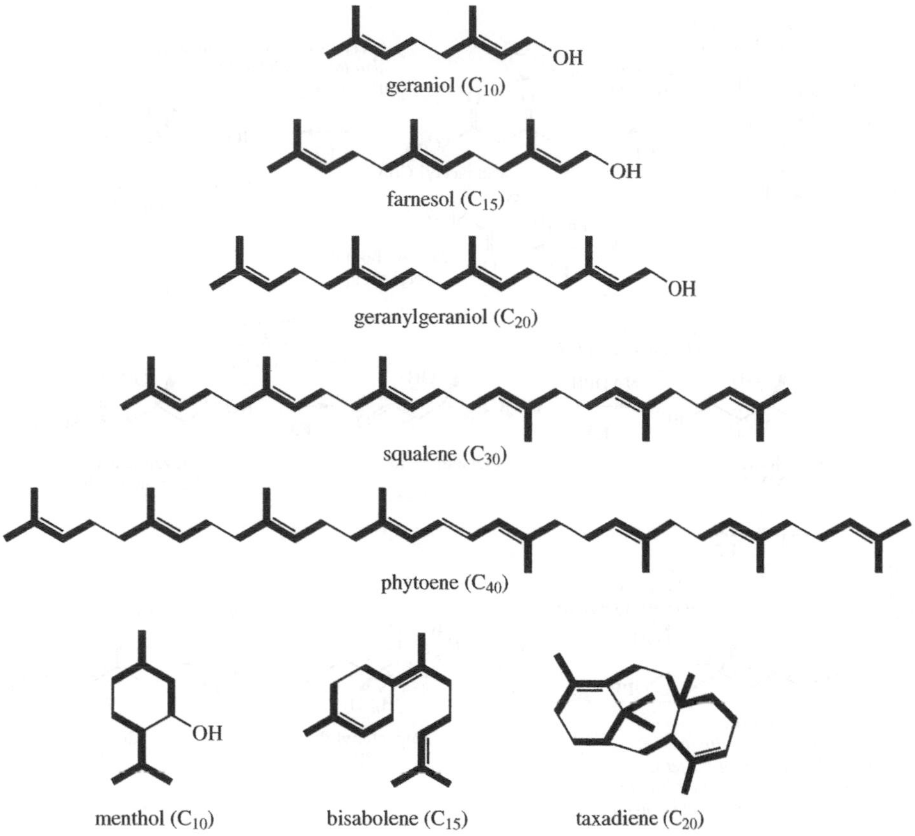

geraniol (C$_{10}$)

farnesol (C$_{15}$)

geranylgeraniol (C$_{20}$)

squalene (C$_{30}$)

phytoene (C$_{40}$)

menthol (C$_{10}$)   bisabolene (C$_{15}$)   taxadiene (C$_{20}$)

Figure 5.3

Three molecules of acetyl-coenzyme A are used to form **MVA**. Two molecules combine initially in a Claisen condensation to give acetoacetyl-CoA, and a third is incorporated via a stereospecific aldol addition giving the branched-chain ester **3-hydroxy-3-methylglutaryl-CoA (HMG-CoA)** (Figure 5.4). Two of the acetyl-CoA molecules appear to be bound to the enzyme via a thiol group. One linkage is broken during the Claisen reaction and the second is subsequently hydrolysed to form the free-acid group of HMG-CoA. Note that the mevalonate pathway does not use malonyl-CoA and it thus diverges from the acetate pathway at the very first step. In the acetate pathway, an equivalent acetoacetyl thioester (bound to the acyl carrier protein, see page 67) would be formed using the thioester of malonic acid as a more nucleophilic species. In the second step of the mevalonate pathway, it should also be noted that, on purely chemical grounds, acetoacetyl-CoA is the more acidic substrate and might be expected to act as the nucleophile rather than the third acetyl-CoA molecule.

The enzyme thus achieves what is a less favourable reaction. The conversion of HMG-CoA into (3*R*)-MVA involves a two-step reduction of the thioester group to a primary alcohol via the aldehyde, and provides an essentially irreversible and rate-limiting transformation. Drug-mediated inhibition of this enzyme (**HMG-CoA reductase**) is an important means of regulating the biosynthesis of mevalonate and ultimately of the steroid cholesterol (see statins, page 98).

The six-carbon compound MVA is transformed into the five-carbon phosphorylated isoprene units in a series of reactions, beginning with phosphorylation of the primary alcohol group. Two different ATP-dependent enzymes are involved, resulting in mevalonic acid diphosphate, and decarboxylation–dehydration then follows to give **IPP**. Whilst a third molecule of ATP is required for this last transformation, there is no evidence for phosphorylation of the tertiary hydroxyl, though this would convert the hydroxyl into a better leaving group. Hydrolysis of ATP may assist decarboxylation, as shown in Figure 5.4. IPP

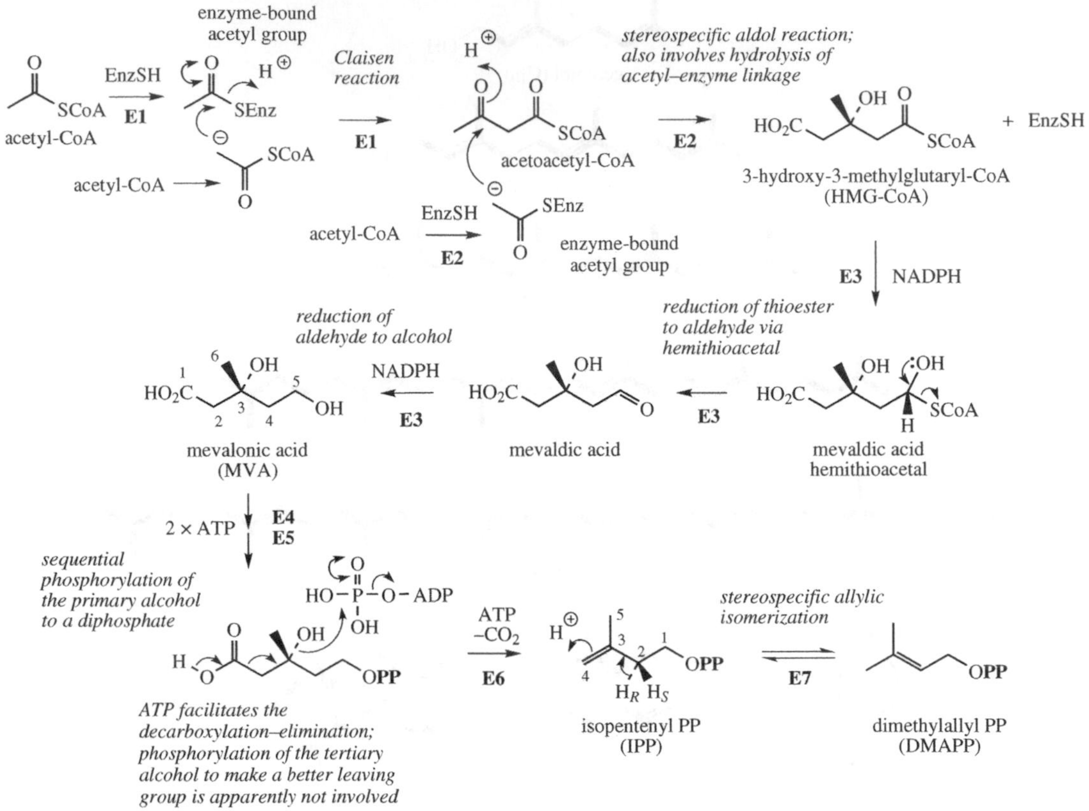

E1: acetoacetyl-CoA synthase
E2: 3-hydroxy-3-methylglutaryl-CoA synthase (HMG-CoA synthase)
E3: 3-hydroxy-3-methylglutaryl-CoA reductase (HMG-CoA reductase)
E4: mevalonate kinase

E5: phosphomevalonate kinase
E6: mevalonate 5-diphosphate decarboxylase
E7: isopentenyl diphosphate isomerase (IPP isomerase)

Figure 5.4

is isomerized to the other isoprene unit, **DMAPP**, by an isomerase enzyme which incorporates a proton from water onto C-4 and stereospecifically removes the *pro-R* proton (H_R) from C-2. This reaction is used to provide the two compounds in the amounts required for further metabolism. Two different types of isomerase enzyme have been distinguished: a type I enzyme requiring a divalent metal ion and a type II enzyme that requires a divalent metal ion together with FMN for activity. Both enzymes appear to employ a protonation–deprotonation mechanism. The conversion of IPP into DMAPP generates a reactive electrophile and, therefore, a good alkylating agent. DMAPP possesses a good leaving group, the diphosphate, and can ionize readily to yield an allylic carbocation which is stabilized by charge delocalization (Figure 5.5). In contrast, IPP with its terminal double bond is more likely to act as a nucleophile, especially towards

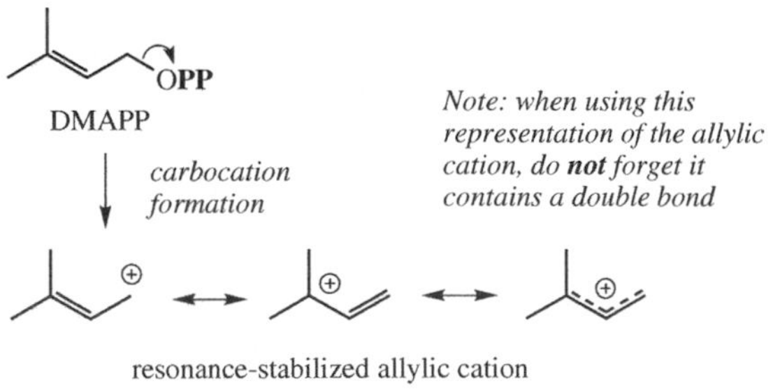

Note: when using this representation of the allylic cation, do **not** forget it contains a double bond

Figure 5.5

the electrophilic DMAPP. These differing reactivities are the basis of terpenoid biosynthesis, and carbocations feature strongly in mechanistic rationalizations of the pathways.

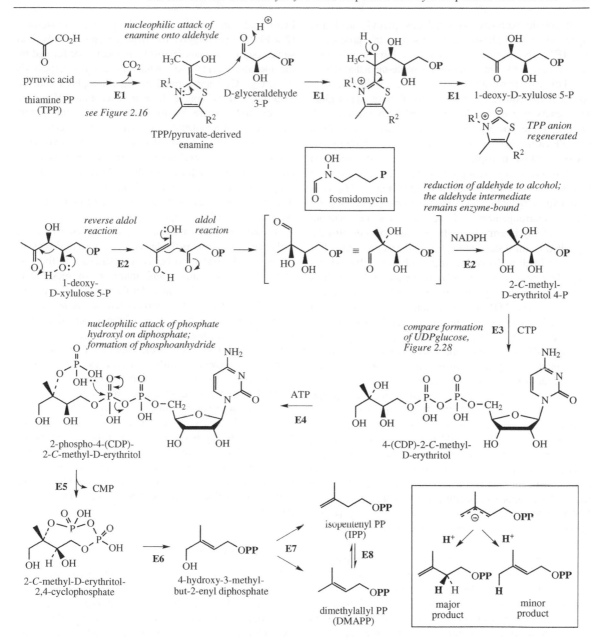

E1: 1-deoxy-D-xylulose 5-phosphate synthase (DXP synthase)
E2: 2-*C*-methyl-D-erythritol 4-phosphate synthase;
    1-deoxy-D-xylulose 5-phosphate reductoisomerase (IspC)
E3: 4-diphosphocytidyl-2-*C*-methyl-D-erythritol synthase (IspD)
E4: 4-diphosphocytidyl-2-*C*-methyl-D-erythritol kinase (IspE)

E5: 2-*C*-methyl-D-erythritol-2,4-cyclodiphosphate synthase (IspF)
E6: 4-hydroxy-3-methylbut-2-enyl diphosphate synthase (IspG)
E7: 4-hydroxy-3-methylbut-2-enyl diphosphate reductase (IspH)
E8: isopentenyl diphosphate isomerase (IPP isomerase)

Figure 5.6

Glycolytic pathway intermediates pyruvic acid and glyceraldehyde 3-phosphate are used in the production of **MEP**; the pyruvate carboxyl is lost in this process (Figure 5.6). Thiamine diphosphate-mediated decarboxylation of pyruvate (compare page 23) produces an acetaldehyde-equivalent bound in the form of an enamine. This reacts as a nucleophile in an addition reaction with the glyceraldehyde 3-phosphate. Subsequent release from the TPP carrier generates 1-deoxy-D-xylulose 5-phosphate (**deoxyxylulose phosphate**), which is transformed into MEP by a rearrangement process. This has been shown to involve a reverse aldol–aldol sequence (Figure 5.6), coupled with a reduction. A single enzyme catalyses these skeletal rearrangement and reduction reactions without release of any intermediate; the product now contains the branched-chain system equivalent to the isoprene unit. Reaction of MEP with cytidine triphosphate (CTP) produces a cytidine diphospho derivative (compare uridine diphosphoglucose in glucosylation, page 31), which is then phosphorylated via ATP. The resultant 2-phosphate is then converted into a cyclic phosphoanhydride with loss of cytidine phosphate. The subsequent steps leading to **IPP** and **DMAPP** are the least understood part of the pathway. Gene methodology has shown that two enzymes are involved, the first producing 4-hydroxy-3-methylbut-2-enyl diphosphate and the second converting this into predominantly IPP, but also DMAPP. Both steps are reductive in nature, but mechanisms are yet to be elucidated. The formation of both IPP and DMAPP (ratios are typically in the region 5:1 to 4:1) is suggested to involve a delocalized allylic system (radical or anion, shown in Figure 5.6 as an anion), with protons being supplied by water. Although this pathway coproduces IPP and DMAPP, isomerism of IPP to DMAPP as in the mevalonate pathway is also possible to balance the pool sizes of these intermediates.

Whether the mevalonate pathway or the MEP pathway supplies isoprene units for the biosynthesis of a particular terpenoid must be established experimentally. This can be determined from the results of feeding [1-^{13}C]-D-glucose as precursor; this leads to different labelling patterns in the isoprene unit according to the pathway operating (Figure 5.7). Animals and fungi appear to lack the MEP pathway, so utilize the mevalonate pathway exclusively. The MEP pathway is present in plants, algae, and most bacteria. Plants and some bacteria are equipped with and employ both pathways, often concurrently. In plants, the two pathways appear to be compartmentalized, so that the mevalonate pathway enzymes are localized in the cytosol, whereas the MEP pathway enzymes are found in chloroplasts. The cytosolic pool of IPP serves as a precursor of C_{15} derivatives (farnesyl PP (FPP) and sesquiterpenes; see

Figure 5.2), and in due course triterpenoids and steroids (2 × FPP). Accordingly, triterpenoids and steroids, and some sesquiterpenoids (cytosolic products) are formed by the mevalonate pathway, whilst most other terpenoids (C_{10}, C_{20}, C_{40}) are formed in the chloroplasts and are MEP derived. Of course there are exceptions. There are also examples where the two pathways can supply different portions of a molecule, or where there is exchange of late-stage common intermediates between the two pathways (cross-talk), resulting in a contribution of isoprene units from each pathway. In the following part of this chapter, these complications will not be considered further, and in most cases there is no need to consider the precise source of the isoprene units.

An area of special pharmacological interest where the early pathway is of particular concern is steroid biosynthesis, which appears to be from mevalonate in the vast majority of organisms. Thus, inhibitors of mevalonate pathway enzymes will reduce steroid production in plants, but will not affect the formation of terpenoids derived via MEP. Equally, it is possible to inhibit terpenoid production without affecting steroid formation by the use of MEP pathway inhibitors, such as the antibiotic **fosmidomycin** from *Streptomyces lavendulae*. This acts as an analogue of the rearrangement intermediate in the reaction catalysed by MEP synthase (Figure 5.6). Enzymes of the MEP pathway are attractive targets for development of drugs against microbial diseases such as malaria or tuberculosis, since the MEP pathway is utilized by the pathogen but is not present in humans. Regulation of cholesterol production in humans is an important health concern (see page 251); the widely used statin drugs are specific inhibitors of the mevalonate pathway enzyme HMG-CoA reductase (see page 98).

HEMITERPENES (C_5)

IPP and DMAPP are reactive hemiterpene intermediates in the pathways leading to more complex terpenoid structures. They are also used as alkylating agents in the formation of meroterpenoids, as indicated above, but examples of these structures are discussed elsewhere under the section appropriate to the major substructure, e.g. alkaloids, shikimate, acetate. Relatively few true hemiterpenes are produced in nature. **Isoprene**, a volatile compound which is released in huge amounts by many species of plants, especially woody trees such as oaks, willows, poplars, and spruce, is the notable example. Isoprene is formed by loss of a proton from the allylic cation. Alternatively, quenching the allylic cation with water leads to methylbutenol, produced by several species of pine (Figure 5.8).

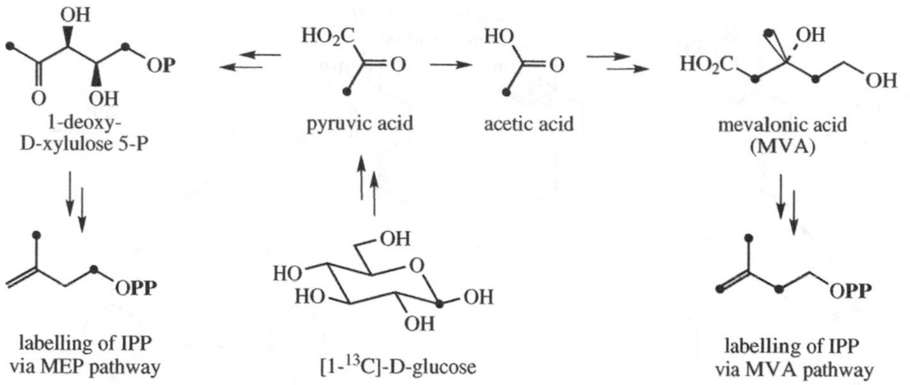

Figure 5.7

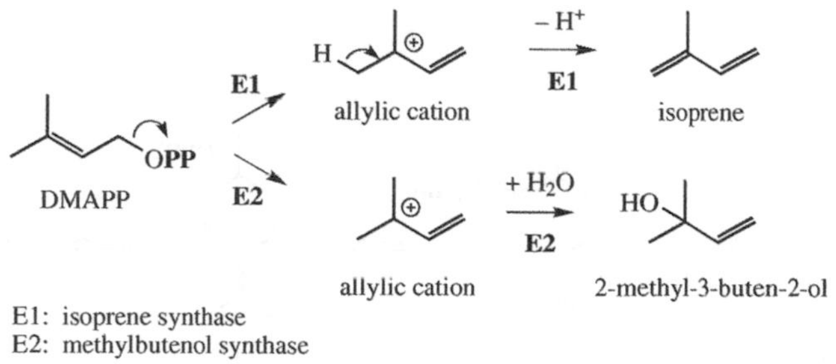

E1: isoprene synthase
E2: methylbutenol synthase

Figure 5.8

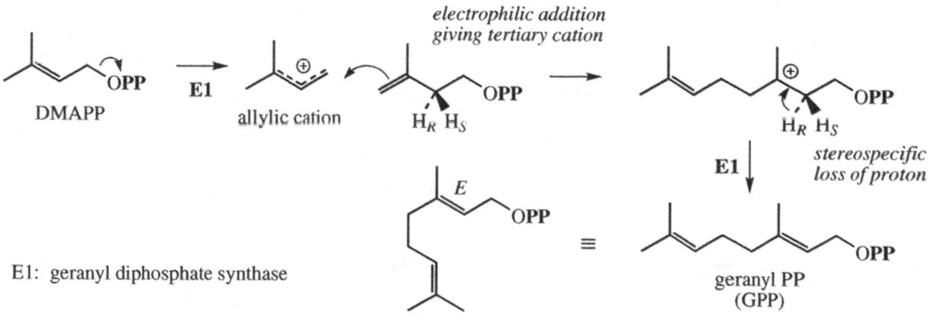

E1: geranyl diphosphate synthase

Figure 5.9

MONOTERPENES (C_{10})

Enzyme-catalysed combination of DMAPP and IPP yields **geranyl diphosphate** (**GPP**; Figure 5.9). This is believed to involve ionization of DMAPP to the allylic cation, addition to the double bond of IPP, followed by loss of a proton. The proton lost (H_R) is stereochemically analogous to that lost on the isomerization of IPP to DMAPP; indeed, the two reaction mechanisms are essentially the same, one involving a proton and the other a carbocation. This produces a monoterpene diphosphate, geranyl PP, in which the new double bond is *trans* (E).

Linalyl PP and **neryl PP** are isomers of GPP, and are likely to be formed from GPP by ionization to the allylic cation, allowing a change in attachment of the diphosphate group (to the tertiary carbon in linalyl PP) or a

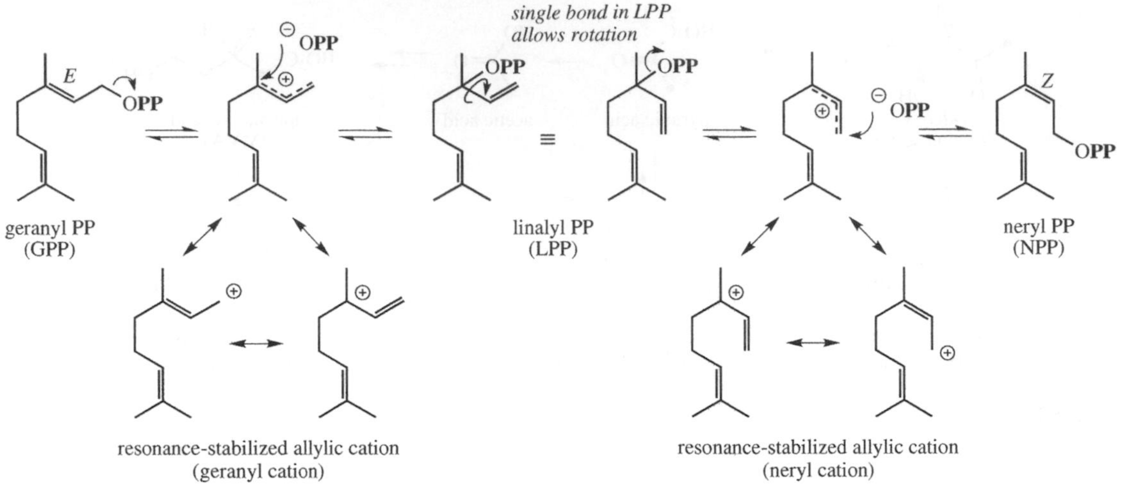

Figure 5.10

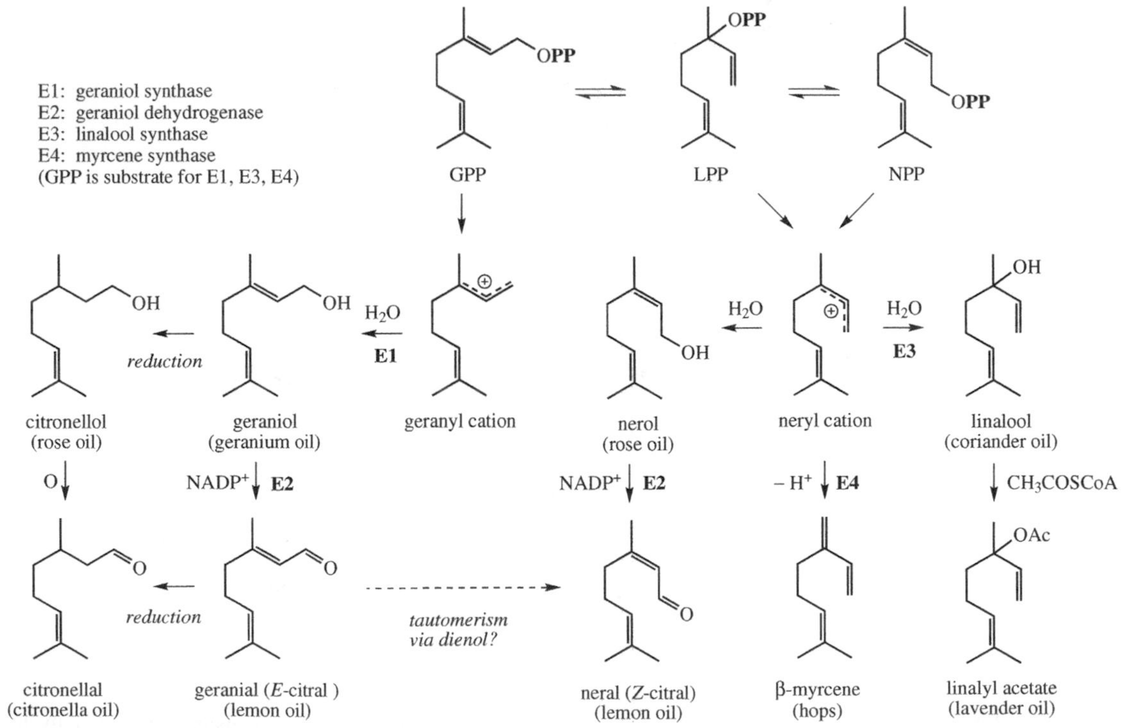

E1: geraniol synthase
E2: geraniol dehydrogenase
E3: linalool synthase
E4: myrcene synthase
(GPP is substrate for E1, E3, E4)

Figure 5.11

change in stereochemistry at the double bond (to Z in neryl PP) (Figure 5.10). These three compounds, by relatively modest changes, can give rise to a range of linear monoterpenes found as components of volatile oils used in flavouring and perfumery (Figure 5.11). The resulting compounds may be hydrocarbons, alcohols, aldehydes, or perhaps esters, especially acetates by reaction with acetyl-CoA. Where enzymes have been characterized, it

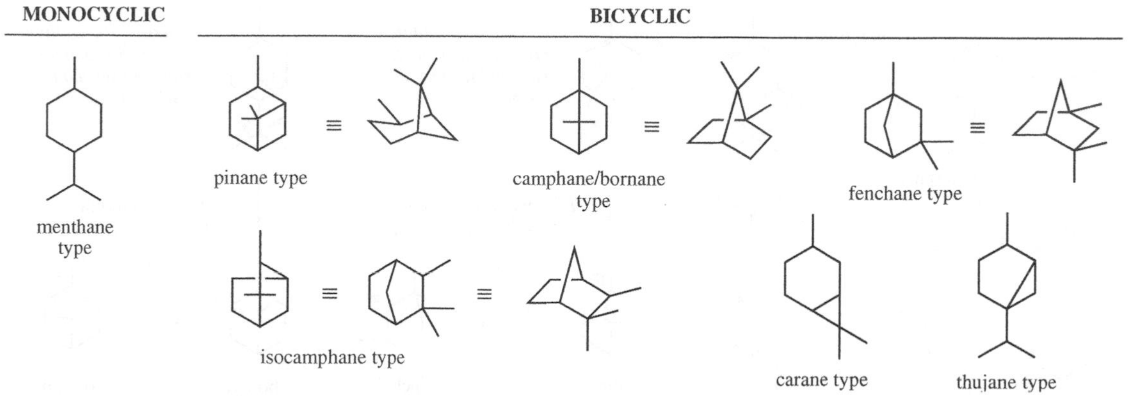

MONOCYCLIC   BICYCLIC

menthane
type

pinane type

camphane/bornane
type

fenchane type

isocamphane type

carane type   thujane type

Figure 5.12

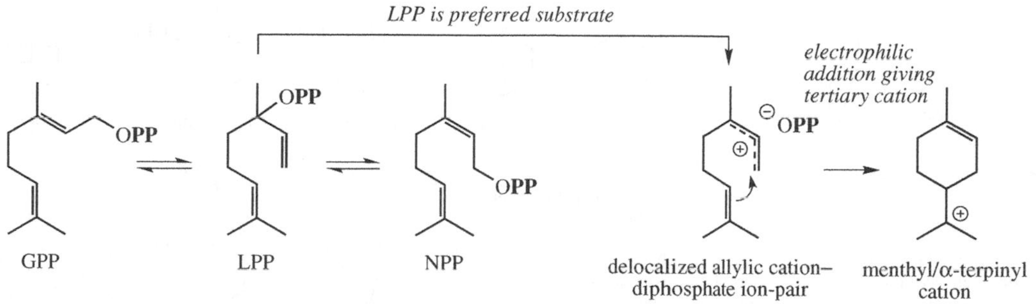

LPP is preferred substrate

*electrophilic
addition giving
tertiary cation*

GPP   LPP   NPP   delocalized allylic cation–   menthyl/α-terpinyl
                                                 diphosphate ion-pair   cation

Figure 5.13

has been demonstrated that the reactions proceed through the carbocation intermediates. Thus, geraniol is the result of addition of water to the geranyl cation, and is not formed by hydrolysis of GPP.

The range of monoterpenes encountered is extended considerably by cyclization reactions, and monocyclic or bicyclic systems can be created. Some of the more important examples of these ring systems are shown in Figure 5.12. A considerable amount of information about the enzymes (terpene cyclases), genes, and cyclization mechanisms is now available, providing a satisfactory and detailed picture of these natural products. Cyclizations would not be expected to occur with the precursor GPP, the *E* stereochemistry of the double bond being unfavourable for ring formation (Figure 5.13). Neryl PP or linalyl PP, however, do have favourable stereochemistry, and either or both of these would be more immediate precursors of the monocyclic menthane system, formation of which could be represented as shown, generating a carbocation (termed menthyl or α-terpinyl) that has the menthane skeleton. It has been found that monoterpene

cyclase enzymes are able to accept all three diphosphates, with linalyl PP being the best substrate, and it appears they have the ability to isomerize the substrates initially as well as to cyclize them. It is convenient, therefore, to consider the species involved in the cyclization as the delocalized allylic cation tightly bound to the diphosphate anion, and bond formation follows due to the proximity of the π-electrons of the double bond (Figure 5.13).

In Chapter 2, the possible fates of carbocations were discussed. These include quenching with nucleophiles (especially water), loss of a proton, cyclization, and the possibility that Wagner–Meerwein rearrangements might occur (see page 15). All of these feature strongly in terpenoid biosynthesis, and examples are shown in Figures 5.14 and 5.15. The newly generated menthyl cation could be quenched by attack of water, in which case the alcohol **α-terpineol** would be formed, or it could lose a proton to give **limonene** (Figure 5.14). Alternatively, folding the cationic side-chain towards the double bond (via the surface characteristics of the enzyme) would allow a repeat of the cyclization mechanism and produce

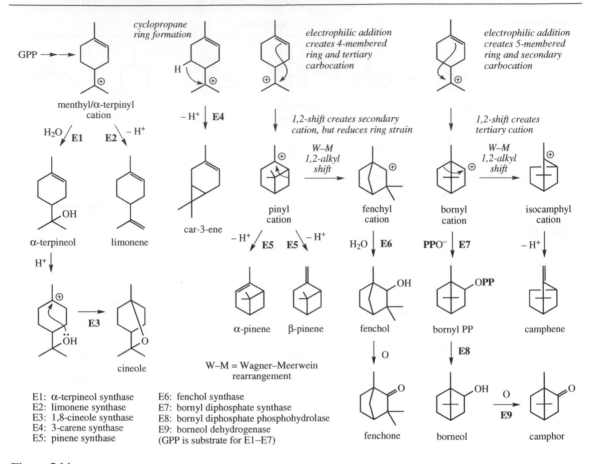

E1: α-terpineol synthase
E2: limonene synthase
E3: 1,8-cineole synthase
E4: 3-carene synthase
E5: pinene synthase

E6: fenchol synthase
E7: bornyl diphosphate synthase
E8: bornyl diphosphate phosphohydrolase
E9: borneol dehydrogenase
(GPP is substrate for E1–E7)

Figure 5.14

bicyclic pinyl and bornyl cations, according to which end of the double bond was involved in forming the new bonds. Thus **α-pinene** and **β-pinene** arise by loss of different protons from the pinyl cation, producing the double bonds as cyclic or exocyclic respectively. **Borneol** could potentially result from quenching of the bornyl cation with water; unusually, though, this alcohol is actually derived by hydrolysis of bornyl diphosphate. Oxidation of the secondary alcohol in borneol then generates the ketone **camphor**. A less common termination step involving loss of a proton is also shown in Figure 5.14. This is the formation of a cyclopropane ring as exemplified by **3-carene** and generation of the carane skeleton.

The chemistry of terpenoid formation is essentially based on the reactivity of carbocations, even though in nature these cations may not exist as discrete species, but rather as tightly bound ion pairs with a counter anion, e.g. diphosphate. The analogy with carbocation

chemistry is justified, however, since a high proportion of natural terpenoids have skeletons which have suffered rearrangement processes. Rearrangements of the Wagner–Meerwein type (see page 15), in which carbon atoms or hydride migrate to achieve enhanced stability for the cation via tertiary against secondary character, or by reduction of ring strain, give a mechanistic rationalization for the biosynthetic pathway. The bicyclic pinyl cation, with a strained four-membered ring, rearranges to the less-strained five-membered fenchyl cation, a change which presumably more than makes up for the unfavourable tertiary to secondary carbocation transformation. This produces the fenchane skeleton, exemplified by **fenchol** and **fenchone** (Figure 5.14). The isocamphyl tertiary carbocation is formed from the bornyl secondary carbocation by a Wagner–Meerwein rearrangement, and so leads to **camphene**.

Examples of Wagner–Meerwein rearrangements involving hydride migrations are featured in Figure 5.15.

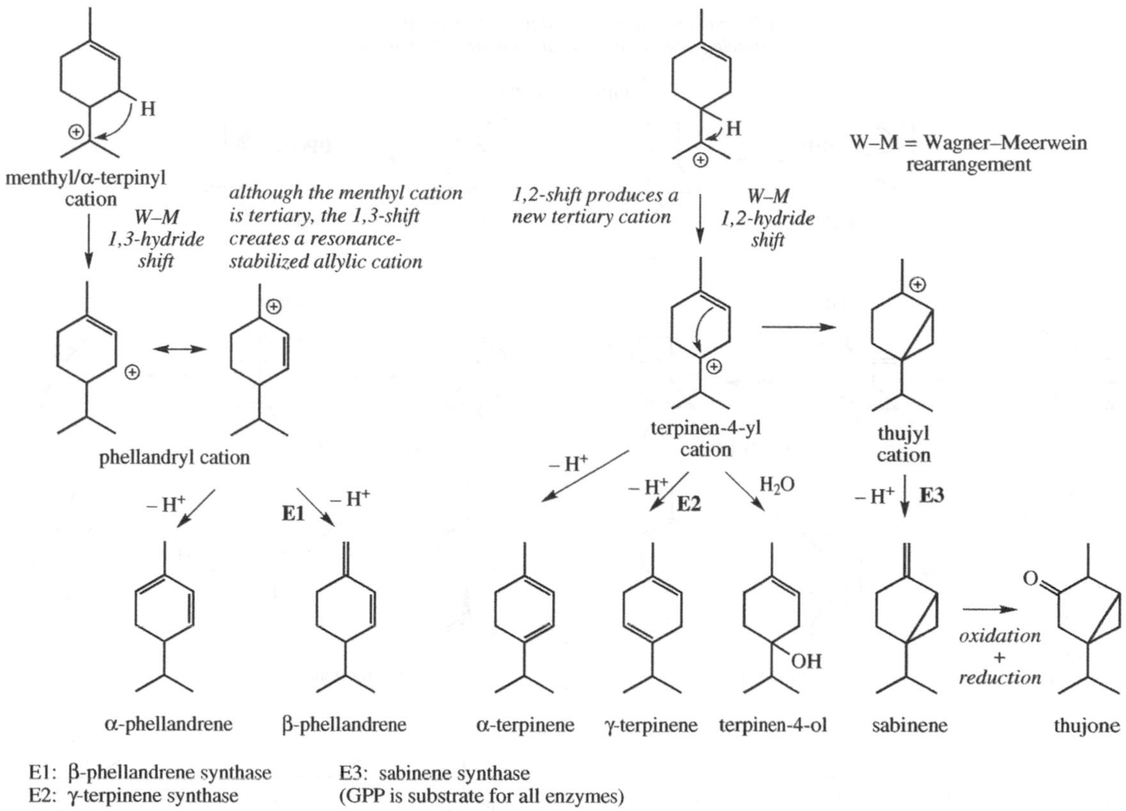

menthyl/α-terpinyl cation

W–M 1,3-hydride shift

although the menthyl cation is tertiary, the 1,3-shift creates a resonance-stabilized allylic cation

phellandryl cation

1,2-shift produces a new tertiary cation

W–M 1,2-hydride shift

W–M = Wagner–Meerwein rearrangement

terpinen-4-yl cation

thujyl cation

– H⁺   E1   – H⁺

– H⁺   – H⁺ E2   H₂O

– H⁺   E3

α-phellandrene   β-phellandrene   α-terpinene   γ-terpinene   terpinen-4-ol   sabinene   *oxidation + reduction*   thujone

E1: β-phellandrene synthase   E3: sabinene synthase
E2: γ-terpinene synthase   (GPP is substrate for all enzymes)

Figure 5.15

The menthyl cation, although it is a tertiary, may be converted by a 1,3-hydride shift into a favourable resonance-stabilized allylic cation. This allows the formation of α- and β-**phellandrenes** by loss of a proton from the phellandryl carbocation. A 1,2-hydride shift converting the menthyl cation into the terpinen-4-yl cation only changes one tertiary carbocation system for another, but allows formation of α-**terpinene**, γ-**terpinene**, and the α-terpineol isomer **terpinen-4-ol**. A further cyclization reaction on the terpinen-4-yl cation generates the thujane skeleton, e.g. **sabinene** and **thujone**. Terpinen-4-ol is the primary antibacterial component of tea tree oil from *Melaleuca alternifolia* (Myrtaceae); thujone has achieved notoriety as the neurotoxic agent in wormwood oil from *Artemisia absinthium* (Compositae/Asteraceae) used in preparation of the drink absinthe, now banned in most countries.

So far, little attention has been given to the stereochemical features of the resultant monoterpene. Individual enzyme systems present in a particular organism will, of course, control the folding of the substrate molecule and, thus, define the stereochemistry of the final product. Most monoterpenes are optically active, and there are many examples known where enantiomeric forms of the same compound can be isolated from different sources, e.g. (+)-**camphor** in sage (*Salvia officinalis*; Labiatae/Lamiaceae) and (−)-camphor in tansy (*Tanacetum vulgare*; Compositae/Asteraceae), or (+)-**carvone** in caraway (*Carum carvi*: Umbelliferae/Apiaceae) and (−)-carvone in spearmint (*Mentha spicata*; Labiatae/Lamiaceae). There are also examples of compounds found in both enantiomeric forms in the same organism, examples being (+)- and (−)-**limonene** in peppermint (*Mentha × piperita*; Labiatae/Lamiaceae) and (+)- and (−)-α-**pinene** in pine (*Pinus* species; Pinaceae). The individual enantiomers can produce different biological responses, especially towards olfactory receptors in the nose. Thus, the characteristic odour of caraway is due to (+)-carvone, whereas (−)-carvone smells of mint. (+)-Limonene smells of oranges, whilst (−)-limonene resembles the smell of lemons. The origins of the different enantiomeric forms of limonene and

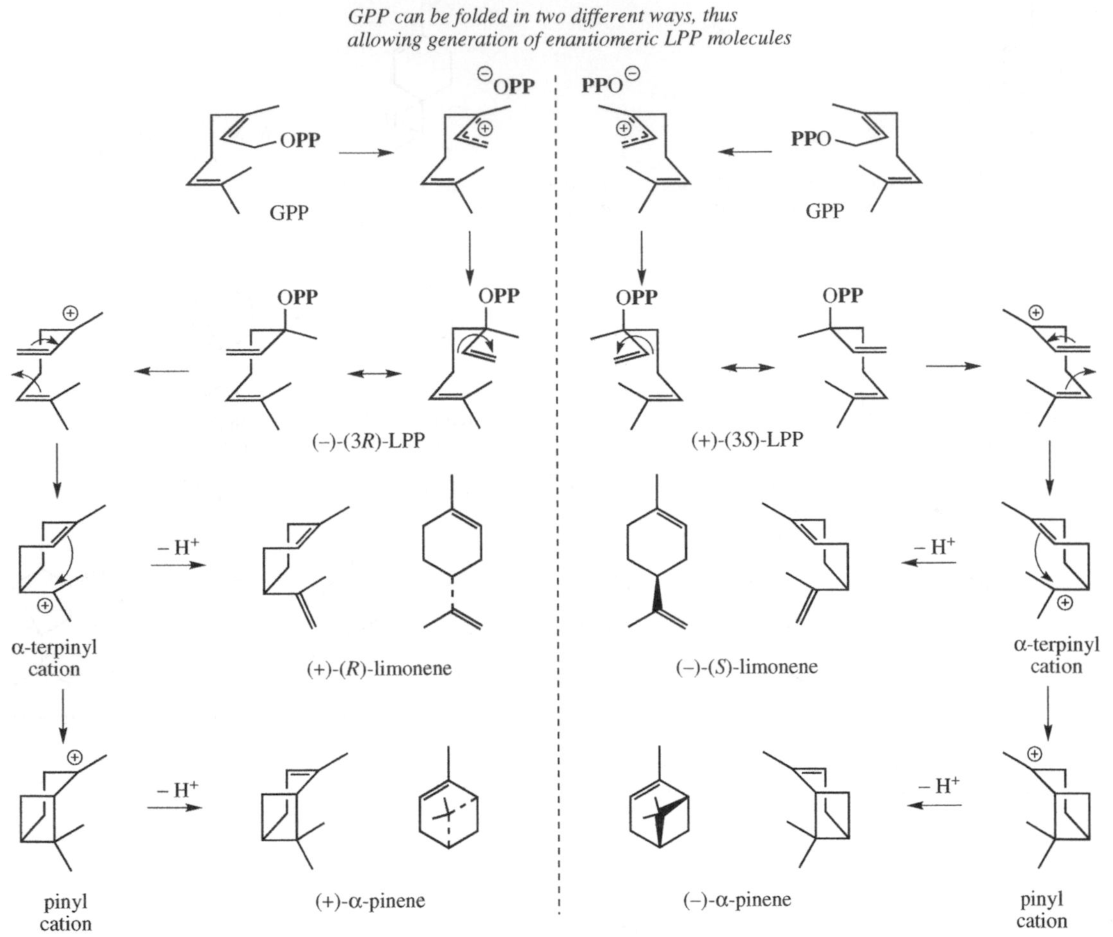

*GPP can be folded in two different ways, thus
allowing generation of enantiomeric LPP molecules*

(−)-(3R)-LPP   (+)-(3S)-LPP

α-terpinyl
cation   (+)-(R)-limonene   (−)-(S)-limonene   α-terpinyl
                                                                    cation

pinyl
cation   (+)-α-pinene   (−)-α-pinene   pinyl
                                                                    cation

Figure 5.16

α-pinene are illustrated in Figure 5.16. This shows the precursor GPP being folded in two mirror-image conformations, leading to formation of the separate enantiomers of linalyl PP. Analogous carbocation reactions will then explain production of the optically active monoterpenes. Where a single plant produces both enantiomers, it appears to contain two separate enzyme systems each capable of elaborating a single enantiomer. Furthermore, a single enzyme typically accepts GPP as substrate, catalyses the isomerization to linalyl PP, and converts this into a final product without the release of free intermediates.

Terpenoid cyclase enzymes, even highly pure proteins obtained by gene expression, rarely convert their substrate into a single product. Sometimes, multiple products in varying amounts, e.g. limonene, myrcene, α-pinene, and β-pinene, are synthesized by a single enzyme,

reflecting the common carbocation chemistry involved in these biosyntheses. This suggests that the enzyme is predominantly providing a suitable environment for the folding and cyclization of the substrate, whilst carbocation chemistry is responsible for product formation. Many of the transformations included in Figures 5.14 and 5.15 are based on analysis of these alternative products and how carbocation chemistry interrelates them. Subsequent reactions, such as oxidation of an alcohol to a ketone, e.g. borneol to **camphor** (Figure 5.14), or heterocyclic ring formation in the conversion of α-terpineol into **cineole** (Figure 5.14), require additional enzyme systems.

In other systems, a particular structure may be found as a mixture of diastereoisomers. Peppermint (*Mentha × piperita*; Labiatae/Lamiaceae) typically produces (−)-**menthol**, with smaller amounts of the stereoisomers (+)-**neomenthol**, (+)-**isomenthol**, and

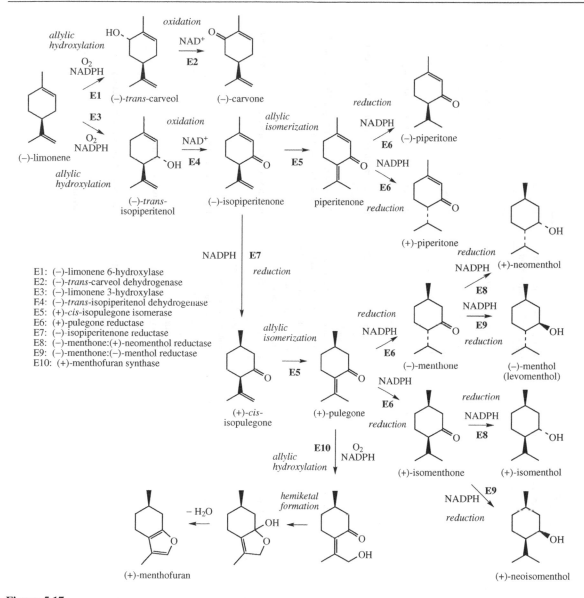

Figure 5.17

E1: (−)-limonene 6-hydroxylase
E2: (−)-*trans*-carveol dehydrogenase
E3: (−)-limonene 3-hydroxylase
E4: (−)-*trans*-isopiperitenol dehydrogenase
E5: (+)-*cis*-isopulegone isomerase
E6: (+)-pulegone reductase
E7: (−)-isopipcritenone reductase
E8: (−)-menthone:(+)-neomenthol reductase
E9: (−)-menthone:(−)-menthol reductase
E10: (+)-menthofuran synthase

(+)-**neoisomenthol**, covering four of the possible eight stereoisomers (Figure 5.17). Oils from various *Mentha* species also contain significant amounts of ketones, e.g. (−)-**menthone**, (+)-**isomenthone**, (−)-**piperitone**, or (+)-**pulegone**. The metabolic relationship of these various compounds has been established as in Figure 5.17, which illustrates how the stereochemistry at each centre can be determined by stereospecific reduction processes on double bonds or carbonyl groups. Note that some of the enzymes involved have rather broad substrate specificity. The pathway also exemplifies that oxygen functions can be introduced into the molecule at positions activated by adjacent double bonds (allylic oxidation), as well as being introduced by quenching of carbocations with water. These reactions are catalysed by cytochrome P-450-dependent monooxygenases. Thus, limonene is a precursor of **carvone** (the main constituent of spearmint oil from *Mentha spicata*) as well as menthone and piperitone, where initial hydroxylation occurs at an alternative allylic site on the ring. **Menthofuran** exemplifies a

Table 5.1   Volatile oils (ii): containing principally terpenoid compounds

Oil	Plant source	Plant part used	Oil content (%)	Major constituents with typical (%) composition	Uses, notes
Bergamot	*Citrus aurantium* ssp. *bergamia* (Rutaceae)	fresh fruit peel (expression)	0.5	limonene (42), linalyl acetate (27), γ-terpinene (8), linalool (7)	flavouring, aromatherapy, perfumery; also contains the furocoumarin bergapten (up to 5%) and may cause severe photosensitization (see page 165)
Camphor oil	*Cinnamomum camphora* (Lauraceae)	wood	1–3	camphor (27–45), cineole (4–21), safrole (1–18)	soaps
Caraway	*Carum carvi* (Umbelliferae /Apiaceae)	ripe fruit	3–7	(+)-carvone (50–70), limonene (47)	flavour, carminative, aromatherapy
Cardamom	*Elettaria cardamomum* (Zingiberaceae)	ripe fruit	3–7	α-terpinyl acetate (25–35), cineole (25–45), linalool (5)	flavour, carminative; ingredient of curries, pickles
Chamomile (Roman chamomile)	*Chamaemelum nobile* (*Anthemis nobilis*) (Compositae/ Asteraceae)	dried flowers	0.4–1.5	aliphatic esters of angelic, tiglic, isovaleric, and isobutyric acids (75–85), small amounts of monoterpenes	flavouring, aromatherapy; blue colour of oil is due to chamazulene
Citronella	*Cymbopogon winterianus, C. nardus* (Graminae/Poaceae)	fresh leaves	0.5–1.2	(+)-citronellal (25–55), geraniol (20–40), (+)-citronellol (10–15), geranyl acetate (8)	perfumery, aromatherapy, insect repellent
Coriander	*Coriandrum sativum* (Umbelliferae /Apiaceae)	ripe fruit	0.3–1.8	(+)-linalool (60–75), γ-terpinene (5), α-pinene (5), camphor (5)	flavour, carminative
Dill	*Anethum graveolens* (Umbelliferae /Apiaceae)	ripe fruit	3–4	(+)-carvone (40–65)	flavour, carminative
Eucalyptus	*Eucalyptus globulus, E. smithii; E. polybractea* (Myrtaceae)	fresh leaves	1–3	cineole (= eucalyptol) (70–85), α-pinene (14)	flavour, antiseptic, aromatherapy
Eucalyptus (lemon-scented)	*Eucalyptus citriodora* (Myrtaceae)	fresh leaves	0.8	citronellal (65–85)	perfumery

Name	Source (Family)	Plant part	%	Constituents (%)	Uses/notes
Ginger	*Zingiber officinale* (Zingiberaceae)	dried rhizome	1.5–3	zingiberene (34), β-sesquiphellandrene (12), β-phellandrene (8), β-bisabolene (6)	flavouring; the main pungent principals (gingerols) in ginger are not volatile (see page 168)
Juniper	*Juniperus communis* (Cupressaceae)	dried ripe berries	0.5–2	α-pinene (45–80), myrcene (10–25), limonene (1–10), sabinene (0–15)	flavouring, antiseptic, diuretic, aromatherapy; juniper berries provide the flavouring for gin
Lavender	*Lavandula angustifolia, L. officinalis* (Labiatae/Lamiaceae)	fresh flowering tops	0.3–1	linalyl acetate (25–45), linalool (25–38)	perfumery, aromatherapy; inhalation produces mild sedation and facilitates sleep
Lemon	*Citrus limon* (Rutaceae)	dried fruit peel (expression)	0.1–3	(+)-limonene (60–80), β-pinene (8–12), γ-terpinene (8–10), citral (= geranial + neral) (2–3)	flavouring, perfumery, aromatherapy; terpeneless lemon oil is obtained by removing much of the terpenes under reduced pressure; this oil is more stable and contains 40–50% citral
Lemon-grass	*Cymbopogon citratus* (Graminae/Poaceae)	fresh leaves	0.1–0.3	citral (= geranial + neral) (50–85)	perfumery, aromatherapy
Matricaria (German chamomile)	*Matricaria chamomilla* (*Chamomilla recutica*) (Compositae/Asteraceae)	dried flowers	0.3–1.5	(−)-α-bisabolol (10–25), bisabolol oxides A and B (10–25), chamazulene (1–15)	flavouring; dark blue colour of oil is due to chamazulene
Orange (bitter)	*Citrus aurantium* ssp. *amara* (Rutaceae)	dried fruit peel (expression)	0.5–2.5	(+)-limonene (92–94), myrcene (2)	flavouring, aromatherapy; the main flavour and odour come from the minor oxygenated components; terpeneless orange oil is obtained by removing much of the terpenes under reduced pressure; this oil contains about 20% aldehydes, mainly decanal
Orange (sweet)	*Citrus sinensis* (Rutaceae)	dried fruit peel (expression)	0.3	(+)-limonene (90–95), myrcene (2)	flavouring, aromatherapy; the main flavour and odour come from the minor oxygenated components; terpeneless orange oil is obtained by removing much of the terpenes under reduced pressure; this oil contains about 20% aldehydes, mainly octanal and decanal

(*continued overleaf*)

Table 5.1   (*continued*)

Oil	Plant source	Plant part used	Oil content (%)	Major constituents with typical (%) composition	Uses, notes
Orange flower (Neroli)	*Citrus aurantium* ssp. *amara* (Rutaceae)	fresh flowers	0.1	linalool (36), β-pinene (16), limonene (12), linalyl acetate (6)	flavour, perfumery, aromatherapy
Peppermint	*Mentha × piperita* (Labiatae/Lamiaceae)	fresh leaf	1–3	menthol (30–50), menthone (15–32), menthyl acetate (2–10), menthofuran (1–9)	flavouring, carminative, aromatherapy
Pine	*Pinus palustris* or other *Pinus* species (Pinaceae)	needles, twigs		α-terpineol (65)	antiseptic, disinfectant, aromatherapy
Pumilio pine	*Pinus mugo* ssp. *pumilio* (Pinaceae)	needles	0.3–0.4	α- and β-phellandrene (60), α- and β-pinene (10–20), bornyl acetate (3–10)	inhalant; the minor components bornyl acetate and borneol are mainly responsible for the aroma
Rose (attar of rose, otto of rose)	*Rosa damascena, R. gallica, R. alba,* and *R. centifolia* (Rosaceae)	fresh flowers	0.02–0.03	citronellol (36), geraniol (17), 2-phenylethanol (3), C_{14}–C_{23} straight-chain hydrocarbons (25)	perfumery, aromatherapy
Rosemary	*Rosmarinus officinalis* (Labiatae/Lamiaceae)	fresh flowering tops	1–2	cineole (15–45), α-pinene (10–25), camphor (10–25), β-pinene (8)	perfumery, aromatherapy
Sage	*Salvia officinalis* (Labiatae/Lamiaceae)	fresh flowering tops	0.7–2.5	thujone (40–60), camphor (5–22), cineole (5–14), β-caryophyllene (10), limonene (6)	aromatherapy, food flavouring
Sandalwood	*Santalum album* (Santalaceae)	heartwood	4.5–6.3	sesquiterpenes: α-santalol (50), β-santalol (21)	perfumery, aromatherapy
Spearmint	*Mentha spicata* (Labiatae/Lamiaceae)	fresh leaf	1–2	(−)-carvone (50–70), (−)-limonene (2–25)	flavouring, carminative, aromatherapy

Tea tree	*Melaleuca alternifolia* (Myrtaceae)	fresh leaf	1.8	terpinen-4-ol (30–45), γ-terpinene (10–28), α-terpinene (5–13), p-cymene (0.5–12), cineole (0.5–10), α-terpineol (1.5–8)	antiseptic, aromatherapy; an effective broad-spectrum antiseptic widely used in creams, cosmetics, toiletries
Thyme	*Thymus vulgaris* (Labiatae/Lamiaceae)	fresh flowering tops	0.5–2.5	thymol (40), p-cymene (30), linalool (7), carvacrol (1)	antiseptic, aromatherapy, food flavouring
Turpentine oil	*Pinus palustris* and other *Pinus* spp. (Pinaceae)	distillation of the resin (turpentine) secreted from bark		(+)- and (−)-α-pinene (35:65) (60–70), β-pinene (20–25)	counter-irritant, important source of industrial chemicals; the residue from distillation is colophony (rosin), composed chiefly of diterpene acids (abietic acids, see page 230)

Note: Major volatile oils have been divided into two groups. Those oils containing principally chemicals which are terpenoid in nature and which are derived by the methylerythritol phosphate or mevalonate pathways are given in this table. Oils which are composed predominantly of aromatic compounds which are derived via the shikimate pathway are listed in Table 4.1 on page 158. The remarks in the footnotes to Table 4.1 are also applicable here. Structures of terpenoids are shown in Figures 5.11–5.18.

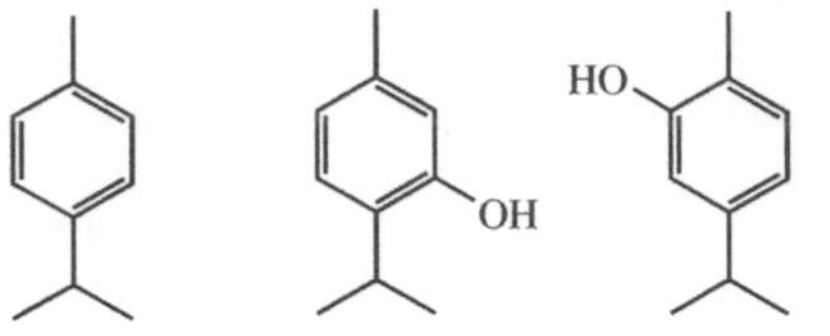

p-cymene   thymol   carvacrol

Figure 5.18

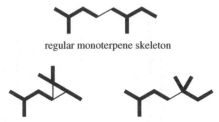

regular monoterpene skeleton

irregular monoterpene skeletons

Figure 5.19

further cytochrome P-450-dependent oxidative modification that leads to generation of a heterocyclic furan ring via the hemiketal. Both pulegone and menthofuran are considered hepatotoxic. Pulegone is a major constituent of oil of pennyroyal from *Mentha pulegium*, which has a folklore history as an abortifacient. Pulegone is metabolized in humans first to menthofuran, by the same mechanism as it is in the plant pathway, and then to electrophilic metabolites that form adducts with cellular proteins (compare pyrrolizidine alkaloids, page 325). High menthofuran levels in peppermint oils are regarded as undesirable. Peppermint plants transformed with an antisense version of the menthofuran synthase gene produce less than half the normal amounts of this metabolite.

p-**Cymene** and the phenol derivatives **thymol** and **carvacrol** (Figure 5.18), found in thyme (*Thymus vulgaris*; Labiatae/Lamiaceae), are representatives of a small group of aromatic compounds that are produced in nature from isoprene units, rather than by the much more common routes to aromatics involving acetate or shikimate (see also cannabinol, page 119, and gossypol, page 220). These compounds all possess the carbon skeleton typical of monocyclic monoterpenes, and their structural relationship to limonene and the more common oxygenated monoterpenes, such as menthone or carvone, suggests pathways in which additional dehydrogenation reactions are involved.

Data on volatile oils containing terpenoid constituents isolated from these and other plant materials are given in Table 5.1. Volatile oils in which the main components are aromatic and derived from the shikimate pathway are listed in Table 4.1, page 158.

IRREGULAR MONOTERPENES

A number of natural monoterpene structures contain carbon skeletons which, although obviously derived from two isoprene C_5 units, do not seem to fit the regular head-to-tail coupling mechanism, e.g. those in Figure 5.19. These structures are termed irregular monoterpenes and seem to be limited almost exclusively to members of the Compositae/Asteraceae plant family. Allowing for possible rearrangements, the two isoprene units appear to have coupled in another manner, and this is borne out by information available on their biosynthesis, though this is far from complete. Thus, although DMAPP and IPP are utilized in their biosynthesis, GPP and neryl PP do not appear to be involved. Pre-eminent amongst these structures are **chrysanthemic acid** and **pyrethric acid** (Figure 5.20), found in ester form as the **pyrethrins** (pyrethrins, cinerins, and jasmolins, Figure 5.20), which are valuable insecticidal components in pyrethrum flowers, the flower heads of *Chrysanthemum cinerariaefolium* (Compositae/Asteraceae) [Box 5.1]. These cyclopropane structures are readily recognizable as derived from two isoprene units, and a mechanism for the derivation of chrysanthemic acid via chrysanthemyl diphosphate is given in Figure 5.21. This invokes two DMAPP units joining by a modification of the standard mechanism, with termination achieved by cyclopropane ring formation (compare carene, page 196). This mechanism is seen again later; it is identical to that involved in the formation of presqualene PP during steroid biosynthesis (see page 235) and of prephytoene PP in carotenoid formation (see page 300). Relatively little is known about the origins of **pyrethrolone, cinerolone**, and **jasmolone** (Figure 5.20), the alcohol portions of the pyrethrins, though it is likely that these are cyclized and modified fatty acid derivatives, the cyclization process resembling the biosynthetic pathway to prostaglandins (see page 60). Thus, α-linolenic acid via 12-oxophytodienoic acid could be the precursor of jasmolone, with β-oxidation and then decarboxylation accounting for the chain shortening (Figure 5.22). Certainly, this type of pathway operates in the formation of **jasmonic acid** (Figure 5.22), which forms part of a general signalling system in plants, particularly the synthesis of secondary metabolites in response to wounding or microbial infection.

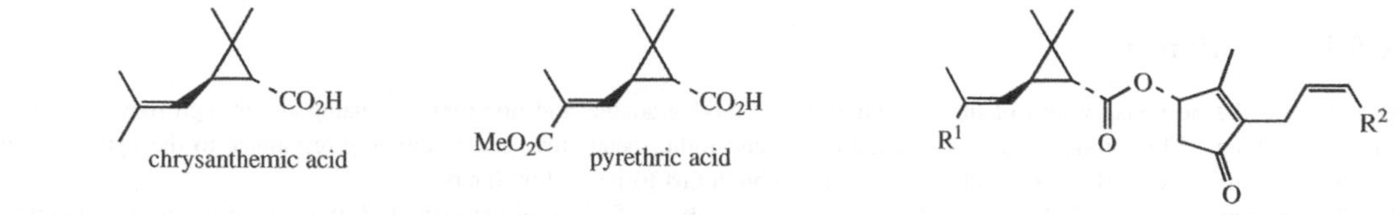
CO2H
chrysanthemic acid
MeO2C
CO2H
pyrethric acid
R1
O
R2

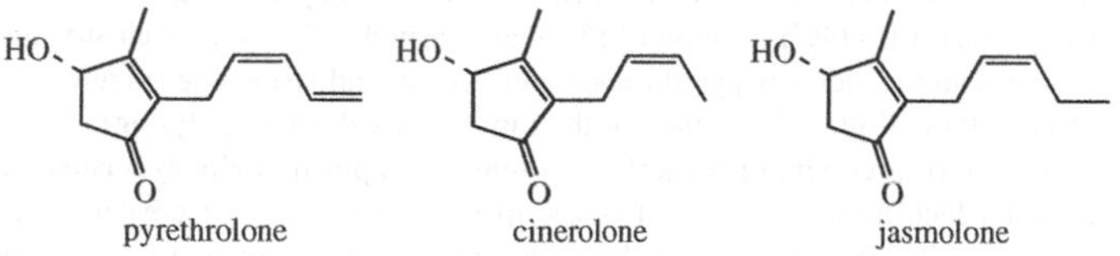
HO
O
pyrethrolone
cinerolone
jasmolone

	R^1	R^2
pyrethrin I	Me	CH=CH$_2$
cinerin I	Me	Me
jasmolin I	Me	Et
pyrethrin II	CO_2Me	CH=CH$_2$
cinerin II	CO_2Me	Me
jasmolin II	CO_2Me	Et

Figure 5.20

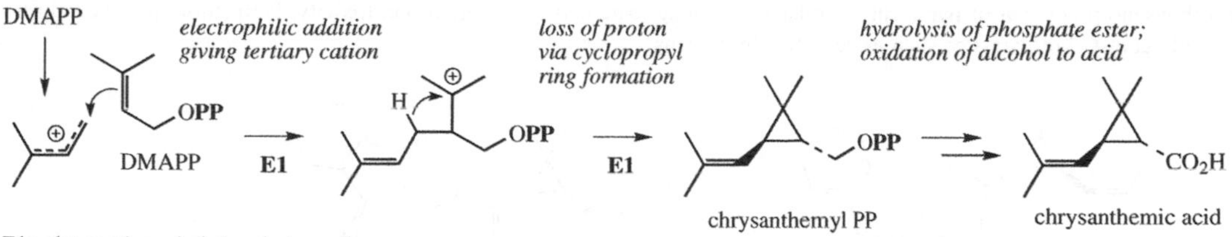
DMAPP
electrophilic addition
giving tertiary cation
OPP
DMAPP
E1
loss of proton
via cyclopropyl
ring formation
H
OPP
E1
hydrolysis of phosphate ester;
oxidation of alcohol to acid
OPP
chrysanthemyl PP
CO2H
chrysanthemic acid

E1: chrysanthemyl diphosphate synthase (substrate DMAPP)

Figure 5.21

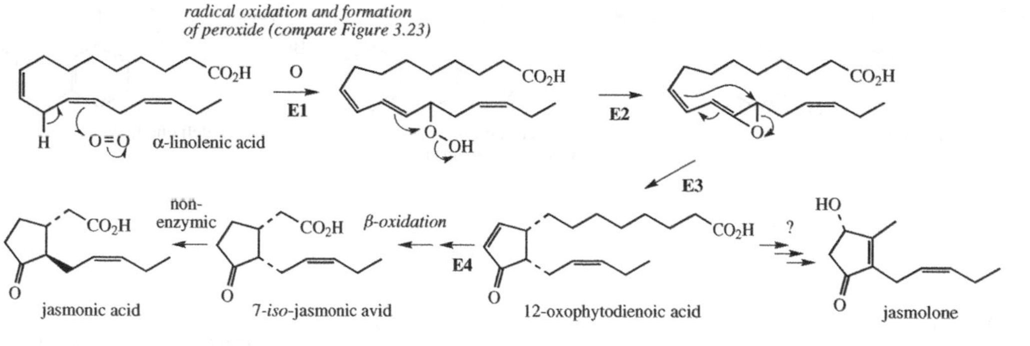
radical oxidation and formation
of peroxide (compare Figure 3.23)
CO2H
O
E1
H
α-linolenic acid
OH
E2
E3
non-
enzymic
β-oxidation
E4
?
HO
jasmonic acid
7-iso-jasmonic avid
12-oxophytodienoic acid
jasmolone

E1: lipoxygenase
E2: allene oxide synthase
E3: allene oxide cyclase
E4: 12-oxophytodienoate reductase

Figure 5.22

Box 5.1

Pyrethrins

The **pyrethrins** are valuable insecticidal components of pyrethrum flowers, *Chrysanthemum cinerariaefolium* (=*Tanacetum cinerariifolium*) (Compositae/Asteraceae). The flowers are harvested just before they are fully expanded, and usually processed to an extract. Pyrethrum cultivation is conducted in East Africa, especially Kenya, and more recently in Ecuador and Australia. The natural pyrethrins are used as a constituent of insect sprays for household use and as a post-harvest insecticides, having a

Box 5.1   (continued)

rapid action on the nervous system of insects, whilst being biodegradable and non-toxic to mammals, though they are toxic to fish and amphibians. This biodegradation, initiated by air and light, means few insects develop resistance to the pyrethrins, but it does limit the lifetime of the insecticide under normal conditions to just a few hours.

The flowers may contain 0.7–2% of pyrethrins, representing about 25–50% of the extract. A typical pyrethrin extract contains pyrethrin I (35%), pyrethrin II (32%), cinerin I (10%), cinerin II (14%), jasmolin I (5%), and jasmolin II (4%), which structures represent esters of chrysanthemic acid or pyrethric acid with the alcohols pyrethrolone, cinerolone, and jasmolone (Figure 5.20). Pyrethrin I is the most insecticidal component, with pyrethrin II providing much of the rapid knock-down (paralysing) effect.

A wide range of synthetic pyrethroid analogues, e.g. **bioresmethrin, tetramethrin, phenothrin, permethrin, cypermethrin**, and **deltamethrin** (Figure 5.23), have been developed which have increased lifetimes up to several days and greater toxicity towards insects. These materials have become widely used household and agricultural insecticides; the commercial insecticides are often a mixture of stereoisomers. Tetramethrin, bioresmethrin, and phenothrin are all esters of chrysanthemic acid, but with a modified alcohol portion, providing improvements in knock-down effect and in insecticidal activity. Replacement of the terminal methyl groups of chrysanthemic acid with halogen atoms, e.g. permethrin, conferred greater stability towards air and light and opened up the use of pyrethroids in agriculture. Inclusion of a cyano group in the alcohol portion, as in cypermethrin and deltamethrin, improved insecticidal activity several-fold. Modern pyrethroids now have insecticidal activities over a thousand times that of pyrethrin I, whilst maintaining extremely low mammalian toxicity. Permethrin and phenothrin are employed against skin parasites such as head lice in humans.

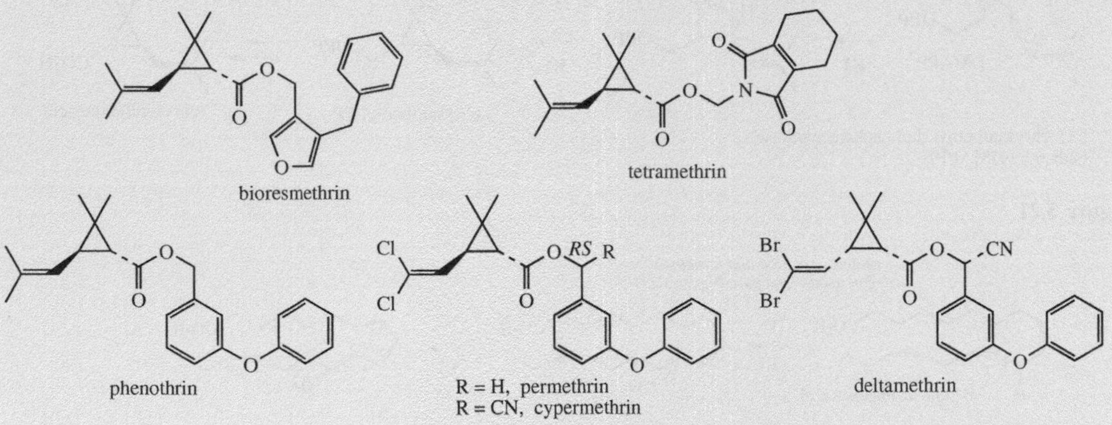

bioresmethrin

tetramethrin

phenothrin

R = H, permethrin
R = CN, cypermethrin

deltamethrin

Figure 5.23

IRIDOIDS (C$_{10}$)

The **iridane** skeleton (Figure 5.24) found in **iridoids** is monoterpenoid in origin and contains a cyclopentane ring which is usually fused to a six-membered oxygen heterocycle. The iridoid system arises from geraniol by a type of folding (Figure 5.25) which is different from that already encountered with monoterpenoids; also markedly different is the lack of phosphorylated intermediates and subsequent carbocation mechanism in its formation. The fundamental cyclization to **iridodial** is formulated as attack of hydride on the dialdehyde, produced by a series of hydroxylation and oxidation reactions on geraniol. Further oxidation gives **iridotrial**, in which hemiacetal formation then leads to production of the heterocyclic ring. In

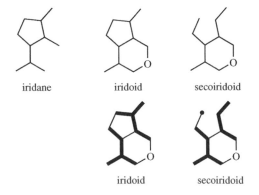

iridane   iridoid   secoiridoid

iridoid   secoiridoid

Figure 5.24

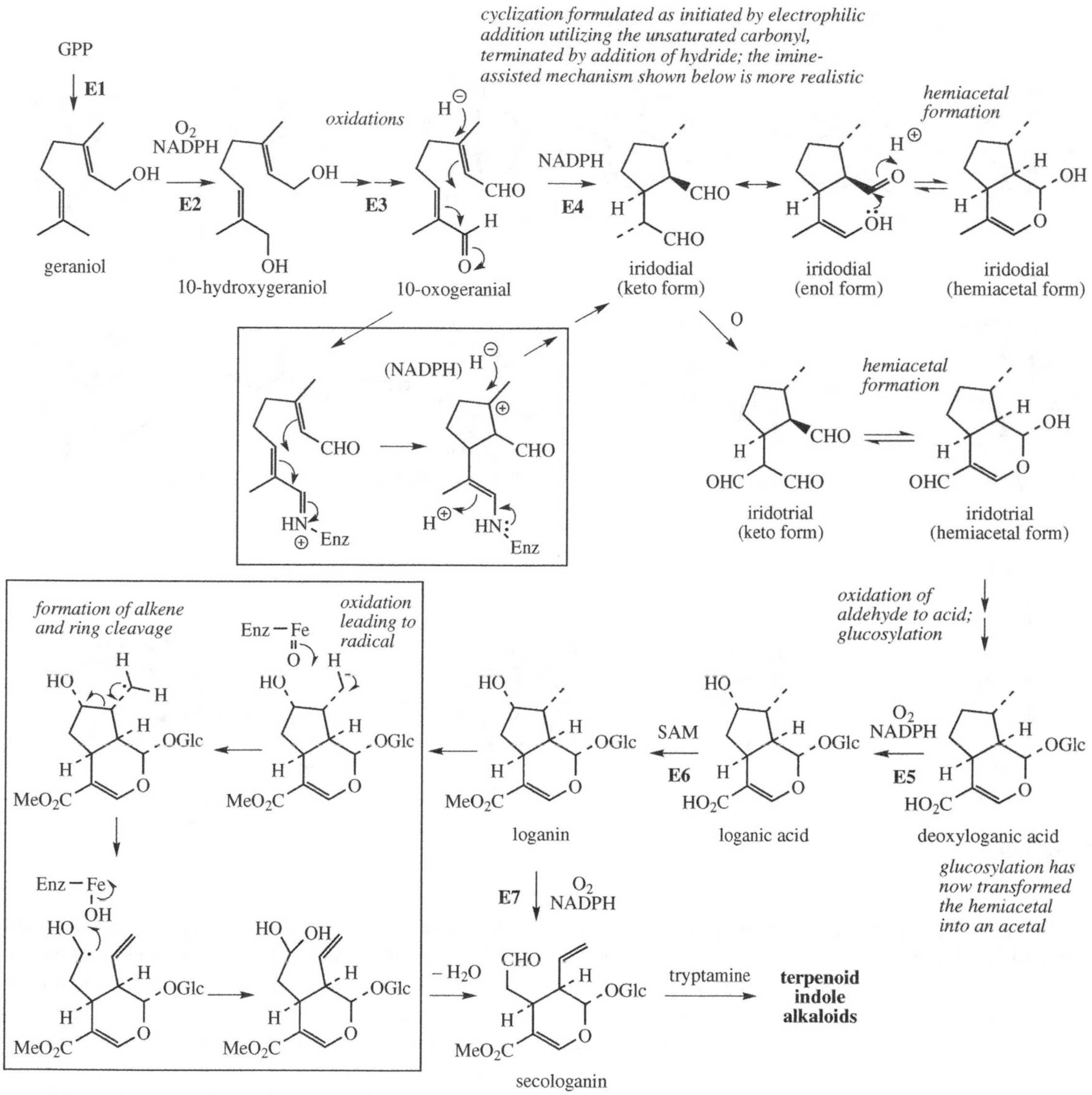

Figure 5.25

E1: geraniol synthase
E2: geraniol 10-hydroxylase
E3: 10-hydroxygeraniol oxidoreductase
(acyclic monoterpene primary alcohol dehydrogenase)

E4: monoterpene cyclase
E5: 7-deoxyloganin 7-hydroxylase (also accepts acid as substrate)
E6: loganic acid methyltransferase
E7: secologanin synthase

iridotrial, there is an equal chance that the original methyl groups from the head of geraniol end up as the aldehyde or in the heterocyclic ring. A large number of iridoids are found as glycosides, e.g. **loganin**; glycosylation effectively transforms the hemiacetal linkage into an acetal. The pathway to loganin involves, in addition, a sequence of reactions in which the remaining aldehyde group is oxidized to the acid and methylated, and the cyclopentane ring is hydroxylated. Loganin is a key intermediate in the pathway to a range of complex terpenoid indole alkaloids (see page 369) and tetrahydroisoquinoline alkaloids (see page 363). Fundamental in this further metabolism is cleavage of the simple monoterpene skeleton, which is still recognizable in

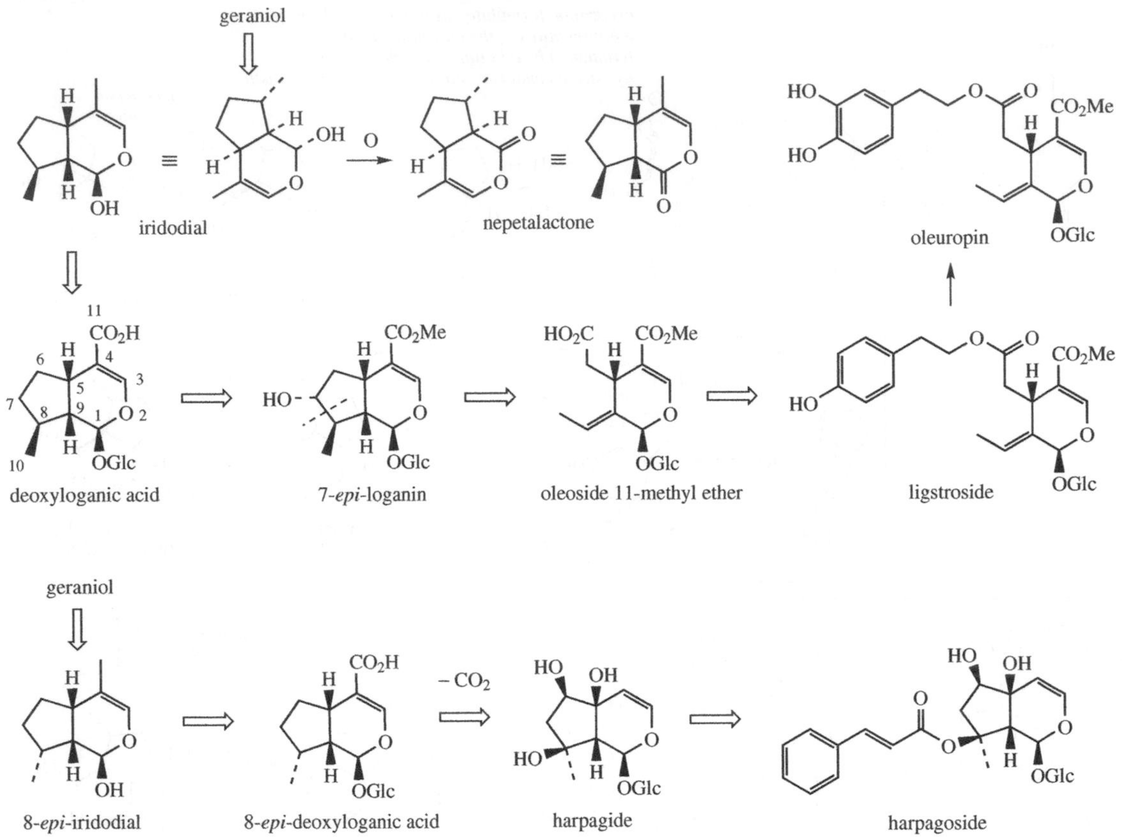

Figure 5.26

loganin, to give **secologanin**, representative of the **secoiridoids** (Figure 5.24). This is catalysed by a cytochrome P-450-dependent monooxygenase, and a radical mechanism is proposed in Figure 5.25. Secologanin now contains a free aldehyde group, together with further aldehyde and enol groups, with these latter two fixed as an acetal by the presence of the glucose. As we shall see with some of the complex alkaloids, these functionalities can be released again by hydrolysing off the glucose and reopening the hemiacetal linkage.

Well over a thousand different natural iridoids and secoiridoids are known. Structural variation arises predominantly from hydroxylations, esterifications, and changes in stereochemistry. Another major change is the loss of a carbon atom by a decarboxylation mechanism. A few examples showing their relationship to the intermediates of Figure 5.25 are presented in Figure 5.26.

Nepetalactone from catmint *Nepeta cataria* (Labiatae/Lamiaceae), a powerful attractant and stimulant for cats, is produced from iridodial by simple oxidation of the hemiacetal to a lactone (Figure 5.26). **Ligstroside** and **oleuropin** are phenolic esters found in olive oil (see page 47) from *Olea europea* (Oleaceae), and are believed to contribute to the health benefits of this oil. These compounds are secoiridoids, but are formed from the 7-epimer of loganin; in this case, the stereochemistry of 7-hydroxylation is different from the normal loganin pathway. **Harpagide** and **harpagoside** are examples of decarboxylated iridoids; these particular examples are formed via 8-*epi*-iridodial, in which geraniol is cyclized by a stereochemically different mechanism. Harpagoside is the cinnamoyl ester of harpagide. These compounds are found in devil's claw (*Harpagophytum procumbens*; Pedaliaceae) and contribute to the anti-inflammatory and analgesic properties associated with this plant drug [Box 5.2].

Box 5.2

Devil's Claw

Devil's claw is the common name for *Harpagophytum procumbens* (Pedaliaceae) from the appearance of its fruit, which have curved, sharp hooks. The plant is a weedy, perennial, tuberous plant with long creeping stems found in southern Africa (South Africa, Namibia, Botswana); commercial material is collected from the wild, but is often a mixture from two species: *H. procumbens* and *H. zeyheri*. Preparations of the secondary roots have gained a reputation as an anti-inflammatory and antirheumatic agent to relieve pain and inflammation in people with arthritis and similar disorders. Clinical studies appear to support its medicinal value as an anti-inflammatory and analgesic, though some findings are less positive.

The main constituents of devil's claw root are a group of decarboxylated iridoid glycosides (about 3%), including harpagoside (at least 1.2%) as the main component and smaller amounts of procumbide, harpagide, and 8-(4-coumaroyl)harpagide (Figure 5.27). The latter compound appears representative of *H. zeyheri* only, so indicates if this species is present in the sample. The secondary roots contain significantly higher levels of iridoids than the primary tubers. Harpagoside and related iridoids have been shown to inhibit thromboxane biosynthesis (see page 65), which may relate to the observed anti-inflammatory activity of devil's claw.

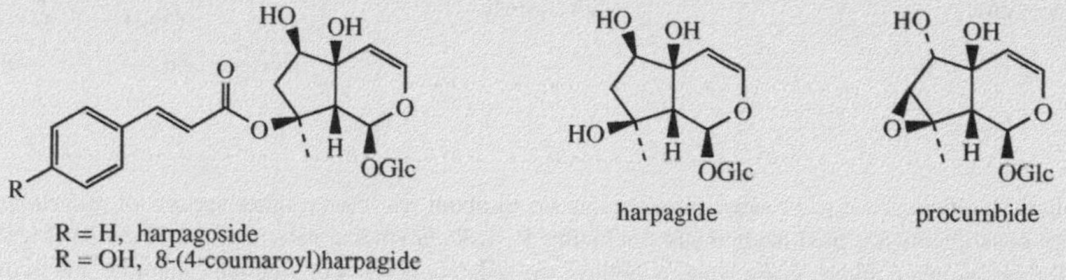

R = H, harpagoside
R = OH, 8-(4-coumaroyl)harpagide

harpagide

procumbide

Figure 5.27

A range of epoxyiridoid esters has been identified in the drug valerian (*Valeriana officinalis*; Valerianaceae) [Box 5.3]. These materials, responsible for the sedative activity of the crude drug, are termed **valepotriates. Valtrate** (Figure 5.28) is a typical example, and illustrates the structural relationship to loganin, though these compounds contain additional ester functions, frequently isovaleryl. The hemiacetal is now fixed as an ester, rather than as a glycoside.

Box 5.3

Valerian

Valerian root consists of the dried underground parts of *Valeriana officinalis* (Valerianaceae), a perennial herb found throughout Europe. Drug material comes from wild and cultivated plants, and is carefully dried at low temperature (less than 40°C) to minimize decomposition of constituents. Valerian preparations are widely used as herbal tranquillizers to relieve nervous tension, anxiety, and as a mild sedative to promote sleep; the drug was especially popular during the First World War, when it was used to treat shell-shock. The drug does possess mild sedative and anxiolytic properties, but the roots need to be freshly harvested and carefully dried for maximum activity. The major active principles are generally held to be a number of epoxyiridoid esters called valepotriates (0.5–1.6%), the principal component of which is valtrate (about 80%) (Figure 5.28). Minor valepotriates have the same parent iridoid alcohol as valtrate, but differ with respect to esterifying acids, e.g. isovaltrate and acevaltrate (Figure 5.28), or are based on the reduced iridoid seen in didrovaltrate, again with various ester functionalities. Acid entities characterized in this group of compounds are mainly isovaleric (3-methylbutyric) and acetic (as in valtrate/isovaltrate/didrovaltrate), though more complex diester groups involving 3-acetoxyisovaleric and isovaleroxyisovaleric acids are encountered. During drying and storage, some of the valepotriate content may decompose by hydrolysis to liberate quantities of isovaleric acid, giving a characteristic odour, and structures such as baldrinal (Figure 5.28) (from valtrate) and homobaldrinal (from isovaltrate). Samples of old or poorly prepared valerian may contain negligible amounts of valepotriates. Standardized mixtures of valepotriates are available in some countries. These materials are usually extracted from the roots of other species of *Valeriana* which produce higher amounts

Box 5.3　(continued)

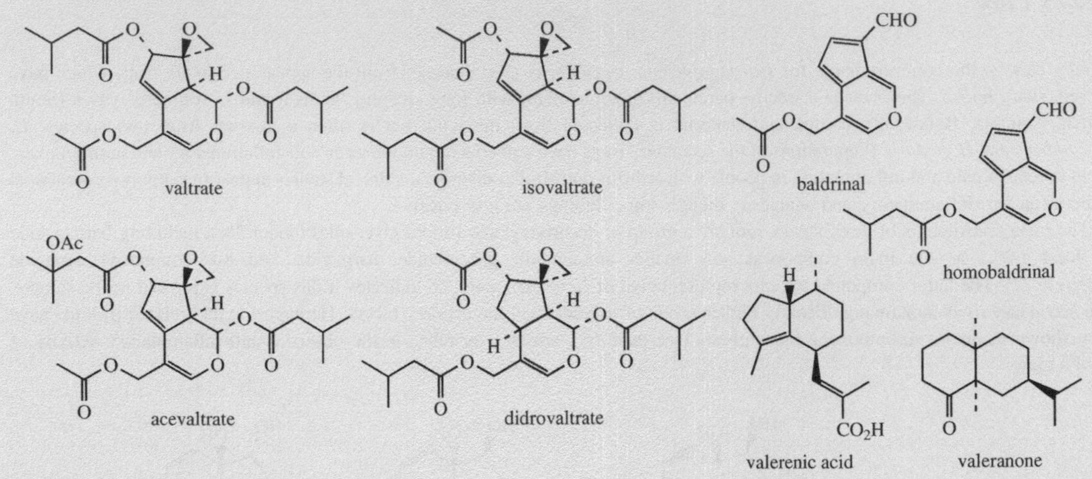

valtrate

isovaltrate

baldrinal

homobaldrinal

acevaltrate

didrovaltrate

valerenic acid

valeranone

Figure 5.28

of valepotriates than *V. officinalis*, e.g. *V. mexicana* contains up to about 8%. Some other species of *Valeriana* which contain similar valepotriate constituents are used medicinally, including *V. wallichi* (Indian valerian) and *V. edulis* (Mexican valerian).

Despite the information given above, many workers believe the sedative activity of valerian cannot be due to the valepotriates, which are rather unstable and not water soluble. Some of the sedative activity is said to arise from sesquiterpene derivatives such as valerenic acid (about 0.3%) and those found in the volatile oil content (0.5–1.3%), e.g. valeranone (Figure 5.28), which have been shown to be physiologically active. GABA and glutamine have also been identified in aqueous extracts of valerian, and these have been suggested to contribute to the sedative properties. The valepotriates valtrate and didrovaltrate are reported to be cytotoxic *in vitro*, and this may restrict future use of valerian. The reactive epoxide group is likely to be responsible for these cytotoxic properties.

SESQUITERPENES (C_{15})

Sesquiterpenes are formed from three C_5 units; the terminology comes from the Latin prefix sesqui: 'one and a half times'. Though the mechanisms involved in their formation closely parallel those seen for the monoterpenes, sesquiterpenes are generally synthesized via the mevalonate pathway (see page 192) rather than from MEP. Addition of a further C_5 IPP unit to GPP in an extension of the GPP synthase reaction (Figure 5.9) leads to the fundamental sesquiterpene precursor **farnesyl diphosphate (FPP)** (Figure 5.29). Again, an initial ionization of GPP seems likely, and the proton lost from C-2 of IPP is stereochemically analogous to that lost in the previous isoprenylation step.

FPP can then give rise to linear and cyclic sesquiterpenes. Because of the increased chain length and an additional double bond, the number of possible cyclization modes is increased, and a huge range of mono-, bi-, and tri-cyclic structures can result; the number of known natural sesquiterpenes greatly exceeds that of known natural

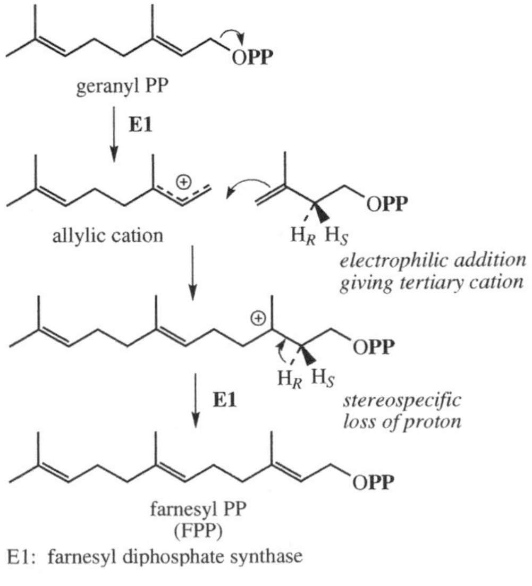

geranyl PP

E1

allylic cation

electrophilic addition giving tertiary cation

H_R H_S

E1

stereospecific loss of proton

H_R H_S

farnesyl PP
(FPP)

E1: farnesyl diphosphate synthase

Figure 5.29

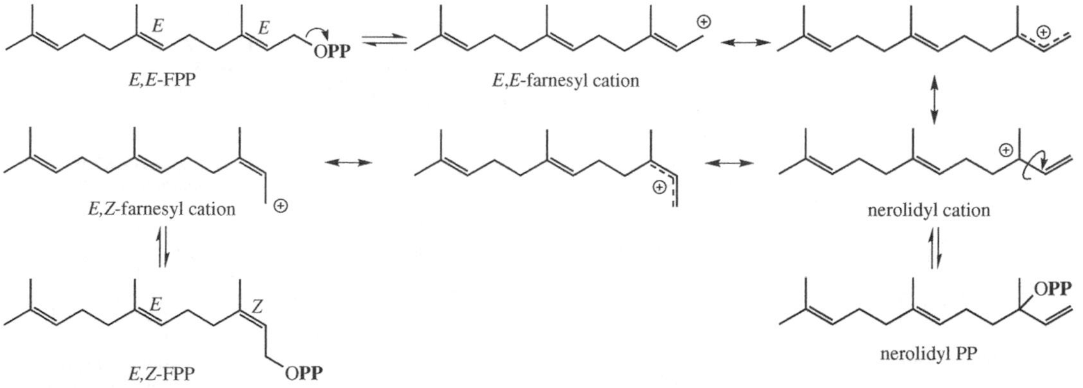

Figure 5.30

monoterpenes. The stereochemistry of the double bond nearest the diphosphate can adopt an *E* configuration (as in FPP) or a *Z* configuration via ionization as found with GPP/neryl PP (Figure 5.30). In some systems, the tertiary diphosphate **nerolidyl PP** (compare linalyl PP, page 194) has been implicated as a more immediate precursor than FPP (Figure 5.30). This allows different possibilities for folding the carbon chain, dictated of course by the enzyme involved, and cyclization by electrophilic attack onto an appropriate double bond.

As with the monoterpenes, standard reactions of carbocations rationally explain most of the common structural skeletons encountered, and a small but representative selection of these is given in Figure 5.31; sesquiterpene cyclase enzymes typically synthesize a major product accompanied by a range of related structures. One of these cyclized systems, the bisabolyl cation, is analogous to the monoterpene menthane system, and further modifications in the six-membered ring can take place to give essentially monoterpene variants with an extended hydrocarbon substituent, e.g. **γ-bisabolene** (Figure 5.32), which contributes to the aroma of ginger (*Zingiber officinale*; Zingiberaceae) along with obviously related structures such as **zingiberene** and **β-sesquiphellandrene** (Figure 5.33). Sesquiterpenes will, in general, be less volatile than monoterpenes. Simple quenching of the bisabolyl cation with water leads to **α-bisabolol** (Figure 5.32), a major component of matricaria (German chamomile) flowers (*Matricaria chamomilla*; Compositae/Asteraceae) [Box 5.5]. So-called **bisabolol oxides** A and B are also present, compounds probably derived from bisabolol by cyclization reactions (Figure 5.32) on an intermediate epoxide (compare Figure 4.34, page 164).

Other cyclizations in Figure 5.31 lead to ring systems larger than six carbon atoms, and 7-, 10-, and

11-membered rings can be formed as shown. The two 10-membered ring systems (germacryl and *cis*-germacryl cations), or the two 11-membered systems (humulyl and *cis*-humulyl cations), differ only in the stereochemistry associated with the double bonds. However, this affects further cyclization processes and is responsible for extending the variety of natural sesquiterpene derivatives. The germacryl cation, without further cyclization, is a precursor of the germacrane class of sesquiterpenes, as exemplified by **costunolide** (Figure 5.34), a bitter principle found in the roots of chicory (*Cichorium intybus*; Compositae/Asteraceae). Costunolide is actually classified as a germacranolide, the suffix 'olide' referring to the lactone group (compare macrolide, page 68). The antimigraine agent in feverfew (*Tanacetum parthenium*; Compositae/Asteraceae) is **parthenolide** [Box 5.4], an epoxide derivative of costunolide (Figure 5.34). At present it is not known whether epoxidation of the double bond in costunolide is involved, or whether an earlier intermediate in the pathway is oxidized.

The α,β-unsaturated carbonyl functionality seen in costunolide and parthenolide is a common feature of many of the biologically active terpenoids. The activity frequently manifests itself as a toxicity, especially cytotoxicity, as seen with the germacranolide **elephantopin** (Figure 5.33) from *Elephantopus elatus* (Compositae/Asteraceae), or skin allergies, as caused by the pseudoguaianolide (a rearranged guaianolide) **parthenin** (Figure 5.33) from *Parthenium hysterophorus* (Compositae/Asteraceae), a highly troublesome weed in India. These compounds can be considered as powerful alkylating agents by a Michael-type addition of a suitable nucleophile, e.g. thiols, onto the α,β-unsaturated carbonyl system. Such alkylation reactions are believed to explain

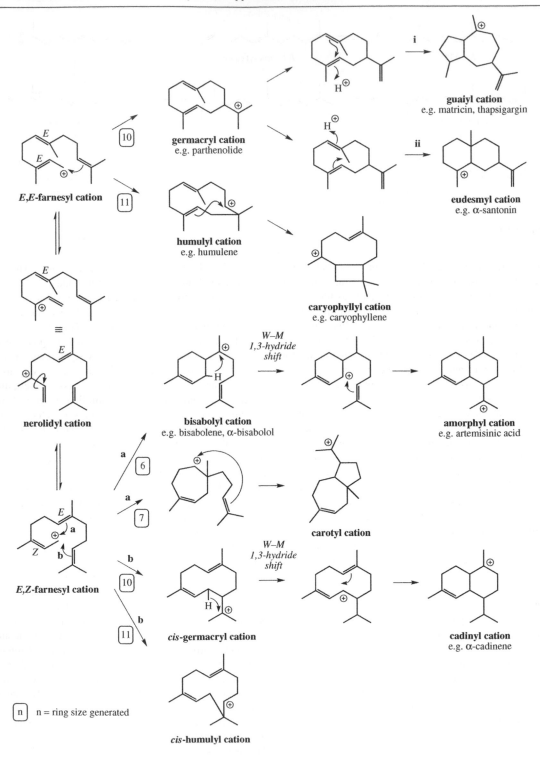

Figure 5.31

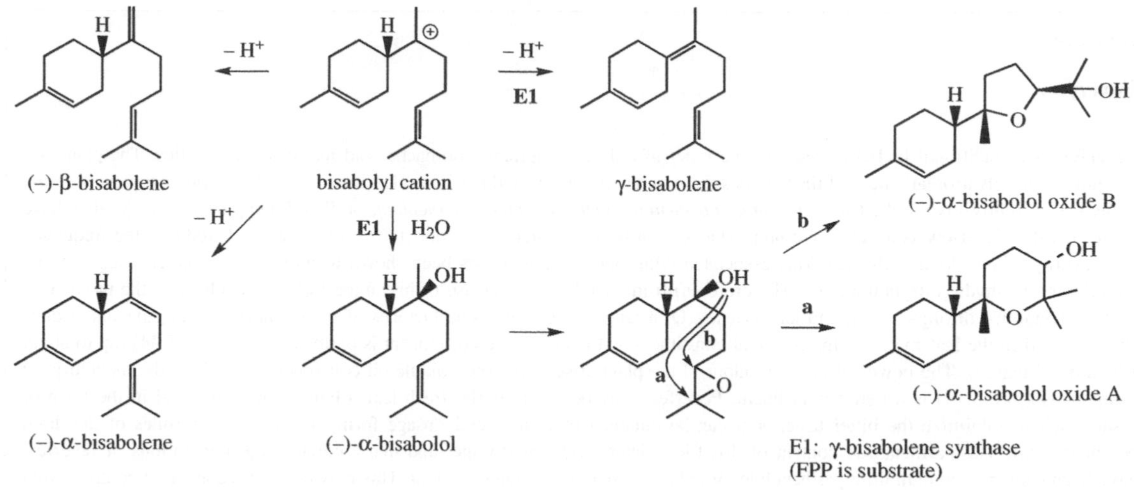

Figure 5.32

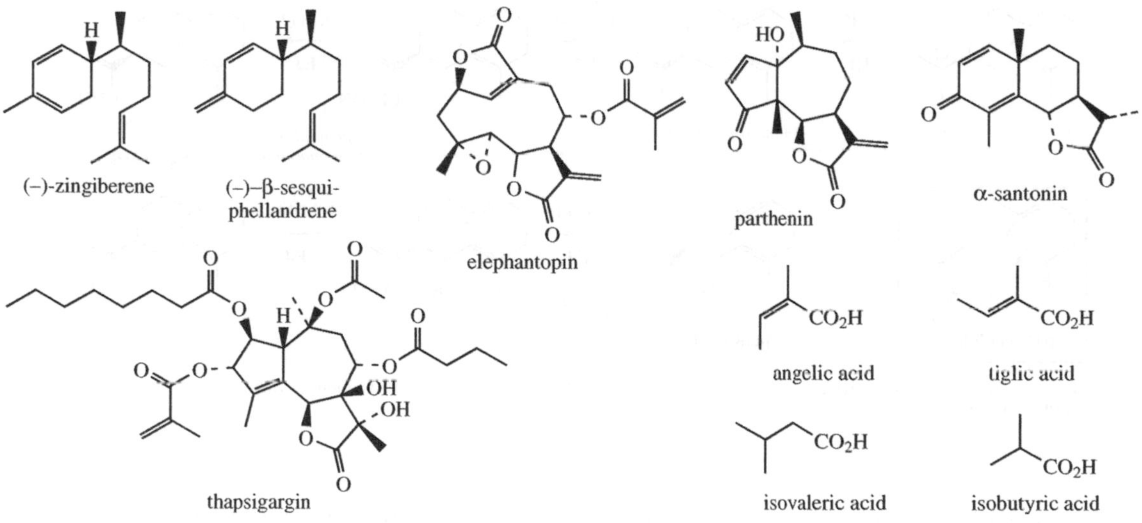

Figure 5.33

biological activity; indeed, activity is typically lost if either the double bond or the carbonyl group is chemically reduced. In some structures, additional electrophilic centres offer further scope for alkylation reactions. In **parthenolide** (Figure 5.35), an electrophilic epoxide group is also present, allowing transannular cyclization and generation of a second alkylation site. Cytotoxic agents may irreversibly alkylate critical enzymes that control cell division, whilst allergenic compounds may conjugate with proteins to form antigens which trigger the allergic response. The beneficial effects of parthenolide and structurally related compounds in feverfew have been demonstrated to relate to alkylation of thiol groups.

Box 5.4

Feverfew

Feverfew is a traditional herbal remedy for the relief of arthritis, migraine, toothache, and menstrual difficulties. The plant is a perennial, strongly aromatic herb of the Compositae/Asteraceae family, and has been classified variously as *Tanacetum parthenium* (which is currently favoured), *Chrysanthemum parthenium, Leucanthemum parthenium*, or *Pyrethrum parthenium*. Studies have confirmed that feverfew is an effective prophylactic treatment for migraine in about 70% of sufferers. It reduces the frequency and severity of attacks and the vomiting associated with them. The herb has been shown to inhibit blood platelet aggregation, the release of 5-hydroxytryptamine (5-HT, serotonin) from platelets, the release of histamine from mast cells, and the production of prostaglandins, thromboxanes, and leukotrienes. Of a range of sesquiterpene lactones of the germacrane and guianane groups characterized in the leaf material, the principal constituent and major active component is parthenolide (Figure 5.34) (up to about 1% in dried leaves). The powerful pungent odour of the plant arises from the volatile oil constituents, of which the monoterpene camphor (Figure 5.14) is a major constituent. Feverfew may be taken as the fresh leaf, often eaten with bread in the form of a sandwich to minimize the bitter taste, or it can be obtained in commercial dosage forms as tablets or capsules of the dried powdered leaf. The parthenolide content of dried leaf deteriorates on storage, and many commercial preparations of feverfew have been shown to contain little parthenolide, or to be well below the stated content. This may be a consequence of complexation with plant thiols via Michael addition. Consumers of fresh leaf can be troubled by sore mouth or mouth ulcers, caused by the sesquiterpenes. Parthenolide is also known to be capable of causing some allergic effects, e.g. contact dermatitis. The proposed mechanism of action of parthenolide via alkylation of thiol groups in proteins is shown in Figure 5.35.

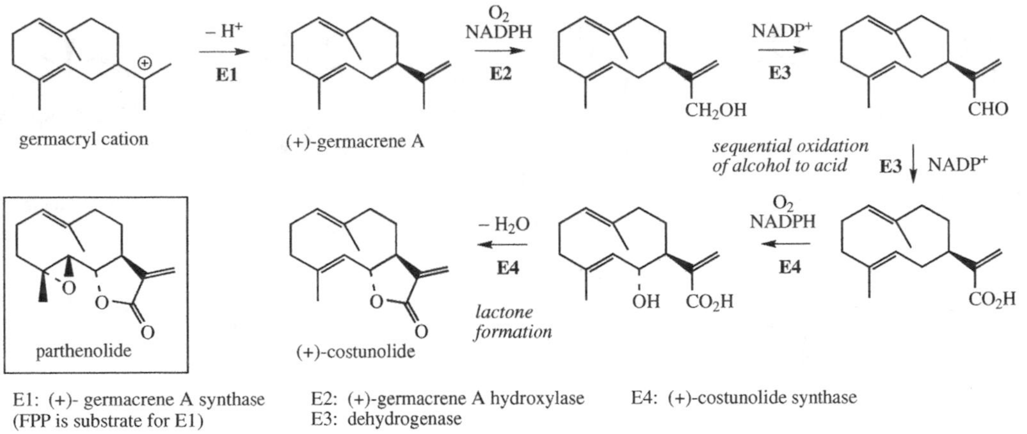

germacryl cation
– H+
E1
(+)-germacrene A
O2
NADPH
E2
CH2OH
NADP+
E3
CHO
sequential oxidation of alcohol to acid
E3
NADP+
O2
NADPH
E4
CO2H
OH
CO2H
– H2O
E4
lactone formation
(+)-costunolide
parthenolide
E1: (+)- germacrene A synthase (FPP is substrate for E1)
E2: (+)-germacrene A hydroxylase
E3: dehydrogenase
E4: (+)-costunolide synthase

Figure 5.34

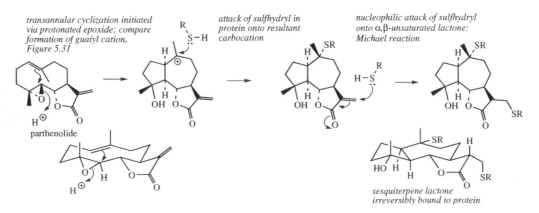

transannular cyclization initiated via protonated epoxide; compare formation of guaiyl cation, Figure 5.31
attack of sulfhydryl in protein onto resultant carbocation
nucleophilic attack of sulfhydryl onto α,β-unsaturated lactone: Michael reaction
parthenolide
sesquiterpene lactone irreversibly bound to protein

Figure 5.35

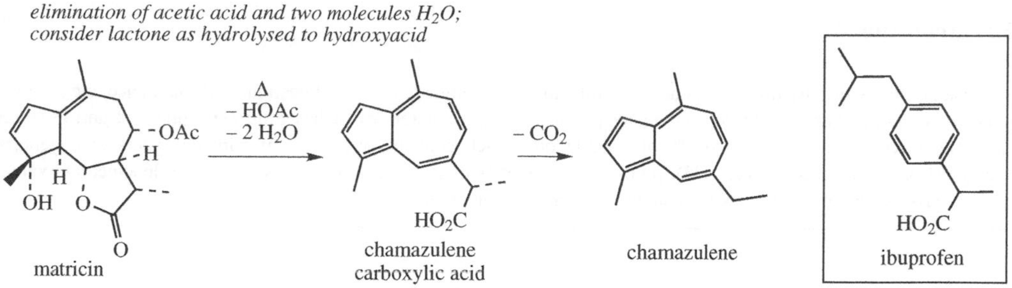

elimination of acetic acid and two molecules H_2O;
consider lactone as hydrolysed to hydroxyacid

Figure 5.36

α-Santonin (Figure 5.33) has been identified as the principal anthelmintic component of various *Artemisia* species, e.g. wormseed (*A. cinia*; Compositae/Asteraceae), and has found considerable use for removal of roundworms, although potential toxicity limits its application. Structurally, α-santonin bears much similarity to parthenolide, and the most marked difference lies in the presence of the bicyclic decalin ring system. This basic skeleton, the eudesmane system, is formed from the germacryl cation by protonation and cyclization via the eudesmyl cation (Figure 5.31, route ii), whereas protonation at the more substituted end of a double bond (anti-Markovnikov addition, route i), could generate the guaiyl cation and guaiane skeleton. This latter skeleton is found in **matricin** (Figure 5.36), again from matricaria flowers [Box 5.5]. This compound degrades on heating, via hydrolysis then elimination of acetic acid and water, followed by decarboxylation to the azulene derivative

chamazulene, which is responsible for the blue coloration of oil distilled from the flowers. Chamazulene carboxylic acid (Figure 5.36) is also found in yarrow (*Achillea millefolium*; Compositae/Asteraceae) and has been shown to be an anti-inflammatory agent by inhibiting cyclooxygenase COX-2, though not COX-1 (see page 62). The structural resemblance to the synthetic analgesic ibuprofen is striking. **Thapsigargin** (Figure 5.33) from seeds and roots of the Mediterranean plant *Thapsia garganica* (Compositae/Asteraceae) provides a further example of a guaianolide, highly oxygenated and esterified with a variety of acid groups. This compound is of considerable pharmacological interest in the study of Ca^{2+} signalling pathways. Thapsigargin is a potent inhibitor of Ca^{2+}-ATPases, and is capable of severely unbalancing cellular Ca^{2+} concentrations, often leading to disrupted cell function and growth, and apoptosis. A thapsigargin pro-drug has been tested in the treatment of prostate cancer.

Box 5.5

Chamomile and Matricaria

Two types of **chamomile** (**camomile**) are commonly employed in herbal medicine: Roman chamomile *Chamaemelum nobile* (formerly *Anthemis nobilis*) (Compositae/Asteraceae) and German chamomile *Matricaria chamomilla* (*Chamomilla recutica*) (Compositae/Asteraceae). German chamomile, an annual plant, is the more important commercially and is often called **matricaria** to distinguish it from the perennial Roman chamomile. Both plants are cultivated in various European countries to produce the flower-heads which are then dried for drug use. Volatile oils obtained by steam distillation or solvent extraction are also available.

Matricaria is also used as a digestive aid, but is mainly employed for its anti-inflammatory and spasmolytic properties. Extracts or the volatile oil find use in creams and ointments to treat inflammatory skin conditions, and as an antibacterial and antifungal agent. Taken internally, matricaria may help in the control of gastric ulcers. The flowers yield 0.5–1.5% volatile oil containing the sesquiterpenes α-bisabolol (10–25%), bisabolol oxides A and B (10–25%) (Figure 5.32), and chamazulene (0–15%) (Figure 5.36). Chamazulene is a thermal decomposition product from matricin, and is responsible for the dark blue coloration of the oil (Roman chamomile oil contains only trace amounts of chamazulene). α-Bisabolol has some anti-inflammatory, antibacterial, and ulcer-protective properties, but chamazulene probably contributes to the anti-inflammatory activity of matricaria preparations. It has been found to block the cyclooxygenase enzyme COX-2, though not COX-1, in prostaglandin biosynthesis (see page 62); the anti-inflammatory activity may result from the subsequent inhibition of leukotriene formation. Matricin itself produces rather greater anti-inflammatory activity than chamazulene and appears to be metabolized in the body to chamazulene carboxylic acid, a natural analogue of the synthetic analgesic ibuprofen (Figure 5.36).

Box 5.5   (continued)

Roman chamomile is usually taken as an aqueous infusion (chamomile tea) to aid digestion, curb flatulence, etc., but extracts also feature in mouthwashes, shampoos, and many pharmaceutical preparations. It has mild antiseptic and anti-inflammatory properties. The flower-heads yield 0.4–1.5% of volatile oil, which contains over 75% of aliphatic esters of angelic, tiglic, isovaleric, and isobutyric acids (Figure 5.33), products from metabolism of the amino acids isoleucine, leucine, and valine (see pages 56, 79, 316), with small amounts of monoterpenes and sesquiterpenes.

The 11-carbon ring of the humulyl carbocation (Figure 5.31) may be retained, as in the formation of **α-humulene** (Figure 5.37), or modified to give the caryophyllyl cation containing a nine-membered ring fused to a four-membered ring, as in **β-caryophyllene** (Figure 5.37). Humulene is found in hops (*Humulus lupulus*; Cannabaceae), and β-caryophyllene is found in a number of plants, e.g. in the oils from cloves (*Syzygium aromaticum*; Myrtaceae) and cinnamon (*Cinnamomum zeylanicum*; Lauraceae).

Another type of decalin-containing sesquiterpene is seen in the structures of cadinenes and amorpha-4,11-diene. **α-Cadinene** (Figure 5.38) is one of the many terpenoids found in juniper berries (*Juniperus communis*; Cupressaceae) used in making gin, and this compound is derived from the 10-carbon ring-containing *cis*-germacryl cation. The double bonds in the *cis*-germacryl cation are unfavourably placed for a cyclization reaction, as observed with the germacryl cation, and available evidence points to an initial 1,3-shift of hydride to the

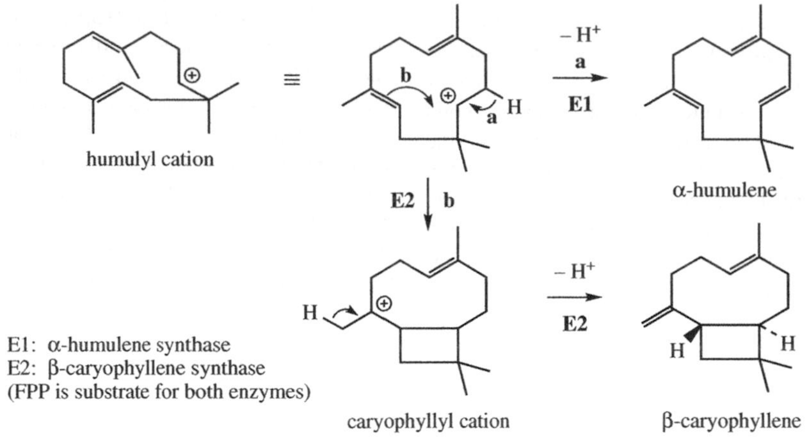

E1: α-humulene synthase
E2: β-caryophyllene synthase
(FPP is substrate for both enzymes)

Figure 5.37

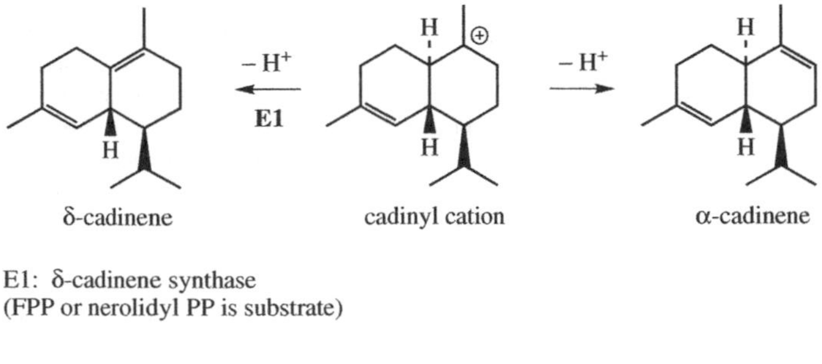

E1: δ-cadinene synthase
(FPP or nerolidyl PP is substrate)

Figure 5.38

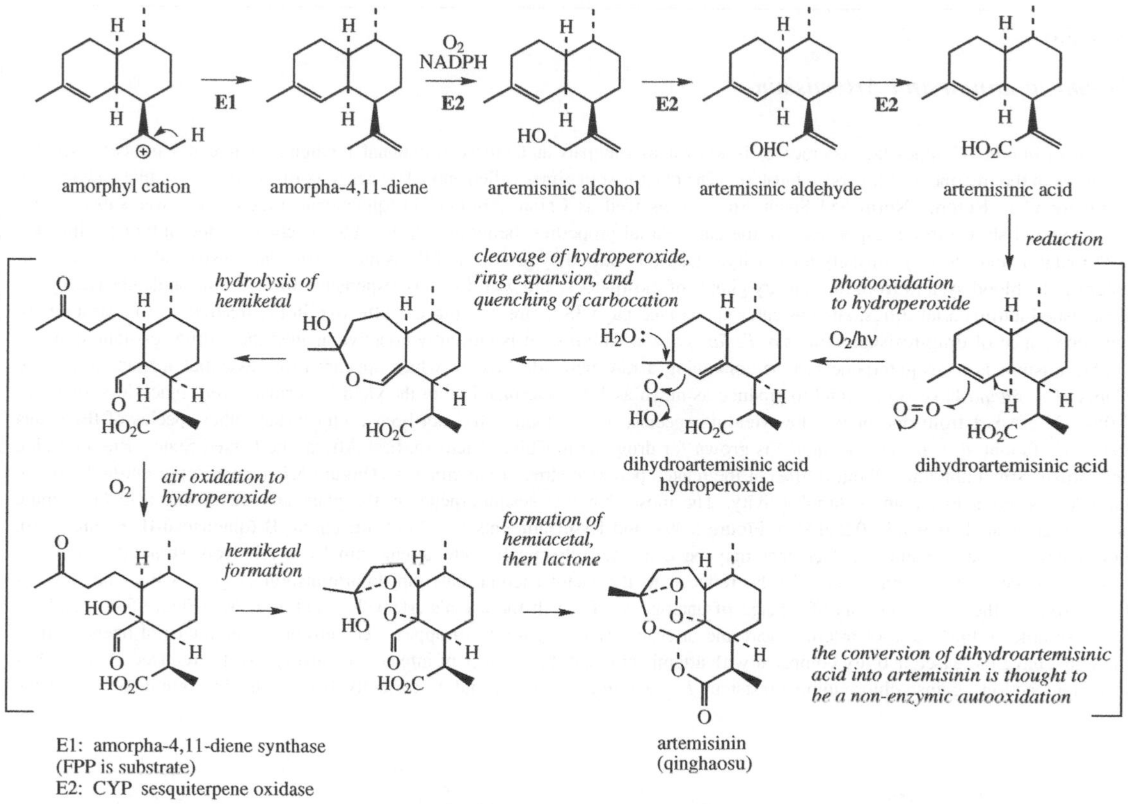

E1: amorpha-4,11-diene synthase
(FPP is substrate)
E2: CYP sesquiterpene oxidase

Figure 5.39

isopropyl side-chain to generate a new cation, and thus allowing cyclization to the cadinyl cation (Figure 5.31). **δ-Cadinene** is an alternative deprotonation product; it is further elaborated to gossypol in cotton (see page 220).

Amorpha-4,11-diene (Figure 5.39) is structurally related to the cadinenes, but the different stereochemistry of ring fusion and the position of the second double bond is a consequence of a different cyclization mechanism operating to produce the initial decalin ring system. In this case, a six-membered ring is formed first to give the bisabolyl cation; again, a 1,3-hydride shift is implicated prior to forming the decalin system of the amorphyl cation (Figure 5.31).

Amorpha-4,11-diene is an intermediate in the pathway leading to **artemisinin** in *Artemisia annua* (Compositae/Asteraceae) (Figure 5.39). **Artemisinic acid** is formed from amorphadiene via modest oxidation processes all catalysed by a single cytochrome P-450-dependent enzyme system. Reduction to **dihydroartemisinic acid** is

followed by transformation into artemisinin by a sequence of reactions that is currently regarded as non-enzymic, and brought about by an oxygen-mediated photochemical oxidation under conditions that might normally be present in the plant. An intermediate in this process, also found naturally in *A. annua*, is the hydroperoxide of dihydroartemisinic acid. The further modifications postulated in Figure 5.39 include ring expansion by cleavage of this hydroperoxide and a second oxygen-mediated hydroperoxidation. The 1,2,4-trioxane system in artemisinin can be viewed more simply as a combination of hemiketal, hemiacetal, and lactone functions, and the later stages of the pathway merely reflect their construction. Artemisinin is an important antimalarial component in *Artemisia annua*, a Chinese herbal drug [Box 5.6]. There is currently strong research effort to produce artemisinin or analogues as new antimalarial drugs, since the malarial parasite has developed resistance to many of the current drugs (see quinine, page 382).

Box 5.6

Artemisia annua and Artemisinin

Artemisia annua (Compositae/Asteraceae) is known as qinghao in Chinese traditional medicine, where it has been used for centuries in the treatment of fevers and malaria. The plant is sometimes called annual or sweet wormwood, and is quite widespread, being found in Europe, North and South America, as well as China. Artemisinin (qinghaosu; Figure 5.40) was subsequently extracted and shown to be responsible for the antimalarial properties, being an effective blood schizontocide in humans infected with malaria, and showing virtually no toxicity. Malaria is caused by protozoa of the genus *Plasmodium*, especially *P. falciparum*, entering the blood system from the salivary glands of mosquitoes, and worldwide is responsible for 2–3 million deaths each year. Established antimalarial drugs, such as chloroquine (see page 382), are proving less effective in the treatment of malaria due to the appearance of drug-resistant strains of *P. falciparum*. Artemisinin is currently effective against these drug-resistant strains.

Artemisinin is a sesquiterpene lactone containing a rare peroxide linkage which appears to be essential for activity. Some plants of *A. annua* have been found to produce as much as 1.4% artemisinin, but the yield is normally very much less, typically 0.05–0.2%. Apart from one or two low-yielding species, the compound has not been found in any other species of the genus *Artemisia* (about 400 species). *A. annua* is grown for drug use in China, Vietnam, East Africa, the United States, Russia, India, and Brazil. Small amounts (about 0.01%) of the related peroxide structure artemisitene (Figure 5.40) are also present in *A. annua*, though this has a lower antimalarial activity. The most abundant sesquiterpenes in the plant are artemisinic acid (arteannuic acid, qinghao acid; typically 0.2–0.8%) (Figure 5.39), and lesser amounts (0.1%) of arteannuin B (qinghaosu-II; Figure 5.40). Fortunately, the artemisinic acid content may be converted chemically into artemisinin by a relatively straightforward and efficient process. Artemisinin may also be reduced to the lactol (hemiacetal) dihydroartemisinin (Figure 5.40), and this has been used for the semi-synthesis of a range of analogues, of which the acetals artemether and arteether (Figure 5.40), and the water-soluble sodium salts of artelinic acid and artesunic acid (Figure 5.40) appear very promising antimalarial agents. These materials have increased activity compared with artemisinin, and the chances of infection recurring are also reduced. **Artemether** has rapid action against chloroquinine-resistant *P. falciparum* malaria, and is currently being used in combination with the

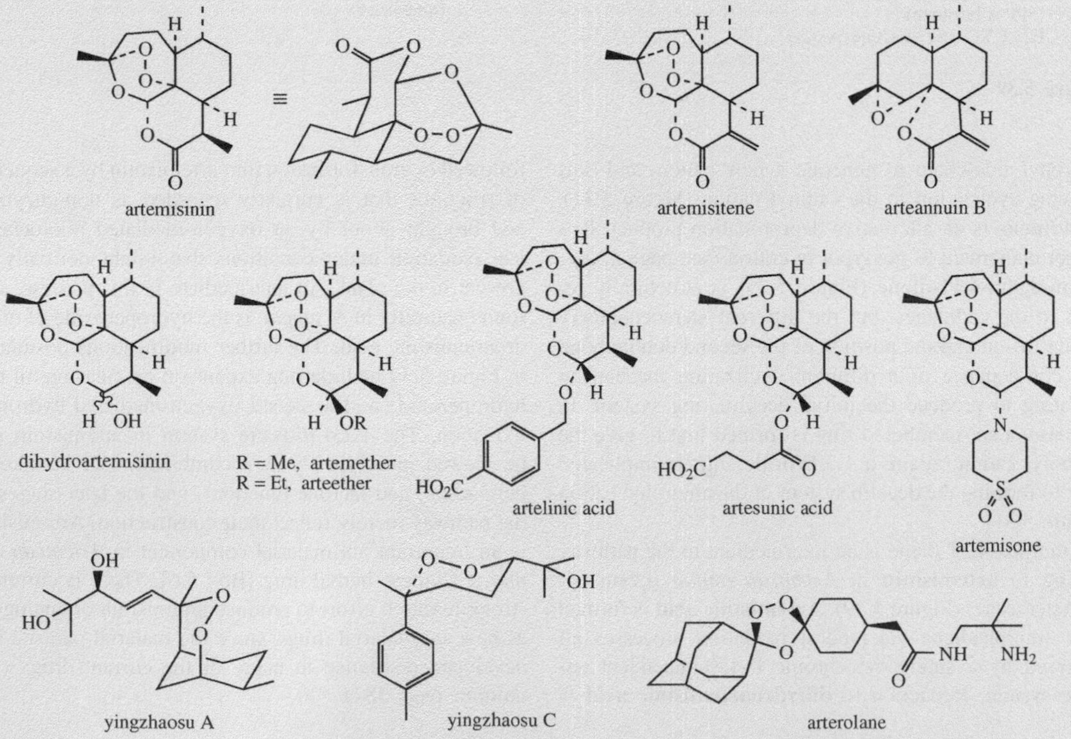

artemisinin   ≡   artemisitene   arteannuin B

dihydroartemisinin   R = Me, artemether   artelinic acid   artesunic acid   artemisone
                      R = Et, arteether HO_2C HO_2C

yingzhaosu A   yingzhaosu C   arterolane

Figure 5.40

Box 5.6 (continued)

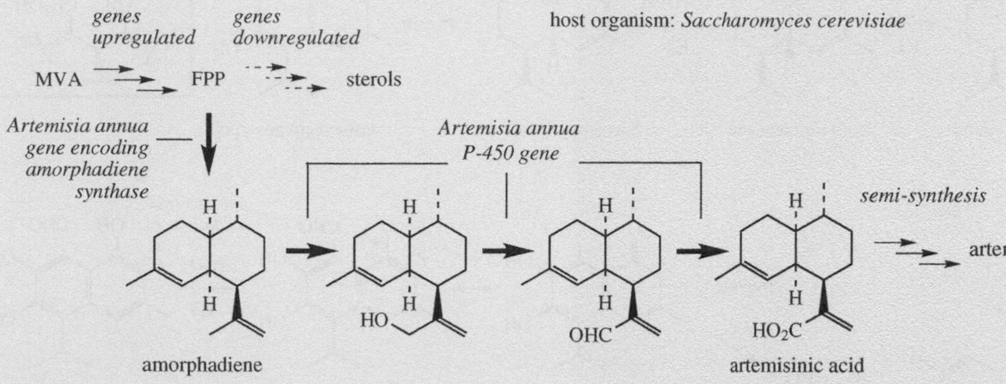

Figure 5.41

synthetic antimalarial lumefantrine. Arteether has similar activity. Being acetals, artemether and arteether are both extensively decomposed in acidic conditions, but are stable in alkali. The ester artesunic acid (**artesunate**) is also used in injection form, but is rather unstable in alkaline solution, hydrolysing to dihydroartemisinin. The ether artelinic acid is considerably more stable. These two compounds have a rapid action and particular application in the treatment of potentially fatal cerebral malaria. Dihydroartemisinin is a more active antimalarial than artemisinin and appears to be the main metabolite of these drugs in the body; unfortunately, dihydroartemisinin displays some neurotoxicity. Nevertheless, these agents rapidly clear the blood of parasites, though they do not have a prophylactic effect. **Artemisone** (Figure 5.40), a new semi-synthetic thiomorpholine dioxide derivative of dihydroartemisinin, is currently in clinical trials. Chemically, these agents are quite unlike any other class of current antimalarial agent, are well tolerated with no major side-effects, and so far no drug resistance is evident. As new analogues are introduced, they may well provide an important group of drugs in the fight against this life-threatening disease.

Currently, there is no shortage of *Artemisia* for drug use. However, as artemisinin derivatives become more widely used, the longer term supply of artemisinin for drug manufacture may need addressing. A significant development then is the use of genetic engineering to produce artemisinic acid in culture based on the biosynthetic pathway deduced for artemisinin (Figure 5.39). Genes from *A. annua* encoding the enzymes amorphadiene synthase and the cytochrome P-450-dependent monooxygenase that performs the three-step oxidation of amorphadiene to artemisinic acid were expressed in yeast *Saccharomyces cerevisiae* (Figure 5.41). To provide sufficient FPP precursor in the yeast host, several genes linked to the MVA → FPP transformation were upregulated; overexpression of the gene for HMGCoA reductase (see page 190) had the most significant effect. To avoid FPP being channelled off into steroid biosynthesis, the gene encoding for squalene synthase (see page 235) was downregulated. The net effect was high production of artemisinic acid (up to 100 mg l^{-1}) in the culture medium. This breakthrough is being commercialized to provide artemisinic acid for semi-synthesis of antimalarial drugs.

The relationship between a peroxide linkage and antimalarial activity is strengthened by the isolation of other sesquiterpene peroxides which have similar levels of activity as artemisinin. Thus, roots of the vine yingzhao (*Artabotrys uncinatus*; Annonaceae) which is also used as a traditional remedy for malaria, contain the bisabolyl derivatives yingzhaosu A and yingzhaosu C (Figure 5.40), the latter structure containing an aromatic ring of isoprenoid origin (compare the monoterpenes thymol and carvacrol, page 204). **Arterolane** (Figure 5.40) is a totally synthetic 1,2,4-trioxolane developed from the artemisinin template that shows potent antimalarial activity and is currently in clinical trials.

Despite intensive studies, there is as yet no generally accepted mechanism of action for artemisinin and derivatives. The malarial parasite utilizes the host's haemoglobin as a food source. However, haem, which is a soluble iron–porphyrin material released from haemoglobin as a result of proteolytic digestion, is toxic to *Plasmodium*, so it is normally converted into the insoluble non-toxic form haemozoin (malarial pigment) by enzymic polymerization. Agents like chloroquine (see page 382) may interfere with this polymerization process, or simply prevent the haemozoin from crystallizing. Of the various mechanisms proposed for the mode of action of artemisinin, initial reductive cleavage of the peroxide function by the Fe^{II} centre of haem to generate an oxygen radical species is widely accepted. This goes on to produce a carbon-centred radical that can interact with and effectively alkylate the porphyrin ring of haem. Alternatively, specific proteins or other biomolecules may suffer alkylation, resulting in death of the malarial parasite. It has also been suggested that artemisinin may act on a single enzymic target, a Ca^{2+}-ATPase; evidence for this comes from the observation that parasites with a mutant enzyme were insensitive to artemisinin.

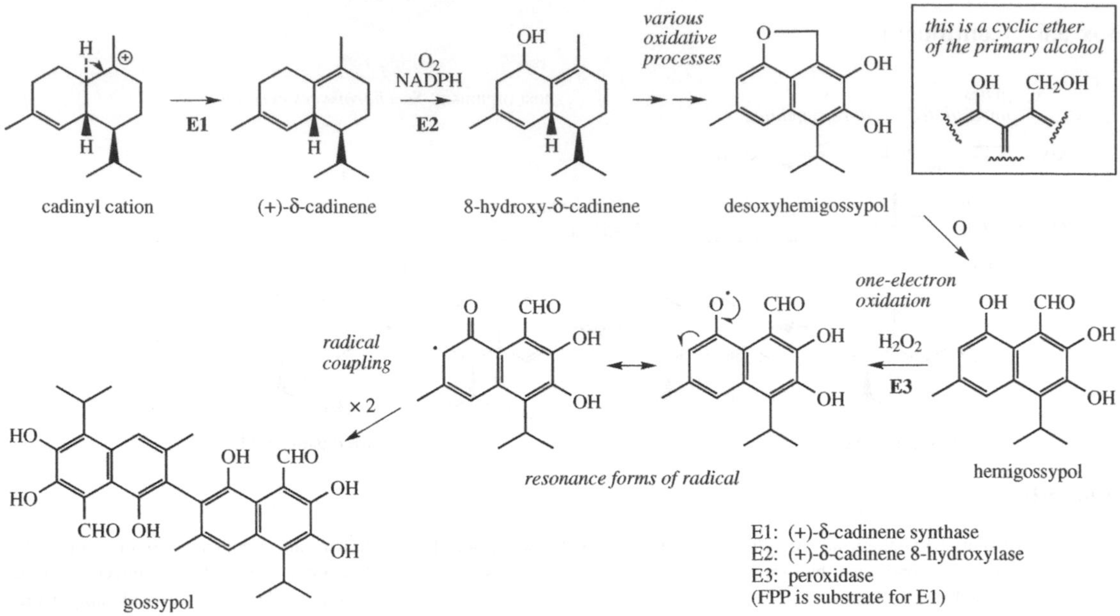

Figure 5.42

Gossypol (Figure 5.42) is an interesting and unusual example of a dimeric sesquiterpene in which loss of hydrogen has led to an aromatic system (compare the phenolic monoterpenes thymol and carvacrol, page 204). The cadinyl carbocation via **δ-cadinene** is involved in generating the basic aromatic sesquiterpene unit **hemigossypol**, and then dimerization is simply an example of phenolic oxidative coupling *ortho* to the phenol groups (Figure 5.42). The coupling is catalysed by an H_2O_2-dependent peroxidase enzyme. Gossypol is found in immature flower buds and seeds of the cotton plant (*Gossypium* species; Malvaceae), though it was originally isolated in small amounts from cottonseed oil. Its toxicity renders cottonseed oil unsafe for human consumption, but it has been used in China as a male infertility agent [Box 5.7].

Box 5.7

Gossypol

Gossypol occurs in the seeds of cotton (*Gossypium* species, e.g. *G. hirsutum, G. herbaceum, G. arboreum, G. barbadense*; Malvaceae) in amounts of 0.1–0.6%. Cotton is a major plant crop, and a considerable amount of cottonseed is produced as a by-product; this is mainly used as a feedstuff for cattle. Cottonseed, though protein-rich, has limited food use because of the cardiotoxic and hepatotoxic effects of gossypol in humans and animals other than ruminants. The contraceptive effects of gossypol were discovered when subnormal fertility in some Chinese rural communities was traced back to the presence of gossypol in dietary cottonseed oil. Gossypol acts as a male contraceptive, altering sperm maturation, spermatozoid motility, and inactivation of sperm enzymes necessary for fertilization. Extensive clinical trials in China have shown the antifertility effect is reversible after stopping the treatment provided that consumption has not been too prolonged. Cases of irreversible infertility have resulted from longer periods of drug use.

The gossypol molecule is chiral due to restricted rotation about the aryl–aryl linkage, and can thus exist as two atropisomers which do not easily racemize (Figure 5.43). Only the (−)-isomer is pharmacologically active as a contraceptive, whereas most of the toxic symptoms appear to be associated with the (+)-isomer. Most species of *Gossypium* (except *G. barbadense*)

Box 5.7 (continued)

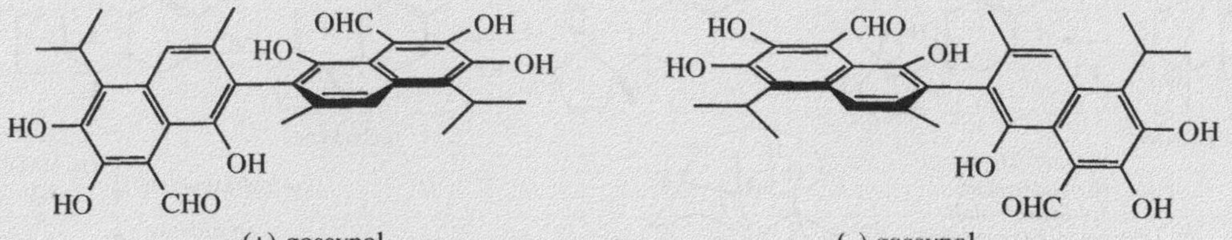

(+)-gossypol   (−)-gossypol

Figure 5.43

produce gossypol where the (+)-isomer predominates over the (−)-isomer, with amounts varying according to species and cultivar. The relative proportions of the two isomers appear to be controlled by regioselective coupling of two hemigossypol molecules by the peroxidase enzyme (Figure 5.42). Racemic (±)-gossypol (but neither of the enantiomers) complexes with acetic acid, so that suitable treatment of cottonseed extracts can separate the racemate from the excess of (+)-isomer. The racemic form can then be resolved. Other plants in the Gossypieae tribe of the Malvaceae also produce gossypol, with the barks of *Thespia populnea* (3.3%) and *Montezuma speciosissima* (6.1%) being particularly rich sources. Unfortunately, gossypol from these sources is almost entirely the (+)-form that lacks contraceptive activity. An exception is *Thespia danis*, where aerial parts contain predominantly (72%) (−)-gossypol; yields are about 0.2%. It is possible to reduce gossypol levels in cottonseed by interfering with the enzyme δ-cadinene synthase in the plant; this has been achieved by expression of the antisense gene, or by gene silencing through RNA interference: supplying a short strand of RNA to target that portion of mRNA responsible for a particular enzyme. Such techniques may lead to production of strains yielding non-toxic cottonseed oil.

The formation of sesquiterpenes by a carbocation mechanism also means that there is considerable scope for rearrangements of the Wagner–Meerwein type. So far, only occasional hydride migrations have been invoked in rationalizing the examples considered. Obviously, fundamental skeletal rearrangements will broaden the range of natural sesquiterpenes even further. That such processes do occur has been proven beyond doubt by appropriate labelling experiments, and a single example will be used as illustration. The **trichothecenes** are a group of fungal toxins found typically in infected grain foodstuffs [Box 5.8]. Their name comes from the fungal genus *Trichothecium*, but most of the known structures are derived from cultures of *Fusarium* species. A typical trichothecene contaminant is **3-acetyldeoxynivalenol**, which is produced from the less-substituted trichothecene **isotrichodermol** by a sequence of oxygenation, esterification, and de-esterification reactions (Figure 5.44). The trichothecenes have their origins in **nerolidyl diphosphate**, and ring closure of the bisabolyl cation derived from it generates a new carbocation with a five-membered ring (Figure 5.44). At this stage, a series of one hydride and two methyl migrations occur to give a cation, which loses a proton to produce the specific trichothecene precursor **trichodiene**. These migrations are fully backed up by experimental data and, although not immediately predictable, can be rationalized satisfactorily by consideration of the cation suitably bound to the enzyme surface, as shown in Figure 5.44. The sequence is initiated by a 1,4-hydride shift, which is spatially allowed by the relative proximity of the centres. Two 1,2-methyl shifts then follow, and it is important to note that each migrating group attacks the opposite side of the centre from which the previous group is departing, i.e. inverting the configuration at these centres. Accordingly, a concerted sequence of migrations is feasible; a more vivid example of this type of concerted process is seen in the formation of triterpenoids and steroids (see page 236). Loss of a proton and generation of a double bond then terminates the process, giving trichodiene. Oxygenation of trichodiene gives, in several steps, **isotrichotriol**. Two of the hydroxylations are at activated allylic positions; hydroxylation on the five-membered ring, therefore, will occur before the epoxidation. Ether formation, involving perhaps protonation, loss of water, and generation of an allylic cation completes the pathway to the basic trichothecene structure as in **isotrichodermol**.

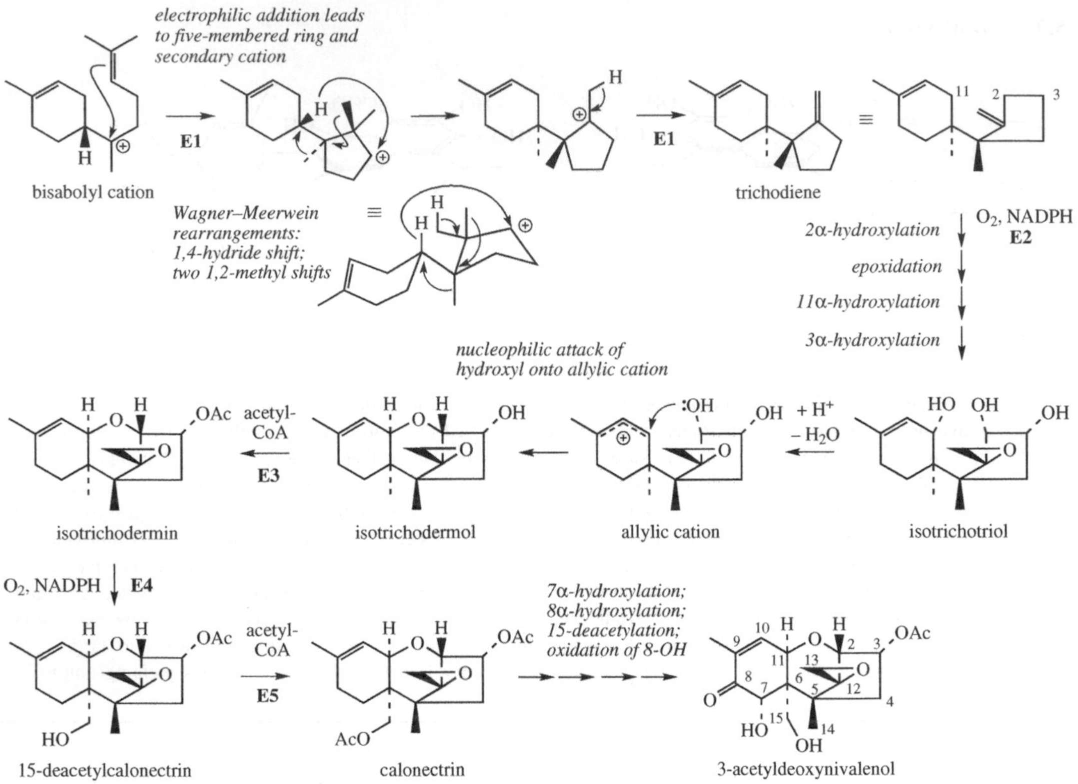

E1: trichodiene synthase (FPP is substrate)
E2: TRI4 (trichodiene 2α-hydroxylase)
E3: TRI101 (isotrichodermol 3-*O*-acetyltransferase)
E4: TRI11 (isotrichodermin 15-hydroxylase)
E5: TRI3 (15-deacetylcalonectrin 15-*O*-acetyltransferase)

Figure 5.44

Box 5.8

Trichothecenes

The trichothecenes are a group of sesquiterpene toxins produced by several fungi of the genera *Fusarium, Myrothecium, Trichothecium*, and *Trichoderma*, which are parasitic on cereals such as maize, wheat, rye, barley, and rice. About 180 different structures have been identified, with some of these being isolated from plants of the genus *Baccharis* (Compositae/Asteraceae), where a symbiotic plant–fungus relationship may account for their production. Examples of trichothecene structures most commonly encountered as food contaminants include deoxynivalenol (DON; vomitoxin), diacetoxyscirpenol, T-2 toxin, and verrucarin A (Figure 5.45). The double bond and the epoxide group in the basic trichothecene skeleton are essential for toxicity, and the number of oxygen substituents and ester functions also contribute. Macrocyclic ester functions as seen in verrucarin A tend to produce the most toxic examples. Although these compounds are more toxic when injected, oral toxicity is relatively high, and lethal amounts can easily be consumed because of the nature of the host plants. They are sufficiently toxic to warrant routine analysis of foodstuffs such as wheat and flour, and also flour-derived products, e.g. bread, since they survive processing and the high temperatures used in baking. DON levels above 1 ppm are considered hazardous for human consumption. It is relevant to note that, when mammals ingest these compounds, a degree of de-epoxidation can occur, ascribed to gut microflora, thus providing a level of detoxification by removing a structural feature necessary for toxicity.

As their main mechanism of action, these compounds inhibit protein biosynthesis by binding to the ribosome and inhibiting peptidyl transferase activity (see page 422). They also inhibit DNA biosynthesis. A major human condition known to be caused by

Box 5.8 (continued)

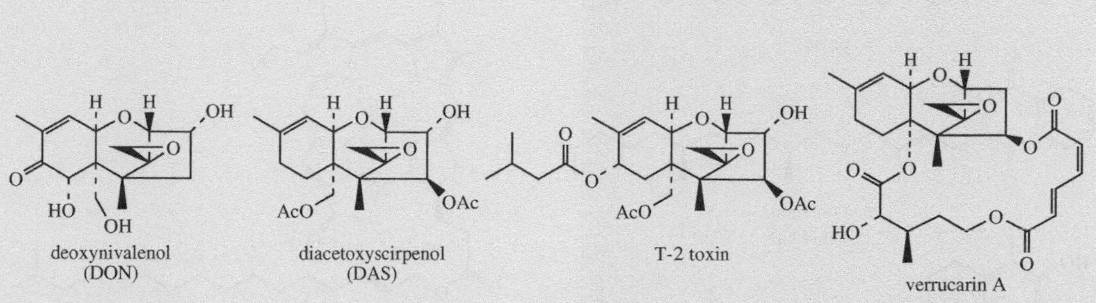

deoxynivalenol
(DON)

diacetoxyscirpenol
(DAS)

T-2 toxin

verrucarin A

Figure 5.45

trichothecenes is alimentary toxic aleukia (ATA), characterized by destruction of the skin, haemorrhaging, inflammation, sepsis, a decrease in red and white blood corpuscles, bone marrow atrophy, and a high mortality rate. A severe outbreak of ATA was recorded in the former Soviet Union shortly after the Second World War when food shortages led to the consumption of grain that had overwintered in the field. This had become badly contaminated with *Fusarium sporotrichioides* and, hence, T-2 toxin. It is estimated that tens of thousands died as a result. Many trichothecene derivatives have been tested as potential anticancer agents, but they have proved too toxic for clinical use.

Finally, it is worth noting how many of the sesquiterpene derivatives described above are found in plants belonging to the daisy family, the Compositae/Asteraceae. Whilst sesquiterpenes are by no means restricted to this family, the Compositae/Asteraceae undoubtedly provides a very rich source.

DITERPENES (C_{20})

The diterpenes arise from **geranylgeranyl diphosphate (GGPP)**, which is formed by addition of a further IPP molecule to FPP in the same manner as described for the lower terpenoids (Figure 5.46). One of the

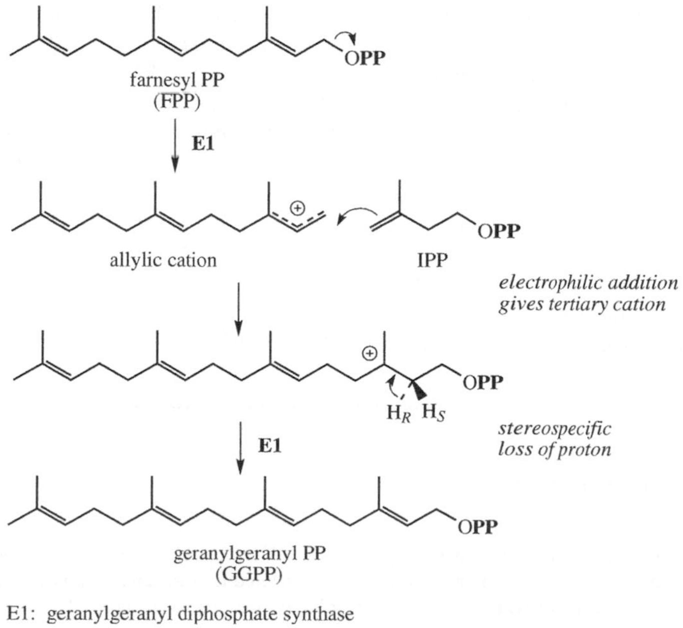

farnesyl PP
(FPP)

E1

allylic cation

IPP

electrophilic addition gives tertiary cation

H_R H_S

stereospecific loss of proton

E1

geranylgeranyl PP
(GGPP)

E1: geranylgeranyl diphosphate synthase

Figure 5.46

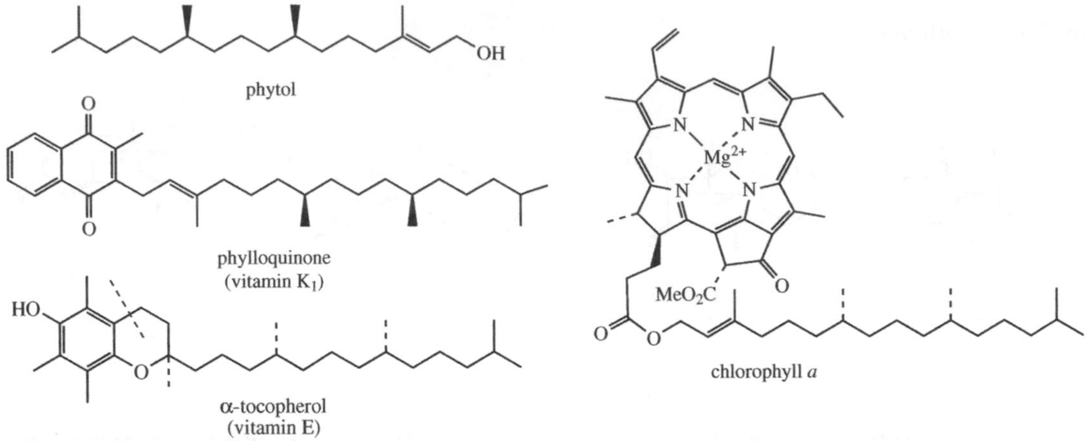

Figure 5.47

simplest and most important of the diterpenes is **phytol** (Figure 5.47), a reduced form of geranylgeraniol, which forms the lipophilic side-chain of the chlorophylls, e.g. **chlorophyll *a*** (Figure 5.47). Related haem molecules, the porphyrin components of haemoglobin, lack such lipophilic side-chains. Available evidence suggests that GGPP is involved in forming the ester linkage, and the three reduction steps necessary to form the phytol ester occur after attachment to the chlorophyll molecule. A phytyl substituent is also found in **vitamin K₁** (**phylloquinone**; Figure 5.47), a naphthoquinone derivative found in plants, though other members of the vitamin K group (**menaquinones**) from bacteria have unsaturated terpenoid side-chains of variable length. The phytyl group of phylloquinone is introduced by alkylation of dihydroxynaphthoic acid with phytyl diphosphate and a similar phytylation of homogentisic acid features in the formation of the E group vitamins (**tocopherols**). These compounds are discussed further under shikimate derivatives (see page 178).

Cyclization reactions of GGPP mediated by carbocation formation, plus the potential for Wagner–Meerwein rearrangements, will allow many structural variants of diterpenoids to be produced. The toxic principle 'taxine' from English yew (*Taxus baccata*; Taxaceae) has been shown to be a mixture of at least 11 compounds based on the **taxadiene** skeleton, which can readily be rationalized as in Figure 5.48, employing the same mechanistic principles as seen with mono- and sesqui-terpenes. A novel feature discovered is the enzyme's ability to utilize the proton released during alkene formation to protonate another double bond; this effectively relocates the cationic centre. Although these compounds are sometimes classified as diterpenoid alkaloids, the nitrogen atom is not incorporated into the diterpene skeleton, as exemplified by **taxol** (**paclitaxel**; Figure 5.48) from Pacific yew (*Taxus brevifolia*) [Box 5.9]. The side-chains in taxol containing aromatic rings are derived from shikimate via phenylalanine. The nitrogen atom derives from the amino acid β-phenylalanine, a rearranged version of the α-amino acid L-phenylalanine. Virtually all of the enzymes and genes for the whole sequence from GGPP to taxol are now characterized. Some enzymes show relatively broad substrate specificity. Taxol is an important anticancer agent, with a broad spectrum of activity against some cancers that do not respond to other agents.

Box 5.9

Taxus brevifolia and Taxol (Paclitaxel)

A note on nomenclature: the name taxol was given to a diterpene ester with anticancer properties when it was first isolated in 1971 from *Taxus brevifolia*. However, the name Taxol had already been registered as a trademark; this was subsequently used when the natural product was exploited commercially as a drug. Accordingly, the generic name paclitaxel has been assigned to the

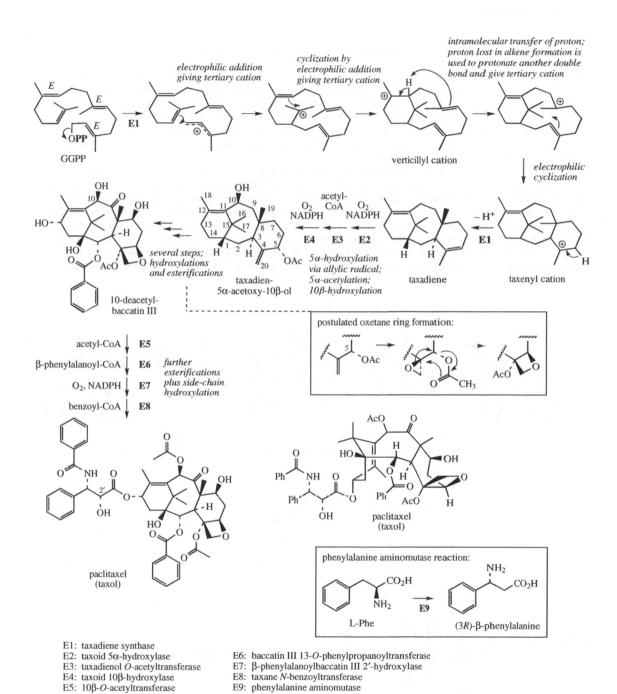

E1: taxadiene synthase
E2: taxoid 5α-hydroxylase
E3: taxadienol *O*-acetyltransferase
E4: taxoid 10β-hydroxylase
E5: 10β-*O*-acetyltransferase

E6: baccatin III 13-*O*-phenylpropanoyltransferase
E7: β-phenylalanoylbaccatin III 2′-hydroxylase
E8: taxane *N*-benzoyltransferase
E9: phenylalanine aminomutase

Figure 5.48

Box 5.9   (continued)

compound, and the literature now contains an unhappy mixture of the two names, though the original name taxol is most often employed.

The anticancer drug taxol (Figure 5.48) is extracted from the bark of the Pacific yew, *T. brevifolia* (Taxaceae), a slow-growing shrub/tree found in the forests of northwest Canada (British Columbia) and the USA (Washington, Oregon, Montana, Idaho, and north California). Although the plant is not rare, it does not form thick populations, and it needs to be mature (about 100 years old) to be large enough for exploitation of its bark. The bark from about three mature trees is required to provide 1 g of taxol, and a course of treatment may need 2 g of the drug. Current demand for taxol is in the region of 250 kg per annum. Harvesting has been strictly regulated, but it was immediately realized that this source would not provide a satisfactory long-term supply of the drug. Taxol is now obtained by alternative means.

All parts of *T. brevifolia* contain a wide range of diterpenoid derivatives termed taxanes, which are structurally related to the toxic constituents found in other *Taxus* species, e.g. the common English yew, *T. baccata*. Nearly 400 taxanes have been characterized from various *Taxus* species, and taxol is a member of a small group of compounds possessing a four-membered oxetane ring and a complex ester side-chain in their structures; both of these features are essential for antitumour activity. Taxol is found predominantly in the bark of *T. brevifolia*, but in relatively low amounts (about 0.01–0.02%). Up to 0.033% of taxol has been recorded in some samples of leaves and twigs, but generally the taxol content is much lower than in the bark. The content of some other taxane derivatives in the bark is considerably higher, e.g. up to 0.2% baccatin III (Figure 5.49). Other taxane derivatives characterized include 10-deacetyltaxol, 10-deacetylbaccatin III, cephalomannine, and 10-deacetylcephalomannine (Figure 5.49).

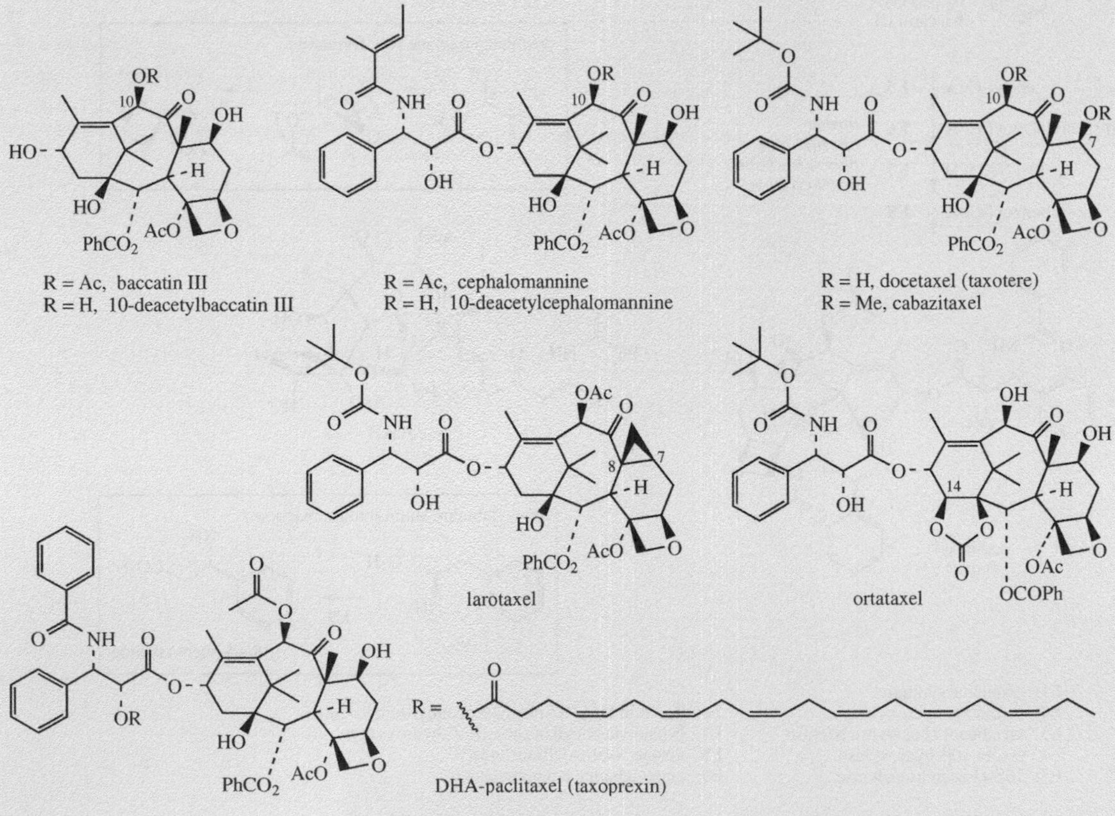

R = Ac, baccatin III
R = H, 10-deacetylbaccatin III

R = Ac, cephalomannine
R = H, 10-deacetylcephalomannine

R = H, docetaxel (taxotere)
R = Me, cabazitaxel

larotaxel

ortataxel

DHA-paclitaxel (taxoprexin)

Figure 5.49

Box 5.9 (continued)

A satisfactory solution currently exploited for the supply of taxol and derivatives for drug use is to produce these compounds by semi-synthesis from more accessible structurally related materials. Both baccatin III and 10-deacetylbaccatin III (Figure 5.49) have been efficiently transformed into taxol, the latter being the preferred substrate. 10-Deacetylbaccatin III is readily extracted from the leaves and twigs of English yew *T. baccata*, and although the content is variable, it is generally present at much higher levels (up to 0.2%) than taxol can be found in *T. brevifolia*. *T. baccata* is widely planted as an ornamental tree in Europe and the USA and is much faster growing than the Pacific yew. Yew trees, typically 8–10 years old, can be harvested by pruning off top growth on a regular basis. In Europe, there are also arrangements whereby unwanted clippings from ornamental yew hedges may be channelled into drug production.

Semi-synthesis provided reliable supplies to establish taxol as an important anticancer drug. However, the chemical conversion of 10-deacetylbaccatin III is still complex enough to consider alternative approaches. Taxol-producing cell cultures of Chinese yew *T. chinensis* have been established and are currently grown in fermentors for commercial drug production. Taxol is isolated and purified from the fermentation broth; taxol yields in the region of 20 mg l^{-1} can be obtained from cultures of various *Taxus* species.

Initial optimism for obtaining taxol by microbial culture was tempered by the very low levels produced. Fungi such as *Taxomyces adreanae* isolated from the inner bark of *T. brevifolia*, and *Pestalotiopsis microspora* from the inner bark of the Himalayan yew (*T. wallachiana*), appear to have inherited the ability to synthesize taxol by gene transfer from the host tree. At best, taxol levels measured were only about 70 μg l^{-1} of culture medium, and thus commercially insignificant. However, the use of microorganisms and enzymes to specifically hydrolyse ester groups from mixtures of structurally related plant-derived taxanes in crude extracts and thus improve the yields of 10-deacetylbaccatin III has been reported.

Paclitaxel (Taxol$^®$) is used clinically in the treatment of ovarian and breast cancers, non-small-cell lung cancer, small-cell lung cancer, and cancers of the head and neck. A disadvantage associated with taxol is its very low water solubility. **Docetaxel (Taxotere$^®$**; Figure 5.49) is a side-chain analogue of taxol; it was one of the intermediates produced during a semi-synthesis of taxol from 10-deacetylbaccatin III. It was found to be more active than taxol, and, crucially, was more water-soluble. It is used against ovarian and breast cancers and non-small-cell lung cancer.

Clinical development of taxol did not proceed until it was realized that it possessed a mode of action different from the drugs available at the time. Taxol acts as an antimitotic by binding to microtubules, promoting their assembly from tubulin, and stabilizing them against depolymerization during cell division. The resultant abnormal tubulin–microtubule equilibrium disrupts the normal mitotic spindle apparatus and blocks cell proliferation. Vincristine and vinblastine (see page 375), and podophyllotoxin (see page 155), also interfere with the tubulin system, but bind to the protein tubulin in the mitotic spindle, preventing polymerization and assembly into microtubules. Taxol thus shares a similar target, but has a different mechanism of action to the other antimitotics. In recent years, other natural products have been discovered that have the same or a similar mode of action; for example, the epothilones (see page 85) and discodermolide (see page 90). Taxol has also been shown to bind to a second target, a protein which normally blocks the process of apoptosis (cell death). Inhibition of this protein allows apoptosis to proceed.

Several taxol analogues are in clinical trials. Considerable effort is being directed to improve solubility and to counter a tendency to induce multiple drug resistance. Four compounds in advanced clinical trials are **DHA–paclitaxel (Taxoprexin$^®$)**, **cabazitaxel, larotaxel**, and **ortataxel** (Figure 5.49). The first is an ester of taxol at the side-chain hydroxyl with the polyunsaturated fatty acid DHA (see page 49); polyunsaturated fatty acids are able to deliver cytotoxic drugs selectively to cancer cells. The other three structures are based on taxotere. Cabazitaxel is the 7,10-dimethyl ether and larotaxel is a 7,8-cyclopropane analogue. Ortataxel is an orally active cyclic carbonate derivative produced from the natural 14β-hydroxy-10-deacetylbaccatin III.

The latex of some plants in the genus *Euphorbia* (Euphorbiaceae) can cause poisoning in humans and animals, skin dermatitis, cell proliferation, and tumour promotion (co-carcinogen activity). Many species of *Euphorbia* are regarded as potentially toxic, and the latex can produce severe irritant effects, especially on mucous membranes and the eye. Most of the biological effects are due to diterpene esters, e.g. esters of **phorbol** (Figure 5.50), which activate protein kinase C, an important and widely distributed enzyme responsible for phosphorylating many biochemical entities. The permanent activation of protein kinase C is thought to lead to the uncontrolled cancerous growth. The most commonly encountered ester of phorbol is 12-*O*-myristoylphorbol 13-acetate (Figure 5.50), one of the most potent tumour promoters known. The origins of phorbol are not known, but may be rationalized as in Figure 5.50. Cyclization of GGPP generates a cation containing a 14-membered ring system. Loss of a proton via cyclopropane ring formation leads to **casbene**, an antifungal metabolite produced by the castor oil plant, *Ricinus communis* (Euphorbiaceae). Casbene, via the ring closures shown in Figure 5.50, is then likely to be the precursor of the phorbol ring system.

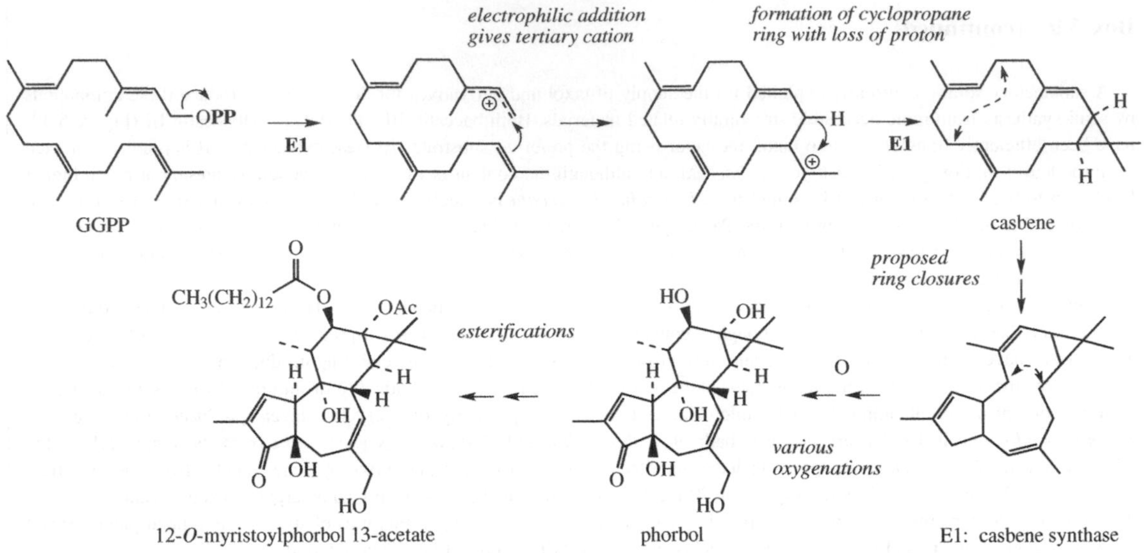

electrophilic addition gives tertiary cation

formation of cyclopropane ring with loss of proton

GGPP

E1

E1

casbene

proposed ring closures

CH₃(CH₂)₁₂

O

OAc

esterifications

H

H

H

OH

H

O

OH

HO

12-*O*-myristoylphorbol 13-acetate

HO

OH

H

H

OH

H

O

OH

HO

phorbol

various oxygenations

O

E1: casbene synthase

Figure 5.50

In contrast to the cyclization mechanisms shown in Figures 5.48 and 5.50, where loss of diphosphate generates the initial carbocation, many of the natural diterpenes have arisen by a different mechanism. Carbocation formation is initiated by protonation of the double bond at the head of the chain, leading to a first cyclization sequence. Loss of the diphosphate later on also produces a carbocation and facilitates further cyclization. The early part of the sequence resembles that involved in hopanoid biosynthesis (see page 242), and to some extent triterpenoid and steroid biosynthesis (see page 238), though in the latter cases it is the opening of the epoxide ring in the precursor squalene oxide rather than protonation of an alkene that is responsible for generation of the cationic intermediates. Protonation of GGPP can initiate a concerted cyclization sequence, terminated by loss of a proton from a methyl, yielding (**–**)-**copalyl PP** (Figure 5.51, a). The stereochemistry in this product is controlled by the folding

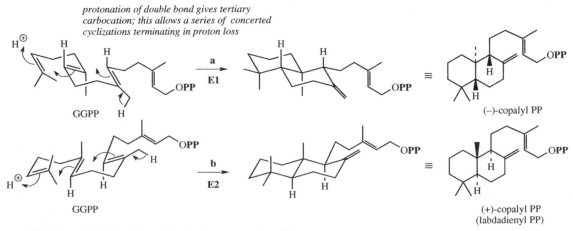

protonation of double bond gives tertiary carbocation; this allows a series of concerted cyclizations terminating in proton loss

GGPP

a
E1

(–)-copalyl PP

GGPP

b
E2

(+)-copalyl PP
(labdadienyl PP)

E1: (–)-copalyl diphosphate synthase (*ent*-kaurene synthase A)
E2: (+)-copalyl diphosphate synthase (part of bifunctional abietadiene synthase)

Figure 5.51

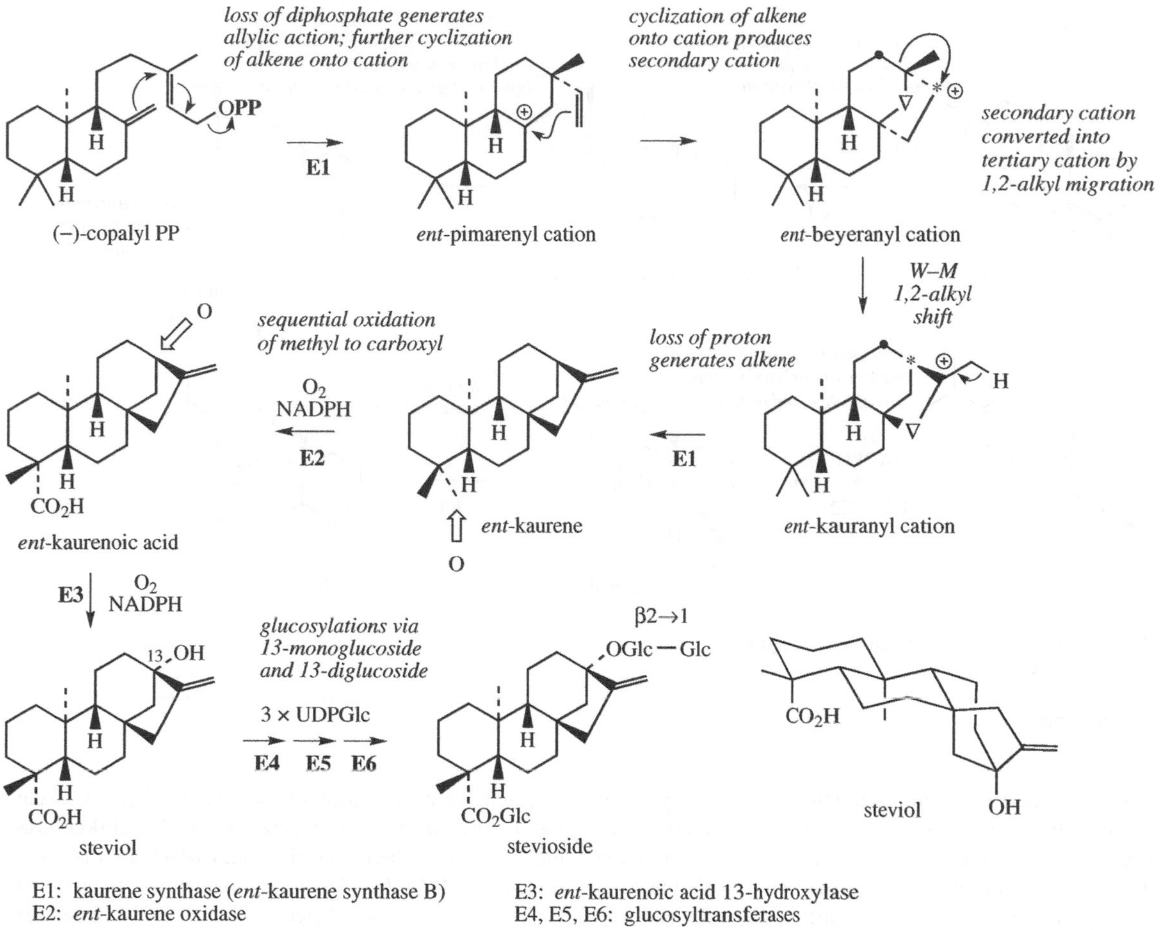

loss of diphosphate generates allylic action; further cyclization of alkene onto cation

(−)-copalyl PP

E1

ent-pimarenyl cation

cyclization of alkene onto cation produces secondary cation

ent-beyeranyl cation

secondary cation converted into tertiary cation by 1,2-alkyl migration

W–M 1,2-alkyl shift

sequential oxidation of methyl to carboxyl

O_2
NADPH

E2

ent-kaurenoic acid

loss of proton generates alkene

E1

ent-kaurene

ent-kauranyl cation

E3 $\begin{array}{c} O_2 \\ NADPH \end{array}$

13 OH

CO_2H

steviol

glucosylations via 13-monoglucoside and 13-diglucoside

3 × UDPGlc

E4 E5 E6

CO_2Glc

stevioside

β2→1

OGlc — Glc

CO_2H

steviol OH

E1: kaurene synthase (*ent*-kaurene synthase B)
E2: *ent*-kaurene oxidase

E3: *ent*-kaurenoic acid 13-hydroxylase
E4, E5, E6: glucosyltransferases

Figure 5.52

of the substrate on the enzyme surface, though an alternative folding can lead to (+)-**copalyl PP** (labdadienyl PP), the enantiomeric product with opposite configurations at the newly generated chiral centres (Figure 5.51, b).

From (−)-copalyl PP, a sequence of cyclizations and a rearrangement, all catalysed by a single enzyme kaurene synthase, leads to ***ent*-kaurene** (Figure 5.52). As shown, this involves loss of the diphosphate leaving group enabling carbocation-mediated formation of the third ring system, and subsequent production of the fourth ring. Then follows a Wagner–Meerwein migration, effectively contracting the original six-membered ring to a five-membered one, whilst expanding the five-membered ring to give a six-membered ring. The driving force is transformation of a secondary carbocation to give a tertiary one, but this also results in the methyl group no longer being at a bridgehead, and what appears at first glance to be merely a confusing change in

stereochemistry. Loss of a proton from this methyl generates the exocyclic double bond of ***ent*-kaurene** and provides an exit from the carbocationic system. The prefix *ent* is used to indicate enantiomeric; the most common stereochemistry is that found in (+)-copalyl PP (Figure 5.51) and derivatives, so the kaurene series is termed 'enantiomeric'.

ent-Kaurene is the precursor of **stevioside** (Figure 5.52) in the plant *Stevia rebaudiana* (Compositae/Asteraceae) by relatively simple oxidation, hydroxylation, and glucosylation reactions. Both glucosyl ester and glucoside linkages are present in stevioside, and these help to confer an intensely sweet taste to this and related compounds. Stevioside is present in the plant leaf in quite large amounts (3–10%), is some 200–300 times as sweet as sucrose, and is being used in many countries as a commercial non-calorific sweetening agent.

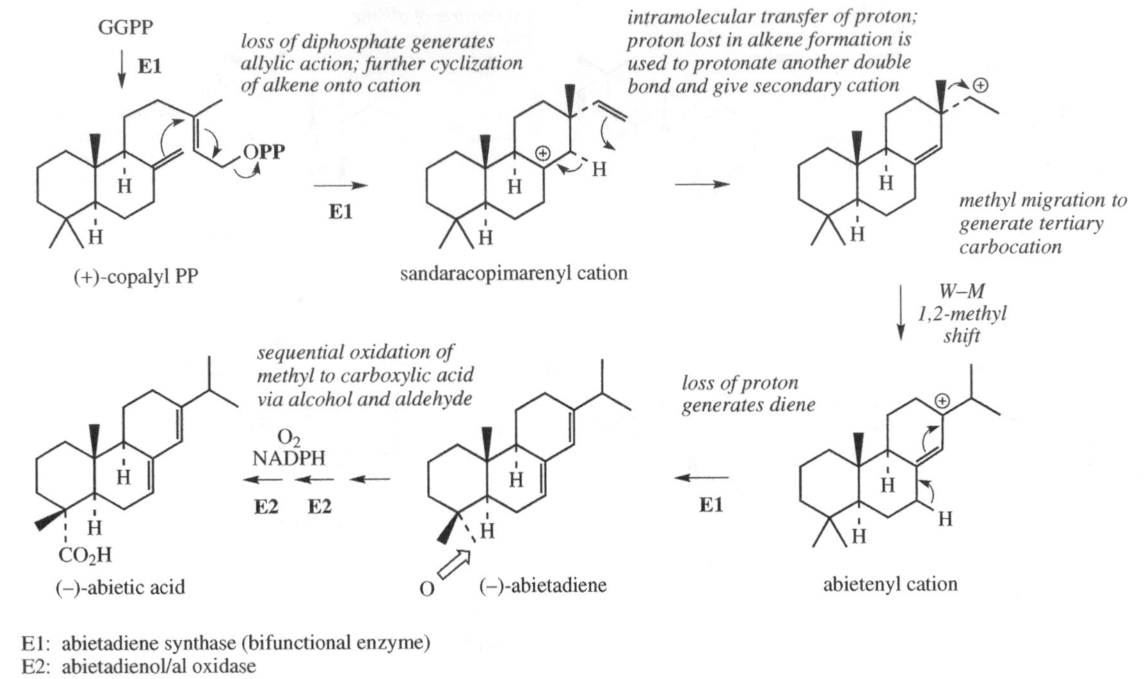

E1: abietadiene synthase (bifunctional enzyme)
E2: abietadienol/al oxidase

Figure 5.53

The alternative stereochemistry typified by (+)-copalyl PP can be seen in the structure of **abietic acid** (Figure 5.53), the major component of the rosin fraction of turpentine from pines and other conifers (Table 5.1). Initially, the tricyclic system is built up as in the pathway to *ent*-kaurene (Figure 5.52), via the same mechanism, but generating the enantiomeric series of compounds. An internal proton transfer (compare taxadiene page 225) relocates the carbocationic centre to the side-chain, which then undergoes a methyl migration (Figure 5.53). Finally, loss of a proton leads to the diene **abietadiene**. The whole sequence from GGPP to abietadiene is catalysed by a single bifunctional enzyme; the first activity is a (+)-copalyl PP synthase, and the product is then transferred to the second active site. In the first step, carbocation formation is proton initiated, whilst the second is ionization initiated. **Abietic acid** results from sequential oxidation of the 4α-methyl. Wounding of pine trees leads to an accumulation at the wound site of both monoterpenes and diterpenes, and fractionation by distillation gives turpentine oil and rosin. The volatile monoterpenes seem to act as a solvent to allow deposition of the rosin layer to seal the wound. The diterpenes in rosin have both antifungal and insecticidal properties.

Extensive modification of the copalyl diterpene skeleton is responsible for generation of the **ginkgolides**, highly oxidized diterpene trilactones which are the active principles of *Ginkgo biloba* (Ginkgoaceae) [Box 5.10]. Apart from the C_{20} skeleton, hardly any typical terpenoid features are recognizable. However, a speculative scheme involving several rearrangements, ring cleavage, and formation of lactone rings can broadly explain its origin and provide a link between these extremely complex ginkgolide structures and the more familiar diterpenes (Figure 5.54). Detailed evidence is lacking. What is known is that levopimaradiene and abietatriene are precursors; levopimaradiene synthase is also a bifunctional enzyme (compare abietadiene synthase) that converts GGPP into levopimaradiene via (+)-copalyl PP. The unusual *tert*-butyl substituent arises as a consequence of the A ring cleavage. **Bilobalide** (Figure 5.54) contains a related C_{15}-skeleton, and is most likely a partially degraded ginkgolide. *Ginkgo* is the world's oldest tree species, and its leaves are a highly popular and currently fashionable health supplement, taken in the anticipation that it can delay some of the degeneration of the faculties normally experienced in old age.

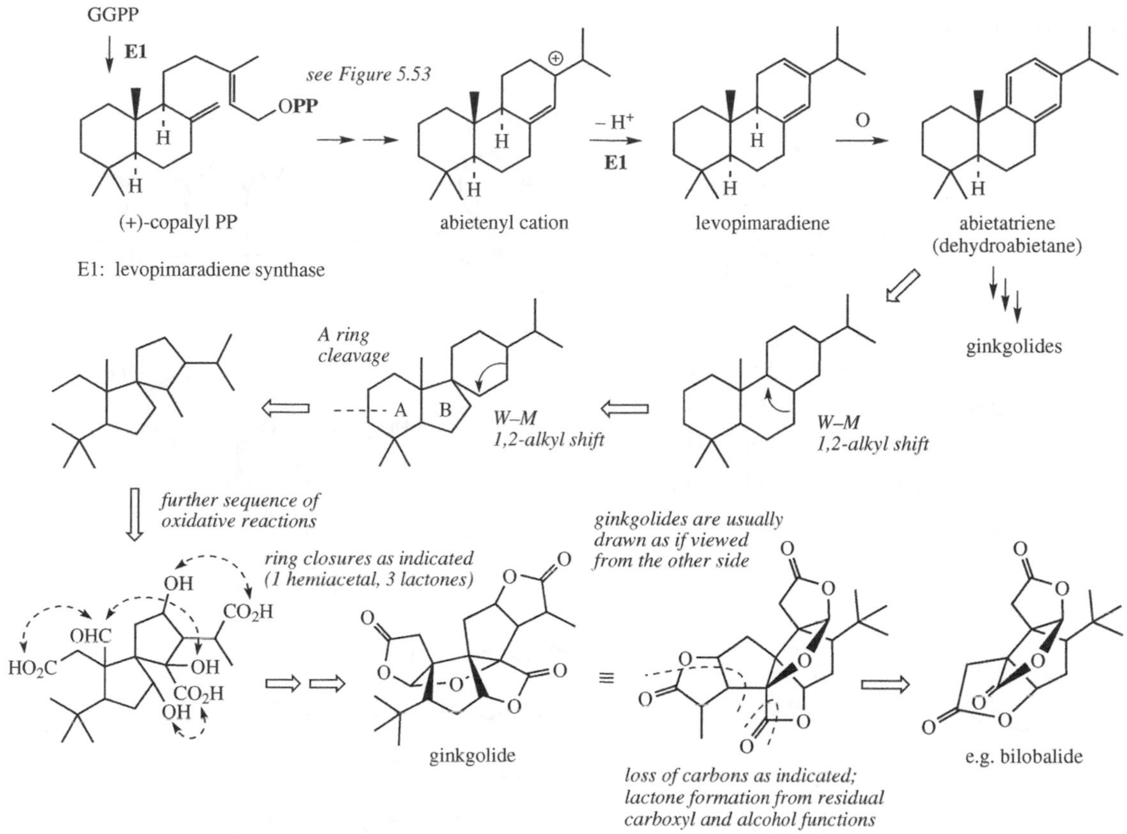

Figure 5.54

Box 5.10

Ginkgo biloba

Ginkgo biloba is a primitive member of the gymnosperms and the only survivor of the Ginkgoaceae, all other species being found only as fossils. It is a small tree native to China, but widely planted as an ornamental, and cultivated for drug use in Korea, France, and the United States. Standardized extracts of the leaves are marketed against cerebral vascular disease and senile dementia. Extracts have been shown to improve peripheral and cerebrovascular circulation. The decline in cognitive function and memory processes in old age can be due to disturbances in brain blood circulation, and thus **ginkgo** may exert beneficial effects by improving this circulation, and assist with other symptoms such as vertigo, tinnitus, and hearing loss. Virtually all clinical studies report positive results regarding cerebral insufficiency.

The active constituents have been characterized as mixtures of terpenoids and flavonoids. The dried leaves contain 0.1–0.25% terpene lactones, comprising predominantly the five ginkgolides (A, B, C, J, and M) and bilobalide (Figure 5.55). Bilobalide accounts for about 30–40% of the mixture, whilst ginkgolide A is the predominant ginkgolide (about 30%). The ginkgolides are diterpenoid in nature, whilst bilobalide is described as sesquiterpenoid. However, bilobalide bears such a structural similarity to the ginkgolides, that it is most probably a degraded ginkgolide. The ginkgolides have been shown to have potent and selective antagonistic activity towards platelet-activating factor (PAF, see page 44), which is implicated in many physiological processes. More recently, ginkgolides have been shown to exert antagonistic effects on glycine receptors; together with GABA receptors, glycine receptors are the main inhibitory receptors in the central nervous system. Bilobalide lacks PAF antagonistic activity but shows neuroprotective effects. The flavonoid content of the dried leaves is 0.5–1.0%, and consists of a mixture of mono-, di-, and tri-glycosides of the flavonols kaempferol and quercetin (Figure 5.55; see also page 171) and some biflavonoids. These probably

Box 5.10   (continued)

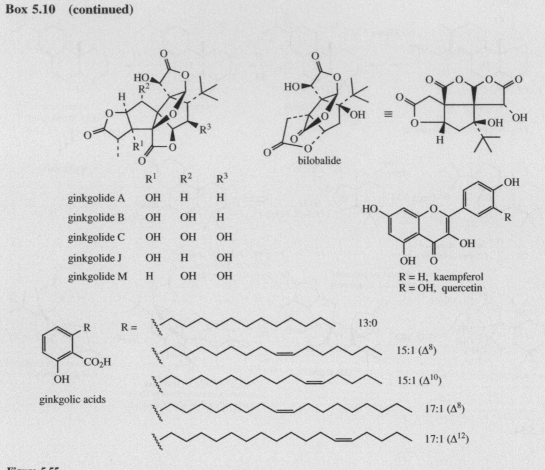

	R¹	R²	R³
ginkgolide A	OH	H	H
ginkgolide B	OH	OH	H
ginkgolide C	OH	OH	OH
ginkgolide J	OH	H	OH
ginkgolide M	H	OH	OH

R = H, kaempferol
R = OH, quercetin

13:0
15:1 (Δ^8)
15:1 (Δ^{10})
17:1 (Δ^8)
17:1 (Δ^{12})

ginkgolic acids

Figure 5.55

also contribute to the activity of ginkgo, and may act as radical scavengers. Ginkgo leaf is also a rich source of shikimic and quinic acids (see page 138).

Extracts of ginkgo for drug use are usually standardized to contain flavonoid glycosides and terpene lactones in a ratio of 24% to 6%, or 27% to 7%. Standardized extracts should also contain less than 5 ppm of alkyl phenols, since these are known to be allergenic, capable of inducing contact dermatitis, as well as being potentially cytotoxic and mutagenic. The ginkgolic acids (also known as anacardic acids, see page 118) form the main group of alkyl phenols found in ginkgo leaf (up to 1.7%). These phenolic acids have long, mainly unsaturated, alkyl chains and are related to the fatty acids (Figure 5.55). Ginkgo may be combined with ginseng (see page 245) in the treatment of geriatric disorders. Ginkgo and the ginkgolides are undergoing extensive investigation in conditions where there are high PAF levels, e.g. shock, burns, ulceration, and inflammatory skin disease.

In **forskolin** (colforsin; Figure 5.56), the third ring is heterocyclic rather than carbocyclic. The basic skeleton of forskolin can be viewed as the result of quenching of the cation by water as opposed to proton loss, followed by S_N2'-type nucleophilic substitution onto the allylic diphosphate (or nucleophilic substitution onto the allylic cation generated by loss of diphosphate) (Figure 5.56). A series of oxidative modifications will

then lead to forskolin. This compound has been isolated from roots of *Coleus forskohlii* (*Plectranthus barbatus*) (Labiatae/Lamiaceae), a plant used in Indian traditional medicine, and has been found to lower blood pressure and have cardioprotective properties. Forskolin has become a valuable pharmacological tool as a potent stimulator of adenylate cyclase activity, and has shown promising potential for the treatment of glaucoma,

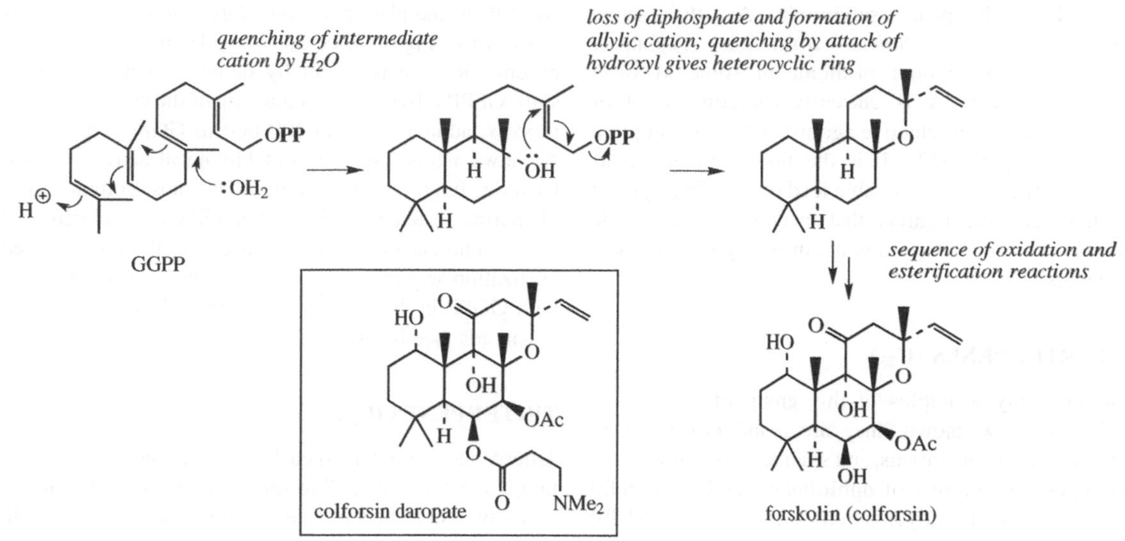

Figure 5.56

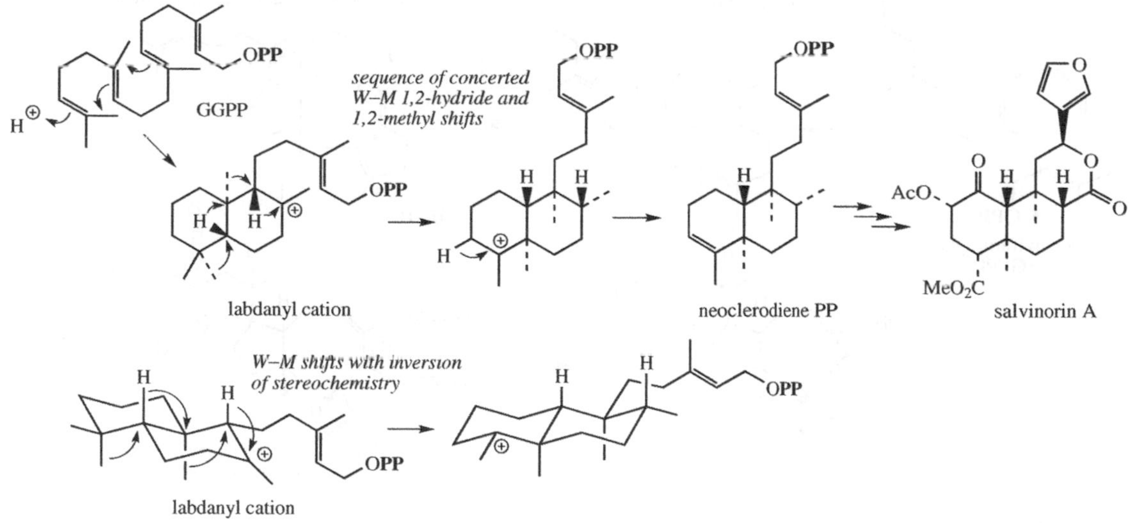

Figure 5.57

congestive heart failure, hypertension, and bronchial asthma. The water-soluble derivative colforsin daropate has been introduced for drug use.

In Figure 5.56, the GGPP-derived cation is quenched by the addition of water; in other cases, it may precipitate a series of concerted Wagner–Meerwein hydride and methyl migrations, resulting in a rearranged skeleton with different stereochemistry (Figure 5.57). Similar processes are encountered in triterpenoid and steroid biosynthesis (see page 238). The first migration is of hydride, generating a new tertiary carbocation, with successive methyl, hydride, and methyl migrations; the sequence is terminated by proton loss, giving **neoclerodiene PP**. The migrating groups are each positioned *anti* to one another, one group entering whilst the other leaves from the opposite side of the stereocentre. This inverts the configuration at each appropriate centre. Undoubtedly arising from this system is the neoclerodane diterpene **salvinorin A**, the active hallucinogen from *Salvia divinorum* (Labiatae/Lamiaceae), a traditional medicine of some Mexican Indians. Salvinorin A is currently attracting a lot of interest, since it is a selective agonist at the kappa-opioid receptor (see page 352). It is the first non-nitrogenous agonist known, and is a valuable lead for the development of other selective ligands that may have therapeutic potential in a range of conditions, including pain, nausea, and depression.

SESTERTERPENES (C$_{25}$)

Although many examples of this group of natural terpenoids are now known, they are found principally in fungi and marine organisms, and span relatively few structural types. The origins of **ophiobolene** and **ophiobolin A** (Figure 5.58) from cyclization of **geranylfarnesyl PP**

(**GFPP**) in the plant pathogen *Helminthosporium maydis* is shown in Figure 5.58, and provides no novel features except for an experimentally demonstrated 1,5-hydride shift. GFPP arises by a continuation of the chain extension process, adding a further IPP unit to GGPP. Ophiobolin A shows a broad spectrum of biological activity against bacteria, fungi, and nematodes. The most common type of marine sesterterpenoid is exemplified by **sclarin**, and this structure can be envisaged as the result of a concerted cyclization sequence (Figure 5.59) analogous to that seen with GGPP in the diterpenoids, and with squalene in the hopanoids (see below).

TRITERPENES (C$_{30}$)

Triterpenes are not formed by an extension of the now familiar process of adding IPP to the growing chain. Instead, two molecules of FPP are joined tail-to-tail to yield

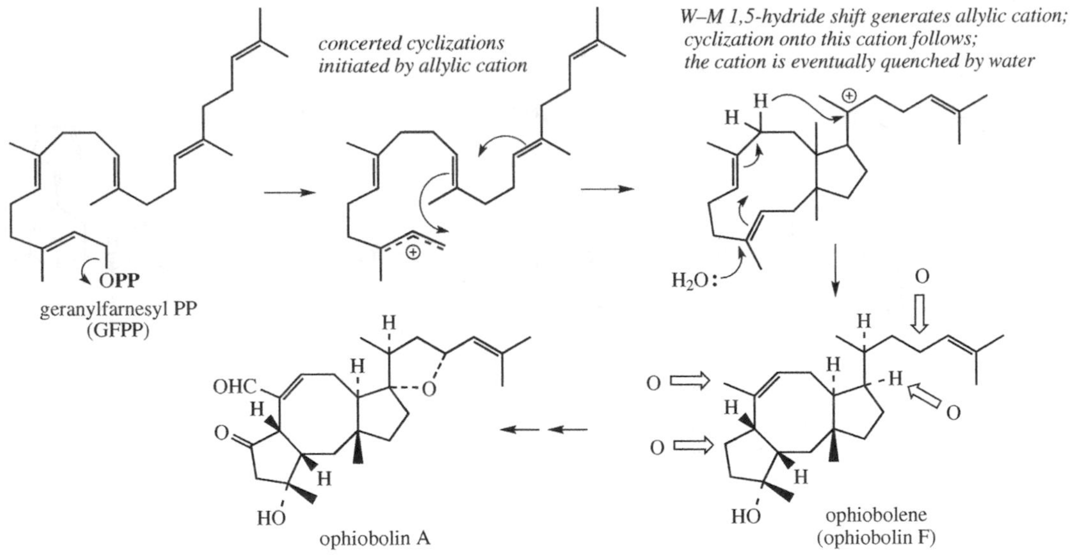

Figure 5.58

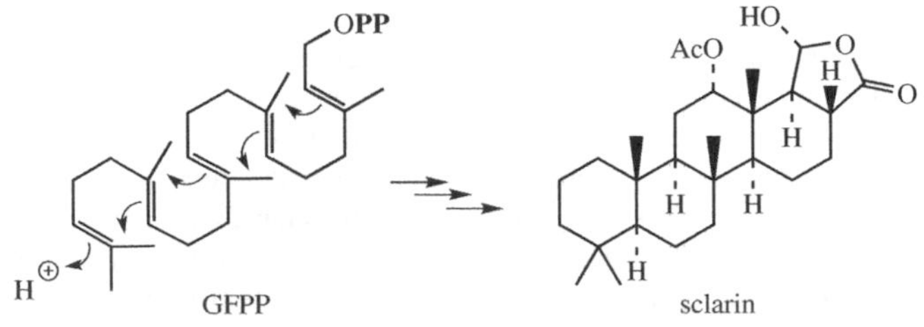

Figure 5.59

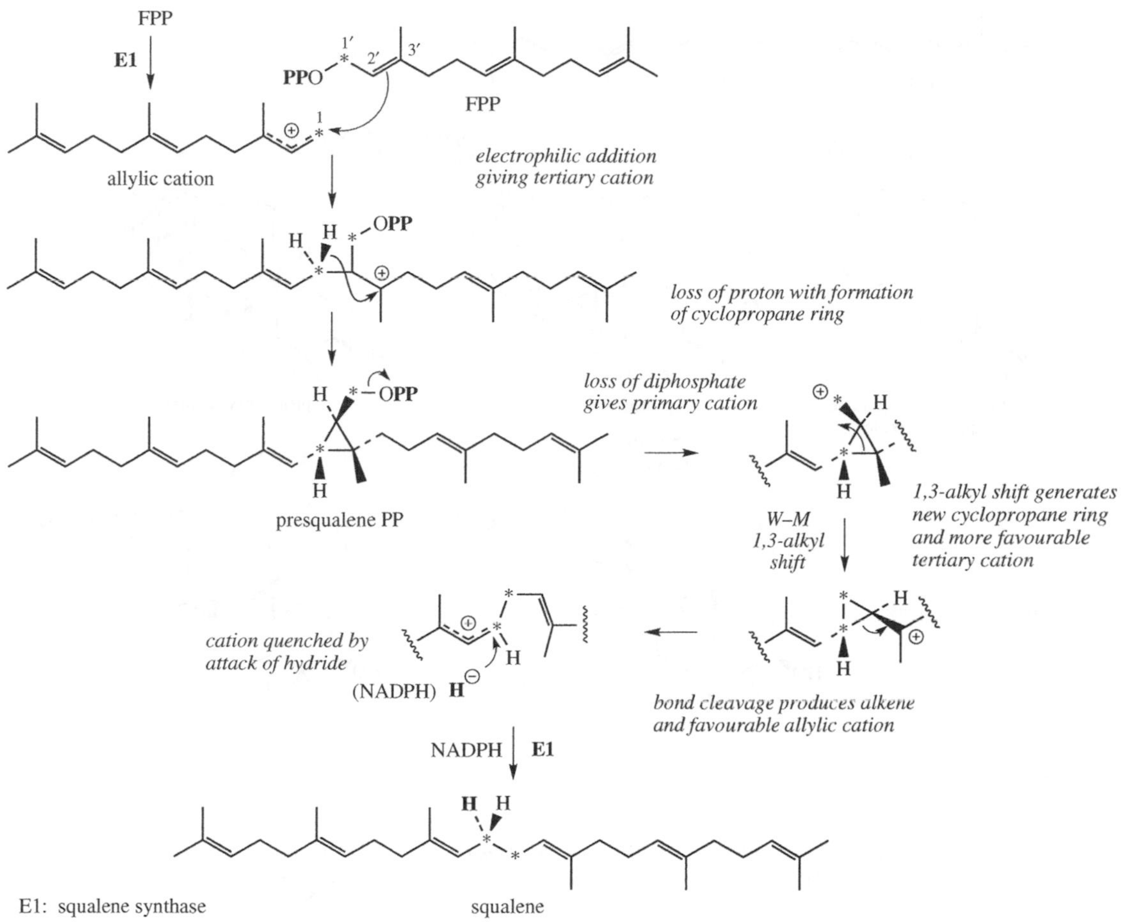

Figure 5.60

squalene (Figure 5.60); in general, FPP is formed from MVA (see page 192). Squalene is a hydrocarbon originally isolated from the liver oil of shark (*Squalus* sp.), but was subsequently found in rat liver and yeast, and these systems were used to study its biosynthetic role as a precursor of triterpenes and steroids. Several seed oils are now recognized as quite rich sources of squalene, e.g. *Amaranthus cruentus* (Amaranthaceae). During the coupling process, which on paper merely requires removal of the two diphosphate groups, a proton from a C-1 position of one molecule of FPP is lost and a proton from NADPH is inserted. Difficulties with formulating a plausible mechanism for this unlikely reaction were resolved when **presqualene diphosphate**, an intermediate in the process, was isolated from rat liver. Its characterization as a cyclopropane derivative immediately ruled out all the hypotheses current at the time.

The formation of presqualene PP in Figure 5.60, is initiated by attack of the 2,3-double bond of FPP onto the farnesyl cation, which is mechanistically equivalent to normal chain extension using IPP. The resultant tertiary cation is discharged by loss of a proton and formation of a cyclopropane ring, giving presqualene PP. An exactly analogous sequence was used for the origins of irregular monoterpenes (see page 205). Obviously, to then form squalene, C-1s of the two FPP units must eventually be coupled, whilst presqualene PP formation has actually joined C-1 of one molecule to C-2 of the other. To account for the subsequent change in bonding of the two FPP units, a further cyclopropane cationic intermediate is proposed. Loss of diphosphate from presqualene PP would give an unfavourable primary cation, which via Wagner–Meerwein rearrangement can generate a tertiary carbocation and achieve the required C-1–C-1′ bond.

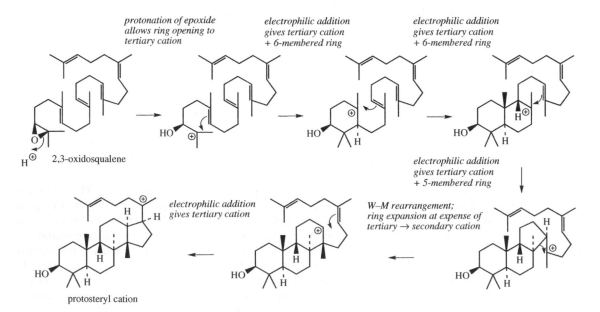

Figure 5.61

Figure 5.62

Breaking the original but now redundant C-1–C-2′ bond can give an allylic cation, and the generation of **squalene** is completed by supply of hydride from NADPH. Squalene synthase catalyses all steps in the sequence, the formation of presqualene PP from FPP, its subsequent rearrangement, and NADPH-dependent reduction.

Cyclization of squalene is via the intermediate **2,3-oxidosqualene** (squalene-2,3-oxide; Figure 5.61), produced in a reaction catalysed by squalene epoxidase, a flavoprotein requiring O_2 and NADPH cofactors. If oxidosqualene is suitably positioned and folded on the enzyme surface, then the polycyclic triterpene structures formed can be rationalized in terms of a series of cyclizations, followed by a sequence of concerted Wagner–Meerwein migrations of hydride and methyl groups (Figure 5.61). The cyclizations are carbocation-mediated and proceed in a stepwise sequence (Figure 5.62). Thus, protonation of the epoxide group will allow opening of this ring and generation of the preferred tertiary carbocation, suitably placed to allow electrophilic addition to a double bond, formation of a six-membered ring, and production of a new tertiary carbocation. This process continues twice more, generating the preferred tertiary carbocation (Markovnikov addition) after each ring formation, though the third ring formed is consequently a five-membered one. This is expanded to a six-membered ring via a Wagner–Meerwein 1,2-alkyl shift, resulting in some relief of ring strain, though sacrificing a tertiary carbocation for a secondary one. A further electrophilic addition generates the tertiary protosteryl cation (Figure 5.62). The stereochemistries in this cation are controlled by the type of folding achieved on the enzyme surface, and this probably also limits the extent of the cyclization process. Thus, if the folded oxidosqualene approximates to a *chair–boat–chair–boat* conformation (Figure 5.63), the transient **protosteryl cation** will be produced with these conformational characteristics. This cation then undergoes a series of Wagner–Meerwein 1,2-shifts, first migrating a hydride and generating a new cation, migrating the next hydride, then a methyl and so on until a proton is lost forming a double bond and thus creating **lanosterol** (Figure 5.63). The stereochemistry of the protosteryl cation in Figure 5.63 shows how favourable this sequence will be, and emphasizes that, in the ring system, the migrating groups are positioned *anti* to each other, one group entering whilst the other leaves from the opposite side of the stereocentre (compare diterpenes, page 233). This, of course, inverts configurations at each appropriate centre. No *anti* group is available to migrate to C-9 (steroid numbering), and the reaction terminates by loss of the proton H-9. Lanosterol is

a typical animal triterpenoid, and the precursor for cholesterol and other sterols in animals (see page 248) and fungi (see page 254). In plants, its intermediate role is taken by **cycloartenol** (Figure 5.63), which contains a cyclopropane ring, generated by inclusion of carbon from the methyl at C-10. For cycloartenol, H-9 is not lost, but migrates to C-8, and the carbocation so formed is quenched by cyclopropane formation and loss of one of the methyl protons. For many plant steroids, this cyclopropane ring has then to be reopened (see page 251). Most natural triterpenoids and steroids contain a 3-hydroxyl group, the original epoxide oxygen from oxidosqualene.

An additional feature of the protosteryl cation is that the C-10 methyl and H-5 also share an *anti*-axial relationship, and are also susceptible to Wagner–Meerwein rearrangements, so that the C-9 cation formed in the cycloartenol sequence may then initiate further migrations. This can be terminated by formation of a 5,6-double bond (Figure 5.64), as in the pathway to the **cucurbitacins**, a group of highly oxygenated triterpenes encountered in the Cucurbitaceae, the cucumber/melon/marrow family. These compounds are characteristically bitter tasting, purgative, and extremely cytotoxic.

Should **oxidosqualene** be folded in a roughly *chair–chair–chair–boat* conformation by binding to another type of cyclase enzyme (Figure 5.65), then an identical carbocation mechanism ensues. However, the transient **dammarenyl cation** formed has different stereochemical features to the protosteryl cation. Whilst a series of Wagner–Meerwein migrations can occur, there is relatively little to be gained on purely chemical grounds, since these would invert stereochemistry and destroy what is already a very favourable conformation. Instead, the dammarenyl cation typically undergoes further carbocation-promoted cyclizations, without any major changes to the ring system already formed. Occasionally, the migrations do occur: **euphol** (Figure 5.65) from *Euphorbia* species (Euphorbiaceae) is a stereoisomer of lanosterol.

If the Wagner–Meerwein rearrangements do not take place, then the dammarenyl cation could be quenched with water (Figure 5.66), giving the epimeric **dammarenediols**, as found in Dammar resin from *Balanocarpus heimii* (Dipterocarpaceae) and ginseng (*Panax ginseng*; Araliaceae) [Box 5.11]. Alternatively, the alkyl migration shown gives the baccharenyl cation, relieving some ring strain by creating a six-membered ring, despite sacrificing a tertiary carbocation for a secondary one. A pentacyclic ring system can now be formed by cyclization onto the double bond, giving a new five-membered ring and the tertiary lupenyl cation. Although this appears to

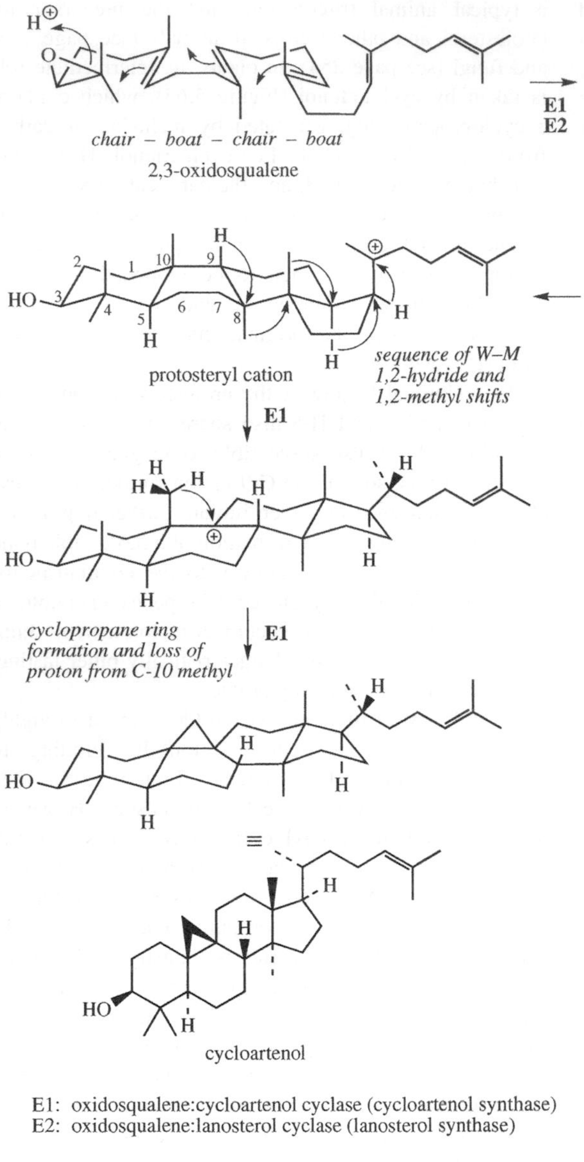

chair – boat – chair – boat
2,3-oxidosqualene

protosteryl cation

sequence of W–M
1,2-hydride and
1,2-methyl shifts

cyclopropane ring
formation and loss of
proton from C-10 methyl

sequence of W–M
1,2-hydride and
1,2-methyl shifts

protosteryl cation

loss of H-9 creates
double bond

cycloartenol

lanosterol

E1: oxidosqualene:cycloartenol cyclase (cycloartenol synthase)
E2: oxidosqualene:lanosterol cyclase (lanosterol synthase)

Figure 5.63

contradict the reasoning used above for the dammarenyl → baccharenyl transformation, the contribution of the enzyme involved must also be considered in each case. A five-membered ring is not highly strained, as evidenced by all the natural examples encountered. Loss of a proton from the lupanyl cation gives **lupeol**, found in lupin (*Lupinus luteus*; Leguminosae/Fabaceae). Ring expansion in the lupanyl cation by bond migration gives the oleanyl system, and labelling studies have demonstrated this ion is discharged by hydride migrations and loss

of a proton, giving the widely distributed **β-amyrin**. Formation of the isomeric **α-amyrin** involves first the migration of a methyl in the oleanyl cation, then discharge of the new taraxasteryl cation by three hydride migrations and loss of a proton. Loss of a proton from the non-migrated methyl in the taraxasteryl cation is an alternative way of achieving a neutral molecule, and yields **taraxasterol** found in dandelion (*Taraxacum officinale*; Compositae/Asteraceae). Comparison with α-amyrin shows the subtly different stereochemistry

first sequence of W–M 1,2-hydride and 1,2-methyl shifts

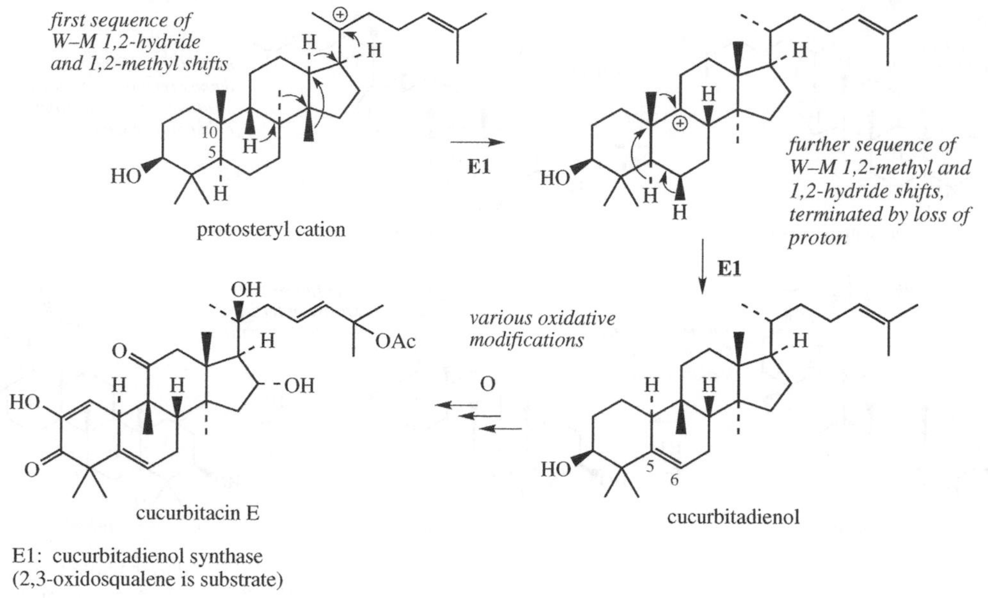

protosteryl cation

E1

further sequence of W–M 1,2-methyl and 1,2-hydride shifts, terminated by loss of proton

E1

various oxidative modifications

O

cucurbitacin E

cucurbitadienol

E1: cucurbitadienol synthase
(2,3-oxidosqualene is substrate)

Figure 5.64

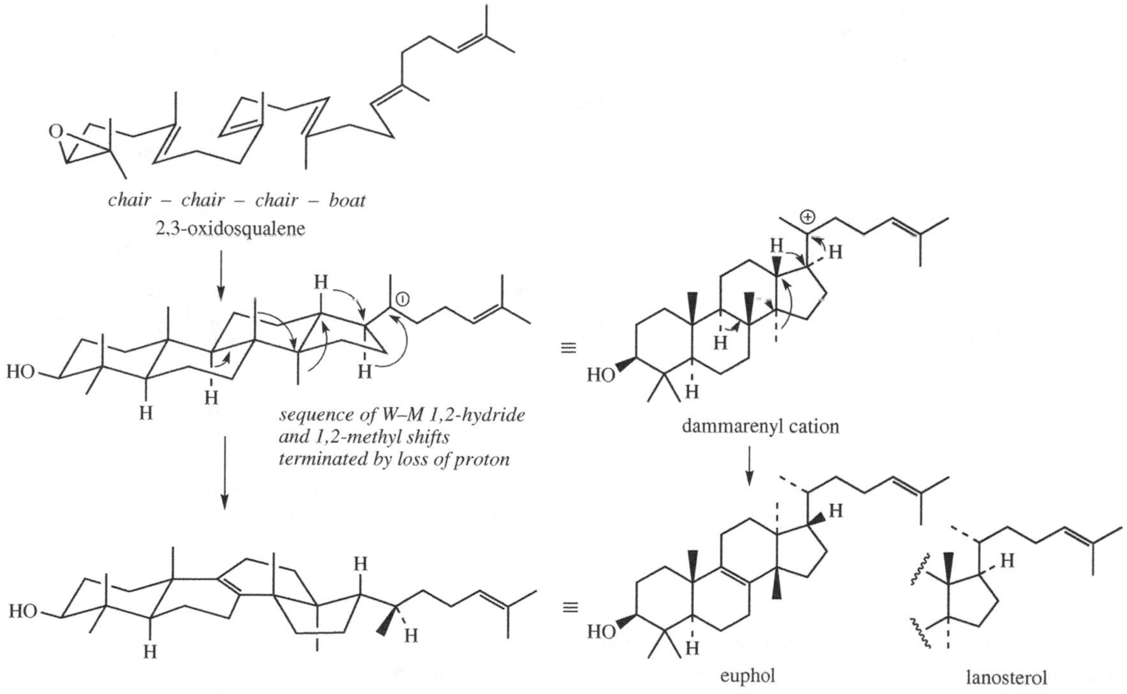

chair – chair – chair – boat
2,3-oxidosqualene

sequence of W–M 1,2-hydride and 1,2-methyl shifts terminated by loss of proton

dammarenyl cation

euphol

lanosterol

Figure 5.65

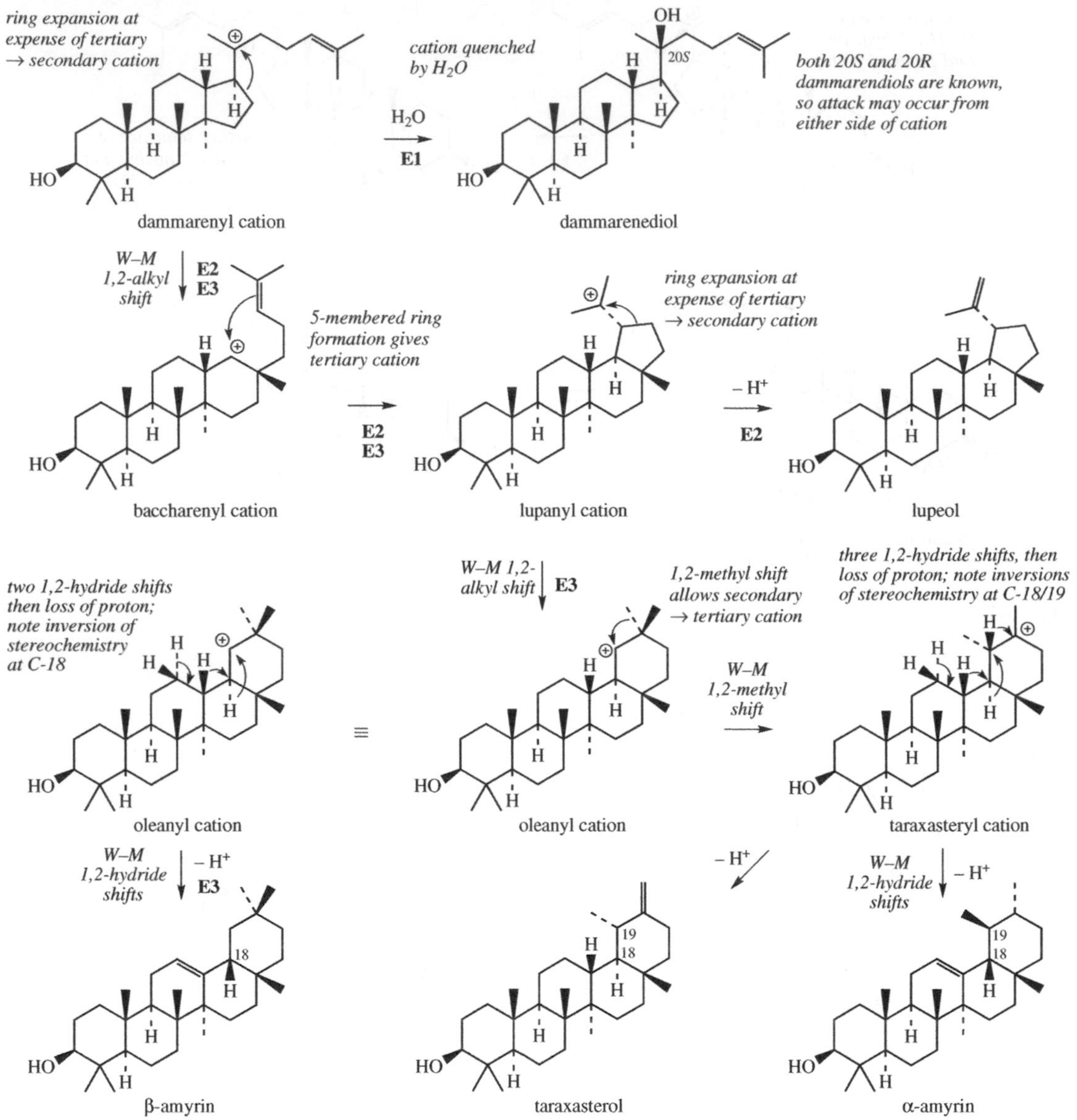

E1: dammarenediol-II synthase
E2: oxidosqualene:lupeol cyclase (lupeol synthase)

E3: oxidosqualene:β-amyrin cyclase (β-amyrin synthase)
(2,3-oxidosqualene is substrate for enzymes)

Figure 5.66

present, because the inversions of configuration caused by hydride migrations have not occurred. Where evidence is available, these extensive series of cyclizations and Wagner–Meerwein rearrangements appear to be catalysed by a single enzyme which converts oxidosqualene into the final product, e.g. lanosterol, cycloartenol, lupeol, or β-amyrin. Enzymes can be monofunctional, synthesizing a single product, or in some cases the triterpenoid cyclase is multifunctional, catalysing formation of a range of structurally related products, e.g. lupeol, α-amyrin, taraxasterol, and others. Remarkably, changing a single amino acid residue in lupeol synthase from olive (*Olea europaea*; Oleaceae) was sufficient to alter the protein's function; it became instead a producer of β-amyrin.

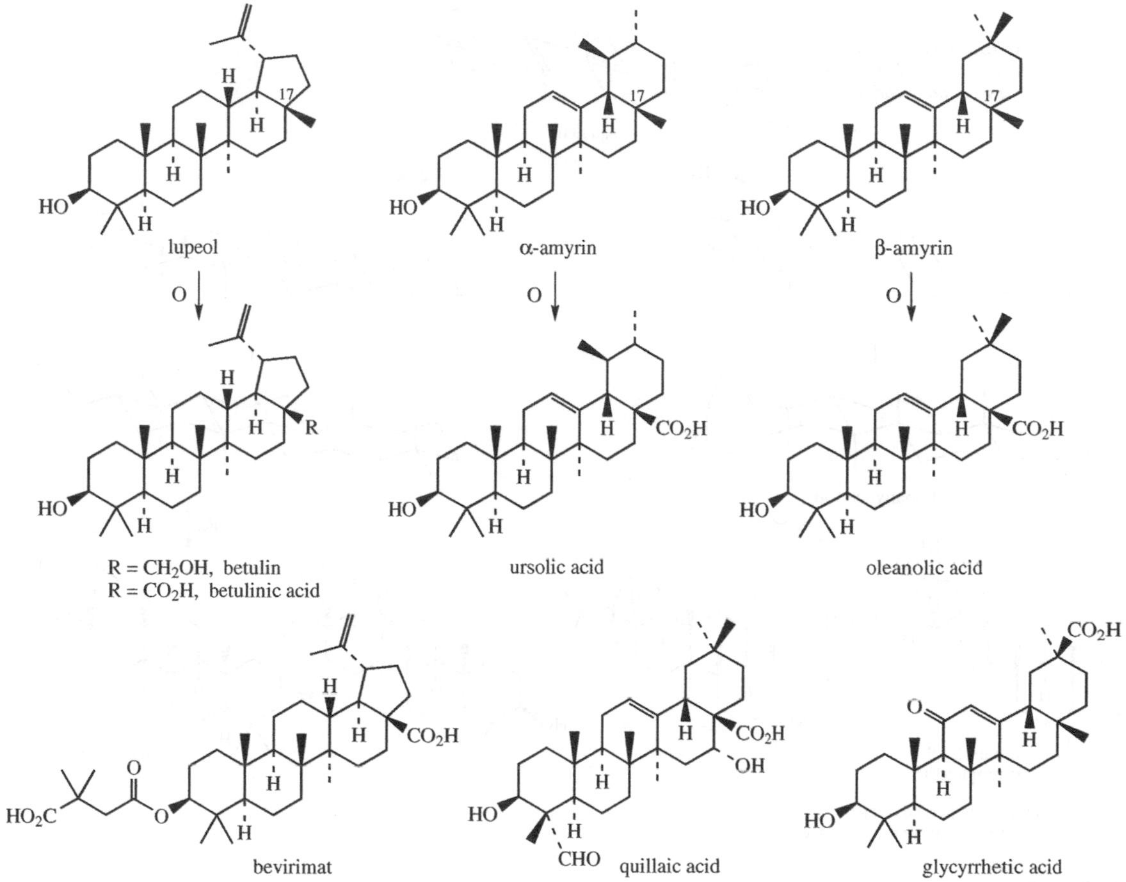

Figure 5.67

Further modification of these hydrocarbons to **triterpene acids** is frequently encountered (Figure 5.67). Olives (*Olea europaea*; Oleaceae) contain large quantities of **oleanolic acid**, in which the C-17 methyl of β-amyrin has been oxidized to a carboxylic acid. The leaves of olive contain a mixture of oleanolic acid and **betulinic acid**, the latter an oxidized version of lupeol. The corresponding derivative of α-amyrin is **ursolic acid**, found in bearberry (*Arctostaphylos uva-ursi*; Ericaceae). These are presumably produced by cytochrome P-450-dependent oxidations, but details are not known. Oxidative transformations at other methyl groups and/or ring carbon atoms are required to produce triterpenoids such as **glycyrrhetic acid** and **quillaic acid** (see triterpenoid saponins below). The alcohol **betulin** is a major component in the bark of white birch (*Betula alba*; Betulaceae), where it comprises up to 24% of the outer bark layer; large amounts of the rarer betulinic acid may be obtained by selective chemical oxidation of betulin. Derivatives of betulinic

acid and betulin are currently attracting considerable interest in HIV therapeutics by inhibiting viral growth in a manner different from that of current HIV drugs. The most promising compound, currently in clinical trials, is the semi-synthetic 3′,3′-dimethylsuccinyl ester **bevirimat** (Figure 5.67).

Bacterial membranes frequently contain **hopanoids** (Figure 5.68), triterpenoid compounds that appear to take the place of the sterols that are typically found in the membranes of higher organisms, helping to maintain structural integrity and to control permeability. Hopanoids are also the characteristic triterpenes in ferns. Hopanoids arise from squalene by a similar carbocation cyclization mechanism, but do not involve the initial epoxidation to oxidosqualene. Instead, the carbocation is produced by protonation (compare the cyclization of GGPP to copalyl PP, page 228), and the resultant compounds tend to lack the characteristic 3-hydroxyl group, e.g. **hopene** from

cyclization is initiated by protonation of double bond to give tertiary cation

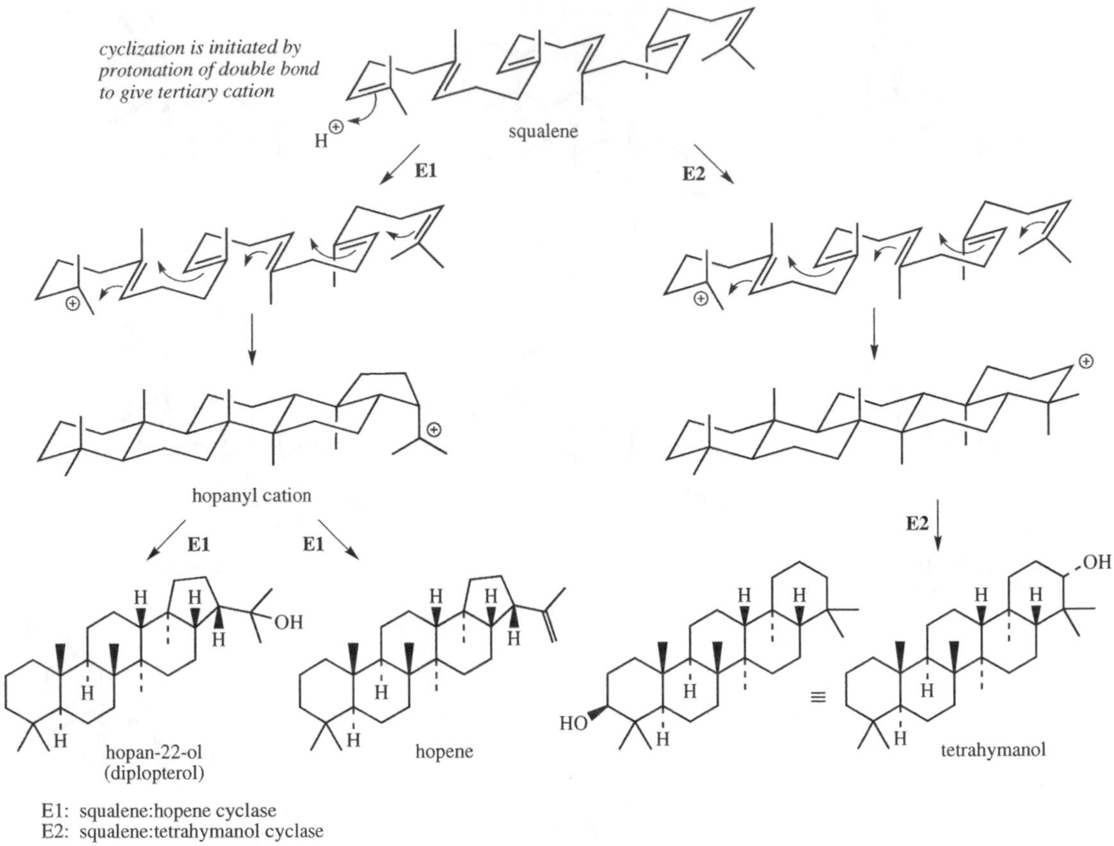

squalene

hopanyl cation

hopan-22-ol
(diplopterol)

hopene

tetrahymanol

E1: squalene:hopene cyclase
E2: squalene:tetrahymanol cyclase

Figure 5.68

Alicyclobacillus acidocaldarius (Figure 5.68). On the other hand, **tetrahymanol** from the protozoan *Tetrahymena pyriformis*, because of its symmetry, might appear to have a 3-hydroxyl group, but this is derived from water and not molecular oxygen, as would be the case if oxidosqualene were involved. As in formation of the protosteryl cation (page 237), Markovnikov additions followed by Wagner–Meerwein ring expansions (rather than anti-Markovnikov additions) may occur during the cyclization mechanisms shown in Figure 5.68.

Triterpenoid Saponins

The pentacyclic triterpenoid skeletons exemplified by lupeol, α-amyrin, and β-amyrin (Figure 5.67) are frequently encountered in the form of triterpenoid saponin structures. Saponins are glycosides which, even at low concentrations, produce a frothing in aqueous solution, because they have surfactant and soap-like properties. The name comes from the Latin *sapo*, meaning soap, and plant materials containing saponins were originally used for cleansing clothes, e.g. soapwort (*Saponaria officinalis*; Caryophyllaceae) and quillaia or soapbark (*Quillaja saponaria*; Rosaceae). These materials also cause haemolysis, lysing red blood cells by increasing the permeability of the plasma membrane, and thus they are highly toxic when injected into the bloodstream. Some saponin-containing plant extracts have been used as arrow poisons. However, saponins are relatively harmless when taken orally, and some of our valuable food materials, e.g. beans, lentils, soybeans, spinach, and oats, contain significant amounts. Sarsaparilla (see page 264) is rich in steroidal saponins, but is widely used in the manufacture of non-alcoholic drinks. Toxicity is

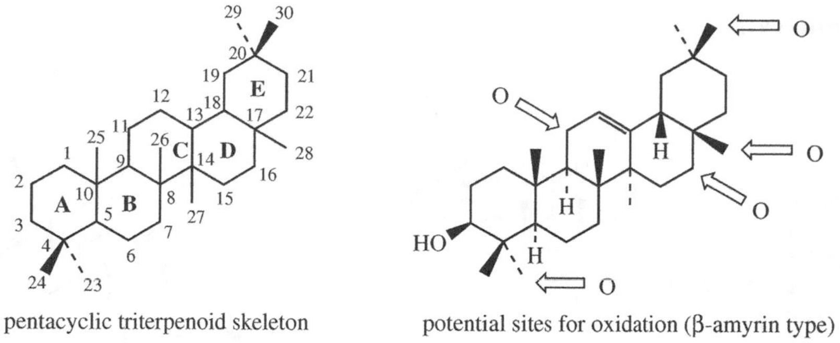

pentacyclic triterpenoid skeleton

potential sites for oxidation (β-amyrin type)

Figure 5.69

minimized during ingestion by low absorption, and by hydrolysis. Acid-catalysed hydrolysis of saponins liberates sugar(s) and an aglycone (sapogenin) which can be either triterpenoid or steroidal (see page 259) in nature. Some plants may contain exceptionally high amounts of saponins, e.g. about 10% in quillaia bark.

Triterpenoid saponins are rare in monocotyledons, but abundant in many dicotyledonous families. Several medicinally useful examples are based on the β-amyrin subgroup (Figure 5.69), and many of these possess carboxylic acid groups derived by oxidation of methyl groups, those at positions 4 (C-23), 17 (C-28) and 20 (C-30) on the aglycone ring system being subject to such oxidation. Less oxidized formyl (–CHO) or hydroxymethyl (–CH$_2$OH) groups may be encountered, and positions 11 and 16 may also be oxygenated. Sugar

residues are usually attached to the 3-hydroxyl, with one to six monosaccharide units, the most common being glucose, galactose, rhamnose, and arabinose, with uronic acid units (glucuronic acid and galacturonic acid) also featuring (see page 488). Thus, quillaia bark contains a saponin mixture with **quillaic acid** (Figure 5.67) as the principal aglycone [Box 5.11]. The medicinally valuable root of liquorice (*Glycyrrhiza glabra*; Leguminosae/Fabaceae) contains **glycyrrhizin**, a mixture of potassium and calcium salts of **glycyrrhizic acid** (Figure 5.70), which is composed of the aglycone **glycyrrhetic acid** and two glucuronic acid units [Box 5.11]. Ginseng is a herbal drug derived from the roots of *Panax ginseng* (Araliaceae) that is widely held to counter stress and improve general well-being. A group of saponins based on the dammarane skeleton and termed **ginsenosides** (see Figure 5.72) are most likely the biologically active components [Box 5.11].

Box 5.11

Liquorice

Liquorice (licorice; glycyrrhiza) is the dried unpeeled rhizome and root of the perennial herb *Glycyrrhiza glabra* (Leguminosae/Fabaceae). A number of different varieties are cultivated commercially, including *G. glabra* var. *typica* (Spanish liquorice) in Spain, Italy, and France, and *G. glabra* var. *glandulifera* (Russian liquorice) in Russia. Russian liquorice is usually peeled before drying. *G. uralensis* (Manchurian liquorice) from China is also commercially important. Much of the liquorice is imported in the form of an extract, prepared by extraction with water, then evaporation to give a dark black solid. Most of the liquorice produced is used in confectionery and for flavouring, including tobacco, beers, and stouts. Its pleasant sweet taste and foaming properties are due to saponins. Liquorice root contains about 20% of water-soluble extractives, and much of this (typically 3–5% of the root, but up to 12% in some varieties) is comprised of glycyrrhizin, a mixture of the potassium and calcium salts of glycyrrhizic (=glycyrrhizinic) acid (Figure 5.70). Glycyrrhizic acid is a diglucuronide of the aglycone glycyrrhetic (=glycyrrhetinic) acid. The bright yellow colour of liquorice root is provided by flavonoids (1–1.5%), including liquiritigenin and isoliquiritigenin and their corresponding glucosides (see page 169). Considerable amounts (5–15%) of sugars (glucose and sucrose) are also present.

Glycyrrhizin is reported to be 50–150 times as sweet as sucrose, and liquorice has thus long been used in pharmacy to mask the taste of bitter drugs. Its surfactant properties have also been exploited in various formulations, as have its demulcent

Box 5.11   (continued)

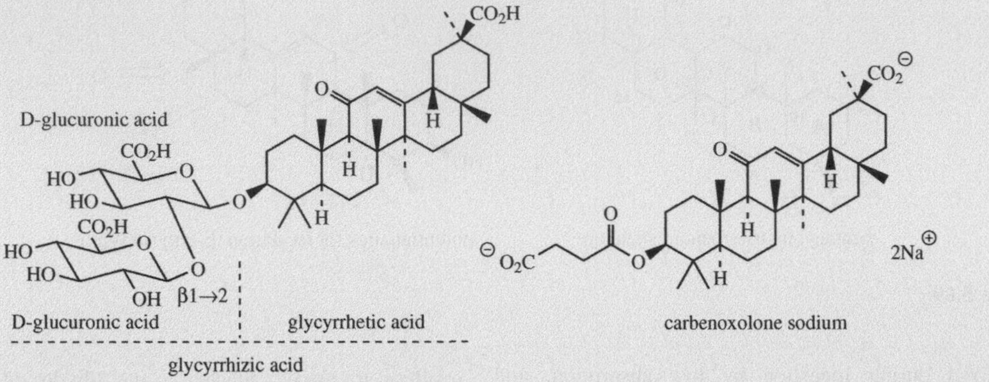

Figure 5.70

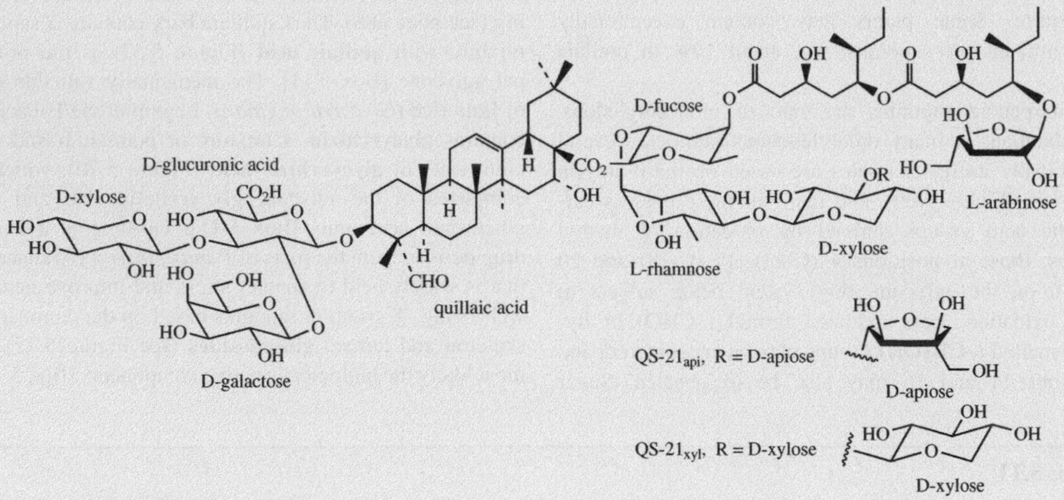

Figure 5.71

and mild expectorant properties. More recently, some corticosteroid-like activity has been recognized, with liquorice extracts displaying mild anti-inflammatory and mineralocorticoid activities. These have been exploited in the treatment of rheumatoid arthritis, Addison's disease (chronic adrenocortical insufficiency), and various inflammatory conditions. Glycyrrhetic acid has been implicated in these activities, and has been found to inhibit enzymes that catalyse the conversion of prostaglandins and glucocorticoids into inactive metabolites. This results in increased levels of prostaglandins, e.g. PGE_2 and $PGF_{2\alpha}$ (see page 60), and of hydrocortisone (see page 278). A semi-synthetic derivative of glycyrrhetic acid, the hemisuccinate **carbenoxolone sodium** (Figure 5.70), has been widely prescribed for the treatment of gastric ulcers, and also duodenal ulcers. Because of side-effects, typically loss of potassium and increase in sodium levels, it has been superseded by newer drugs.

Quillaia

Quillaia bark or soapbark is derived from the tree *Quillaja saponaria* (Rosaceae) and other *Quillaja* species found in Chile, Peru, and Bolivia. The bark contains up to 10% saponins, a mixture known as 'commercial saponin' which is used as a foaming agent

Box 5.11 (continued)

in beverages and emulsifier in foods. Quillaia's surfactant properties are occasionally exploited in pharmaceutical preparations, where in the form of quillaia tincture it is used as an emulsifying agent, particularly for fats, tars, and volatile oils. The bark contains a mixture of saponins which on hydrolysis liberates quillaic acid (Figure 5.67) as the aglycone, together with sugars, uronic acids, and acids from ester functions.

Saponins from quillaia are also showing great promise as immunoadjuvants, substances added to vaccines and other immunotherapies designed to enhance the body's immune response to the antigen. One such agent, QS-21A (Figure 5.71), is a mixture of the two saponins $QS-21_{api}$ and $QS-21_{xyl}$, each incorporating a quillaic acid triterpenoid core, flanked on either side by complex oligosaccharides, one of which includes a chain comprised of two fatty acyl components. These compounds were isolated from the 21st chromatographic fraction of the *Q. saponaria* extract.

Ginseng

The roots of the herbaceous plants *Panax ginseng* (Araliaceae) from China, Korea and Russia, and related *Panax* species, e.g. *P. quinquefolium* (American ginseng) from the USA and Canada and *P. notoginseng* (Sanchi-ginseng) from China, have been widely used in China and Russia for the treatment of a number of diseases, including anaemia, diabetes, gastritis, insomnia, sexual impotence, and as a general restorative, promoting health and longevity. Interest in the drug has increased considerably in recent years, and ginseng is widely available as a health food in the form of powders, extracts, and teas. The dried and usually peeled root provides white ginseng, whereas red ginseng is obtained by steaming the root, this process generating a reddish-brown caramel-like colour, and reputedly enhancing biological activity. **Ginseng** is classified as an 'adaptogen', an agent that helps the body to adapt to stress, improving stamina and concentration, and providing a normalizing and restorative effect. It is also widely promoted as an aphrodisiac. The Korean root is highly prized and the most expensive. Long-term use of ginseng can lead to symptoms similar to those of corticosteroid poisoning, including hypertension, nervousness, and sleeplessness in some subjects, yet hypotension and tranquillizing effects in others.

The benefits of ginseng treatment are by no means confirmed at the pharmacological level, though ginseng does possess antioxidant activity, can affect both central nervous system and neuroendocrine functions, can alter carbohydrate and lipid metabolism, and modulates immune function. Many of the secondary metabolites present in the root have now been identified. It contains a large number of triterpenoid saponins based on the dammarane subgroup, saponins that have been termed ginsenosides by Japanese investigators, or panaxosides by Russian researchers. These are derivatives of two main aglycones, protopanaxadiol and protopanaxatriol (Figure 5.72), though the aglycones liberated on acid hydrolysis are panaxadiol and panaxatriol respectively. Acid-catalysed cyclization in the side-chain produces an ether ring (Figure 5.72). Sugars are present in the saponins on the 3- and 20-hydroxyl groups in the diol series, and the 6- and 20-hydroxyl groups in the triol series. Over 30 ginsenosides have been characterized from the different varieties of ginseng, with ginsenoside Rb_1 (Figure 5.72) of the diol series typically being the most abundant constituent. Ginsenoside Rg_1 (Figure 5.72) is usually the major component representative of the triol series. Some other variants are shown in Figure 5.72. Particularly in white ginseng, many of the ginsenosides are also present as esters with malonic acid. Steaming to prepare red ginseng causes partial hydrolysis of esters and glycosides; there is undoubtedly greater antioxidant activity in the steamed product, though this is derived mainly from various phenolic constituents. Ginsenosides Rb_1 and Rg_1 appear to be the main representatives in *P. ginseng*, ginsenosides Rb_1, Rg_1, and Rd in *P. notoginseng*, and ginsenosides Rb_1, Re, and malonylated Rb_1 in *P. quinquefolium*. The pentacyclic triterpenoid sapogenin oleanolic acid (Figure 5.67) is also produced by hydrolysis of the total saponins of *P. ginseng*, and is present in some saponin structures (chikusetsusaponins). The saponin contents of *P. notoginseng* (about 12%) and *P. quinquefolium* (about 6%) are generally higher than that of *P. ginseng* (1.5–2%). Pharmacological studies on individual ginsenosides are being facilitated by development of enzymic procedures for interconverting the structures, e.g. by selective hydrolysis of sugar groups, or by selective glycosylations. The aglycone mixture of protopanaxadiol and protopanaxatriol (**pandimex**®) is being tested clinically as an anticancer agent.

The root of *Eleutherococcus senticosus* (*Acanthopanax senticosus*) (Araliaceae) is used as an inexpensive substitute for ginseng, and is known as **Russian** or **Siberian ginseng**. This material is held to have adaptogenic properties similar to *P. ginseng*, and a number of eleutherosides have been isolated. However, the term eleutheroside has been applied to compounds of different chemical classes, and the main active anti-stress constituents appear to be lignan glycosides, e.g. eleutheroside E (≡syringaresinol diglucoside; Figure 5.72) (compare page 152) and phenylpropane glycosides, e.g. eleutheroside B (≡syringin). The leaves of Russian ginseng contain a number of saponins based on oleanolic acid (Figure 5.67), but these are quite different to the ginsenosides/panaxosides found in *Panax*. Whilst there is sufficient evidence to support the beneficial adaptogen properties for *E. senticosus*, detailed pharmacological confirmation is not available.

Box 5.11   (continued)

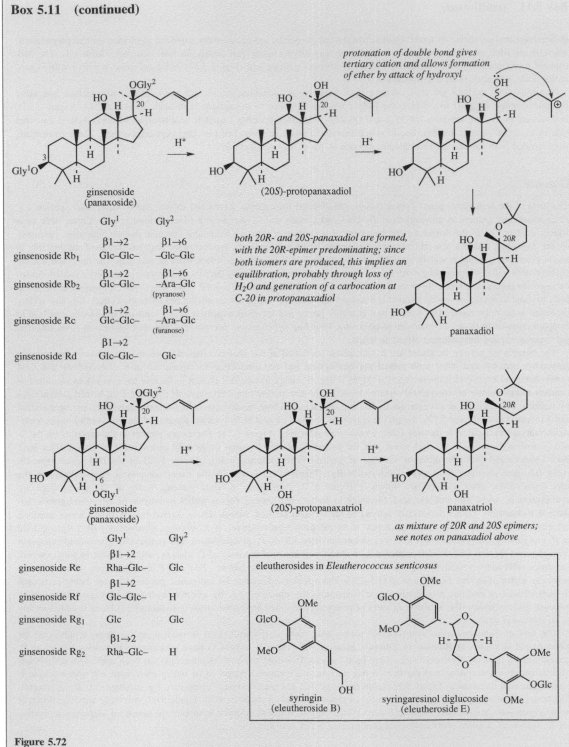

protonation of double bond gives tertiary cation and allows formation of ether by attack of hydroxyl

ginsenoside
(panaxoside)

(20S)-protopanaxadiol

both 20R- and 20S-panaxadiol are formed, with the 20R-epimer predominating; since both isomers are produced, this implies an equilibration, probably through loss of H_2O and generation of a carbocation at C-20 in protopanaxadiol

panaxadiol

	Gly[1]	Gly[2]
	β1→2	β1→6
ginsenoside Rb₁	Glc–Glc–	–Glc–Glc
	β1→2	β1→6
ginsenoside Rb₂	Glc–Glc–	–Ara–Glc (pyranose)
	β1→2	β1→6
ginsenoside Rc	Glc–Glc–	–Ara–Glc (furanose)
	β1→2	
ginsenoside Rd	Glc–Glc–	Glc

ginsenoside
(panaxoside)

(20S)-protopanaxatriol

panaxatriol

as mixture of 20R and 20S epimers; see notes on panaxadiol above

	Gly[1]	Gly[2]
	β1→2	
ginsenoside Re	Rha–Glc–	Glc
	β1→2	
ginsenoside Rf	Glc–Glc–	H
ginsenoside Rg₁	Glc	Glc
	β1→2	
ginsenoside Rg₂	Rha–Glc–	H

eleutherosides in *Eleutherococcus senticosus*

syringin
(eleutheroside B)

syringaresinol diglucoside
(eleutheroside E)

Figure 5.72

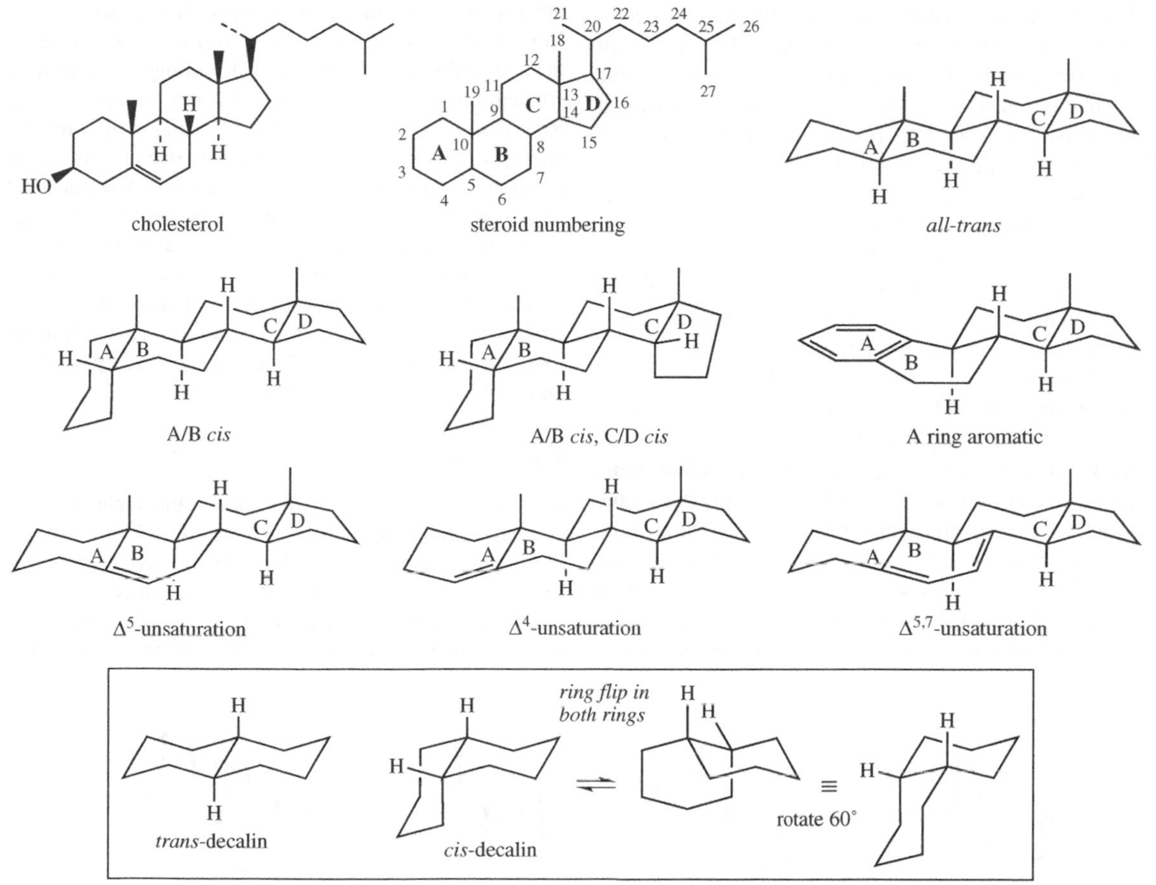

cholesterol

steroid numbering

all-trans

A/B *cis*

A/B *cis*, C/D *cis*

A ring aromatic

Δ^5-unsaturation

Δ^4-unsaturation

$\Delta^{5,7}$-unsaturation

trans-decalin

cis-decalin

ring flip in both rings

rotate 60°

Figure 5.73

STEROIDS

The steroids are modified triterpenoids containing the tetracyclic ring system of lanosterol (Figure 5.61), but lacking the three methyl groups at C-4 and C-14. **Cholesterol** (Figure 5.73) typifies the fundamental structure, but further modifications, especially to the side-chain, help to create a wide range of biologically important natural products, e.g. sterols, steroidal saponins, cardioactive glycosides, bile acids, corticosteroids, and mammalian sex hormones. Because of the profound biological activities encountered, many natural steroids and a considerable number of synthetic and semi-synthetic steroidal compounds are routinely employed in medicine. The markedly different biological activities observed emanating from compounds containing a common structural skeleton are in part ascribed to the functional groups attached to the steroid nucleus, and in part to the overall shape conferred on this nucleus by the stereochemistry of ring fusions.

Stereochemistry and Nomenclature

Ring systems containing six-membered or five-membered rings can be *trans*-fused as exemplified by *trans*-decalin or *cis*-fused as in *cis*-decalin (Figure 5.73). The *trans*-fusion produces a flattish molecule when two chair conformations are present. The only conformational mobility allowed is to less favourable boat forms. Bridgehead hydrogen atoms (or other substituents) are axial to both of the rings. In contrast, the *cis*-fused decalin is basically a bent molecule; it is found to be flexible, in that alternative conformers are possible, with both rings still being in chair form. Bridgehead substituents are axial to one ring, whilst being equatorial to the other, in each conformer. However, this conformational flexibility will be lost if either ring is then fused to a third ring, so is not encountered in steroid structures

In natural steroids, there are examples of the A/B ring fusion being *trans* or *cis*, or having unsaturation, either

Δ^4 or Δ^5. In some compounds, notably the oestrogens, ring A can even be aromatic; clearly, there can then be no bridgehead substituent at C-10 and, therefore, the normal C-10 methyl (C-19) must be lost. All natural steroids have a *trans* B/C fusion, though *cis* forms can be made synthetically. The C/D fusion is also usually *trans*, though there are notable exceptions, such as the cardioactive glycosides. Of course, such comments apply equally to some of the triterpenoid structures already considered. However, it is in the steroid field where the relationship between stereochemistry and biological activity is most marked. The overall shapes of some typical steroid skeletons are shown in Figure 5.73.

Systematic **steroid nomenclature** is based on a series of parent hydrocarbons, including **gonane**, **estrane**, **androstane**, **pregnane**, **cholane**, **cholestane**, **ergostane**, **campestane**, **stigmastane**, and **poriferastane** (Figure 5.74). The triterpenoid hydrocarbons **lanostane** and **cycloartane** are similarly used in systematic nomenclature and are also included in Figure 5.74. It is usual to add only unsaturation (ene/yne) and the highest priority functional group as suffixes to the root name;

other groups are added as prefixes. Stereochemistry of substituents is represented by α (on the lower face of the molecule when it is drawn according to customary conventions as in Figure 5.74), or β (on the upper face). Ring fusions may be designated by using α or β for the appropriate bridgehead hydrogen, particularly those at positions 5 and 14, which will define the A/B and C/D fusions respectively, e.g. 5β-cholestane has the A/B rings *cis*-fused. Since the parent hydrocarbon assumes that ring fusions are *trans*, the stereochemistry for ring fusions is usually only specified where it is *cis*. Cholesterol is thus cholest-5-en-3β-ol. The term *nor* is affixed to indicate loss of a carbon atom, e.g. 19-norsteroids (see page 288) lack C-19 (the methyl at C-10).

Cholesterol

In animals, the triterpenoid alcohol **lanosterol** (C_{30}) is converted into **cholesterol** (C_{27}) (Figure 5.75), a process that, as well as the loss of three methyl groups, requires reduction of the side-chain double bond, and generation of a $\Delta^{5,6}$ double bond in place of the $\Delta^{8,9}$ double bond. The sequence of these steps is, to some extent, variable and

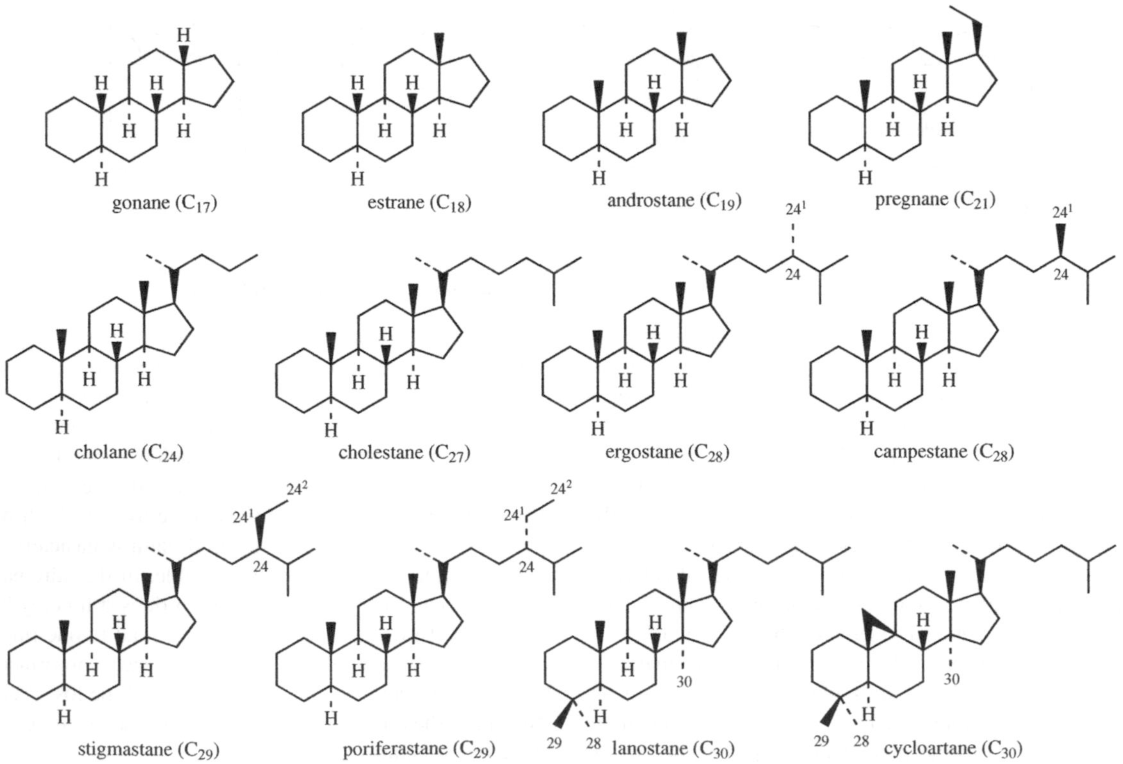

Figure 5.74

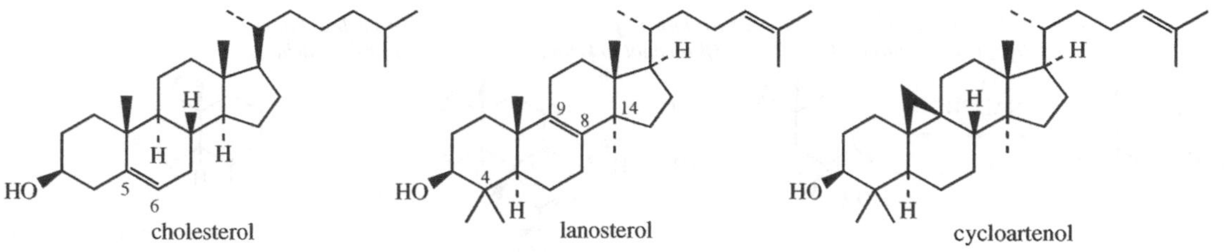

Figure 5.75

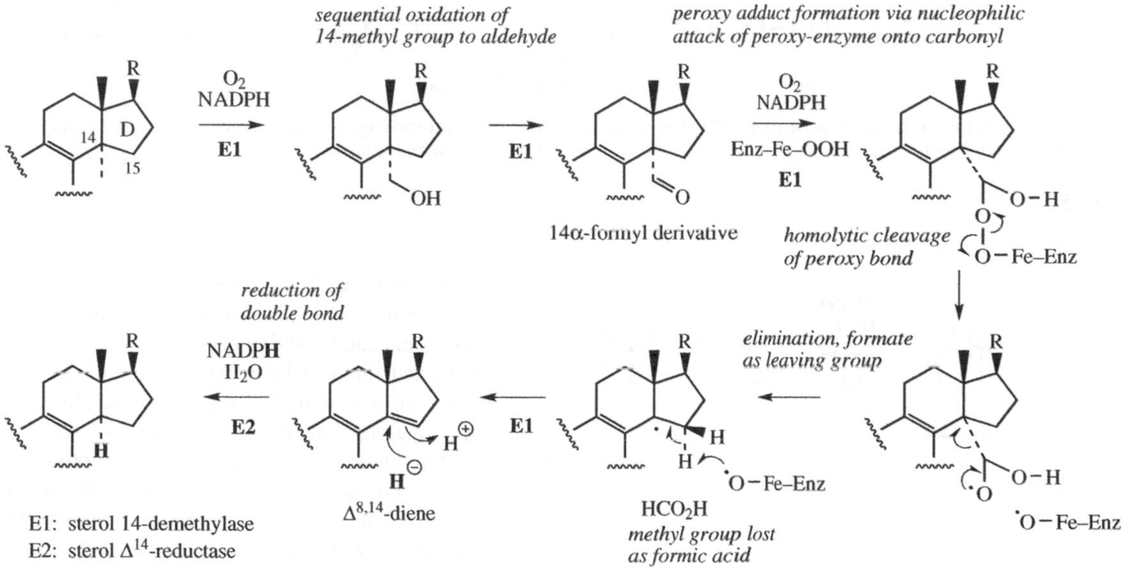

E1: sterol 14-demethylase
E2: sterol Δ^{14}-reductase

Figure 5.76

dependent on the organism involved. Accordingly, these individual transformations are considered rather than the overall pathway.

The methyl at C-14 is usually the one lost first, and this is removed as formic acid. The reaction is catalysed by a cytochrome P-450 monooxygenase which achieves two oxidation reactions to give the 14α-formyl derivative (Figure 5.76), and loss of this formyl group giving the $\Delta^{8,14}$ diene, most probably via homolytic cleavage of the peroxy adduct as indicated (compare similar peroxy adducts and mechanisms involved in side-chain cleavage from ring D, page 292, and in A ring aromatization, page 292). The 14-demethyl sterol is then obtained by an NADPH-dependent reduction step, the 15-proton being derived from water.

Loss of the C-4 methyl groups occurs sequentially, usually after removal of the 14α-methyl. Both carbon atoms are oxidized to carboxyl groups, then cleaved off

via a decarboxylation mechanism (Figure 5.77). This is facilitated by oxidizing the 3-hydroxyl to a ketone, thus generating intermediate β-keto acids. In this sequence, the enolate is restored to a ketone in which the remaining C-4 methyl takes up the more favourable equatorial (4α) orientation. In animals and yeasts, the two methyl groups are removed by the same enzyme systems; plants are found to employ a different set of enzymes for the removal of the α- and β-methyl groups.

The side-chain Δ^{24} double bond is reduced by an NADPH-dependent reductase, hydride from the coenzyme being added at C-25, with H-24 being derived from water (Figure 5.78). The Δ^8 double bond is effectively migrated to Δ^5 via Δ^7 and the $\Delta^{5,7}$ diene (Figure 5.79). This sequence involves an allylic isomerization, a dehydrogenation, and a reduction. Newly introduced protons at C-9 and C-8 originate from water, and that at C-7 from NADPH.

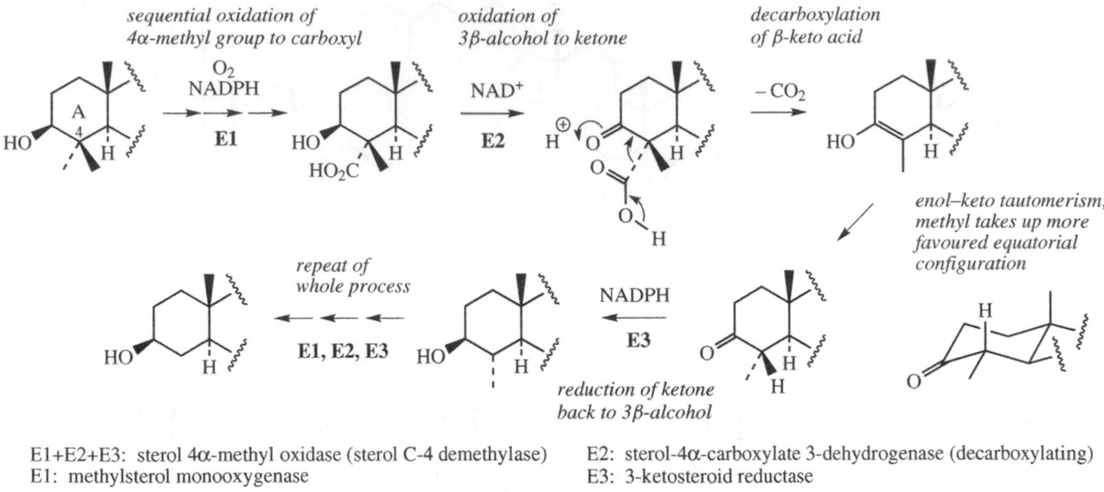

sequential oxidation of
4α-methyl group to carboxyl

oxidation of
3β-alcohol to ketone

decarboxylation
of β-keto acid

enol–keto tautomerism;
methyl takes up more
favoured equatorial
configuration

repeat of
whole process

reduction of ketone
back to 3β-alcohol

E1+E2+E3: sterol 4α-methyl oxidase (sterol C-4 demethylase)   E2: sterol-4α-carboxylate 3-dehydrogenase (decarboxylating)
E1: methylsterol monooxygenase   E3: 3-ketosteroid reductase

Figure 5.77

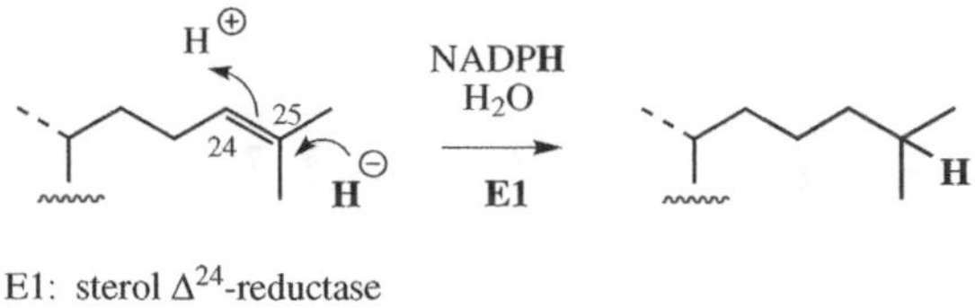

E1: sterol Δ^{24}-reductase

Figure 5.78

The role of lanosterol in non-photosynthetic organisms (animals, fungi) is taken in photosynthetic organisms (plants, algae) by the cyclopropane triterpenoid **cycloartenol** (Figure 5.75). This cyclopropane feature is found in a number of plant sterols, though the majority of plant steroids contain the typical C-10 methyl group. This means that, in addition to the lanosterol → cholesterol

modifications outlined above, a further mechanism to re-open the cyclopropane ring is necessary. This is shown in Figure 5.80. Acid-catalysed ring opening is the most likely mechanism, and this can be achieved non-enzymically, but under severe conditions. It is suggested, therefore, that a nucleophilic group from the enzyme attacks C-9, opening the cyclopropane ring and incorporating a proton from water. A *trans* elimination then generates the Δ^8 double bond. The stereochemistry at C-8 (Hβ) is unfavourable for a concerted mechanism involving loss of H-8 with cyclopropane ring opening. The cyclopropane ring-opening process seems to occur only with 4α-monomethyl sterols. In plants, removal of the first 4-methyl group (4α; note the remaining 4β-methyl group then takes up the α-orientation as in Figure 5.77) is also known to precede loss of the 14α-methyl. Accordingly, the general substrate shown in Figure 5.80 has both 4α- and 14α-methyl groups;

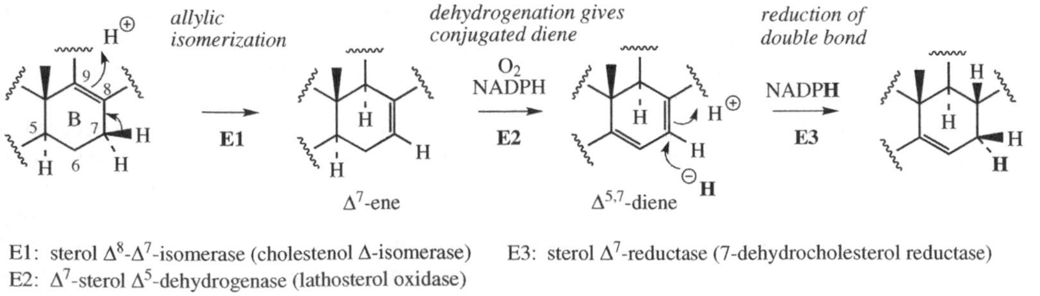

allylic
isomerization

dehydrogenation gives
conjugated diene

reduction of
double bond

Δ^7-ene $\Delta^{5,7}$-diene

E1: sterol Δ^8-Δ^7-isomerase (cholestenol Δ-isomerase)   E3: sterol Δ^7-reductase (7-dehydrocholesterol reductase)
E2: Δ^7-sterol Δ^5-dehydrogenase (lathosterol oxidase)

Figure 5.79

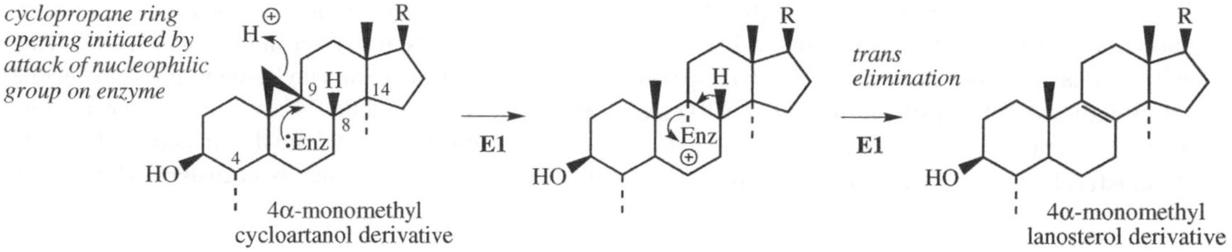

4α-monomethyl
cycloartanol derivative

E1

4α-monomethyl
lanosterol derivative

E1: cycloeucalenol cycloisomerase (cycloeucalenol-obtusifoliol isomerase)

Figure 5.80

side-chain alkylation (see page 252) also commonly pre-cedes these other modifications. The specificity of the cyclopropane ring-opening enzyme means cycloartenol is not converted into lanosterol, and lanosterol is thus absent from virtually all plant tissues. Cholesterol [Box 5.12] is almost always present in plants, though in only trace amounts, and is formed via cycloartenol.

Phytosterols

The major sterol found in mammals is the C_{27} compound cholesterol, which acts as a precursor for other steroid structures such as sex hormones and corticosteroids. The main sterols in plants, fungi, and algae are character-ized by extra one-carbon or two-carbon substituents on

---

Box 5.12

Cholesterol

Cholesterol (Figure 5.75) is the principal animal sterol, and since it is a constituent of cell membranes has been found in all animal tissues. It maintains membrane fluidity, microdomain structure, and permeability. It functions as a precursor for steroid hormones (pages 279, 291), bile acids (page 276), vitamin D (page 257), and lipoproteins, but is also correlated with cardiovascular disease, atherosclerosis, hypercholesterolaemia, and gallstone disease. Human gallstones are almost entirely composed of cholesterol precipitated from the bile.

Although the processes involved are quite complex, there appears to be a clear correlation between human blood cholesterol levels and heart disease. Atherosclerosis is a hardening of the arteries caused by deposition of cholesterol, cholesterol esters, and other lipids in the artery wall, causing a narrowing of the artery and, thus, an increased risk of forming blood clots (thrombosis). Normally, most of the cholesterol serves a structural element in cell walls, whilst the remainder is transported via the blood and is used for synthesis of steroid hormones, vitamin D, or bile acids. Transport of cholesterol is facilitated by formation of lipoprotein carriers, comprising protein and phospholipid shells surrounding a core of cholesterol, in both free and esterified forms. Low density lipoproteins (LDLs) carry larger amounts of cholesterol and deposit it around the body. High-density lipoproteins (HDLs) pick up free cholesterol and return it to the liver to be degraded. Risk of atherosclerosis increases with increasing levels of LDL cholesterol, and is reduced with increasing levels of HDL cholesterol. Blood LDL cholesterol levels are thus a good statistical indicator of the potential risk of a heart attack. Current recommendations are that total cholesterol levels in blood should be lower than 4 mmol/l, of which LDL cholesterol concentration should be less than 2 mmol/l. The risks can be lessened by avoiding foods rich in cholesterol, e.g. eggs, reducing the intake of foods containing high amounts of saturated fatty acids such as animal fats, and replacing these with vegetable oils and fish that are rich in polyunsaturated fatty acids (see page 49). Blood LDL cholesterol levels may also be reduced by incorporating into the diet plant sterol esters or plant stanol esters, which reduce the absorption of cholesterol (see page 255). In humans, dietary cholesterol is actually a smaller contributor to LDL cholesterol levels than is dietary saturated fat. Cholesterol biosynthesis may also be inhibited by drug therapy using specific inhibitors of HMG-CoA reductase in the mevalonate pathway, e.g. lovastatin and related compounds (see page 98).

Cholesterol is one of the primary sources for the semi-synthesis of medicinal steroids. It is currently available in quantity via the brains and spinal cords of cattle as a by-product of meat production. Large quantities are also extractable from lanolin, the fatty material coating sheep's wool. This is a complex mixture of esters of long-chain fatty acids (including straight-chain, branched-chain, and hydroxy acids) with long-chain aliphatic alcohols and sterols. Cholesterol is a major sterol component. Saponification of crude lanolin gives an alcohol fraction (lanolin alcohols or wool alcohols) containing about 34% cholesterol and 38% lanosterol/dihydrolanosterol. Wool alcohols are also used as an ointment base.

the side-chain, attached at C-24. These substituent carbon atoms are numbered 24^1 and 24^2 (Figure 5.74); some older publications may use 28 and 29. The widespread plant sterols **campesterol** and **sitosterol** (Figure 5.81) are respectively 24-methyl and 24-ethyl analogues of cholesterol. **Stigmasterol** contains additional unsaturation in the side-chain, a *trans-*Δ^{22} double bond, a feature seen in many plant sterols, but never in mammalian ones. The introduction of methyl and ethyl groups at C-24 generates a new chiral centre, and the 24-alkyl groups in campesterol, sitosterol, and stigmasterol are designated α. The predominant sterol found in fungi is **ergosterol** (Figure 5.81),

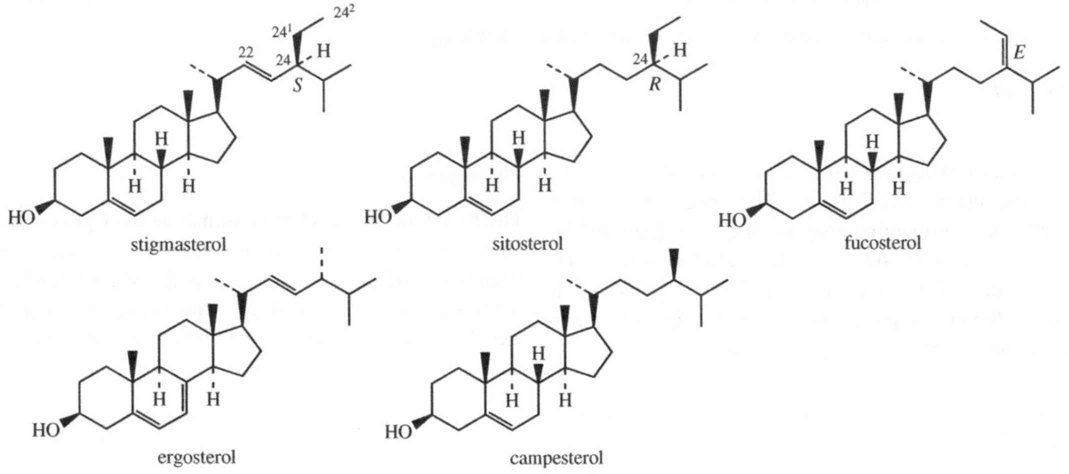

Figure 5.81

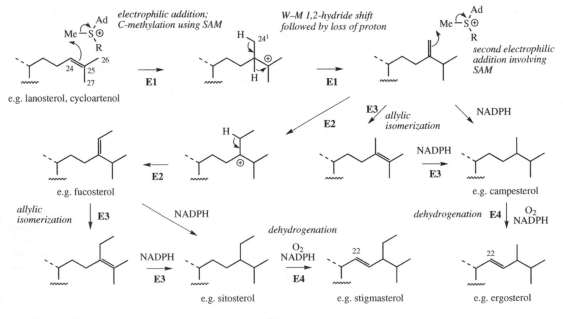

E1: sterol C-24 methyltransferase
E2: 24-methylenesterol *C*-methyltransferase
E3: Δ^{24} sterol reductase
E4: sterol C-22 desaturase

Figure 5.82

which has a β-oriented 24-methyl, as well as a *trans*-Δ^{22} double bond and additional Δ^7 unsaturation. The descriptors α and β unfortunately, and also confusingly, do not relate to similar terms used with the steroid ring system, but are derived from consideration of Fischer projections for the side-chain, substituents to the left being designated α and those to the right as β. Systematic *RS* nomenclature is thus preferred, but note that this defines sitosterol as 24*R*, whilst stigmasterol, because of its extra double bond, is 24*S*. The majority of plant sterols have a 24α-methyl or 24α-ethyl substituent, whilst algal sterols tend to have 24β-ethyl groups, and fungi 24β-methyl groups. The most abundant sterol in brown algae (*Fucus* spp.; Fucaceae) is **fucosterol** (Figure 5.81), which demonstrates a further variant, a 24-ethylidene substituent. Such groups can have *E*-configurations, as in fucosterol, or the alternative *Z*-configuration. Sterols are found predominantly in free alcohol form, but also as esters with long-chain fatty acids (e.g. palmitic, oleic, linoleic, and α-linolenic acids), as glycosides, and as glycosides acylated with fatty acids. These sterols, termed phytosterols, are structural components of membranes in plants, algae, and fungi and affect the permeability of these membranes. They also appear to play a role in cell proliferation.

The source of the extra methyl or ethyl side-chain carbons in both cases is SAM, and to achieve alkylation, the side-chain must have a Δ^{24} double bond, i.e. the side-chains as seen in lanosterol and cycloartenol. The precise mechanisms involved have been found to vary according to organism, but some of the demonstrated sequences are given in Figure 5.82. Methylation of the Δ^{24} double bond at C-24 via SAM yields a carbocation which undergoes a hydride shift and loss of a proton from C-24[1] to generate the 24-methylene side-chain. This can be

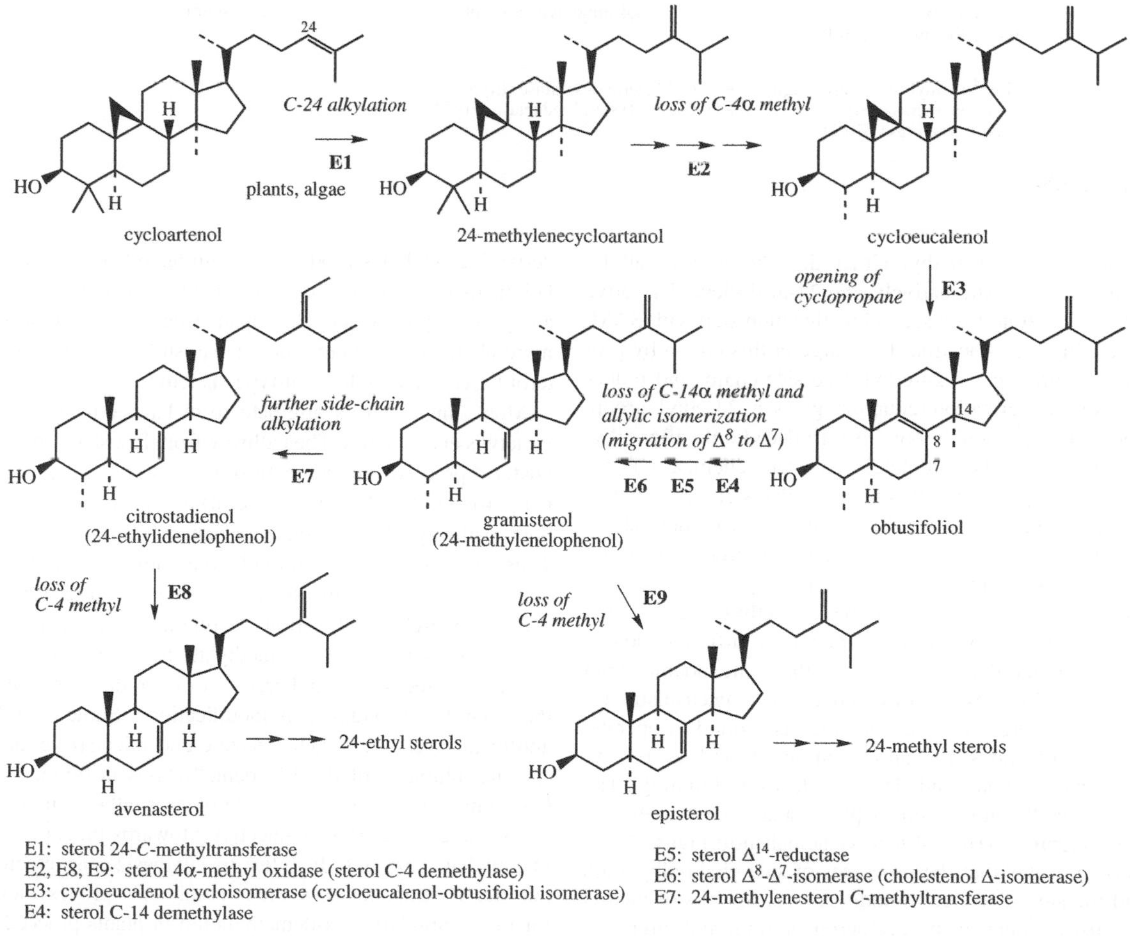

E1: sterol 24-*C*-methyltransferase
E2, E8, E9: sterol 4α-methyl oxidase (sterol C-4 demethylase)
E3: cycloeucalenol cycloisomerase (cycloeucalenol-obtusifoliol isomerase)
E4: sterol C-14 demethylase
E5: sterol Δ^{14}-reductase
E6: sterol Δ^8-Δ^7-isomerase (cholestenol Δ-isomerase)
E7: 24-methylenesterol *C*-methyltransferase

Figure 5.83

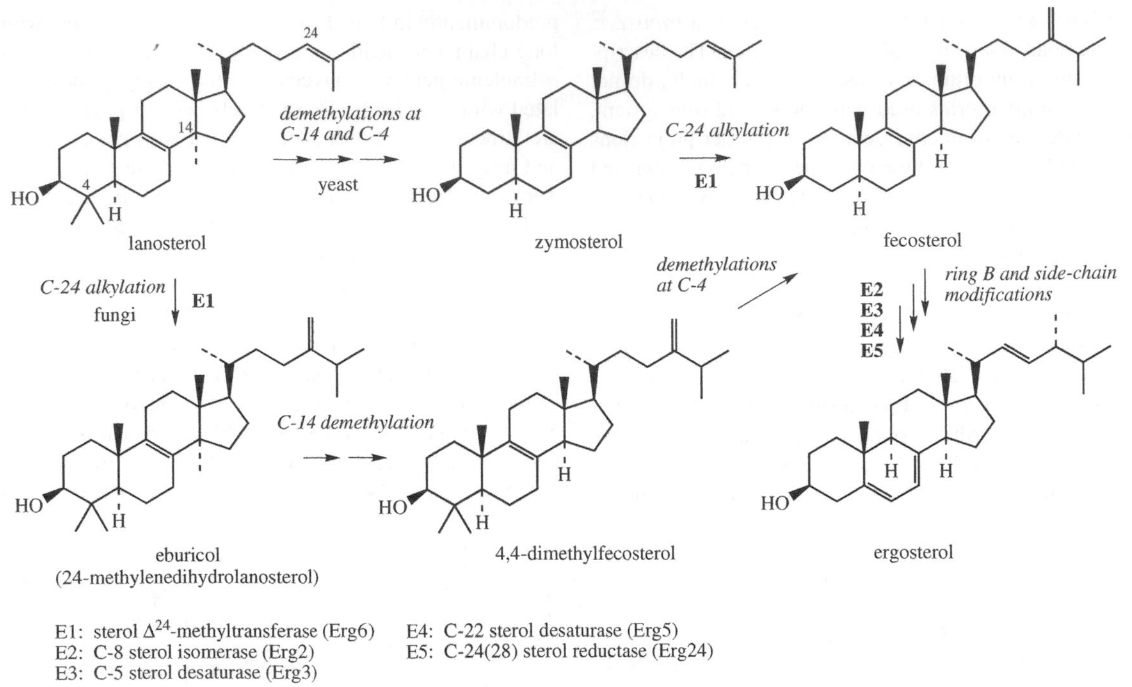

E1: sterol Δ^{24}-methyltransferase (Erg6)   E4: C-22 sterol desaturase (Erg5)
E2: C-8 sterol isomerase (Erg2)   E5: C-24(28) sterol reductase (Erg24)
E3: C-5 sterol desaturase (Erg3)

Figure 5.84

reduced to a 24-methyl either directly or after allylic isomerization. Alternatively, the 24-methylene derivative acts as substrate for a second methylation step with SAM, producing a carbocation. Discharge of this cation by proton loss produces a 24-ethylidene side-chain, and reduction or isomerization/reduction gives a 24-ethyl group. The *trans*-Δ^{22} double bond is introduced only after alkylation at C-24 is completed. No stereochemistry is intended in Figure 5.82. It is apparent that stereochemistries in the 24-methyl, 24-ethyl, and 24-ethylidene derivatives could be controlled by the reduction processes or by proton loss as appropriate. It is more plausible for different stereochemistries in the 24-methyl and 24-ethyl side-chains to arise from reduction of different double bonds, rather than reduction of the same double bond in two different ways. In practice, other mechanisms involving a 25(26)-double bond are also found to operate.

The substrates for alkylation are found to be cycloartenol in plants and algae, and lanosterol in fungi. The second methylation step in plants and algae usually involves **gramisterol** (24-methylenelophenol; Figure 5.83). This indicates that the processes of side-chain alkylation and the steroid skeleton modifications, i.e. loss of methyl groups, opening of the cyclopropane ring, and migration of double bond, tend to run concurrently rather than sequentially. Accordingly, the range of plant and algal sterol

derivatives includes products containing side-chain alkylation, that retain one or more skeletal methyl groups, and perhaps possess a cyclopropane ring, as well as those more abundant examples such as sitosterol and stigmasterol based on a cholesterol-type skeleton.

Most fungal sterols originate from lanosterol, so less variety is encountered. The pathway from lanosterol to **ergosterol** proceeds via **zymosterol** in yeast, but via **eburicol** (24-methylenedihydrolanosterol) in filamentous fungi; both pathways lead through **fecosterol** (Figure 5.84). Thus, in yeast, there is initially the same sequence of demethylations as in mammals, but then zymosterol undergoes side-chain alkylation. In fungi, side-chain alkylation occurs before the demethylations. The pathways converge at fecosterol, and ergosterol is produced by further ring B and side-chain modifications. Some useful antifungal agents, e.g. ketoconazole and miconazole, are specific inhibitors of the 14α-demethylases in fungi, but have minimal effect on cholesterol biosynthesis in humans; there is a degree of selectivity towards fungal over mammalian enzymes. Inability to synthesize the essential sterol components of their membranes proves fatal for fungi. Similarly, 14-demethylation in plants proceeds via **obtusifoliol** (Figure 5.83) and plants are unaffected by azole derivatives developed as agricultural fungicides.

The antifungal effect of polyene antibiotics, such as amphotericin and nystatin, depends on their ability to bind strongly to ergosterol in fungal membranes and not to cholesterol in mammalian cells (see page 81).

Sitosterol and **stigmasterol** (Figure 5.81) are produced commercially from soya beans (*Glycine max*; Leguminosae/Fabaceae) as raw materials for the semi-synthesis

of medicinal steroids (see pages 282, 294) [Box 5.13]. For many years, only stigmasterol was utilized, since the Δ^{22} double bond allowed chemical degradation of the side-chain to be effected with ease. The utilization of sitosterol was not realistic until microbiological processes for removal of the saturated side-chain became available.

Box 5.13

Soya Bean Sterols

Soya beans or soybeans (*Glycine max*; Leguminosae/Fabaceae) are grown extensively in the United States, China, Japan, and Malaysia as a food plant. They are used as a vegetable, and provide a high protein flour, an important edible oil (Table 3.1), and an acceptable non-dairy soybean milk. The flour is increasingly used as a meat substitute. Soy sauce is obtained from fermented soybeans and is an indispensible ingredient in Chinese cookery. The seeds also contain substantial amounts (about 0.2%) of sterols. These include stigmasterol (about 20%), sitosterol (about 50%), and campesterol (about 20%) (Figure 5.81), the first two of which are used for the semi-synthesis of medicinal steroids. In the seed, about 40% of the sterol content is in the free form, the remainder being combined in the form of glycosides or as esters with fatty acids. The oil is usually solvent extracted from the dried flaked seed using hexane. The sterols can be isolated from the oil after basic hydrolysis as a by-product of soap manufacture, and form part of the unsaponifiable matter.

The efficacy of dietary plant sterols in reducing cholesterol levels in laboratory animals has been known for many years. This has more recently led to the introduction of **plant sterol esters** as food additives, particularly in margarines and dairy products, as an aid to reducing blood levels of low density lipoprotein (LDL) cholesterol, known to be a contributory factor in atherosclerosis and the incidence of heart attacks (see page 251). Plant sterol esters are usually obtained by esterifying sitosterol from soya beans with fatty acids to produce a fat-soluble product. Regular consumption of this material (recommended 1.3 g per day) is shown to reduce blood LDL cholesterol levels by 10–15%. The plant sterols are more hydrophobic than cholesterol and have a higher affinity for micelles involved in fat digestion, effectively decreasing intestinal cholesterol absorption. The plant sterols themselves are poorly absorbed from the gastrointestinal tract. Of course, the average diet will naturally include small amounts of plant sterol esters.

Related materials used in a similar way are **plant stanol esters**. Stanols are obtained by hydrogenation of plant sterols, and will consist mainly of sitostanol (from sitosterol and stigmasterol) and campestanol (from campesterol) (Figure 5.85); these are then esterified with fatty acids. Regular consumption of plant stanol esters (recommended 3.4 g per day) is shown to reduce blood LDL cholesterol levels by an average of 14%. Much of the material used in preparation of plant stanol esters originates from tall oil, a by-product of the wood pulping industry. This contains campesterol, sitosterol, and also sitostanol. The stanols are usually transesterified with rapeseed oil, which is rich in unsaturated fatty acids (see page 47). Stanols tend to be more effective in reducing cholesterol levels than sterols. It has proved possible to engineer food plants such as rapeseed and soya to produce the reduced sterols, i.e. stanols, so that the chemical reduction steps may be rendered unnecessary. This has been achieved by incorporating a *Streptomyces*-derived 3-hydroxysteroid oxidase gene into the plant host. This enzyme is an FAD-dependent

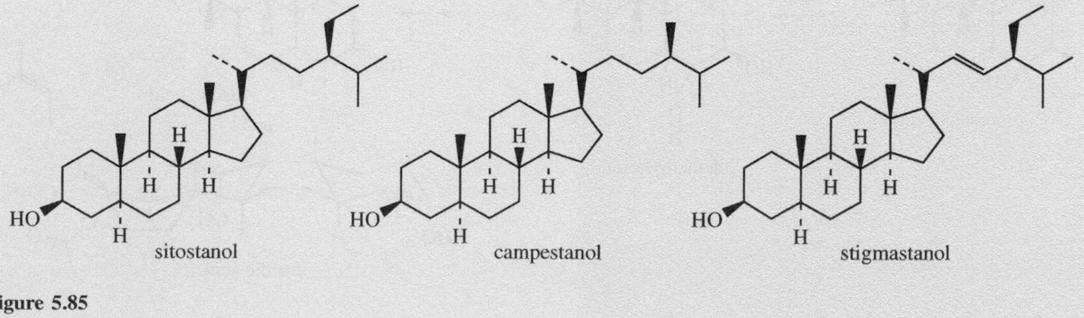

Figure 5.85

Box 5.13   (continued)

bifunctional enzyme that catalyses oxidation and isomerization in ring A of cholesterol and other steroids (compare page 264); additional reductive steps are presumably achieved by endogenous plant enzymes (Figure 5.86). Genetically modified soya seeds contained sitostanol, stigmastanol, and campestanol (Figure 5.85), with up to 80% of each normal sterol content being replaced by the reduced compound. Side-chain double bonds were not reduced.

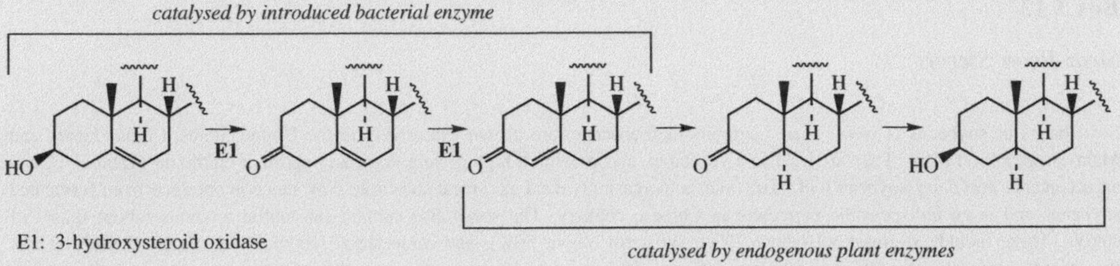

E1: 3-hydroxysteroid oxidase

Figure 5.86

Fusidic acid (Figure 5.87), an antibacterial agent from *Fusidium coccineum* [Box 5.14], has no additional side-chain alkylation, but has lost one C-4 methyl and undergone hydroxylation and oxidation of a side-chain methyl. The stereochemistry in fusidic acid is not typical of most steroids, and ring B adopts a boat conformation; the *chair–boat–chair–chair* molecular shape is comparable to the protosteryl cation (Figure 5.63, page 238). Its relationship to the protosteryl cation is shown in Figure 5.87, and its formation probably involves initial proton loss, which thus prevents the normal carbocation-mediated migrations.

Vitamin D

Vitamin D$_3$ (colecalciferol, cholecalciferol) [Box 5.15] is a sterol metabolite formed photochemically in animals from **7-dehydrocholesterol** by the sun's irradiation of the skin (Figure 5.88). 7-Dehydrocholesterol is the immediate $\Delta^{5,7}$ diene precursor of cholesterol (see

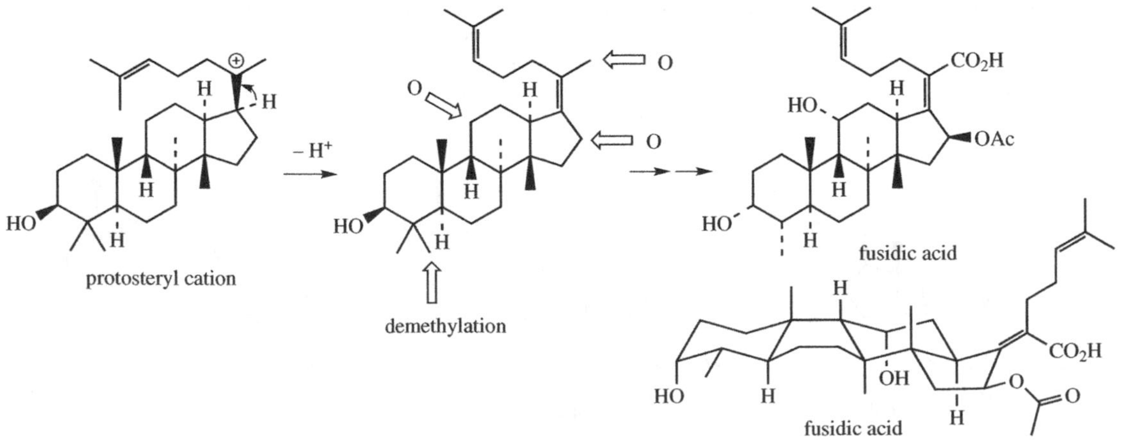

protosteryl cation

demethylation

fusidic acid

fusidic acid

Figure 5.87

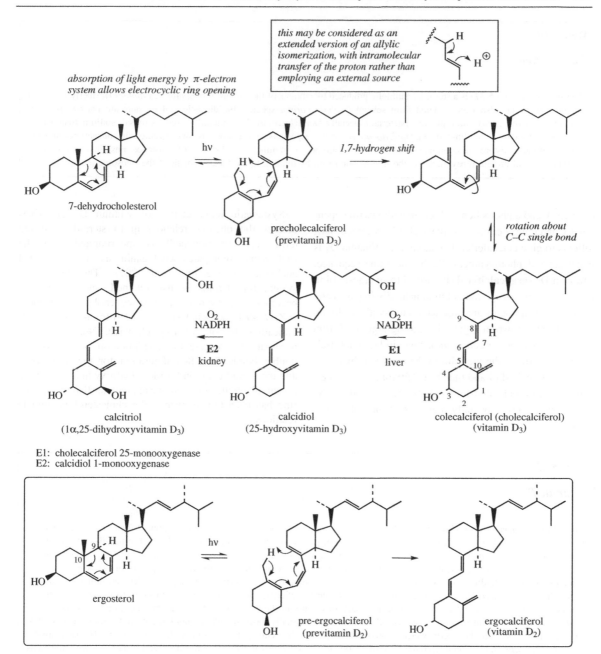

Figure 5.88

Box 5.14

Fusidic Acid

Fusidic acid (Figure 5.87) is a steroidal antibiotic produced by cultures of the fungus *Acremonium fusidioides* (formerly *Fusidium coccineum*). It has also been isolated from several *Cephalosporium* species. Fusidic acid and its salts are narrow-spectrum antibiotics active against Gram-positive bacteria, especially *Staphylococcus*. It is primarily used, as its salt **sodium fusidate**, in infections caused by penicillin-resistant *Staphylococcus* species, especially in osteomyelitis, since fusidic acid concentrates in bone. It is usually administered in combination with another antibiotic to minimize development of resistance. Fusidic acid reversibly inhibits bacterial protein biosynthesis at the translocation step by binding to the larger subunit of the ribosome (see page 422).

---

page 250), and a photochemical reaction allows ring opening to precholecalciferol. A thermal 1,7-hydrogen shift follows to give colecalciferol (vitamin D_3). Vitamin D_3 is manufactured photosynthetically by the same sequence. **Vitamin D_2 (ergocalciferol)** is formed from ergosterol in exactly the same way, and is found naturally in plants and yeasts. Large amounts are produced semi-synthetically by the sequence shown in Figure 5.88, using ergosterol from yeast (*Saccharomyces cerevisiae*). Vitamin D_3 is not itself the active form of the vitamin; in the body it is hydroxylated first to **calcidiol** and then to **calcitriol** (Figure 5.88). Colecalciferol and calcitriol have also been found in several plant species, especially members of the Solanaceae family.

Systematic nomenclature of vitamin D derivatives utilizes the obvious relationship to steroids, and the term *seco* (ring-opened) is incorporated into the root name (compare secologanin as a ring-opened analogue of loganin, page 207). The numbering system for steroids is also retained, and vitamin D_3 becomes a derivative of 9,10-secocholestane, namely (5Z,7E)-9,10-secocholesta-5,7,10(19)-trien-3β-ol, '9,10' indicating the site of ring cleavage. Note that it is necessary to indicate the configuration of two of the double bonds. The β-configuration for the 3-hydroxyl and the α-configuration for the 1-hydroxy in calcitriol are potentially confusing until one appreciates that the bottom ring has been turned over in the cholesterol–vitamin D relationship.

Box 5.15

Vitamin D

Vitamin D_3 (colecalciferol, cholecalciferol; Figure 5.88) is the main form of the fat-soluble vitamin D found in animals, though **vitamin D_2 (ergocalciferol**; Figure 5.88) is a constituent of plants and yeasts. Vitamin D_3 is obtained in the diet from liver and dairy products such as butter, cream, and milk, whilst large amounts can be found in oily fish and fish liver oils, e.g. cod liver oil and halibut liver oil (Table 3.1, page 46). A major source for normal requirements is the casual exposure of the skin to sunlight, when the sterol 7-dehydrocholesterol is converted into colecalciferol by UV irradiation. With a proper diet, and sufficient exposure to sunshine, vitamin D deficiency should not occur. Vitamin D deficiency leads to rickets, an inability to calcify the collagen matrix of growing bone, and is characterized by a lack of rigidity in the bones, particularly in children. In adults, osteoporosis may occur. In most countries, foods such as milk and cereals are usually fortified with vitamin D_3, obtained commercially by UV irradiation of 7-dehydrocholesterol which is produced in quantity by semi-synthesis from cholesterol. Vitamin D_2 has a similar activity in humans and is manufactured by UV irradiation of yeast, thereby transforming the ergosterol content. Other compounds with vitamin D activity have also been produced: vitamin D_4 from 22,23-dihydroergosterol, vitamin D_5 from 7-dehydrositosterol, vitamin D_6 from 7-dehydrostigmasterol, and vitamin D_7 from 7-dehydrocampesterol. Vitamin D_1 was an early preparation later shown to be a mixture of vitamin D_2 and a photochemical by-product lumisterol. Lumisterol (9β,10α-ergosterol) is formed by recyclization of pre-ergocalciferol (Figure 5.88), giving the ergosterol stereoisomer. Vitamin D is unstable to heat, light, and air.

Vitamin D_3 is not itself the active form of the vitamin, and in the body it is hydroxylated firstly to 25-hydroxyvitamin D_3 (calcidiol; Figure 5.88) by an enzyme in the liver, and then to 1α,25-dihydroxyvitamin D_3 (calcitriol) by a kidney enzyme; both enzymes are cytochrome P-450 systems. Calcitriol is then transported to the bones, intestine, and other organs. It stimulates calcium absorption from the intestine, reabsorption from the kidney, and mobilization from bone. **Calcitriol** and other analogues, e.g. **alfacalcidol, dihydrotachysterol**, and **paricalcitol** (Figure 5.89), are available for use where chronic vitamin D deficiency is due to liver or kidney malfunction. The long-term use of calcitriol and alfacalcidol (1α-hydroxyvitamin D_3) in the treatment of osteoporosis may lead to toxic effects arising from elevated serum calcium levels.

Box 5.15 (continued)

Vitamin D is also known to have other physiological functions, including a role in immune suppression, hormone secretion, and the differentiation of both normal and malignant cells. Vitamin D derivatives, including **calcipotriol, tacalcitol,** and **maxacalcitol** (Figure 5.89), are widely used in the topical treatment of psoriasis, to inhibit the cell proliferation characteristic of this condition. In maxacalcitol, the side-chain modification (22-methylene replaced by an ether link) appears to retain the non-calcaemic action whilst reducing calcaemic activity, the latter probably because of more rapid oxidative metabolism in the liver.

Vitamin D_2 is also employed as a rodenticide. High doses are toxic to rats and mice, since the vitamin causes fatal hypercalcaemia.

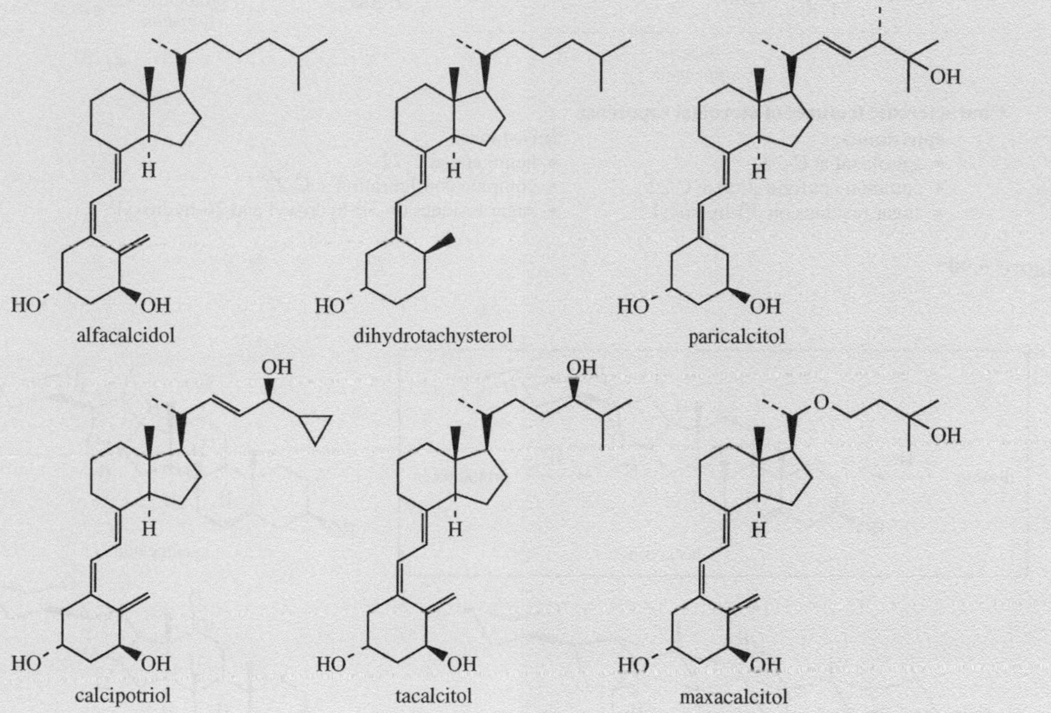

alfacalcidol   dihydrotachysterol   paricalcitol

calcipotriol   tacalcitol   maxacalcitol

Figure 5.89

Steroidal Saponins

Steroidal saponins have similar biological properties to the triterpenoid saponins, e.g. surfactant and haemolytic activities (see page 242), but are less widely distributed in nature. They are found in many monocot families, especially the Dioscoreaceae (e.g. *Dioscorea*), the Agavaceae (e.g. *Agave, Yucca*) and the Liliaceae (e.g. *Smilax, Trillium*). Their sapogenins are C_{27} sterols in which the side-chain of cholesterol has undergone modification to produce either a spiroketal (spirostane saponins), e.g. **dioscin**, or a hemiketal (furostane saponins), e.g. **protodioscin** (Figure 5.90). As described below, the furostanes feature

in the biosynthetic sequence to the spirostanes. All the steroidal saponins have the same configuration at the centre C-22, but stereoisomers at C-25 exist, e.g. **yamogenin** (Figure 5.91), and often mixtures of the C-25 stereoisomers co-occur in the same plant. Sugars are found at position 3, with a second glycoside function at C-26 in the furostanes. The sugar moiety at position 3 typically contains fewer monosaccharide units than are found with triterpenoid saponins; one to three monosaccharide units are most common. In general, the more sugar residues there are attached, the greater is the haemolytic activity. The sugar at C-26 in furostanes is usually glucose.

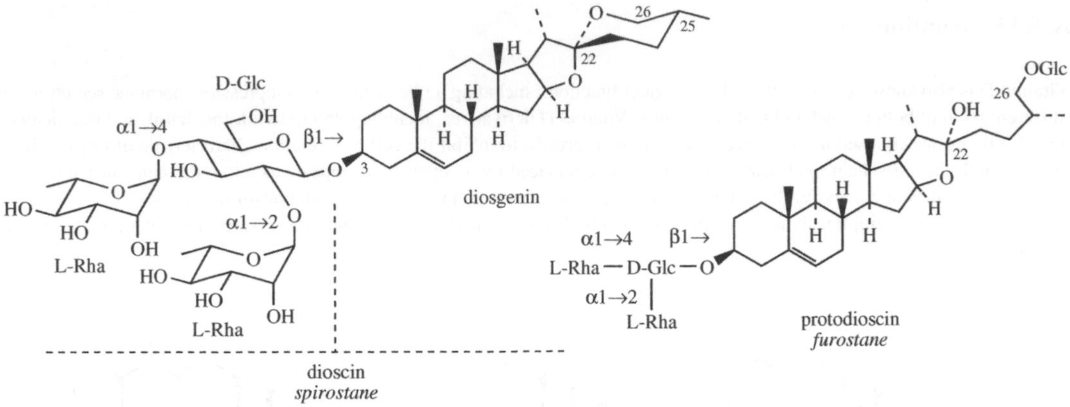

Characteristic features of steroidal saponins:

spirostanes:
- spiroketal at C-22
- common configuration at C-22
- sugar residues on 3β-hydroxyl

furostanes:
- hemiketal at C-22
- common configuration at C-22
- sugar residues on 3β-hydroxyl and 26-hydroxyl

Figure 5.90

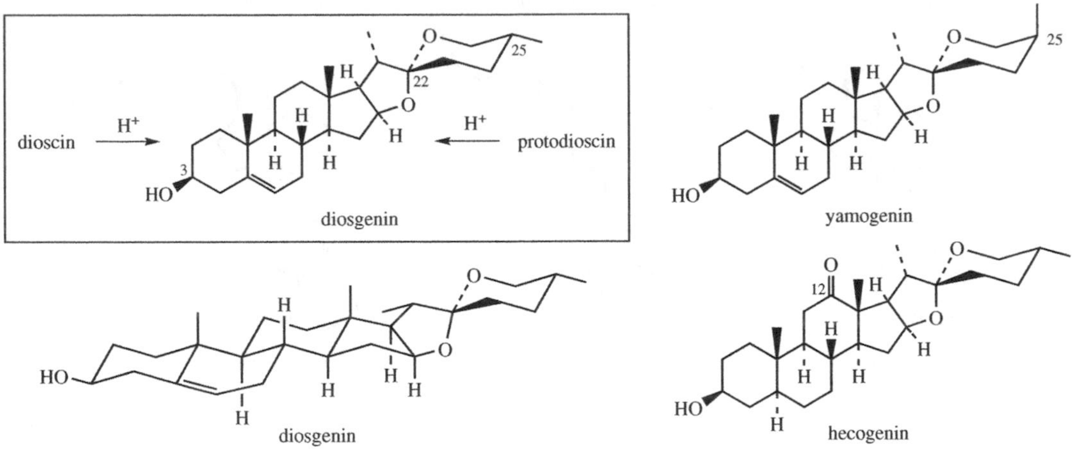

Figure 5.91

Acid hydrolysis of either dioscin or protodioscin liberates the aglycone **diosgenin** (Figure 5.91); the hydrolytic conversion of protodioscin into diosgenin is analogous to the biosynthetic sequence. The three-dimensional shape of diosgenin is indicated in Figure 5.91.

The spiroketal function is derived from the cholesterol side-chain by a series of oxygenation reactions, hydroxylating one of the terminal methyl groups and at C-16, and then producing a ketone function at C-22 (Figure 5.92). This proposed intermediate is transformed into the hemiketal and then the spiroketal. The

chirality at C-22 is fixed by the stereospecificity in the formation of the ketal, whilst the different possible stereochemistries at C-25 are dictated by whether C-26 or C-27 is hydroxylated in the earlier step. Enzymic glycosylation at the 3-hydroxyl of spirostane sapogenins has been reported, but knowledge of other steps at the enzymic level is lacking. Furostane derivatives, e.g. **protodioscin** (Figure 5.90), can co-occur with spirostanes, and undoubtedly represent glycosylation of the intermediate hemiketal at the 26-hydroxyl. These compounds are readily hydrolysed and then

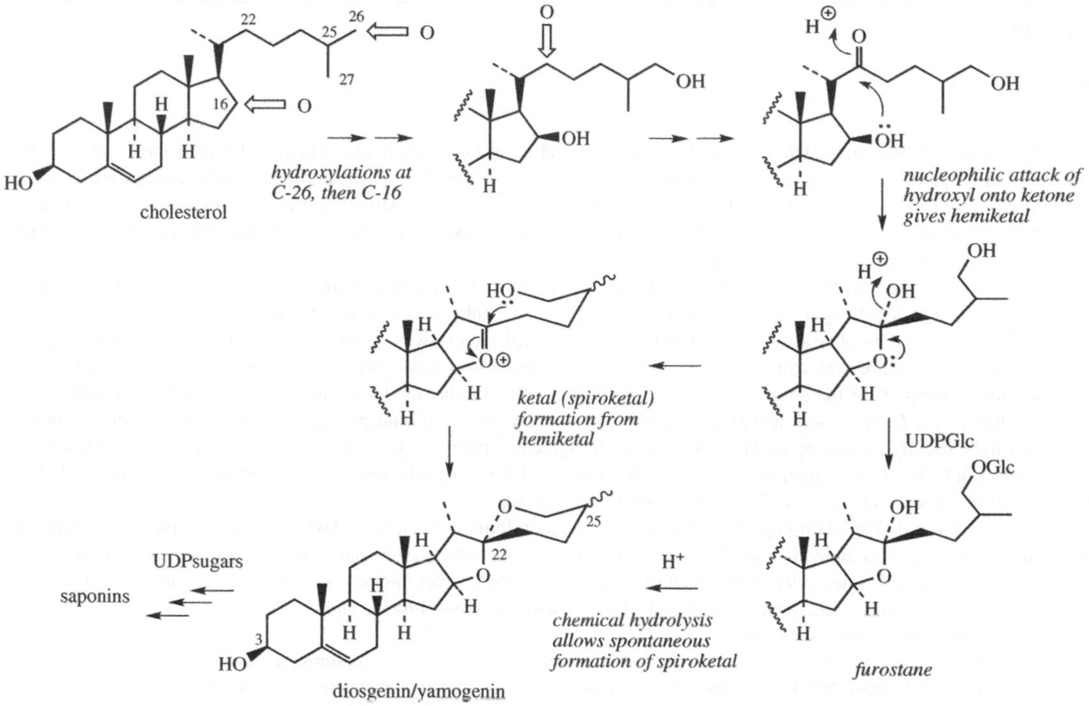

Figure 5.92

spontaneously cyclize to the spiroketal. Allowing homogenized fresh plant tissues to stand and autolyse through the action of endogenous glycosidase enzymes not only achieves cyclization of such open-chain saponins, but also can hydrolyse off the sugar units at C-3, thus yielding the aglycone or sapogenin. This is a standard approach employed in commercial production of steroidal sapogenins, which are important starting materials for the semi-synthesis of steroidal drugs [Box 5.16]. **Diosgenin** is the principal example and is obtained from Mexican yams (*Dioscorea* spp.; Dioscoreaceae). Fenugreek (*Trigonella foenum-graecum*; Leguminosae/Fabaceae) is another potentially useful commercial source of diosgenin. Sisal (*Agave sisalana*; Agavaceae) is also used commercially, yielding **hecogenin** (Figure 5.91), a 12-keto derivative with *trans*-fused A/B rings, the result of reduction of the Δ^5 double bond.

Undoubtedly related to the furostans and spirostans is the sterol glycoside **OSW-1** (Figure 5.93) isolated from the bulbs of *Ornithogalum saundersiae* (Liliaceae). This material has shown quite remarkable cytotoxicity, inhibiting the growth of various tumour cells, and being 10–100 times more potent than some clinical anticancer agents. It appears to damage the mitichondrial membrane in

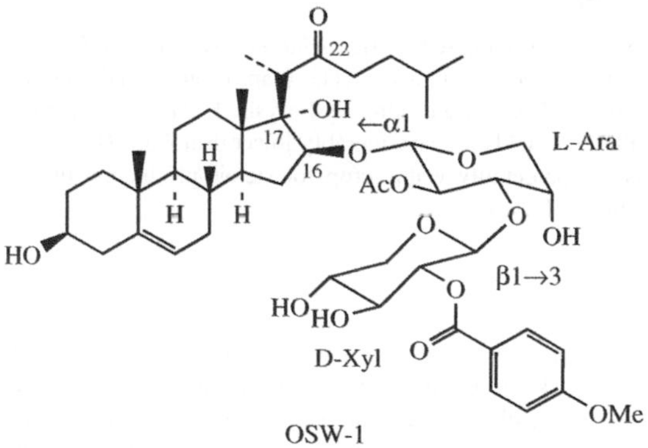

Figure 5.93

cancer cells, so its mechanism of action is also different to current anticancer agents. In addition, it shows little toxicity towards normal cells. The aglycone part of OSW-1 is structurally similar to early intermediates in the spirostanol pathway, in that it possesses the 22-keto and 16-hydroxyl functions, with an additional hydroxyl at C-17. 17α-Hydroxylation is characteristic of corticosteroids (see page 278) and the oestrogen/androgen biosynthetic pathway (see page 291).

Box 5.16

Dioscorea

About 600 species of *Dioscorea* (Dioscoreaceae) are known, and a number of these are cultivated for their large starchy tubers, commonly called yams, which are an important food crop in many parts of the world. Important edible species are *Dioscorea alata* and *D. esculenta* (Southeast Asia), *D. rotundata* and *D. cayenensis* (West Africa) and *D. trifida* (America). A number of species accumulate quite high levels of saponins in their tubers, which make them bitter and inedible, but these provide suitable sources of steroidal material for drug manufacture.

Dioscorea spp. are herbaceous, climbing, vine-like plants, the tuber being totally buried, or sometimes protruding from the ground. Tubers weigh anything up to 5 kg, but in some species the tubers have been recorded to reach weights as high as 40–50 kg. Drug material is obtained from both wild and cultivated plants, with plants collected from the wild having been exploited considerably more than cultivated ones. Commercial cultivation is less economic, requiring a 4–5-year growing period and some form of support for the climbing stems. Much of the world's production has come from Mexico, where tubers from *D. composita* (barbasco), *D. mexicana*, and *D. floribunda*, mainly harvested from wild plants, are utilized. The saponin content of the tubers varies, usually increasing as tubers become older. Typically, tubers of *D. composita* may contain 4–6% total saponins, and *D. floribunda* 6–8%. Other important sources of *Dioscorea* used commercially now include India (*D. deltoidea*), South Africa (*D. sylvatica*) and China (*D. collettii*, *D. pathaica*, and *D. nipponica*).

Sapogenins are isolated by chopping the tubers, allowing them to ferment for several days, and then completing the hydrolysis of saponins by heating with aqueous acid. The sapogenins can then be solvent extracted. The principal sapogenin in the species given above is diosgenin (Figure 5.91), with small quantities of the 25β-epimer yamogenin (Figure 5.91). Demand for diosgenin for pharmaceuticals is huge, equivalent to 10 000 t of *Dioscorea* tuber per annum, and it is estimated that about 60% of all steroidal drugs are derived from diosgenin.

Powdered *Dioscorea* (wild yam) root or extract is also marketed to treat the symptoms of menopause as an alternative to hormone replacement therapy (HRT; see page 294). Although there is a belief that this increases levels of progesterone, which is then used as a biosynthetic precursor of other hormones, there is little definitive evidence that diosgenin is metabolized in the human body to progesterone, and any beneficial effects may arise from diosgenin itself.

Fenugreek

The seeds of fenugreek (*Trigonella foenum-graecum*; Leguminosae/Fabaceae) are an important spice material, and are ingredients in curries and other dishes. The plant is an annual and is grown widely, especially in India, both as a spice and as a forage crop. Seeds can yield, after hydrolysis, 1–2% of sapogenins, principally diosgenin (Figure 5.91) and yamogenin (Figure 5.91). Although yields are considerably lower than from *Dioscorea*, the ease of cultivation of fenugreek and its rapid growth make the plant a potentially viable crop for steroid production in temperate countries. Field trials of selected high-yielding strains have been conducted.

Sisal

Sisal (*Agave sisalana*; Agavaceae) has long been cultivated for fibre production, being the source of sisal hemp, used for making ropes, sacking, and matting. The plant is a large, rosette-forming succulent with long, tough, spine-tipped leaves containing the very strong fibres. The main area of sisal cultivation is East Africa (Tanzania, Kenya), with smaller plantations in other parts of the world. The sapogenin hecogenin (Figure 5.91) was initially produced from the leaf waste (0.6–1.3% hecogenin) after the fibres had been stripped out. The leaf waste was concentrated, allowed to ferment for several days, and then treated with steam under pressure to complete hydrolysis of the saponins. Filtration then produced a material containing about 12% hecogenin, plus other sapogenins. This was refined further in the pharmaceutical industry. Other sapogenins present include tigogenin and neotigogenin (Figure 5.94).

As the demand for natural fibres declined due to the availability of synthetics, so did the supply of sisal waste and, thus, hecogenin. In due course, hecogenin became a more valuable commodity than sisal, and efforts were directed specifically towards hecogenin production. This has resulted in the cultivation of *Agave* hybrids with much improved hecogenin content. The highest levels (2.5%) of hecogenin recorded have been found in a Mexican species *Agave sobria* var *roseana*.

The fermented sap of several species of Mexican *Agave*, especially *Agave tequilana*, provides the alcoholic beverage pulque. Distillation of the fermented sap produces tequila.

Box 5.16   (continued)

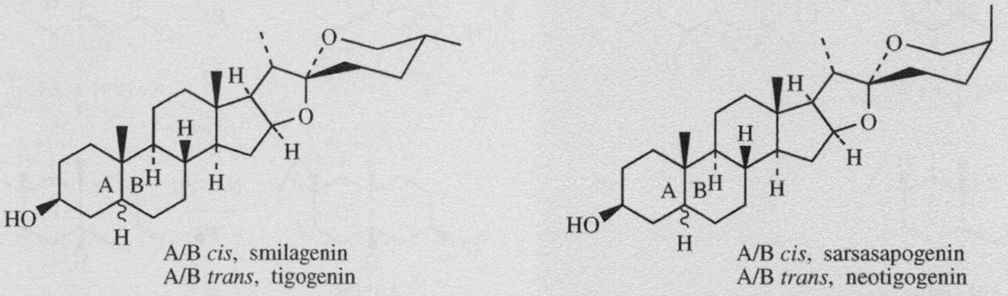

A/B *cis*, smilagenin
A/B *trans*, tigogenin

A/B *cis*, sarsasapogenin
A/B *trans*, neotigogenin

Figure 5.94

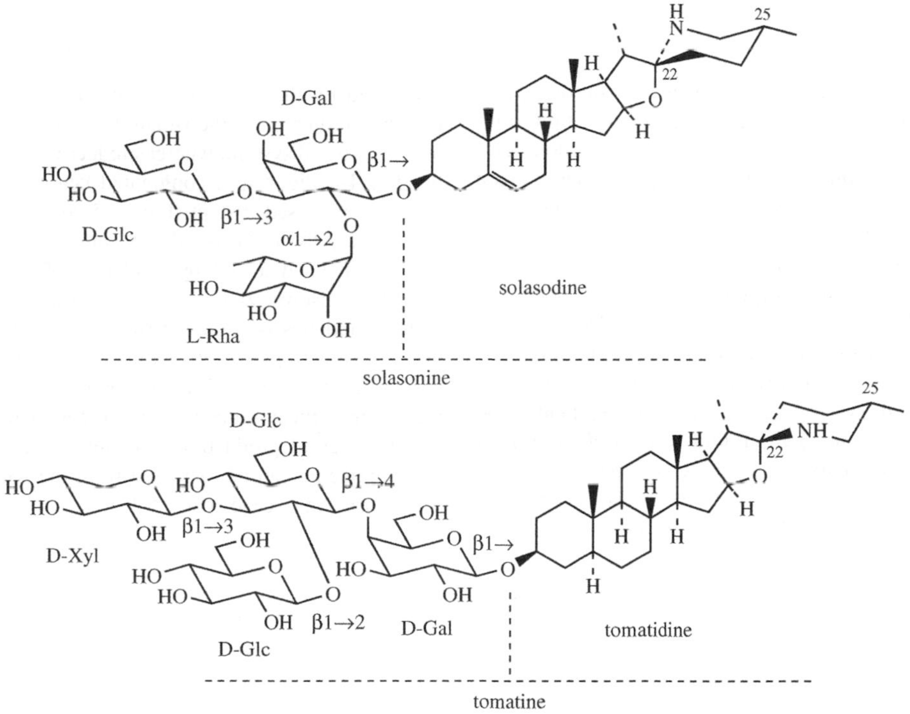

Figure 5.95

Some steroidal alkaloids are nitrogen analogues of steroidal saponins and display similar properties, such as surface activity and haemolytic activity, but these compounds *are* toxic when ingested. These types of compound, e.g. **solasonine** (Figure 5.95; aglycone **solasodine**), are found in many plants of the genus *Solanum* (Solanaceae), and such plants must thus be regarded as potentially toxic. In contrast to the oxygen analogues, all compounds have the same stereochemistry at C-25 (methyl always equatorial), whilst isomers at C-22 do exist, e.g. **tomatine** (Figure 5.95; aglycone **tomatidine**) from tomato (*Lycopersicon esculente*; Solanaceae). The nitrogen atom is introduced by a transamination reaction, typically employing an amino acid as donor (see page

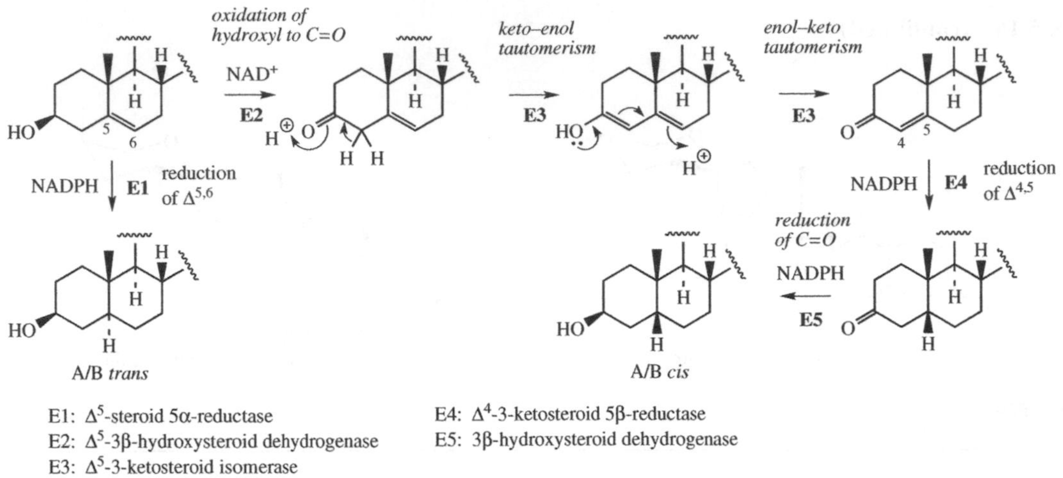

E1: Δ^5-steroid 5α-reductase
E2: Δ^5-3β-hydroxysteroid dehydrogenase
E3: Δ^5-3-ketosteroid isomerase

E4: Δ^4-3-ketosteroid 5β-reductase
E5: 3β-hydroxysteroid dehydrogenase

Figure 5.96

409). Since the production of medicinal steroids from steroidal saponins requires preliminary degradation to remove the ring systems containing the original cholesterol side-chain, it is immaterial whether these rings contain oxygen or nitrogen. Thus, plants rich in solasodine or tomatidine can also be employed for commercial steroid production (see page 410).

Smilagenin and **sarsasapogenin** (Figure 5.94) found in sarsaparilla (*Smilax* spp.; Liliaceae/Smilacaceae) are reduced forms of diosgenin and yamogenin respectively. These contain *cis*-fused A/B rings, whilst the corresponding *trans*-fused systems are present in **tigogenin** and **neotigogenin** (Figure 5.94) found in *Digitalis purpurea* along with cardioactive glycosides (see page 269). All four stereoisomers are derived from cholesterol, and the stereochemistry of the A/B ring fusion appears to be controlled by the nature of the substrate being reduced. Enzymic reduction of the isolated Δ^5 double bond yields the *trans*-fused system, whereas reduction of a Δ^4 double bond (1,4-addition to a conjugated ketone) gives the alternative *cis*-fused system (Figure 5.96). Accordingly, to obtain the A/B *cis* fusion, the Δ^5 unsaturation of cholesterol is changed to Δ^4 by oxidation of the 3-hydroxyl and allylic isomerization to the conjugated 4-ene-3-one system, and this is followed by reduction of both functional groups (Figure 5.96) (compare biosynthesis of progesterone, page 267). The sarsaparilla saponins are not present in sufficient quantities to be commercially important for steroid production, but quite large amounts of sarsasapogenin can be extracted from the seeds of *Yucca brevifolia* (Agavaceae) [Box 5.17].

Box 5.17

Sarsaparilla

Sarsaparilla consists of the dried roots of various *Smilax* species (Liliaceae/Smilacaceae), including *S. aristolochiaefolia*, *S. regelii*, and *S. febrifuga*, known respectively as Mexican, Honduran, and Ecuadorian sarsaparilla. The plants are woody climbers indigenous to Central America. Sarsaparilla has a history of use in the treatment of syphilis, rheumatism, and skin diseases, but is now mainly employed as a flavouring in the manufacture of non-alcoholic drinks. It has some potential as a raw material for the semi-synthesis of medicinal steroids, being a source of sarsasapogenin and smilagenin (Figure 5.94). The roots contain 1.8–2.4% steroidal saponins, including parillin (Figure 5.97).

Yucca

Yucca brevifolia (Agavaceae) has been explored as a potential source of sarsasapogenin for steroid production, especially at times when market prices of diosgenin from *Dioscorea* became prohibitively expensive. The plant grows extensively in the

Box 5.17   (continued)

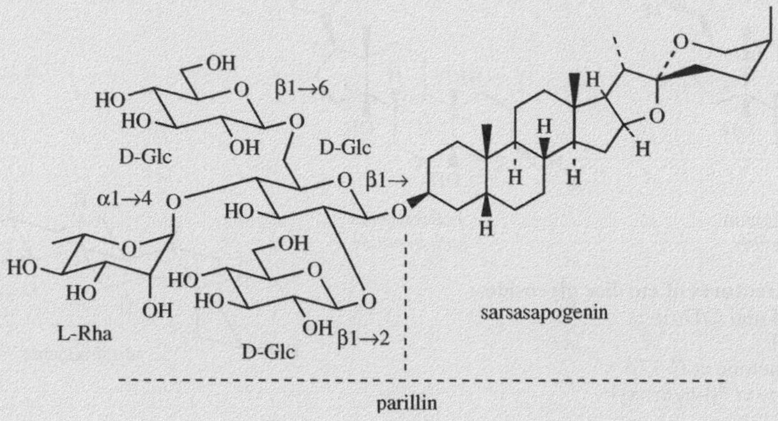

Figure 5.97

Mojave Desert in California, and high levels of sarsasapogenin (8–13%) are present in the seeds. This means the plants can be harvested regularly without damage. The subsequent stabilization of *Dioscorea* prices in the 1970s stopped any further commercial exploitation.

Cardioactive Glycosides

Many of the plants known to contain cardiac or cardiotonic glycosides have long been used as arrow poisons (e.g. *Strophanthus*) or as heart drugs (e.g. *Digitalis*). They are used medicinally to strengthen a weakened heart and allow it to function more efficiently, though the dosage must be controlled very carefully, since the therapeutic dose is so close to the toxic dose. The cardioactive effects of *Digitalis* were discovered as a result of its application in the treatment of dropsy, an accumulation of water in the body tissues. *Digitalis* alleviated dropsy indirectly by its effect on the heart, improving the blood supply to the kidneys and so removing excess fluid.

The therapeutic action of cardioactive glycosides depends on the structure of the aglycone, and on the type and number of sugar units attached. Two types of aglycone are recognized, **cardenolides** (e.g. **digitoxigenin** from *D. purpurea*), which are C_{23} compounds, and **bufadienolides** (e.g. **hellebrigenin** from *Helleborus niger*), which are C_{24} structures (Figure 5.98). Stereochemistry is very important for activity, and these compounds have *cis* fusions for both the A/B and C/D ring junctions, 3β- and 14β-hydroxyl groups with the glycoside function at C-3, and an α,β-unsaturated lactone grouping at C-17β. This lactone ring is five-membered

in the cardenolides and six-membered in the bufadienolides. The hellebrigenin structure shows two other modifications not found in the basic steroid skeleton, namely a hydroxyl at the bridgehead carbon C-5 and a formyl group at C-10, being an oxidized form of the normal methyl. The three-dimensional shape of digitoxigenin is shown in Figure 5.98. These basic structures arise biosynthetically by metabolism of cholesterol, in which the side-chain is cleaved to a two-carbon acetyl group, followed by incorporation of either two carbonatoms for cardenolides or three carbonatoms for bufadienolides (Figure 5.99).

Shortening of the cholesterol side-chain is accomplished by stepwise hydroxylation at C-22 and then C-20, then cleavage of the C-20/22 bond to give **pregnenolone**, a sequence catalysed by the 'side-chain cleaving enzyme'. Pregnenolone is then oxidized in ring A to give **progesterone** (Figure 5.99). This can be reduced to give the *cis*-fused A/B system as in 5β-pregnan-3,20-dione (compare Figure 5.96) which is the substrate for 14β-hydroxylation, i.e. inverting the stereochemistry at this centre. Inversion is atypical for hydroxylation by monooxygenases, which are found to hydroxylate with retention of configuration. Whatever the mechanism of this hydroxylation, no Δ^8 or Δ^{15} double-bond intermediates are involved. Hydroxylation in the side-chain at C-21 follows. The lactone ring is

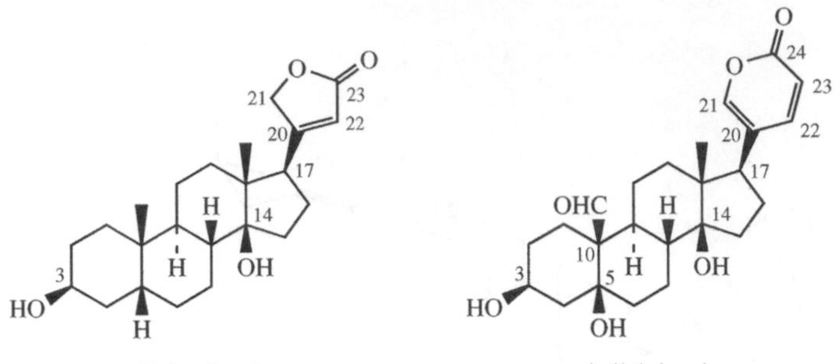

digitoxigenin
cardenolide

hellebrigenin
bufadienolide

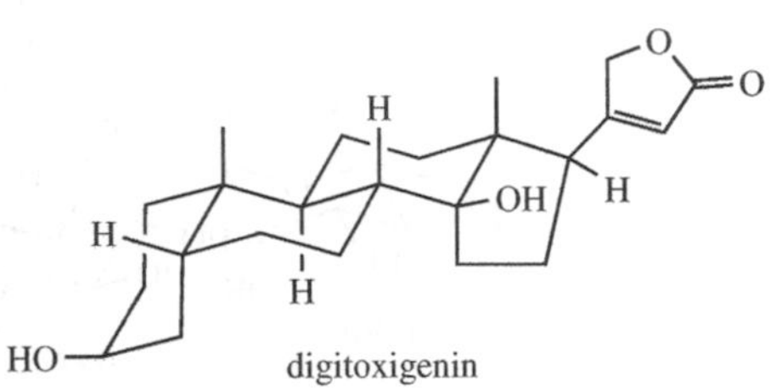

digitoxigenin

Characteristic features of cardiac glycosides:
- *cis*-fused A/B and C/D rings
- 14β-hydroxyl
- unsaturated lactone at C-17β
- sugar residues on 3β-hydroxyl

Figure 5.98

created at this stage. An intermediate malonate ester is involved, and ring formation probably occurs via the aldol addition process shown in Figure 5.100 to give the cardenolide **digitoxigenin**, the carboxyl carbon of the malonate ester being lost by decarboxylation during the process (compare the role of malonate in the acetate pathway). Digitoxigenin is the precursor of **digoxigenin** and **gitoxigenin** by specific hydroxylations (Figure 5.99). Oxaloacetate is the source of the extra carbon atoms in bufadienolides; a similar esterification/aldol reaction sequence may be proposed (Figure 5.100). This would produce **bufalin** (Figure 5.99), a bufadienolide structure found in the skin of toad (*Bufo* spp.), from which this class of compound was originally isolated and has subsequently taken the general name. Note that, in the subsequent formation of **hellebrigenin** (Figure 5.99), hydroxylation at C-5 occurs with the expected retention of stereochemistry, not with inversion as seen at C-14.

The fundamental pharmacological activity of the cardioactive glycosides resides in the aglycone portion, but is considerably modified by the nature of the sugar at C-3. This increases water solubility and binding to heart muscle. The sugar unit may have one to four monosaccharides; many (e.g. D-**digitoxose** and D-**digitalose**; Figure 5.101) are unique to this group of compounds. About 20 different sugars have been characterized, and with the exception of D-glucose, they are 6-deoxy- (e.g. L-rhamnose, D-digitalose) or 2,6-dideoxy- (e.g. D-digitoxose, D-cymarose) hexoses, some of which are also 3-methyl ethers (e.g. D-digitalose

and D-**cymarose**; Figure 5.101). In plants, cardiac glycosides are confined to the angiosperms, but are found in both monocotyledons and dicotyledons. The cardenolides are more common, and the plant families the Apocynaceae (e.g. *Strophanthus*), Liliaceae (e.g. *Convallaria*), and Scrophulariaceae (e.g. *Digitalis*) yield medicinal agents [Box 5.18]. The rarer bufadienolides are found in some members of the Liliaceae (e.g. *Urginea*) [Box 5.18] and Ranunculaceae (e.g. *Helleborus*), as well as in toads. Monarch butterflies and their larvae are known to accumulate in their bodies a range of cardenolides which they ingest from their food plant, the common milkweed (*Asclepias syriaca*; Asclepiadaceae). This makes them unpalatable to predators such as birds. Endogenous *Digitalis*-like compounds have also been detected, albeit in very small quantities, in mammalian tissues. **Ouabain** (see below, Figure 5.106), first isolated from the bark of the African ouabio tree (*Acokanthera ouabio*; Apocynaceae), is now known to occur naturally in blood plasma, adrenal glands, and the hypothalamus of mammals; it is a major component in the seeds of *Strophanthus gratus* [Box 5.18]. **19-Norbufalin** (Figure 5.101) is found in the lens of human eyes, at higher levels if these are cataract-afflicted, and it is believed to regulate ATPase activity under some physiological and pathological conditions.

Spirostane saponins are known to co-occur with cardenolide glycosides in *D. purpurea* (see page 264), indicating the plant is able to synthesize structures in which A/B and C/D ring fusions may be *trans–trans* or

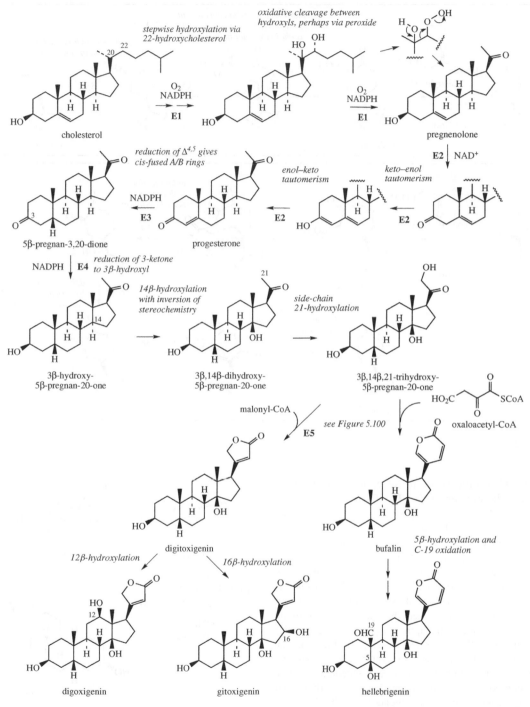

E1: cholesterol monooxygenase (side-chain cleaving)
(side-chain cleaving enzyme)
E2: Δ⁵-3β-hydroxysteroid dehydrogenase/Δ⁵-Δ⁴-ketosteroid isomerase

E3: progesterone 5β-reductase
E4: 3β-hydroxysteroid dehydrogenase
E5: 21-hydroxypregnane 21-*O*-malonyltransferase

Figure 5.99

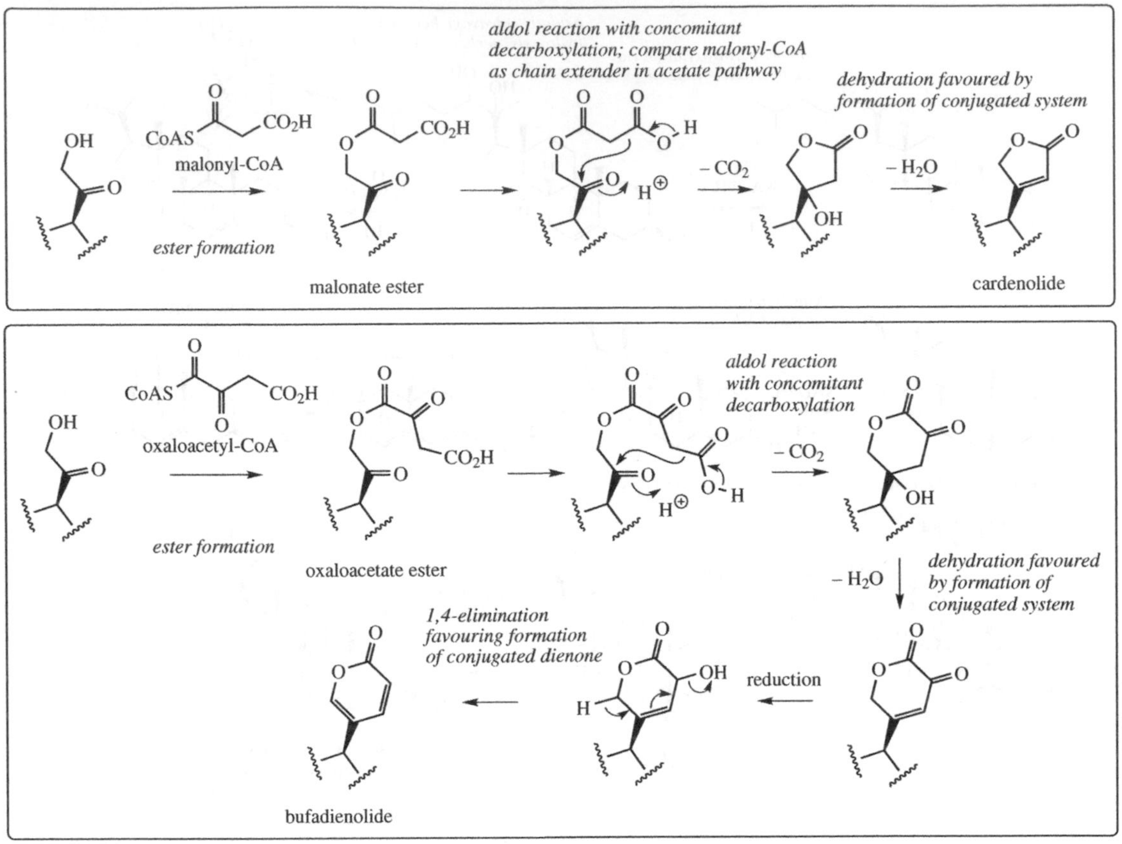

Figure 5.100

cis–cis. Other variants are possible, in that compounds such as **digipurpurogenin II** (Figure 5.102) have also been found in *D. purpurea.* This compound has the C/D ring fusion *cis*, with a 14β-hydroxyl group, characteristic of the cardiac glycosides, yet the A/B rings are typical of more common sterols. The glycoside **P57A3** of this compound found in the South African succulents *Hoodia pilifera* and *H. gordonii* is attracting considerable interest as an appetite suppressant. Indigenous people have used the plants as a substitute for food and water, and appetite-suppressant properties were subsequently confirmed, leading to the isolation of P57A3 and its glycoside (a cymaroside). A *Hoodia* herbal appetite suppressant has been patented. Though these compounds are pregnane derivatives, they bear a strong relationship to cardenolides via the 12β- and 14β-oxygenation, and the chain of 2,6-dideoxy sugar units.

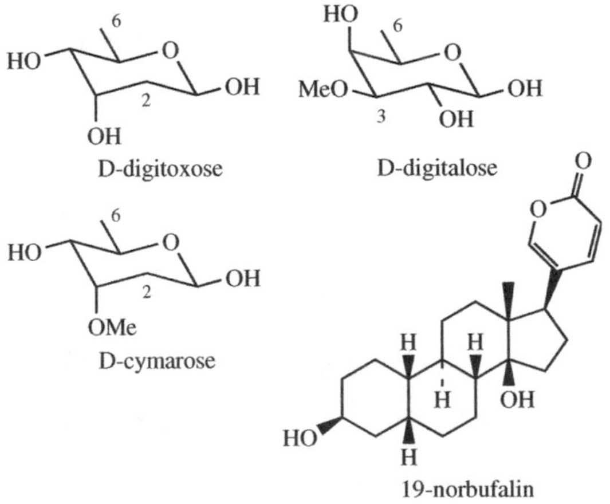

Figure 5.101

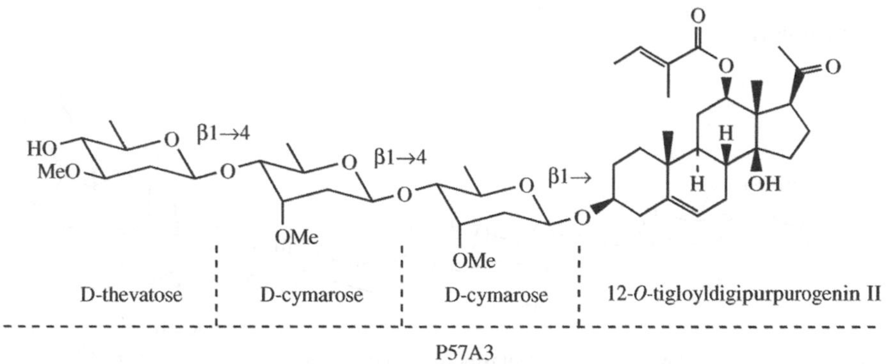

Figure 5.102

Box 5.18

Digitalis purpurea

Digitalis leaf consists of the dried leaf of the red foxglove *Digitalis purpurea* (Scrophulariaceae). The plant is a biennial herb, common in Europe and North America, which forms a low rosette of leaves in the first year and its characteristic spike of purple (occasionally white) bell-shaped flowers in the second year. It is potentially very toxic, but the leaf is unlikely to be ingested by humans. *D. purpurea* is cultivated for drug production, principally in Europe, the first year leaves being harvested then rapidly dried at 60°C as soon as possible after collection. This procedure is necessary to inactivate hydrolytic enzymes which would hydrolyse glycoside linkages in the cardioactive glycosides, giving rise to less active derivatives. Even so, some partial hydrolysis does occur. Excess heat may also cause dehydration in the aglycone to biologically inactive Δ^{14}-anhydro compounds.

Because of the pronounced cardiac effects of digitalis, the variability in the cardiac glycoside content, and also differences in the range of structures present due to the effects of enzymic hydrolysis, the crude leaf drug is usually assayed biologically rather than chemically. Prepared digitalis is a biologically standardized preparation of powdered leaf, its activity being assessed on cardiac muscle of guinea pig or pigeon and compared against a standard preparation. It may be diluted to the required activity by mixing in powdered digitalis of lower potency, or inactive materials such as lucerne (*Medicago sativa*) or grass. The crude drug is hardly ever used now, having been replaced by the pure isolated glycosides.

The cardioactive glycoside content of *D. purpurea* leaf is 0.15–0.4%, consisting of about 30 different structures. The major components are based on the aglycones digitoxigenin, gitoxigenin, and gitaloxigenin (Figure 5.103), the latter being a formate ester. The glycosides comprise two series of compounds, those with a tetrasaccharide *glucose–(digitoxose)*$_3$– unit and those with a trisaccharide (*digitoxose*)$_3$– unit. The latter group (the secondary glycosides) is produced by partial hydrolysis from the former group (the primary glycosides) during drying by the enzymic action of a β-glucosidase which removes the terminal glucose. Thus, the principal glycosides in the fresh leaves, namely purpureaglycoside A and purpureaglycoside B (Figure 5.103), are partially converted into digitoxin and gitoxin respectively (Figure 5.103), which normally predominate in the dried leaf. These transformations are indicated schematically in Figure 5.104. In the fresh leaf, purpureaglycoside A can constitute about 50% of the glycoside mixture, whilst in the dried leaf the amounts could be negligible if the plant material is old or poorly stored. The gitaloxigenin-based glycosides are relatively unstable, and the formyl group on the aglycone is readily lost by hydrolysis. Other minor glycosides are present, but neither the fresh nor dried leaf contains any significant quantities of the free aglycones.

Glycosides of the gitoxigenin series are less active than the corresponding members of the digitoxigenin-derived series. **Digitoxin** is the only compound routinely used as a drug, and it is employed in congestive heart failure and treatment of cardiac arrhymias, particularly atrial fibrillation.

Digitalis lanata

Digitalis lanata (Scrophulariaceae), the Grecian foxglove, is a perennial or biennial herb from southern and central Europe. It differs in appearance from the red foxglove by its long narrow smoother leaves and its smaller flowers of a yellow–brown colour. It is cultivated in Europe, the United States, and South America and is harvested and dried in a similar manner to *D. purpurea*. It has not featured as a crude drug, but is used exclusively for the isolation of individual cardiac glycosides, principally digoxin and lanatoside C (Figure 5.105).

Box 5.18   (continued)

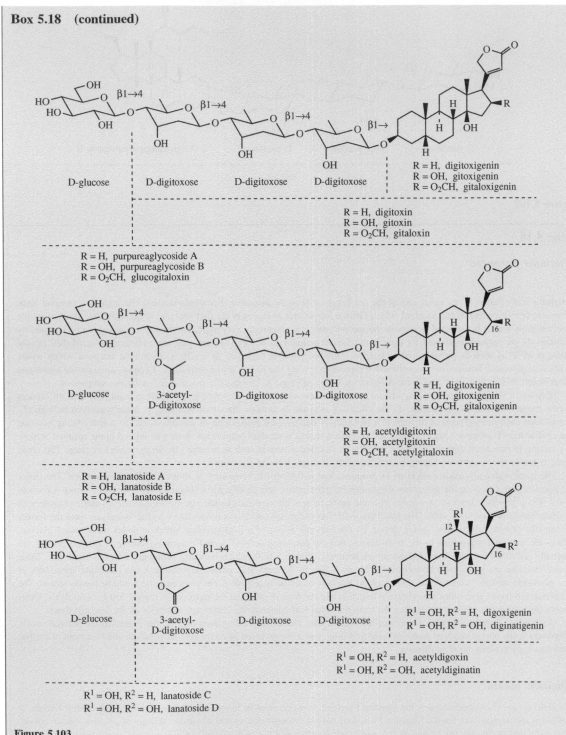

Figure 5.103

Box 5.18   (continued)

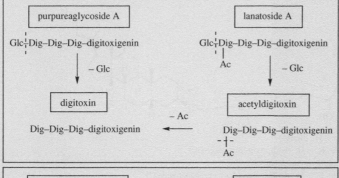

purpureaglycoside A

Glc–Dig–Dig–Dig–digitoxigenin

↓ – Glc

digitoxin

Dig–Dig–Dig–digitoxigenin

lanatoside A

Glc–Dig–Dig–Dig–digitoxigenin
|
Ac

↓ – Glc

acetyldigitoxin

← – Ac

Dig–Dig–Dig–digitoxigenin
-|-
Ac

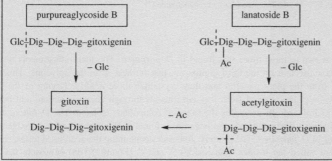

purpureaglycoside B

Glc–Dig–Dig–Dig–gitoxigenin

↓ – Glc

gitoxin

Dig–Dig–Dig–gitoxigenin

lanatoside B

Glc–Dig–Dig–Dig–gitoxigenin
|
Ac

↓ – Glc

acetylgitoxin

← – Ac

Dig–Dig–Dig–gitoxigenin
-|-
Ac

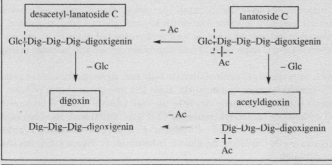

desacetyl-lanatoside C

Glc–Dig–Dig–Dig–digoxigenin

↓ – Glc

digoxin

Dig–Dig–Dig–digoxigenin

← – Ac

lanatoside C

Glc–Dig–Dig–Dig–digoxigenin
-|-
Ac

↓ – Glc

acetyldigoxin

← – Ac

Dig–Dig–Dig–digoxigenin
-|-
Ac

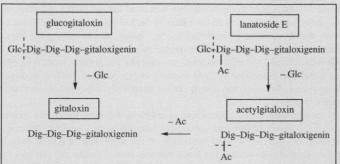

glucogitaloxin

Glc–Dig–Dig–Dig–gitaloxigenin

↓ – Glc

gitaloxin

Dig–Dig–Dig–gitaloxigenin

lanatoside E

Glc–Dig–Dig–Dig–gitaloxigenin
|
Ac

↓ – Glc

acetylgitaloxin

← – Ac

Dig–Dig–Dig–gitaloxigenin
-|-
Ac

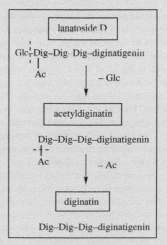

lanatoside D

Glc–Dig–Dig–Dig–diginatigenin
|
Ac

↓ – Glc

acetyldiginatin

Dig–Dig–Dig–diginatigenin
-|-
Ac

↓ – Ac

diginatin

Dig–Dig–Dig–diginatigenin

Figure 5.104

Box 5.18   (continued)

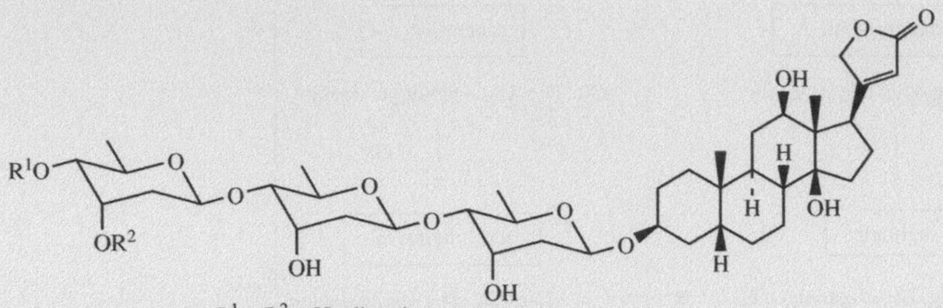

R^1 = R^2 = H, digoxin
R^1 = Me, R^2 = H, medigoxin (metildigoxin)
R^1 = Glc, R^2 = Ac, lanatoside C
R^1 = Glc, R^2 = H, deslanoside (desacetyl-lanatoside C)

Figure 5.105

The total cardenolide content of up to 1% is two to three times that found in *D. purpurea*. The main constituents resemble those of *D. purpurea*, but contain an acetyl ester function on the third digitoxose, that furthest from the aglycone. This acetyl group makes the compounds easier to isolate from the plant material and they crystallize more readily. Drying of the leaf is similarly accompanied by some partial hydrolysis of the original fresh leaf constituents through enzymic action, and both the terminal glucose and the acetyl group may be hydrolysed off, extending the range of compounds isolated. The *D. lanata* cardiac glycosides are based on five aglycones: digitoxigenin, gitoxigenin and gitaloxigenin, as found in *D. purpurea*, plus digoxigenin and diginatigenin (Figure 5.103), which do not occur in *D. purpurea*. The primary glycosides containing the acetylated tetrasaccharide unit *glucose–acetyldigitoxose–(digitoxose)$_2$–* are called lanatosides. Lanatosides A and C (Figure 5.103) constitute the major components in the fresh leaf (about 50–70%) and are based on the aglycones digitoxigenin and digoxigenin respectively. Lanatosides B, D, and E (Figure 5.103) are minor components derived from gitoxigenin, diginatigenin, and gitaloxigenin respectively. Enzymic hydrolysis of the lanatosides generally involves loss of the terminal glucose prior to removal of the acetyl function, so that compounds like acetyldigitoxin and acetyldigoxin, as well as digitoxin and digoxin, are present in the dried leaf as decomposition products from lanatosides A and C respectively. These transformations are also indicated in simplified form in Figure 5.104.

Digoxin (Figure 5.105) has a rapid action and is more quickly eliminated from the body than digitoxin; therefore, it is the most widely used of the cardioactive glycosides. It is more hydrophilic than digitoxin, binds less strongly to plasma proteins, and is mainly eliminated by the kidneys, whereas digitoxin is metabolized more slowly by the liver. Digoxin is used in congestive heart failure and atrial fibrillation. **Lanatoside C** and **deslanoside** (**desacetyl-lanatoside C**) (Figure 5.105) have also been employed, though not to the same extent. They have very rapid action and are suited for treatment of cardiac emergencies by injection. The semi-synthetic derivative **medigoxin** or **metildigoxin** (methyl replacing the glucose in lanatoside C; Figure 5.105) has also been available, being more active through better bioavailability.

The cardioactive glycosides increase the force of contractions in the heart, thus increasing cardiac output and allowing more rest between contractions. The primary effect on the heart appears to be inhibition of the ion transport activity of the enzyme Na$^+$/K$^+$-ATPase in the cell membranes of heart muscle, specifically inhibiting the Na$^+$ pump, thereby raising the intracellular Na$^+$ concentration. The resultant decrease in the Na$^+$ gradient across the cell membrane reduces the energy available for transport of Ca^{2+} out of the cell, leads to an increase in intracellular Ca^{2+} concentration, and provides the positive ionotropic effect and increased force of contractions. The improved blood circulation also tends to improve kidney function, leading to diuresis and loss of oedema fluid often associated with heart disease. However, the diuretic effect, historically important in the treatment of dropsy, is more safely controlled by other diuretic drugs.

To treat congestive heart failure, an initial loading dose of the cardioactive glycoside is followed by regular maintenance doses, the amounts administered depending on drug bioavailability and subsequent metabolism or excretion. Because of the extreme toxicity associated with these compounds (the therapeutic level is 50–60% of the toxic dose; a typical daily dose is only about 1 mg), dosage must be controlled very carefully. Bioavailability has sometimes proved erratic and can vary between different manufacturers' formulations, so patients should not be provided with different preparations during their treatment. Individual patients also excrete the glycosides or metabolize them by hydrolysis to the aglycone at different rates, and ideally these processes should be monitored. Levels of the drug in blood plasma can be measured quite rapidly by radioimmunoassay using a specific antibody. A preparation of **digoxin-specific antibody fragments** derived from sheep is available both for assay

Box 5.18   (continued)

and also as a means of reversing life-threatening digoxin overdose. It has also successfully reversed digitoxin overdose, thus demonstrating a somewhat broader specificity. The value of digoxin treatment for heart failure where the heartbeat remains regular has recently been called into question. It still remains a recognized treatment for atrial fibrillation.

Many other species of *Digitalis*, e.g. *D. dubia, D. ferruginea, D. grandiflora, D. lutea, D. mertonensis, D. nervosa, D. subalpina,* and *D. thaspi* contain cardioactive glycosides in their leaves, and some have been evaluated and cultivated for drug use. Gene engineering may be able to increase the amount of cardioactive glycosides obtainable. Thus, transgenic *D. minor* plants expressing an additional gene for the mevalonate pathway enzyme HMG-CoA reductase from *Arabidopsis thaliana* have shown increased sterol and cardenolide production.

Strophanthus

Strophanthus comprises the dried ripe seeds of *Strophanthus kombé* or *S. gratus* (Apocynaceae), which are tall vines from equatorial Africa. *S. kombé* has a history of use by African tribes as an arrow poison, and the seeds contain 5–10% cardenolides, a mixture known as K-strophanthin. This has little drug use today, though was formerly used medicinally as a cardiac stimulant. The main glycoside (about 80%) is K-strophanthoside (Figure 5.106) with smaller amounts of K-strophanthin-β and cymarin, related to K-strophanthoside as shown. These are derivatives of the aglycone strophanthidin. *S. gratus* contains 4–8% of **ouabain** (G-strophanthin; Figure 5.106), the rhamnoside of ouabigenin. Ouabigenin is rather unusual in having additional hydroxylation

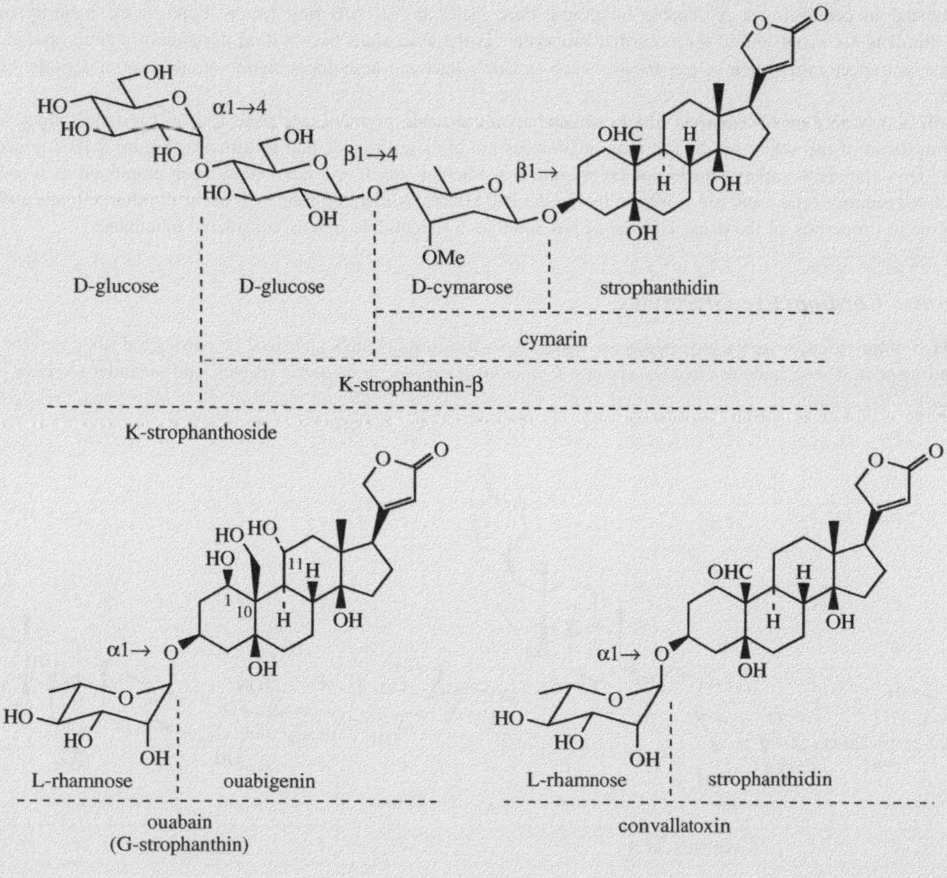

Figure 5.106

Box 5.18   (continued)

at 1β and 11α, as well as a hydroxymethyl at C-10. Ouabain is a stable, crystalline material which is often employed as the biological standard in assays for cardiac activity. It is a potent cardiac glycoside, acts quickly, but wears off rapidly. It is very polar, with rapid renal elimination and must be injected because it is so poorly absorbed orally. It has been used for emergency treatment in cases of acute heart failure.

Convallaria

The dried roots and tops of lily of the valley *Convallaria majalis* (Liliaceae/Convallariaceae) contain cardioactive glycosides (0.2–0.3%) and in the past have been used in some European countries rather than digitalis. The effects are similar, but the drug is less cumulative. This plant is widely cultivated as an ornamental, particularly for its intensely perfumed small white flowers, and must be considered potentially toxic. The major glycoside (40–50%) is convallatoxin (Figure 5.106), the rhamnoside of strophanthidin.

Squill

Squill (white squill) consists of the dried sliced bulbs of the white variety of *Urginea maritima* (formerly *Scilla maritima*; also known as *Drimia maritima*) (Liliaceae/Hyacinthaceae) which grows on seashores around the Mediterranean. The plant contains bufadienolides (up to 4%), principally scillaren A and proscillaridin A (Figure 5.107). The aglycone of scillaren A is scillarenin, which is unusual in containing a Δ^4 double bond and, thus, lacks the *cis* A/B ring fusion found in the majority of cardiac glycosides. Squill is not usually used for its cardiac properties, as the glycosides have a short duration of action. Instead, squill is employed for its expectorant action in preparations such as Gee's linctus. Large doses cause vomiting and a digitalis-like action on the heart.

Red squill is a variety of *U. maritima* which contains an anthocyanin pigment (see page 170) and bufadienolides which are different from those of the white squill. The main glycosides are glucoscilliroside and scilliroside (Figure 5.107), glucosides of scillirosidin. This chemical variety should not be present in medicinal squill, and has mainly been employed as a rodenticide. Rodents lack a vomiting reflex and are poisoned by the cardiac effects, whilst vomiting will occur in other animals and humans due to the emetic properties of the drug. The use of red squill as a rodenticide is now considered inhumane.

Toxic Plants: Cardioactive Glycosides

Many plants containing cardioactive glycosides are widely grown as ornamentals and must be considered toxic and treated with due care and respect. These include *Digitalis* species, *Convallaria majalis*, *Helleborus* species, and oleander (*Nerium oleander*; Apocynaceae).

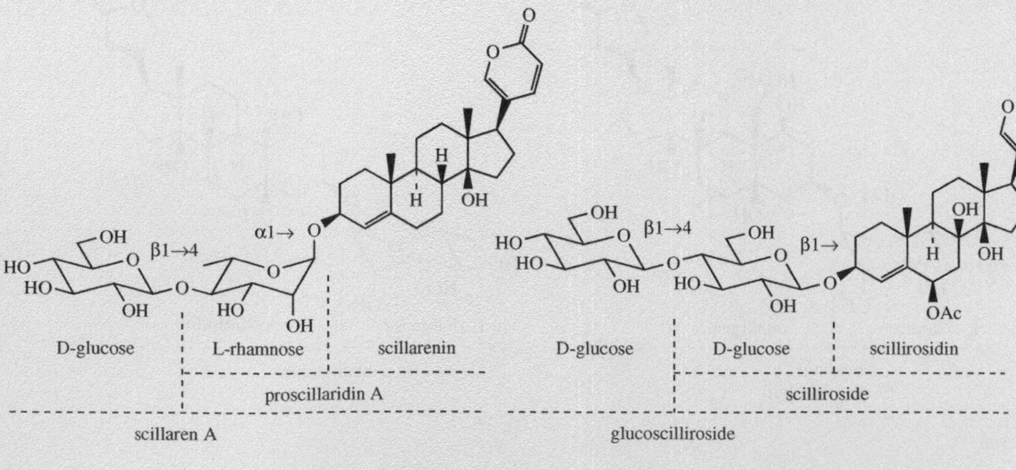

Figure 5.107

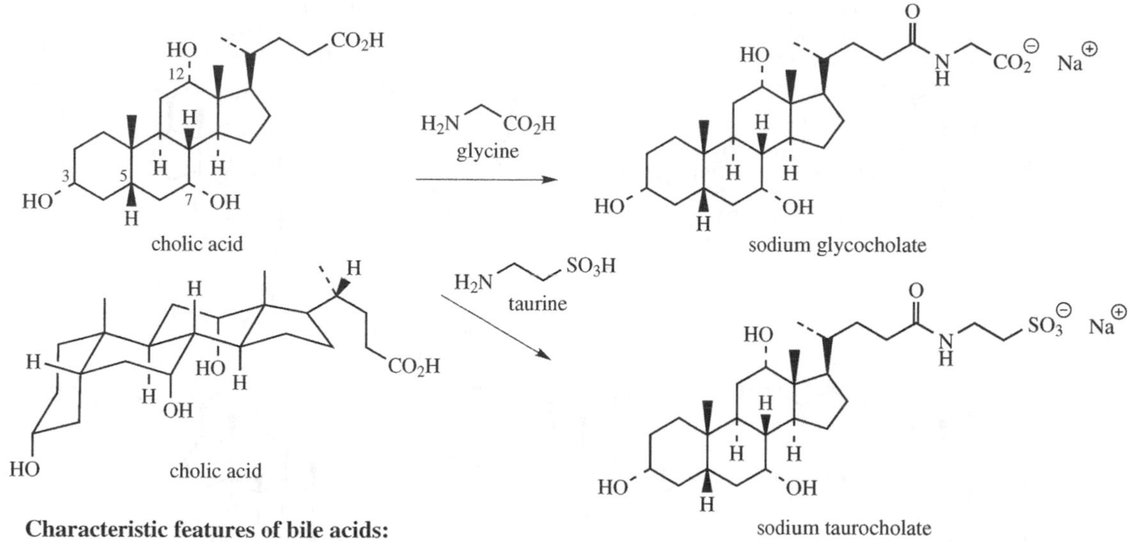

Characteristic features of bile acids:

- C_{24} cholane skeleton
- *cis*-fusion of A/B rings
- C_5-carboxylic acid side-chain
- 3α- and 7α-hydroxyls

Figure 5.108

Bile Acids

The **bile acids** are C_{24} steroidal acids, e.g. **cholic acid** (Figure 5.108), which occur in salt form in bile, secreted into the gut to emulsify fats and encourage digestion [Box 5.19]. The carboxyl group is typically bound via an amide linkage to glycine (about 75%) or taurine (about 25%), e.g. cholic acid is found as **sodium glycocholate** and **sodium taurocholate**. Conjugation with glycine or taurine increases the water solubility of bile salts under physiological conditions. Taurine (2-aminoethanesulphonic acid) was first isolated from ox bile, but is now known to be widely distributed in animal tissues. Metabolism to bile acids is the principal way in which mammals degrade cholesterol absorbed from the diet. Cholesterol is extremely hydrophobic; its removal is dependent upon increasing hydrophilicity, achieved by the introduction of several polar groups into the molecule. The *cis* fusion of rings A and B confers a curvature to the steroidal skeleton, and the polar hydroxyl groups are all positioned on the lower α face, contrasting with the non-polar upper β face. Because of this ambiphilicity, they can form micelles and act as detergents.

The bile acids are formed in the liver from cholesterol by a sequence which also removes three carbon atoms from the side-chain (Figure 5.109). This is achieved by initial oxidation of one of the side-chain methyl groups to an acid, followed by a β-oxidation sequence as seen with fatty acids (see Figure 2.11), removing the three-carbon unit as propionyl-CoA. Other essential features of the molecule are introduced earlier. The A/B ring system is *cis*-fused, and this is achieved by reduction of a Δ^4 rather than a Δ^5 double bond (see page 264). Migration of the double bond is accomplished via the 3-ketone; when this is reduced back to a hydroxyl, the configuration at C-3 is changed to 3α. The pathway to **chenodeoxycholic acid** (Figure 5.110) is essentially the same, though diverges early on by omitting the 12α-hydroxylation step. Alternative pathways ('acidic' pathways) to bile acids are known, though the one described (the 'classical' or 'neutral' pathway) predominates. These other pathways are characterized by initial 27-hydroxylation in the side-chain, and side-chain modifications tend to precede modifications in the sterol nucleus.

Both cholic acid and chenodeoxycholic acid are formed in the liver, stored in the gall bladder, and released into the intestine; they are termed primary bile acids. However, the 7α-hydroxyl functions of these compounds can be removed by intestinal microflora, so that mammalian bile also contains **deoxycholic acid** and **lithocholic acid** (Figure 5.110), which are termed secondary bile acids. The bile salts are then usually reabsorbed and stored in

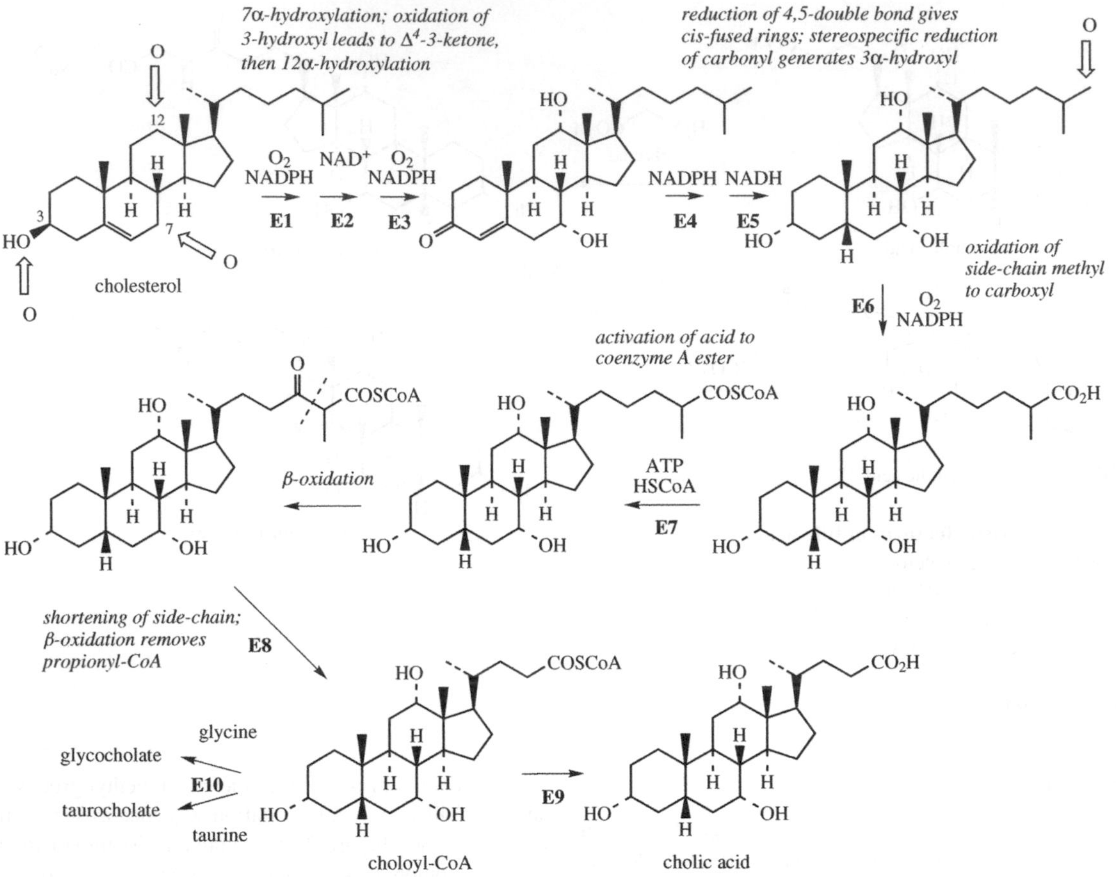

*7α-hydroxylation; oxidation of
3-hydroxyl leads to Δ⁴-3-ketone,
then 12α-hydroxylation*

*reduction of 4,5-double bond gives
cis-fused rings; stereospecific reduction
of carbonyl generates 3α-hydroxyl*

cholesterol

O₂   NAD⁺   O₂
NADPH   NADPH
E1   **E2**   **E3**

NADPH NADH
E4   **E5**

*oxidation of
side-chain methyl
to carboxyl*

E6 ↓ O₂
NADPH

*activation of acid to
coenzyme A ester*

β-oxidation

ATP
HSCoA
E7

*shortening of side-chain;
β-oxidation removes
propionyl-CoA*   **E8**

glycine

glycocholate

E10

taurocholate

taurine

E9

choloyl-CoA

cholic acid

E1: cholesterol 7α-hydroxylase (cholesterol 7α-monooxygenase)
E2: cholest-5-ene-3β,7α-diol 3β-dehydrogenase
E3: 7α-hydroxycholest-4-ene-3-one 12α-hydroxylase
E4: Δ⁴-3-ketosteroid 5β-reductase
E5: 3α-hydroxysteroid dehydrogenase

E6: cholestanetriol 26-monooxygenase
E7: cholestanate-CoA ligase
E8: propanoyl-CoA C-acyltransferase
E9: cholate-CoA ligase
E10: bile acid-CoA: amino acid N-acyltransferase

Figure 5.109

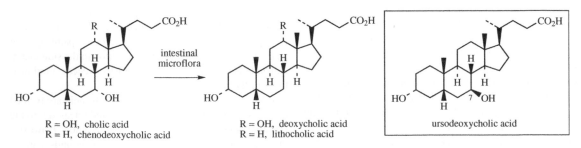

intestinal
microflora

R = OH, cholic acid
R = H, chenodeoxycholic acid

R = OH, deoxycholic acid
R = H, lithocholic acid

ursodeoxycholic acid

Figure 5.110

the gall bladder, although they are also excreted as the body's main means of eliminating excess cholesterol. Inability to remove cholesterol by bile acid synthesis and excretion may contribute to atherosclerosis and gallstone disease; gallstones often contain more than 70% of cholesterol (see page 251). 7α-Hydroxylation of cholesterol is the critical and rate-limiting step. This is catalysed by the cytochrome P-450-dependent monooxygenase cholesterol 7α-hydroxylase, and any deficiency of this enzyme leads to high levels of total serum and LDL cholesterol (see page 251).

Box 5.19

Bile Acids

Bile acids are obtained by purification from fresh ox bile taken from beef carcasses as a by-product of the meat trade. Bile acids are still important as starting materials for the semi-synthesis of other medicinal steroids, being a cheap and readily accessible raw material. The 7-epimer of chenodeoxycholic acid, **ursodeoxycholic acid** (ursodiol; Figure 5.110), is a minor secondary bile acid in humans, but a major component in bear bile. It is produced semi-synthetically from cholic acid or chenodeoxycholic acid and used medicinally to dissolve cholesterol gallstones as an alternative to surgery. By suppressing synthesis of both cholesterol and cholic acid, they contribute to removal of biliary cholesterol and, consequently, a gradual dissolution of gallstones which can have formed due to supersaturation. Partial or complete dissolution requires treatment over a period of many months, and is not effective for radio-opaque gallstones, which contain appreciable levels of calcium salts. Anion-exchange resins such as colestyramine (cholestyramine) and colestipol are used as cholesterol-lowering drugs to bind bile acids and prevent their reabsorption. This promotes hepatic conversion of cholesterol into bile acids, thus increasing breakdown of LDL cholesterol, and is of value in treating high-risk coronary patients.

Adrenocortical Hormones/Corticosteroids

A large number of steroid hormones have been isolated and characterized from the adrenal glands. Since they are produced by the adrenal cortex, the outer part of the adrenal glands near the kidneys, they are termed **adrenocortical hormones** or **corticosteroids** [Box 5.20]. They contain a pregnane C_{21} skeleton and fall into two main activity groups, the **glucocorticoids** and the **mineralocorticoids**, although it is difficult to separate entirely the two types of activity in one molecule. Glucocorticoids are concerned with the synthesis of carbohydrate from protein and the deposition of glycogen in the liver. They also play an important role in inflammatory processes. Mineralocorticoids are concerned with the control of electrolyte balance, active compounds promoting the retention of Na^+ and Cl^-, and the excretion of K^+.

Examples of natural glucocorticoids include **hydrocortisone (cortisol)** and **corticosterone**, whilst **aldosterone** and **deoxycorticosterone (cortexone)** typify mineralocorticoids (Figure 5.111). Deoxycorticosterone has also been found in plants. Some common features of these molecules are the β-CO.CH$_2$OH side-chain at C-17 and, frequently, an α-hydroxy also at this position. Ring A usually contains a Δ^4-3-keto functionality. The 11β-hydroxy is essential for glucocorticoid activity. In aldosterone, the principal mineralocorticoid hormone, the methyl group (C-18) has been oxidized to an aldehyde, and this is able to react with the 11β-hydroxyl, so that aldosterone exists predominantly in the hemiacetal form. This essentially eliminates the glucocorticoid activity.

The corticosteroids are produced from cholesterol via **pregnenolone** and **progesterone**. This involves side-chain cleavage as seen in the biosynthesis of cardioactive glycosides (see page 267), and the same sequence of reactions is operative. From progesterone, the formation of **deoxycorticosterone, corticosterone**, and **hydrocortisone (cortisol)** (Figure 5.112) requires only a series of hydroxylation steps, catalysed by cytochrome P-450-dependent hydroxylases with NADPH and O_2 cofactors. Thus, positions 17, 21, and 11 may be hydroxylated, and the exact order can in fact vary from that shown in Figure 5.112, according to species. It can be seen that production of hydrocortisone from cholesterol actually utilizes cytochrome P-450-dependent enzymes in four of the five steps. The further oxidation of C-18 to an aldehyde via the alcohol allows formation of **aldosterone** from corticosterone, again involving a P-450 system.

Semi-Synthesis of Corticosteroids

The medicinal use of corticosteroids was stimulated by reports of the dramatic effects of **cortisone** (Figure 5.113) on patients suffering from rheumatoid arthritis in the late 1940s and early 1950s. The cortisone employed

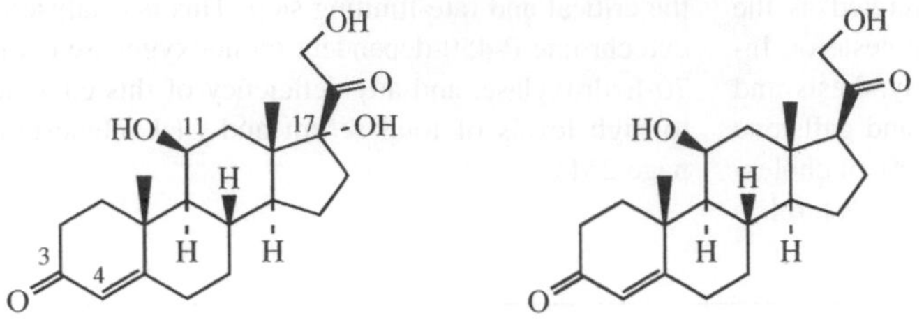

hydrocortisone
(cortisol)

corticosterone

Characteristic features of glucocorticoids:
- C_{21} pregnane skeleton
- 17β-CO.CH$_2$OH side-chain
- 11β-hydroxyl
- Δ^4-3-keto (usually)
- 17α-hydroxyl (usually)

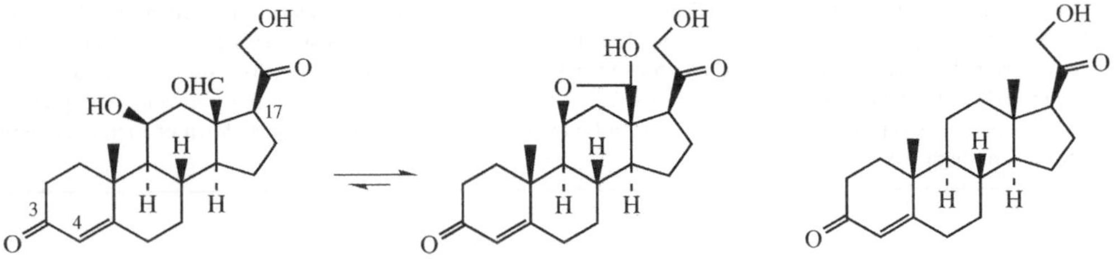

aldosterone

aldosterone
(hemiacetal form)

deoxycorticosterone
(cortexone)

Characteristic features of mineralocorticoids:
- C_{21} pregnane skeleton
- 17β-CO.CH$_2$OH side-chain
- Δ^4-3-keto (usually)

Figure 5.111

was isolated from the adrenal glands of cattle, and later was produced semi-synthetically by a laborious process from **deoxycholic acid** (see page 276) isolated from ox bile, a sequence necessitating over 30 chemical steps. In due course, it was shown that cortisone itself was not the active agent; it was reduced in the liver to **hydrocortisone** as the active agent (Figure 5.113). Two dehydrogenase enzymes regulate hydrocortisone levels, both of them acting in a unidirectional sense. One of these activates cortisone by reduction, and though prevalent in the liver is found in a wide range of tissues. The other enzyme is found predominantly in mineralocorticoid target tissues, e.g. kidney and colon, where it protects receptors against excess hydrocortisone by converting it to inactive cortisone. Increased demand for cortisone and hydrocortisone (cortisol) led to exploitation of alternative raw materials, particularly plant sterols and saponins. A major difficulty in any semi-synthetic conversion was

the need to provide the 11β-hydroxyl group, which was essential for glucocorticoid activity.

Sarmentogenin (Figure 5.114) had been identified as a natural 11-hydroxy cardenolide in *Strophanthus sarmentosus*, but it was soon appreciated that the amounts present in the seeds, and the limited quantity of plant material available, would not allow commercial exploitation of this compound. As an alternative to using a natural 11-oxygenated substrate, compounds containing a 12-oxygen substituent might be used instead, in that this group provides activation and allows chemical modification at the adjacent site. Indeed, this was a feature of the semi-synthesis of cortisone from deoxycholic acid, which contains a 12α-hydroxyl. However, it was the 12-keto steroidal sapogenin **hecogenin** (Figure 5.114) from sisal (*Agave sisalana*; Agavaceae) (see page 262) that made possible the economic production of cortisone on a commercial scale. This material is still used

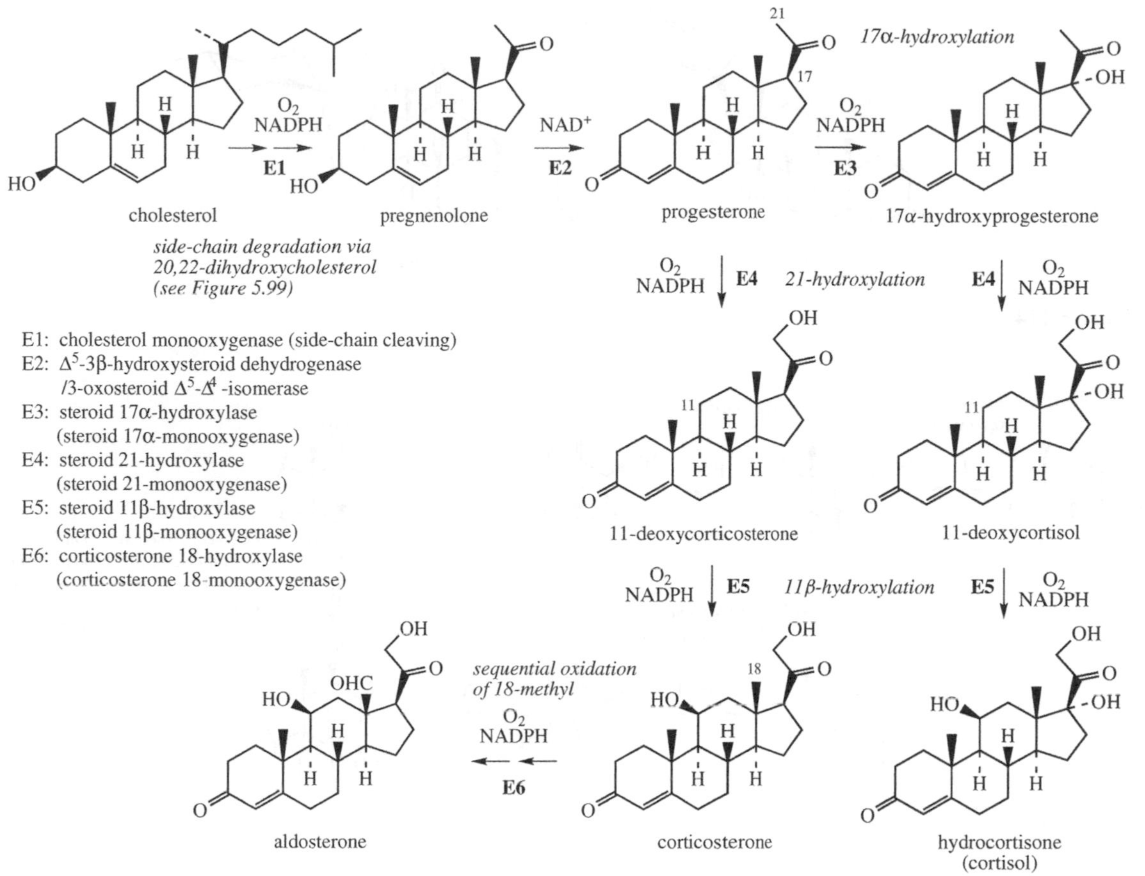
cholesterol
pregnenolone
progesterone
17α-hydroxyprogesterone
11-deoxycorticosterone
11-deoxycortisol
aldosterone
corticosterone
hydrocortisone (cortisol)
side-chain degradation via 20,22-dihydroxycholesterol (see Figure 5.99)
17α-hydroxylation
21-hydroxylation
11β-hydroxylation
sequential oxidation of 18-methyl
O2 NADPH
NAD+
E1
E2
E3
E4
E5
E6
E1: cholesterol monooxygenase (side-chain cleaving)
E2: Δ5-3β-hydroxysteroid dehydrogenase /3-oxosteroid Δ5-Δ4-isomerase
E3: steroid 17α-hydroxylase (steroid 17α-monooxygenase)
E4: steroid 21-hydroxylase (steroid 21-monooxygenase)
E5: steroid 11β-hydroxylase (steroid 11β-monooxygenase)
E6: corticosterone 18-hydroxylase (corticosterone 18-monooxygenase)

Figure 5.112

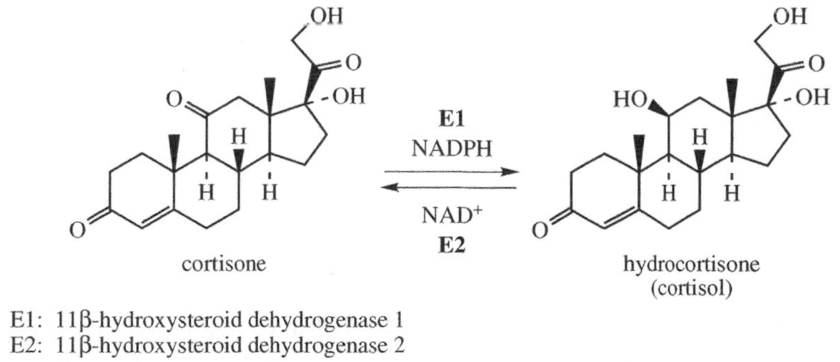
cortisone
hydrocortisone (cortisol)
E1
NADPH
NAD+
E2
E1: 11β-hydroxysteroid dehydrogenase 1
E2: 11β-hydroxysteroid dehydrogenase 2

Figure 5.113

in the semi-synthesis of steroidal drugs, and the critical modifications in ring C are shown in Figure 5.115. Bromination α to the 12-keto function generates the 11α-bromo derivative, which on treatment with base gives the 12-hydroxy-11-ketone by a base-catalysed keto–enol tautomerism mechanism. The 12-hydroxyl is then removed by hydride displacement of the acetate using calcium in liquid ammonia. The 11-keto sapogenin derived

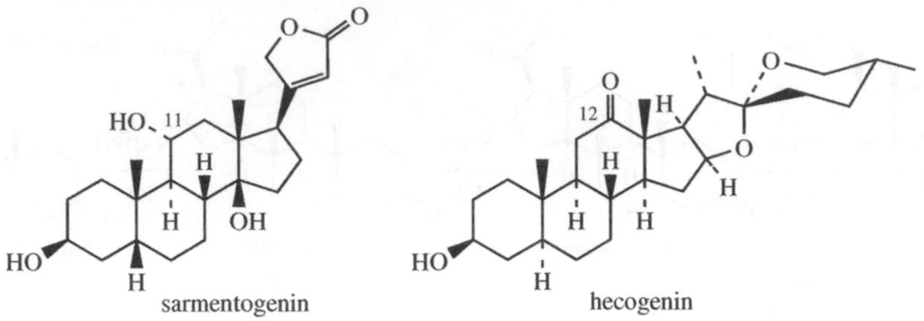

Figure 5.114

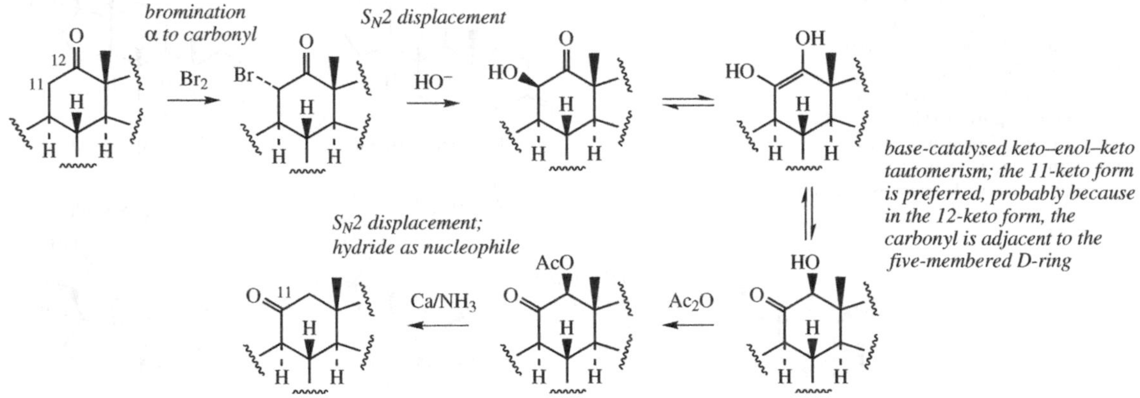

Figure 5.115

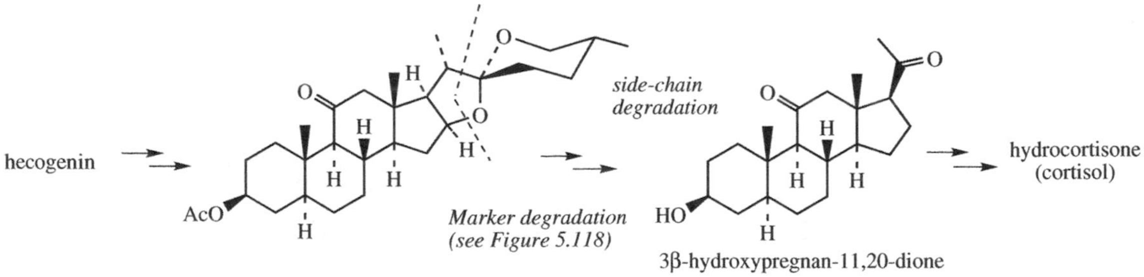

Figure 5.116

by this sequence is subjected to the side-chain degradation used with other sapogenins, e.g. diosgenin (see later, Figure 5.118), giving the 11-ketopregnane (Figure 5.116). This compound can then be used for conversion into cortisone, hydrocortisone, and other steroid drugs.

Of much greater importance was the discovery in the mid 1950s that hydroxylation at C-11 could be achieved via a microbial fermentation. **Progesterone** was transformed by *Rhizopus arrhizus* into

11α-hydroxyprogesterone (Figure 5.117) in yields of up to 85%. More recently, *R. nigricans* has been employed to give even higher yields. 11α-Hydroxyprogesterone is then converted into hydrocortisone by chemical means, the 11β configuration being introduced via oxidation to the 11-ketone followed by a stereospecific reduction step.

Progesterone could be obtained in good yields (about 50%) from **diosgenin** extracted from Mexican yams (*Dioscorea* species; Dioscoreaceae; see page

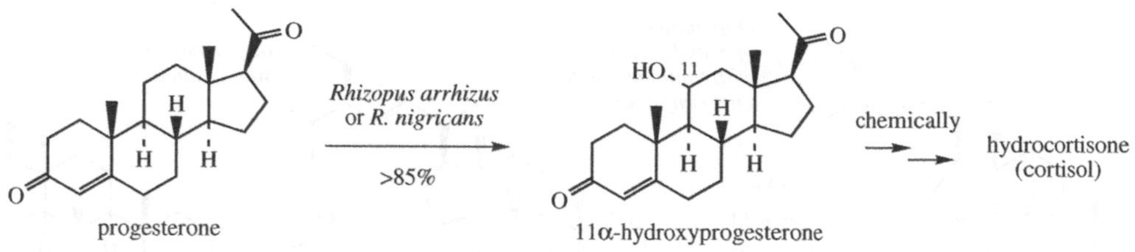

Figure 5.117

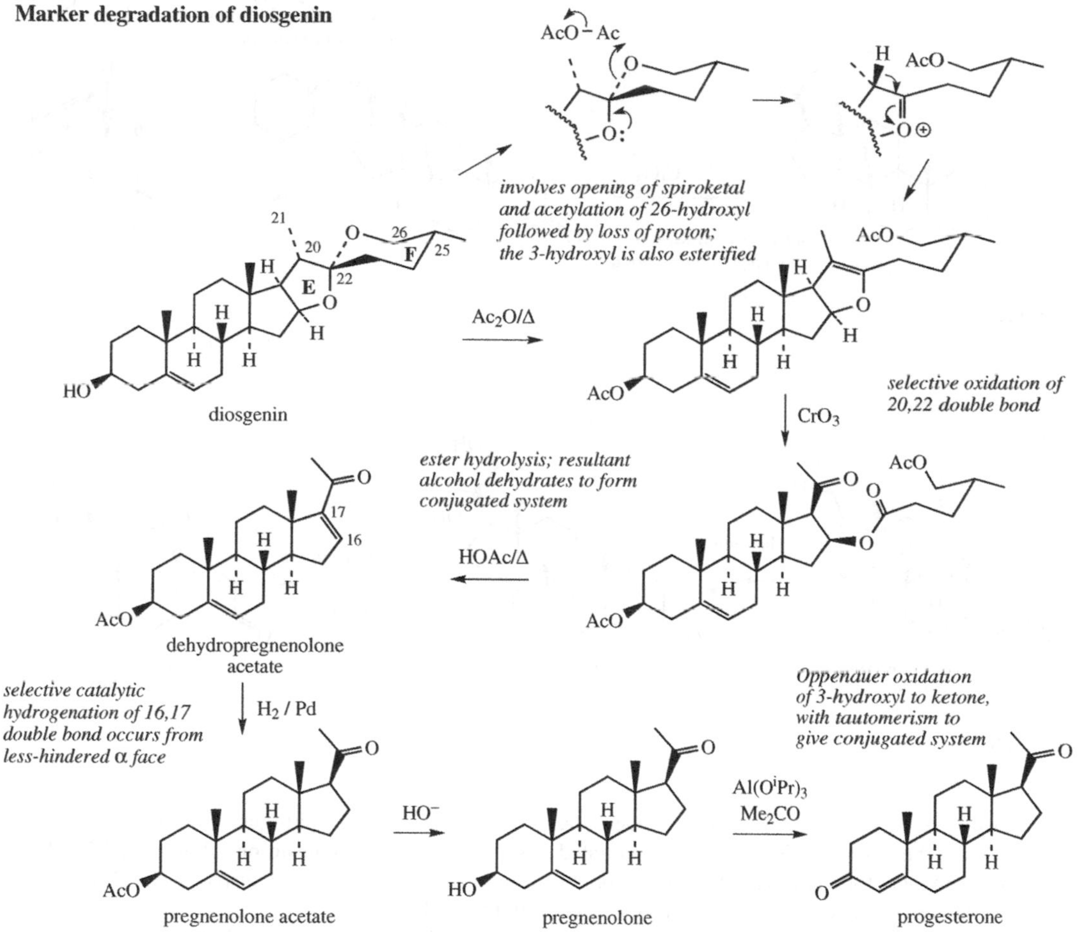

Marker degradation of diosgenin

Figure 5.118

262) or **stigmasterol** from soya beans (*Glycine max*; Leguminosae/Fabaceae; see page 255). Steroidal sapogenins such as diosgenin may be degraded by the **Marker degradation** (Figure 5.118), which removes the spiroketal portion, leaving carbon atoms C-20 and C-21 still attached to contribute to the pregnane

system. Initial treatment with acetic anhydride produces the 3,26-diacetate, by opening the ketal, dehydrating in ring E and acetylating the remaining hydroxyl groups. The double bond in ring E is then selectively oxidized to give a product, which now contains the unwanted side-chain carbon atoms as an ester

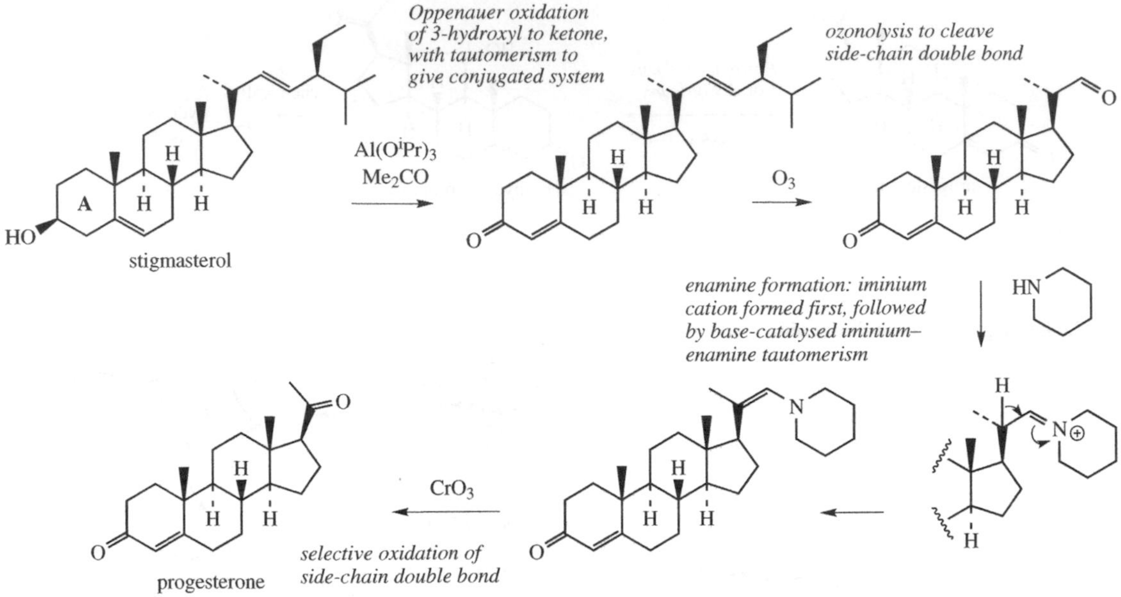

Figure 5.119

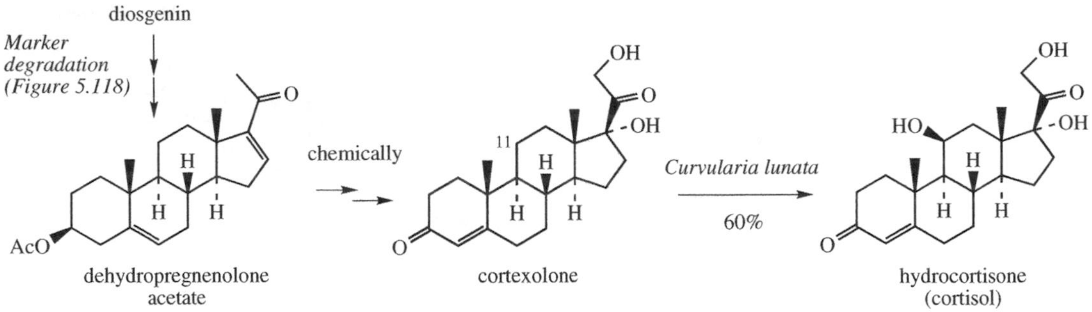

Figure 5.120

function, easily removed by hydrolysis. Under the conditions used, the product is the α,β-unsaturated ketone dehydropregnenolone acetate. Hydrogenation of the double bond is achieved in a regioselective and stereoselective manner, addition of hydrogen being from the less-hindered α-face to give pregnenolone acetate. **Progesterone** is obtained by hydrolysis of the ester function and Oppenauer oxidation to give the preferred α,β-unsaturated ketone (compare page 264). It is immediately obvious from Figure 5.119 that, since the objective is to remove the unwanted ring F part of the sapogenin, features like the stereochemistry at C-25 are irrelevant, and the same general degradation procedure can be used for other sapogenins. It is equally

applicable to the nitrogen-containing analogues of sapogenins, e.g. **solasodine** (Figure 5.95). In such compounds, the stereochemistry at C-22 is also quite immaterial.

Degradation of the sterol **stigmasterol** to progesterone is achieved by the sequence shown in Figure 5.119. The double bond in the side-chain allows cleavage by ozonolysis, and the resultant aldehyde is chain-shortened via formation of an enamine with piperidine. This can be selectively oxidized to progesterone. In this sequence, the ring A transformations are more conveniently carried out as the first reaction. A similar route can be used for the fungal sterol **ergosterol**, though an additional step is required for reduction of the Δ^7 double bond.

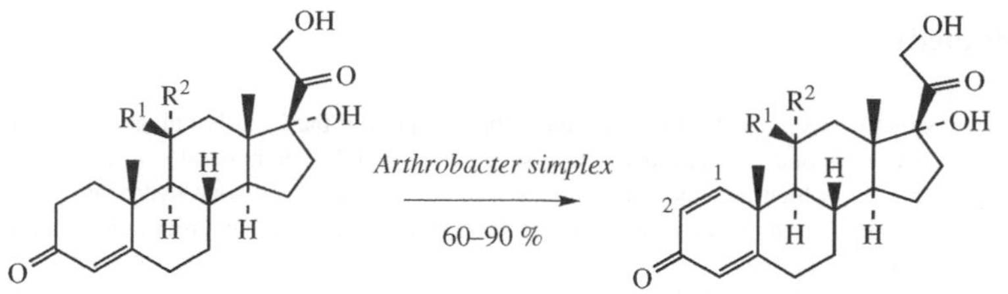

$R^1R^2 = O$, cortisone
$R^1 = OH$, $R^2 = H$, hydrocortisone (cortisol)

$R^1R^2 = O$, prednisone
$R^1 = OH$, $R^2 = H$, prednisolone

Figure 5.121

An alternative sequence from diosgenin to hydrocortisone has been devised, making use of another microbiological hydroxylation, this time a direct 11β-hydroxylation of the steroid ring system (Figure 5.120). The fungus *Curvularia lunata* is able to 11β-hydroxylate **cortexolone** to **hydrocortisone** in yields of about 60%. Although a natural corticosteroid, cortexolone may be obtained in large amounts by chemical transformation from 16-dehydropregnenolone acetate, an intermediate in the Marker degradation of diosgenin (Figure 5.118).

Some steroid drugs are produced by total synthesis, but, in general, the ability of microorganisms to biotransform steroid substrates has proved invaluable in exploiting inexpensive natural steroids as sources of drug materials.

It is now possible via microbial fermentation to hydroxylate the steroid nucleus at virtually any position and with defined stereochemistry. These processes are, in general, more expensive than chemical transformations, and are only used commercially when some significant advantage is achieved, e.g. replacement of several chemical steps. For example, the therapeutic properties of cortisone and hydrocortisone can be further improved by the microbial introduction of a 1,2-double bond, giving **prednisone** and **prednisolone** respectively (Figure 5.121). These agents surpass the parent hormones in antirheumatic and antiallergic activity with fewer side-effects. As with cortisone, prednisone is converted in the body by reduction into the active agent, in this case prednisolone.

Box 5.20

Corticosteroid Drugs

Glucocorticoids are primarily used for their antirheumatic and anti-inflammatory activities. They give valuable relief to sufferers of rheumatoid arthritis and osteoarthritis, and find considerable use for the treatment of inflammatory conditions by suppressing the characteristic development of swelling, redness, heat, and tenderness. One of the key ways in which they exert their action is by interfering with prostaglandin biosynthesis, via production of a peptide that inhibits the phospholipase enzyme responsible for release of arachidonic acid from phospholipids (see page 62). However, these agents merely suppress symptoms; they do not provide a cure for the disease. Long-term usage may result in serious side-effects, including adrenal suppression, osteoporosis, ulcers, fluid retention, and increased susceptibility to infections. Because of these problems, steroid drugs are rarely the first choice for inflammatory treatment, and other therapies are usually tried first. Nevertheless, corticosteroids are widely used for inflammatory conditions affecting the ears, eyes, and skin, and in the treatment of burns. Some have valuable antiallergic properties, helping in reducing the effects of hay fever and asthma. In some disease states, e.g. Addison's disease, the adrenal cortex is no longer able to produce these hormones, and replacement therapy becomes necessary. The most common genetic deficiency is lack of the 21-hydroxylase enzyme in the biosynthetic pathway, necessary for both hydrocortisone and aldosterone biosynthesis (Figure 5.112). This can then lead to increased synthesis of androgens (see Figure 5.141).

Mineralocorticoids are primarily of value in maintaining electrolyte balance where there is adrenal insufficiency.

Natural corticosteroid drugs **cortisone** (as **cortisone acetate**) and **hydrocortisone** (**cortisol**) (Figure 5.113) are valuable in replacement therapies, and hydrocortisone is one of the most widely used agents for topical application in the treatment of inflammatory skin conditions. The early use of the natural corticosteroids for anti-inflammatory activity tended to show up some serious side-effects on water, mineral, carbohydrate, protein, and fat metabolism. In particular, the mineralocorticoid activity is usually considered an undesirable effect. In an effort to optimize anti-inflammatory activity, many thousands of chemical

Box 5.20   (continued)

modifications to the basic structure were tried. Three structural changes proved particularly valuable. Introduction of a Δ^1 double bond modifies the shape of ring A and was found to increase glucocorticoid over mineralocorticoid activity, e.g. **prednisone** and **prednisolone** (Figure 5.121), which are about four times more potent than cortisone/hydrocortisone as anti-inflammatory agents. A 9α-fluoro substituent increased all activities, whereas 16α- or 16β-methyl groups reduced the mineralocorticoid activity without affecting the glucocorticoid activity.

The discovery that 9α-fluoro analogues had increased activity arose indirectly from attempts to epimerize 11α-hydroxy compounds into the active 11β-hydroxy derivatives (Figure 5.122). Thus, when an 11α-tosylate ester was treated with acetate, a base-catalysed elimination was observed rather than the hoped-for substitution, which is hindered by the axial methyl groups. This *syn* elimination suggests that an E1 mechanism is involved. The same $\Delta^{9(11)}$-ene can also be obtained by dehydration of the 11β-alcohol using thionyl chloride. The alkene was then treated with aqueous bromine to generate an 11-hydroxy compound. Addition of HOBr to the 9(11)-double bond proceeds via electrophilic attack from the less-hindered α-face, giving the cyclic bromonium ion, and then ring opening involves β-attack of hydroxide at C-11. Attack at C-9 is sterically hindered by the methyl at C-10. The 9α-bromocortisol (as its 21-acetate) produced in this way was less active as an anti-inflammatory agent than cortisol 21-acetate by a factor of three, and 9α-iodocortisol acetate was also less active by a factor of 10. For fluoro compounds, fluorine has to be introduced indirectly; this was achieved from the β-epoxide formed by base treatment of the 9α-bromo-11β-hydroxy analogue (Figure 5.122). The derivative 9α-fluorocortisol 21-acetate (*fluorohydrocortisone* acetate; **fludrocortisone acetate**; Figure 5.123) was found to be about 11 times more active than cortisol acetate. However, its mineralocorticoid activity was also increased some 300-fold, so that its potential anti-inflammatory activity has no clinical relevance, and this drug is only employed for its mineralocorticoid activity; in practice, it is the only mineralocorticoid agent routinely used. The introduction of a 9α-fluoro substituent into prednisolone causes powerful Na$^+$ retention. These effects can be reduced (though usually not eliminated entirely) by introducing a substituent at C-16, either a 16α-hydroxy or a 16α/16β-methyl.

The 16α-hydroxyl can be introduced microbiologically, e.g. as in the conversion of 9α-fluoroprednisolone into **triamcinolone** (Figure 5.124). The ketal formed from triamcinolone and acetone, **triamcinolone acetonide** (Figure 5.124), provides the most satisfactory means of administering this anti-inflammatory. **Methylprednisolone** (Figure 5.123) is a 6α-methyl derivative of prednisolone showing a modest increase in activity over the parent compound. A 6-methyl group can be supplied via reaction

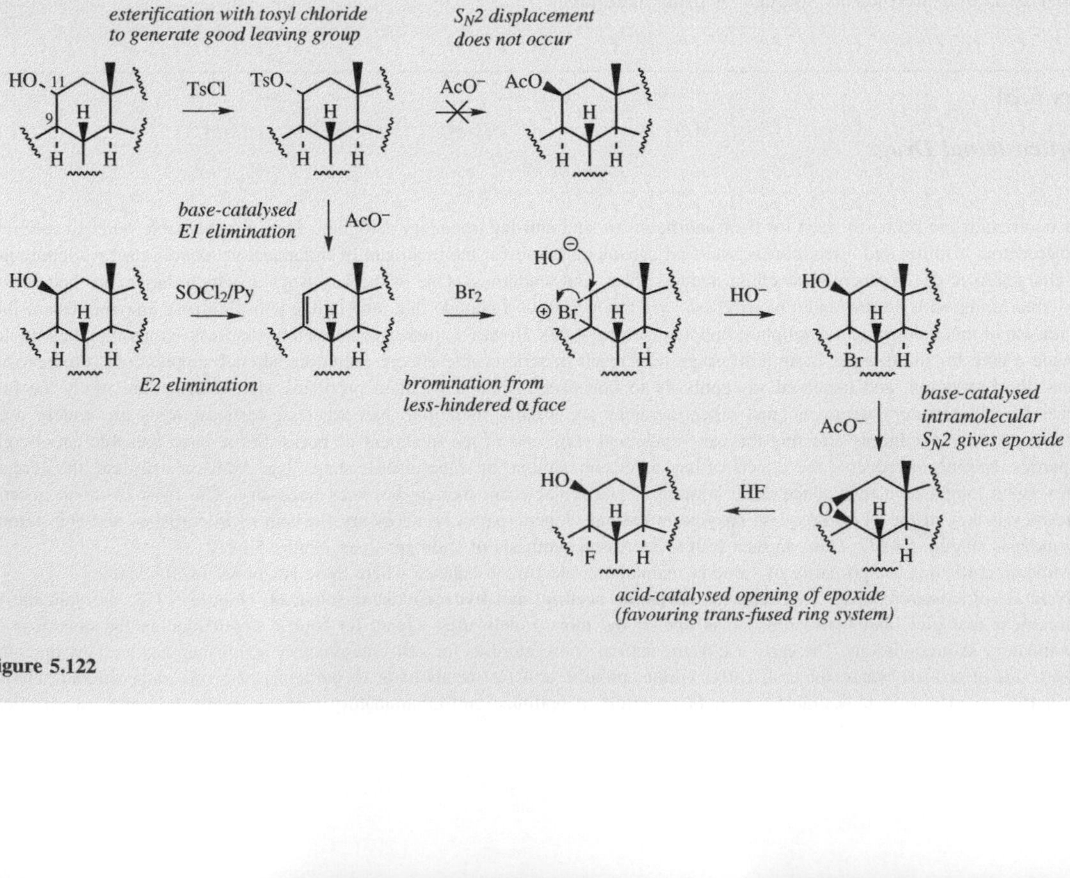

Figure 5.122

Box 5.20 (continued)

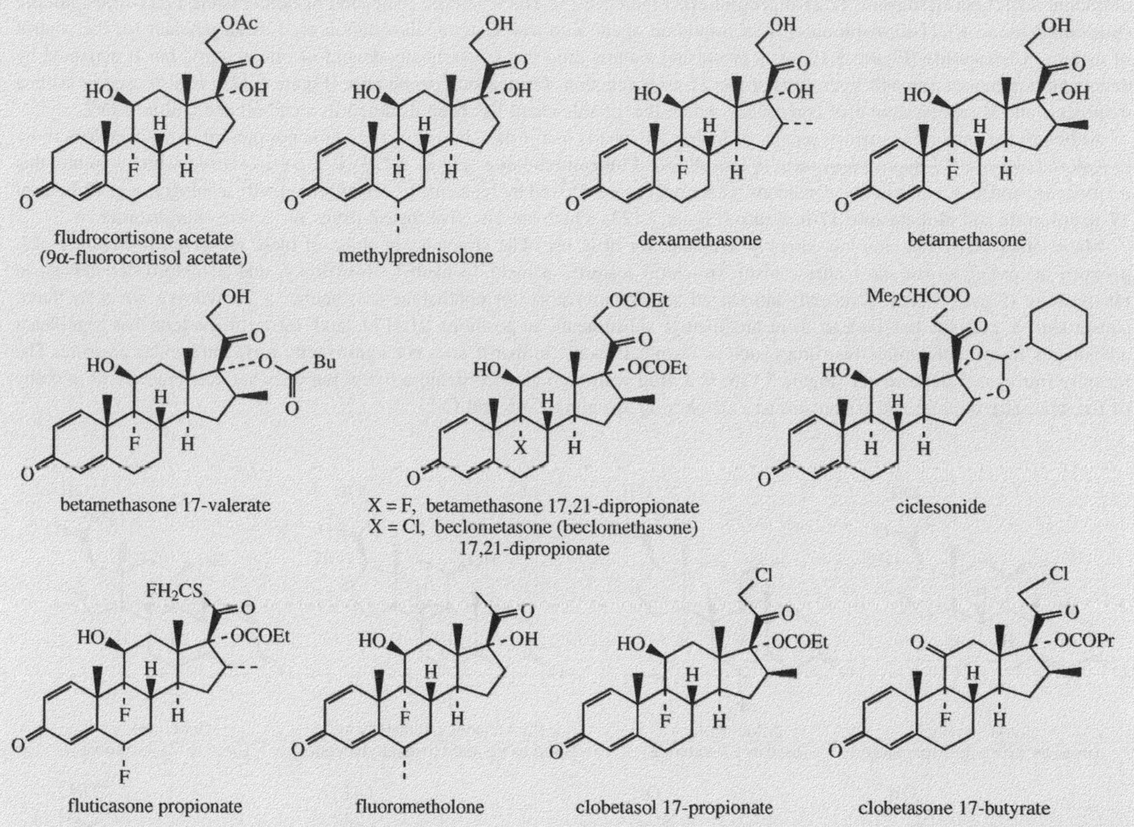

fludrocortisone acetate
(9α-fluorocortisol acetate)

methylprednisolone

dexamethasone

betamethasone

betamethasone 17-valerate

X = F, betamethasone 17,21-dipropionate
X = Cl, beclometasone (beclomethasone)
17,21-dipropionate

ciclesonide

fluticasone propionate

fluorometholone

clobetasol 17-propionate

clobetasone 17-butyrate

Figure 5.123

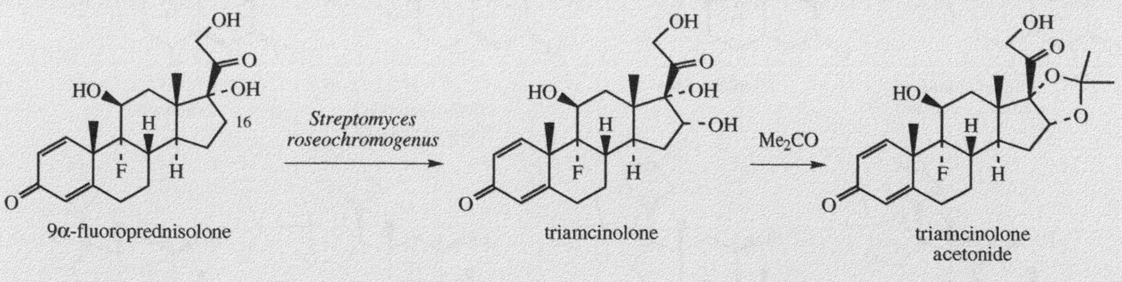

9α-fluoroprednisolone

Streptomyces roseochromogenus

triamcinolone

Me₂CO

triamcinolone
acetonide

Figure 5.124

of the Grignard reagent MeMgBr with a suitable 5,6-epoxide derivative. **Dexamethasone** and **betamethasone** (Figure 5.123) exemplify respectively 16α- and 16β-methyl derivatives in drugs with little, if any, mineralocorticoid activity. The 16-methyl group is easily introduced by a Grignard reaction with an appropriate α,β-unsaturated Δ¹⁶-20-ketone. Betamethasone, for topical application, is typically formulated as a C-17 ester with valeric acid (**betamethasone 17-valerate**), or as the 17,21-diester with

Box 5.20   (continued)

propionic acid (**betamethasone 17,21-dipropionate**; Figure 5.123). The 9α-chloro compound **beclometasone 17,21-dipropionate** (**beclomethasone 17,21-dipropionate**) is an important agent with low systemic distribution used as an inhalant for the control of asthma. **Ciclesonide** (Figure 5.123) is a promising asthma drug that is essentially devoid of oral activity, but is activated by endogenous esterases through hydrolysis of the 21-ester function. **Fluticasone propionate** (Figure 5.123) is also used in asthma treatment, and is representative of compounds where the 17-side-chain has been modified to a carbothiate (sulfur ester).

Although the anti-inflammatory activity of hydrocortisone is lost if the 21-hydroxyl group is not present, considerable activity is restored when a 9α-fluoro substituent is introduced. **Fluorometholone** (Figure 5.123) is a corticosteroid which exploits this relationship and is of value in eye conditions. Other agents are derived by replacing the 21-hydroxyl with a halogen, e.g. **clobetasol 17-propionate** and **clobetasone 17-butyrate** (Figure 5.123) which are effective topical drugs for severe skin disorders.

Many other corticosteroids are currently available for drug use. The structures of some of these are given in Figure 5.125, grouped according to the most characteristic structural features, namely 16-methyl, 16-hydroxy, and 21-chloro derivatives. In **rixemolone** (Figure 5.126), a recently introduced anti-inflammatory for ophthalmic use, neither a 21-hydroxy nor a 9α-fluoro substituent is present, but instead there are methyl substituents at positions 21, 17α, and 16α. Rimexolone has significant advantages in eye conditions over drugs such as dexamethasone, in that it does not significantly raise intraocular pressure. The recently introduced **deflazacort** (Figure 5.126) is a drug with high glucocorticoid activity, but does not conveniently fit into any of the general groups in that it contains an oxazole ring spanning C-16 and C-17.

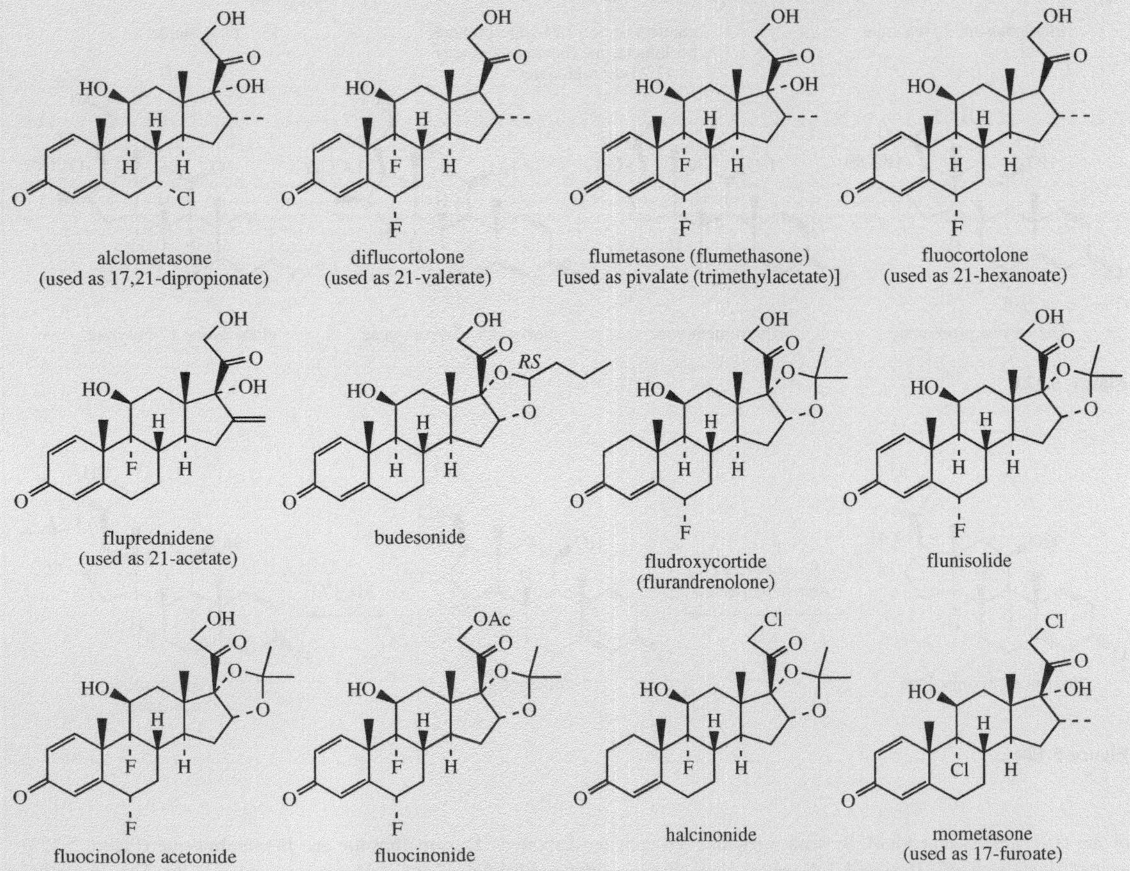

alclometasone
(used as 17,21-dipropionate)

diflucortolone
(used as 21-valerate)

flumetasone (flumethasone)
[used as pivalate (trimethylacetate)]

fluocortolone
(used as 21-hexanoate)

fluprednidene
(used as 21-acetate)

budesonide

fludroxycortide
(flurandrenolone)

flunisolide

fluocinolone acetonide

fluocinonide

halcinonide

mometasone
(used as 17-furoate)

Figure 5.125

Box 5.20 (continued)

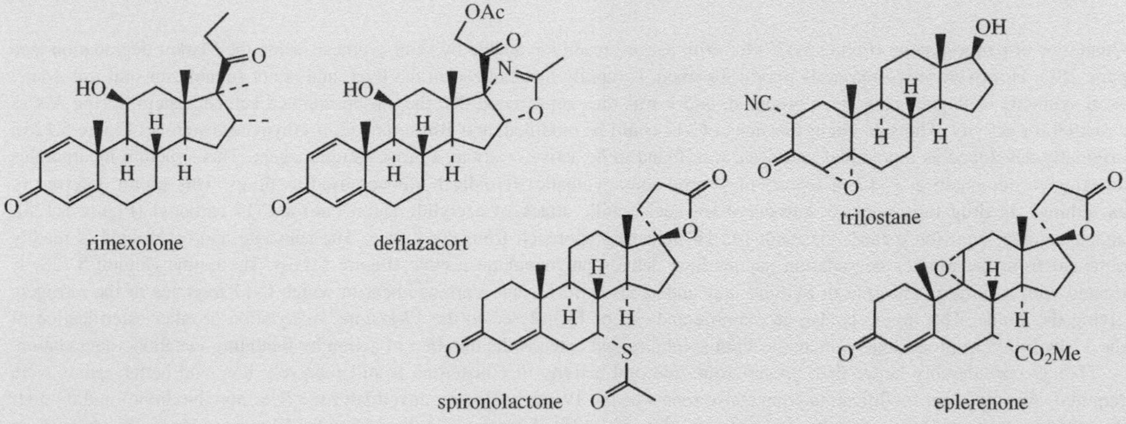

rimexolone

deflazacort

trilostane

spironolactone

eplerenone

Figure 5.126

Trilostane (Figure 5.126) is an adrenocortical suppressant which inhibits synthesis of glucocorticoids and mineralocorticoids and has value in treating Cushing's syndrome, a condition characterized by a moon-shaped face and caused by excessive glucocorticoids. This drug is an inhibitor of the dehydrogenase–isomerase which transforms pregnenolone into progesterone (Figure 5.112).

Spironolactone (Figure 5.126) is an antagonist of the endogenous mineralocorticoid aldosterone and inhibits the sodium-retaining action of aldosterone whilst also decreasing the potassium-secreting effect. Classified as a potassium-sparing diuretic, it is employed in combination with other diuretic drugs to prevent excessive potassium loss. Low doses of spironolactone are beneficial in severe heart failure. Progesterone (page 288) is also an aldosterone antagonist; the spironolactone structure differs from progesterone in its 7α-thioester substituent, and replacement of the 17β side-chain with a 17α-spirolactone. **Eplerenone** (Figure 5.126), a 9,10-epoxy spironolactone analogue, is a newer aldosterone antagonist that has fewer side-effects than spironolactone due to more selective binding to the mineralocorticoid receptor. It is used for the treatment of hypertension and heart failure.

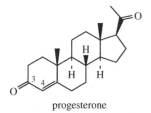

progesterone

Characteristic features of progestogens:

- C_{21} pregnane skeleton
- Δ^4-3-keto

Figure 5.127

Progestogens

Progestogens (progestins; gestogens) are female sex hormones, concerned with preparing the uterus for pregnancy, and then maintaining the necessary conditions [Box 5.21]. There is only one naturally occurring progestational steroid and that is **progesterone** (Figure 5.127), which is secreted by the corpus luteum following release of an ovum. Progesterone is also an intermediate in the biosynthesis of the corticosteroids, e.g. hydrocortisone and aldosterone (see page 279), and its derivation from cholesterol via pregnenolone has also been seen in the formation of cardioactive glycosides (see page 267).

Box 5.21

Progestogen Drugs

Quantities of **progesterone** (Figure 5.127) for drug use are readily available by semi-synthesis using the Marker degradation (see page 281). However, progesterone is poorly absorbed, is rapidly metabolized in the liver, and is not suitable for oral use. Many semi-synthetic analogues have been produced, and it was thus appreciated that the α,β-unsaturated ketone system in ring A was essential for activity. The side-chain function at C-17 could be modified, and ethisterone (17α-ethynyltestosterone; Figure 5.128), originally developed as a potential androgen, was found to be active orally as a progestational agent. This structure incorporates an ethynyl side-chain at C-17, a feature of several semi-synthetic steroidal hormones used as drugs. This group, referred to as 'ethinyl' in drug molecules, is introduced by nucleophilic attack of acetylide anion onto a C-17 carbonyl (Figure 5.128), attack coming from the α-face, the methyl C-18 hindering approach from the β-face. The substrate androstenolone is readily obtained from the Marker degradation intermediate dehydropregnenolone acetate (Figure 5.118). The oxime (Figure 5.128) is treated with a sulfonyl chloride in pyridine and undergoes a Beckmann rearrangement in which C-17 migrates to the nitrogen, giving the amide. This amide is also an enamine and can be hydrolysed to the 17-ketone. Acetylation or other esterification of the 17-hydroxyl in progestogens increases lipid solubility and extends the duration of action by inhibiting metabolic degradation.

Though considerably better than progesterone, the oral activity of ethisterone is still relatively low, and better agents were required. An important modification from ethisterone was the 19-nor analogue, **norethisterone** (US: norethindrone) and its ester **norethisterone acetate** (US: norethindrone acetate) (Figure 5.129). Attention was directed to the 19-norsteroids by the observation that 19-nor-14β,17α-progesterone (Figure 5.129), obtained by degradation of the cardioactive glycoside strophanthidin (see page 273), displayed eight times higher progestational activity than progesterone, despite lacking the methyl C-19, and having the unnatural configurations at the two centres C-14 (C/D rings *cis*-fused) and C-17. Norethisterone can be synthesized from the oestrogen estrone (see page 291), which already lacks the C-9 methyl, or from androstenolone (Figure 5.128) by a sequence which allows oxidation of C-19 to a carboxyl, which is readily lost by decarboxylation when adjacent to the α,β-unsaturated ketone system.

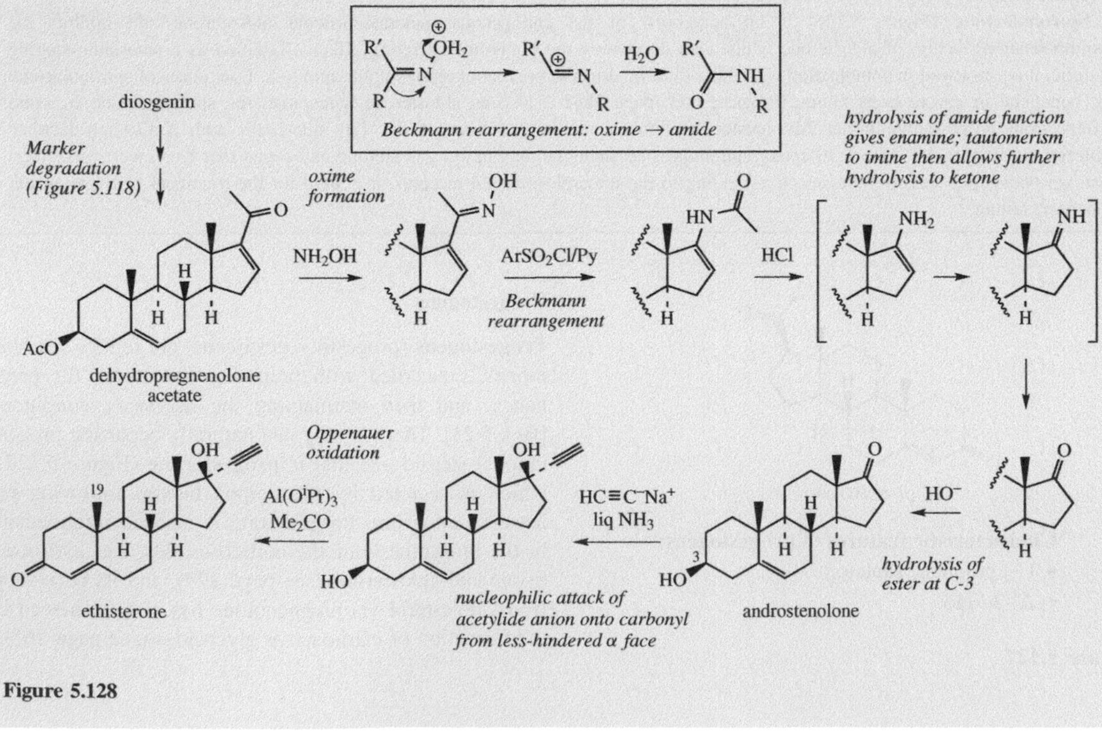

Figure 5.128

Box 5.21 (continued)

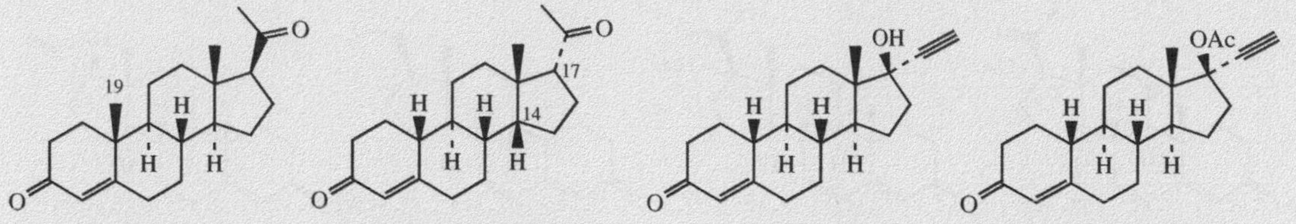

| progesterone | 19-nor-14β,17α-progesterone | norethisterone (US: norethindrone) | norethisterone acetate (US: norethindrone acetate) |

Figure 5.129

Although ethisterone and norethisterone are structurally C_{21} pregnane derivatives, they may also be regarded as 17-ethynyl derivatives of testosterone (see page 296), the male sex hormone, and 19-nortestosterone respectively. It is thus convenient to classify the progestogens as either progesterone derivatives or 17-ethynyl-testosterone derivatives, with the latter group then being subdivided into estranes or 13-ethylgonanes. Structures of some currently available progestogen drugs are shown in Figure 5.130. Semi-synthetic progesterone structures, still containing the 17-acetyl side-chain, tend to be derivatives of 17α-hydroxyprogesterone, a biosynthetic intermediate on the way to hydrocortisone (Figure 5.112), which also has progesterone-like activity. Relatively few examples of this class remain in use. **Medroxyprogesterone acetate** contains an additional 6α-methyl, introduced to block potential deactivation by metabolic hydroxylation, and is 100–300 times as potent as ethisterone on oral administration. **Megestrol acetate** contains a 6-methyl group and an additional Δ^6 double bond, whilst **dydrogesterone** unusually has different stereochemistries at C-9 and C-10. Norethisterone and norethisterone acetate (Figure 5.129), along with **etynodiol (ethynodiol) diacetate** (Figure 5.130), are 17-ethynylestranes. **Norgestrel** (Figure 5.130) is representative of 17-ethynyl progestogens with an ethyl group replacing the 13-methyl. Although these can be obtained by semi-synthesis from natural 13-methyl compounds, norgestrel is produced by total synthesis as the racemic compound. Since only the laevorotatory enantiomer which has the natural configuration is biologically active, this enantiomer, **levonorgestrel**, is now replacing the racemic form for drug use. In **desogestrel**, further features are the modification of an 11-oxo function to an 11-methylene, and removal of the 3-ketone. **Norgestimate** and **norelgestromin** are characterized by conversion of the 3-keto into an oxime group.

During pregnancy, the corpus luteum continues to secrete progesterone for the first 3 months, after which the placenta becomes the supplier of both progesterone and oestrogen. Progesterone prevents further ovulation and relaxes the uterus to prevent the fertilized egg being dislodged. In the absence of pregnancy, a decline in progesterone levels results in shedding of the uterine endometrium and menstruation. Progestogens are useful in many menstrual disorders and as **oral contraceptives**, either alone at low dosage (progestogen-only contraceptives, e.g. norethisterone, levonorgestrel) or in combination with oestrogens (combined oral contraceptives, e.g. ethinylestradiol + norethisterone, ethinylestradiol + levonorgestrel). The combined oestrogen–progestogen preparation inhibits ovulation, but normal menstruation occurs when the drug is withdrawn for several days each month. The low-dosage progestogen-only pill appears to interfere with the endometrial lining to inhibit fertilized egg implantation and thickens cervical mucus, making a barrier to sperm movement. The progestogen-only formulation is less likely to cause thrombosis, a serious side-effect sometimes experienced from the use of oral contraceptives. There appears to be a slightly higher risk of thrombosis in patients using the so-called 'third-generation' oral contraceptive pills containing the newer progestogens **desogestrel** and **gestodene**. Current oral contraceptives have a much lower hormone content than the early formulations of the 1960s and 1970s, typically about 10% of the progestogen and 50% of the oestrogen content. Deep muscular injections of medroxyprogesterone or norethisterone esters and implants of **etonogestrel** can be administered to provide long-acting contraception. A high dose of **levonorgestrel** is the drug of choice for emergency contraception after unprotected intercourse, i.e. the 'morning-after' pill. Hormone replacement therapy (HRT) in non-hysterectomized women also involves progestogen–oestrogen combinations (see page 294), whilst progestogens such as norethisterone, megestrol acetate, and medroxyprogesterone acetate also find limited application in the treatment of breast cancers.

The structure of **drospirenone** (Figure 5.130) is quite unlike other progestogens, but is based on that of the aldosterone antagonist spironolactone (Figure 5.126, page 287). This new agent is a progestogen with antimineralocorticoid activity. The oestrogen content in a combined contraceptive pill tends to increases aldosterone levels, causing side-effects such as sodium and water retention, leading to swelling and increased blood pressure; these side-effects are thus reduced. Drospirenone also displays antiandrogenic activity, and this may be of value in patients who suffer effects such as androgen-related skin disorders.

Box 5.21   (continued)

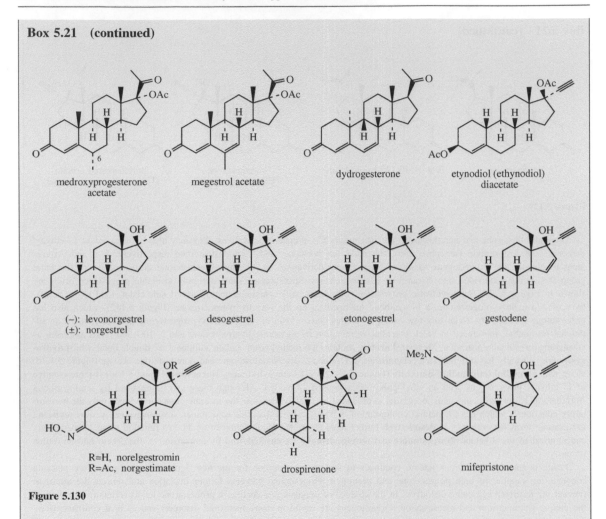

medroxyprogesterone acetate

megestrol acetate

dydrogesterone

etynodiol (ethynodiol) diacetate

(–): levonorgestrel
(±): norgestrel

desogestrel

etonogestrel

gestodene

R=H, norelgestromin
R=Ac, norgestimate

drospirenone

mifepristone

Figure 5.130

Mifepristone (Figure 5.130) is a progestogen antagonist used orally as an abortifacient to terminate pregnancy. This drug has a higher affinity for the progesterone receptor than does the natural hormone and prevents normal responses. This leads to loss of integrity of the uterine endometrial lining and detachment of the implanted fertilized egg.

Oestrogens

The **oestrogens** (US spelling: **estrogens**) are female sex hormones produced in the ovaries, and also in the placenta during pregnancy. They are responsible for the female sex characteristics and, together with progesterone, control the menstrual cycle [Box 5.22]. Oestrogens were first isolated from the urine of pregnant women, in which levels increase some 50-fold during the pregnancy. In horses, levels rise by as much as 500 times during pregnancy. Oestrogens occur both in free form and as glucuronides and sulfates at position 3; they are not restricted to females, since small amounts are produced in the male testis.

The principal and most potent example is **estradiol** (also **oestradiol**, but US spelling has been generally adopted), though only low levels are found in urine, and larger amounts of the less-active metabolites **estrone** (**oestrone**) and the 16α-hydroxylated derivative **estriol** (**oestriol**) are present (Figure 5.131). Estrone has also been found in significant quantities in some plant seeds, e.g. pomegranate and date palm. These compounds have an aromatic A ring, a consequence of which is that C-19, the methyl on C-10, is absent. There is no carbon side-chain at C-17, and the basic C_{18} skeleton is termed estrane.

The biosynthetic pathway to estradiol and estrone (Figure 5.132) proceeds from cholesterol via

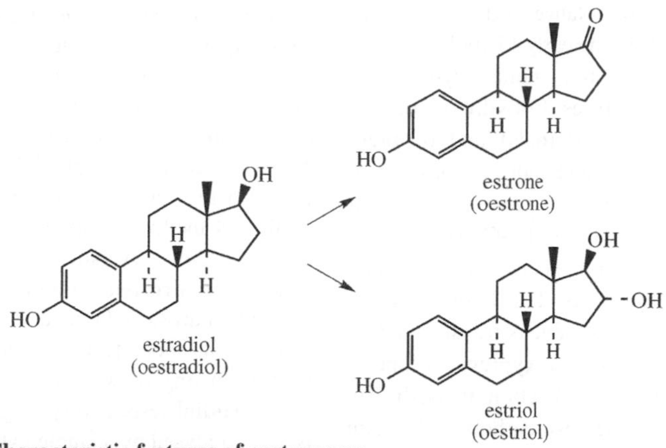

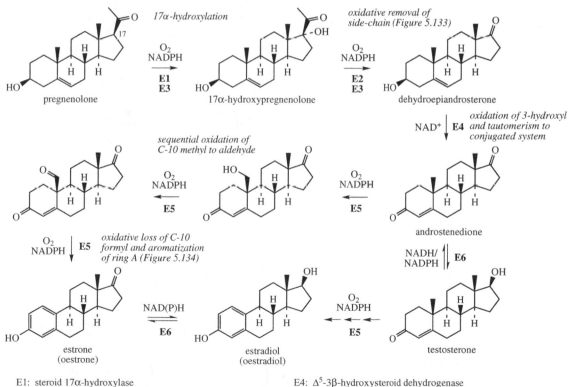

estrone
(oestrone)

estradiol
(oestradiol)

estriol
(oestriol)

Characteristic features of oestrogens:

- C_{18} estrane skeleton
- aromatic A ring (consequently no methyl at C-10)
- no side-chain

Figure 5.131

pregnenolone

17α-hydroxylation

O_2
NADPH

E1
E3

17α-hydroxypregnenolone

*oxidative removal of
side-chain (Figure 5.133)*

O_2
NADPH

E2
E3

dehydroepiandrosterone

NAD^+ **E4** *oxidation of 3-hydroxyl
and tautomerism to
conjugated system*

*sequential oxidation of
C-10 methyl to aldehyde*

O_2
NADPH

E5

O_2
NADPH

E5

androstenedione

O_2
NADPH **E5** *oxidative loss of C-10
formyl and aromatization
of ring A (Figure 5.134)*

NADH/
NADPH **E6**

estrone
(oestrone)

NAD(P)H

E6

estradiol
(oestradiol)

O_2
NADPH

E5

testosterone

E1: steroid 17α-hydroxylase
E2: steroid C-17/C-20 lyase
E3: steroid 17α-hydroxylase-17,20-lyase (CPY17) (bifunctional)

E4: Δ^5-3β-hydroxysteroid dehydrogenase
   /3-oxosteroid Δ^5-Δ^4-isomerase
E5: cytochrome P-450 aromatase (P450arom; CPY19)
E6: 17β-hydroxysteroid dehydrogenase

Figure 5.132

pregnenolone and bears a resemblance to the hydrocortisone pathway (Figure 5.112) in the early 17-hydroxylation step. Indeed, in humans, the same cytochrome P-450-dependent enzyme catalyses 17-hydroxylation of progesterone (leading to corticosteroids) and of pregnenolone (leading to oestrogens and androgens). Further, in the presence of cytochrome b_5, it catalyses the next step in oestrogen biosynthesis, the C-17–C-20 cleavage. Overall, this enzyme plays a significant role in controlling the direction of steroid synthesis. For hydrocortisone biosynthesis, 17α-hydroxyprogesterone is transformed by 21-hydroxylation, whereas 17α-hydroxypregnenolone is oxidized in the α-hydroxyketone function in oestrogen biosynthesis, cleaving off the two-carbon side-chain as acetic acid. The product is the 17-ketone **dehydroepiandrosterone**, which is the most abundant steroid in the blood of young adult humans, with levels peaking

at about 20 years of age, then declining as the person ages. Apart from its role as a precursor of hormones, it presumably has other physiological functions, though these still remain to be clarified (see page 297). A mechanism for the side-chain cleavage reaction, initiated by attack of an enzyme-linked peroxide, is shown in Figure 5.133 and is analogous to that proposed for loss of the 14-methyl group during cholesterol biosynthesis (see page 249). Oxidation and tautomerism in rings A/B then give **androstenedione** (Figure 5.133).

Both androstenedione and its reduction product the male sex hormone **testosterone** are substrates for aromatization in ring A, with loss of C-19, leading to **estrone** and **estradiol** respectively (Figure 5.132). This sequence is also catalysed by a single cytochrome P-450-dependent enzyme, called **aromatase**, and the reaction proceeds via sequential oxidation of the methyl, with its final elimination as formic acid (Figure 5.134). The mechanism

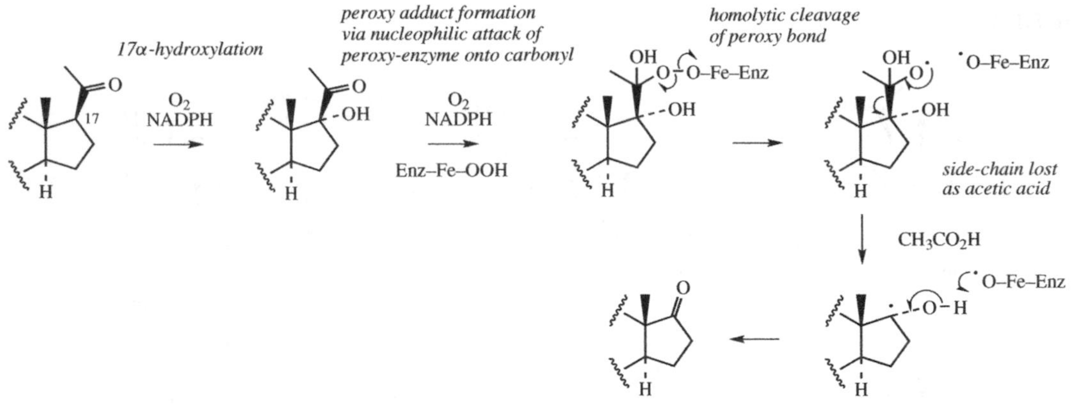

Figure 5.133

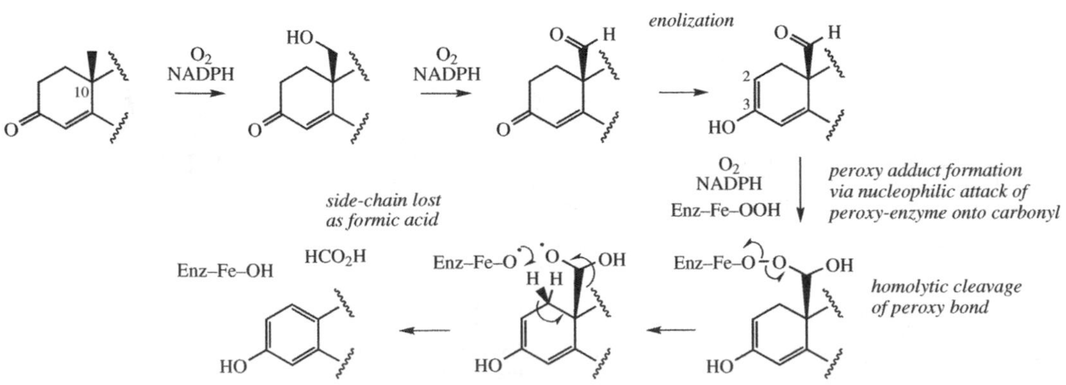

Figure 5.134

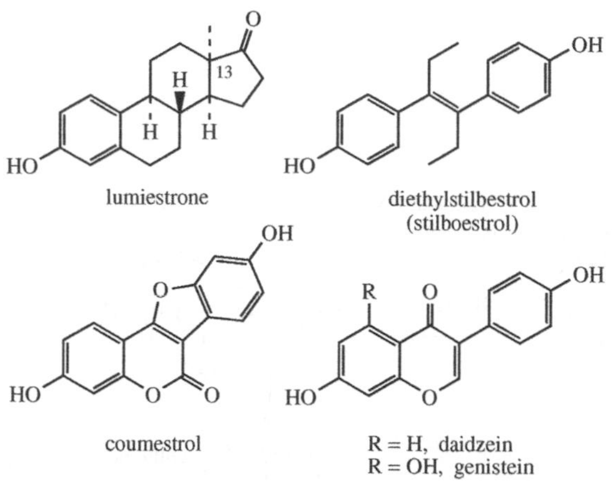

lumiestrone

diethylstilbestrol
(stilboestrol)

coumestrol

R = H, daidzein
R = OH, genistein

Figure 5.135

suggested is analogous to that of the side-chain cleavage reaction. The 2,3-enolization is also enzyme-catalysed and is a prerequisite for aromatization. As with other steroid hormones, the exact order of some of the steps, including formation of the Δ^4-3-keto function, 17-hydroxylation, reduction of the 17-keto, and aromatization in ring A, can vary according to the organism or the site of

synthesis in the body. Since many breast tumours require oestrogens for growth, the design of **aromatase inhibitors** has become an important target for anticancer drug research [Box 5.22].

The aromatic ring makes the oestrogen molecule almost planar (see page 247) and is essential for activity. Changes which remove the aromaticity, e.g. partial reduction, or alter stereochemistry, give analogues with reduced or no activity. Thus, exposure of estrone to UV light leads to inversion of configuration at C-13 adjacent to the carbonyl function and, consequently, to formation of a *cis*-fused C/D ring system. The product, **lumiestrone** (Figure 5.135), is no longer biologically active. Some planar non-steroidal structures can also demonstrate oestrogenic activity as a result of a similar shape and relative spacing of oxygen functions. Thus, the synthetic **diethylstilbestrol** (**stilboestrol**; Figure 5.136) has been widely used as an oestrogen drug, and **coumestrol, daidzein**, and **genistein** (Figure 5.136) are naturally occurring isoflavonoids with oestrogenic properties from plants such as alfalfa, clovers, and soya beans and are termed phyto-oestrogens (see page 177). Dietary natural isoflavonoids are believed to give some protection against breast cancers and are also recommended to alleviate the symptoms of menopause.

Box 5.22

Oestrogen Drugs

Oestrogens suppress ovulation, and with progestogens they form the basis of combined oral contraceptives (see page 289) and hormone replacement therapy (HRT). They are also used to supplement natural oestrogen levels where these are insufficient, as in some menstrual disorders, and to suppress androgen formation and, thus, tumour growth of cancers dependent on androgens, e.g. prostate cancers. Oestrogens appear to offer a number of beneficial effects to women, including protection against osteoporosis, heart attacks, and possibly Alzheimer's disease. However, some cancers, e.g. breast and uterine cancers, are dependent on a supply of oestrogen for growth, especially during the early stages, so high oestrogen levels are detrimental.

Steroidal oestrogens for drug use were originally obtained by processing pregnancy urines, but the dramatic increase in demand resulting from the introduction of oral contraceptives required development of semi-synthetic procedures. Androstenolone formed via the Marker degradation of diosgenin plus side-chain removal (Figure 5.128) may be transformed to a dione by catalytic reduction of the Δ^5 double bond and oxidation of the 3-hydroxyl (Figure 5.136). This then allows production of androstadienedione by dibromination and base-catalysed elimination of HBr. Alternatively, it is now possible to achieve the synthesis of androstadienedione in a single step by a microbiological fermentation of either sitosterol obtained from soya beans (see page 255), or of cholesterol obtained in large quantities from the woolfat of sheep or from the spinal cord of cattle (see page 251). These materials lack unsaturation in the side-chain and were not amenable to simple chemical oxidation processes, e.g. as with stigmasterol (see page 282). Their exploitation required the development of suitable biotransformations, and this objective has now achieved through fermentation with *Mycobacterium phlei* (Figure 5.136). The aromatization step to estrone can be carried out in low yields by vapour-phase free-radical-initiated thermolysis, or more recently with considerably better yields using a dissolving-metal reductive thermolysis. In both processes, the methyl at C-10 is lost. This sequence gives estrone, from which **estradiol** (**oestradiol**) may be obtained by reduction of the 17-carbonyl. However, by far the most commonly used medicinal oestrogen is **ethinylestradiol** (**ethinyloestradiol**; Figure 5.136), which is 12 times as effective as estradiol when administered orally. This analogue can be synthesized from estrone by treatment with sodium acetylide in liquid ammonia, which attacks from the less-hindered α-face (compare page 282). The ethynyl substituent prevents oxidation at C-17; metabolism of estradiol leads to

Box 5.22   (continued)

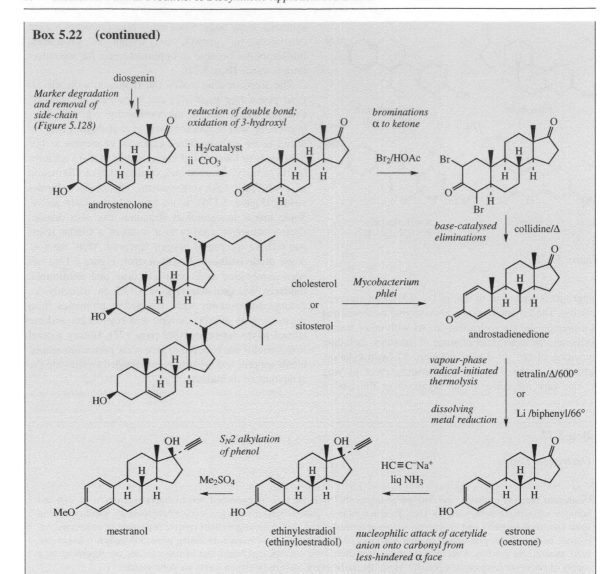

Figure 5.136

the less-active estrone. To retain oestrogenic activity, structural modifications appear effectively limited to the addition of the 17α-ethynyl group and to substitution on the 3-hydroxyl. The phenol group allows synthesis of other derivatives, for example, the 3-methyl ether **mestranol** (Figure 5.136); this acts as a pro-drug, being oxidized in the liver to ethinylestradiol. The ester **estradiol valerate (oestradiol valerate)** facilitates prolonged action through slower absorption and metabolism.

The lower activity metabolites **estriol (oestriol**; about 2% activity of estradiol) and **estrone (oestrone**; about 33% activity) (Figure 5.131) are sometimes used in **hormone replacement therapy (HRT)**. Oestrogen and progesterone levels decline naturally at menopause when the menstrual cycle ceases. The sudden reduction in oestrogen levels can lead to a number of unpleasant symptoms, including tiredness, hot flushes, vaginal dryness, and mood changes. HRT reduces these symptoms and delays other

Box 5.22 (continued)

long-term consequences of reduced oestrogen levels, including osteoporosis and atherosclerosis. HRT currently provides the best therapy for preventing osteoporosis, a common disease in post-menopausal women. Osteoporosis is characterized by a generalized loss of bone mass, leading to increased risk of fracture, and a sharp reduction in endogenous oestrogen levels is recognized as a critical factor. However, it has been discovered that HRT also increases the risk of developing some types of cancer, especially breast, ovarian, and endometrial cancers. Oestrogen and progestogen combinations are used in HRT unless the woman has had a hysterectomy, in which case oestrogen alone is prescribed. Natural oestrogen structures are preferred to the synthetic structures, such as ethinylestradiol or mestranol. Before the availability of plant-derived semi-synthetic oestrogens, extraction of urine from pregnant women and pregnant horses allowed production of oestrogen mixtures for drug use. **Conjugated equine oestrogens** are still widely prescribed for HRT and are obtained by extraction from the urine of pregnant mares and subsequent purification, predominantly in Canada and the USA. Animal welfare groups voice concern over the conditions the animals endure and urge women to reject these drug preparations in favour of plant-derived alternatives. Conjugated equine oestrogens provide a profile of natural oestrogens based principally on estrone and equilin (Figure 5.137). They consist of a mixture of oestrogens in the form of sodium salts of their sulfate esters, comprising mainly estrone (50–60%) and equilin (20–30%), with smaller amounts of 17α- and 17β-dihydroequilin, and 17α-estradiol. The semi-synthetic **estropipate** (Figure 5.137) is also a conjugated oestrogen, the piperazine salt of estrone sulfate.

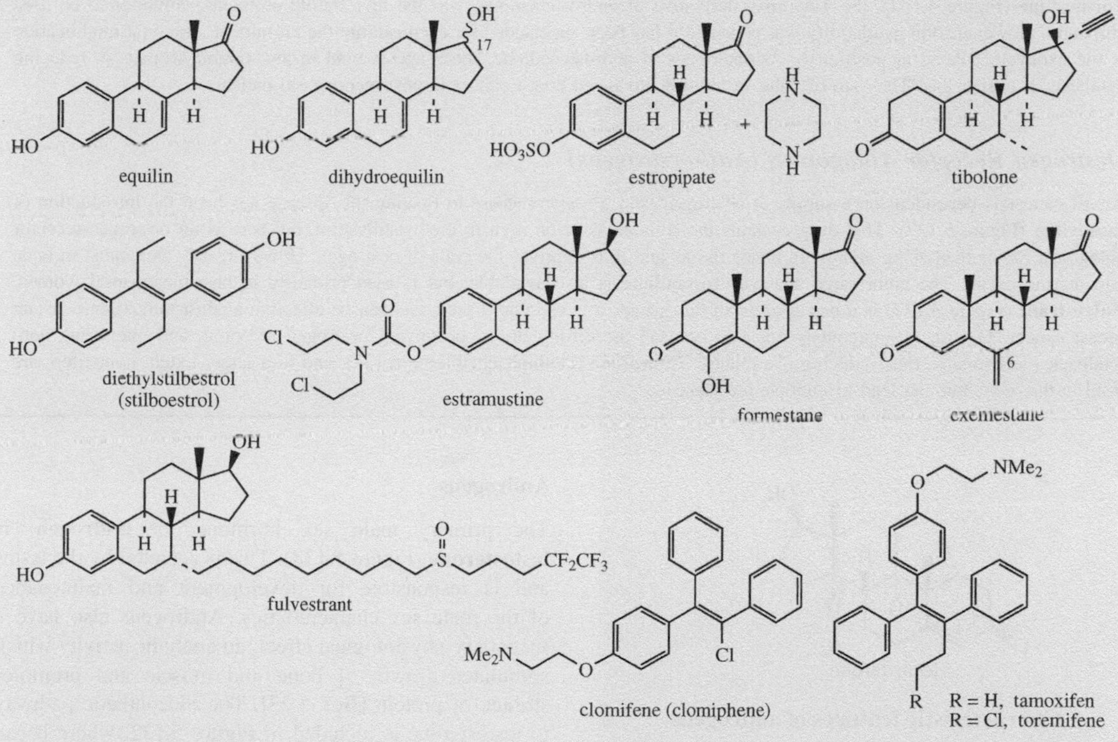

Figure 5.137

Box 5.22   (continued)

Phyto-oestrogens are predominantly isoflavonoid derivatives found in food plants and are used as dietary supplements to provide similar benefits to HRT, especially in countering some of the side-effects of the menopause in women. These compounds are discussed under isoflavonoids (see page 177). **Dioscorea (wild yam)** root or extract (see page 262) is also marketed to treat the symptoms of menopause as an alternative to HRT. Although there is a belief that this increases levels of progesterone, which is then used as a biosynthetic precursor of other hormones, there is little definitive evidence that diosgenin is metabolized in the human body to progesterone.

The structure of **tibolone** (Figure 5.137) appears to resemble that of a progestogen more than it does an oestrogen. Although it does not contain an aromatic A ring, the 5(10)-double bond ensures a degree of planarity. This agent combines both oestrogenic and progestogenic activity, and also has weak androgenic activity; it has been introduced for short-term treatment of symptoms of oestrogen deficiency.

Diethylstilbestrol (stilboestrol; Figure 5.137) is the principal non-steroidal oestrogen drug, but now finds only occasional use, since there are safer and better agents available. It may sometimes be used in treating breast cancer in postmenopausal women.

In **estramustine** (Figure 5.137), estradiol is combined with a cytotoxic alkylating agent of the nitrogen mustard class via a carbamate linkage. This drug has a dual function, a hormonal effect by suppressing androgen (testosterone) formation and an antimitotic effect from the mustine residue. It is used for treating prostate cancers.

Aromatase Inhibitors

Formestane (Figure 5.137), the 4-hydroxy derivative of androstenedione, was the first steroid aromatase inhibitor to be used clinically. However, oral availability was poor and it has been superseded by **exemestane**; the prominent structural modification is the exocyclic alkene at position 6. As analogues of androstenedione, these agents bind to and inhibit aromatase, reducing synthesis of oestrogens. They are of value in treating advanced breast cancer in post-menopausal patients.

Oestrogen Receptor Antagonists (Anti-oestrogens)

Breast cancer is dependent on a supply of oestrogen, and a major success in treating this disease has been the introduction of **tamoxifen** (Figure 5.137). This drug contains the stilbene skeleton seen in diethylstilbestrol, but acts as an oestrogen-receptor antagonist rather than as an agonist in breast tissue, and thus deprives the cells of oestrogen. However, it is an agonist in bone and uterine tissue. The chlorinated analogue **toremifene** is also available, but is used primarily in post-menopausal women. **Fulvestrant** (Figure 5.137) is a newer agent in this group; it is estradiol-based, and can be effective against tamoxifen-resistant breast cancer. Oestrogen antagonists can also be used as fertility drugs, occupying oestrogen receptors and interfering with feedback mechanisms, thus inducing ova release. **Clomifene** (clomiphene; Figure 5.137), and to a lesser extent tamoxifen, are used in this way, but can lead to multiple pregnancies.

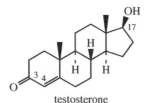

testosterone

Characteristic features of androgens:

- C_{19} androstane skeleton
- no side-chain
- Δ^4-3-keto
- 17β-hydroxyl

Figure 5.138

Androgens

The primary male sex hormone, or **androgen**, is **testosterone** (Figure 5.138). This is secreted by the testes and is responsible for development and maintenance of the male sex characteristics. Androgens also have a secondary physiological effect, an anabolic activity which stimulates growth of bone and muscle and promotes storage of protein [Box 5.23]. The biosynthetic pathway to testosterone is included in Figure 5.132, where it can feature as an intermediate in the pathway to oestrogens. Low levels of testosterone are also synthesized in females in the ovary. Testosterone lacks any side-chain and has a 17β-hydroxyl as in estradiol, but still contains the methyl C-19 and the Δ^4-3-one system in ring A. This C_{19} skeleton is designated androstane.

Box 5.23

Androgen Drugs

Testosterone can be produced from androstenolone (Figure 5.128) by chemical routes, requiring reduction of the 17-carbonyl and oxidation of the 3-hydroxyl, with the use of appropriate protecting groups. A simple high-yielding process (Figure 5.139) exploits microbiological conversion with yeast, in which fermentation first under aerobic conditions oxidizes the 3-hydroxyl and then in the absence of air reduces the 17-keto group.

Testosterone is not active orally, since it is easily metabolized in the liver; it has to be implanted or injected in the form of esters. Transdermal administration from impregnated patches has also proved successful and is now the method of choice for treating male sexual impotence caused by low levels of sex hormones (hypogonadism). Testosterone may also be prescribed for menopausal women as an adjunct to HRT (see page 294) to improve sex drive, and occasionally in the treatment of oestrogen-dependent breast cancer. The ester testosterone undecanoate is orally active, as is **mesterolone** (Figure 5.140), which features introduction of a 1α-methyl group and reduction of the Δ^4 double bond.

The ratio of androgenic to anabolic activity can vary in different molecules. Considerable effort has been put into producing steroids with low androgenic activity but high anabolic activity to use for various metabolic and endocrine disorders. However, it is difficult to remove the androgenic activity completely from anabolic steroids. **Nandrolone** (19-nortestosterone; Figure 5.140) is probably the only example currently in clinical use. Abuse of these materials by athletes wishing to promote muscle development and strength is considerably more frequent. Androgenic activity can affect the sexual characteristics of women, making them more masculine, whilst prolonged use of these drugs can lower fertility in either sex and endanger long-term health by increasing the risks of heart and liver disease or cancer.

The progestogen **cyproterone acetate** (Figure 5.140) is a competitive androgen antagonist or anti-androgen that reduces male libido and fertility, and finds use in the treatment of severe hypersexuality and sexual deviation in the male, as well as in prostate cancer. **Finasteride** and the structurally similar **dutasteride** (Figure 5.140) are also anti-androgens that have value in prostate conditions. These are 4-aza-steroids and specific inhibitors of the 5α-reductase involved in testosterone metabolism. This enzyme reduces the 4,5-double bond and converts testosterone into dihydrotestosterone, which is actually a more potent androgen. High levels of dihydrotestosterone are implicated in prostate cancer and benign prostatic hyperplasia; inhibition of 5α-reductase helps to reduce prostate tissue growth. Finasteride has also been noted to prevent hair loss in men, and is marketed to treat male-pattern baldness. Continuous use for 3–6 months is necessary; unfortunately, the effects are reversed 6–12 months after the treatment is discontinued. **Abiraterone** (Figure 5.140) is an inhibitor of the steroid 17α-hydroxylase-17,20-lyase (CYP17) and thus blocks formation of testosterone (and also oestrogens). This agent is proving very successful in clinical trials for prostate cancer treatment. Structurally, this is a side-chain modification of pregnenolone, the normal substrate for the enzyme.

Dehydroepiandrosterone (DHEA; Figure 5.132) is a precursor of androgens and oestrogens; it is the most abundant steroid in the blood of young adult humans, levels peaking at about 20 years of age and then declining as the person ages. Whilst this hormone has a number of demonstrated biological activities, its precise physiological functions remain to be clarified. This material has become popular in the hope that it will maintain youthful vigour and health, and counter the normal symptoms of ageing. These claims are as yet unsubstantiated, but taking large amounts of this androgen and oestrogen precursor can lead to side-effects associated with high levels of these hormones, e.g. increased risk of prostate cancer in men or of breast cancer in women, who may also develop acne and facial hair. DHEA is not a precursor of glucocorticoids, mineralocorticoids, or of progestogens.

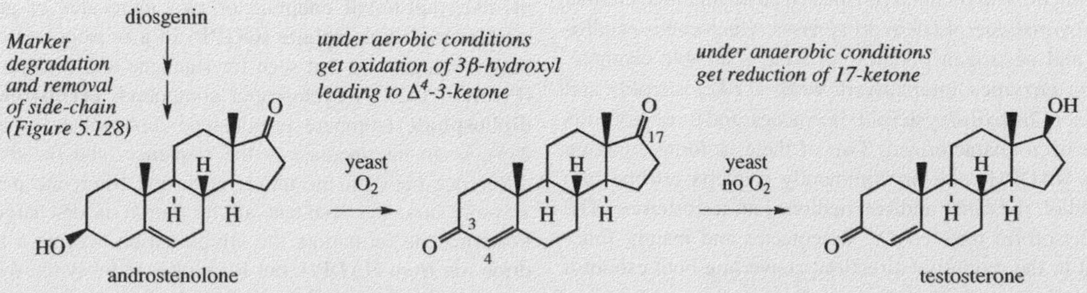

Figure 5.139

Box 5.23   (continued)

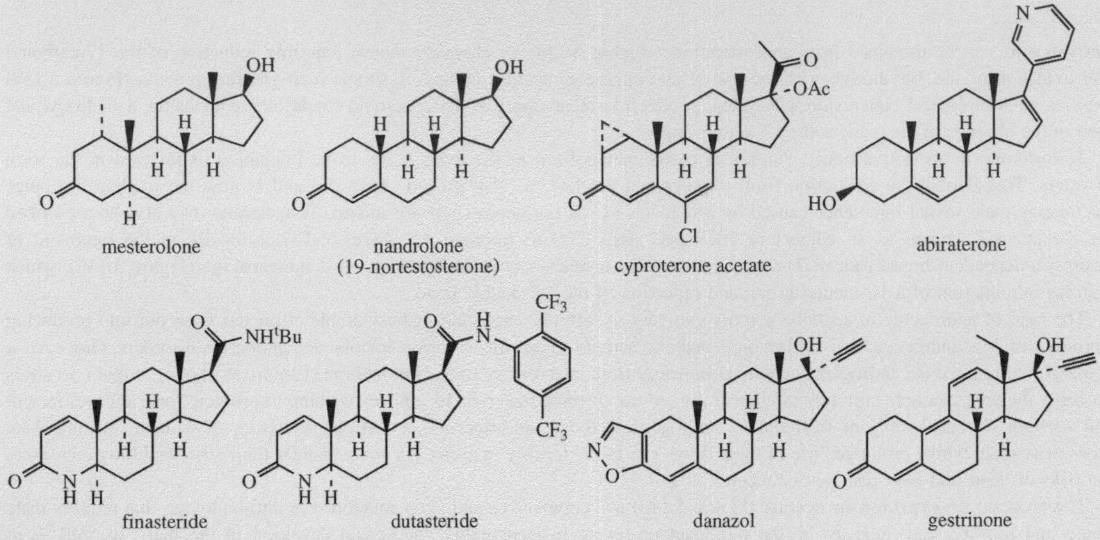

mesterolone

nandrolone
(19-nortestosterone)

cyproterone acetate

abiraterone

finasteride

dutasteride

danazol

gestrinone

Figure 5.140

Danazol and **gestrinone** (Figure 5.140) are inhibitors of pituitary gonadotrophin release, combining weak androgenic activity with anti-oestrogenic and antiprogestogenic activity. These highly modified structures bear one or more of the features we have already noted in discussions of androgens, progestogens, and oestrogens, possibly accounting for their complex activity. These compounds are used particularly to treat endometriosis, where endometrial tissue grows outside the uterus.

It is particularly worthy of note that the routes to corticosteroids, progestogens, oestrogens, and androgens involve common precursors or partial pathways, and some enzymes display multiple functions or perhaps broad specificity. The main relationships are summarized in Figure 5.141. Consequently, these processes need to be under very tight control for a person's normal physiological functions and characteristics to be maintained. Human 17β-hydroxysteroid dehydrogenases, which control androgen and oestrogen potency, provide a suitable example. These enzymes interconvert weak (17-ketosteroid) and potent (17β-hydroxysteroid) hormones, and three isoforms have been characterized. Two of these isoforms function with NADPH; one predominantly converts estrone into estradiol, the other androstenedione into testosterone. The third isoform uses NAD^+ as cofactor and mainly functions in the oxidative direction, converting both estradiol and testosterone into the less-active ketones. According to their activities and tissue localization, these enzyme isoforms help to achieve the required functional balance of the various hormones.

TETRATERPENES (C_{40})

The tetraterpenes are represented by only one group of compounds, the **carotenoids**, though several hundred natural structural variants are known. These compounds play a role in photosynthesis, but they are also found in non-photosynthetic plant tissues, in fungi, and in bacteria. Formation of the tetraterpene skeleton, e.g. **phytoene**, involves tail-to-tail coupling of two molecules of **geranylgeranyl diphosphate (GGPP)** in a sequence essentially analogous to that seen for squalene and triterpenes (Figure 5.142). A cyclopropyl compound, **prephytoene diphosphate** (compare presqualene diphosphate, page 235), is an intermediate in the sequence, and the main difference between the tetraterpene and triterpene pathways is how the resultant allylic cation is discharged. For squalene formation, the allylic cation accepts a hydride ion from NADPH, but for phytoene biosynthesis, a proton is lost, generating a double bond in the centre of the molecule, and thus a short conjugated chain is developed. In plants and fungi, this new double bond has the Z (*cis*) configuration, whilst in bacteria it is E (*trans*). This

Steroid hormone biosynthetic interrelationships

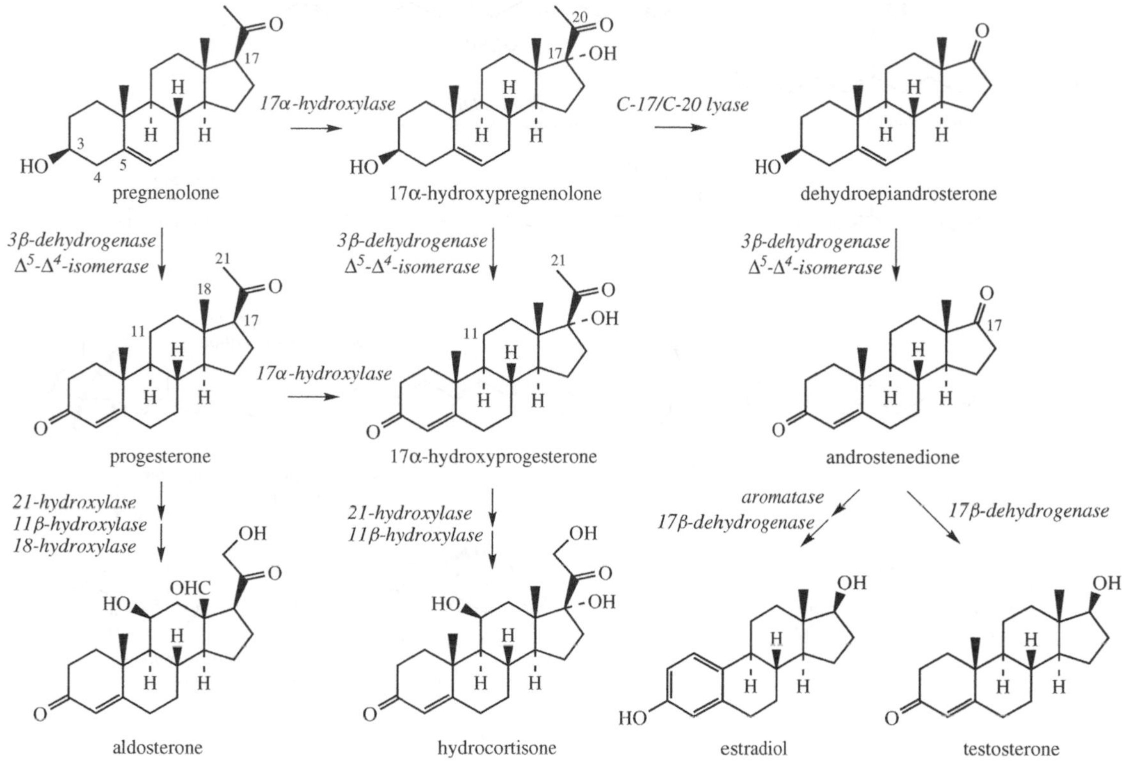

Figure 5.141

triene system prevents the types of cyclization seen with squalene. Conjugation is extended then by a sequence of desaturation reactions, removing pairs of hydrogen atoms alternately from each side of the triene system, giving eventually **lycopene** (Figure 5.142), which, in common with the majority of carotenoids, has the *all-trans* configuration. In bacteria and fungi, a single phytoene desaturase enzyme converts phytoene into lycopene. In plants, two desaturase enzymes and an isomerase are involved, the isomerase being responsible for changing the configurations of those double bonds introduced initially in *cis* form. The central double bond appears to be isomerized as part of the first dehydrogenation step.

The extended π-electron system confers colour to the carotenoids, and accordingly they contribute yellow, orange, and red pigmentations to plant tissues; phytoene is colourless. Lycopene is the characteristic carotenoid pigment in ripe tomato fruit (*Lycopersicon esculente*; Solanaceae). The orange colour of carrots (*Daucus carota*; Umbelliferae/Apiaceae) is caused by β-**carotene** (Figure 5.143), though this compound is widespread in higher plants. β-Carotene and other natural carotenoids (Figure 5.143) are widely employed as colouring agents for foods, drinks, confectionery, and drugs. β-Carotene displays additional cyclization of the chain ends, which can be rationalized by the carbocation mechanism shown in Figure 5.144. The new proton introduced is derived from water; depending on which proton is then lost from the cyclized cation, three different cyclic alkene systems can arise at the end of the chain, described as β-, γ-, or ε-ring systems. The γ-ring system is the least common. α-**Carotene** (Figure 5.143) has a β-ring at one end of the chain and an ε-type at the other, and is representative of carotenoids lacking symmetry. γ-**Carotene** (a precursor of β-carotene) and δ-**carotene** (a precursor of α-carotene) illustrate carotenoids where only one end of the chain has become cyclized. Oxygenated carotenoids (termed xanthophylls) are also widely distributed, and the biosynthetic origins of the oxygenated rings found in some of these, such as **zeaxanthin, lutein**, and **violaxanthin** (Figure 5.143), all common green-leaf carotenoids, are

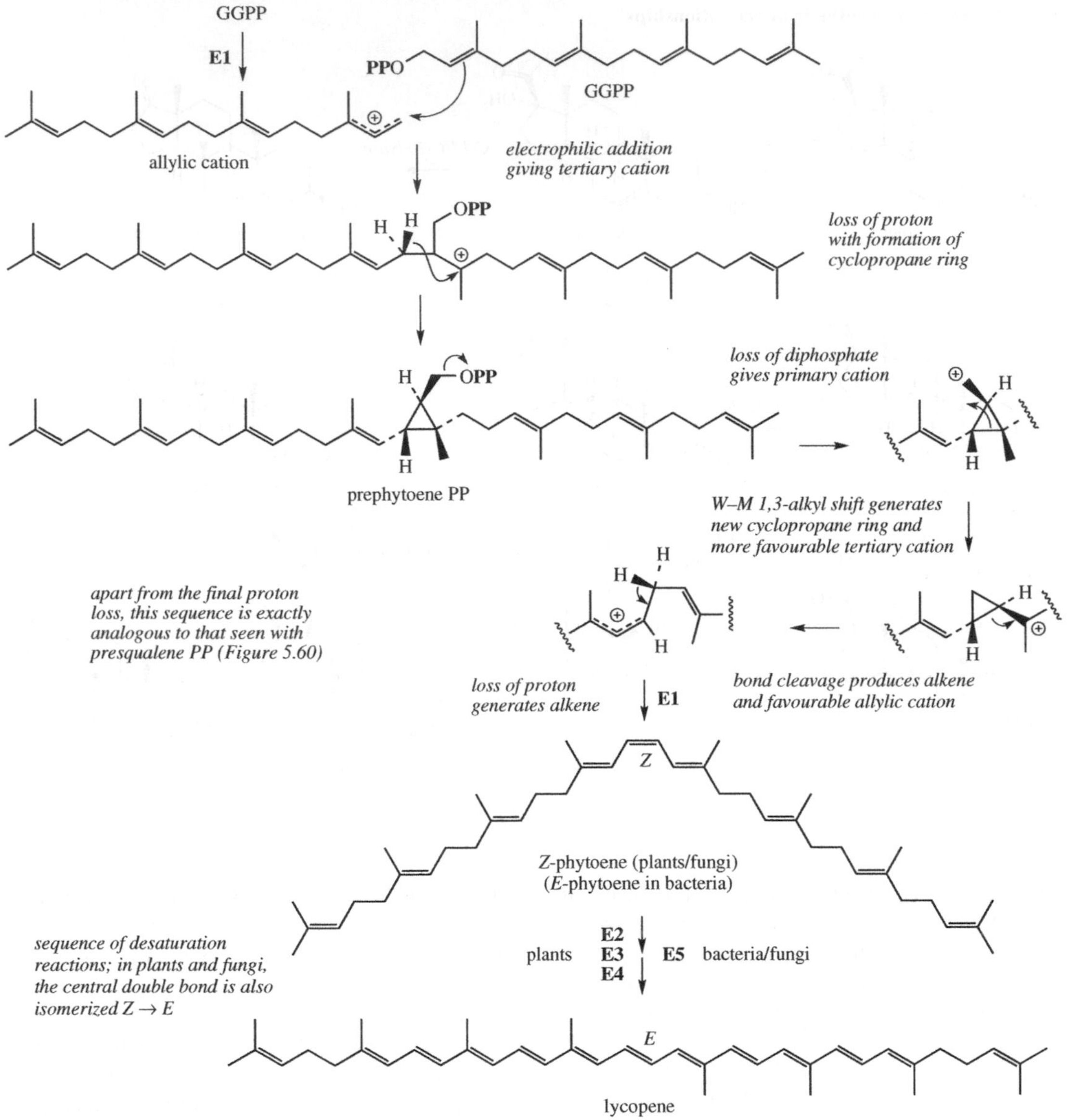

E1: phytoene synthase
E2, E5: phytoene desaturase
E3: ζ-carotene desaturase
E4: carotene isomerase

Figure 5.142

shown in Figure 5.144. The epoxide grouping in violaxanthin allows further chemical modifications, such as ring contraction to a cyclopentane, exemplified by **capsanthin** (Figure 5.143), the brilliant red pigment of sweet peppers (*Capsicum annuum*; Solanaceae), or formation of an allene as in **fucoxanthin**, an abundant carotenoid in brown algae (*Fucus* species; Fucaceae). **Astaxanthin** (Figure 5.143) is commonly found in marine animals and is responsible for the pink/red coloration of crustaceans, shellfish, and fish such as salmon. These animals are unable to synthesize carotenoids, and astaxanthin is produced by modification of plant carotenoids, e.g. β-carotene, obtained in the diet.

Carotenoids function along with chlorophylls in photosynthesis as accessory light-harvesting pigments, effectively extending the range of light absorbed by the

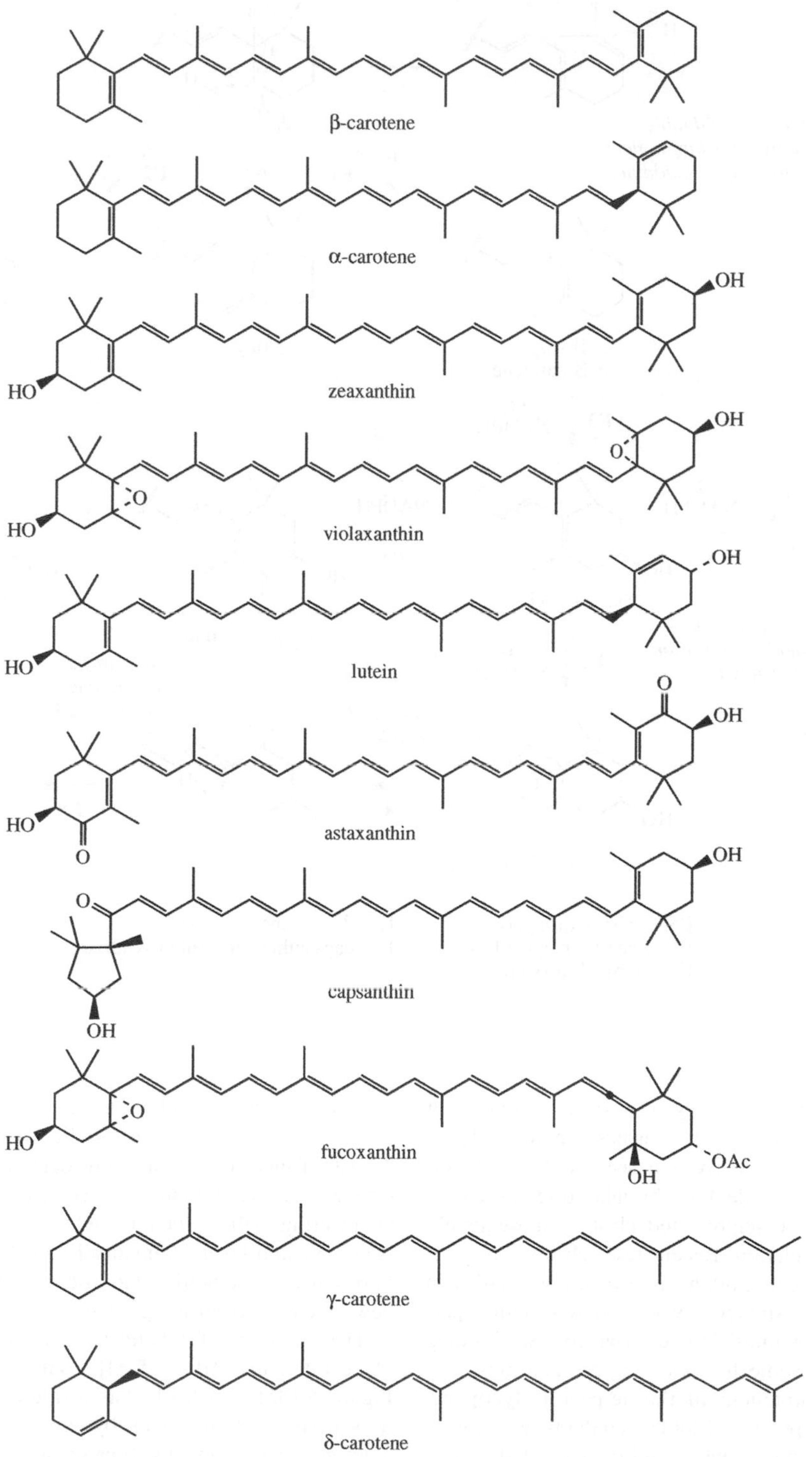
β-carotene
α-carotene
OH
HO
zeaxanthin
OH
O
O
HO
violaxanthin
OH
HO
lutein
O
OH
HO
O
astaxanthin
OH
O
OH
capsanthin
O
HO
fucoxanthin
OH
OAc
γ-carotene
δ-carotene

Figure 5.143

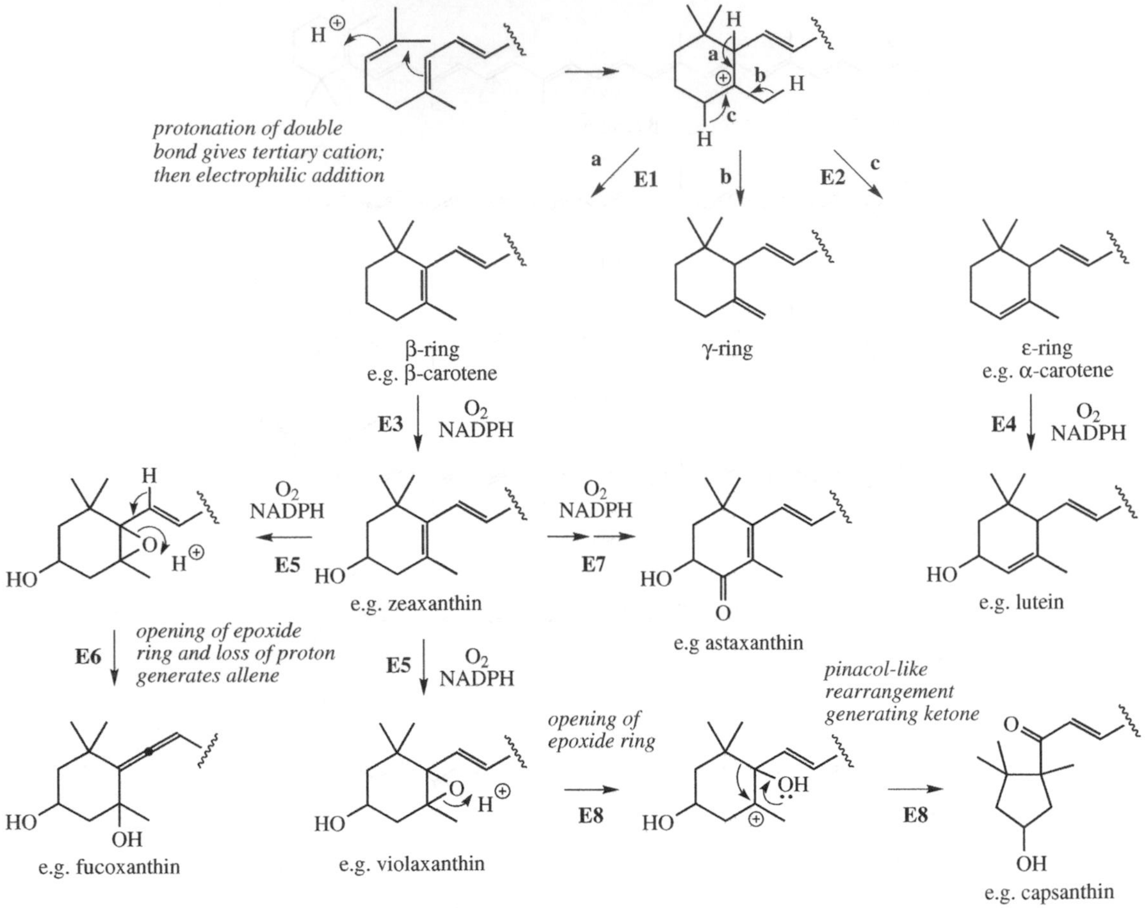

Figure 5.144

E1: β-cyclase
E2: ε-cyclase
E3: β-ring hydroxylase

E4: ε-ring hydroxylase
E5: zeaxanthin epoxidase
E6: neoxanthin synthase

E7: β-carotene oxygenase
E8: capsanthin/capsorubin synthase

photosynthetic apparatus. They also serve as important protectants for plants and algae against photo-oxidative damage, quenching toxic oxygen species. Some herbicides (bleaching herbicides) act by inhibiting carotenoid biosynthesis, and the unprotected plant is subsequently killed by photo-oxidation. Recent research also suggests that carotenoids are important antioxidant molecules in humans, quenching singlet oxygen and scavenging peroxyl radicals, thus minimizing cell damage and affording protection against some forms of cancer. The most significant dietary carotenoid in this respect is **lycopene**, with tomatoes and processed tomato products featuring as the predominant source. The extended conjugated system

allows free-radical addition reactions and hydrogen abstraction from positions allylic to this conjugation. Tomato fruits accumulate considerable amounts of lycopene, and thus provide an excellent vehicle for genetic engineering in the carotenoid field. It is possible to diminish carotenoid levels dramatically, or divert the lycopene into other carotenoids, e.g. β-carotene, by expressing new genes controlling the cyclization reactions.

The A group of vitamins are important metabolites of carotenoids [Box 5.24]. **Vitamin A_1** (**retinol**; Figure 5.145) effectively has a diterpene structure, but it is derived in mammals by oxidative metabolism of a tetraterpenoid, mainly β-carotene, taken in the diet.

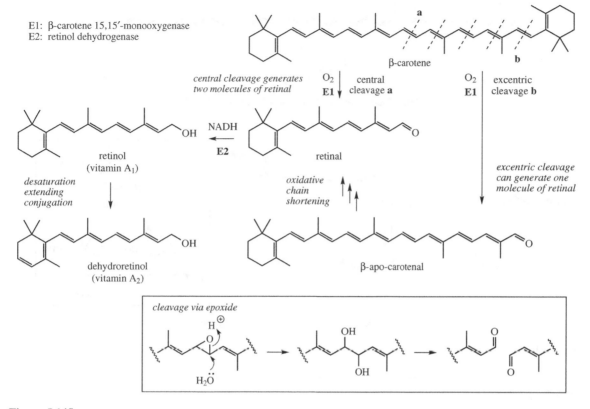

E1: β-carotene 15,15′-monooxygenase
E2: retinol dehydrogenase

β-carotene

central cleavage generates two molecules of retinal

O_2 | central cleavage **a**
E1

O_2 | excentric cleavage **b**
E1

NADH
E2

retinal

retinol (vitamin A₁)

desaturation extending conjugation

oxidative chain shortening

excentric cleavage can generate one molecule of retinal

dehydroretinol (vitamin A₂)

β-apo-carotenal

cleavage via epoxide

Figure 5.145

Cleavage occurs in the mucosal cells of the intestine and is catalysed by an O_2-dependent monooxygenase, most probably via an intermediate epoxide. This can theoretically yield two molecules of the intermediate aldehyde **retinal**, which is subsequently reduced to the alcohol retinol (Figure 5.145). Although β-carotene cleaved at the central double bond is capable of giving rise to two molecules of retinol, there is evidence that cleavage can also occur at other double bonds, so-called excentric cleavage (Figure 5.145). Further chain shortening from the larger cleavage product, an apocarotenal, then produces retinal, but only one molecule can be produced per molecule of β-carotene. **Vitamin A₂ (dehydroretinol**; Figure 5.145) is an analogue of retinol containing a cyclohexadiene ring system; the corresponding aldehyde and retinal are also included in the A group of vitamins. Retinol and its derivatives are found only in animal products, and these provide some of our dietary needs. Cod liver oil and halibut liver oil are rich sources that are used as dietary supplements.

However, carotenoid sources are equally important. These need to have at least one non-hydroxylated ring system of the β-type, e.g. β-carotene, α-carotene, and γ-carotene; about 50 natural carotenoids fall in this group.

Cleavage of carotenoid precursors also explains the formation of **bixin** and **crocetin** (Figure 5.146); indeed, these compounds are classified as apocarotenoids. Large amounts (up to 10%) of the red pigment bixin are found in the seed coats of annatto (*Bixa orellana*; Bixaceae), and bixin is widely used as a natural food colorant, especially for cheese and other dairy products. Bixin originates from cleavage of the non-cyclic carotenoid lycopene. Crocetin, in the form of esters with gentiobiose [D-Glc(β1→6)D-Glc], is the major pigment in stigmas of *Crocus sativus* (Iridaceae) which comprise the extremely expensive spice saffron. Crocetin is known to be produced from the cyclic carotenoid zeaxanthin; the monoterpenoid **safranal** (Figure 5.146) also contributes to the aroma and taste of saffron, and is formed from the remaining portion of the carotenoid structure.

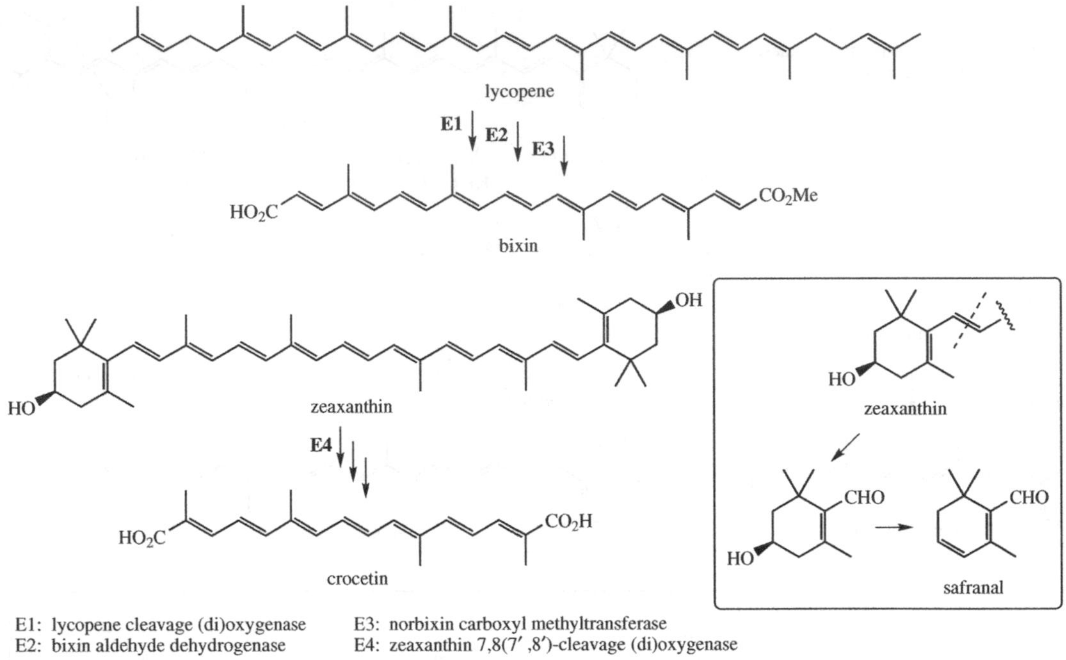
lycopene
E1
E2
E3
HO2C
CO2Me
bixin
OH
HO
zeaxanthin
E4
HO2C
CO2H
crocetin
HO
zeaxanthin
CHO
HO
CHO
safranal
E1: lycopene cleavage (di)oxygenase
E2: bixin aldehyde dehydrogenase
E3: norbixin carboxyl methyltransferase
E4: zeaxanthin 7,8(7′,8′)-cleavage (di)oxygenase

Figure 5.146

Box 5.24

Vitamin A

Vitamin A_1 (retinol) and **vitamin A_2** (dehydroretinol) (Figure 5.145) are fat-soluble vitamins found only in animal products, particularly eggs, dairy products, and animal livers and kidneys. Fish liver oils, e.g. cod liver oil, halibut liver oil (see Table 3.1) are particularly rich sources. They exist as the free alcohols, or as esters with acetic and palmitic acid. Vitamin A_2 has about 40% of the activity of vitamin A_1. Carotenoid precursors (provitamins) are widely distributed in plants, and after ingestion, these are subsequently transformed into vitamin A in the liver. Green vegetables and plant sources rich in carotenoids, such as carrots, help to provide adequate levels. A deficiency of vitamin A leads to vision defects, including impairment at low light levels (night blindness) and a drying and degenerative disease of the cornea. It is also necessary for normal growth of young animals. Retinoids (vitamin A and analogues) are now known to act as signalling molecules which regulate diverse aspects of cell differentiation, embryonic development, growth, and vision.

For the processes of vision, *all-trans*-retinol needs to be converted into 11-*cis*-retinol and then 11-*cis*-retinal. The isomerization is accomplished via esterification, which creates a better leaving group, and a likely S_N2' addition–elimination mechanism that allows rotation about a single bond is shown in Figure 5.147. 11-*cis*-Retinal is then bound to the protein opsin in the retina via an imine linkage to give the red visual pigment rhodopsin; its sensitivity to light involves isomerization of the *cis*-retinal portion back to the *all-trans* form, thus translating the light energy into a molecular change which triggers a nerve impulse to the brain. The absorption of light energy promotes an electron from a π- to a π^*-orbital, thus temporarily destroying the double-bond character and allowing rotation. A similar *cis–trans* isomerization affecting cinnamic acids was discussed under coumarins (see page 161). *all-trans*-Retinal is then subsequently released from the protein by hydrolysis, and the process can continue.

Vitamin A is relatively unstable, and sensitive to oxidation and light. Antioxidant stabilizers such as vitamin E and vitamin C are sometimes added. It is more stable in oils such as the fish liver oils, which are thus good vehicles for administering the vitamin. Synthetic material is also used. Excessive intake of vitamin A can lead to toxic effects, including pathological changes in the skin, hair loss, blurred vision, and headaches.

all-trans-Retinoic acid, the biologically most active metabolite of vitamin A, has been found to play a major role in the regulation of gene expression, in cellular differentiation, and in the proliferation of epithelial cells. Synthetic retinoic acid **(tretinoin)** and **isotretinoin** (13-*cis*-retinoic acid) (Figure 5.148) are used as topical or oral treatments for acne vulgaris, reducing levels of dehydroretinol and modifying skin keratinization. Dehydroretinol levels in the skin become markedly elevated in

Box 5.24   (continued)

conditions such as eczema and psoriasis. **Acitretin** (Figure 5.148) is an aromatic analogue which can give relief in severe cases of psoriasis. All these materials can produce toxic side-effects, including increased sensitivity to UV light. Tretinoin has also proven useful in cancer chemotherapy, particularly in acute promyelocytic leukaemia.

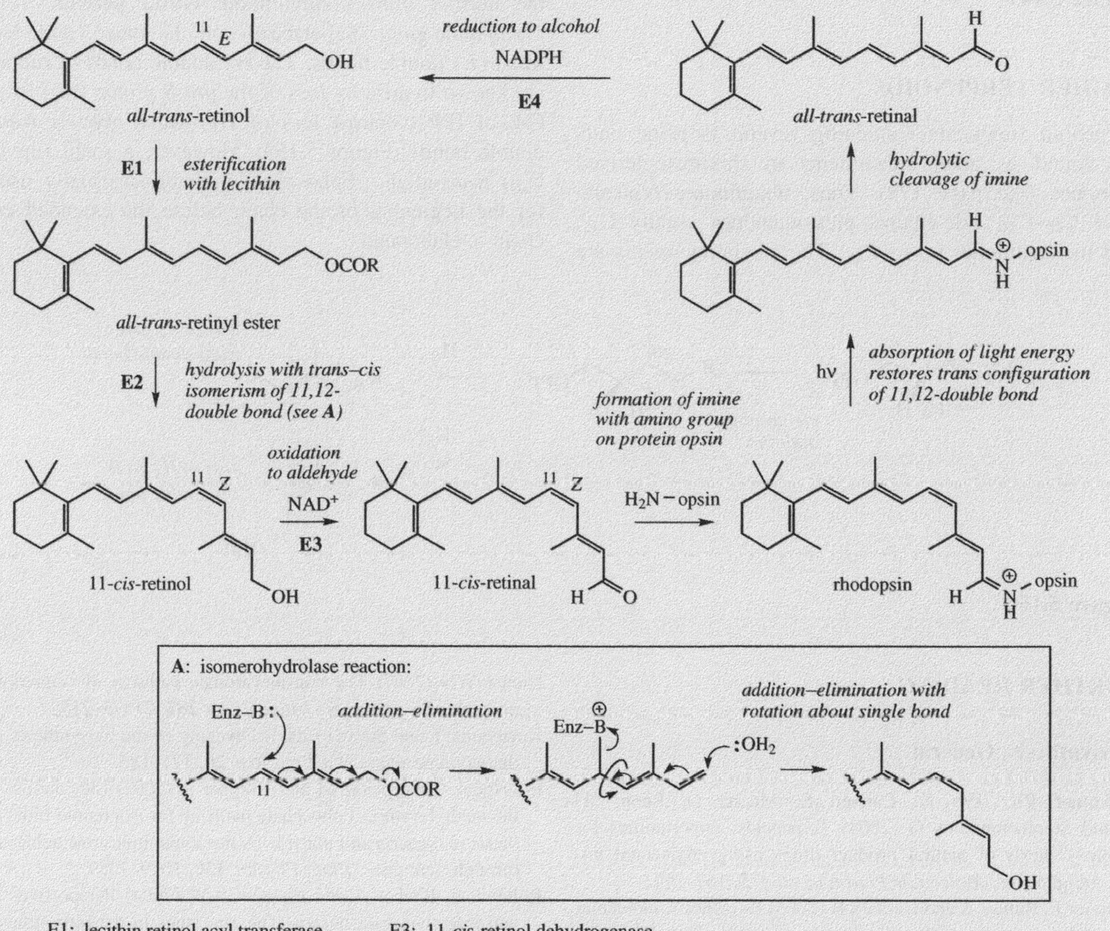

E1: lecithin retinol acyl transferase   E3: 11-*cis*-retinol dehydrogenase
E2: isomerohydrolase   E4: *all-trans*-retinol dehydrogenase

Figure 5.147

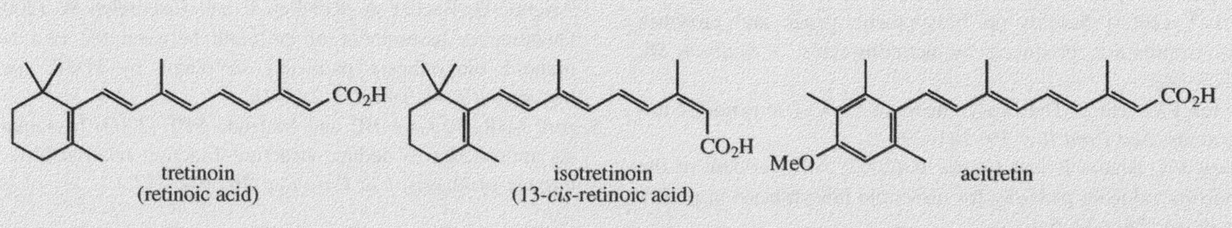

Figure 5.148

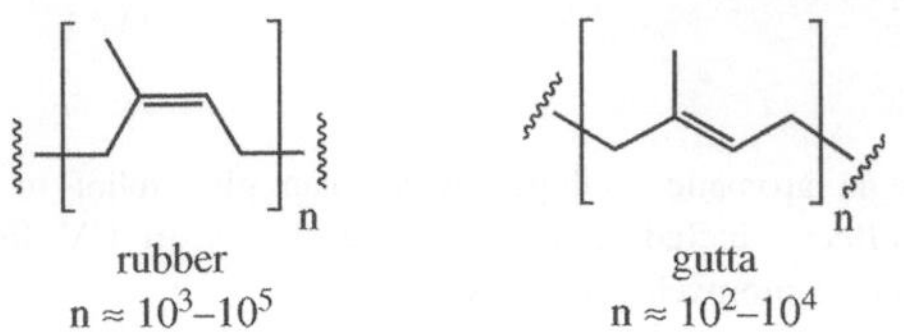

rubber
$n \approx 10^3 - 10^5$

gutta
$n \approx 10^2 - 10^4$

Figure 5.149

HIGHER TERPENOIDS

Terpenoid fragments containing several isoprene units are found as alkyl substituents in shikimate-derived quinones (see page 178). Thus, ubiquinones typically have C_{40}–C_{50} side-chains, plastoquinones usually C_{45}, and menaquinones up to C_{65}. The alkylating agents are polyprenyl diphosphates, formed simply by increasing the chain length with further addition of IPP residues. Even longer polyisoprene chains are encountered in some natural polymers, especially rubber and gutta percha. **Rubber** (Figure 5.149), from the rubber tree *Hevea brasiliensis* (Euphorbiaceae), is unusual in possessing an extended array of *cis* (*Z*) double bonds rather than the normal *trans* configuration. **Gutta percha**, from *Palaquium gutta* (Sapotaceae), on the other hand, has *trans* (*E*) double bonds. The *cis* double bonds in rubber are known to arise by loss of the *pro-S* proton (H_S) from C-2 of IPP (contrast loss of H_R which gives a *trans* double bond) (Figure 5.150). However, a small (up to C_{20}) *trans*-allylic diphosphate initiator is actually used for the beginning of the chain before the extended *cis* chain is elaborated.

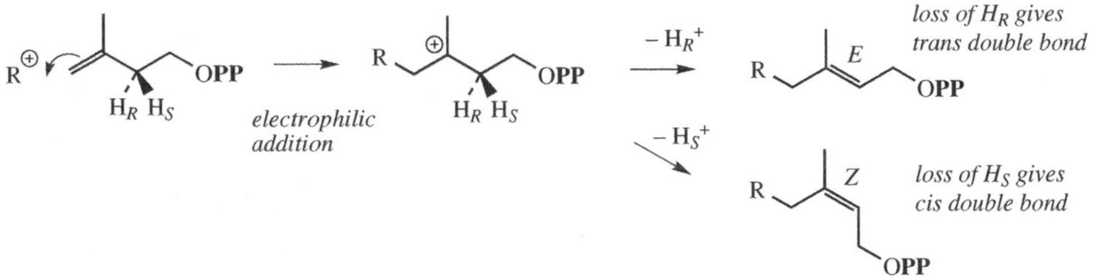

Figure 5.150

FURTHER READING

Biosynthesis, General

Ajikumar PK, Tyo K, Carlsen S, Mucha O, Phon TH and Stephanopoulos G (2008) Terpenoids: opportunities for biosynthesis of natural product drugs using engineered microorganisms. *Molecular Pharmaceutics* **5**, 167–190.

Bouvier F, Rahier A and Camara B (2005) Biogenesis, molecular regulation and function of plant isoprenoids. *Prog Lipid Res* **44**, 357–429.

Christianson DW (2006) Structural biology and chemistry of the terpenoid cyclases. *Chem Rev* **106**, 3412–3442.

Dairi T (2005) Studies on biosynthetic genes and enzymes of isoprenoids produced by actinomycetes. *J Antibiot* **58**, 227–243.

Dewick PM (2002) The biosynthesis of C_5–C_{25} terpenoid compounds. *Nat Prod Rep* **19**, 181–222.

Dubey VS, Bhalla R and Luthra R (2003) An overview of the non-mevalonate pathway for terpenoid biosynthesis in plants. *J Biosci* **28**, 637–646.

Eisenreich W, Bacher A, Arigoni D and Rohdich F (2004) Biosynthesis of isoprenoids via the non-mevalonate pathway. *Cell Mol Life Sci* **61**, 1401–1426.

Hunter WN (2007) The non-mevalonate pathway of isoprenoid precursor biosynthesis. *J Biol Chem* **282**, 21573–21577.

Kuzuyama T and Seto H (2003) Diversity of the biosynthesis of the isoprene units. *Nat Prod Rep* **20**, 171–183.

Rodríguez-Concepción M and Boronat A (2002) Elucidation of the methylerythritol phosphate pathway for isoprenoid biosynthesis in bacteria and plastids. A metabolic milestone achieved through genomics. *Plant Physiol* **130**, 1079–1089.

Rohdich F, Bacher A and Eisenreich W (2004) Perspectives in anti-infective drug design. The late steps in the biosynthesis of the universal terpenoid precursors, isopentenyl diphosphate and dimethylallyl diphosphate. *Bioorg Chem* **32**, 292–308.

Schuhr CA, Radykewicz T, Sagner S, Latzel C, Zenk MH, Arigoni D, Bacher A, Rohdich F and Eisenreich W (2003) Quantitative assessment of crosstalk between the two isoprenoid biosynthesis pathways in plants by NMR spectroscopy. *Phytochem Rev* **2**, 3–16.

Segura MJR, Jackson BE and Matsuda SPT (2003) Mutagenesis approaches to deduce structure–function relationships in terpene synthases. *Nat Prod Rep* **20**, 304–317.

Isoprene

Sharkey TD and Yeh S (2001) Isoprene emission from plants. *Annu Rev Plant Physiol Plant Mol Biol* **52**, 407–36.

Volatile Oils

Bentley R (2006) The nose as a stereochemist. Enantiomers and odor. *Chem Rev* **106**, 4099–4112.

Dudareva N, Pichersky E and Gershenzon J (2004) Biochemistry of plant volatiles. *Plant Physiol* **135**, 1893–1902.

Edris AE (2007) Pharmaceutical and therapeutic potentials of essential oils and their individual volatile constituents: a review. *Phytother Res* **21**, 308–323.

Kalemba D and Kunicka A (2003) Antibacterial and antifungal properties of essential oils. *Curr Med Chem* **10**, 813–829.

Pichersky E and Gershenzon J (2002) The formation and function of plant volatiles: perfumes for pollinator attraction and defense. *Curr Opin Plant Biol* **5**, 237–243.

Pyrethrins

Khambay BPS (2002) Pyrethroid insecticides. *Pesticide Outlook* 49–54.

Iridoids

Birkett MA and Pickett JA (2003) Aphid sex pheromones: from discovery to commercial production. *Phytochemistry* **62**, 651–656.

Dinda B, Debnath S and Harigaya Y (2007) Naturally occurring iridoids. A review, part 1. *Chem Pharm Bull* **55**, 159–222; Naturally occurring secoiridoids and bioactivity of naturally occurring iridoids and secoiridoids. A review, part 2. *Chem Pharm Bull* **55**, 689–728.

Grant L, McBean DE, Fyfe L and Warnock AM (2007) A review of the biological and potential therapeutic actions of *Harpagophytum procumbens*. *Phytother Res* **21**, 199–209.

Jensen SR, Franzyka H and Wallander E (2002) Chemotaxonomy of the Oleaceae: iridoids as taxonomic markers. *Phytochemistry* **60**, 213–231.

Oudin A, Courtois M, Rideau M and Clastre M (2007) The iridoid pathway in *Catharanthus roseus* alkaloid biosynthesis. *Phytochem Rev* **6**, 259–276.

Sesquiterpenoids

Staneva JD, Todorova MN and Evstatieva LN (2008) Sesquiterpene lactones as chemotaxonomic markers in genus *Anthemis*. *Phytochemistry* **69**, 607–618.

Artemisinin

Covello PS, Teoh KH, Polichuk DR, Reed DW and Nowak G (2007) Functional genomics and the biosynthesis of artemisinin. *Phytochemistry* **68**, 1864–1871.

Efferth T (2007) Antiplasmodial and antitumor activity of artemisinin – from bench to bedside. *Planta Med* **73**, 299–309.

Gelb MH (2007) Drug discovery for malaria: a very challenging and timely endeavor. *Curr Opin Chem Biol* **11**, 440–445.

Jansen FH and Soomro SA (2007) Chemical instability determines the biological action of the artemisinins. *Curr Med Chem* **14**, 3243–3259.

Jefford CW (2001) Why artemisinin and certain synthetic peroxides are potent antimalarials. Implications for the mode of action. *Curr Med Chem* **8**, 1803–1826.

Li Y and Wu Y-L (2003) An over four millennium story behind qinghaosu (artemisinin) – a fantastic antimalarial drug from a traditional Chinese herb. *Curr Med Chem* **10**, 2197–2230.

O'Neill PM and Posner GH (2004) A medicinal chemistry perspective on artemisinin and related endoperoxides. *J Med Chem* **47**, 2945–2964.

Ro D-K, Paradise EM, Ouellet M, Fisher KJ, Newman KL, Ndungu JM, Ho KA, Eachus RA, Ham TS, Kirby J, Chang MCY, Withers ST, Shiba Y, Sarpong R and Keasling JD (2006) Production of the antimalarial drug precursor artemisinic acid in engineered yeast. *Nature* **440**, 940–943.

Robert A, Dechy-Cabaret O, Cazelles J and Meunier B (2002) From mechanistic studies on artemisinin derivatives to new modular antimalarial drugs. *Acc Chem Res* **35**, 167–174.

Utzinger J, Shuhuac X, Keiser J, Minggan C, Jiang Z and Tanner M (2001) Current progress in the development and use of artemether for chemoprophylaxis of major human schistosome parasites. *Curr Med Chem* **8**, 1841–1859.

Wright CW (2005) Plant derived antimalarial agents: new leads and challenges. *Phytochem Rev* **4**, 55–61.

Wu Y (2002) How might qinghaosu (artemisinin) and related compounds kill the intraerythrocytic malaria parasite? A chemist's view. *Acc Chem Res* **35**, 255–259.

Gossypol

Kovacic P (2003) Mechanism of drug and toxic actions of gossypol: focus on reactive oxygen species and electron transfer. *Curr Med Chem* **10**, 2711–2718.

Trichothecenes

Grove JF (2007) The trichothecenes and their biosynthesis. *Prog Chem Org Nat Prod* **88**, 63–130.

Diterpenoids

Chinou I (2005) Labdanes of natural origin – biological activities (1981–2004). *Curr Med Chem* **12**, 1295–1317.

Keeling CI and Bohlmann J (2006) Diterpene resin acids in conifers. *Phytochemistry* **67**, 2415–2423.

Peters RJ (2006) Uncovering the complex metabolic network underlying diterpenoid phytoalexin biosynthesis in rice and other cereal crop plants. *Phytochemistry* **67**, 2307–2317.

Trapp S and Croteau R (2001) Defensive resin biosynthesis in conifers. *Annu Rev Plant Physiol Plant Mol Biol* **52**, 689–724.

Taxol

Altmann K-H and Gertsch J (2007) Anticancer drugs from nature – natural products as a unique source of new microtubule-stabilizing agents. *Nat Prod Rep* **24**, 327–357.

Croteau R, Ketchum REB, Long RM, Kaspera R and Wildung MR (2006) Taxol biosynthesis and molecular genetics. *Phytochem Rev* **5**, 75–97.

Fang W-S and Liang X-T (2005) Recent progress in structure activity relationship and mechanistic studies of taxol analogues. *Mini Rev Med Chem* **5**, 1–12.

Ganem B and Franke RR (2007) Paclitaxel from primary taxanes: a perspective on creative invention in organozirconium chemistry. *J Org Chem* **72**, 3981–3987.

Ganesh T (2007) Improved biochemical strategies for targeted delivery of taxoids. *Bioorg Med Chem* **15**, 3597–3623.

Kingston DGI (2001) Taxol, a molecule for all seasons. *Chem Commun* 867–880.

Kingston DGI (2007) The shape of things to come: structural and synthetic studies of taxol and related compounds. *Phytochemistry* **68**, 1844–1854.

Kingston DGI, Jagtap PG, Yuan H and Samala L (2002) The chemistry of taxol and related taxoids. *Prog Chem Org Nat Prod* **84**, 53–225.

Ojima I (2008) Guided molecular missiles for tumor-targeting chemotherapy – case studies using the second-generation taxoids as warheads. *Acc Chem Res* **41**, 108–119.

Skwarczynski M, Hayashi Y and Kiso Y (2006) Paclitaxel prodrugs: toward smarter delivery of anticancer agents. *J Med Chem* **49**, 7253–7269.

Walker K and Croteau R (2001) Taxol biosynthetic genes. *Phytochemistry* **58**, 1–7.

Zhao CF, Yu LJ, Li LQ and Xiang F (2007) Simultaneous identification and determination of major taxoids from extracts of *Taxus chinensis* cell cultures. *Z Naturforsch Teil C* **62**, 1–10.

Ginkgo

Nakanishi K (2005) Terpene trilactones from *Gingko biloba*: from ancient times to the 21st century. *Bioorg Med Chem* **13**, 4987–5000.

Strømgaard K and Nakanishi K (2004) Chemistry and biology of terpene trilactones from *Ginkgo biloba*. *Angew Chem Int Ed* **43**, 1640–1658.

Stevioside

Brandle JE and Telmer PG (2007) Steviol glycoside biosynthesis. *Phytochemistry* **68**, 1855–1863.

Geuns JMC (2003) Molecules of interest. Stevioside. *Phytochemistry* **64**, 913–921.

Salvinorin

Grundmann O, Phipps SM, Zadezensky I and Butterweck V (2007) *Salvia divinorum* and salvinorin A: and update on pharmacology and analytical methodology. *Planta Med* **73**, 1039–1046.

Imanshahidi M and Hosseinzadeh H (2006) The pharmacological effects of *Salvia* species on the central nervous system. *Phytother Res* **20**, 427–437.

Prisinzano TE and Rothman RB (2008) Salvinorin A analogs as probes in opioid pharmacology. *Chem Rev* **108**, 1732–1743.

Triterpenoids

Abe I (2007) Enzymatic synthesis of cyclic triterpenes. *Nat Prod Rep* **24**, 1311–1331.

Charlton-Menys V and Durrington PN (2007) Squalene synthase inhibitors: clinical pharmacology and cholesterol-lowering potential. *Drugs* **67**, 11–16.

Chen JC, Chiu MH, Nie RL, Cordell GA and Qiu SX (2005) Cucurbitacins and cucurbitane glycosides: structures and biological activities. *Nat Prod Rep* **22**, 386–399; errata 794–795.

Dzubak P, Hajduch M, Vydra D, Hustova A, Kvasnica M, Biedermann D, Markova L, Urban M and Sarek J (2006) Pharmacological activities of natural triterpenoids and their therapeutic implications. *Nat Prod Rep* **23**, 394–411.

Hoshino T and Sato T (2002) Squalene–hopene cyclase: catalytic mechanism and substrate recognition. *Chem Commun* 291–301.

Krasutsky PA (2006) Birch bark research and development. *Nat Prod Rep* **23**, 919–942.

Phillips DR, Rasbery JM, Bartel B and Matsuda SPT (2006) Biosynthetic diversity in plant triterpene cyclization. *Curr Opin Plant Biol* **9**, 305–314.

Wendt KU (2005) Enzyme mechanisms for triterpene cyclization: new pieces of the puzzle. *Angew Chem Int Ed* **44**, 3966–3971.

Wendt KU, Schultz GE, Corey EJ and Liu DR (2000) Enzyme mechanisms for polycyclic triterpene formation. *Angew Chem Int Ed* **39**, 2812–2833.

Yogeeswari P and Sriram D (2005) Betulinic acid and its derivatives: a review on their biological properties. *Curr Med Chem* **12**, 657–666.

Triterpenoid Saponins

Kalinowska M, Zimowski J, Plczkowski C and Wojciechowski ZA (2005) The formation of sugar chains in triterpenoid saponins and glycoalkaloids. *Phytochem Rev* **4**, 237–257.

Osbourn AE (2003) Saponins in cereals. *Phytochemistry* **62**, 1–4.

Sparg SG, Light ME and van Staden J (2004) Biological activities and distribution of plant saponins. *J Ethnopharmacol* **94**, 219–243.

Townsend B, Jenner H and Osbourn A (2006) Saponin glycosylation in cereals. *Phytochem Rev* **5**, 109–114.

Xu R, Fazio GC and Matsuda SPT (2004) On the origins of triterpenoid skeletal diversity *Phytochemistry* **65**, 261–291.

Liquorice

Asl MN and Hosseinzadeh H (2008) Review of pharmacological effects of *Glycyrrhiza* sp. and its bioactive compounds. *Phytother Res* **22**, 709–724.

Baltina LA (2003) Chemical modification of glycyrrhizic acid as a route to new bioactive compounds for medicine. *Curr Med Chem* **10**, 155–171.

Fiore C, Eisenhut M, Ragazzi E, Zanchin G and Armanini D (2005) A history of the therapeutic use of liquorice in Europe. *J Ethnopharmacol* **99**, 317–324.

Ginseng

Davydov M and Krikorian AD (2000) *Eleutherococcus senticosus* (Araliaceae) as an adaptogen: a closer look. *J Ethnopharmacol* **72**, 345–393.

Leung KW, Yung KKL, Mak NK, Yue PYK, Luo H-B, Cheng Y-K, Fan TPD, Yeung HW, Ng TB and Wong RNS (2007) Angiomodulatory and neurological effects of ginsenosides. *Curr Med Chem* **14**, 1371–1380.

Panossian A and Wagner H (2005) Stimulating effect of adaptogens: an overview with particular reference to their efficacy following single dose administration. *Phytother Res* **19**, 819–838.

Park JD, Rhee DK and Lee YH (2005) Biological activities and chemistry of saponins from *Panax ginseng*. *Phytochem Rev* **4**, 159–175.

Cholesterol

Jain KS, Kathiravan MK, Somani RS and Shishoo CJ (2007) The biology and chemistry of hyperlipidemia. *Bioorg Med Chem* **15**, 4674–4699.

Libby P, Aikawa M and Schönbeck U (2000) Cholesterol and atherosclerosis. *Biochim Biophys Acta* **1529**, 299–309.

Risley JM (2002) Cholesterol biosynthesis: lanosterol to cholesterol. *J Chem Educ* **79**, 377–384.

Vance DE and Van den Bosch H (2000) Cholesterol in the year 2000. *Biochim Biophys Acta* **1529**, 1–8.

Wang M and Briggs MR (2004) HDL: The metabolism, function, and therapeutic importance. *Chem Rev* **104**, 119–137.

Waterham HR (2006) Defects of cholesterol biosynthesis. *FEBS Lett* **580**, 5442–5449.

Sterols

Benveniste P (2004) Biosynthesis and accumulation of sterols. *Annu Rev Plant Biol* **55**, 429–457.

Moreaua RA, Whitaker BD and Hicks KB (2002) Phytosterols, phytostanols, and their conjugates in foods: structural diversity, quantitative analysis, and health-promoting uses. *Prog Lipid Res* **41**, 457–500.

Nes WD (2000) Sterol methyl transferase: enzymology and inhibition. *Biochim Biophys Acta* **1529**, 63–88.

Nes WD (2003) Enzyme mechanisms for sterol *C*-methylations. *Phytochemistry* **64**, 75–95.

Schaller H (2003) The role of sterols in plant growth and development. *Prog Lipid Res* **42**, 163–175.

Vitamin D

Boland R, Skliar M, Curino A and Milanesi L (2003) Vitamin D compounds in plants. *Plant Sci* **164**, 357–369.

Chen TC, Chimeh F, Lu Z, Mathieu J, Person KS, Zhang A, Kohn N, Martinello S, Berkowitz R and Holick MF (2007) Factors that influence the cutaneous synthesis and dietary sources of vitamin D. *Arch Biochem Biophys* **460**, 213–217.

Nagpal S, Lu J and Boehm MF (2001) Vitamin D analogs: mechanism of action and therapeutic applications. *Curr Med Chem* **8**, 1661–1679.

Cardioactive Glycosides

Gobbini M and Cerri A (2005) *Digitalis*-like compounds: the discovery of the *o*-aminoalkyloxime group as a very powerful substitute for the unsaturated γ-butyrolactone moiety. *Curr Med Chem* **12**, 2343–2355.

Paula S, Tabet MR and Ball WJ (2005) Interactions between cardiac glycosides and sodium/potassium-ATPase: three-dimensional structure–activity relationship models for ligand binding to the E_2–P_i form of the enzyme versus activity inhibition. *Biochemistry* **44**, 498–510.

Schoner W (2002) Endogenous cardiac glycosides, a new class of steroid hormones. *Eur J Biochem* **269**, 2440–2448.

Bile Acids

Nonappaa and Maitra U (2008) Unlocking the potential of bile acids in synthesis, supramolecular/materials chemistry and nanoscience. *Org Biomol Chem* **6**, 657–669.

Corticosteroids

Gupta R, Jindal DP and Kumar G (2004) Corticosteroids: the mainstay in asthma therapy. *Bioorg Med Chem* **12**, 6331–6342.

Whitehouse MW (2005) Drugs to treat inflammation: a historical introduction. *Curr Med Chem* **12**, 2031–2942.

Steroid Hormones

Ahmed S, Owen CP, James K, Sampson L and Patel CK (2002) Review of estrone sulfatase and its inhibitors – an important new target against hormone dependent breast cancer. *Curr Med Chem* **9**, 263–273.

Brožic P, Lanišnik Rižner T and Gobec S (2008) Inhibitors of 17β-hydroxysteroid dehydrogenase type 1. *Curr Med Chem* **15**, 137–150.

Brunoa RD and Njar VCO (2007) Targeting cytochrome P450 enzymes: a new approach in anti-cancer drug development. *Bioorg Med Chem* **15**, 5047–5060.

Gao W, Bohl CE and Dalton JT (2005) Chemistry and structural biology of androgen receptor. *Chem Rev* **105**, 3352–3370.

Jordan VC (2003) Antiestrogens and selective estrogen receptor modulators as multifunctional medicines. 1. Receptor interactions. *J Med Chem* **46**, 883–908.

Moreira VM, Salvador JAR, Vasaitis TS and Njar VCO (2008) CYP17 inhibitors for prostate cancer treatment – an update. *Curr Med Chem* **15**, 868–899.

Poirier D (2003) Inhibitors of 17β-hydroxysteroid dehydrogenases. *Curr Med Chem* **10**, 453–477.

Saiah E (2008) The role of 11β-hydroxysteroid dehydrogenase in metabolic disease and therapeutic potential of 11β HSD1 inhibitors. *Curr Med Chem* **15**, 642–649.

Stanczyk FZ (2003) All progestins are not created equal. *Steroids* **68**, 879–890.

Yen SSC (2001) Dehydroepiandrosterone sulfate and longevity: new clues for an old friend. *Proc Natl Acad Sci USA* **98**, 8167–8169.

Carotenoids

Auldridge ME, McCarty DR and Klee HJ (2006) Plant carotenoid cleavage oxygenases and their apocarotenoid products. *Curr Opin Plant Biol* **9**, 315–321.

DellaPenna D and Pogson BJ (2006) Vitamin synthesis in plants: tocopherols and carotenoids. *Annu Rev Plant Biol* **57**, 711–738.

Enfissi EMA, Fraser PD and Bramley PM (2006) Genetic engineering of carotenoid formation in tomato. *Phytochem Rev* **5**, 59–65.

Fraser PD and Bramley PM (2004) The biosynthesis and nutritional uses of carotenoids. *Prog Lipid Res* **43**, 228–265.

Grotewold E (2006) The genetics and biochemistry of floral pigments. *Annu Rev Plant Biol* **57**, 761–780.

Herbers K (2003) Vitamin production in transgenic plants. *Plant Physiol* **160**, 821–829.

Leuenberger MG, Engeloch-Jarret C and Woggon W-D (2001) The reaction mechanism of the enzyme-catalyzed central cleavage of β-carotene to retinal. *Angew Chem Int Ed* **40**, 2614–2617.

Römer S and Fraser PD (2005) Recent advances in carotenoid biosynthesis, regulation and manipulation. *Planta* **221**, 305–308.

Vitamin A

Kagechika H and Shudo K (2005) Synthetic retinoids: recent developments concerning structure and clinical utility. *J Med Chem* **48**, 5875–5883.

Lidén M and Eriksson U (2006) Understanding retinol metabolism: structure and function of retinol dehydrogenases. *J Biol Chem* **281**, 13001–13004.

Njar VCO, Gediya L, Purushottamachar P, Chopra P, Vasaitis TS, Khandelwal A, Mehta J, Huynh C, Belosay A and Patel J (2006) Retinoic acid metabolism blocking agents (RAMBAs) for treatment of cancer and dermatological diseases. *Bioorg Med Chem* **14**, 4323–4340.

Higher Terpenoids

Swiezewska E and Danikiewicz W (2005) Polyisoprenoids: structure, biosynthesis and function. *Prog Lipid Res* **44**, 235–258.

6

ALKALOIDS

The alkaloids are low molecular weight nitrogen-containing compounds found mainly in plants, but also to a lesser extent in microorganisms and animals; over 27 000 different alkaloid structures have been characterized, with 21 000 from plants. They contain one or more nitrogen atoms, typically as primary, secondary, or tertiary amines, and this usually confers basicity on the alkaloid, facilitating isolation and purification, since water-soluble salts can be formed in the presence of mineral acids. The name alkaloid is in fact derived from alkali. However, the degree of basicity varies greatly, depending on the structure of the alkaloid molecule and on the presence and location of other functional groups. Indeed, some alkaloids, e.g. where the nitrogen is part of an amide function, are essentially neutral. Alkaloids containing quaternary amines are also found in nature. The biological activity of many alkaloids is often dependent on the amine function being transformed into a quaternary system by protonation at physiological pH values.

Alkaloids are often classified according to the nature of the nitrogen-containing structure (e.g. pyrrolidine, piperidine, quinoline, isoquinoline, indole), though the structural complexity of some examples rapidly expands the number of subdivisions. The nitrogen atoms in alkaloids originate from an amino acid, and, in general, the carbon skeleton of the particular amino acid precursor is also largely retained intact in the alkaloid structure, though the carboxylic acid carbon is often lost through decarboxylation. Accordingly, subdivision of alkaloids into groups based on amino acid precursors forms a rational and often illuminating approach to classification. Relatively few amino acid precursors are actually involved in alkaloid biosynthesis, the principal ones being ornithine, lysine, nicotinic acid, tyrosine, tryptophan, anthranilic acid,

and histidine. Building blocks from the acetate, shikimate, or methylerythritol phosphate pathways are also frequently incorporated into the alkaloid structures. However, a large group of alkaloids are found to acquire their nitrogen atoms via transamination reactions, incorporating only the nitrogen from an amino acid, whilst the rest of the molecule may be derived from acetate or shikimate; others may be terpenoid or steroid in origin. The term 'pseudoalkaloid' is sometimes used to distinguish this group.

ALKALOIDS DERIVED FROM ORNITHINE

L-**Ornithine** (Figure 6.1) is a non-protein amino acid forming part of the urea cycle in animals, where it is produced from L-arginine in a reaction catalysed by the enzyme arginase. In plants it is formed mainly from L-glutamic acid (Figure 6.2). Ornithine contains both δ- and α-amino groups, and it is the nitrogen from the former group which is incorporated into alkaloid structures along with the carbon chain, except for the carboxyl group. Thus, ornithine supplies a C_4N building block to the alkaloid, principally as a pyrrolidine ring system, but also as part of the tropane alkaloids (Figure 6.1). Most of the other amino acid alkaloid precursors typically supply nitrogen from their solitary α-amino group. However, the reactions of ornithine are almost exactly paralleled by those of L-lysine, which incorporates a C_5N unit containing its ε-amino group (see page 326).

Polyamines

Simple polyamines were first isolated from human semen over 300 years ago, though another 250 years

Medicinal Natural Products: A Biosynthetic Approach. 3rd Edition Paul Dewick
© 2009 John Wiley & Sons, Ltd

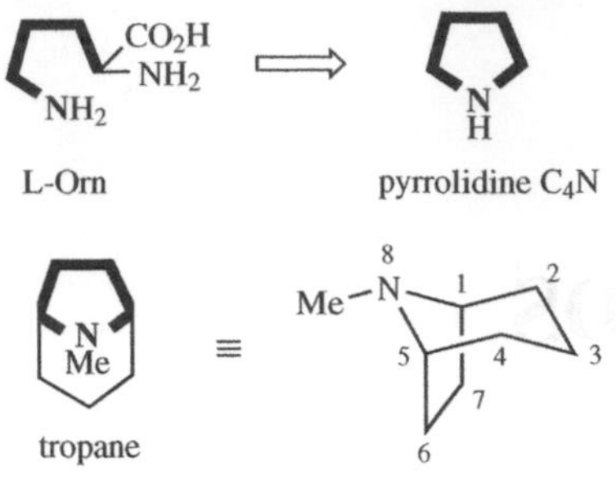

Figure 6.1

passed before their chemical characterization as **spermidine** and **spermine** (Figure 6.2). The simpler compound **putrescine** (1,4-diaminobutane) was first isolated from *Vibrio cholerae*, but its common name relates to its presence in decomposing animal flesh.

We now know polyamines are found in virtually all living species and are critical regulators of cell growth, differentiation, and cell death. In eukaryotic cells, the three polyamines are synthesized from L-ornithine and L-methionine. In animals, PLP-dependent decarboxylation (see page 22) of ornithine gives putrescine. In plants and microorganisms, an alternative sequence to putrescine starting from arginine also operates concurrently as indicated in Figure 6.2. The arginine pathway also involves decarboxylation, but requires additional hydrolysis reactions to cleave the guanidine portion. Aminopropyl groups are then transferred from a decarboxylated SAM (dcSAM) to putrescine. The alkylation reaction is mechanistically analogous to SAM-mediated methylation, though an alternative alkyl group is transferred. These reactions give firstly spermidine and then spermine. The polyamine pathway is an important target for chemotherapy, since depletion of polyamines can lead to the disruption of a variety of cellular functions. Spermidine synthase is currently considered a promising drug target in the malaria parasite *Plasmodium falciparum* (see page 218).

Pyrrolidine and Tropane Alkaloids

Simple pyrrolidine-containing alkaloid structures are exemplified by hygrine and cuscohygrine found in those plants of the Solanaceae which accumulate medicinally valuable tropane alkaloids such as hyoscyamine or cocaine (see Figure 6.4). The pyrrolidine ring system is formed initially as a Δ^1-pyrrolinium cation (Figure 6.3). **Putrescine** is methylated to *N*-methylputrescine, then oxidative deamination of *N*-methylputrescine by the

action of a diamine oxidase (see page 27) gives the aldehyde. The *N*-methyl-Δ^1-pyrrolinium cation is then generated via imine formation. Indeed, the aminoaldehyde in aqueous solution is known to exist as an equilibrium mixture with the imine.

The extra carbon atoms required for hygrine formation are derived from acetate via acetyl-CoA, and the sequence appears to involve stepwise addition of two acetyl-CoA units (Figure 6.4). In the first step, the enolate anion from acetyl-CoA acts as nucleophile towards the pyrrolinium ion in a Mannich-like reaction, which could yield products with either *R* or *S* stereochemistry. The second addition is then a Claisen condensation extending the side-chain, and the product is the 2-substituted pyrrolidine, retaining the thioester group of the second acetyl-CoA. **Hygrine** and most of the natural tropane alkaloids lack this particular carbon atom, which can be lost by suitable hydrolysis/decarboxylation reactions.

The bicyclic structure of the tropane skeleton in **hyoscyamine** and **cocaine** is achieved by a repeat of the Mannich-like reaction just observed. This requires an oxidation step to generate a new Δ^1-pyrrolinium cation, and removal of a proton α to the carbonyl. The *intramolecular* Mannich reaction on the *R* enantiomer accompanied by decarboxylation generates **tropinone**, and stereospecific reduction of the carbonyl yields **tropine** with a 3α-hydroxyl, or the isomeric **ψ-tropine**, the precursor of the calystegines (see page 321). Hyoscyamine is the ester of tropine with (*S*)-tropic acid, which is derived from L-phenylalanine via phenyl-lactic acid (see Figure 6.5).

Should the carboxyl carbon from the acetoacetyl side-chain not be lost as it was in the formation of tropine, then the subsequent intramolecular Mannich reaction will generate a tropane skeleton with an additional carboxyl substituent (Figure 6.4). However, this event is rare, and is only exemplified by the formation of ecgonine derivatives such as **cocaine** in *Erythroxylum coca* (Erythroxylaceae). The pathway is in most aspects analogous to that already described for hyoscyamine, but must proceed through the *S*-enantiomer of the *N*-methylpyrrolidineacetoacetyl-CoA. The ester function is then modified from a coenzyme A thioester to a simple methyl oxygen ester, and **methylecgonine** is subsequently obtained from the methoxycarbonyltropinone by stereospecific reduction of the carbonyl. Note that, in this case, reduction of the carbonyl occurs from the opposite face to that noted with the tropinone → tropine conversion and, thus, yields the 3β configuration in ecgonine. **Cocaine** is a diester of ecgonine, the benzoyl moiety arising from phenylalanine via cinnamic acid and benzoyl-CoA (see page 157).

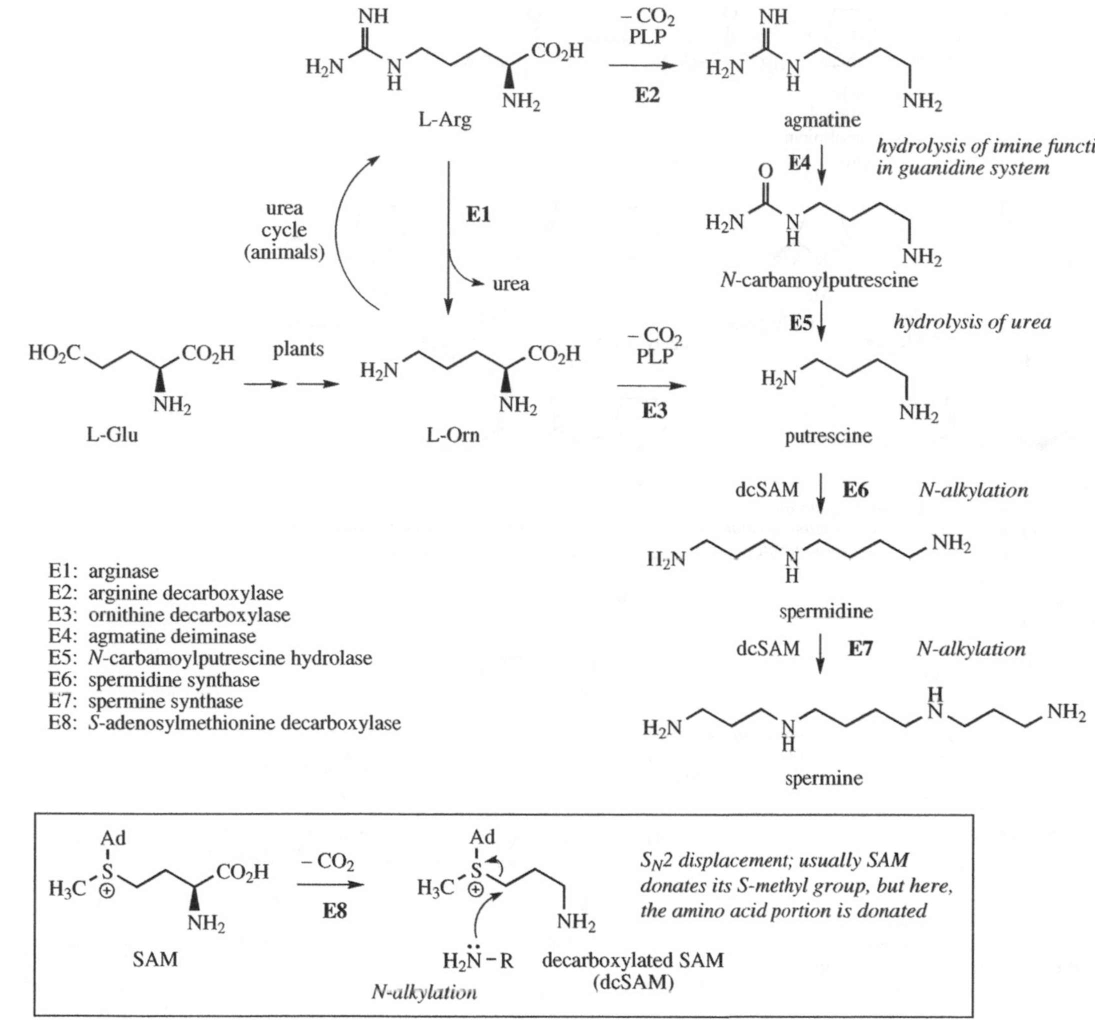

Figure 6.2

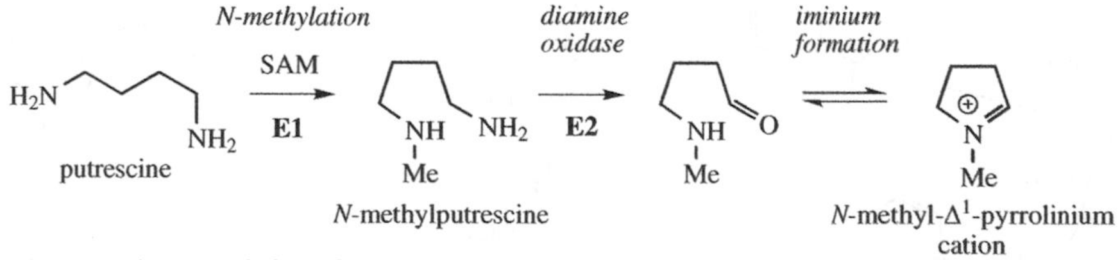

E1: arginase
E2: arginine decarboxylase
E3: ornithine decarboxylase
E4: agmatine deiminase
E5: *N*-carbamoylputrescine hydrolase
E6: spermidine synthase
E7: spermine synthase
E8: *S*-adenosylmethionine decarboxylase

Figure 6.3

E1: putrescine *N*-methyltransferase
E2: methylputrescine oxidase

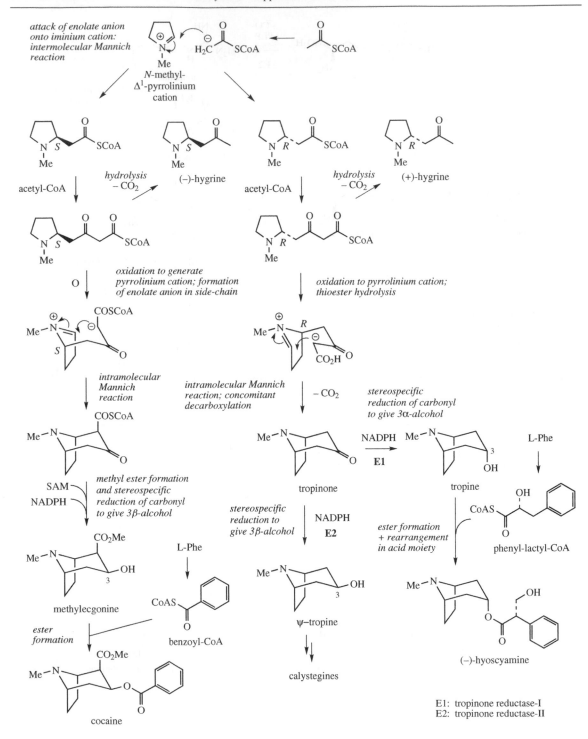

Figure 6.4

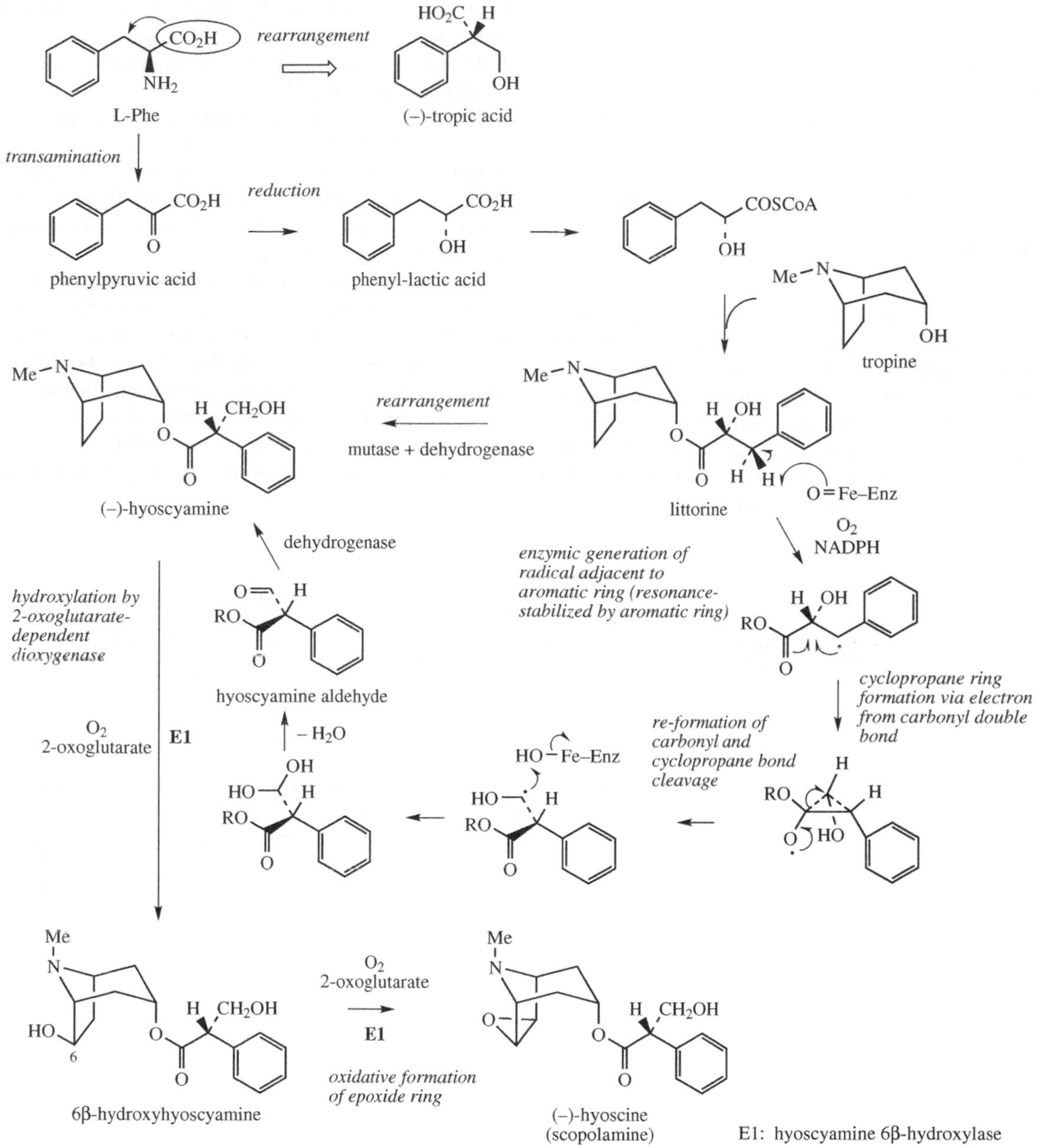

Figure 6.5

A novel rearrangement process occurs in the phenyl-alanine → tropic acid transformation in which the carboxyl group apparently migrates to the adjacent carbon (Figure 6.5). Phenylpyruvic acid and phenyl-lactic acid have been shown to be involved, and tropine becomes esterified with phenyl-lactic acid (most likely via the coenzyme-A ester) to form **littorine** before the rearrangement occurs. A cytochrome P-450-dependent enzyme transforms littorine to hyoscyamine aldehyde, and a radical process (Figure 6.5) with an intermediate cyclopropane-containing radical would accommodate the available data. Further modifications to the tropane

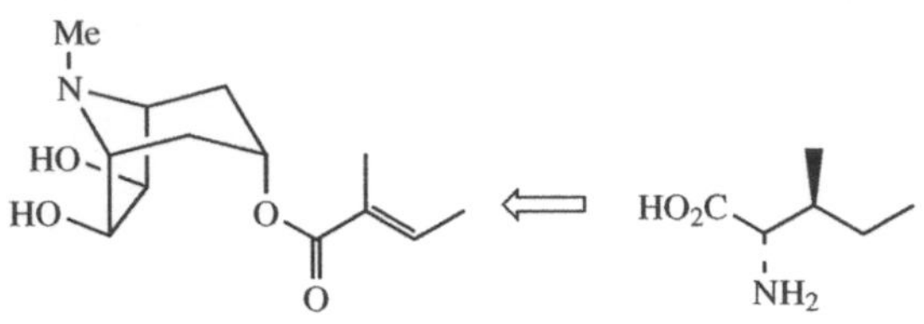

meteloidine   L-Ile

Figure 6.6

skeleton then occur on the ester, not on the free alcohol. These include hydroxylation to 6β-hydroxyhyoscyamine and additional oxidation, allowing formation of an epoxide grouping as in **hyoscine (scopolamine)**. Both of these reactions are catalysed by a single 2-oxoglutarate-dependent dioxygenase (see page 27). Other esterifying acids may be encountered in tropane alkaloid structures, e.g. tiglic acid in **meteloidine** (Figure 6.6)

from *Datura meteloides* and phenyl-lactic acid in littorine above, which is a major alkaloid in *Anthocercis littorea*. Tiglic acid is known to be derived from the amino acid L-isoleucine (see page 216).

The structure of **cuscohygrine** arises by an *intermolecular* Mannich reaction involving a second N-methyl-Δ^1-pyrrolinium cation (Figure 6.7). Cuscohygrine (and also hygrine) racemize rapidly in solution via a self-catalysed (i.e. base-catalysed) Michael-type retroconjugate addition process.

The tropane alkaloids (−)-hyoscyamine and (−)-hyoscine are among the most important of the natural alkaloids used in medicine. They are found in a variety of solanaceous plants, including *Atropa belladonna* (deadly nightshade), *Datura stramonium* (thornapple) and other *Datura* species, *Hyoscyamus niger* (henbane), and *Duboisia* species [Box 6.1]. These alkaloids are also responsible for the pronounced toxic properties of these plants.

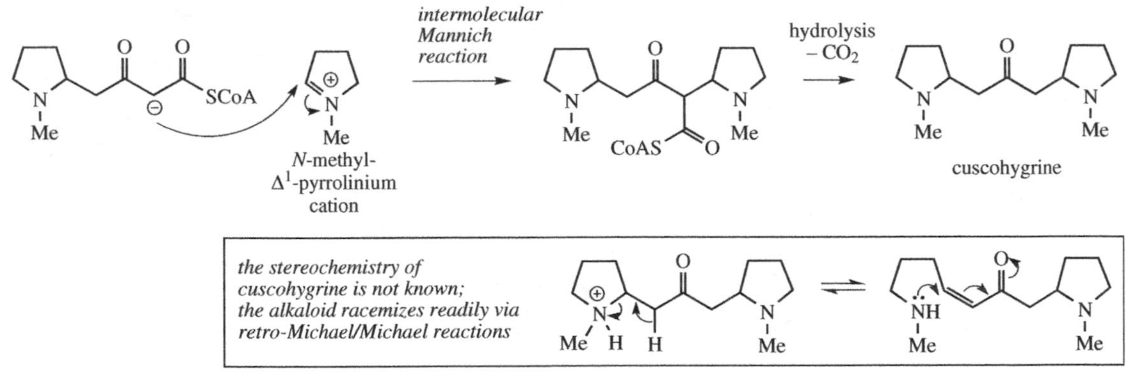

Figure 6.7

Box 6.1

Belladonna

The deadly nightshade *Atropa belladonna* (Solanaceae) has a long history as a highly poisonous plant. The generic name is derived from *Atropos*, in Greek mythology the Fate who cut the thread of life. The berries are particularly dangerous, but all parts of the plant contain toxic alkaloids, and even handling of the plant can lead to toxic effects, since the alkaloids are readily absorbed through the skin. Although humans are sensitive to the toxins, some animals, including sheep, pigs, goats, and rabbits, are less susceptible. Cases are known where the consumption of rabbits or birds that have ingested belladonna has led to human poisoning. The plant is a tall perennial herb producing dull-purple bell-shaped flowers followed by conspicuous shiny black fruits, the size of a small cherry. *Atropa belladonna* is indigenous to central and southern Europe, though it is not especially common. It is cultivated for drug use in Europe and the United States. The tops of the plant are harvested two or three times per year and dried to give **belladonna** herb. Roots from plants some 3–4 years old are less commonly employed as a source of alkaloids.

Box 6.1   (continued)

Belladonna herb typically contains 0.3–0.6% of alkaloids, mainly (−)-hyoscyamine (Figure 6.4). Belladonna root has only slightly higher alkaloid content at 0.4–0.8%, again mainly (−)-hyoscyamine. Minor alkaloids, including (−)-hyoscine (Figure 6.5) and cuscohygrine (Figure 6.7), are also found in the root, though these are not usually significant in the leaf. The mixed alkaloid extract from belladonna herb is still used as a gastrointestinal sedative, usually in combination with antacids. Root preparations can be used for external pain relief, e.g. in belladonna plasters.

Stramonium

Datura stramonium (Solanaceae) is commonly referred to as thornapple on account of its spikey fruit. It is a tall bushy annual plant widely distributed in Europe and North America, and because of its alkaloid content is potentially very toxic. Indeed, a further common name, Jimson or Jamestown weed, originates from the poisoning of early settlers near Jamestown, Virginia. At subtoxic levels, the alkaloids can provide mild sedative action and a feeling of well-being. In the Middle Ages, stramonium was employed to drug victims prior to robbing them. During this event, the victim appeared normal and was cooperative, though afterwards could usually not remember what had happened. For drug use, the plant is cultivated in Europe and South America. The leaves and tops are harvested when the plant is in flower. **Stramonium** leaf usually contains 0.2–0.45% of alkaloids, principally (−)-hyoscyamine and (−)-hyoscine in a ratio of about 2:1. In young plants, (−)-hyoscine can predominate.

The generic name *Datura* is derived from dhat, an Indian poison used by the Thugs. The narcotic properties of *Datura* species, especially *Datura metel*, have been known and valued in India for centuries. The plant material was usually absorbed by smoking. Most species of *Datura* contain similar tropane alkaloids and are potential sources of medicinal alkaloids. In particular, *Datura sanguinea*, a perennial of tree-like stature with blood-red flowers, is cultivated in Ecuador and yields leaf material with a high (0.8%) alkaloid content in which the principal component is (−)-hyoscine. The plants can be harvested several times a year. *Datura sanguinea* and several other species of the tree-daturas (now classified as a separate genus *Brugmansia*) are widely cultivated as ornamentals, especially for conservatories, because of their attractive large tubular flowers. The toxic potential of these plants is not always recognized.

Hyoscyamus

Hyoscyamus niger (Solanaceae), or henbane, is a European native with a long history as a medicinal plant. Its inclusion in mediaeval concoctions and its power to induce hallucinations with visions of flight may well have contributed to our imaginary view of witches on broomsticks. The plant has both annual and biennial forms, and is cultivated in Europe and North America for drug use, the tops being collected when the plant is in flower and then dried rapidly. The alkaloid content of **hyoscyamus** is relatively low at 0.045–0.14%, but this can be composed of similar proportions of (−)-hyoscine and (−)-hyoscyamine. Egyptian henbane, *Hyoscyamus muticus*, has a much higher alkaloid content than *Hyoscyamus niger*, and although it has mainly been collected from the wild, especially from Egypt, it functions as a major commercial source for alkaloid production. Some commercial cultivation occurs in California. The alkaloid content of the leaf is from 0.35 to 1.4%, of which about 90% is (−)-hyoscyamine.

Duboisia

Duboisia is a small genus of Australian trees, containing only three species, again from the family Solanaceae. Two of these, *Duboisia myoporoides* and *Duboisia leichhardtii* are grown commercially in Australia for tropane alkaloid production. The small trees are kept as bushes to allow frequent harvesting, with up to 70–80% of the leaves being removed every 7–8 months. The alkaloid content of the leaf is high (up to 3% has been recorded), and it includes (−)-hyoscyamine, (−)-hyoscine, and a number of related structures. The proportion of hyoscyamine to hyoscine varies according to the species used and the area in which the trees are grown. The hyoscine content is frequently much higher than that of hyoscyamine. Indeed, interest in *Duboisia* was very much stimulated by the demand for hyoscine as a treatment for motion sickness in military personnel during the Second World War. Even higher levels of alkaloids, and higher proportions of hyoscine, can be obtained from selected *Duboisia myoporoides* × *Duboisia leichhardtii* hybrids, which are currently cultivated. The hybrid is superior to either parent and can yield 1–2.5% hyoscine and 0–1% hyoscyamine. *Duboisia* leaf is an important commercial source of medicinal tropane alkaloids.

Duboisia hopwoodi, the third species of *Duboisia*, contains little tropane alkaloid content, but produces mainly nicotine and related alkaloids, e.g. nornicotine (see page 331). Leaves of this plant were chewed by aborigines for their stimulating effects.

Box 6.1 (continued)

Allied Drugs

Tropane alkaloids, principally hyoscyamine and hyoscine, are also found in two other medicinal plants, namely scopolia and mandrake, but these plants find little current use. Scopolia (*Scopolia carniolica*; Solanaceae) resembles belladonna in appearance, though it is considerably smaller. Both root and leaf materials have been employed medicinally. The European mandrake (*Mandragora officinarum*; Solanaceae) has a complex history as a hypnotic, a general panacea, and an aphrodisiac. Its collection has been surrounded by much folklore and superstition, in that pulling it from the ground was said to drive its collector mad due to the unearthly shrieks emitted. The roots are frequently forked and are loosely likened to a man or woman. Despite the Doctrine of Signatures, which teaches that the appearance of an object indicates its special properties, from a pharmacological point of view this plant would be much more efficient as a pain reliever than as an aphrodisiac.

Hyoscyamine, Hyoscine, and Atropine

All the above solanaceous plants contain as main alkaloidal constituents the tropane esters (−)-**hyoscyamine** and (−)-**hyoscine**, together with other minor tropane alkaloids. The piperidine ring in the bicyclic tropane system has a chair-like conformation, and there is a ready inversion of configuration at the nitrogen atom so that the *N*-methyl group can equilibrate between equatorial and axial positions (Figure 6.8). An equatorial methyl is strongly favoured provided there are no substituents on the two-carbon bridge, in which case the axial form may predominate. (−)-Hyoscyamine is the ester of tropine (Figure 6.4) with (−)-(*S*)-tropic acid, whilst (−)-hyoscine contains scopine (see Figure 6.10) esterified with (−)-(*S*)-tropic acid. The optical activity of both hyoscyamine and hyoscine stems from the chiral centre in the acid portion, (*S*)-tropic acid. Tropine itself, although containing chiral centres, is a symmetrical molecule and is optically inactive; it can be regarded as a *meso* structure. The chiral centre in the tropic acid portion is adjacent to a carbonyl and the aromatic ring, and racemization can be achieved under mild conditions by heating or treating with base. This will involve an intermediate enol (or enolate) which is additionally favoured by conjugation with the aromatic ring (Figure 6.9). Indeed, normal base-assisted fractionation of plant extracts to isolate the alkaloids can sometimes result in production of significant amounts of racemic alkaloids. The plant material itself generally contains only the enantiomerically pure alkaloids. Hyoscyamine appears to be much more easily racemized than hyoscine. Hydrolysis of the esters using acid or base usually gives racemic tropic acid. Note that littorine (Figure 6.5), in which the chiral centre is not adjacent to the phenyl ring, is not readily racemized, and base hydrolysis gives optically pure phenyl-lactic acid. The racemic form of hyoscyamine is

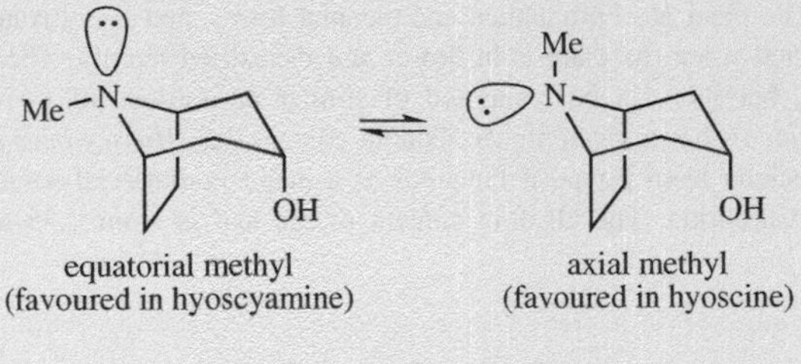

equatorial methyl
(favoured in hyoscyamine)

axial methyl
(favoured in hyoscine)

Figure 6.8

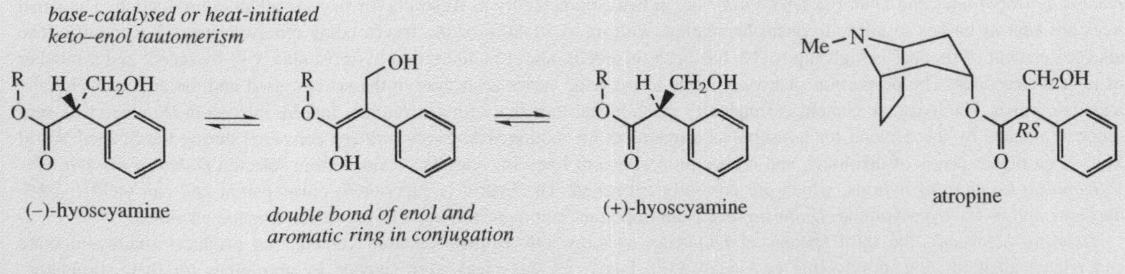

*base-catalysed or heat-initiated
keto–enol tautomerism*

(−)-hyoscyamine

*double bond of enol and
aromatic ring in conjugation*

(+)-hyoscyamine

atropine

Figure 6.9

Box 6.1   (continued)

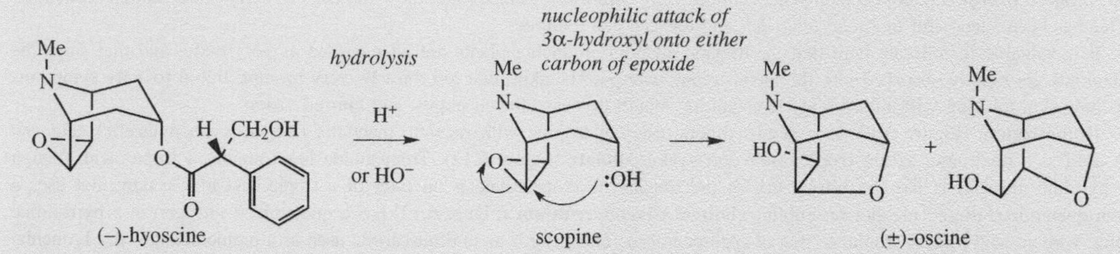

Figure 6.10

called **atropine** (Figure 6.9), whilst that of hyoscine is called atroscine. In each case, the biological activity of the (+)-enantiomer is some 20–30 times less than that of the natural (−)-form. Chemical hydrolysis of hyoscine in an attempt to obtain the alcohol scopine is not feasible. Instead, the alcohol oscine is generated because of the proximity of the 3α-hydroxyl group to the reactive epoxide function (Figure 6.10).

Probably for traditional reasons, salts of both (−)-hyoscyamine and (±)-hyoscyamine (atropine) are used medicinally, whereas usage of hyoscine is restricted to the natural laevorotatory form. These alkaloids compete with acetylcholine for the muscarinic site of the parasympathetic nervous system, thus preventing the passage of nerve impulses, and are classified as anticholinergics. Acetylcholine binds to two types of receptor site, described as muscarinic or nicotinic. These are triggered specifically by the alkaloid muscarine from the fly agaric fungus *Amanita muscaria* or by the tobacco alkaloid nicotine (see page 334) respectively. The structural similarity between acetylcholine and muscarine (Figure 6.11) can readily be appreciated, and hyoscyamine is able to occupy the same receptor site by virtue of the spatial relationship between the nitrogen atom and the ester linkage (Figure 6.11). The side-chain also plays a role in the binding, explaining the marked difference in activities between the two enantiomeric forms. The agonist properties of hyoscyamine and hyoscine give rise to a number of useful effects, including antispasmodic action on the gastrointestinal tract, antisecretory effect controlling salivary secretions during surgical operations, and as mydriatics to dilate the pupil of the eye. Hyoscine has a depressant action on the central nervous system and finds particular use as a sedative to control motion sickness. One of the side-effects from oral administration of tropane alkaloids is dry mouth (the antisecretory effect), but this can be much reduced by transdermal administration. In motion sickness treatment, hyoscine can be supplied via an impregnated patch worn behind the ear. Hyoscine under its synonym **scopolamine** is also well known, especially in fiction, as a 'truth drug'. This combination of sedation, lack of will, and amnesia was first employed in childbirth to give what was termed 'twilight sleep', and may be compared with the mediaeval use of stramonium. The mydriatic use also has a very long history. Indeed, the specific name *belladonna* for deadly nightshade means 'beautiful lady' and refers to the practice of ladies at court who applied the juice of the fruit to the eyes, giving widely dilated pupils and a striking appearance, though at the expense of blurred vision through an inability to focus. Atropine also has useful antidote action in cases of poisoning caused by acetylcholinesterase inhibitors e.g. physostigmine and neostigmine (see page 386) and organophosphate insecticides.

Hyoscine is commercially more valuable than hyoscyamine, but most of the plant sources described produce considerably more hyoscyamine. Researchers are thus trying to establish plants or plant systems that accumulate predominantly hyoscine. Outside of conventional plant breeding and selection, genetic manipulation offers considerable scope. For example, it has been demonstrated that introducing the hyoscyamine 6β-hydroxylase gene from *Hyoscyamus niger* into *Atropa belladonna* increases hyoscine content, whilst overexpressing the *Hyoscyamus niger* gene in *Hyoscyamus muticus* root cultures can increase hyoscine production up to 100-fold. An alternative approach is to express the hyoscyamine 6β-hydroxylase gene in an organism that does

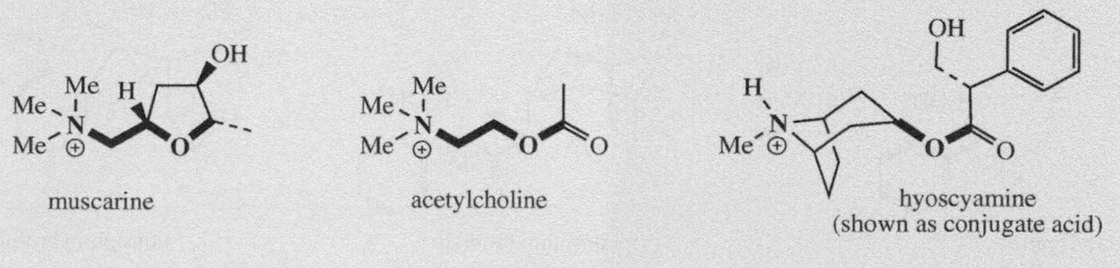

muscarine   acetylcholine   hyoscyamine (shown as conjugate acid)

Figure 6.11

Box 6.1   (continued)

not produce tropane alkaloids, supply the cultured cells with hyoscyamine, and allow the cells to carry out a biotransformation. This has been successful in *Escherichia coli* and tobacco cell cultures.

It is valuable to reiterate here that the tropane alkaloid-producing plants are all regarded as very toxic, and that since the alkaloids are rapidly absorbed into the bloodstream, even via the skin, first aid must be very prompt. Initial toxicity symptoms include skin flushing with raised body temperature, mouth dryness, dilated pupils, and blurred vision.

Homatropine (Figure 6.12) is a semi-synthetic ester of tropine with racemic mandelic (2-hydroxyphenylacetic) acid and is used as a mydriatic, as are **tropicamide** and **cyclopentolate** (Figure 6.12). Tropicamide is an amide of tropic acid, though a pyridine nitrogen is used to mimic that of the tropane. Cyclopentolate is an ester of a tropic acid-like system, but uses a non-quaternized amino alcohol resembling choline. **Glycopyrronium** (Figure 6.12) has a quaternized nitrogen in a pyrrolidine ring, with an acid moiety similar to that of cyclopentolate. This drug is an antimuscarinic used as a premedicant to dry bronchial and salivary secretions. **Hyoscine butylbromide** (Figure 6.13) is a gastrointestinal antispasmodic synthesized from (−)-hyoscine by quaternization of the amine function with butyl bromide. The quaternization of tropane alkaloids by *N*-alkylation proceeds such that the incoming alkyl group always approaches from the equatorial position. The potent bronchodilator **ipratropium bromide** (Figure 6.13) is thus synthesized from noratropine by successive isopropyl and methyl alkylations. This drug is used in inhalers for the treatment of chronic bronchitis. **Tiotropium bromide** (Figure 6.13) is a newer, longer-acting agent.

Benzatropine (benztropine; Figure 6.12) is an ether of tropine used as an antimuscarinic drug in the treatment of Parkinson's disease. It is able to inhibit dopamine reuptake, helping to correct the deficiency which is characteristic of Parkinsonism.

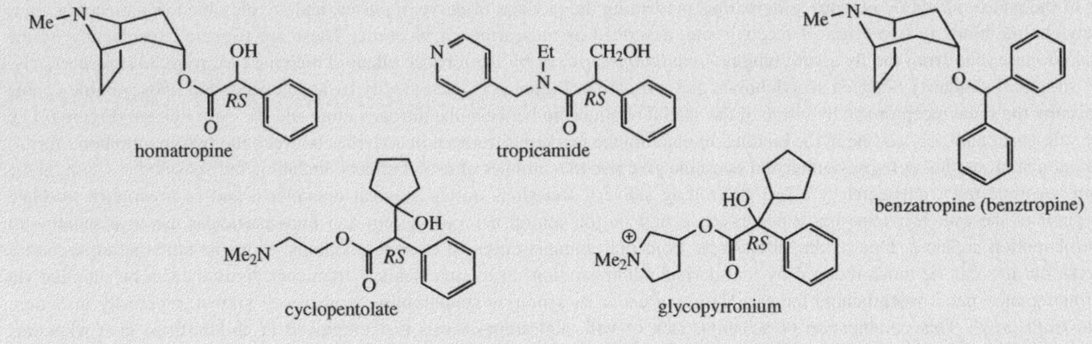

Figure 6.12

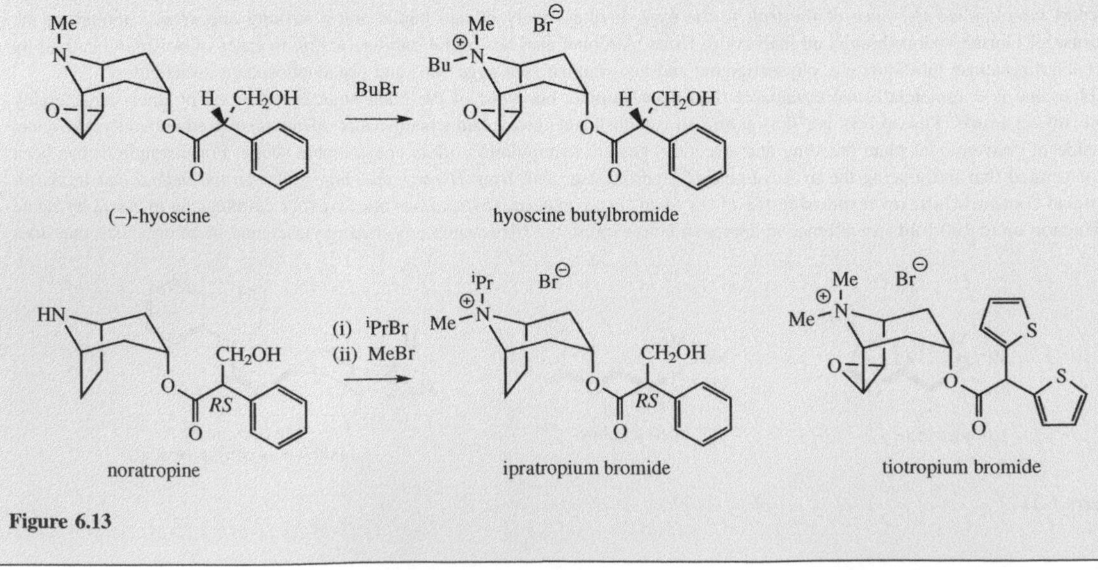

Figure 6.13

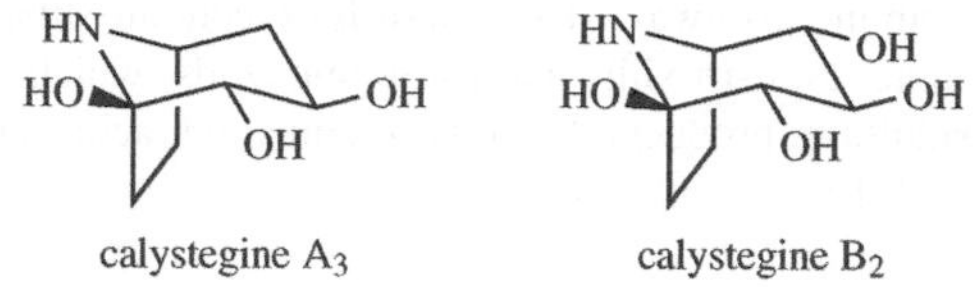

calystegine A₃   calystegine B₂

Figure 6.14

The **calystegines** are a group of recently discovered, water-soluble, polyhydroxy nortropane derivatives that are found in the leaves and roots of many of the solanaceous plants, including *Atropa, Datura, Duboisia, Hyoscyamus, Mandragora, Scopolia* and *Solanum*. They were first isolated from *Calystegia sepium* (Convolvulaceae). These compounds, e.g. calystegin A₃ and calystegin B₂ (Figure 6.14), are currently of great interest as glycosidase inhibitors. They have similar potential for the development of drugs with activity against the

AIDS virus HIV as the polyhydroxyindolizidines such as castanospermine (see page 330), and the aminosugars such as deoxynojirimycin (see page 498). Examples of tri-, tetra-, and penta-hydroxy calystegines are currently known. These alkaloids appear to be produced from the 3β-alcohol ψ-tropine (Figure 6.4) by a sequence of *N*-demethylation, followed by further hydroxylation steps. *N*-Demethylation is also a feature in the formation of nornicotine from nicotine (see page 333).

Cocaine (Figure 6.4) is a rare alkaloid restricted to some species of *Erythroxylum* (Erythroxylaceae). *Erythroxylum coca* (coca) is the most prominent as a source of cocaine, used medicinally as a local anaesthetic, and as an illicit drug for its euphoric properties [Box 6.2]. Coca also contains significant amounts of **cinnamoylcocaine** (**cinnamylcocaine**; Figure 6.15), where cinnamic acid rather than benzoic acid is the esterifying acid, together with some typical tropine derivatives without

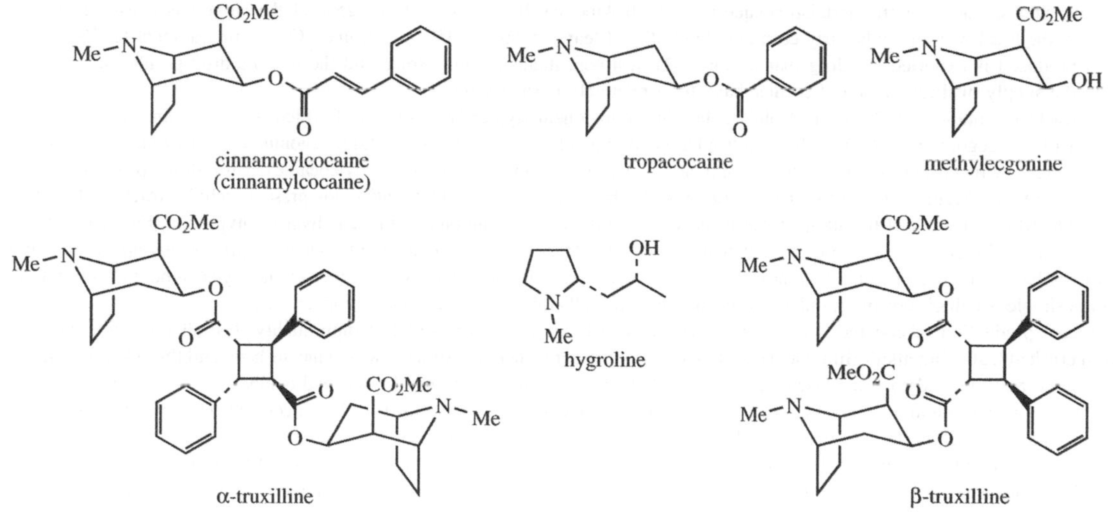

Figure 6.15

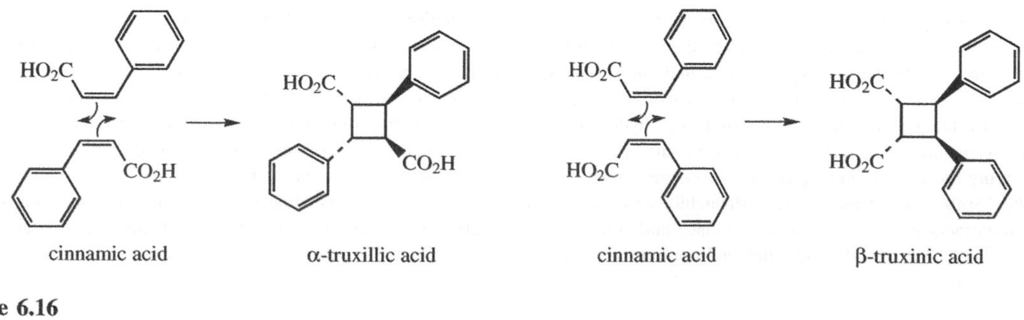

Figure 6.16

the extra carboxyl, e.g. **tropacocaine** (Figure 6.15). Tropacocaine still retains the 3β-configuration, showing that the stereospecific carbonyl reduction is the same as with the cocaine route, and not as with the hyoscyamine pathway. The **truxillines** contain dibasic acid moieties, α-truxillic and β-truxinic acids, which are cycloaddition products from two cinnamic acid units (Figure 6.16).

Box 6.2

Coca

Coca leaves are obtained from species of *Erythroxylum* (Erythroxylaceae), small shrubs native to the Andes region of South America, namely Colombia, Ecuador, Peru, and Bolivia. Peru is the only producer of medicinal coca; illicit supplies originate from Colombia, Peru, and Bolivia. Two main species provide drug materials, *Erythroxylum coca* and *Erythroxylum novogranatense*, though each species exists in two distinguishable varieties. *Erythroxylum coca* var. *coca* provides Peruvian or Huanaco coca, *Erythroxylum coca* var. *ipadu* Amazonian coca, *Erythroxylum novogranatense* var. *novogranatense* Colombian coca, and *Erythroxylum novogranatense* var. *truxillense* gives Trujillo coca. Cultivated plants are kept small by pruning, and a quantity of leaves is harvested from each plant three or more times per year.

Coca-leaf chewing has been practised by South American Indians for many years and has been an integral part of the native culture pattern. Leaf is mixed with lime, thus liberating the principal alkaloid cocaine as the free base, and the combination is then chewed. Cocaine acts as a potent antifatigue agent, and allows labourers to ignore hunger, fatigue, and cold, enhancing physical activity and endurance. Originally, the practice was limited to the Inca high priests and favoured individuals, but it became widespread after the Spanish conquest of South America. It is estimated that 25% of the harvest is consumed in this way by the local workers, who may each use about 50 g of leaf per day ($\equiv$ 350 mg cocaine). Only a tiny amount (1–2%) of the coca produced is exported for drug manufacture. The rest contributes to illicit trade and the world's drug problems. Efforts to stem the supply of illicit coca and cocaine have been relatively unsuccessful.

Coca leaf contains 0.7–2.5% of alkaloids, the chief component (typically 40–50%) of which is (−)-cocaine (Figure 6.4), a diester of (−)-ecgonine. Note that although tropine is an optically inactive *meso* structure, ecgonine contains four chiral centres, is no longer symmetrical, and is, therefore, optically active. Cinnamoylcocaine (cinnamylcocaine), α-truxilline, β-truxilline, and methylecgonine (Figure 6.15) are minor constituents also based on ecgonine. Other alkaloids present include structures based on φ-tropine (the 3β-isomer of tropine), such as tropacocaine (Figure 6.15), and on hygrine, e.g. hygrine, hygroline (Figure 6.15), and cuscohygrine (Figure 6.7). Cuscohygrine typically accounts for 20–30% of the alkaloid content. *Erythroxylum novogranatense* varieties have a higher cocaine content than *Erythroxylum coca* varieties, but also a higher cinnamoylcocaine content and are less desirable for illicit cocaine production, in that the latter alkaloid hinders crystallization of cocaine.

Illegal production of cocaine is fairly unsophisticated, but can result in material of high quality. The alkaloids are extracted from crushed leaf using alkali (lime) and petrol. The petrol extract is then re-extracted with aqueous acid, and this alkaloid fraction is basified and allowed to stand, yielding the free alkaloid as a paste. Alternatively, the hydrochloride or sulfate salts may be prepared. The coca alkaloids are often diluted with carrier to give a preparation with 10–12% of cocaine. The illicit use of cocaine and cocaine hydrochloride is a major problem worldwide. The powder is usually sniffed into the nostrils, where it is rapidly absorbed by the mucosa, giving stimulation and short-lived euphoria through inhibiting reuptake of neurotransmitters dopamine, noradrenaline, and serotonin, so prolonging and augmenting their effects. Regular usage induces depression, dependence, and damage to the nasal membranes. The drug may also be injected intravenously or the vapour inhaled. For inhalation, the free base or 'crack' is employed to increase volatility. The vaporized cocaine is absorbed extremely rapidly and carried to the brain within seconds, speeding up and enhancing the euphoric lift. Taken in this form, cocaine has proved highly addictive and dangerous. Cocaine abuse is currently regarded as a greater problem than heroin addiction, and despite intensive efforts, there is no useful antagonist drug available to treat cocaine craving and addiction. Considerable research is being directed towards an alternative strategy, namely to enhance enzymic hydrolysis of cocaine and to accelerate its clearance from the body. The major cocaine-metabolizing enzymes in humans are a non-specific serum cholinesterase (termed butyrylcholinesterase), which hydrolyses the benzoyl ester, and two liver carboxylesterases, one of which catalyses hydrolysis of the methyl ester, the other the benzoyl ester. Enhancement of benzoyl ester hydrolysis is the more desirable, since the alternative product benzoylecgonine is itself psychoactive. Further, when users consume cocaine and alcohol concurrently, transesterification of benzoylecgonine may occur, giving the corresponding ethyl ester cocaethylene, which has increased toxicity and half-life.

In the 1800s, coca drinks were fashionable, and one in particular, Coca-Cola®, became very popular. This was originally based on extracts of coca (providing cocaine) and cola (supplying caffeine) (see page 415), but although the coca content was omitted from 1906 onwards, the name and popularity continue.

Box 6.2   (continued)

Medicinally, **cocaine** is of value as a local anaesthetic for topical application. It is rapidly absorbed by mucous membranes and paralyses peripheral ends of sensory nerves. This is achieved by blocking ion channels in neural membranes. It was widely used in dentistry, but has been replaced by safer drugs, though it still has applications in ophthalmic and in ear, nose and throat surgery.

The essential functionalities of cocaine required for activity were eventually assessed to be the aromatic carboxylic acid ester and the basic amino group, separated by a lipophilic hydrocarbon chain. Synthetic drugs developed from the cocaine structure have been introduced to provide safer, less toxic local anaesthetics (Figure 6.17). **Procaine**, though little used now, was the first major analogue employed. **Benzocaine** is used topically, but has a short duration of action. **Tetracaine (amethocaine)**, **oxybuprocaine**, and **proxymetacaine** are valuable local anaesthetics employed principally in ophthalmic work. The ester function can be replaced by an amide, and this gives better stability towards hydrolysis in aqueous solution or by esterases. **Lidocaine (lignocaine)** is an example of an amino amide analogue and is perhaps the most widely used local anaesthetic, having rapid action, effective absorption, good stability, and may be used by injection or topically. Other amino amide local anaesthetic structures include **prilocaine**, with similar properties to lidocaine and very low toxicity, and **bupivacaine**, which has a long duration of action and is currently the most widely used local anaesthetic agent in both surgery and obstetrics. The (*S*)-enantiomer **levobupivacaine** has considerably fewer side-effects than the (*R*)-isomer, and is thus preferred over the racemic drug. **Ropivacaine, mepivacaine,** and **articaine (carticaine)** are some recently inroduced amide-type local anaesthetics, the latter two being used predominantly in dentistry. **Cinchocaine** is often incorporated into preparations to soothe haemorrhoids.

Lidocaine, although introduced as a local anaesthetic, was subsequently found to be a potent antiarrhythmic agent, and it now finds further use as an antiarrhythmic drug, for treatment of ventricular arrhythmias especially after myocardial infarction. Other cocaine-related structures also find application in the same way, including **procainamide** and **flecainide** (Figure 6.17). Procainamide is an amide analogue of procaine, whilst in **mexiletene**, a congener of lidocaine, the amide group has been replaced by a simple ether linkage.

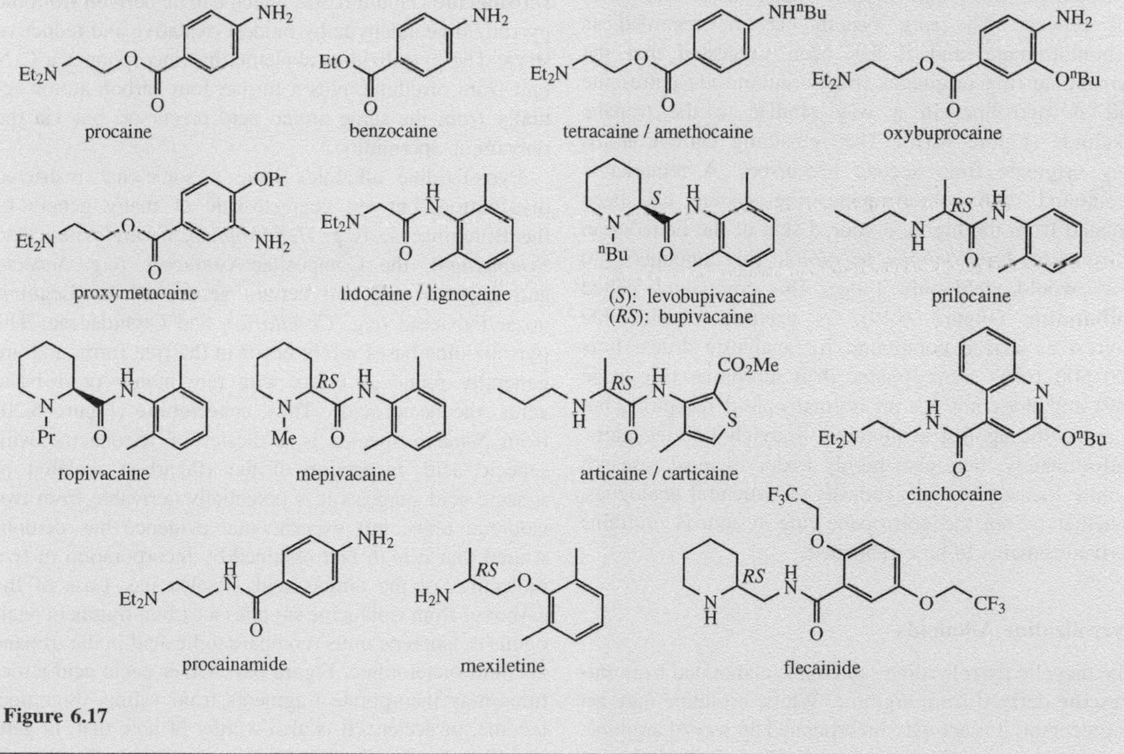

Figure 6.17

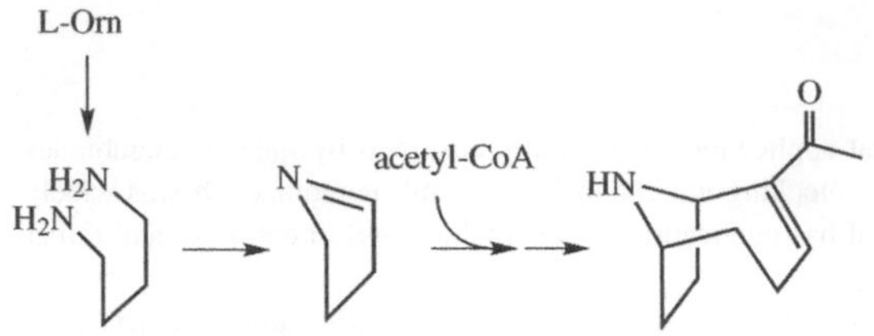

Figure 6.18

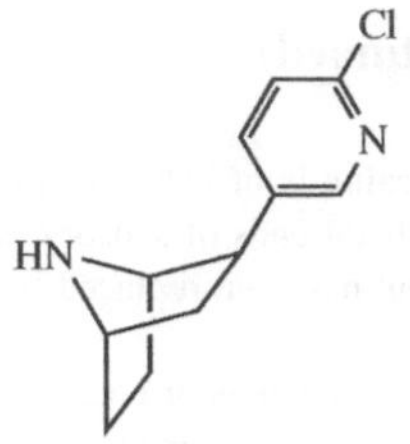

Figure 6.19

Anatoxin-*a* (Figure 6.18) is a toxic tropane-related alkaloid produced by a number of cyanobacteria, e.g. *Anabaena flos-aquae* and *Aphanizomenon flos-aquae*, species which proliferate in lakes and reservoirs during periods of hot, calm weather. A number of animal deaths have been traced back to consumption of water containing the cyanobacteria and ingestion of the highly potent neurotoxin anatoxin-*a*, which has been termed Very Fast Death Factor. Anatoxin-*a* is one of the most powerful nicotinic acetylcholine receptor agonists known, and has become a useful pharmacological probe for elucidating the mechanism of acetylcholine-mediated neurotransmission and disease states associated with this process. The ring system may be regarded as a homotropane, and it has been suggested that the pyrrolidine ring originates from ornithine via putrescine and Δ¹-pyrroline, in a way similar to the tropane alkaloids (Figure 6.18). The remaining carbon atoms may originate from acetate precursors. A remarkable compound with a nortropane ring system has been isolated from the highly coloured skin of the Ecuadorian poison frog *Epipedobates tricolor* in tiny amounts (750 frogs would yield only 1 mg). This compound, called **epibatidine** (Figure 6.19), is exciting considerable interest as a lead compound for analgesic drugs. It is 200–500 times more potent than morphine (see page 349) and does not act on normal opioid receptors, but is a specific agonist at nicotinic acetylcholine receptors. Unfortunately, it is also highly toxic, so most research is now focused on the synthesis of structural analogues. Whether or not the nortropane ring system is ornithine derived remains to be established.

Pyrrolizidine Alkaloids

The bicyclic pyrrolizidine skeleton is elaborated from **putrescine** derived from arginine. Whilst ornithine may act as precursor, it is actually incorporated by way of arginine, because plants synthesizing pyrrolizidine alkaloids appear to lack the decarboxylase enzyme transforming ornithine

into putrescine (Figure 6.2). Putrescine is converted into the polyamine **homospermidine** by a process that transfers an aminopropyl group from spermidine (Figure 6.2) in an NAD^+-dependent reaction (Figure 6.20). The cofactor requirement suggests this may involve imine intermediates. The pyrrolizidine skeleton is built up from homospermidine by a sequence of oxidative deamination, imine formation, and an intramolecular Mannich reaction which exploits the enolate anion generated from the aldehyde. This latter reaction is analogous to that proposed in formation of the tropane ring system (see page 314). A typical simple natural pyrrolizidine structure is that of **retronecine** (Figure 6.20), which can be derived from the pyrrolizidine aldehyde by modest oxidative and reductive steps. The pyrrolizidine skeleton thus incorporates a C_4N unit from ornithine, plus a further four carbon atoms actually from the same amino acid precursor, but via the polyamine spermidine.

Pyrrolizidine alkaloids have a somewhat restricted distribution, but are characteristic of many genera of the Boraginaceae (e.g. *Heliotropium, Cynoglossum,* and *Symphytum*), the Compositae/Asteraceae (e.g. *Senecio* and *Eupatorium*), and certain genera of the Leguminosae/Fabaceae (e.g. *Crotalaria*), and Orchidaceae. The pyrrolizidine bases rarely occur in the free form, and are generally found as esters with rare mono- or di-basic acids, the necic acids. Thus, **senecionine** (Figure 6.20) from *Senecio* species is a diester of retronecine with senecic acid. Inspection of the 10-carbon skeleton of senecic acid suggests it is potentially derivable from two isoprene units, but experimental evidence has demonstrated that it is in fact obtained by incorporation of two molecules of the amino acid L-isoleucine. Loss of the carboxyl from isoleucine supplies a carbon fragment analogous to isoprene units (compare tiglic acid in the tropane alkaloid meteloidine, Figure 6.6). Other necic acid structures may incorporate fragments from valine, threonine, leucine, or acetate. It is also worthy of note that, in general, the pyrrolizidine alkaloids accumulate in the plant as polar, salt-like *N*-oxides, facilitating their transport and,

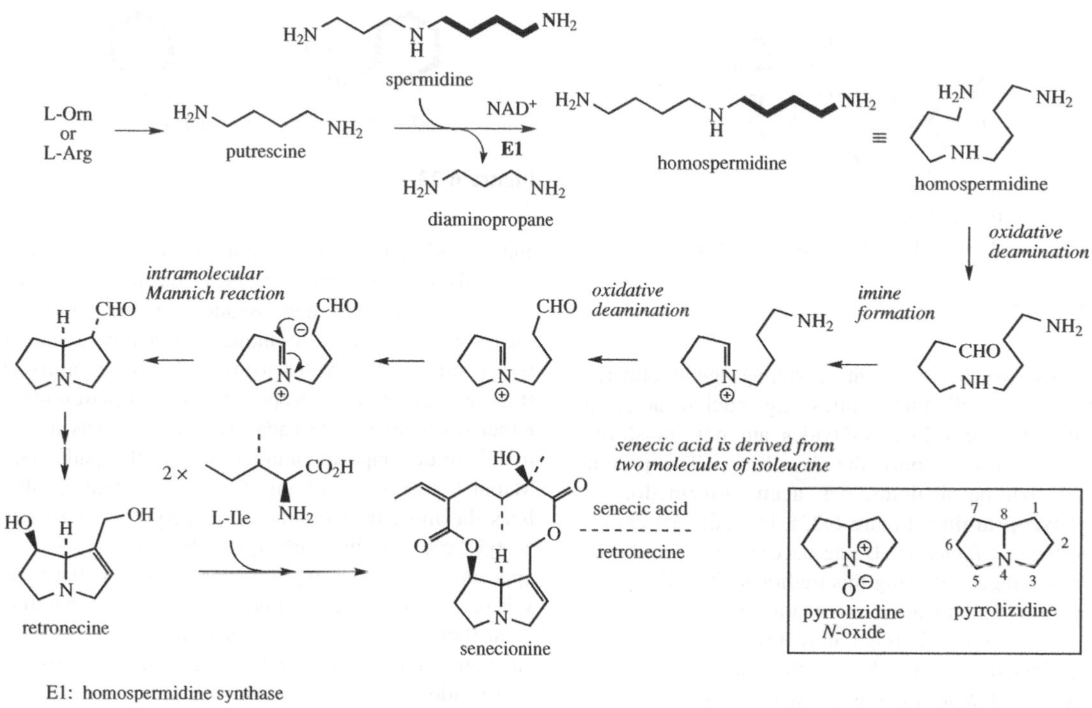

E1: homospermidine synthase

Figure 6.20

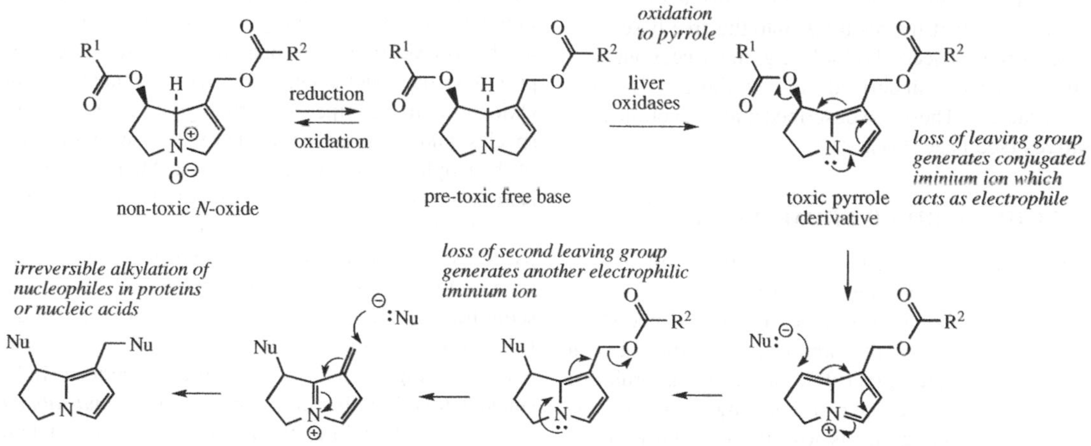

Figure 6.21

above all, maintaining them in a non-toxic form. The *N*-oxides are easily changed back to the tertiary amines by mild reduction, as will occur in the gut of a herbivore (Figure 6.21).

Many pyrrolizidine alkaloids are known to produce pronounced hepatic toxicity, and there are many recorded cases of livestock poisoning, particularly from ingestion of *Senecio* species (ragworts). Potentially toxic structures have 1,2-unsaturation in the pyrrolizidine ring and an ester function on the side-chain. Although themselves non-toxic, these alkaloids are transformed by mammalian liver oxidases into reactive pyrrole

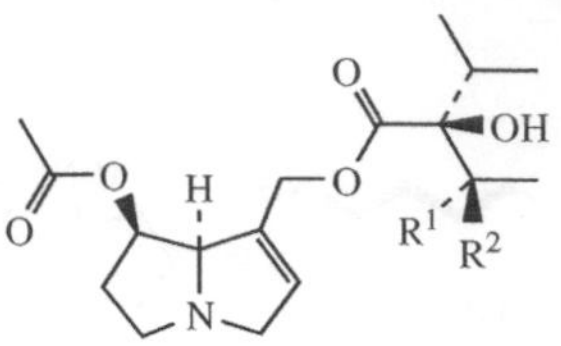

R^1 = H, R^2 = OH, acetyl-intermedine
R^1 = OH, R^2 = H, acetyl-lycopsamine

Figure 6.22

Figure 6.23

structures, which are potent alkylating agents and react with suitable cell nucleophiles, e.g. nucleic acids and proteins (Figure 6.21). *N*-Oxides are not transformed by these oxidases, only the free bases. The presence of pyrrolizidine alkaloids, e.g. **acetyl-intermedine** and **acetyl-lycopsamine** (Figure 6.22), in medicinal comfrey (*Symphytum officinale*; Boraginaceae) has emphasized potential dangers of using this traditional herbal drug as a remedy for inflammatory, rheumatic, and gastrointestinal disorders. Prolonged usage may lead to liver damage. Caterpillars of the cinnabar moth *Tyria jacobaeae* feed on species of *Senecio* (e.g. ragwort, *Senecio jacobaea*, and groundsel, *Senecio vulgaris*) with impunity, building up levels of pyrrolizidine alkaloids in their bodies (in the form of non-toxic *N*-oxides). This makes them distasteful to predators and also potentially toxic should the predator convert the alkaloids into the free bases.

Some of the tobacco alkaloids, e.g. nicotine, contain a pyrrolidine ring system derived from ornithine as a portion of their structure. These are described under nicotinic acid derivatives (see page 331).

ALKALOIDS DERIVED FROM LYSINE

L-**Lysine** is the homologue of L-ornithine, and it, too, functions as an alkaloid precursor, using pathways analogous to those noted for ornithine. The extra methylene group in lysine means this amino acid participates in forming six-membered piperidine rings, just as ornithine provided five-membered pyrrolidine rings. As with ornithine, the carboxyl group is lost, the ε-amino nitrogen rather than the α-amino nitrogen is retained, and lysine thus supplies a C$_5$N building block (Figure 6.23).

Piperidine Alkaloids

N-**Methylpelletierine** (Figure 6.24) is an alkaloidal constituent of the bark of pomegranate (*Punica granatum*; Punicacae), where it cooccurs with **pelletierine**

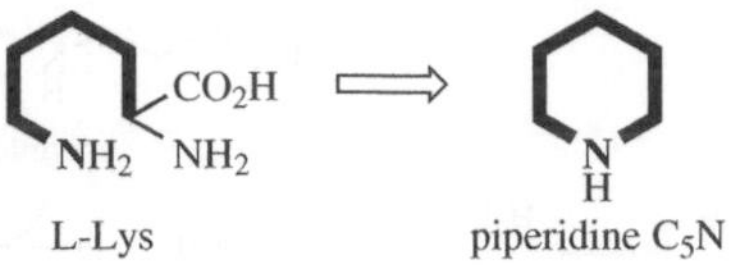

and **pseudopelletierine** (Figure 6.24), the mixture of alkaloids having activity against intestinal tapeworms. *N*-Methylpelletierine and pseudopelletierine are homologues of hygrine and tropinone respectively, and a pathway similar to Figure 6.4 using the diamine **cadaverine** (Figure 6.24) may be proposed. As with putrescine, the rather distinctive name cadaverine also reflects its isolation from decomposing animal flesh. In this pathway, the Mannich reaction involving the Δ^1-piperidinium salt utilizes the more nucleophilic acetoacetyl-CoA rather than acetyl-CoA, and the carboxyl carbon from acetoacetate appears to be lost during the reaction by suitable hydrolysis/decarboxylation reactions (Figure 6.24). **Anaferine** from *Withania somnifera* (Solanaceae; Figure 6.24) is an analogue of cuscohygrine in which a further piperidine ring is added via an intermolecular Mannich reaction.

The alkaloids found in the antiasthmatic plant *Lobelia inflata* (Campanulaceae) contain piperidine rings with C$_6$C$_2$ side-chains derived from phenylalanine via cinnamic acid. These alkaloids are produced as in Figure 6.25, in which benzoylacetyl-CoA, an intermediate in the β-oxidation of cinnamic acid (see page 160), provides the nucleophile for the Mannich reaction. Oxidation in the piperidine ring gives a new iminium species, and this can react further with a second molecule of benzoylacetyl-CoA, again via a Mannich reaction. Naturally, because of the nature of the side-chain, the second intramolecular Mannich reaction as involved in pseudopelletierine biosynthesis is not feasible. Alkaloids such as **lobeline** and **lobelanine** from *Lobelia inflata*, or **sedamine** from *Sedum acre* (Crassulaceae), are products from further *N*-methylation and/or carbonyl reduction reactions (Figure 6.25). The North American Indians smoked lobelia rather like tobacco (*Nicotiana tabacum*; Solanaceae). **Lobeline** stimulates nicotinic acetylcholine receptor sites in a similar way to nicotine, but with a weaker effect. It has been employed in smoking cessation preparations, and more recently to treat methamphetamine abuse (see page 404). Ketopiperidines such as lobeline, *N*-methylpelletierine, and related structures epimerize readily at the centre adjacent to the nitrogen via retro-Michael/Michael conjugate addition reactions (compare cuscohygrine, page 316).

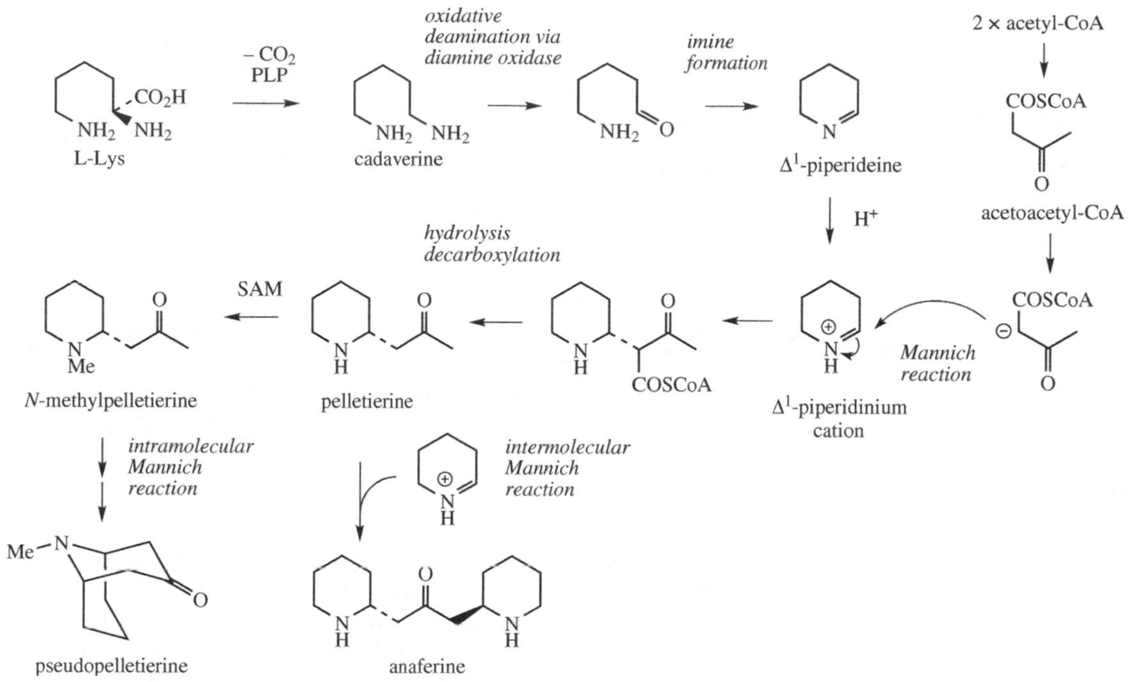

Figure 6.24

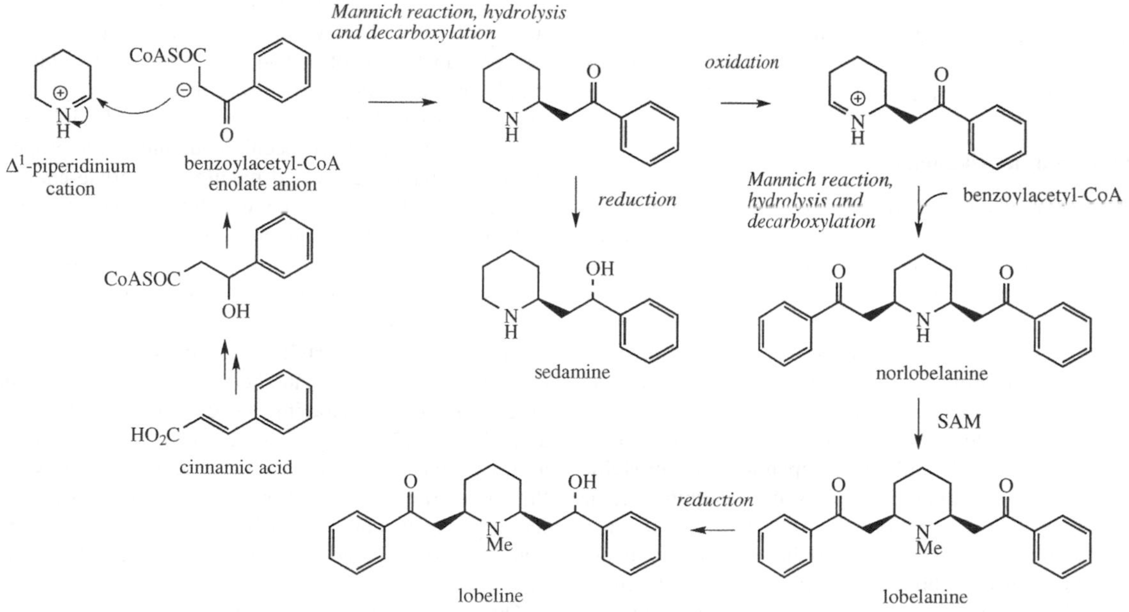

Figure 6.25

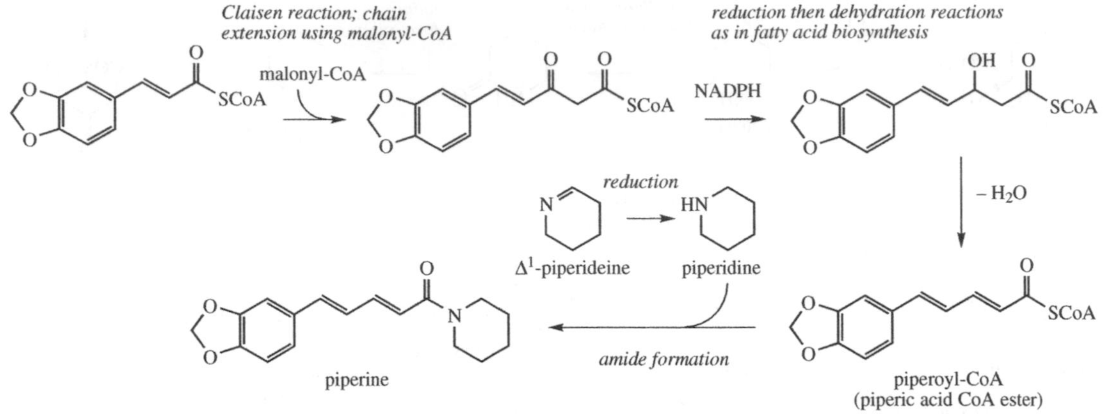

Figure 6.26

The simple piperidine alkaloid coniine from poison hemlock is not derived from lysine, but originates by an amination process and is discussed on page 401.

The pungency of the fruits of black pepper (*Piper nigrum*; Piperaceae), a widely used condiment, is mainly due to the piperidine alkaloid **piperine** (Figure 6.26). In this structure, the piperidine ring forms part of a tertiary amide structure and is incorporated via piperidine itself, the reduction product of Δ^1-piperideine (Figure 6.24). The piperic acid portion is derived from a cinnamoyl-CoA precursor, with chain extension using acetate/malonate (see page 168), and combines as its CoA ester with piperidine.

Quinolizidine Alkaloids

The lupin alkaloids, found in species of *Lupinus* (Leguminosae/Fabaceae) and responsible for the toxic properties associated with lupins, are characterized by a quinolizidine skeleton (Figure 6.27). This bicyclic ring system is closely related to the ornithine-derived pyrrolizidine system, but is formed from two molecules of lysine. **Lupinine** from *Lupinus luteus* is a relatively simple structure, very comparable to the basic ring system of the pyrrolizidine alkaloid retronecine (see page 325), but other lupin alkaloids, e.g. **lupanine** and **sparteine** (Figure 6.27) contain a tetracyclic bis-quinolizidine ring system and are formed by incorporation of a third lysine molecule. (−)-Sparteine is also the major alkaloid in Scotch broom (*Cytisus scoparius*; Leguminosae/Fabaceae); both enantiomeric forms of sparteine are found in nature. The alkaloid (−)-**cytisine**, a toxic component of *Laburnum* species (Leguminosae/Fabaceae), contains a modified tricyclic ring system, and comparison with the structures of lupanine or sparteine shows its

likely relationship by loss of four carbon atoms from the tetracyclic system of (+)-sparteine (Figure 6.27). The structural similarity of lupinine and retronecine is not fully reflected in the biosynthetic pathways. Experimental evidence shows lysine to be incorporated into lupinine via **cadaverine**, but no intermediate corresponding to homospermidine is implicated. Δ^1-Piperideine seems to be an important intermediate after cadaverine, and the pathway proposed (Figure 6.27) invokes coupling of two such molecules. The two tautomers of Δ^1-piperideine, nitrogen analogues of keto–enol systems, are able to couple by an aldol-type mechanism (see page 20). Indeed, this coupling occurs in solution at physiological pH values, though stereospecific coupling to the product shown in Figure 6.27 would require appropriate enzyme participation. Following the coupling, it is suggested that the imine system is hydrolysed, the primary amine group then oxidized, and formation of the quinolizidine ring is achieved by imine formation. **Lupinine** is then synthesized by a reductive step.

The pathway to **sparteine** and **lupanine** undoubtedly requires participation of another molecule of cadaverine or Δ^1-piperideine. Experimental data are not clear cut, and Figure 6.27 merely indicates how incorporation of a further piperidine ring might be envisaged. Loss of the fourth ring and oxidation to a pyridone system offers a potential route to **cytisine**. Note that either of the outermost rings could be lost to yield the same product.

Quinolizidine alkaloids are mainly found in plants of the Leguminosae/Fabaceae family. They deter or repel the feeding of herbivores and are toxic to them by a variety of mechanisms. A number of plants (*Laburnum, Cytisus, Lupinus*) containing significant quantities of these

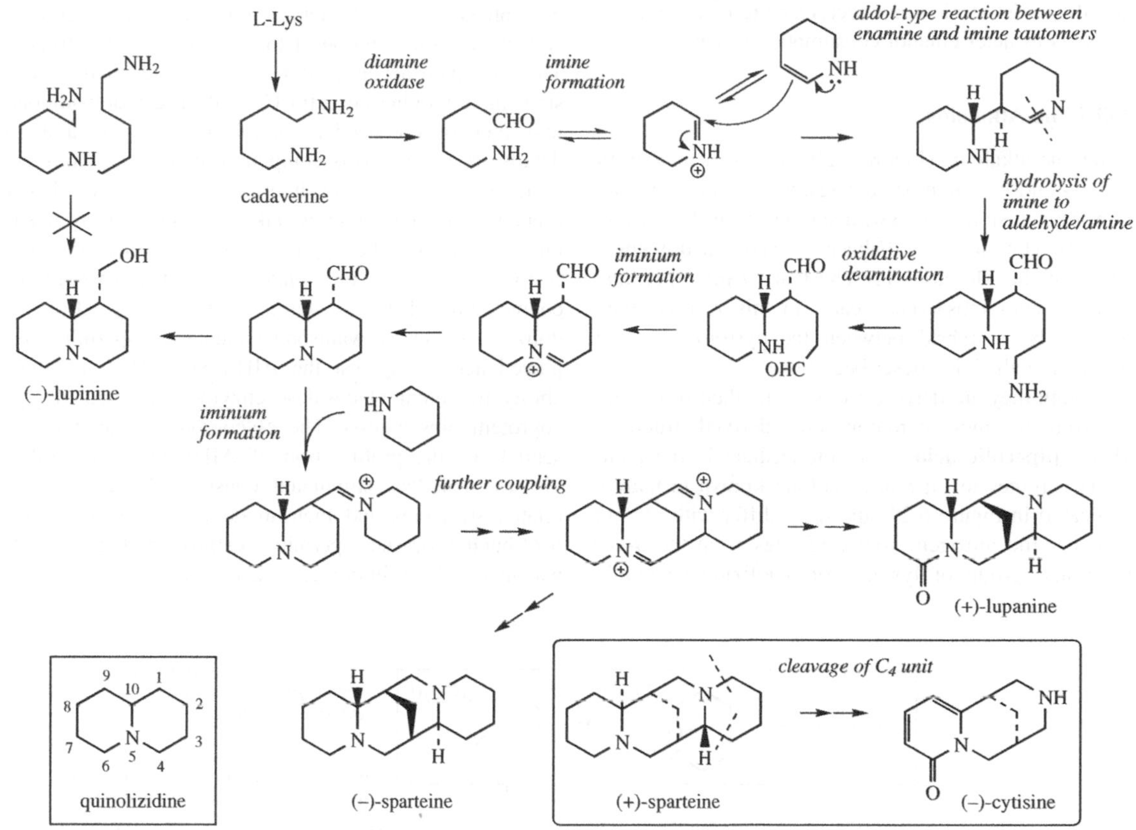

Figure 6.27

alkaloids must be regarded as potentially toxic to humans, and are known to be responsible for human poisoning. The widely planted and ornamental laburnum trees offer a particular risk, since all parts, including the pea-like seeds, contain dangerously high amounts of alkaloids. So-called 'sweet lupins' are selected strains with an acceptably low alkaloid content (typically about a quarter of the total alkaloids of 'bitter' strains), and are cultivated as a high-protein crop. Natural (−)-sparteine, a chiral diamine,

is a useful chiral ligand in asymmetric synthesis; it can be extracted readily from branches of *Cytisus scoparius.* (−)-Cytisine can be obtained in large amounts from the seeds of *Laburnum anagyroides* and is a potent agonist for nicotinic acetylcholine receptors, more so than nicotine itself. It has been used in eastern Europe for many years as a smoking cessation aid. The synthetic **varenicline** (Figure 6.28) is loosely based on the cytisine structure and has also been introduced to help stop smoking; it is

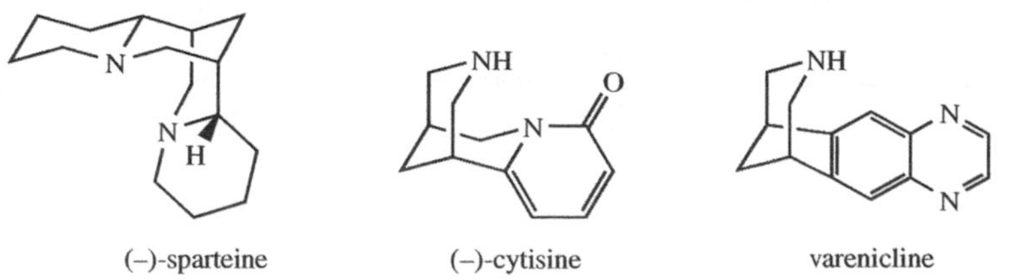

(−)-sparteine   (−)-cytisine   varenicline

Figure 6.28

a partial agonist at nicotinic acetylcholine receptors, and also seems to deter ethanol consumption in patients.

Indolizidine Alkaloids

Indolizidine alkaloids (Figure 6.29) are characterized by fused six- and five-membered rings, with a nitrogen atom at the ring fusion, e.g. **swainsonine** from *Swainsona canescens* (Leguminosae/Fabaceae) and **castanosper-mine** from the Moreton Bay chestnut *Castanospermum australe* (Leguminosae/Fabaceae). In this respect, they appear to be a hybrid between the pyrrolizidine and quinolizidine alkaloids described above.

Although they are derived from lysine, their origin deviates from the more common lysine-derived structures in that L-**pipecolic acid** is an intermediate in the pathway. Two routes to pipecolic acid are known in nature, as indicated in Figure 6.29, and these differ with respect to whether the nitrogen atom originates from the α- or the ε-amino group of lysine. For indolizidine alkaloid

biosynthesis, pipecolic acid is formed via the aldehyde and imine with retention of the α-amino group nitrogen. This can then act as starter for malonate chain extension, incorporating an extra C_2 unit. The indolizidinone may then be produced by simple reactions, though no details are known. This compound leads to castanospermine by a sequence of hydroxylations, but is also a branch-point compound to alkaloids such as swainsonine which have the opposite configuration at the ring fusion. Involvement of a planar iminium ion would account for the change in stereochemistry. Polyhydroxyindolizidines such as swainsonine and castanospermine displayed activity against the AIDS virus HIV, by their ability to inhibit glycosidase enzymes involved in glycoprotein biosynthesis. The glycoprotein coating is essential for the proliferation of AIDS and some other viruses. This has stimulated considerable research on related structures and their mode of action. The ester 6-*O*-butanoyl-castanospermine (**celgosivir**; Figure 6.30) was unsuccessful in clinical trials as an anti-AIDS agent,

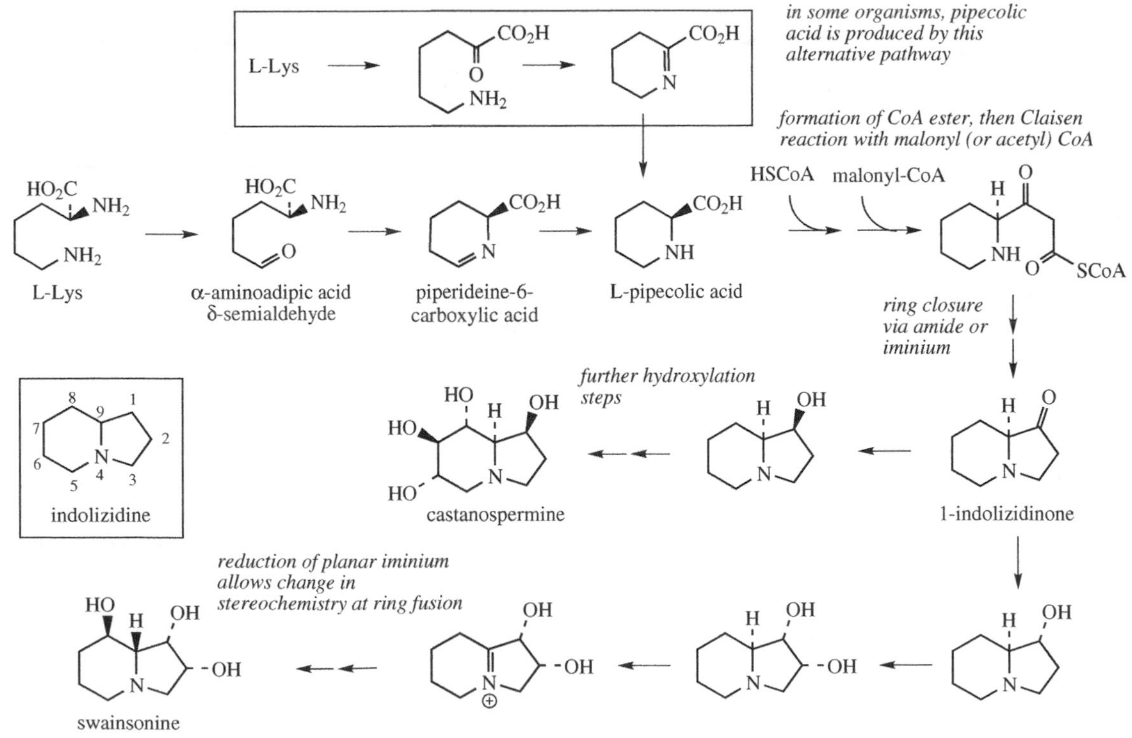

Figure 6.29

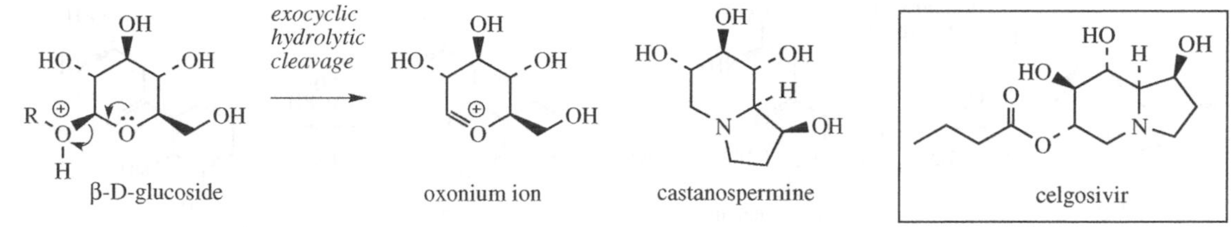

Figure 6.30

but is still being evaluated against hepatitis C. There is a strong similarity between castanospermine and the oxonium ion formed by hydrolytic cleavage of a glucoside (Figure 6.30) (see page 30), but there appears to be little stereochemical relationship with some other sugars, whose hydrolytic enzymes are also strongly inhibited. These alkaloids are also toxic to animals, causing severe gastrointestinal upset and malnutrition by severely affecting intestinal hydrolases. Indolizidine alkaloids are found in many plants in the Leguminosae/Fabaceae (e.g. *Swainsona, Astragalus, Oxytropis*) and also in some fungi (e.g. *Rhizoctonia leguminicola*, which produces swainsonine).

ALKALOIDS DERIVED FROM NICOTINIC ACID

Pyridine Alkaloids

The alkaloids found in tobacco (*Nicotiana tabacum*; Solanaceae) [Box 6.3] include **nicotine** and **anabasine** (Figure 6.31). The structures contain a pyridine system together with a pyrrolidine ring (in nicotine) or a piperidine ring (in anabasine), the latter rings arising from ornithine

and lysine respectively. The pyridine unit has its origins in **nicotinic acid** (**vitamin B$_3$**; Figure 6.31), the vitamin sometimes called **niacin** (see page 31). The amide nicotinamide forms an essential component of coenzymes such as NAD$^+$ and NADP$^+$ (see page 25). The nicotinic acid component of nicotinamide is synthesized in animals by degradation of L-tryptophan through the **kynurenine** pathway and **3-hydroxyanthranilic acid** (Figure 6.32) (see also dactinomycin, page 452), the pyridine ring being formed by oxidative cleavage of the benzene ring and subsequent inclusion of the amine nitrogen (Figure 6.32). However, plants such as *Nicotiana* use a different pathway employing dihydroxyacetone phosphate and L-aspartic acid precursors (Figure 6.33). The dibasic acid **quinolinic acid** features in both pathways. Decarboxylation to nicotinic acid involves assimilation of the quinolinic acid into the pyridine nucleotide cycle, and the subsequent side-chain modifications take place whilst the pyridine system is part of a nucleotide complex. Nicotinamide and nicotinic acid are subsequently released from nucleotide carriers. The pyridine nucleotide cycle is responsible for the biosynthesis of NAD$^+$, the degradation of NAD$^+$ to nicotinic acid, and the recycling of nicotinic acid to NAD$^+$. **Picolinic acid** (Figure 6.32) is an isomer of nicotinic acid also originating from the kynurenine pathway.

In the formation of **nicotine**, a pyrrolidine ring derived from ornithine, most likely as the *N*-methyl-Δ^1-pyrrolinium cation (see Figure 6.3), is attached to the pyridine ring of nicotinic acid, displacing the carboxyl during the sequence (Figure 6.34). A dihydronicotinic acid intermediate is likely to be involved allowing decarboxylation to the enamine 1,2-dihydropyridine. This allows an aldol-type reaction with the *N*-methylpyrrolinium cation, and finally dehydrogenation of the dihydropyridine ring back to a pyridine gives nicotine. **Nornicotine** is derived by oxidative demethylation of nicotine involving a cytochrome P-450-dependent enzyme. **Anabasine**

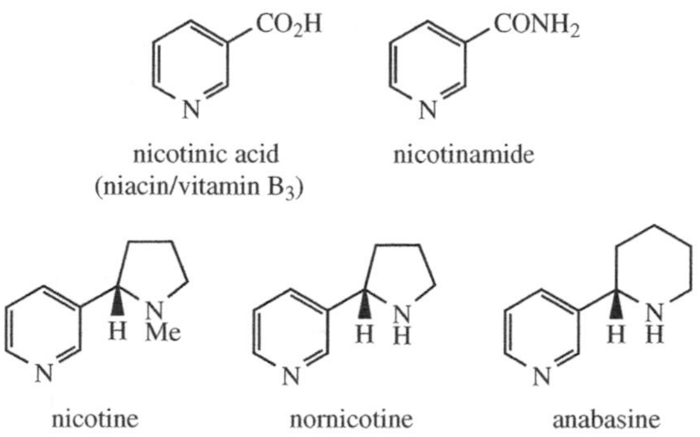

Figure 6.31

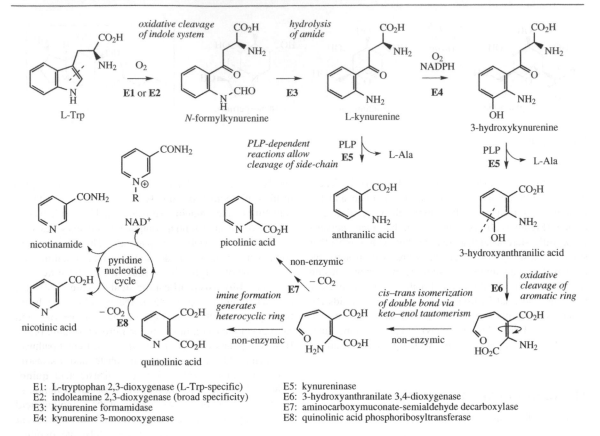

E1: L-tryptophan 2,3-dioxygenase (L-Trp-specific)
E2: indoleamine 2,3-dioxygenase (broad specificity)
E3: kynurenine formamidase
E4: kynurenine 3-monooxygenase

E5: kynureninase
E6: 3-hydroxyanthranilate 3,4-dioxygenase
E7: aminocarboxymuconate-semialdehyde decarboxylase
E8: quinolinic acid phosphoribosyltransferase

Figure 6.32

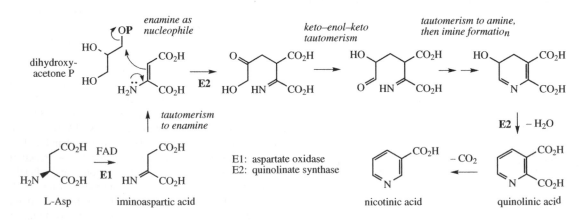

E1: aspartate oxidase
E2: quinolinate synthase

Figure 6.33

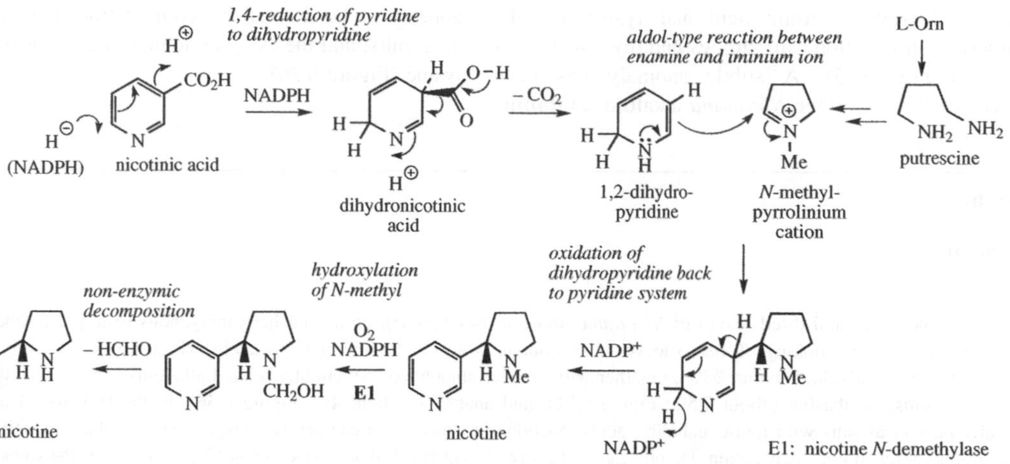

Figure 6.34

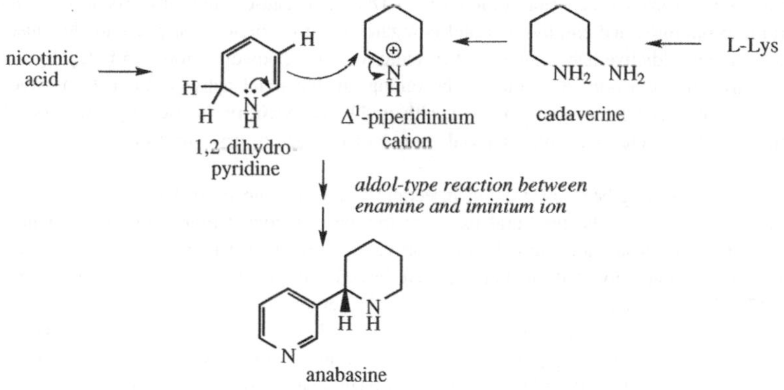

Figure 6.35

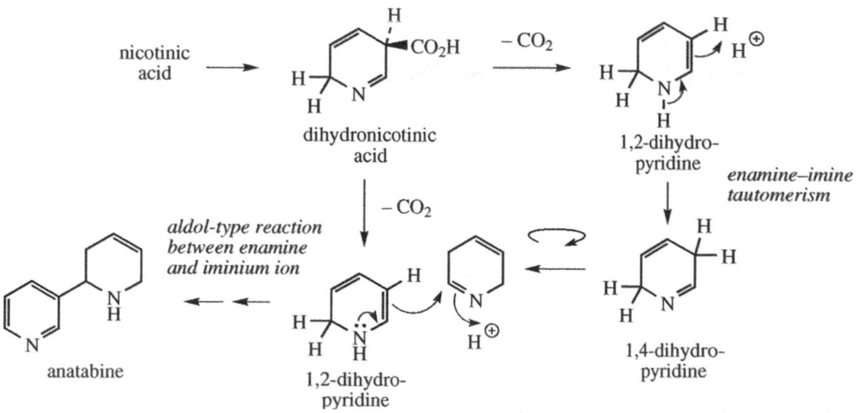

Figure 6.36

is produced from nicotinic acid and lysine via the Δ^1-piperidinium cation in an essentially analogous manner (Figure 6.35). A subtle anomaly has been exposed, in that a further *Nicotiana* alkaloid **anatabine**

appears to be derived by combination of two nicotinic acid units, and the Δ^3-piperideine ring is *not* supplied by lysine (Figure 6.36).

Box 6.3

Tobacco

Tobacco is the cured and dried leaves of *Nicotiana tabacum* (Solanaceae), an annual herb indigenous to tropical America, but cultivated widely for smoking. Tobacco leaves may contain from 0.6 to 9% of (−)-nicotine (Figure 6.31), an oily, volatile liquid as the major alkaloid (about 93%), together with smaller amounts of structurally related alkaloids, e.g. nornicotine (about 3% of alkaloids), anabasine (about 0.5%; Figure 6.31), and anatabine (about 4%; Figure 6.36). In the leaf, the alkaloids are typically present as salts with malic and citric acids. Nicotine in small doses can act as a respiratory stimulant, though in larger doses it causes respiratory depression. Despite the vast array of evidence linking tobacco smoking and cancer, the smoking habit continues throughout the world, and tobacco remains a major crop plant. Tobacco smoke contains over 4000 compounds, including more than 60 known carcinogens formed by incomplete combustion. Amongst these are polycyclic aromatic hydrocarbons, e.g. benzopyrene, nitrosamines, aromatic amines, aldehydes, and other volatile compounds. Metabolism by the body's P-450 system leads to further reactive intermediates which can combine with DNA and cause mutations. Tobacco smoking also contributes to atherosclerosis, chronic bronchitis, and emphysema and is regarded as the single most preventable cause of death in modern society. Smoking tobacco is an addictive habit; unlike other addictive drugs, however, tobacco is legally and widely accessible. Nevertheless, in many parts of the world, it is rapidly becoming an antisocial activity, in that 'passive smoking' (inhaling smoke from users in confined spaces) can also lead to health problems. **Nicotine** is used by smokers who wish to stop the habit. It is available in the form of chewing gum or nasal sprays, or can be absorbed transdermally from nicotine-impregnated patches.

Powdered tobacco leaves have long been used as an insecticide, and nicotine from *Nicotiana tabacum* or *Nicotiana rustica* has been formulated for agricultural and horticultural use. The free base is considerably more toxic than salts, and soaps may be included in the formulations to ensure a basic pH and to provide a surfactant. Other *Nicotiana* alkaloids, e.g. anabasine and nornicotine, share this insecticidal activity. Although an effective insecticide, nicotine has generally been replaced by other agents considered to be safer. Nicotine is toxic to man due to its effect on the nervous system, interacting with the nicotinic acetylcholine receptors, though the tight binding observed is only partially accounted for by the structural similarity between acetylcholine and nicotine (Figure 6.37). Recent studies suggest that nicotine can improve memory by stimulating the transmission of nerve impulses, and this finding may account for the lower incidence of Alzheimer's disease in smokers. Any health benefits conferred by smoking are more than outweighed by the increased risk of heart, lung, and respiratory diseases.

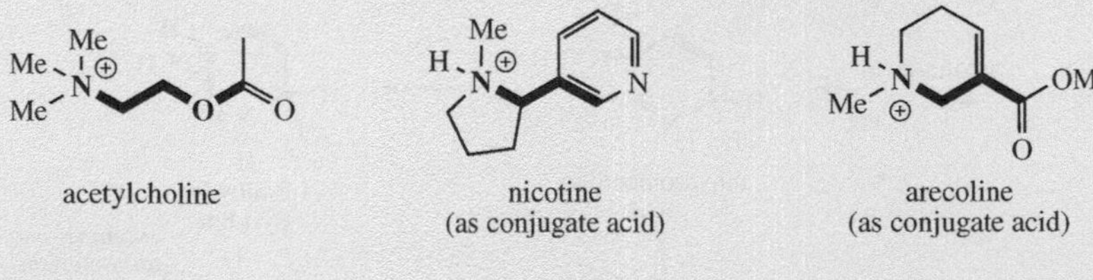

acetylcholine

nicotine
(as conjugate acid)

arecoline
(as conjugate acid)

Figure 6.37

Arecoline (Figure 6.38) is a tetrahydronicotinic acid derivative found in betel-nuts (*Areca catechu*: Palmae/Arecaceae); no biosynthetic information has been

reported. Betel-nuts are chewed in India and Asia for the stimulant effect of arecoline [Box 6.4].

Box 6.4

Areca

Areca nuts (betel-nuts) are the seeds of *Areca catechu* (Palmae/Arecaceae), a tall palm cultivated in the Indian and Asian continents. These nuts are mixed with lime, wrapped in leaves of the betel pepper (*Piper betle*) and then chewed for their stimulant effect, and subsequent feeling of well-being and mild intoxication. The teeth and saliva of chewers stain bright red. The major stimulant alkaloid is arecoline (up to 0.2%; Figure 6.38), the remainder of the alkaloid content (total about 0.45%) being composed of structurally related reduced pyridine structures, e.g. arecaidine, guvacine (tetrahydronicotinic acid), and guvacoline (Figure 6.38). Arecoline is an agonist for muscarinic acetylcholine receptors (see Figure 6.37), although the ester function is reversed compared with acetylcholine. **Arecoline** has been employed in veterinary practice as a vermicide to eradicate worms.

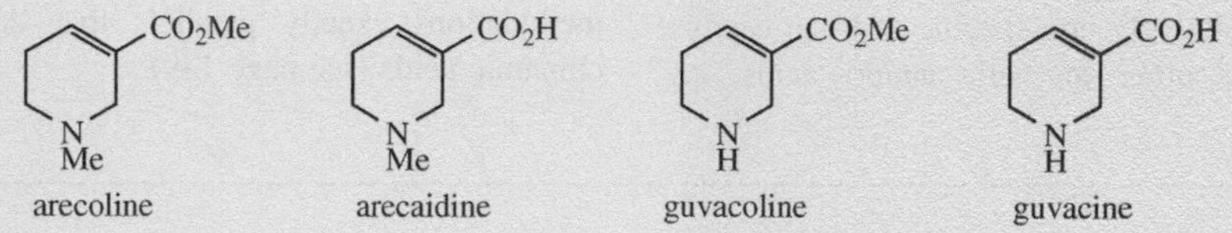

| arecoline | arecaidine | guvacoline | guvacine |

Figure 6.38

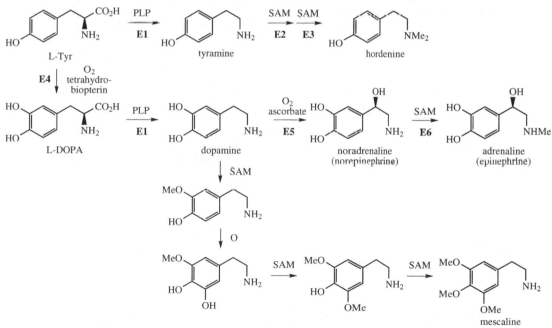

E1: aromatic L-amino acid decarboxylase
      (tyrosine decarboxylase; DOPA decarboxylase)
E2: tyramine *N*-methyltransferase
E3: *N*-methyltyramine *N*-methyltransferase

E4: tyrosine hydroxylase
E5: dopamine β-monooxygenase
E6: phenylethanolamine *N*-methyltransferase

Figure 6.39

ALKALOIDS DERIVED FROM TYROSINE

Phenylethylamines and Simple Tetrahydroisoquinoline Alkaloids

PLP-dependent decarboxylation of L-**tyrosine** gives the simple phenylethylamine derivative **tyramine**, which on di-*N*-methylation yields **hordenine**, a germination inhibitory alkaloid from barley (*Hordeum vulgare*; Graminae/Poaceae; Figure 6.39). More commonly, phenylethylamine derivatives possess 3,4-di- or 3,4,5-tri-hydroxylation patterns and are derived via **dopamine** (Figure 6.39), the decarboxylation product from L-**DOPA** (see page 147). The enzyme aromatic amino acid decarboxylase is relatively non-specific and can catalyse decarboxylation of other aromatic amino acids, e.g.

tryptophan and histidine. Pre-eminent amongst the simple phenylethylamine derivatives are the catecholamines **noradrenaline** (**norepinephrine**), a mammalian neurotransmitter, and **adrenaline** (**epinephrine**), the 'fight or flight' hormone released in animals from the adrenal gland as a result of stress [Box 6.5]. These compounds are synthesized by sucessive β-hydroxylation and *N*-methylation reactions on dopamine (Figure 6.39). Aromatic hydroxylation and *O*-methylation reactions in the cactus *Lophophora williamsii* (Cactaceae) convert dopamine into **mescaline** (Figure 6.39), an alkaloid with pyschoactive and hallucinogenic properties [Box 6.5]. Note that the sequence of hydroxylations and methylations exactly parallels that described for the cinnamic acids (see page 149).

Box 6.5

Catecholamines

The catecholamines dopamine, noradrenaline (norepinephrine), and adrenaline (epinephrine) are produced in the adrenal glands and nervous tissue and act as neurotransmitters in mammals. Several adrenergic receptors have been identified. α-Receptors are usually excitatory and produce a constricting effect on vascular, uterine, and intestinal muscles. β-Receptors are usually inhibitory on smooth muscle, but stimulatory on heart muscles. **Dopamine** (Figure 6.39) can act on both vascular α_1 and cardiac β_1 receptors, but also has its own receptors in several other structures. In Parkinson's disease, there is a deficiency of dopamine due to neural degeneration, affecting the balance between excitatory and inhibitory transmitters. Treatment with L-**DOPA** (levodopa; Figure 6.39) helps to increase the dopamine levels in the brain. Unlike dopamine, DOPA can cross the blood–brain barrier, but needs to be administered with a DOPA-decarboxylase inhibitor, e.g. **carbidopa** (Figure 6.40), to prevent rapid decarboxylation in the bloodstream. Injections of dopamine or **dobutamine** (Figure 6.40) are valuable as cardiac stimulants in cases of cardiogenic shock. These agents act on β_1 receptors; **dopexamine** (Figure 6.40) is also used for chronic heart failure, but acts on β_2 receptors in cardiac muscle.

   Noradrenaline (**norepinephrine**) (Figure 6.39) is a powerful peripheral vasoconstrictor predominantly acting on α-adrenergic receptors, and is useful in restoring blood pressure in cases of acute hypotension. The structurally related alkaloid **ephedrine** (see page 403) may be used in the same way, and synthetic analogues, e.g. **phenylephrine** and **metaraminol** (Figure 6.40), have also been developed. **Methyldopa** is used to treat hypertension; it is a centrally acting agent that becomes decarboxylated and hydroxylated to form the false transmitter α-methylnoradrenaline which competes with noradrenaline.

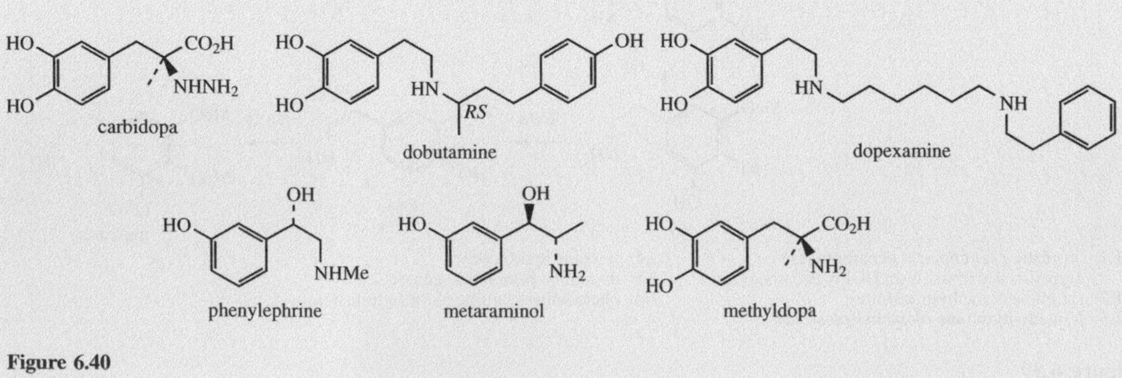

Figure 6.40

Box 6.5 (continued)

Adrenaline (epinephrine; Figure 6.39) is released from the adrenal glands when an animal is confronted with an emergency situation, markedly stimulating glycogen breakdown in muscle, increasing respiration, and triggering catabolic processes that result in energy release. Adrenaline interacts with both α- and β-receptors, an α-response being vasoconstriction of smooth muscle in the skin. β-Responses include mediation of cardiac muscle contractions and the relaxation of smooth muscle in the bronchioles of the lung. Injection of adrenaline is thus of value in cases of cardiac arrest, or in allergic emergencies such as bronchospasm or severe allergy (anaphylactic shock). It is not effective orally. A wide range of cardioactive β-adrenoceptor blocking agents (**beta-blockers**) has been developed to selectively bind to β-receptors to control the rate and force of cardiac contractions in the management of hypertension and other heart conditions. The prototype of the beta-blocker drugs is **propranolol** (Figure 6.41), in which the catechol ring system has been modified to a naphthalene ether and a bulky *N*-alkyl substituent has been incorporated. Many structural variants have been produced, and there is now a huge, perhaps bewildering, variety of beta-blockers in regular use, with subtle differences in properties and action affecting the choice of drug for a particular condition or individual patient. These are shown in Figure 6.41. **Atenolol, bisoprolol, metoprolol, nebivolol**, and, to a lesser extent, **acebutolol** have less effect on the β_2 bronchial receptors and are thus relatively cardioselective, but not cardiospecific. Most other agents are non-cardioselective, and could also provoke breathing difficulties. **Esmolol** and **sotalol** are used only in the management of arrhythmias.

Other β-agonists are mainly selective towards the β_2-receptors and are valuable as antiasthmatic drugs. Important examples include **salbutamol (albuterol)** and **terbutaline**, which are very widely prescribed principally for administration by inhalation at the onset of an asthma attack, but as with cardioactive beta-blockers, a range of agents is in current use (Figure 6.41).

non-selective beta-adrenoceptor blockers

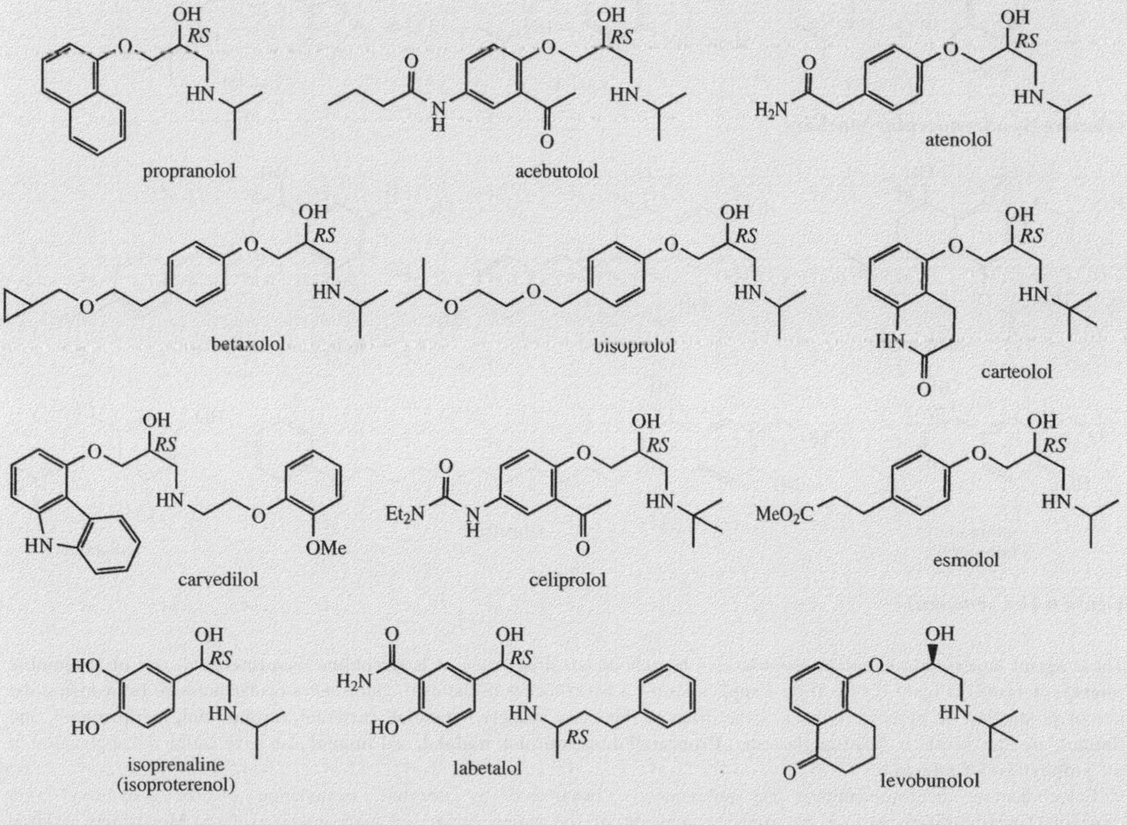

Figure 6.41 (*continued overleaf*)

Box 6.5 (continued)

non-selective beta-adrenoceptor blockers

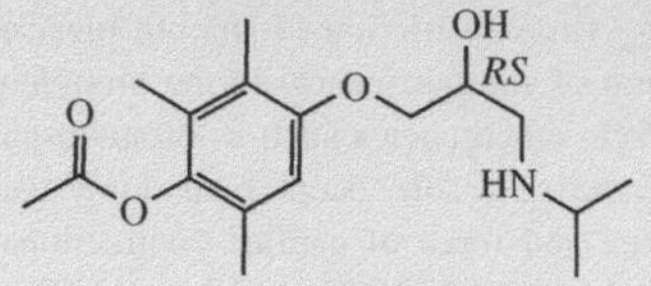

metipranolol

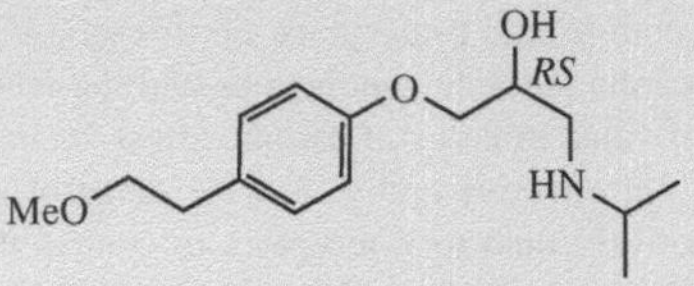

metoprolol

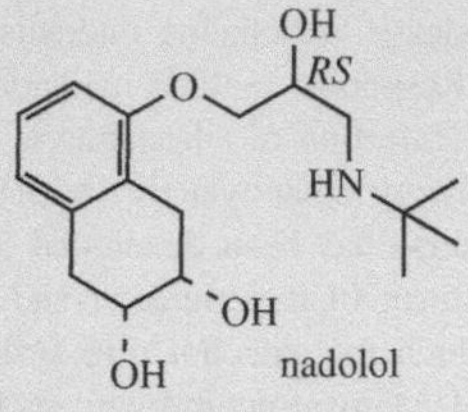

nadolol

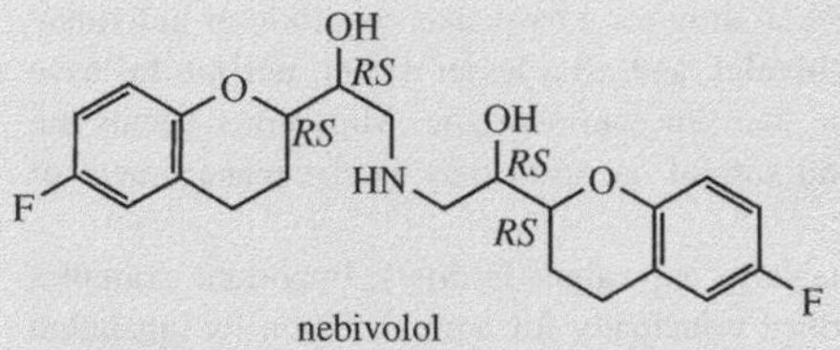

nebivolol

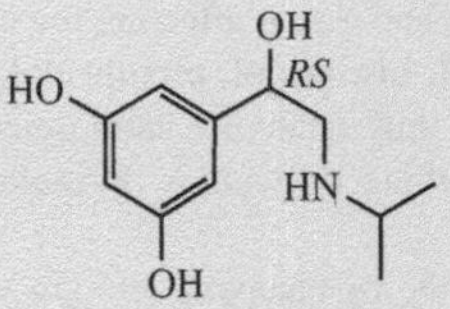

orciprenaline
(metaproterenol)

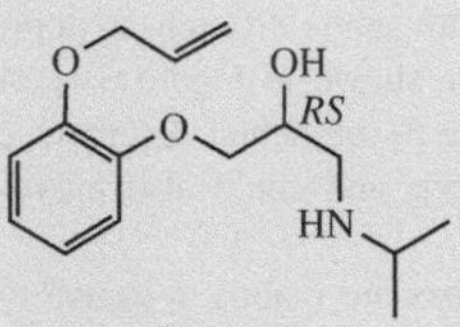

oxprenolol

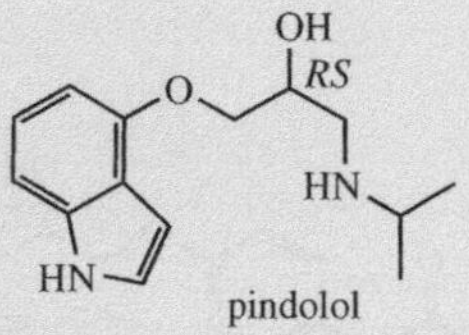

pindolol

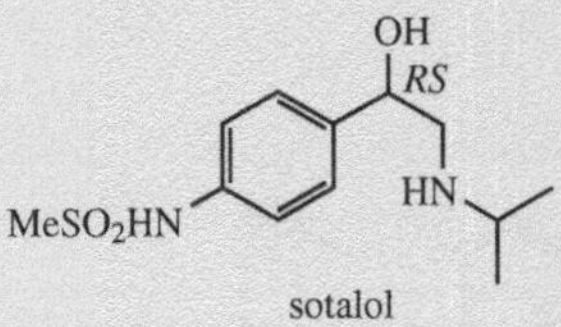

sotalol

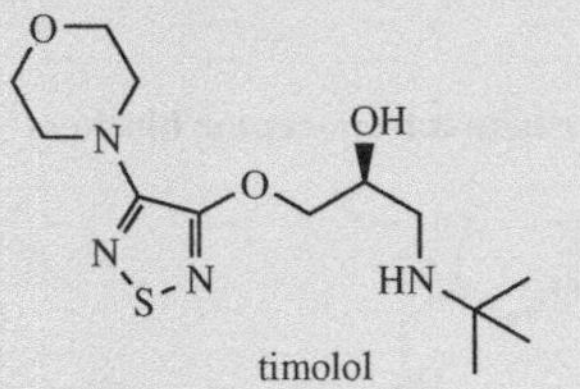

timolol

selective β₂-adrenoceptor blockers

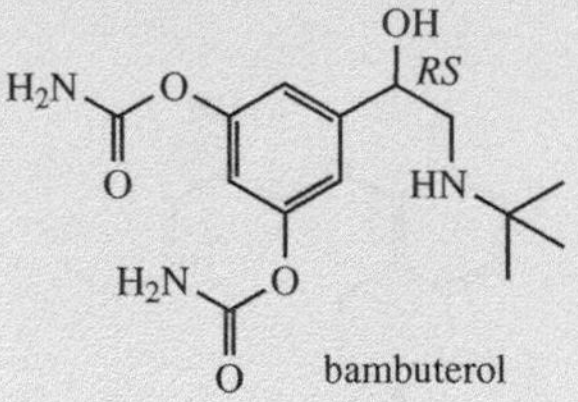

bambuterol

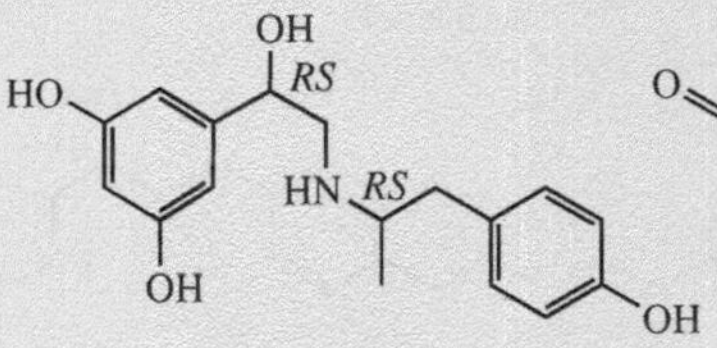

fenoterol

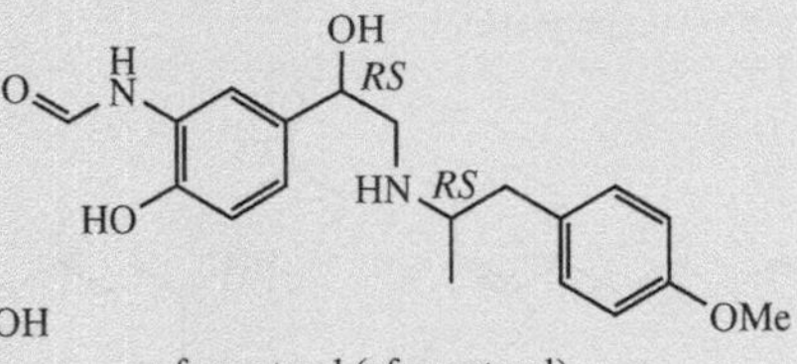

formoterol (eformoterol)

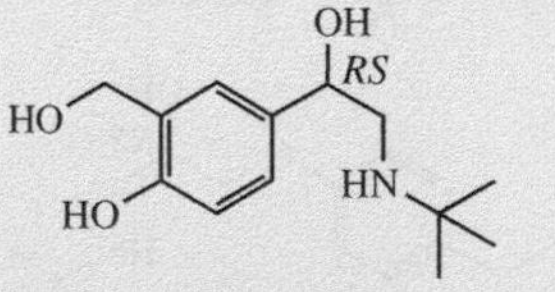

salbutamol
(albuterol)

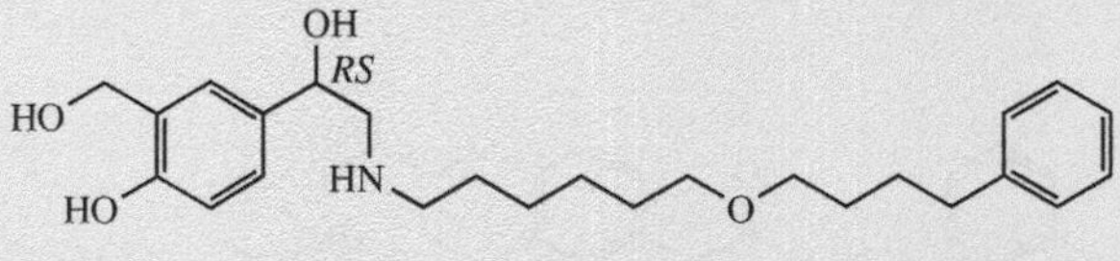

salmeterol

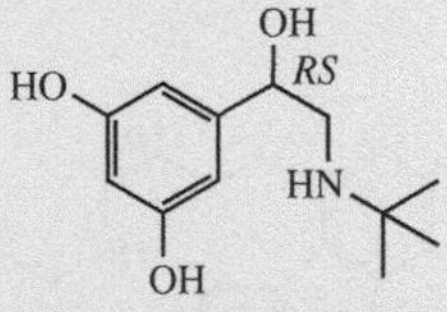

terbutaline

Figure 6.41 (*continued*)

These agents supersede the earlier less-selective bronchodilator drugs such as **isoprenaline (isoproterenol)** and **orciprenaline (metaproterenol)** (Figure 6.41). Topical application of a beta-blocker to the eye reduces intra-ocular pressure by reducing the rate of production of aqueous humour. Some drugs in this class, namely **betaxolol, carteolol, levobunolol, metipranolol**, and **timolol**, are thus useful in treating glaucoma. **Propranolol, metoprolol, nadolol**, and **timolol** also have additional application in the prophylaxis of migraine.

Catecholamine neurotransmitters are subsequently inactivated by enzymic methylation of the 3-hydroxyl (via catechol-*O*-methyltransferase) or by oxidative removal of the amine group via monoamine oxidase. Monoamine oxidase

Box 6.5   (continued)

inhibitors are sometimes used to treat depression, and these drugs cause an accumulation of amine neurotransmitters. Under such drug treatment, simple amines such as tyramine (in cheese, beans, fish, and yeast extract) are also not metabolized and can cause dangerous potentiation of neurotransmitter activity.

Though generally thought of as animal neurotransmitters, catecholamines are also fairly widespread in plants. Significant levels of dopamine accumulate in the flesh of banana (*Musa* species; Araceae), and adrenaline and noradrenaline occur in the peyote cactus (*Lophophora williamsii*; Cactaceae) (see below).

Lophophora

Lophophora or peyote consists of the dried sliced tops of *Lophophora williamsii* (Cactaceae), a small cactus from Mexico and the southwestern United States. The plant has been used by the Aztecs and then by the Mexican Indians for many years, especially in religious ceremonies to produce hallucinations and establish contact with the gods. The so-called mescal buttons were ingested, and this caused unusual and bizarre coloured images. The plant is still used by people seeking drug-induced experiences. The most active of the range of alkaloids found in lophophora (total 8–9% alkaloids in the dried mescal buttons) is mescaline (Figure 6.39), a simple phenylethylamine derivative. Other constituents include anhalamine, anhalonidine, and anhalonine (Figure 6.42). **Mescaline** has been used as a hallucinogen in experimental psychiatry. The dosage required is quite large (300–500 mg), but the alkaloid can readily be obtained by total synthesis, which is relatively uncomplicated. Mescaline is also found in other species of cactus, e.g. *Trichocereus pachanoi*, a substantially larger columnar plant that can grow up to 20 feet tall and found mainly in the Andes.

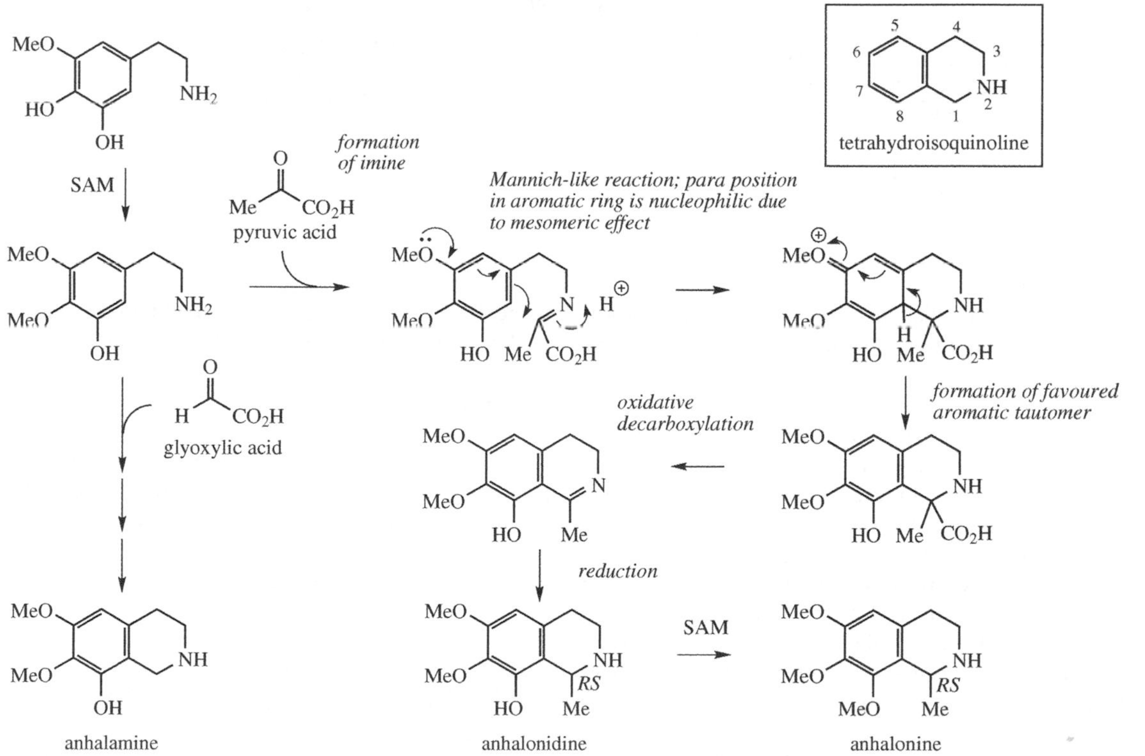

Figure 6.42

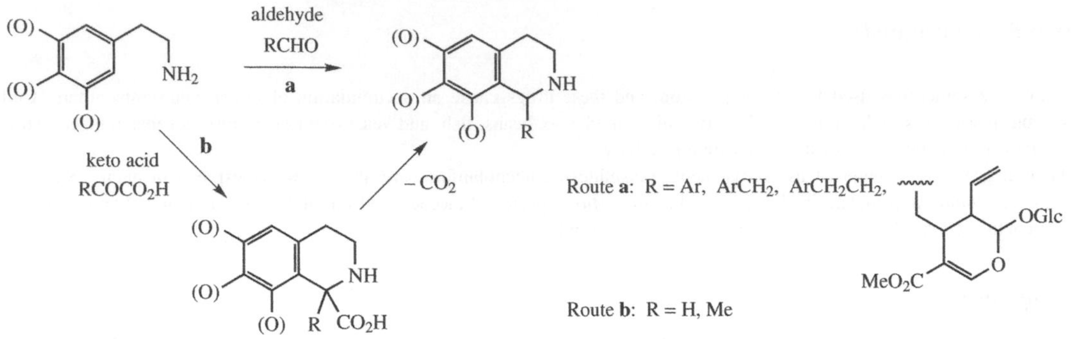

Figure 6.43

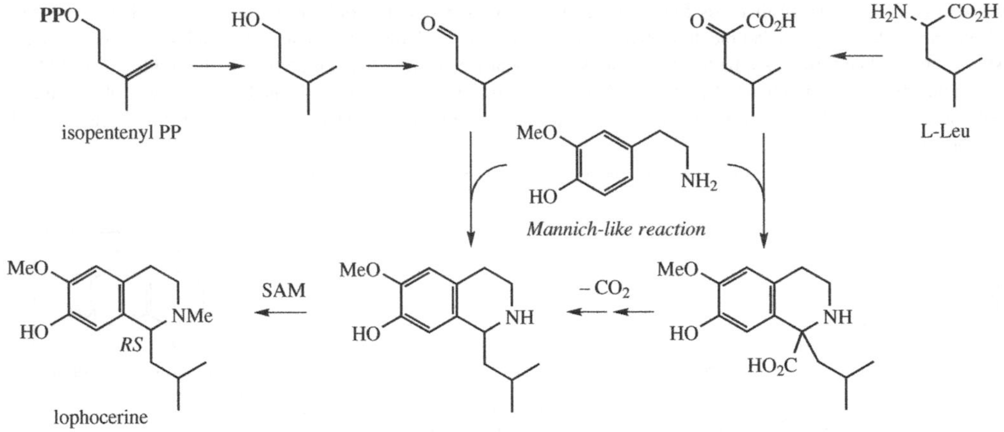

Figure 6.44

Closely related alkaloids co-occurring with mescaline are **anhalamine, anhalonine**, and **anhalonidine** (Figure 6.42), which are representatives of simple tetrahydroisoquinoline derivatives. The additional carbon atoms, two in the case of anhalonidine and anhalonine, and one for anhalamine, are supplied by pyruvate and glyoxylate respectively. In each case, a carboxyl group is lost from this additional precursor. The keto acid pyruvate reacts with a suitable phenylethylamine, in this case the dimethoxy-hydroxy derivative, giving an imine (Figure 6.42). In a Mannich-like mechanism, cyclization occurs to generate the isoquinoline system, the mesomeric effect from an oxygen substituent providing the nucleophilic site on the aromatic ring. Restoration of aromaticity via proton loss gives the tetrahydroisoquinoline, overall a biosynthetic equivalent of the Pictet–Spengler synthesis. The carboxyl group is then removed, not by a simple decarboxylation, but via an unusual oxidative decarboxylation first generating an intermediate

imine. Reduction then leads to **anhalonidine**, with further methylation giving **anhalonine. Anhalamine** is derived from the same phenylethylamine precursor, but utilizing glyoxylic acid (Figure 6.42).

Chemical synthesis of tetrahydroisoquinolines by the Pictet–Spengler reaction does not usually employ keto acids like pyruvate or aldehyde acids like glyoxylate. Instead, simple aldehydes, e.g. acetaldehyde

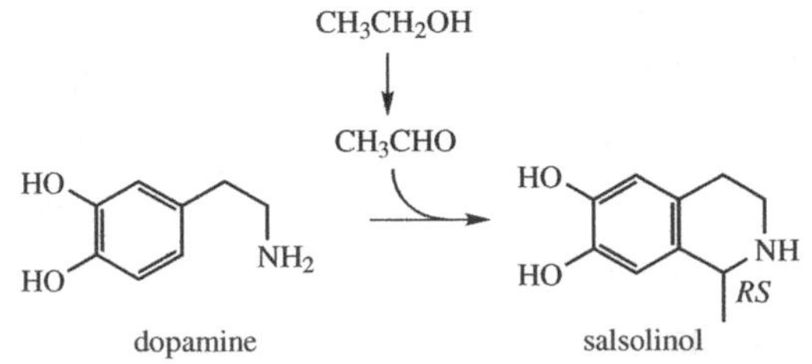

Figure 6.45

or formaldehyde, could be used (Figure 6.43, route **a**), giving the same product directly without the need for a decarboxylation step to convert the intermediate tetrahydroisoquinolinecarboxylic acid (Figure 6.43, route **b**). In nature, both routes are in fact found to operate, depending on the complexity of the R group. Thus, the keto acid (route **b**) is used for relatively simple substrates (R = H, Me), whilst more complex precursors (R = ArCH$_2$, ArCH$_2$CH$_2$, etc.) are incorporated via the corresponding aldehydes (route **a**). The stereochemistry in the product is thus controlled by the condensation/Mannich reactions

(route **a**) or by the final reduction reaction (route **b**). Occasionally, both types of transformation have been demonstrated in the production of a single compound, an example being the *Lophophora schotti* alkaloid **lophocerine** (Figure 6.44). This requires utilization of a C$_5$ isoprene unit, incorporated via an aldehyde. However, a second route using the keto acid derived from the amino acid L-leucine by transamination has also been demonstrated. The alkaloid **salsolinol** (Figure 6.45) is found in plants, e.g. *Corydalis* spp. (Papaveraceae), but can also be detected in the urine of humans, where it is

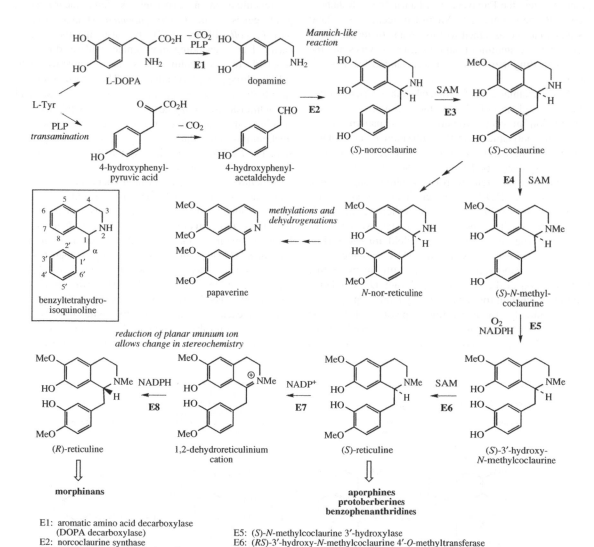

E1: aromatic amino acid decarboxylase
  (DOPA decarboxylase)
E2: norcoclaurine synthase
E3: norcoclaurine 6-*O*-methyltransferase
E4: (*RS*)-coclaurine *N*-methyltransferase

E5: (*S*)-*N*-methylcoclaurine 3′-hydroxylase
E6: (*RS*)-3′-hydroxy-*N*-methylcoclaurine 4′-*O*-methyltransferase
E7: 1,2-dehydroreticuline synthase
E8: 1,2-dehydroreticuline reductase

Figure 6.46

a product from dopamine and acetaldehyde formed via a Pictet–Spengler reaction. Acetaldehyde is typically a metabolite produced after ingestion of ethanol.

Incorporation of a phenylethyl unit into the phenylethylamine gives rise to a benzyltetrahydroisoquinoline skeleton (Figure 6.46), which can undergo further modifications to produce a wide range of plant alkaloids, many of which feature as important drug materials. Fundamental changes to the basic skeleton increase the diversity of structural types, as described under 'modified benzyltetrahydroisoquinolines'. Though found mainly in five plant families (the Papaveraceae, Fumariaceae, Ranunculaceae, Berberidaceae, and Menispermaceae), over 2500 alkaloids can be assigned to this group. In recent years, a considerable amount of data covering enzymes and the genes encoding them has been accumulated for these alkaloids.

Most examples of benzyltetrahydroisoquinoline alkaloids and modified structures contain *ortho* dioxygenation in each aromatic ring, a pattern that is potentially derivable from the utilization of two DOPA molecules. Although two tyrosine molecules are used in the biosynthetic pathway, only the phenylethylamine fragment of the tetrahydroisoquinoline ring system is formed via DOPA, the remaining carbon atoms coming from tyrosine via 4-hydroxyphenylacetaldehyde (Figure 6.46). The product from the Mannich-like reaction is thus the trihydroxy alkaloid **norcoclaurine**, formed stereospecifically as the (*S*)-enantiomer. The tetrahydroxy substitution pattern is built up by further hydroxylation in the benzyl ring, though *O*-methylation (giving (*S*)-**coclaurine**) and *N*-methylation steps precede this. Eventually, (*S*)-**reticuline**, a pivotal intermediate to other alkaloids, is attained by *N*-methylation. Surprisingly, some alkaloids, such as the opium alkaloids

morphine, codeine, and thebaine (see page 348), are elaborated from (*R*)-reticuline rather than the first-formed (*S*)-isomer. The change in configuration is known to be achieved by an oxidation–reduction process through the intermediate 1,2-dehydroreticulinium cation, as shown in Figure 6.46. **Papaverine**, a benzylisoquinoline alkaloid found in opium (see page 350), is formed from *N*-nor-reticuline by successive *O*-methylations and oxidation in the heterocyclic ring (Figure 6.46).

The potential of genetic engineering has been demonstrated by producing substantial quantities of reticuline in a transgenic system, incorporating plant genes from *Coptis japonica* (Ranunculaceae) into *E. coli* (Figure 6.47). By also incorporating a bacterial gene encoding monoamine oxidase, dopamine supplied to the culture was also converted into 3,4-dihydroxyphenylacetaldehyde. The broad substrate specificity of the plant norcoclaurine synthase and methyltransferase enzymes was then exploited by creating an alternative pathway to reticuline via **norlaudanosoline**, thus avoiding the late hydroxylation step. The *Escherichia coli* host system naturally provided the methylating agent SAM.

Structures in which two (or more) benzyltetrahydroisoquinoline units are linked together are readily explained by a phenolic oxidative coupling mechanism (see page 28). Thus, **tetrandrine** (Figure 6.48), a bis-benzyltetrahydroisoquinoline alkaloid isolated from *Stephania tetrandra* (Menispermaceae), is easily recognized as a coupling product from two molecules of (*S*)-*N*-methylcoclaurine (Figure 6.48). The two diradicals, formed by one-electron oxidations of a free phenol group in each ring, couple to give ether bridges, and the product is then methylated to tetrandrine. The pathway is much more likely to follow a stepwise

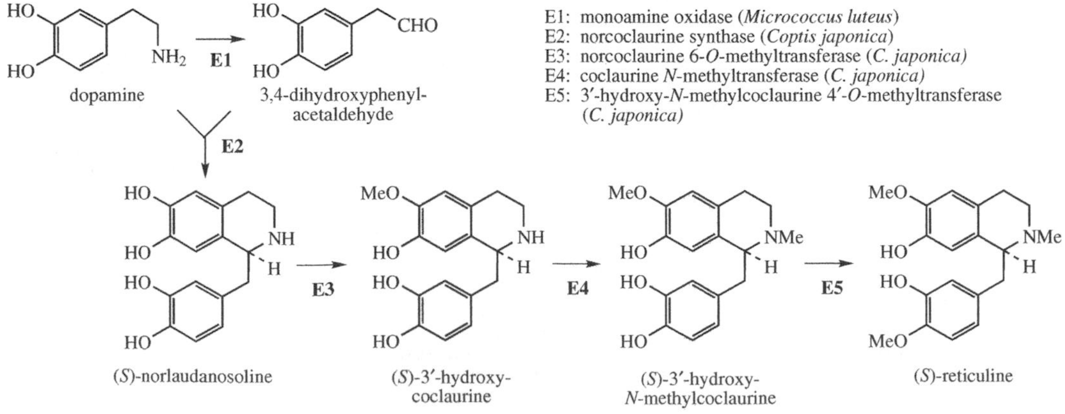

E1: monoamine oxidase (*Micrococcus luteus*)
E2: norcoclaurine synthase (*Coptis japonica*)
E3: norcoclaurine 6-*O*-methyltransferase (*C. japonica*)
E4: coclaurine *N*-methyltransferase (*C. japonica*)
E5: 3′-hydroxy-*N*-methylcoclaurine 4′-*O*-methyltransferase (*C. japonica*)

dopamine
3,4-dihydroxyphenyl-acetaldehyde
(*S*)-norlaudanosoline
(*S*)-3′-hydroxy-coclaurine
(*S*)-3′-hydroxy-*N*-methylcoclaurine
(*S*)-reticuline

Figure 6.47

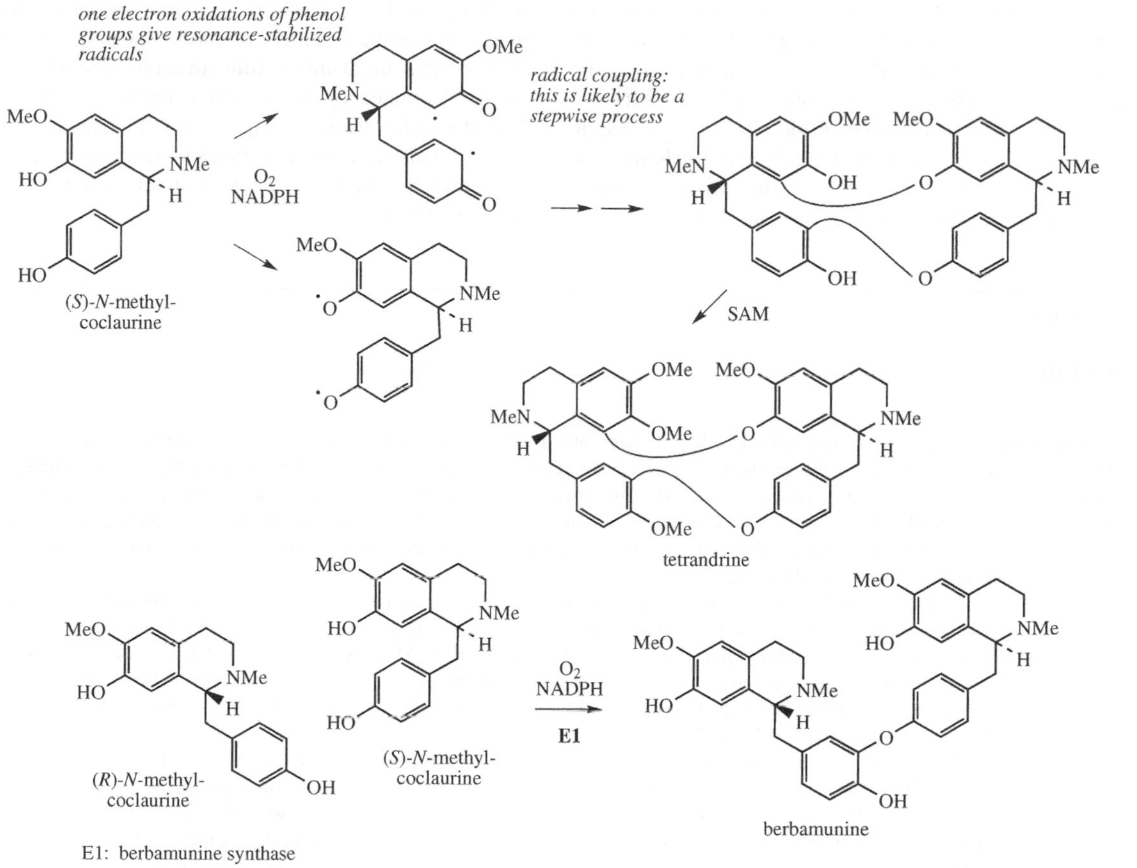

Figure 6.48

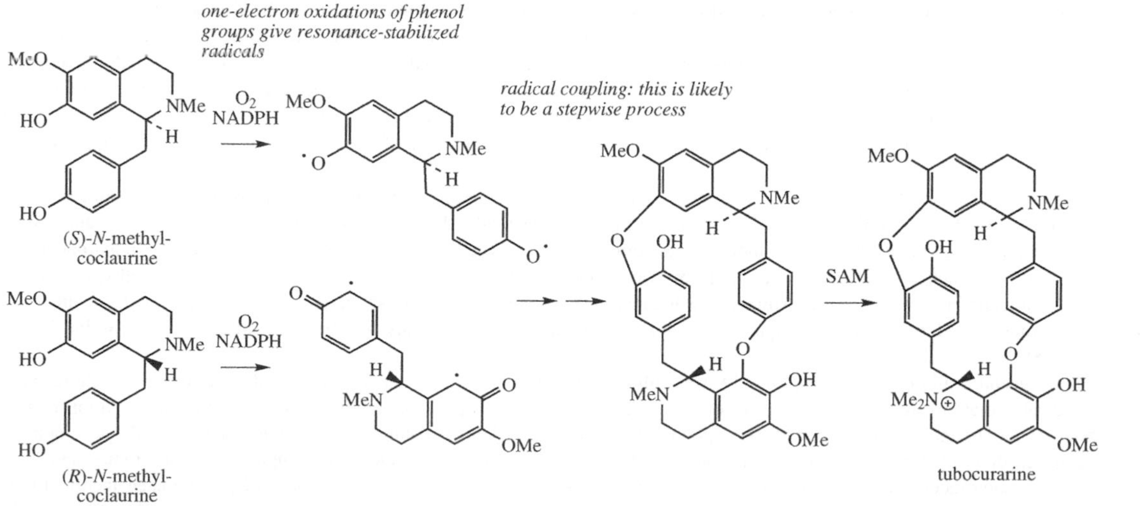

Figure 6.49

coupling process requiring two oxidative enzymes rather than the combined one suggested in Figure 6.48. Indeed, a cytochrome P-450-dependent enzyme from *Berberis stolonifera* couples one molecule each of (*S*)- and (*R*)-*N*-methylcoclaurine in a regiospecific and stereospecific manner to produce **berbamunine** (Figure 6.48) containing a single ether linkage between the two units. Tetrandrine is currently of interest for its ability to block calcium channels, and may have applications in the treatment of cardiovascular disorders. By a similar mechanism, **tubocurarine** (Figure 6.49) can be elaborated by a different coupling of (*S*)- and (*R*)-*N*-methylcoclaurine (Figure 6.49). Tubocurarine from *Chondrodendron tomentosum* (Menispermaceae) is the principal active component in the arrow poison curare [Box 6.6].

Box 6.6

Curare

Curare is the arrow poison of the South American Indians, and it may contain as many as 30 different plant ingredients, which may vary widely from tribe to tribe according to local custom. Curare is prepared in the rain forests of the Amazon and Orinoco and represents the crude dried extract from the bark and stems of various plants. The young bark is scraped off, pounded, and the fibrous mass percolated with water in a leaf funnel. The liquor so obtained is then concentrated by evaporation over a fire. Further vegetable material may be added to make the preparation more glutinous so that it will stick to the arrows or darts. The product is dark brown or black, and tar-like.

In the 1880s, it was found that the traditional container used for curare was fairly indicative of the main ingredients that had gone into its preparation. Three main types were distinguished. Tube curare was packed in hollow bamboo canes, and its principal ingredient was the climbing plant *Chondrodendron tomentosum* (Menispermaceae). Calabash curare was packed in gourds, and was derived from *Strychnos toxifera* (Loganiaceae). Pot curare was almost always derived from a mixture of loganiaceous and menispermaceous plants, and was packed in small earthenware pots. Current supplies of curare are mainly of the menispermaceous type, i.e. derived from *Chondrodendron*.

The potency of curare as an arrow poison is variable and consequently needs testing. A frequently quoted description of this testing is as follows: 'If a monkey hit by a dart is only able to get from one tree to the next before it falls dead, this is 'one-tree curare', the superior grade. 'Two-tree curare' is less satisfactory, and 'three-tree curare' is so weak that it can be used to bring down live animals that the Indians wish to keep in captivity.' Thus, the poison does not necessarily cause death; it depends on the potency. Curare is only effective if it enters the bloodstream, and small amounts taken orally give no ill effects provided there are no open sores in the mouth or throat. Animals killed by the poison could still be safely eaten.

Curare kills by producing paralysis, a limp relaxation of voluntary muscles. It achieves this by competing with acetylcholine at nicotinic receptor sites (see page 334), thus blocking nerve impulses at the neuromuscular junction. Death occurs because the muscles of respiration cease to operate, and artificial respiration is an effective treatment prior to the effects gradually wearing off through normal metabolism of the drug. Anti-acetylcholinesterase drugs, such as physostigmine and neostigmine (see page 386), are specific antidotes for moderate curare poisoning. Curare thus found medicinal use as a muscle relaxant, especially in surgical operations such as abdominal surgery, tonsillectomy, etc., where tense muscles needed to be relaxed. Curare was also found to be of value in certain neurological conditions, e.g. multiple sclerosis, tetanus, and Parkinson's disease, to temporarily relax rigid muscles and control convulsions, but was not a curative. However, the potency of curare varied markedly, and supplies were sometimes limited.

The alkaloid content of curare is from 4 to 7%. The most important constituent in menispermaceous curare is the bis-benzyltetrahydroisoquinoline alkaloid (+)-tubocurarine (Figure 6.50). This is a monoquaternary ammonium salt, and is water-soluble. Other main alkaloids include non-quaternary dimeric structures, e.g. curine (bebeerine) and isochondrodendrine (Figure 6.50), which appear to be derived from two molecules of (*R*)-*N*-methylcoclaurine, with the latter also displaying a different coupling mode. The constituents in loganiaceous curare (from calabash curare, i.e. *Strychnos toxifera*) are even more complex, and a series of 12 quaternary dimeric strychnine-like alkaloids has been identified, e.g. toxiferine-1 (see page 378).

Until recently, **tubocurarine** (Figure 6.50) was still extracted from menispermaceous curare and injected as a muscle relaxant in surgical operations, reducing the need for deep anaesthesia. Artificial respiration is required until the drug has been inactivated (about 30 min) or antagonized (e.g. with neostigmine). However, the limited availability of tubocurarine has led to the development of a series of synthetic analogues as neuromuscular blocking drugs, some of which have improved characteristics and have effectively superseded the natural product. Interestingly, the structure of tubocurarine was originally formulated incorrectly as a diquaternary salt, rather than a monoquaternary salt, and analogues were based on the pretext that curare-like effects might be obtained from compounds containing two quaternary nitrogens separated by a polymethylene chain. This was borne out in

Box 6.6   (continued)

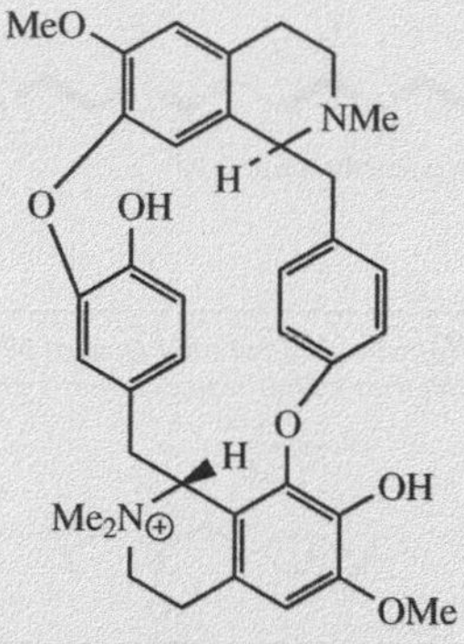

(+)-tubocurarine

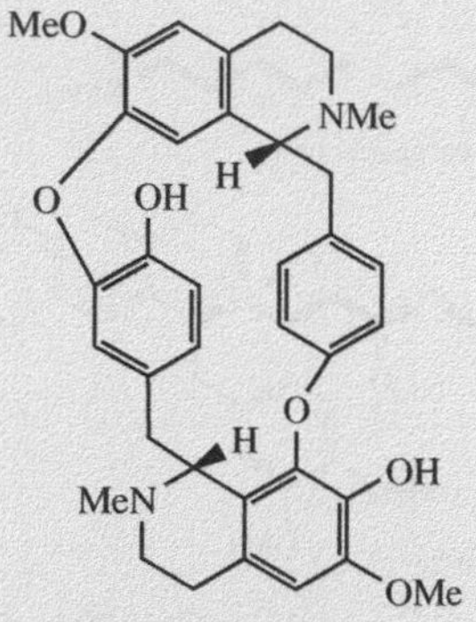

(–)-curine
[(–)-(R,R)-bebeerine]

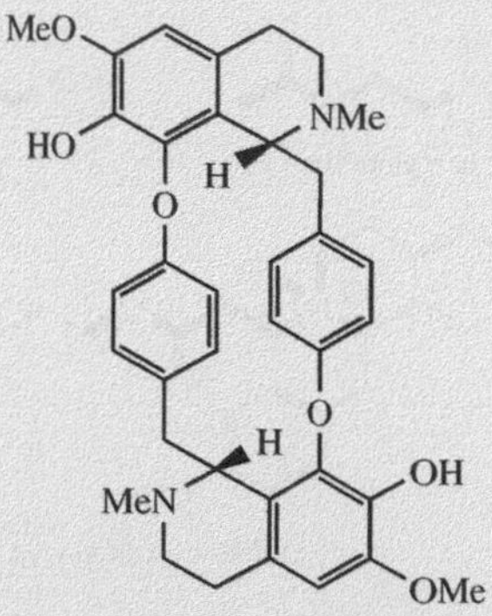

(+)-(R,R)-isochondrodendrine

Figure 6.50

practice, and the separation of the quaternary centres was found to be optimal at about 10 carbon atoms. **Decamethonium** (Figure 6.51) was the first synthetic curare-like muscle relaxant, but it has been superseded too. In tubocurarine, the two nitrogen atoms are also separated by 10 atoms, and at physiological pH values it is likely that both centres will be positively charged. Obviously, the interatomic distance (1.4 nm in tubocurarine) is very dependent on the structure and stereochemistry rather than just the number of atoms separating the centres, but an extended conformation of decamethonium approximates to this distance. **Suxamethonium** (Figure 6.51) is an effective agent with a very short duration of action, due to the two ester functions which are rapidly metabolized in the body by an esterase (pseudocholinesterase: a non-specific serum cholinesterase), and this means the period during which artificial respiration is required is considerably reduced. It also has a 10-atom separation between the quaternary nitrogen atoms.

   Atracurium (Figure 6.51) is a more recent development, containing two quaternary nitrogen atoms in benzyltetrahydroisoquinoline structures separated by 13 atoms. In addition to enzymic ester hydrolysis, atrocurium is also degraded in the body by non-enzymic Hofmann elimination (Figure 6.51), which is independent of liver or kidney function. Normally, this elimination would require strongly alkaline conditions and a high temperature, but the presence of the carbonyl group increases the acidity of the proton and thus facilitates its loss. The elimination proceeds readily under physiological conditions. This is particularly valuable in patients with low or atypical pseudocholinesterase enzymes. Atracurium contains four chiral centres (including the quaternary nitrogen atoms) and is supplied as a mixture of stereoisomers; the single isomer **cisatracurium** has now been introduced. This isomer is more potent than the mixture, has a slightly longer duration of action, and produces less cardiovascular side-effects. **Mivacurium** (Figure 6.51) has similar benzyltetrahydroisoquinoline structures to provide the quaternary centres, but their separation has now been increased to 16 atoms. In **pancuronium**, separation of the two quaternary centres is achieved by a steroidal skeleton. This agent is about five times as potent as tubocurarine. **Vecuronium** is the equivalent monoquaternary structure and has the fewest side-effects. **Rocuronium** is also based on an aminosteroid skeleton and provides rapid action with no cardiovascular effects. Neuromuscular blocking drugs are classified according to their duration of action as ultra-short (8 min, e.g. suxamethonium), short acting (15–30 min, e.g. mivacurium), intermediate (30–40 min, e.g. atracurium), and long acting (60–120 min, e.g. pancuronium).

   The *Strychnos*-derived toxiferines (see Figure 6.89, page 378) also share the diquaternary character. **Alcuronium** is a semi-synthetic skeletal muscle relaxant containing a dimeric strychnine-like structure and is produced chemically from toxiferine-1 (see page 378).

   These neuromuscular blocking agents are competitive antagonists at nicotinic acetylcholine (Figure 6.51) receptor sites. All the structures have two acetylcholine-like portions which can interact with the receptor. Where these are built into a rigid framework, e.g. tubocurarine and pancuronium, the molecule probably spans and effectively blocks several receptor sites without activating them. Tubocurarine and the heterocyclic analogues are termed non-depolarizing or competitive muscle relaxants; their action may be reversed with anticholinesterase agents such as neostigmine that increase acetylcholine concentration at the neuromuscular junction by inhibiting its breakdown. The straight-chain structures, e.g. decamethonium and suxamethonium, initially mimic the action of acetylcholine but then persist at the receptor site, and are termed depolarizing blocking agents. Thus, they trigger a

Box 6.6   (continued)

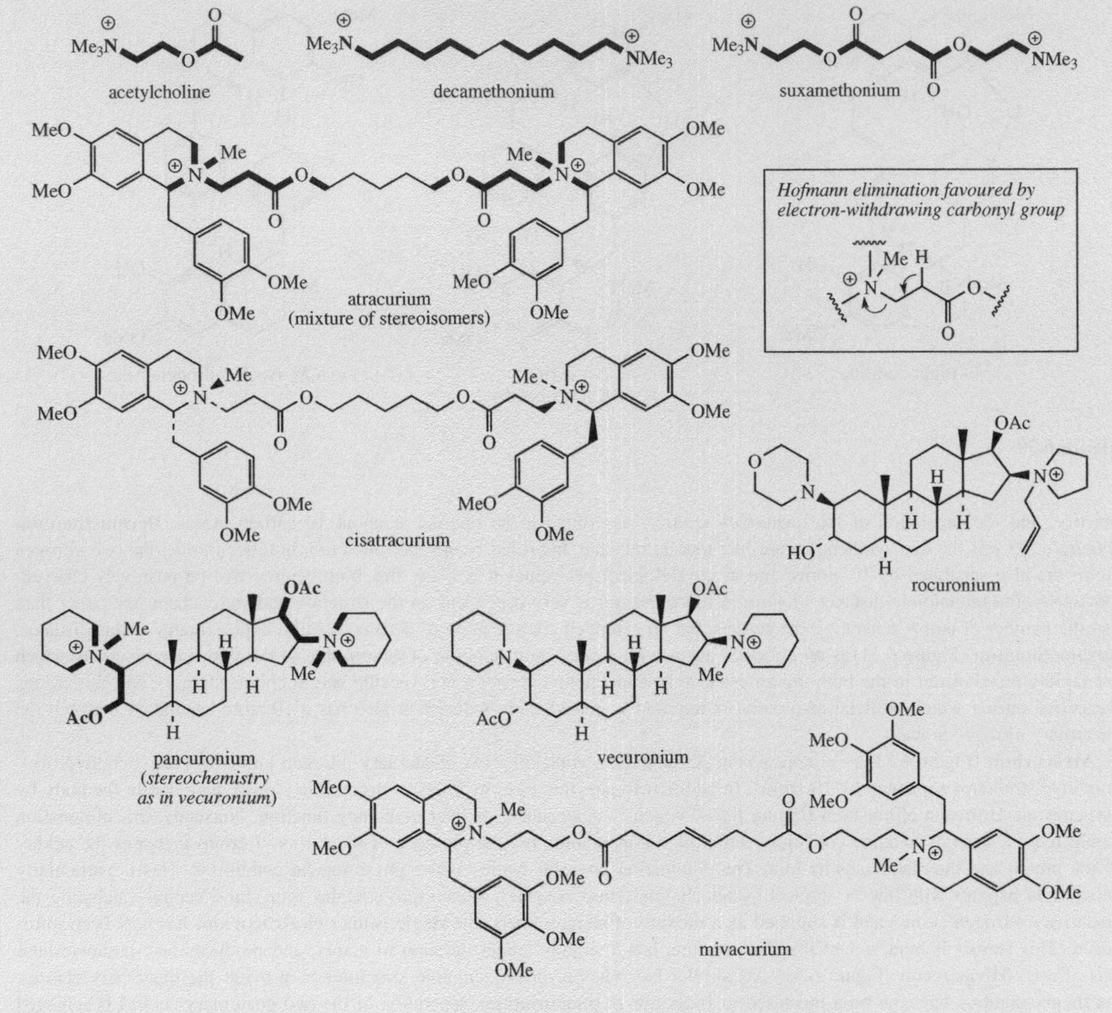

Figure 6.51

response, a brief contraction of the muscle, which is then followed by a prolonged period of muscular paralysis until the compound is metabolized. Their action cannot be reversed by anticholinesterase drugs. Because of their mode of action, depolarizing muscle relaxants may produce more side-effects than non-depolarizing agents, but no alternative ultra-short-acting agents are currently available.

Modified Benzyltetrahydroisoquinoline Alkaloids

The concept of phenolic oxidative coupling is a crucial theme in modifying the basic benzyltetrahydroisoquinoline skeleton to many other types of alkaloids. Tetrandrine (Figure 6.48) and tubocurarine (Figure 6.49) represent coupling of two benzyltetrahydroisoquinoline molecules by ether bridges, but this form of coupling is perhaps less frequent than that involving intramolecular carbon–carbon bonding between aromatic rings. **Morphine, codeine** and

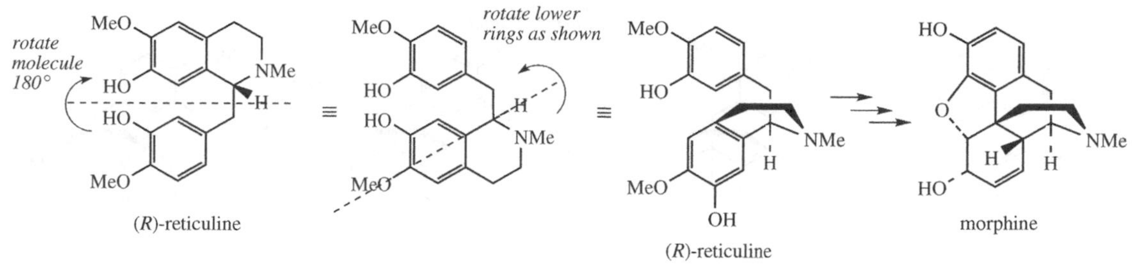

Figure 6.52

thebaine (see Figure 6.53), the principal opium alkaloids [Box 6.7], are derived by this type of coupling, though the subsequent reduction of one aromatic ring to some extent disguises their benzyltetrahydroisoquinoline origins. (*R*)-**Reticuline** (Figure 6.47) is firmly established as the precursor of these morphinan alkaloids; the structural relationship between these groups can be appreciated by careful manipulation of the precursor molecule (Figure 6.52).

(*R*)-Reticuline, redrawn as in Figure 6.53, is the substrate for one-electron oxidations of the phenol group in each ring, giving the diradical. Coupling *ortho* to the phenol group in the tetrahydroisoquinoline and *para* to the phenol in the benzyl substituent then yields the dienone **salutaridine**, found as a minor alkaloid constituent in the opium poppy *Papaver somniferum* (Papaveraceae). The coupling enzyme salutaridine synthase is a cytochrome P-450-dependent monooxygenase. Only the original benzyl aromatic ring can be restored to aromaticity, since the tetrahydroisoquinoline fragment is coupled *para* to the phenol function, a position which is already substituted. The alkaloid **thebaine** is obtained by way of **salutaridinol**, formed from salutaridine by stereospecific reduction of the carbonyl group. Ring closure to form the ether linkage in thebaine would be the result of nucleophilic attack of the phenol group onto the dienol system and subsequent displacement of the hydroxyl. This cyclization step can be demonstrated chemically by treatment of salutaridinol with acid. *In vivo*, however, an additional reaction is used to improve the nature of the leaving group, and this is achieved by acetylation with acetyl-CoA. The cyclization then occurs readily, and without any enzyme participation.

Subsequent reactions involve conversion of thebaine into **morphine** by way of **codeine**, a process which modifies the oxidation state of the diene ring, but most significantly removes two *O*-methyl groups. One is present as an enol ether, removal generating **neopinone**, which gives **codeinone** and then codeine by non-enzymic keto–enol tautomerism and NADPH-dependent reduction respectively. The final step, demethylation of the phenol ether codeine to the phenol morphine, is the type of reaction only achievable in the laboratory by the use of powerful and reactive demethylating agents, e.g. HBr or BBr₃. Because of the other functional groups present, chemical conversion of codeine into morphine is not usually a satisfactory process. However, the enzyme-mediated conversion in *Papaver somniferum* proceeds smoothly and efficiently. The enzymic demethylations of both the enol ether (in thebaine) and the phenol ether (in codeine) most probably involve initial cytochrome P-450-dependent hydroxylation followed by loss of the methyl groups as formaldehyde (Figure 6.53).

The involvement of these *O*-demethylation reactions is rather unusual; secondary metabolic pathways tend to increase the complexity of the product by adding methyl groups rather than removing them. In this pathway, it is convenient to view the methyl groups in reticuline as protecting groups, which reduce the possible coupling modes available during the oxidative coupling process, and these groups are then removed towards the end of the synthetic sequence. There is also some evidence that the later stages of the pathway in Figure 6.53 are modified in some strains of opium poppy. In such strains, thebaine is converted by way of **oripavine** and **morphinone**, this pathway removing the phenolic *O*-methyl before that of the enol ether, i.e. carrying out the same steps but in a different order. The enzymic transformation of thebaine into morphine and the conversion of (*R*)-reticuline into salutaridinol have also been observed in mammalian tissues, giving strong evidence that the trace amounts of morphine and related alkaloids which can sometimes be found in mammals are actually of endogenous origin rather than dietary.

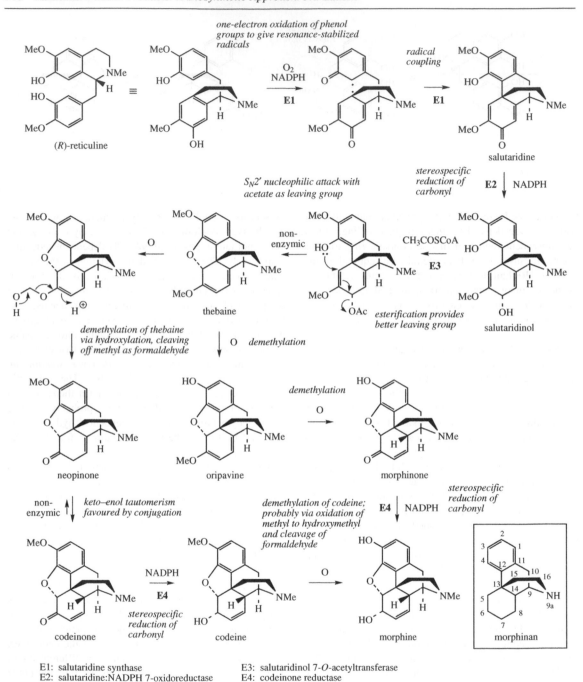

E1: salutaridine synthase
E2: salutaridine:NADPH 7-oxidoreductase

E3: salutaridinol 7-*O*-acetyltransferase
E4: codeinone reductase

Figure 6.53

Box 6.7

Opium

Opium is the air-dried milky exudate, or latex, obtained by incising the unripe capsules of the opium poppy *Papaver somniferum* (Papaveraceae). The plant is an annual herb with large solitary flowers, of white, pink, or dull red–purple colour. For opium production, the ripening capsules, just changing colour from blue–green to yellow, are carefully incised with a knife to open the latex tubes, but not to cut through to the interior of the capsule. These latex tubes open into one another, so it is not necessary to incise them all. Cuts are made transversely or longitudinally according to custom. The initially white milky latex quickly oozes out, but rapidly turns brown and coagulates. This material, the raw opium, is then removed early the following morning, being collected by scraping from the capsule. Further incising and collection may be carried out over a period of about a week. The raw opium is moulded into balls or blocks, and typically these are wrapped in poppy leaves and shade-dried. The blocks may be dusted with various plant materials to prevent cohering. Fresh opium is pale to dark brown and plastic, but it becomes hard and brittle when stored.

Opium has been known and used for 4000 years or more. In recent times, attempts have been made at governmental and international levels to control the cultivation of the opium poppy, but with only limited success. In endeavours to reduce drug problems involving opium-derived materials, especially heroin, where extremely large profits can be made from smuggling relatively small amounts of opium, much pharmaceutical production has been replaced by the processing of the bulkier 'poppy straw'. The entire plant tops are harvested and dried, then extracted for their alkaloid content in the pharmaceutical industry. Poppy straw now accounts for most of the medicinal opium alkaloid production, but there is still considerable trade in illicit opium. In addition to opium, the opium poppy yields seeds which are used in baking and are also pressed to give poppy seed oil. The remaining seed cake is used as cattle feed, and it is held that these poppy seed products cover all the growing expenses, with opium providing the profit. Poppy seeds do not contain any significant amounts of alkaloids.

The main producer of medicinal opium for the world market is India, with China producing supplies for its own domestic use. Poppy straw is cultivated in Australia, France, Hungary, Spain, and Turkey, and more recently in the United Kingdom. Almost all (more than 90%) of the opium destined for the black market now originates from Afghanistan; other sources include Southeast Asia, (mainly in Myanmar (Burma) and Laos) and Latin America (principally Mexico and Colombia).

Crude opium has been used since antiquity as an analgesic, sleep-inducer (narcotic), and for the treatment of coughs. It has been formulated in a number of simple preparations for general use, though these are now uncommon. Laudanum, or opium tincture, was once a standard analgesic and narcotic mixture. Paregoric, or camphorated opium tincture, was used in the treatment of severe diarrhoea and dysentery, but is still an ingredient in the cough and cold preparation Gee's linctus. In Dover's powder, powdered opium was combined with powdered ipecacuanha (see page 363) to give a popular sedative and diaphoretic (promotes perspiration) to take at the onset of colds and influenza. Opium has traditionally been smoked for pleasure, but habitual users developed a craving for the drug followed by addiction. An unpleasant abstinence syndrome was experienced if the drug was withdrawn.

In modern medicine, only the purified opium alkaloids and their derivatives are commonly employed. Indeed, the analgesic preparation '**papaveretum**' (see below) which once contained the hydrochlorides of total opium alkaloids is now formulated from selected purified alkaloids, in the proportions likely to be found in opium. Although the ripe poppy capsule can contain up to 0.5% total alkaloids, opium represents a much concentrated form, and up to 25% of its mass is composed of alkaloids. Of the many (>40) alkaloids identified, some six represent almost all of the total alkaloid content. Actual amounts vary widely, as shown by the following figures: morphine (4–21%; Figure 6.53); codeine (0.8–2.5%; Figure 6.53); thebaine (0.5–2.0%; Figure 6.53); papaverine (0.5–2.5%; Figure 6.47); noscapine (narcotine; 4–8%; Figure 6.54); narceine (0.1–2%; Figure 6.54, and see also Figure 6.65, page 359). A typical commercial sample of opium would probably have a morphine content of about 12%. **Powdered opium** is standardized to contain 10% of anhydrous morphine, usually by dilution with an approved diluent, e.g. lactose or cocoa husk powder. The alkaloids are largely combined in salt form with meconic acid (poppy acid; Figure 6.54), opium containing some 3–5% of this material. Meconic acid is invariably found in opium, but apart from its presence in other *Papaver* species it has not been detected elsewhere. It gives a deep red-coloured complex with ferric chloride, and this has thus been used as a rapid and reasonably specific test for opium. In the past, the urine of suspected opium smokers could also be tested in this way. Of the main opium alkaloids, only morphine and narceine display acidic properties as well as the basic properties due to the tertiary amine. Narceine has a carboxylic acid function, whilst morphine is acidic due to its phenolic hydroxyl. This acidity can be exploited for the preferential extraction of these alkaloids (principally morphine) from an organic solvent by partitioning with aqueous base.

Morphine (Figure 6.53) is a powerful analgesic and narcotic, and remains one of the most valuable analgesics for relief of severe pain. It also induces a state of euphoria and mental detachment, together with nausea, vomiting, constipation, tolerance, and addiction. Regular users experience withdrawal symptoms, including agitation, severe abdominal cramps, diarrhoea, nausea, and vomiting, which may last for 10–14 days unless a further dose of morphine is taken. This leads to physical dependence

Box 6.7   (continued)

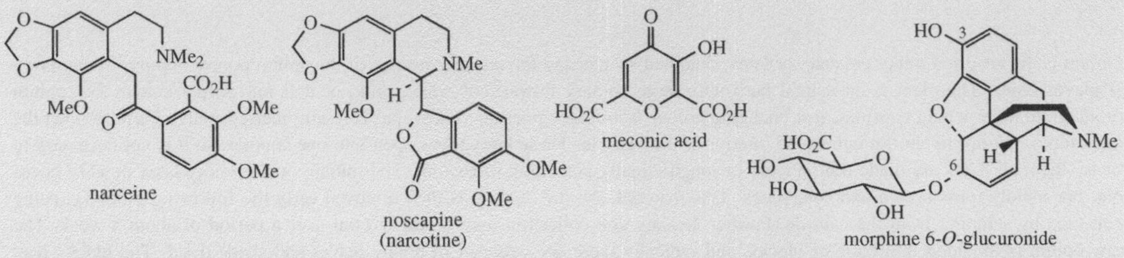

Figure 6.54

which is difficult to overcome, so that the major current use of morphine is thus in the relief of terminal pain. Although orally active, to obtain rapid relief of acute pain it is usually injected. The side-effect of constipation is utilized in some anti-diarrhoea preparations, e.g. kaolin and morphine. Morphine is metabolized in the body to glucuronides which are readily excreted. Whilst morphine 3-*O*-glucuronide is antagonistic to the analgesic effects of morphine, **morphine 6-*O*-glucuronide** (Figure 6.54) is actually a more effective and longer lasting analgesic than morphine, with fewer side-effects, such as nausea and vomiting. This agent is in clinical trials for the treatment of cancer-related pain. Since it is significantly hydrolysed in the gut, it is much less effective taken orally than when administered by injection.

Codeine (Figure 6.53) is the 3-*O*-methyl ether of morphine and is the most widely used of the opium alkaloids. Because of the relatively small amounts found in opium, almost all of the material prescribed is manufactured by semi-synthesis from morphine. Its action is dependent on partial demethylation in the liver to produce morphine, so it produces morphine-like analgesic effects, but little if any euphoria. As an analgesic, codeine has about one-tenth the potency of morphine. Codeine is almost always taken orally and is a component of many compound analgesic preparations. It is a relatively safe non-addictive medium analgesic, but is still too constipating for long-term use. Codeine also has valuable antitussive action, helping to relieve and prevent coughing. It effectively depresses the cough centre, raising the threshold for sensory cough impulses.

Thebaine (Figure 6.53) differs structurally from morphine/codeine mainly by its possession of a conjugated diene ring system. It is almost devoid of analgesic activity, but may be used as a morphine antagonist. Its main value is as substrate for the semi-synthesis of other drugs (see below).

Papaverine (Figure 6.47) is a benzylisoquinoline alkaloid, and is structurally very different from the morphine, codeine, thebaine group of alkaloids (morphinans). It has little or no analgesic or hypnotic properties, but it relaxes smooth muscle in blood vessels. It is sometimes used as an effective treatment for male impotence, being administered by direct injection to achieve erection of the penis. The advent of orally active agents such as sildenafil (Viagra®) has presumably diminished this application.

Noscapine (Figure 6.54) is a member of the phthalideisoquinoline alkaloids and provides a further structural variant in the opium alkaloids. Noscapine has good antitussive and cough suppressant activity comparable to that of codeine, but no analgesic or narcotic action. Its original name 'narcotine' was changed to reflect this lack of narcotic action. Despite many years of use as a cough suppressant, the finding that noscapine may have teratogenic properties (i.e. may deform a fetus) has resulted in noscapine preparations being deleted. In recent studies, antitumour activity has been noted from noscapine, which binds to tubulin as do podophyllotoxin and colchicine (see pages 155 and 361), thus arresting cells at mitosis. The chemotherapeutic potential of this orally effective agent merits further evaluation.

Papaveretum is a mixture of purified opium alkaloids, as their hydrochlorides, and is now formulated to contain only morphine (85.5%), codeine (7.8%) and papaverine (6.7%). It is used for pain relief during operations. It may be combined with the antisecretory tropane alkaloid hyoscine (see page 318).

A vast range of semi-synthetic or totally synthetic morphine-like derivatives have been produced. These are collectively referred to as 'opioids'. Many have similar narcotic and pain-relieving properties as morphine, but are less habit forming. Others possess the cough-relieving activity of codeine, but without the analgesic effect. More than 90% of the morphine extracted from opium (or poppy straw) is currently processed to give other derivatives (Figure 6.55). Most of the codeine is obtained by semi-synthesis from morphine, mono-*O*-methylation occurring at the acidic phenolic hydroxyl. Similarly, **pholcodine** (Figure 6.55), an effective and reliable antitussive, can be obtained by alkylation with *N*-(chloroethyl)morpholine. **Dihydrocodeine** (Figure 6.55) is a reduced form of codeine with similar analgesic properties, the double bond not being essential for activity. In **hydromorphone** (Figure 6.55), the double bond of morphine has been reduced, and in addition the 6-hydroxyl has been oxidized to a ketone. This increases the analgesic effects, but also the side-effects; the drug is used for severe pain associated with cancer. **Diamorphine** (or

Box 6.7   (continued)

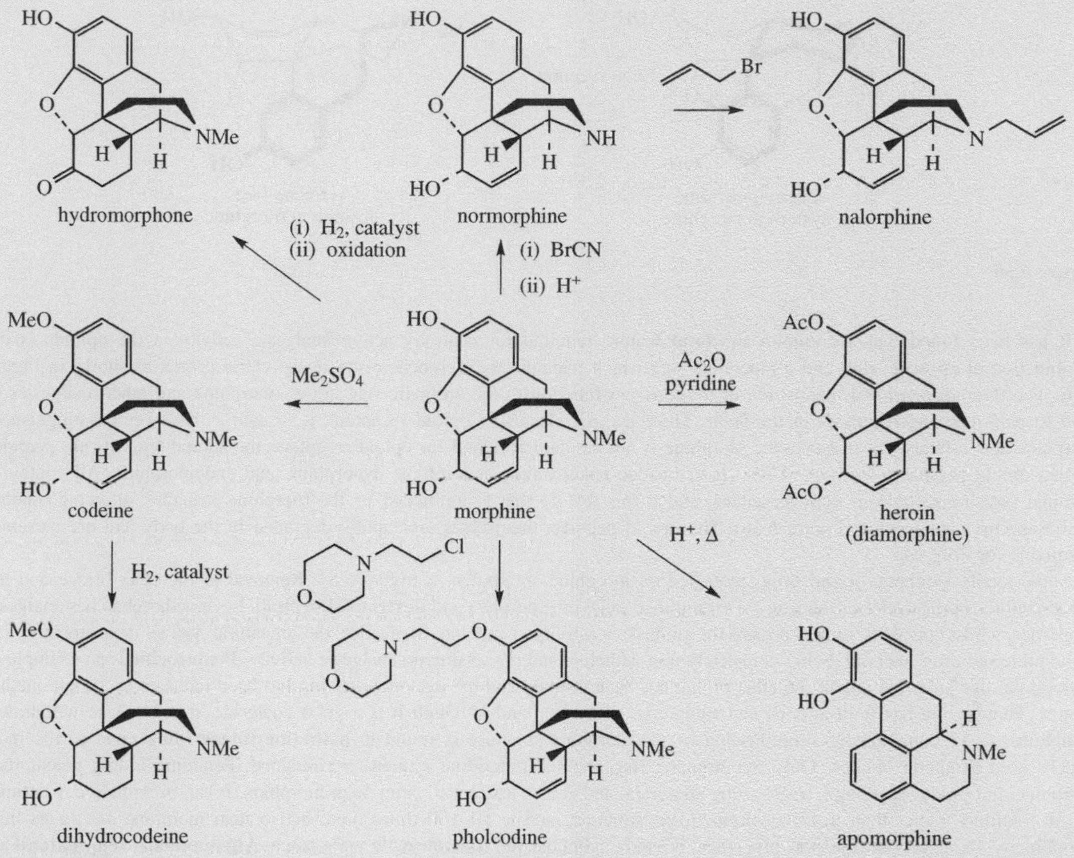

Figure 6.55

heroin; Figure 6.55), is merely the diacetate of morphine; it is a highly addictive analgesic and hypnotic. The increased lipophilic character of heroin over morphine results in improved solubility, with better transport and absorption, though the active agent is probably the 6-acetate, the 3-acetate group being hydrolysed by esterases in the brain. Heroin was synthesized originally as a cough suppressant; and though most effective in this role, it has unpleasant addictive properties, with users developing a psychological craving for the drug. It is widely used for terminal care, e.g. cancer sufferers, both as an analgesic and cough suppressant. The euphoria induced by injection of heroin has resulted in much abuse of the drug and creation of a worldwide major drug problem.

The *N*-methyl group of morphine can be removed by treatment with cyanogen bromide, then hydrolysis. A variety of *N*-alkyl derivatives, e.g. *N*-allyl-normorphine (**nalorphine**; Figure 6.55) may be produced by use of appropriate alkyl bromides. Nalorphine has some analgesic activity, but it was also found to counter the effects of morphine and is thus a mixed agonist–antagonist. It has been used as a narcotic antagonist, but is principally regarded as the forerunner of pure opiate antagonists such as naloxone (see below). Treatment of morphine with hot acid induces a rearrangement process, resulting in a highly modified structural skeleton, a representative of the aporphine group of alkaloids (see page 355). The product **apomorphine** (Figure 6.55) has no analgesic properties, but morphine's side-effects of nausea and vomiting are highly emphasized. Apomorphine is a powerful emetic and can be injected for emergency treatment of poisoning. This is now regarded as dangerous, but apomorphine is currently valuable to control the symptoms of Parkinson's disease, being a stimulator of D_1 and D_2 dopamine receptors. Apomorphine's structure contains a dihydroxyphenylethylamine (dopamine) fragment conferring potent dopamine agonist properties to this agent (see page 336).

Box 6.7   (continued)

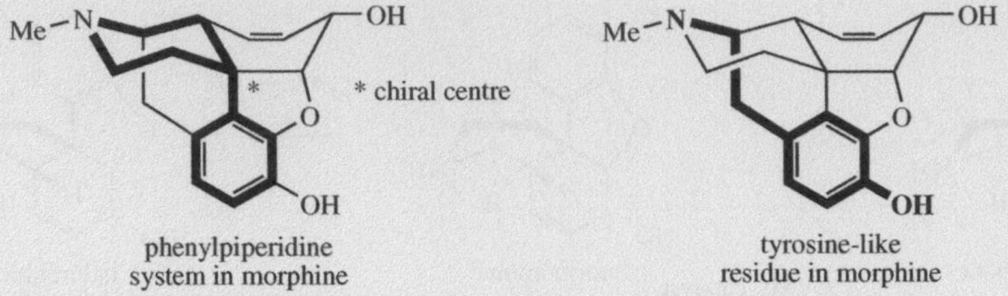

phenylpiperidine
system in morphine

tyrosine-like
residue in morphine

Figure 6.56

It has been found that a common structural feature required for centrally acting analgesic activity in the opioids is the combination of aromatic ring and a piperidine ring which maintain the stereochemistry at the chiral centre, as shown in Figure 6.56. The three-dimensional disposition of the nitrogen function to the aromatic ring allows morphine and other analgesics to bind to pain-reducing receptors in the brain. Three distinct classes of opioid receptors, μ, δ, and κ, have been distinguished; morphine acts primarily at μ-receptors. Morphine is not the natural ligand for opioid receptors; the natural agonists are peptides termed opioid peptides (see page 434). These include enkephalins, endorphins, dynorphins, and endomorphins. All contain a terminal tyrosine residue in their structures, and it this feature that is mimicked by the morphine structure, allowing binding to the appropriate receptor (Figure 6.56). The opioid peptides themselves are rapidly degraded in the body and are currently unsuitable for drug use.

Some totally synthetic opioid drugs modelled on morphine are shown in Figure 6.57. Removal of the ether bridge and the functionalities in the cyclohexene ring are exemplified in levomethorphan and **dextromethorphan**. Levomethorphan has analgesic properties, whilst both enantiomers possess the antitussive activity of codeine. In practice, the 'unnatural' isomer dextromethorphan is the preferred drug material, being completely non-addictive and possessing no analgesic activity. **Pentazocine** is an example of a morphine-like structure where the ether bridge has been omitted and the cyclohexene ring has been replaced by simple methyl groups. Pentazocine has both agonist and antagonist properties, and although it is a good analgesic, it can induce withdrawal symptoms. Even more drastic simplification of the morphine structure is found in **pethidine** (**meperidine**), one of the most widely used synthetic opiates. Only the aromatic ring and the piperidine systems are retained. Pethidine is less potent than morphine, but produces prompt, short-acting analgesia, and is also less constipating than morphine. It can be addictive. **Fentanyl** has a 4-anilino- rather than a 4-phenyl-piperidine structure, and is 50–100 times more active than morphine due to its high lipophilicity and excellent transport properties; it can be administered transdermally via a patch. **Alfentanil** and **remifentanil** are further variants on the fentanyl structure; all three drugs are rapid-acting and used during operative procedures. The piperidine ring system is no longer present in **methadone**, though this diphenylpropylamine derivative can be drawn in such a way as to mimic the piperidine ring conformation. Methadone is orally active, has similar activity to morphine, but is less euphorigenic and has a longer duration of action. Although it is potentially addictive as morphine, the withdrawal symptoms are different and much less severe than with other drugs such as heroin, so that methadone is widely used for the treatment and rehabilitation of heroin addicts. However, it only replaces one addiction with another, albeit a less dangerous one. **Dipipanone** is a structural variant on methadone, and is used for moderate to severe pain; it is usually administered in combination with an anti-emetic. **Diphenoxylate** is used as an antidiarrhoeal; to minimize its habit-forming properties and potential abuse it is combined with a sub-therapeutic amount of the anticholinergic atropine (see page 318). **Dextropropoxyphene** contains an ester function but mimics the piperidine ring in a rather similar manner. This agent has only low analgesic activity, about half that of codeine, and finds application in combination formulations with aspirin or paracetamol. **Meptazinol** is structurally unlike the other opiate analgesics, in that it contains a seven-membered nitrogen heterocycle. It is an effective analgesic, and it produces relatively few side-effects with a low incidence of respiratory depression. **Tramadol** is a recent drug claimed to produce analgesia by two mechanisms, an opioid mechanism and also by enhancement of serotoninergic and adrenergic pathways. It produces few typical opioid side-effects.

Thebaine, for many years regarded as an unwanted by-product from opium, is now the raw material for semi-synthesis of several useful drugs. On treatment with hydrogen peroxide, the conjugated diene undergoes 1,4-addition, and hydrolysis results in formation of a 4-hydroxy cyclohexenone system (Figure 6.58). Reduction and demethylation lead respectively to **oxycodone** and **oxymorphone**, which are potent analgesics. The conjugated diene system in thebaine can also be exploited in a Diels–Alder reaction, building on yet another ring system (Figure 6.59). Some of these adducts have quite remarkable levels of analgesic activity, but are too powerful for human use. Some, e.g. **etorphine** (Figure 6.59), are used in veterinary practice to sedate large

Box 6.7 (continued)

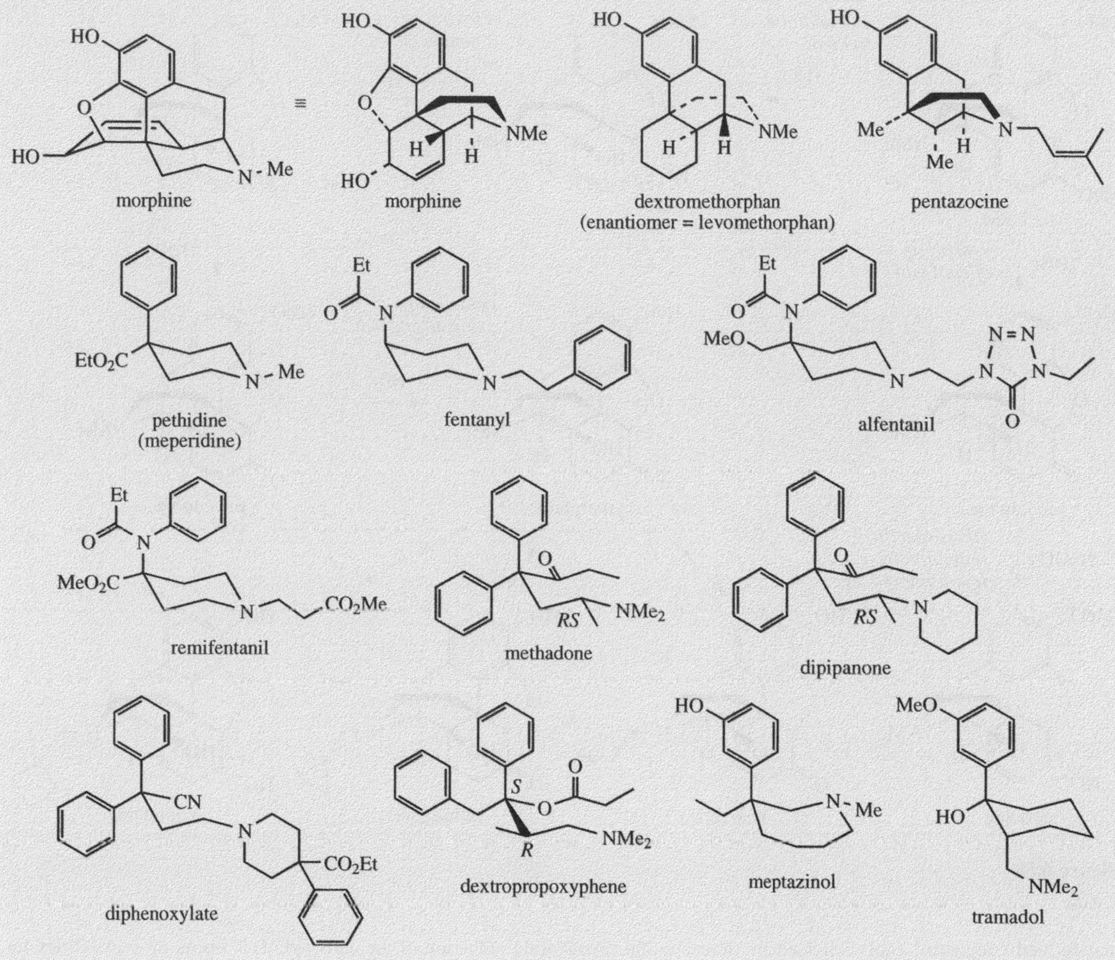

Figure 6.57

animals (elephants, rhinos) by means of tranquillizer darts. Etorphine is some 5000–10 000 times more potent than morphine. **Buprenorphine** (Figure 6.59) is an etorphine analogue with an *N*-cyclopropylmethyl substituent and *tert*-butyl instead of *n*-propyl in the side-chain. This material has both opioid agonist and antagonist properties. Mixed agonist–antagonist properties offer scope for producing analgesia whilst negating the effects of other opioids to which a patient may be addicted. Buprenorphine has a long duration of action and only low dependence potential, but it may precipitate withdrawal symptoms in patients dependent on other opioids. In addition to use as an analgesic, it is now being used as an alternative to methadone in the treatment of opioid dependence. **Nalbuphine** (Figure 6.58), produced semi-sythetically from thebaine, also displays mixed agonist–antagonist properties and has similar agonist activity as morphine, but it produces less side-effects and has less abuse potential. **Naloxone** (Figure 6.58) shows hardly any agonist activity, but it is a potent antagonist at all opioid receptors and is used to treat opiate poisoning, including that in children born to heroin addicts. **Naltrexone** (Figure 6.58) also has antagonist activity similar to naloxone. These agents are *N*-alkyl derivatives related to oxymorphone/oxycodone.

Thebaine may also be transformed very efficiently into codeine in about 75% yield (Figure 6.58). The two-stage synthesis involves acid-catalysed hydrolysis of the enol ether function to give codeinone (this being the more favoured tautomer of the

Box 6.7   (continued)

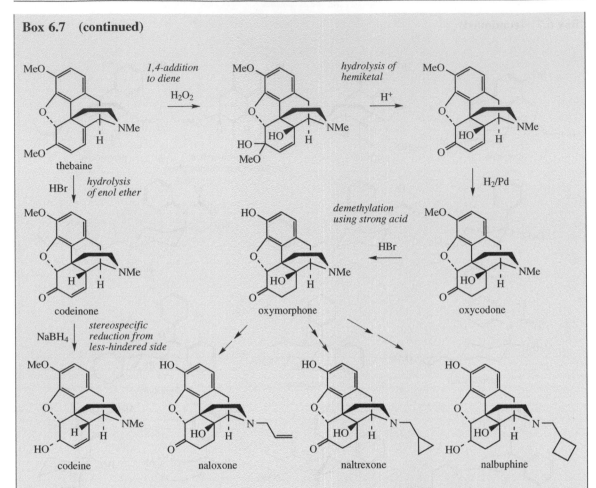

Figure 6.58

first-formed conjugated enol) followed by stereospecific borohydride reduction of the carbonyl. This opens up possibilities for producing codeine (the most widely used of the opium alkaloids) without using morphine. At present, almost all of the codeine used is synthesized by methylation of morphine. The advantage of using thebaine is that the raw material for the pharmaceutical industry could be shifted away from morphine and opium. This might then help in the battle to eliminate illicit morphine production and its subsequent conversion into heroin. Conversion of thebaine into morphine and heroin is much more difficult and low yielding. Thus, considerable effort has been put into selecting thebaine-rich varieties of *Papaver somniferum* and cultivating these for alkaloid production. A significant proportion of the poppy crop in Australia and France is now composed of thebaine-rich varieties. These strains produce thebaine and oripavine (Figurer 6.53) as main alkaloids, and appear to lack enzymes that carry out the late demethylation steps in Figure 6.53. Most species of *Papaver* seem to lack the enzyme that reduces salutaridine to salutaridinol (Figure 6.53) and, thus, they do not synthesize morphine-like alkaloids. Oripavine (3-demethylthebaine) may be used in the same way as thebaine in the synthesis of drugs. Metabolic engineering would seem to offer scope for modifying the alkaloid patterns in *Papaver somniferum*. In one study, the late step catalysed by codeinone reductase has been blocked by gene silencing through RNA interference: supplying a short strand of RNA to target that portion of mRNA responsible for a particular enzyme. Morphine and codeine production was markedly diminished, but other morphinan alkaloids, such as thebaine and oripavine, were also suppressed. Instead, the major alkaloids accumulating were reticuline and some of its methyl ethers. The metabolic block appeared to be several steps further back in the pathway than predicted.

Box 6.7 (continued)

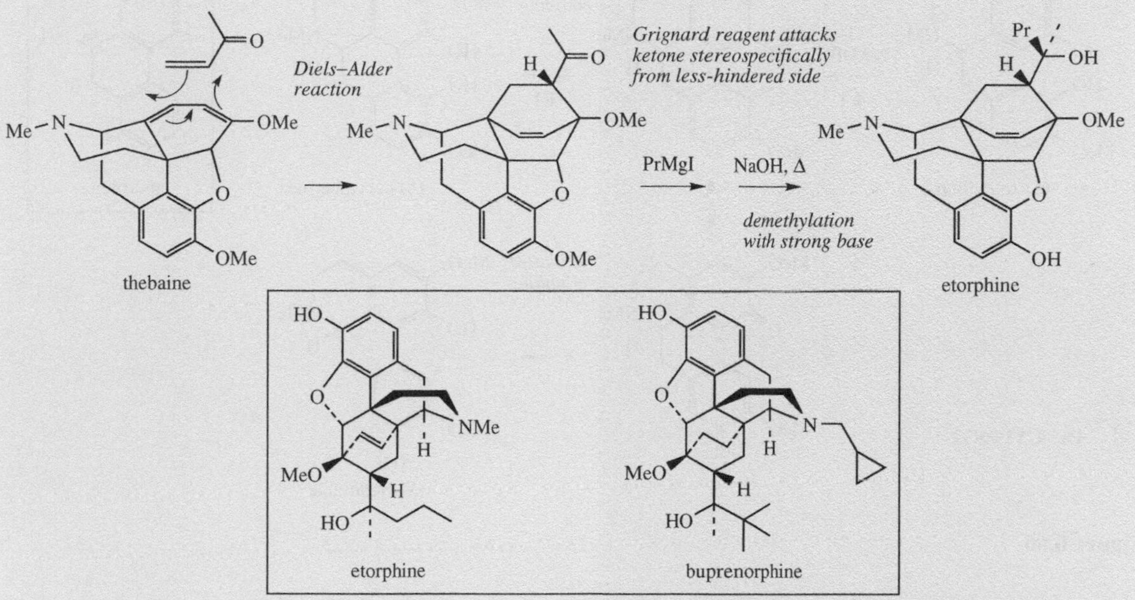

Figure 6.59

Remarkably, there is now considerable evidence that various animals, including humans and other mammals, are also able to synthesize morphine and related alkaloids in small amounts. These compounds have been detected in various tissues, including brain, liver, spleen, adrenal glands, and skin, and endogenous morphine may thus play a role in pain relief, combining its effects with those provided by the enkephalin peptides.

A minor constituent of *Papaver somniferum* is the aporphine alkaloid **isoboldine** (Figure 6.60). Other species of poppy, e.g. *Papaver orientale* and *Papaver pseudoorientale*, are known to synthesize aporphine alkaloids rather than morphinan structures as their principal constituents. Aporphines constitute one of the largest groups of isoquinoline alkaloids, with more than 500 representatives known. Apomorphine (Figure 6.55), the acid rearrangement product from morphine, is a member of this group, though is not a natural product. A cytochrome P-450-dependent enzyme catalysing the *ortho–ortho* oxidative coupling of (*S*)-**reticuline** to (*S*)-**corytuberine** (Figure 6.60) has been characterized from *Coptis japonica* (Ranunculaceae), a plant that also synthesizes berberine (see below). (*S*)-Isoboldine is readily appreciated to be the product of a similar oxidative coupling of (*S*)-reticuline, though coupling *ortho* to the phenol group in the tetrahydroisoquinoline portion and *para* to the

phenol of the benzyl substituent (Figure 6.60). Some structures, e.g. **isothebaine** (Figure 6.60) from *Papaver orientale*, are not as easily rationalized. (*S*)-**Orientaline** is a precursor of isothebaine (Figure 6.61). This benzyltetrahydroisoquinoline, with a different methylation pattern to reticuline, is able to participate in oxidative coupling, but inspection of the structures indicates a phenol group appears to be lost in the transformation. The pathway (Figure 6.61) involves an unexpected rearrangement process, however. Thus, oxidative coupling *ortho–para* to the phenol groups gives a dienone **orientalinone** (compare the structure of salutaridine in Figure 6.53). After reduction of the carbonyl group, a rearrangement occurs, restoring aromaticity and expelling the hydroxyl (originally a phenol group) to produce **isothebaine**. This type of rearrangement, for which good chemical analogies are available, is a feature of many other alkaloid biosynthetic pathways and occurs because normal keto–enol

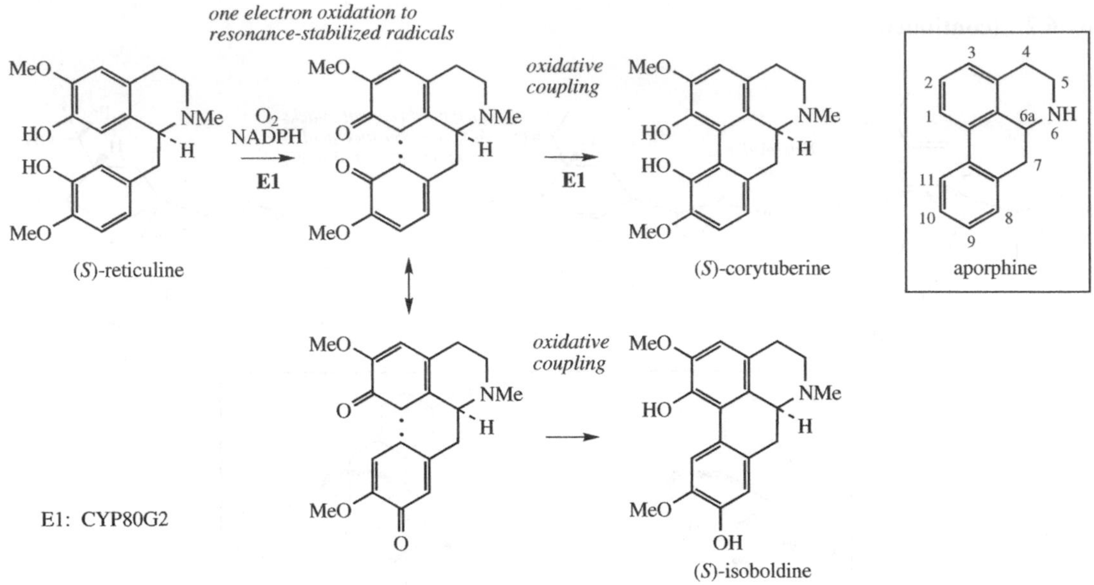

Figure 6.60

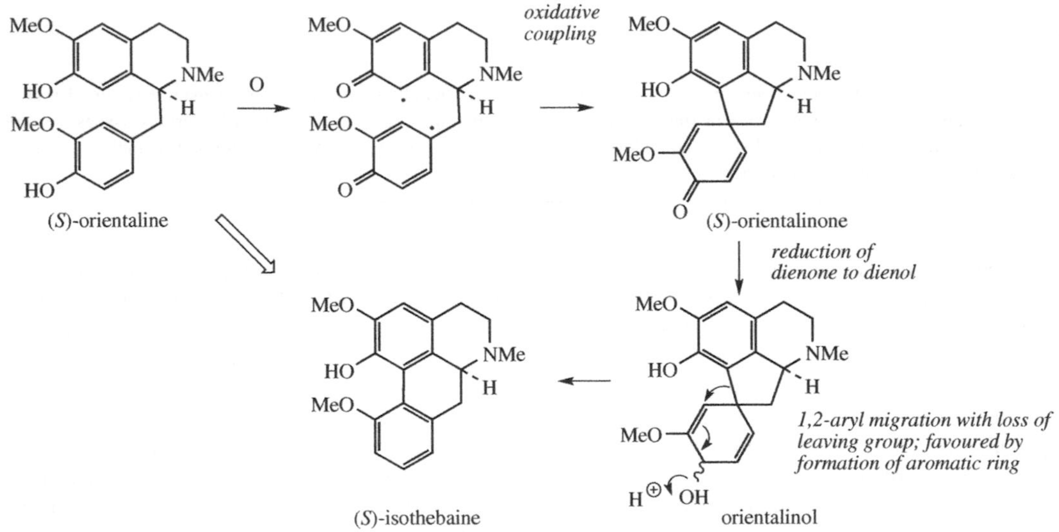

Figure 6.61

tautomerism is not possible for rearomatization when coupling positions are already substituted. The process is fully borne out by experimental evidence, including the subsequent isolation of orientalinone and orientalinol from *Papaver orientale*.

Stephanine (Figure 6.62) from *Stephania* species (Menispermaceae) is analogous to isothebaine and shares

a similar pathway, though this time from (*R*)-**orientaline**. The different substitution pattern in stephanine compared with isothebaine is a consequence of the intermediate dienol suffering migration of the alkyl rather than aryl group (Figure 6.62). **Aristolochic acid** is a novel modified aporphine containing a nitro group and is produced from stephanine by oxidative reactions leading

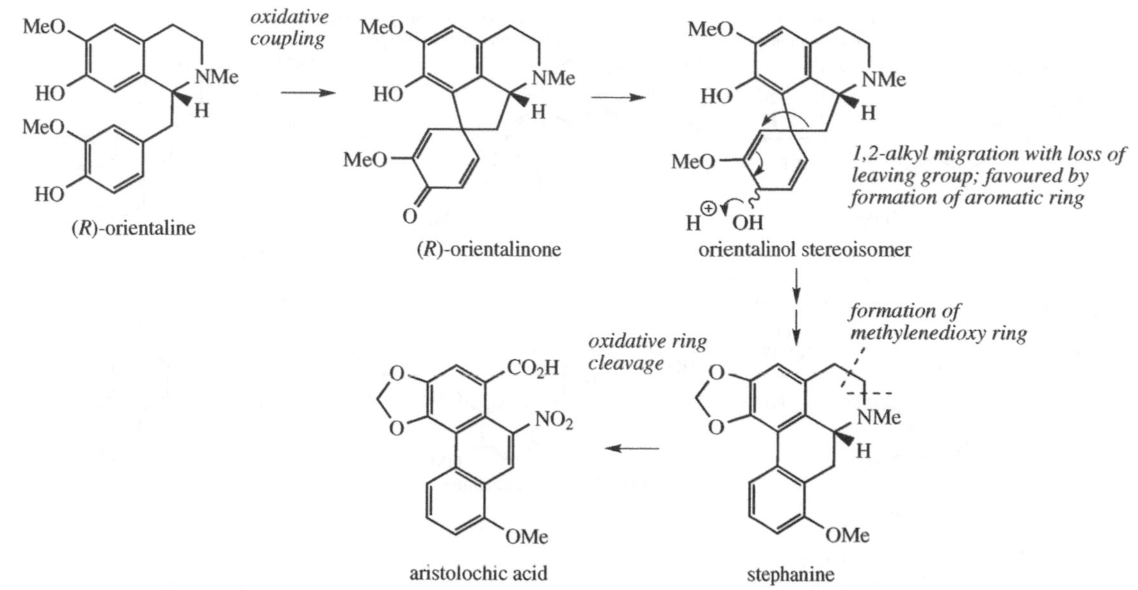

Figure 6.62

to ring cleavage (Figure 6.62). Aristolochic acid is present in many species of *Aristolochia* (Aristolochiaceae) used in traditional medicine, e.g. snake-root *Aristolochia serpentina*. However, because aristolochic acid is now known to be nephrotoxic and to cause acute kidney failure, the use of *Aristolochia* species in herbal medicines, especially Chinese remedies, has been banned in several countries.

The alkaloid **berberine** (Figure 6.63) is found in many members of the Berberidaceae (e.g. *Berberis, Mahonia*), the Ranunculaceae (e.g. *Hydrastis*), and other families. Berberine has antiamoebic, antibacterial, and anti-inflammatory properties, and plants containing berberine have long been used in traditional medicine. Its tetracyclic skeleton is derived from a benzyltetrahydroisoquinoline system with the incorporation of an extra

carbon atom, supplied from SAM via an *N*-methyl group (Figure 6.63). This extra skeletal carbon is known as a 'berberine bridge'. Formation of the berberine bridge is readily rationalized as an oxidative process in which the *N*-methyl group is oxidized to an iminium ion, and a cyclization to the aromatic ring occurs by virtue of the phenolic group (Figure 6.64).

The oxidative cyclization process resembles formation of a methylenedioxy group (see page 27), whilst the mechanism of cyclization is exactly the same as that invoked in formation of a tetrahydroisoquinoline ring, i.e. a Mannich-like reaction (see page 19). The product from the enzymic transformation of (*S*)-**reticuline** is the protoberberine alkaloid (*S*)-**scoulerine**, the berberine bridge enzyme requiring molecular oxygen as oxidant and releasing H_2O_2 as by-product (Figure 6.64). Its role in

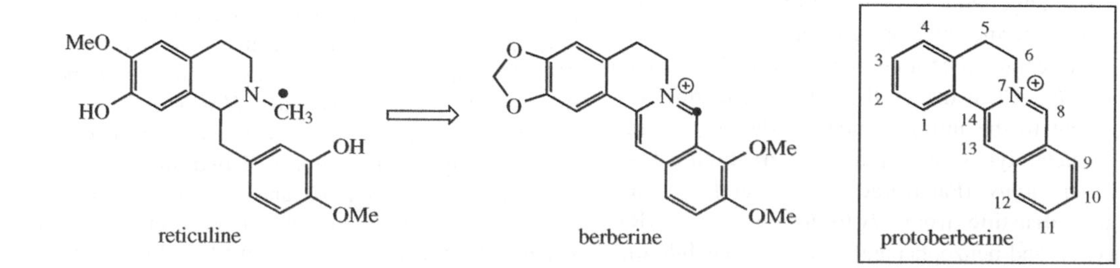

Figure 6.63

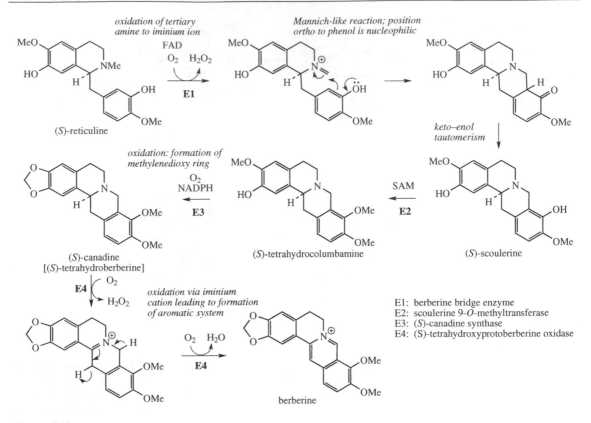

Figure 6.64

the cyclization reaction completed, the phenol group in scoulerine is then methylated to give **tetrahydrocolumb-amine**, and this step is followed by construction of the methylenedioxy group from the *ortho*-methoxyphenol, via an O_2-, NADPH- and cytochrome P-450-dependent enzyme. **Canadine** is oxidized to give the quaternary isoquinolinium system of **berberine**. This appears to involve two separate oxidation steps, both requiring molecular oxygen, with H_2O_2 and H_2O produced in the successive processes. The mechanistic sequence through an iminium ion has been suggested to account for these observations.

The protoberberine skeleton of scoulerine may be subjected to further modifications, some of which are given in Figure 6.65. Cleavage of the heterocyclic ring systems adjacent to the nitrogen atom as shown give rise to new skeletal types: protopine, e.g. **protopine** from *Chelidonium majus* (Papaveraceae), phthalideisoquinoline, e.g. **hydrastine** from *Hydrastis canadensis* (Ranunculaceae), and benzophenanthridine, e.g. **chelidonine** also from *Chelidonium majus*. The non-heterocyclic system seen in the opium alkaloid **narceine** from *Papaver*

somniferum can be visualized as the result of cleavage of two of these bonds. Some enzymes implicated in these modifications have been characterized. These include an *N*-methyltransferase, yielding the quaternary amine with defined stereochemistry at the new chiral centre (Figure 6.65), and a cytochrome P-450-dependent monooxygenase that hydroxylates at position 14. This initiates ring opening to the protopine-type systems. A similar process can be formulated to rationalize cleavage of other rings. Some alkaloids of the phthalide-type are medicinally important. **Noscapine** (Figure 6.66) is one of the opium alkaloids and although it lacks any analgesic activity it is an effective cough suppressant (see page 350). **Hydrastine** is beneficial as a traditional remedy in the control of uterine bleeding. *Hydrastis* also contains berberine, indicating the close biosynthetic relationship of the two types of alkaloid. **Bicuculline** (Figure 6.66) from species of *Corydalis* and *Dicentra* (Fumariaceae) and its quaternary methiodide have been identified as potent GABA antagonists and have found widespread application as pharmacological probes for convulsants acting at GABA neuroreceptors.

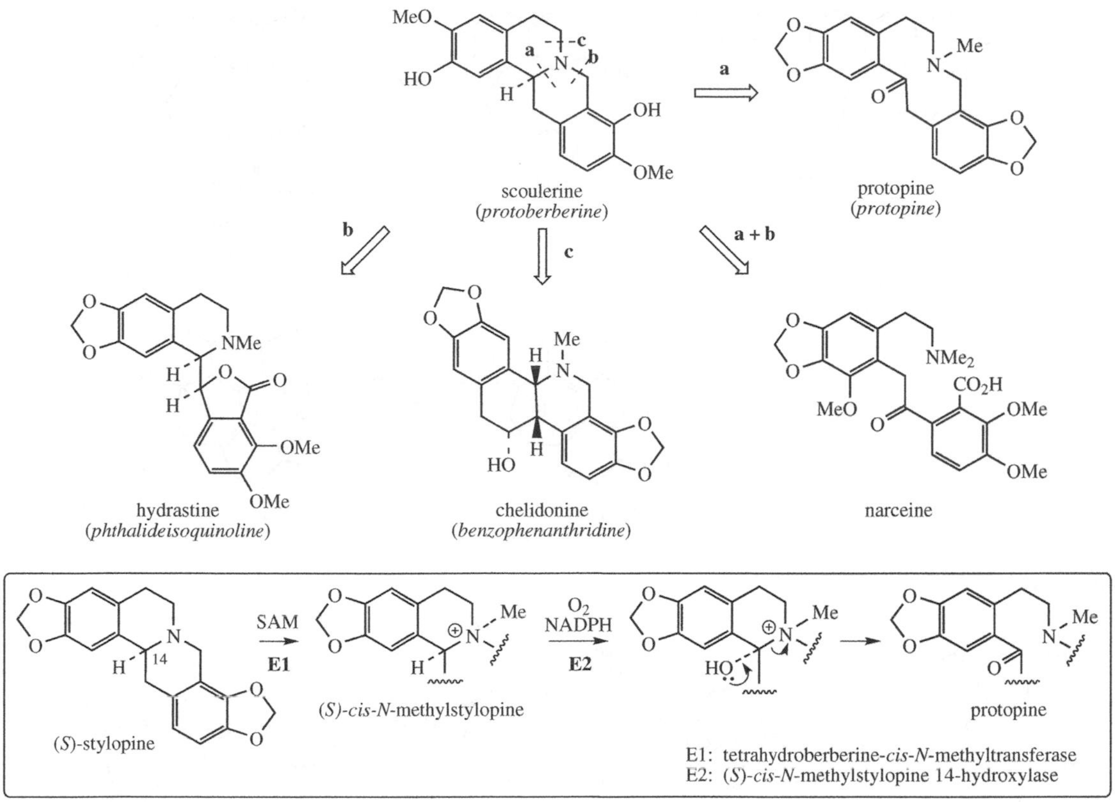

Figure 6.65

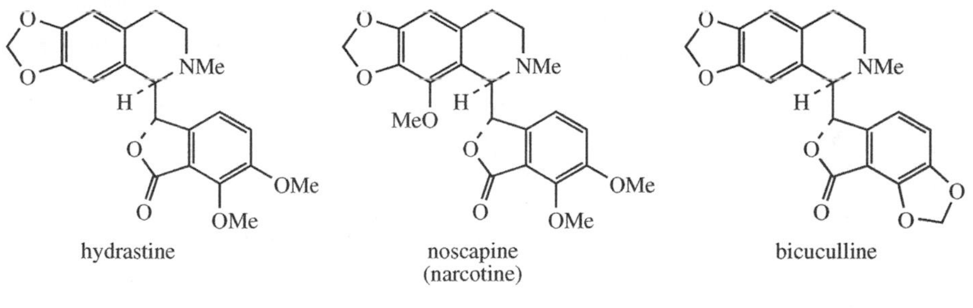

Figure 6.66

Phenethylisoquinoline Alkaloids

Several genera in the lily family (Liliaceae) are found to synthesize analogues of the benzyltetrahydroisoquinoline alkaloids, e.g. **autumnaline** (Figure 6.67), which contain an extra carbon between the tetrahydroisoquinoline and the pendant aromatic rings. This skeleton is formed in a similar way to that in the benzyltetrahydroisoquinolines

from a phenylethylamine and an aldehyde (Figure 6.67), but a whole C_6C_3 unit rather than a C_6C_2 fragment functions as the reacting aldehyde. Typically, dopamine (from tyrosine) and 4-hydroxydihydrocinnamaldehyde (from phenylalanine) are involved in the initial condensation, and further hydroxylation and methylation steps then build up the substitution pattern to that of autumnaline. Phenolic oxidative coupling accounts for

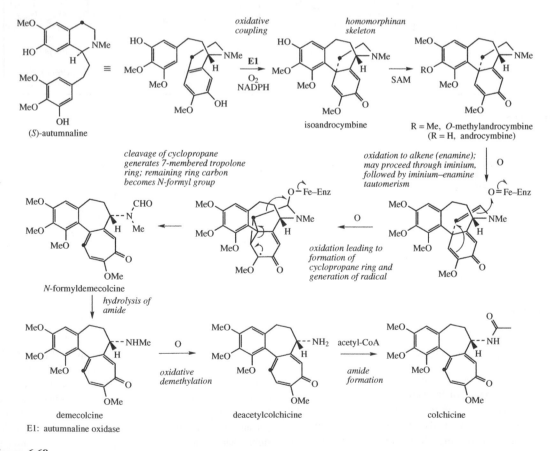

Figure 6.67

Figure 6.68

the occurrence of homoaporphine alkaloids such as **floramultine** and **kreysigine** in *Kreysigia multiflora* (Liliaceae/Convallariaceae).

(*S*)-**Autumnaline** has also been found to act as a precursor for **colchicine** (Figure 6.68), an alkaloid containing an unusual tropolone ring. Colchicine is found in species of *Colchicum*, e.g. *Colchicum autumnale* (Liliaceae/Colchicaceae), as well as many other plants in the Liliaceae [Box 6.8]. Colchicine no longer has its nitrogen atom in a ring system, and extensive reorganization of the autumnaline structure has occurred. The seven-membered tropolone ring was shown by labelling experiments to originate by ring expansion of the tyrosine-derived aromatic ring taking in the adjacent benzylic carbon (Figure 6.68). Prior to these remarkable rearrangements, oxidative coupling of autumnaline in the *para–para* sense features in the pathway giving the dienone **isoandrocymbine**, which has a homomorphinan skeleton (compare salutaridine, Figure 6.53). The isomer **androcymbine** (Figure 6.68) had been

isolated earlier from *Androcymbium melanthioides* (Liliaceae/Colchicaceae), thus giving a clue to the biosynthetic pathway. Methylation follows, giving *O*-methylandrocymbine, and it is then proposed that enzymic oxidation to an enamine yields the substrate for ring modification. Experimental labelling studies are then best explained by a cytochrome P-450-dependent process in which formation of a cyclopropane ring is followed by ring opening to generate the 6π electron aromatic tropolone system. This incorporates the original tyrosine benzylic carbon into the seven-membered ring, and also breaks the original phenylethylamine side-chain between the carbon atoms. One carbon is left on the nitrogen as a formyl group, and this can be lost by hydrolysis. **Colchicine** is produced by exchanging the *N*-methyl group for an *N*-acetyl group, by way of an oxidative demethylation followed by acetylation using acetyl-CoA. **Demecolcine** and **deacetylcolchicine** are intermediates in the process.

Box 6.8

Colchicum

Colchicum seed and corm are obtained from *Colchicum autumnale* (Liliaceae/Colchicaceae), the autumn crocus or meadow saffron. The plant, though not a crocus, produces crocus-like flowers in the autumn, the leaves not emerging until the spring. It is a native of Europe, is widely cultivated as an ornamental garden plant, and is grown for drug use, mainly in Europe and North Africa. The principal alkaloid is colchicine (Figure 6.68), which occurs to the level of about 0.8% in the seed, and 0.6% in the corm. The nitrogen in colchicine is part of an amide function, so colchicine does not display any significant basicity and does not form well-defined salts. Demecolcine (*N*-deacetyl-*N*-methylcolchicine; Figure 6.68) is a minor constituent in both corm and seeds.

Extracts of *Colchicum autumnale*, and later **colchicine** itself, have been used in the treatment of gout, a painful condition in which impaired purine metabolism leads to a build-up of uric acid crystals in the joints. Colchicine is an effective treatment for acute attacks, but it is very toxic, and this restricts its general use. It appears to act primarily as an anti-inflammatory agent and does not itself affect uric acid metabolism, which needs to be treated with other agents, e.g. a xanthine oxidase inhibitor such as allopurinol. The cytotoxic properties of colchicine and related alkaloid structures from *Colchicum autumnale* led to their being tested as potential anticancer agents. Colchicine binds to tubulin in the mitotic spindle, preventing polymerization and assembly into microtubules, as do podophyllotoxin (see page 155) and vincristine (see page 375), and provides a useful biochemical probe for this process. A feature of colchicine's structure is that the two aromatic rings are not coplanar, but are twisted relative to each other with a dihedral angle of 54° (Figure 6.69). This is essential for binding to tubulin. Colchicine and most related compounds are too toxic for medicinal use as anticancer agents, though research still progresses. One derivative under active development

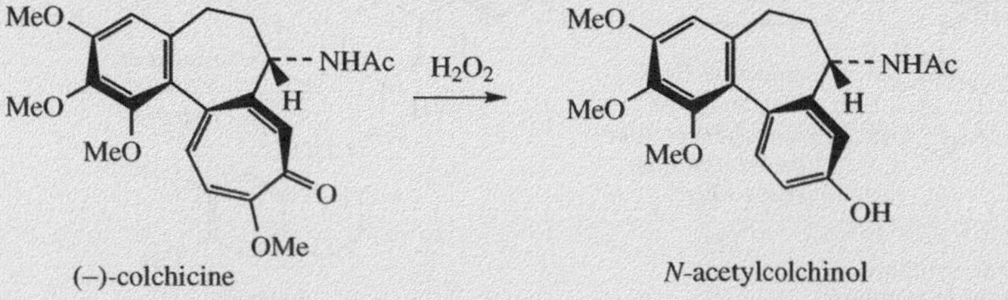

(−)-colchicine H_2O_2 *N*-acetylcolchinol

Figure 6.69

Box 6.8   (continued)

is a water-soluble phosphate pro-drug of **N-acetylcolchinol**, a compound which was initially synthesized from colchicine by an oxidation reaction that prompts a ring contraction process (Figure 6.69).

The ability of colchicine to act as a mitotic poison is exploited in plant breeding, since the interference with mitosis results in multiplication of chromosomes in the cell nucleus without the process of cell division. Cell division recommences on cessation of treatment. This allows generation of mutations (polyploids) and possible new varieties of plant. Colchicine is also found in other species of *Colchicum*, as well as in many other plants in the Liliaceae (e.g. *Bulbocodium, Gloriosa, Merendera*, and *Sandersonia*), a group of plants now classified as the family Colchicaceae. *Gloriosa superba* is currently a commercial source of colchicine.

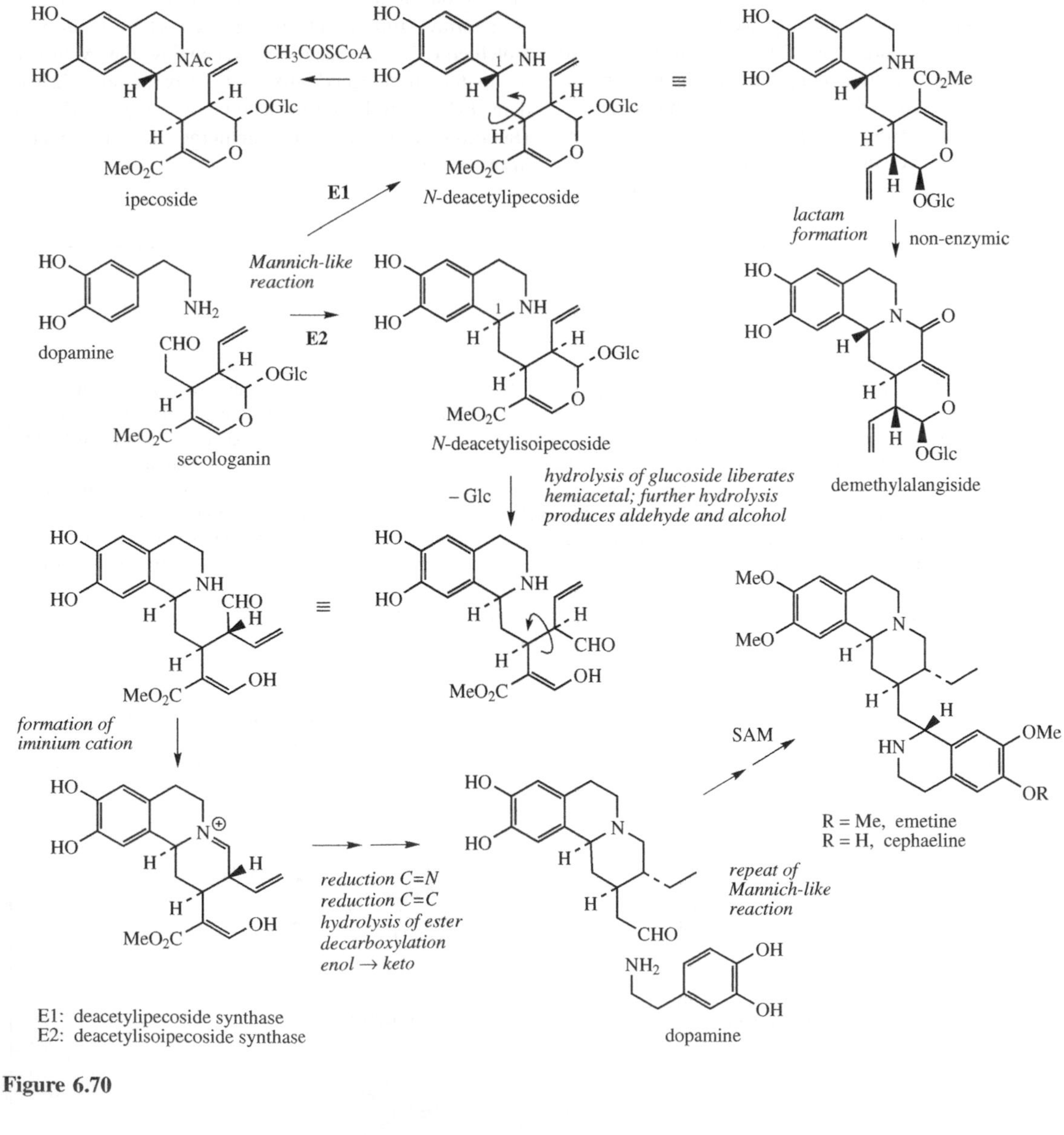

E1: deacetylipecoside synthase
E2: deacetylisoipecoside synthase

Figure 6.70

Terpenoid Tetrahydroisoquinoline Alkaloids

The alkaloids found in ipecacuanha [Box 6.9], the dried rhizome and roots of *Cepahaelis ipecacuanha* (Rubiaceae), have a long history of use in the treatment of amoebic dysentery, and provide unusual examples of tetrahydroisoquinoline structures. The principal alkaloids, e.g. **emetine** and **cephaeline** (Figure 6.70), possess a skeleton with two tetrahydroisoquinoline ring systems plus a further fragment which has its origin in a terpenoid-derived molecule. This terpenoid substrate is the secoiridoid **secologanin** (see page 207), a compound which also features in the biosynthesis of many complex indole alkaloids (see page 370).

Secologanin is an aldehyde and can condense with dopamine in a Mannich-like reaction to give the tetrahydroisoquinoline alkaloids *N*-deacetylisoipecoside or *N*-deacetylipecoside with different configurations at C-1 (Figure 6.70). Indeed, dopamine and secologanin react readily under mildly acidic conditions to give a mixture of these two alkaloids; in nature, two different enzymes are involved. The *N*-acetate **ipecoside** is also

---

Box 6.9

Ipecacuanha

Ipecacuanha or **ipecac** is derived from the dried rhizome and roots of *Cephaelis ipecacuanha* or *Cephaelis acuminata* (Rubiaceae). These are low, straggling shrubs possessing horizontal rhizomes with prominently ridged roots. *Cephaelis ipecacuanha* yields what is termed Rio or Brazilian ipecac, and is cultivated mainly in Brazil, whilst *Cephaelis acuminata* gives Cartagena, Nicaragua, or Panama ipecac, and comes principally from Colombia and Nicaragua. Most of the commercial ipecac now derives from *Cephaelis acuminata*. Ipecac is an age-old remedy of the South American Indians, who used it for the treatment of dysentery. More recently it was mixed with powdered opium to give Dover's powder (see page 349), where the ipecac content functioned as a diaphoretic (promotes perspiration).

Ipecac contains 2–2.5% of alkaloids, the principal ones being emetine and cephaeline (Figure 6.70). Typically, in *Cephaelis ipecacuanha* the emetine to cephaeline ratio might be about 2:1, whereas in *Cephaelis acuminata* the ratio ranges from about 1:2 to 1:1. Minor alkaloids characterized include psychotrine and *O*-methylpsychotrine (Figure 6.71), which are dehydro variants of cephaeline and emetine respectively.

Both **emetine** and the synthetic **dehydroemetine** (Figure 6.71) have been useful as anti-amoebics, particularly in the treatment of amoebic dysentery. However, they also cause nausea, and this has now made other drugs preferable. The emetic action of the alkaloids is particularly valuable though, and the crude drug extract in the form of **ipecacuanha emetic mixture** is an important preparation used for drug overdose or poisoning. The emetic mixture is often a standard component in poison antidote kits. Ipecacuanha also has expectorant activity, and extracts are still components of a number of compound expectorant preparations. Emetine has more expectorant and less emetic action than cephaeline; thus, the Brazilian drug is preferred for such mixtures. If required, emetine may be obtained in larger amounts by methylating the cephaeline component of the plant material.

Emetine and cephaeline are both potent inhibitors of protein synthesis, inhibiting at the translocation stage. They display antitumour, antiviral, and antiamoebic activity, but they are too toxic for therapeutic use. In recent studies, *O*-methylpsychotrine has displayed fairly low effects on protein synthesis, but a quite potent ability to curb viral replication through inhibition of HIV reverse transcriptase. This may give it potential in the treatment of AIDS.

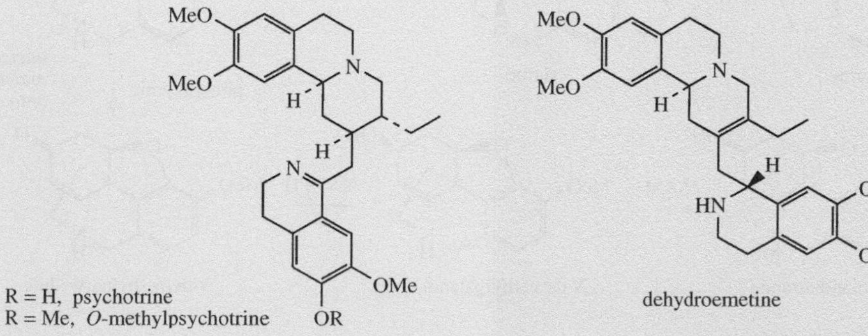

R = H, psychotrine
R = Me, *O*-methylpsychotrine

dehydroemetine

Figure 6.71

found in ipecacuanha; however, this has the opposite stereochemistry at C-1 to the biosynthetic intermediates leading to emetine. In the absence of the *N*-acetyl function, lactam (amide) formation between the amine and the carboxylic ester may occur spontaneously to give **demethylalangiside**; the C-1 epimer is also known. The pathway to emetine from *N*-deacetylisoipecoside may be postulated as follows. The secologanin fragment in *N*-deacetylisoipecoside contains an acetal function, which can be restored to its component aldehyde and alcohol fragments by hydrolysis of the glucosidic bond. The newly liberated aldehyde can then bond with the secondary amine to give the quaternary iminium cation. This

intermediate is converted into an aldehyde by a sequence of reactions: reduction of iminium, reduction of alkene, plus hydrolysis of ester and subsequent decarboxylation, though not necessarily in that order. The decarboxylation step is facilitated by the β-aldehyde function, shown as an enol in Figure 6.71. Most of the reactions taking place in the secologanin-derived part of the structure are also met in discussions of terpenoid indole alkaloids (see page 370). The resultant aldehyde is now able to participate in formation of a second tetrahydroisoquinoline ring system, by reaction with a second dopamine molecule. Methylation gives **cephaeline** and **emetine**, though these methylation steps may well occur earlier in the pathway.

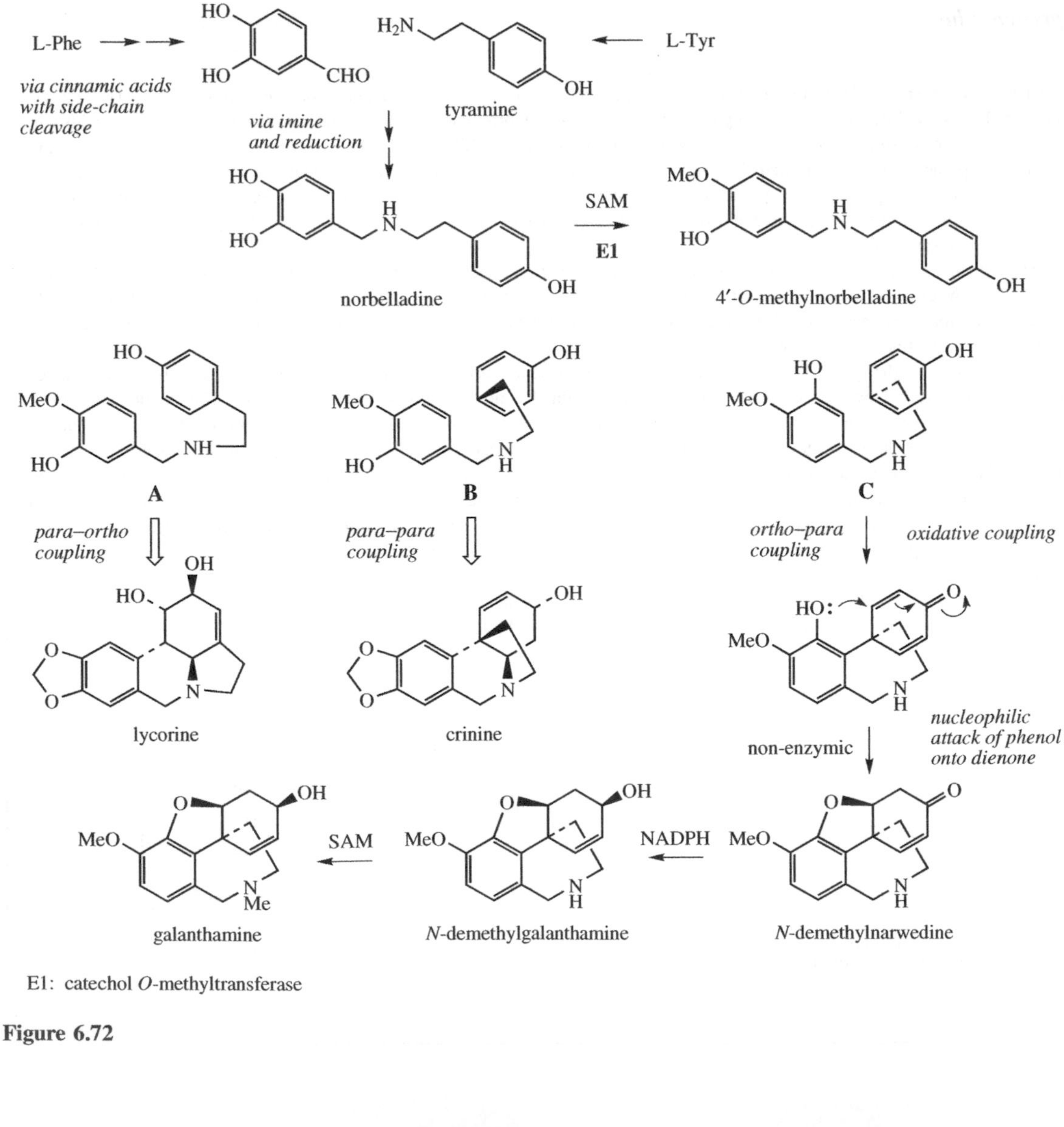

E1: catechol *O*-methyltransferase

Figure 6.72

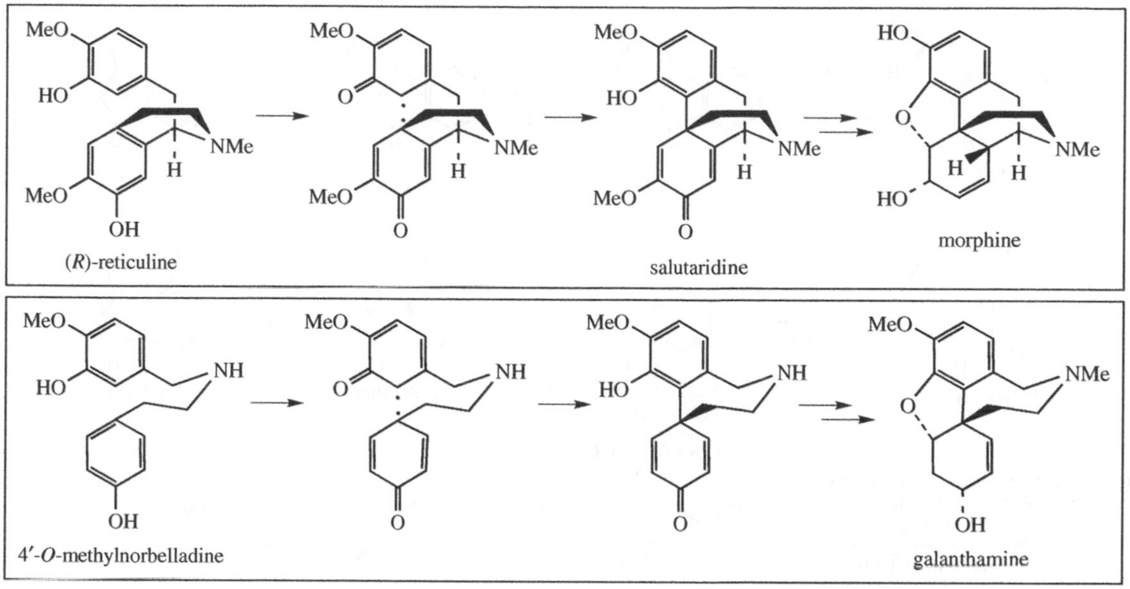

(R)-reticuline → salutaridine → morphine

4'-O-methylnorbelladine → galanthamine

Figure 6.73

Amaryllidaceae Alkaloids

Various types of alkaloid structure are encountered in the daffodil/narcissus family, the Amaryllidaceae, and they can be rationalized better through biosynthesis than by structural type. The alkaloids arise by alternative modes of oxidative coupling of precursors related to **norbelladine** (Figure 6.72), which is formed through combination of 3,4-dihydroxybenzaldehyde with tyramine, these two precursors arising from phenylalanine and tyrosine respectively. Three structural types of alkaloid can be related to **4'-O-methylnorbelladine** by different alignments of the phenol rings allowing couplings *para–ortho* (A), *para–para* (B), or *ortho–para* (C), as shown in Figure 6.72.

For **galanthamine**, the dienone formed via oxidative coupling (C) undergoes a spontaneous nucleophilic addition from the phenol group, forming an ether linkage (compare opium alkaloids, page 348), and the sequence is completed by reduction and methylation reactions. The analogy with morphine biosynthesis is quite striking and can be appreciated from Figure 6.73. For **lycorine** and **crinine**, although details are not given in Figure 6.72,

---

Box 6.10

Galanthamine

Galantamine (galanthamine; Figure 6.72) can be isolated from a number of species of the Amaryllidaceae, including snowdrops (*Galanthus* species), daffodils (*Narcissus pseudonarcissus*), and snowflakes (*Leucojum* species), where typical content varies from about 0.05 to 0.2% in the bulbs. Several structurally related alkaloids are also present. It is currently isolated for drug use from the bulbs of wild *Leucojum aestivum* and *Galanthus* species; longer term supplies will require development of cheaper synthetic procedures. Galantamine acts as a centrally acting competitive and reversible inhibitor of acetylcholinesterase, and significantly enhances cognitive function in the treatment of Alzheimer's disease by raising acetylcholine levels in brain areas lacking cholinergic neurones. It is less toxic than other acetylcholinesterase inhibitors, such as physostigmine (see page 386). There is also evidence that galantamine displays an increased beneficial effect due to a sensitizing action on nicotinic acetylcholine receptors in the central nervous system. In common with other treatments for Alzheimer's disease, it does not cure the condition, but merely slows the rate of cognitive decline.

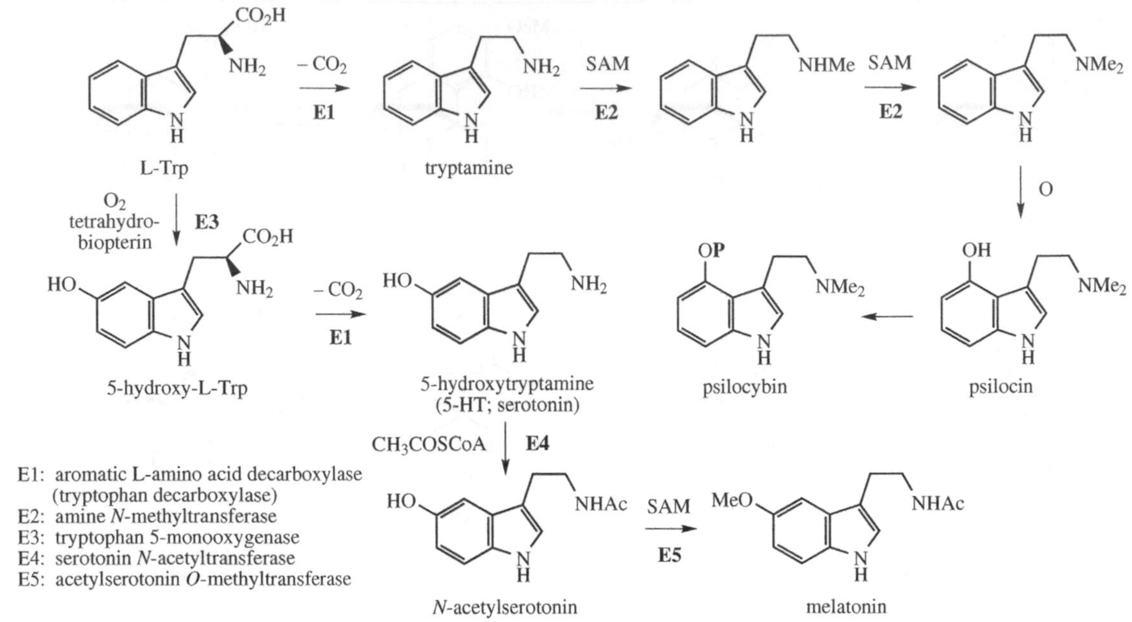

Figure 6.74

it is apparent that the nitrogen atom acts as a nucleophile towards the dienone system, generating the new heterocyclic ring systems. Alkaloids such as lycorine, crinine, and galanthamine can undergo further modifications, which include ring cleavage reactions, generating many more variations than can be considered here. The Amaryllidaceae family includes *Amaryllis, Narcissus,* and *Galanthus,* and the alkaloid content of bulbs from most members makes these toxic. Lycorine was first isolated from *Lycorus radiata,* but is common and found throughout the family. **Galanthamine** from snowdrops (*Galanthus* species) is currently an important drug material of value in treating Alzheimer's disease [Box 6.10].

ALKALOIDS DERIVED FROM TRYPTOPHAN

L-Tryptophan is an aromatic amino acid containing an indole ring system, having its origins in the shikimate pathway (Chapter 4) via anthranilic acid. It acts as a precursor of a wide range of indole alkaloids, but there is also definite proof that major rearrangment reactions can convert the indole ring system into a quinoline ring, thus increasing further the ability of this amino acid to act as an alkaloid precursor (see page 380).

Simple Indole Alkaloids

Tryptamine and its *N*-methyl and *N,N*-dimethyl derivatives (Figure 6.74) are widely distributed in plants, as are simple hydroxylated derivatives such as **5-hydroxytryptamine (5-HT, serotonin)**. These are formed (Figure 6.74) by a series of decarboxylation, methylation, and hydroxylation reactions, though the sequences of these reactions are found to vary according to final product and/or organism involved. 5-HT is also found in mammalian tissue, where it acts as a neurotransmitter in the central nervous system [Box 6.11]. It is formed from tryptophan by hydroxylation and then decarboxylation, paralleling the tyrosine → dopamine pathway (see page 335). *N*-Acetylation of serotonin followed by *O*-methylation leads to **melatonin**, an animal hormone that regulates daily (circadian) rhythms: melatonin levels inform the organism about the time of day [Box 6.11]. Melatonin is also found in plants, however. In the formation of **psilocin** (Figure 6.74), decarboxylation precedes *N*-methylation, and hydroxylation occurs last. Phosphorylation of the hydroxyl in psilocin gives **psilocybin**. These two compounds are responsible for the hallucinogenic properties of so-called magic mushrooms, which include species of *Psilocybe, Panaeolus,* etc. [Box 6.11].

Box 6.11

5-Hydroxytryptamine (Serotonin)

5-Hydroxytryptamine (5-HT, serotonin) is a monoamine neurotransmitter found in cardiovascular tissue, the peripheral nervous system, blood cells, and the central nervous system. It mediates many central and peripheral physiological functions, including contraction of smooth muscle, vasoconstriction, food intake, sleep, pain perception, and memory, a consequence of it acting on several distinct receptor types; currently, seven types have been identified. Although 5-HT may be metabolized by monoamine oxidase, platelets and neurons possess a high affinity 5-HT reuptake mechanism. This mechanism may be inhibited, thereby increasing levels of 5-HT in the central nervous system, by widely prescribed antidepressant drugs termed selective serotonin re-uptake inhibitors (SSRIs), e.g. fluoxetine (Prozac®).

Migraine headaches that do not respond to analgesics may be relieved by the use of an agonist of the 5-HT_1 receptor, since these receptors are known to mediate vasoconstriction. Though the causes of migraine are still not clear, they are characterized by dilation of cerebral blood vessels. 5-HT_1 agonists based on the 5-HT structure in current use include the sulfonamide derivative **sumatriptan**, and the more recent agents **almotriptan, eletriptan, frovatriptan, naratriptan, rizatriptan**, and **zolmitriptan** (Figure 6.75). These are of considerable value in treating acute attacks. Several of the ergot alkaloids (page 390) also interact with 5-HT receptors.

Melatonin

In animals, **melatonin** (Figure 6.74) is a hormone synthesized by the pineal gland in the brain. It plays a key role in the circadian rhythm, sleep regulation, and seasonal photoperiodic regulation. The duration of elevated melatonin levels is usually proportional to night length in vertebrates. Melatonin concentration and its daily rhythm can thus inform the organism about the time of day and about the season. It is also found in invertebrates and plants, though less is known about its role. Melatonin is claimed to be effective in helping to regulate disrupted circadian rhythms and sleep disorders. A slow-release formulation is available for treating insomnia in older patients; melatonin production is found to decrease with age. It is currently also popular to reduce the effects of jet-lag by resetting the internal body clock.

Psilocybe

The genus *Psilocybe* constitutes a group of small mushrooms with worldwide distribution. It has achieved notoriety on account of the hallucinogenic effects experienced following ingestion of several species, particularly those from Mexico, and has led to the collective description 'magic mushrooms'. Over 80 species of *Psilocybe* have been found to be psychoactive, whereas over

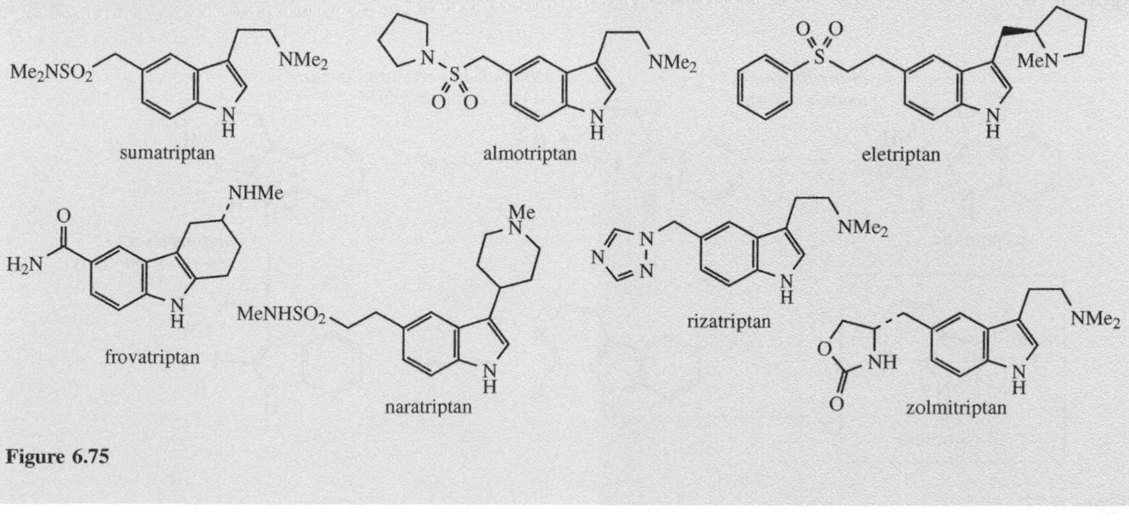

Figure 6.75

Box 6.11   (continued)

50 species are inactive. More than 30 of the hallucinogenic species have been identified in Mexico, but active species may be found in all areas of the world. *Psilocybe mexicana* has been used by the Mexican Indians in ancient ceremonies for many years, and its history can be traced back to the Aztecs. In temperate regions, *Psilocybe semilanceata*, the liberty cap, is a common species with similar activity. All the psychoactive members of the genus are said to stain blue when the fresh tissue, particularly that near the base of the stalk, is damaged, though the converse is not true. Ingestion of the fungus causes visual hallucinations with rapidly changing shapes and colours, and different perceptions of space and time, the effects gradually wearing off and causing no lasting damage or addiction.

The active hallucinogens, present at about 0.3%, are the tryptamine derivatives psilocybin and psilocin (Figure 6.74), which are structurally related to the neurotransmitter 5-HT, thus explaining their neurological effects. Psilocybin is probably the main active ingredient, though to produce hallucinations a dose of some 6–20 mg is required. In addition to species of *Psilocybe*, these compounds may be found in some fungi from other genera, including *Conocybe, Panaeolus*, and *Stropharia*. Misidentification of fungi can lead to the consumers experiencing possible unwanted toxic effects, especially gastrointestinal upsets, instead of the desired psychedelic visions.

Gramine (Figure 6.76) is a simple amine found in barley (*Hordeum vulgare*; Graminae/Poaceae) and is derived from tryptophan by a biosynthetic pathway which cleaves off two carbon atoms, yet surprisingly retains the tryptophan nitrogen atom. It has been suggested that the nitrogen reacts with a cofactor, e.g. pyridoxal phosphate, and is subsequently transferred back to the indolemethyl group after the chain shortening. Only the *N*-methyltransferases have been characterized.

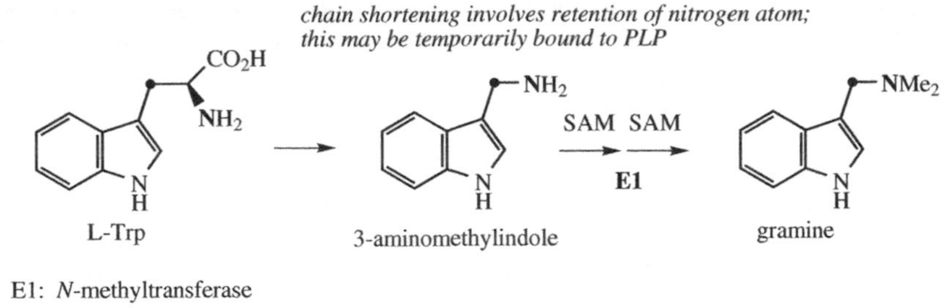

Figure 6.76

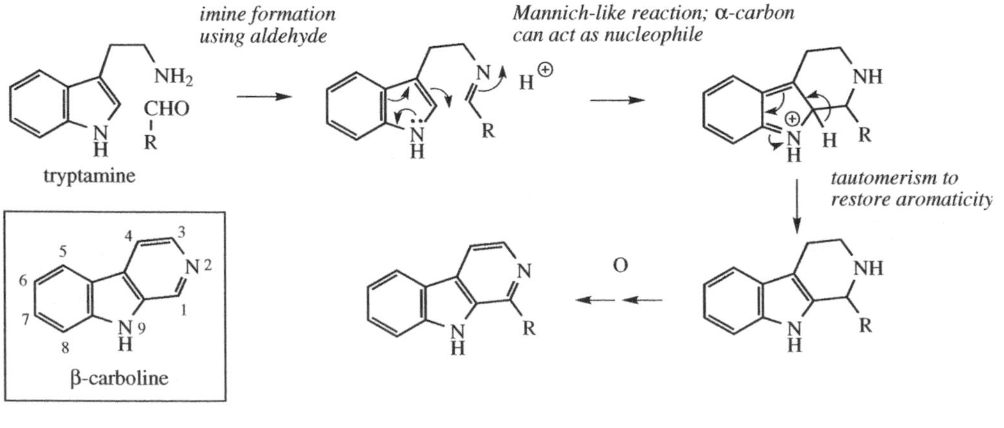

Figure 6.77

Simple β-Carboline Alkaloids

Alkaloids based on a β-carboline system (Figure 6.77) exemplify the formation of a new six-membered heterocyclic ring using the ethylamine side-chain of tryptamine in a process analogous to generation of tetrahydroisoquinoline alkaloids (see page 340). Position 2 of the indole system is nucleophilic, due to the adjacent nitrogen, and can participate in a Mannich/Pictet–Spengler-type reaction, attacking an imine generated from tryptamine and an aldehyde (or keto acid) (Figure 6.77). Aromaticity is restored by subsequent loss of the C-2 proton. (It should be noted that the analogous chemical reaction actually involves nucleophilic attack from C-3, and then a subsequent rearrangement occurs to give bonding at C-2; this type of process does not appear to participate in biosynthetic pathways.)

Extra carbon atoms are supplied by aldehyes or keto acids, according to the complexity of the substrate (compare tetrahydroisoquinoline alkaloids, page 340). Thus, complex β-carbolines, e.g. the terpenoid indole alkaloid ajmalicine (see page 371), are produced by a pathway using an aldehyde such as secologanin. Simpler structures employ keto acids; for example, **harmine** (Figure 6.78) incorporates two extra carbon atoms from pyruvate. In such cases, an intermediate acid is involved, and oxidative decarboxylation gives the dihydro-β-carboline, from which reduced tetrahydro-β-carboline structures, e.g. **elaeagnine** from *Elaeagnus angustifolia* (Elaeagnaceae), or fully aromatic β-carboline structures, e.g. **harman** and **harmine** from *Peganum harmala* (Zygophyllaceae), are derived (Figure 6.78). The methoxy substitution in the indole system of harmine is introduced at some stage in the pathway by successive hydroxylation and methylation reactions. A sequence from 6-hydroxytryptamine is also feasible. The reported psychoactive properties of the plants *Peganum harmala* and *Banisteriopsis caapi* (Malpighiaceae) is due to β-carboline alkaloids such as harmine, harmaline, and tetrahydroharmine (Figure 6.78), which are potent serotonin antagonists.

Terpenoid Indole Alkaloids

More than 3000 terpenoid indole alkaloids are recognized, making this one of the major groups of alkaloids in plants. They are found mainly in eight plant families, of which the Apocynaceae, the Loganiaceae, and the Rubiaceae provide the best sources. In terms of structural complexity, many of these alkaloids are quite outstanding, and it is a tribute to the painstaking experimental studies of various groups of workers that we are able to rationalize these structures in terms of their biochemical origins. Many of the steps have now been characterized at the enzymic level, and appropriate genes have been identified.

In virtually all structures, a tryptamine portion can be recognized. The remaining fragment is usually a C_9 or C_{10} residue, and three main structural types are discernible according to the arrangement of atoms in this fragment. These are termed the *Corynanthe* type, as in **ajmalicine** and **akuammicine**, the *Aspidosperma* type, as in **tabersonine**, and the *Iboga* type, exemplified by **catharanthine** (Figure 6.79). The C_9 or C_{10} fragment was shown to be of terpenoid origin, and the secoiridoid **secologanin** (see page 207) was identified as the terpenoid derivative which initially combined with the tryptamine portion of the molecule. Furthermore, the *Corynanthe, Aspidosperma* and *Iboga* groups of alkaloids could then be related and rationalized in terms of rearrangements occurring in the terpenoid part of the structures (Figure 6.79). Secologanin itself contains the 10-carbon framework that is typical of the *Corynanthe* group. The *Aspidosperma* and *Iboga* groups could then arise by rearrangement of

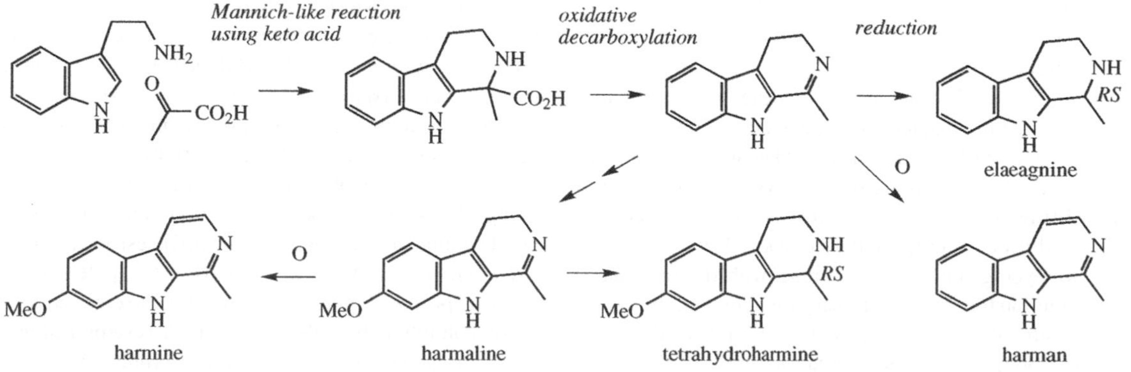

Figure 6.78

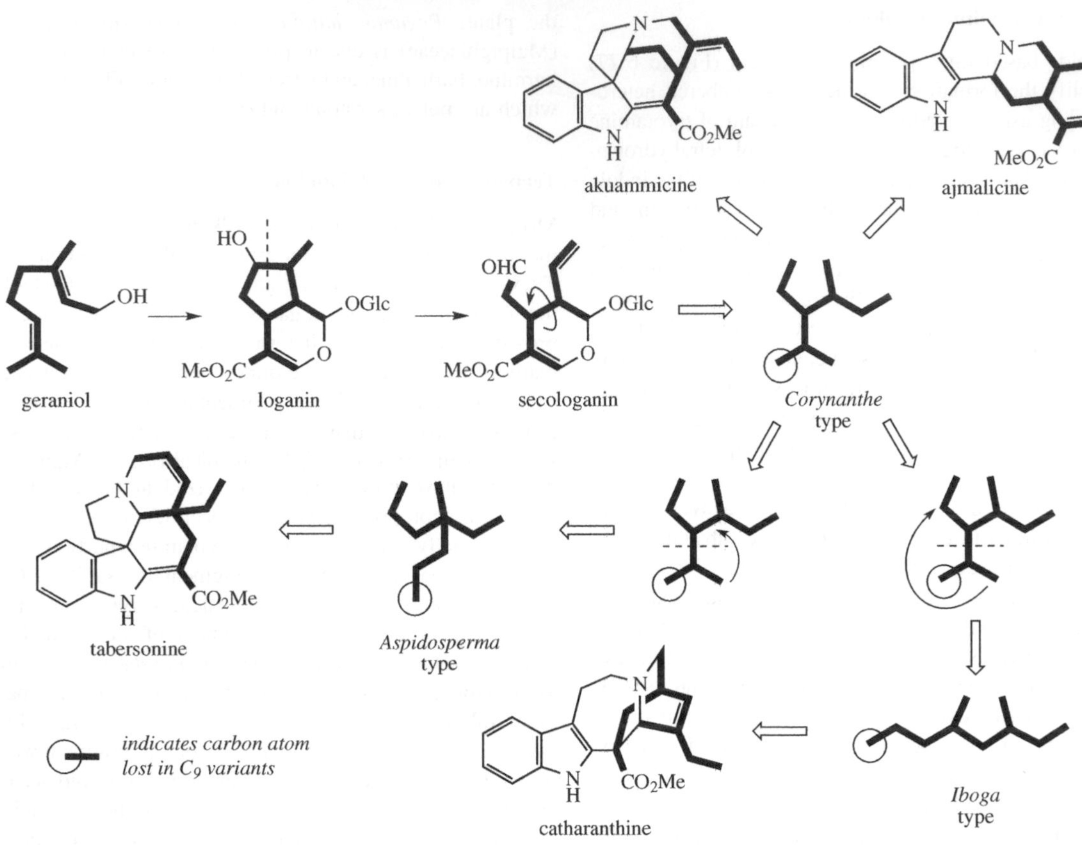

Figure 6.79

the *Corynanthe* skeleton as shown. This is represented by detachment of a three-carbon unit which is then rejoined to the remaining C_7 fragment in one of two different ways. Where C_9 terpenoid units are observed, the alkaloids normally appear to have lost the carbon atom indicated in the circle. This corresponds to the carboxylate function of secologanin and its loss by hydrolysis/decarboxylation is now understandable.

The origins of loganin and secologanin have already been discussed in Chapter 5 (see page 207). Condensation of secologanin with tryptamine in a Mannich-like reaction generates the tetrahydro-β-carboline system and produces **strictosidine** (Figure 6.80). Hydrolysis of the glycoside function allows opening of the hemiacetal and exposure of an aldehyde group which can react with the secondary amine function to give a quaternary iminium cation. These reactions are also seen in the pathway to ipecac alkaloids (see page 362). Allylic isomerization, moving the vinyl double bond into conjugation with the iminium, generates **dehydrogeissoschizine**, and cyclization to **cathenamine**

follows. Cathenamine is reduced to **ajmalicine** in the presence of NADPH. Oxidation of ajmalicine to **serpentine** is catalysed by a peroxidase.

Although details are not confirmed, carbocyclic variants related to ajmalicine, such as **yohimbine**, are likely to arise from dehydrogeissoschizine by the mechanism indicated in Figure 6.81. Yohimbine is found in yohimbe bark (*Pausinystalia yohimbe*; Rubiaceae) and also aspidosperma bark (*Aspidosperma* species; Apocynaceae) and has been used in folk medicine as an aphrodisiac. It does have some pharmacological activity and is known to dilate blood vessels. More important examples containing the same carbocyclic ring system are the alkaloids found in species of *Rauwolfia*, especially *Rauwolfia serpentina* (Apocynaceae) [Box 6.12]. **Reserpine** and **deserpidine** (Figure 6.82) are trimethoxybenzoyl esters of yohimbine-like alkaloids, whilst **rescinnamine** is a trimethoxycinnamoyl ester. Both reserpine and rescinnamine contain an additional methoxyl substituent on the indole system at position 11, the result of hydroxylation

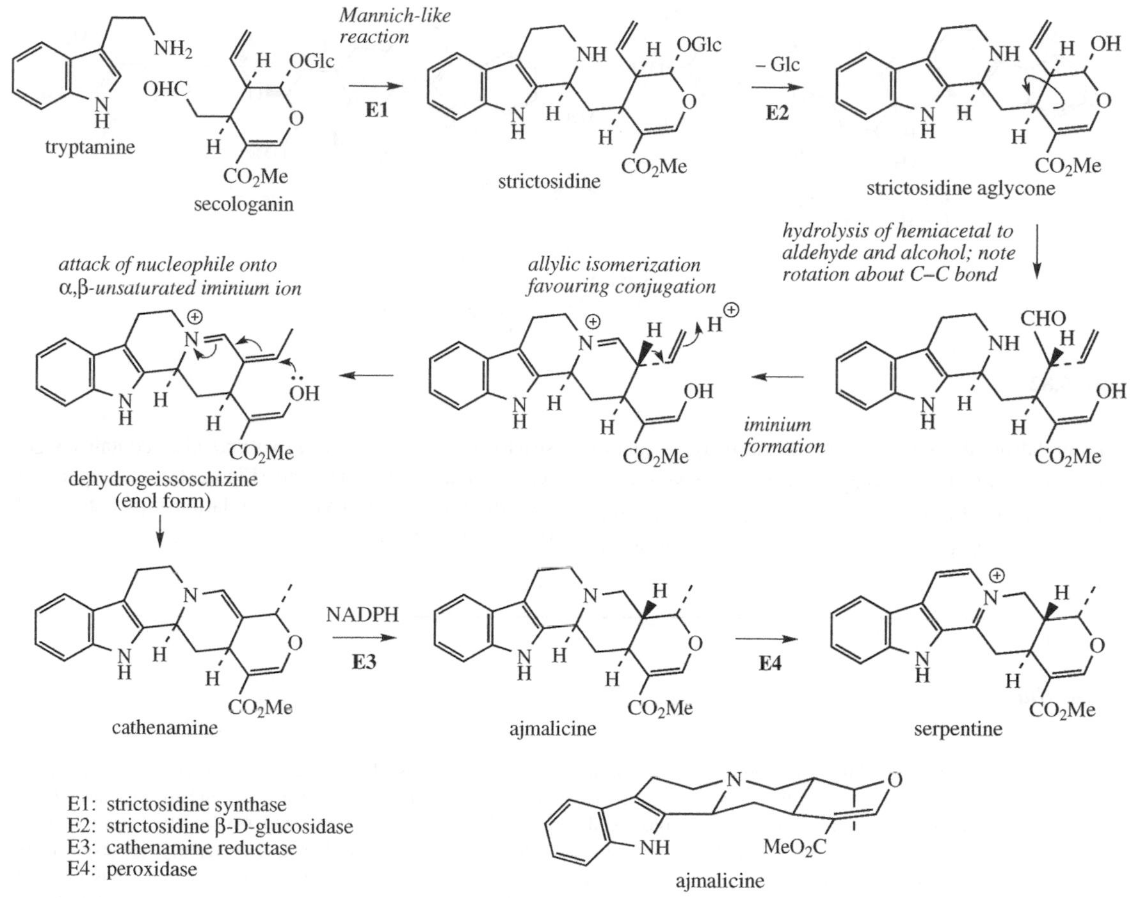

E1: strictosidine synthase
E2: strictosidine β-D-glucosidase
E3: cathenamine reductase
E4: peroxidase

Figure 6.80

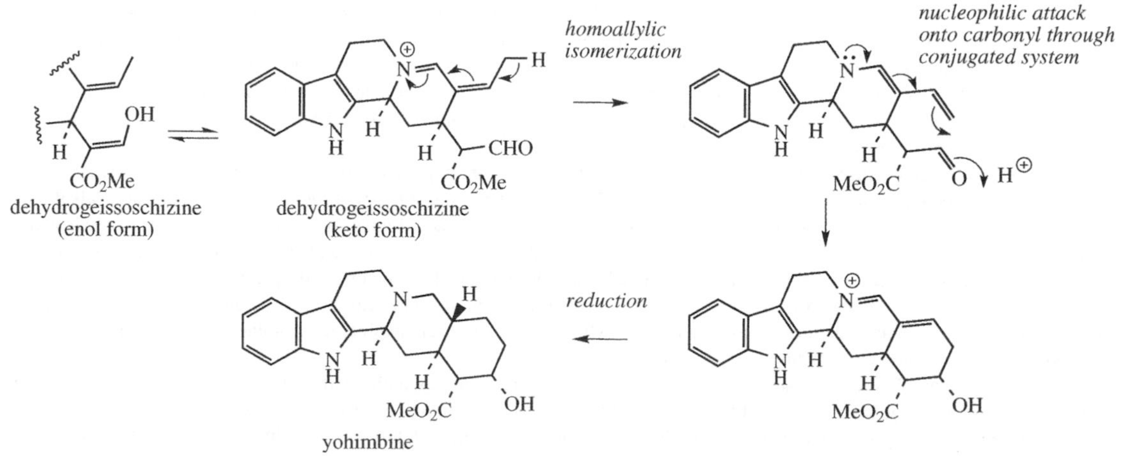

Figure 6.81

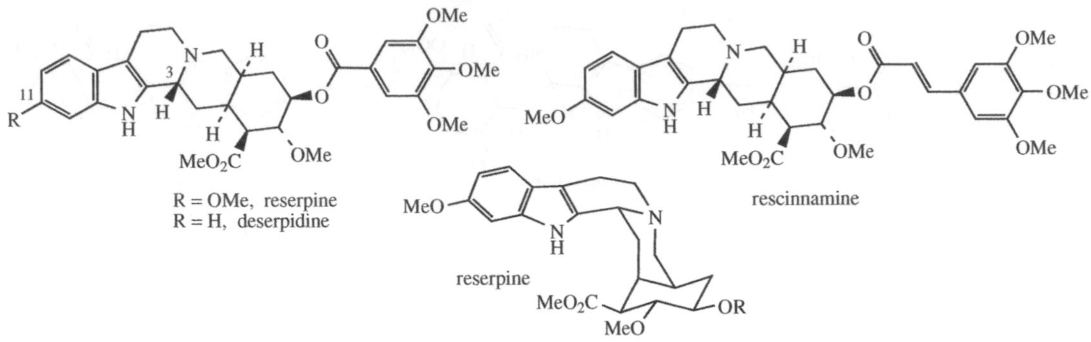

Figure 6.82

and methylation at a late stage in the pathway. A feature of these alkaloids is that they have the opposite stereochemistry at position 3 to yohimbine and strictosidine, and it is likely they are formed from the C-3 epimer of strictosidine. *Rauwolfia serpentina* also contains significant amounts of ajmalicine (Figure 6.80), emphasizing the structural and biosynthetic relationships between the two types of alkaloid.

Box 6.12

Rauwolfia

Rauwolfia has been used in Africa for hundreds of years, and in India for at least 3000 years. It was used as an antidote to snake-bite, to remove white spots in the eyes, against stomach pains, fever, vomiting, and headache, and to treat insanity. It appeared to be a universal panacea, and was not considered seriously by Western scientists until the late 1940s/early 1950s. Clinical tests showed the drug to have excellent antihypertensive and sedative activity. It was then rapidly and extensively employed in treating high blood pressure and to help mental conditions, relieving anxiety and restlessness, and thus initiated the tranquillizer era. The 'cure for insanity' was thus partially justified, and rauwolfia was instrumental in showing that mental disturbance has a chemical basis and may be helped by the administration of drugs.

Rauwolfia is the dried rhizome and roots of *Rauwolfia* (sometimes *Rauvolfia*) *serpentina* (Apocynaceae) or snakeroot, a small shrub from India, Pakistan, Burma, and Thailand. Other species used in commerce include *Rauwolfia vomitoria* from tropical Africa, a small tree whose leaves after ingestion cause violent vomiting, and *Rauwolfia canescens* (= *Rauwolfia tetraphylla*) from India and the Caribbean. Most of the drug material has been collected from the wild. *Rauwolfia serpentina* contains a wide range of indole alkaloids, totalling 0.7–2.4%, though only 0.15–0.2% consists of desirable therapeutically active compounds, principally reserpine, rescinnamine, and deserpidine (Figure 6.82). Other alkaloids of note are serpentine (Figure 6.80), ajmalicine (Figure 6.80), and ajmaline (see Figure 6.86). Reserpine and deserpidine are major alkaloids in *Rauwolfia canescens*, and *Rauwolfia vomitoria* contains large amounts of rescinnamine and reserpine.

Reserpine and **deserpidine** (Figure 6.82) have been widely used as antihypertensives and mild tranquillizers. They act by interfering with catecholamine storage, depleting levels of available neurotransmitters. Prolonged use of the pure alkaloids, reserpine in particular, has been shown to lead to severe depression in some patients, a feature not so prevalent when the powdered root was employed. The complex nature of the alkaloidal mixture means the medicinal action is somewhat different from that of reserpine alone. Accordingly, crude powdered rauwolfia remained an important drug for many years, and selected alkaloid fractions from the crude extract have also been widely used. The alkaloids can be fractionated according to basicity. Thus, serpentine and similar structures are strongly basic, whilst reserpine, rescinnamine, deserpidine, and ajmalicine are weak bases. Ajmaline and related compounds have intermediate basicity.

The rauwolfia alkaloids are now hardly ever prescribed in the UK, either as antihypertensives or as tranquillizers. Over a period of a few years, they have been rapidly superseded by synthetic alternatives. Reserpine has also been suggested to play a role in the promotion of breast cancers. Both **ajmalicine** (= raubasine) (Figure 6.80) and **ajmaline** (see Figure 6.86) are used clinically in Europe, though not in the UK. Ajmalicine is employed as an antihypertensive, whilst ajmaline is of value in the treatment of cardiac arrhythmias. Ajmalicine is also extracted commercially from *Catharanthus roseus* (see page 376).

The structural changes involved in converting the *Corynanthe*-type skeleton into those of the *Aspidosperma* and *Iboga* groups are quite complex, and are summarized in Figure 6.83. In parts, the pathways have yet to be confirmed. Only the tabersonine to vindoline conversion has been characterized in any detail. Early intermediates are alkaloids such as **preakuammicine**, which, although clearly of the *Corynanthe* type, is sometimes designated as *Strychnos* type (compare strychnine, page 377). This is because the *Corynanthe* terpenoid unit, originally attached to the indole α-carbon (as in ajmalicine), is now bonded to the β-carbon, and a new bonding

between the rearrangeable C_3 unit and C-α is in place (see transformation **b**). **Stemmadenine** arises through fission of the bond to C-β, and then further fission yields **dehydrosecodine**, the importance of which is that the rearrangeable C_3 unit has been cleaved from the rest of the terpenoid carbon atoms. Hypothetically, this compound could undergo Diels–Alder-type coupling (see page 96) in two different ways. Alkaloids of the *Aspidosperma* type, e.g. **tabersonine** and **vindoline**, and *Iboga* type, e.g. **catharanthine**, emerge from the different bonding modes (Figure 6.83).

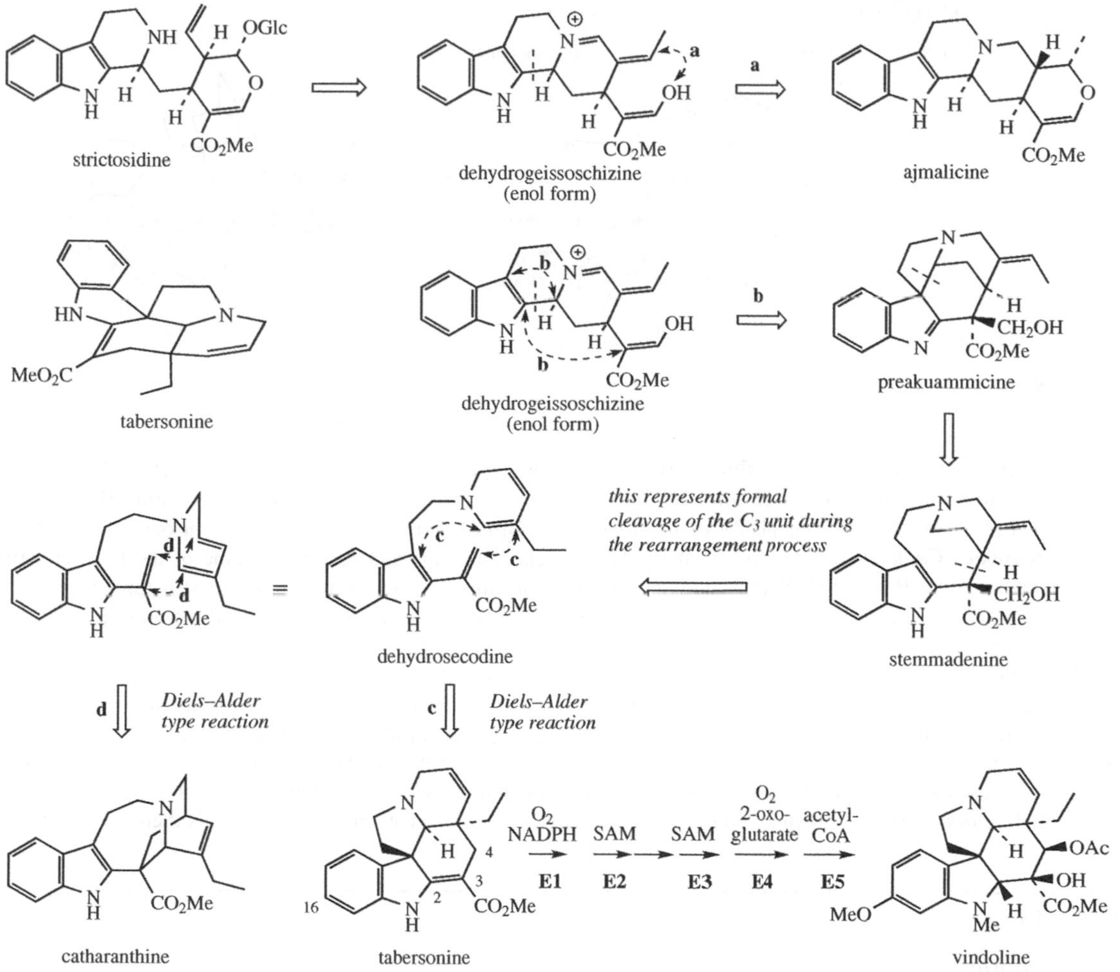

strictosidine

dehydrogeissoschizine
(enol form)

ajmalicine

tabersonine

dehydrogeissoschizine
(enol form)

preakuammicine

this represents formal cleavage of the C_3 unit during the rearrangement process

stemmadenine

dehydrosecodine

d Diels–Alder type reaction

c Diels–Alder type reaction

catharanthine

tabersonine

O_2 NADPH SAM SAM O_2 2-oxo-glutarate acetyl-CoA

E1 **E2** **E3** **E4** **E5**

vindoline

E1: tabersonine 16-hydroxylase
E2: 16-hydroxytabersonine 16-*O*-methyltransferase
E3: 16-methoxy-2,3-dihydro-3-hydroxytabersonine *N*-methyltransferase

E4: deacetoxyvindoline 4-hydroxylase
E5: deacetylvindoline *O*-acetyltransferase

Figure 6.82

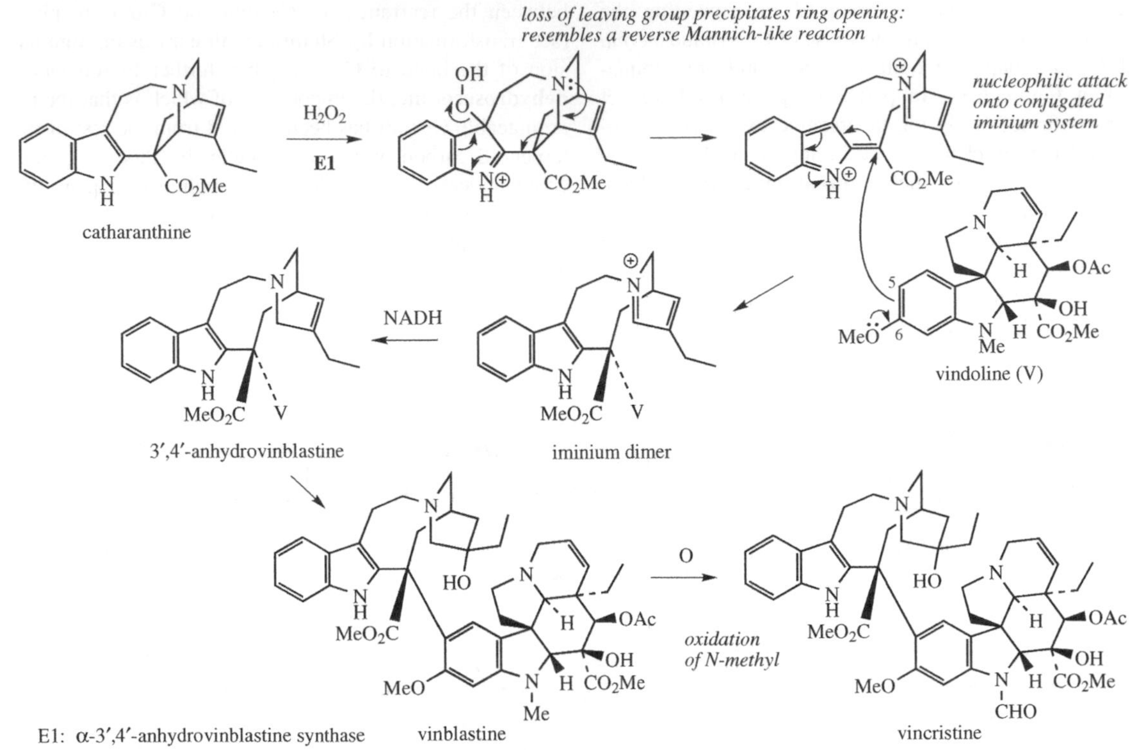

Figure 6.82

Many of the experimental studies which have led to an understanding of terpenoid indole alkaloid biosynthesis have been carried out using plants of the Madagascar periwinkle (*Catharanthus roseus*, formerly *Vinca rosea*; Apocynaceae) [Box 6.13]. Representatives of all the main classes of these alkaloids are produced, including **ajmalicine** (*Corynanthe*), **catharanthine** (*Iboga*), and **vindoline** (*Aspidosperma*). The sequence of alkaloid formation has been established initially by noting which alkaloids become labelled as a feeding experiment progresses, and more recently by appropriate enzyme and gene studies. However, the extensive investigations of the *Catharanthus roseus* alkaloids have also been prompted by the anticancer activity detected in a group of bisindole alkaloids. Two of these, **vinblastine** and **vincristine** (Figure 6.84), have been introduced into cancer chemotherapy and feature as some of the most effective anticancer agents available. These structures are seen to contain the elements of catharanthine and vindoline; indeed, they are derived by coupling of these two alkaloids. A peroxidase enzyme catalyses the coupling, and the product generated is **anhydrovinblastine**. It is proposed that **catharanthine** is oxidized to an iminium cation via a peroxide which loses the peroxide as a leaving group, breaking a carbon–carbon bond as shown (Figure 6.84). This intermediate electrophilic ion is then attacked by the

---

Box 6.13

Catharanthus

The Madagascar periwinkle *Catharanthus roseus* (= *Vinca rosea*) (Apocynaceae) is a small herb or shrub originating in Madagascar, but now common in the tropics and widely cultivated as an ornamental for its shiny dark green leaves and pleasant five-lobed flowers. Drug material is now cultivated in many parts of the world, including the USA, Europe, India, Australia, and South America.

Box 6.13 (continued)

The plant was originally investigated for potential hypoglycaemic activity because of folklore usage as a tea for diabetics. Although plant extracts had no effects on blood sugar levels in rabbits, test animals succumbed to bacterial infection due to depleted white blood cell levels (leukopenia), though no other adverse effects were apparent. The selective action suggested anticancer potential for the plant, and an exhaustive study of the constituents was initiated. The activity was found in the alkaloid fraction, and more than 150 alkaloids have since been characterized in the plant. These are principally terpenoid indole alkaloids, many of which are known in other plants, especially from the same family. Useful antitumour activity was demonstrated in a number of dimeric indole alkaloid structures (more correctly, bisindole alkaloids, since the 'monomers' are different), including vincaleukoblastine, leurosine, leurosidine, and leurocristine. These compounds became known as vinblastine, vinleurosine, vinrosidine, and vincristine respectively, the vin- prefix being a consequence of the earlier botanical nomenclature *Vinca rosea*, in common use at that time. The alkaloids vinblastine and vincristine (Figure 6.85) were introduced into cancer chemotherapy and have proved to be extremely valuable drugs.

Despite the minor difference in structure between vinblastine and vincristine, a significant difference exists in the spectrum of human cancers which respond to the drugs. **Vinblastine** (Figure 6.85) is used mainly in the treatment of Hodgkin's disease, a cancer affecting the lymph glands, spleen, and liver. **Vincristine** (Figure 6.85) has superior antitumour activity compared with vinblastine but is more neurotoxic. It is clinically more important than vinblastine, and is especially useful in the treatment of childhood leukaemia, giving a high rate of remission. Some other cancer conditions, including lymphomas, small-cell lung cancer, and cervical and breast cancers, also respond favourably. The alkaloids need to be injected, and both generally form part of a combination regimen with other anticancer drugs.

Vindesine (Figure 6.85) is a semi-synthetic amide derivative of vinblastine which has been introduced for the treatment of acute lymphoid leukaemia in children. **Vinorelbine** (Figure 6.85) is a newer semi-synthetic modification obtained from anhydrovinblastine (see below). In this structure, the indole.C_2N bridge in the catharanthine-derived unit has been shortened by one carbon; other agents feature structural modifications in the vindoline unit. It is orally active and has a broader anticancer activity, yet with lower neurotoxic side-effects than either vinblastine or vincristine. It is given intravenously for the treatment of advanced breast cancer and non-small-cell lung cancer, or orally to treat small-cell lung cancer. These compounds all inhibit cell mitosis, acting by binding to the protein tubulin in the mitotic spindle, preventing polymerization into microtubules, a mode of action shared with other natural agents, e.g. colchicine (see page 361) and podophyllotoxin (see page 155).

A major problem associated with the clinical use of vinblastine and vincristine is that only very small amounts of these desirable alkaloids are present in the plant. Although the total alkaloid content of the leaf can reach 1% or more, over 500 kg

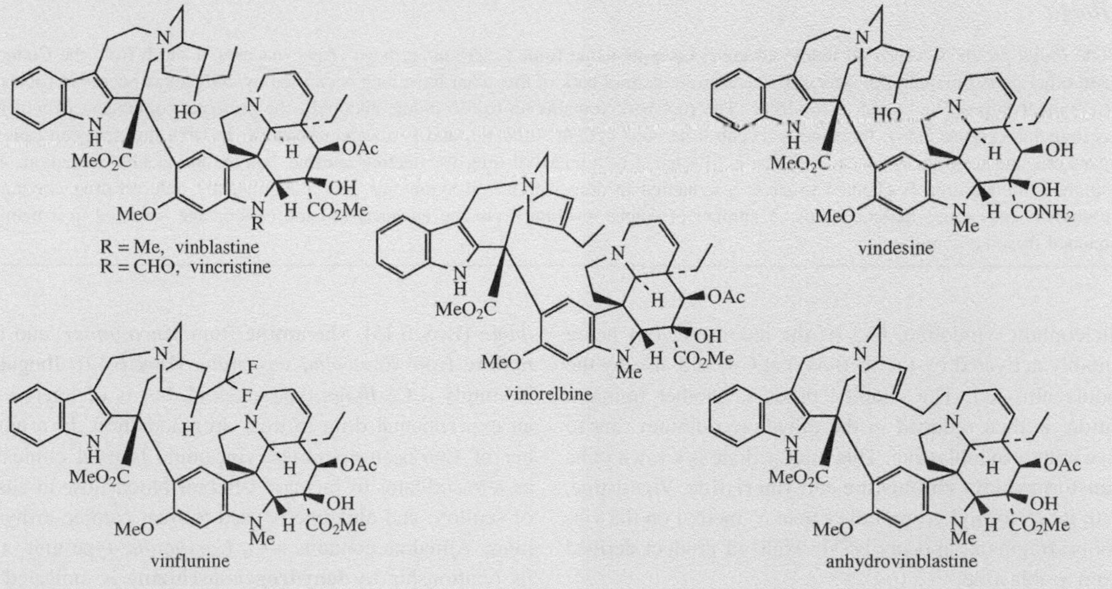

R = Me, vinblastine
R = CHO, vincristine

vindesine

vinorelbine

vinflunine

anhydrovinblastine

Figure 6.85

Box 6.13   (continued)

of catharanthus is needed to yield 1 g of vincristine. This yield (0.0002%) is the lowest of any medicinally important alkaloid isolated on a commercial basis. Extraction is both costly and tedious, requiring large quantities of raw material and extensive use of chromatographic fractionations. The plant also produces a much higher proportion of vinblastine than vincristine, and the latter drug is medicinally more valuable. Fortunately, it is possible to convert vinblastine into vincristine by controlled chromic acid oxidation, or by chemical formylation of demethylvinblastine. The latter compound occurs naturally or can be obtained from vinblastine via a microbiological *N*-demethylation using *Streptomyces albogriseolus*. Considerable effort has been expended on the semi-synthesis of the 'dimeric' alkaloids from 'monomers' such as catharanthine and vindoline, which are produced in *Catharanthus roseus* in much larger amounts. Efficient, stereospecific coupling has eventually been achieved, and it is now possible to convert catharanthine and vindoline into vinblastine in about 40% yield, though reliance on the natural monomers restricts its commercial application. Excellent yields of anhydrovinblastine (Figure 6.85) can also be obtained by electrochemical oxidation of a catharanthine–vindoline mixture; coupling can also be accomplished enzymically using commercial horseradish peroxidase. This is the starting material for vinorelbine production. This drug (noranhydrovinblastine) was the unexpected product from the attempted conversion of anhydrovinblastine into vinblastine, which resulted in loss of a carbon atom. Further testing demonstrated its drug potential.

Total synthesis is beginning to improve the supply of these alkaloids and derivatives, and also allow more detailed studies of structure–activity relationships to be undertaken. Synthesis is currently the approach being used to maintain supplies. So far, attempts to manipulate alkaloid levels in the plant, or in plant cell cultures, have had limited success, and metabolic engineering has yet to make a significant impact. Though several *Catharanthus roseus* biosynthetic genes have been overexpressed in other plants, overexpression in the host plant has only increased alkaloid production marginally. Heterologous constructs may provide greatest potential. Thus, *Catharanthus roseus* genes encoding strictosidine synthase and strictosidine glucosidase (Figure 6.80) have been expressed in yeast, and the enzymes were found in the culture medium and the cells respectively. When the culture was supplied with secologanin and tryptamine, yeast produced high levels of strictosidine in the medium; upon breaking the cells, strictosidine was then converted into cathenamine by the action of the released strictosidine glucosidase enzyme. In due course it may prove possible to construct even more of the plant metabolic pathway in a microbial host.

This group of compounds is still of very high interest, and development programmes for analogues continue. Indeed, **vinflunine** (Figure 6.85), a fluorinated analogue produced from vinorelbine, is undergoing clinical trials.

Ajmalicine (see rauwolfia, page 372) is present in the roots of *Catharanthus roseus* at a level of about 0.4%, and this plant is used as a commercial source in addition to *Rauwolfia serpentina*.

Iboga

The *Iboga* group of terpenoid indole alkaloids takes its name from *Tabernanthe iboga* (Apocynaceae), a shrub from the Congo and other parts of equatorial Africa. Extracts from the root bark of this plant have long been used by indigenous people in rituals, to combat fatigue, and as an aphrodisiac. The root bark contains up to 6% indole alkaloids, the principal component of which is ibogaine (Figure 6.86). Ibogaine is a central nervous system stimulant, and is also psychoactive. In large doses, it can cause paralysis and respiratory arrest. Ibogaine is of interest as a potential drug for treating cocaine, heroin, and alcohol addiction. A single administration is claimed to cause a reduction in drug withdrawal symptoms, and a significantly reduced drug craving, though it does cause hallucinations. A number of deaths resulting from the unsupervised use of ibogaine have led to it being banned in some countries.

---

nucleophilic vindoline, C-5 of the indole nucleus being suitably activated by the methoxyl at C-6 and also by the indole nitrogen. The coupled product, another iminium cation, is then reduced in the dihydropyridinium ring to give anhydrovinblastine. This intermediate is known to be transformed into **vinblastine** and **vincristine**. Vincristine, with its *N*-formyl group rather than *N*-methyl on the vindoline fragment, appears be an oxidized product derived from vinblastine.

Further variants on the terpenoid indole alkaloid skeleton (Figure 6.86) are found in **ibogaine** from *Tabernanthe*

iboga [Box 6.13], **vincamine** from *Vinca minor*, and **ajmaline** from *Rauwolfia serpentina* [Box 6.12]. Ibogaine is simply a C_9 *Iboga*-type alkaloid, but is of interest as an experimental drug to treat drug addiction. In a number of European countries, vincamine is used clinically as a vasodilator to increase cerebral blood flow in cases of senility, and ajmaline is used to treat cardiac arrhythmias. Ajmaline contains a C_9 *Corynanthe*-type unit, and its relationship to **dehydrogeissoschizine** is indicated in Figure 6.86; most enzyme and gene aspects of the pathway have now been delineated. Vincamine still retains a

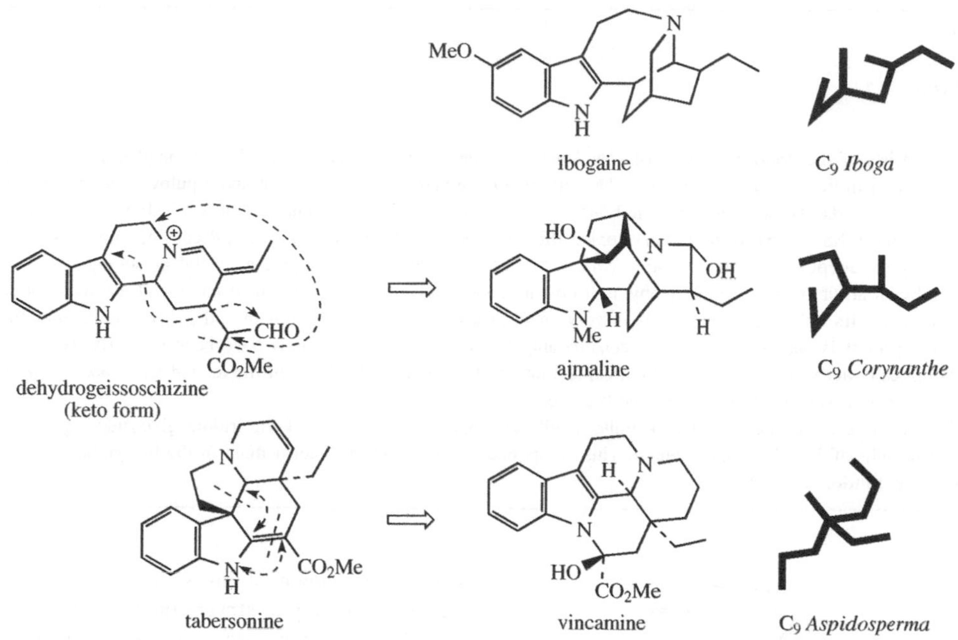

Figure 6.86

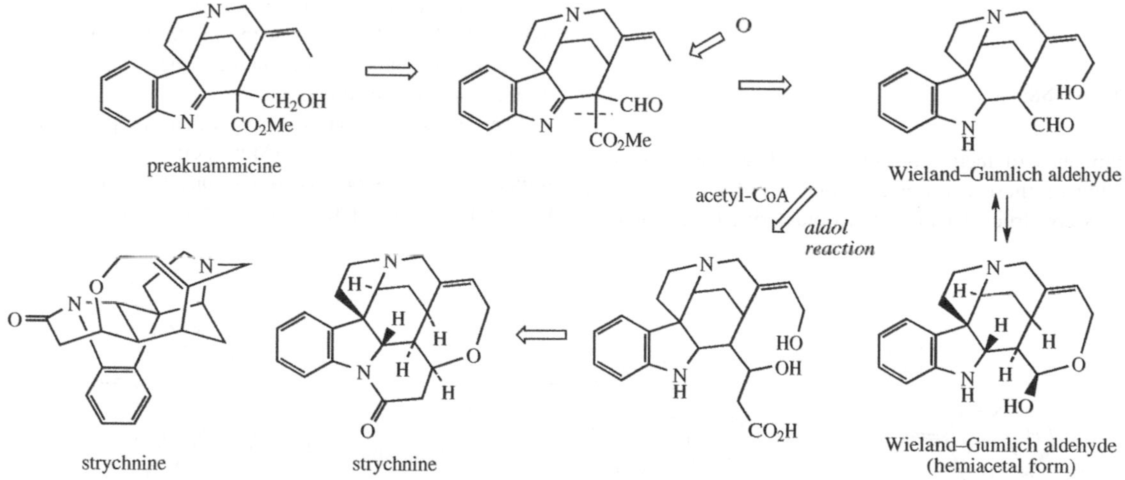

Figure 6.87

C_{10} *Aspidosperma* unit, and it originates from **taberson-ine** by a series of reactions that involve cleavage of bonds to both α and β positions of the indole (Figure 6.86).

Alkaloids like **preakuammicine** (Figure 6.83) and **akuammicine** (Figure 6.79) contain the C_{10} and C_9 *Corynanthe*-type terpenoid units respectively. They are, however, representatives of a subgroup of *Corynanthe*

alkaloids termed the *Strychnos* type because of their structural similarity to many of the alkaloids found in *Strychnos* species (Loganiaceae), e.g. *Strychnos nux-vomica* [Box 6.14], noteworthy examples being the extremely poisonous **strychnine** (Figure 6.87) and its dimethoxy analogue **brucine** (Figure 6.88). The structures of these natural products are regarded as the most complex possible for

brucine

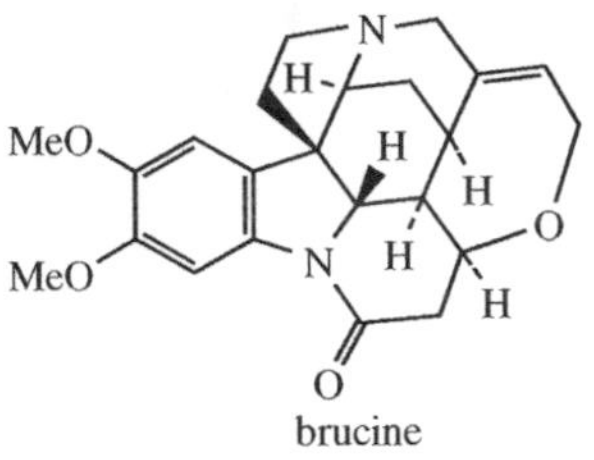

Figure 6.88

compounds of their molecular size. The non-tryptamine portion of these compounds contains 11 carbon atoms, and is constructed from an iridoid-derived C_9 unit plus two further carbon atoms supplied from acetate. The suggested pathway to **strychnine** in Figure 6.87 involves loss of one carbon from a preakuammicine-like structure via hydrolysis/decarboxylation to give the so-called Wieland–Gumlich aldehyde, a demonstrated precursor. The Wieland–Gumlich aldehyde normally exists as a hemiacetal. Addition of the extra two carbon atoms requires aldol condensation with the formyl group, and subsequent formation of strychnine is merely construction of ether and amide linkages.

The arrow poison curare, when prepared from *Chondrodendron* species (Menispermaceae), contains principally the bis-benzyltetrahydroisoquinoline alkaloid tubocurarine (see page 343). Species of *Strychnos*,

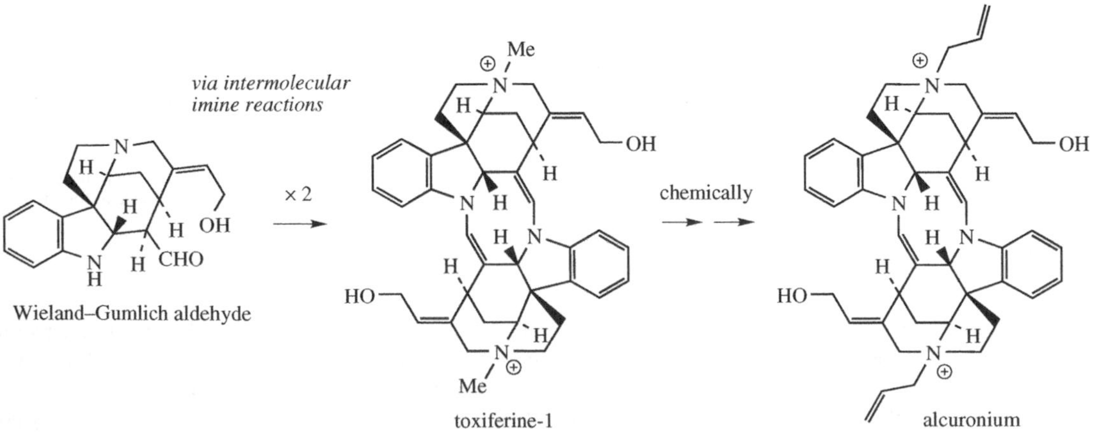

Wieland–Gumlich aldehyde

via intermolecular imine reactions

toxiferine-1

alcuronium

Figure 6.89

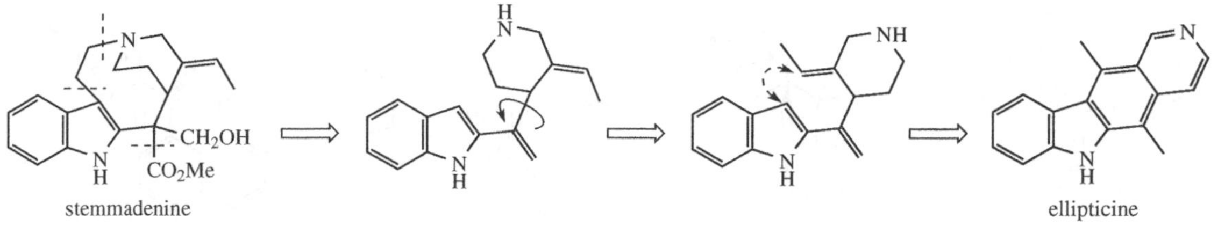

Figure 6.90

especially *Strychnos toxifera*, are employed in making loganiaceous curare, and biologically active alkaloids isolated from such preparations have been identified as a series of toxiferines, e.g. **toxiferine-1** (Figure 6.89). The structures appear remarkably complex, but may be envisaged as a combination of two Wieland–Gumlich aldehyde-like molecules (Figure 6.89). The presence of two quaternary nitrogen atoms, separated by an appropriate distance, is responsible for the curare-like activity (compare tubocurarine and synthetic analogues, page 344). **Alcuronium** (Figure 6.89) is a semi-synthetic skeletal muscle relaxant produced from toxiferine-1 (see curare, page 345).

Ellipticine (Figure 6.90) contains a pyridocarbazole skeleton which is also likely to be formed from a tryptamine–terpenoid precursor. Although little evidence is available, it is suggested that a precursor like **stemmadenine** may undergo transformations that effectively remove the two-carbon bridge originally linking the indole and the nitrogen in tryptamine (Figure 6.90). The remaining C_9 terpenoid fragment now containing the tryptamine nitrogen can then be used to generate the rest of the skeleton. Ellipticine is found in *Ochrosia elliptica* (Apocynaceae) and related species and has useful anticancer properties [Box 6.15].

Box 6.15

Ellipticine

Ellipticine (Figure 6.90) and related alkaloids, e.g. 9-methoxyellipticine (Figure 6.91), are found in the bark of *Ochrosia elliptica* (Apocynaceae) and other *Ochrosia* species. Clinical trials with these alkaloids and a number of synthetic analogues showed them to be potent inhibitors of several cancerous disorders, but preclinical toxicology indicated a number of side-effects, including haemolysis and cardiovascular effects. Ellipticines are planar molecules that intercalate between the base pairs of DNA and cause a partial unwinding of the helical array. Recent research suggests they also inhibit the enzyme topoisomerase II (see page 155). Ellipticine is oxidized *in vivo* mainly to 9-hydroxyellipticine, which has increased activity, and it is believed that this may in fact be the active agent. Poor water solubility of ellipticine and derivatives gave problems in formulation for clinical use, but quaternization of 9-hydroxyellipticine to give the water-soluble 9-hydroxy-2-*N*-methylellipticinium acetate (**elliptinium acetate**; Figure 6.91) has produced a highly active material, of value in some forms of breast cancer and perhaps also in renal cell cancer. Several other quaternized derivatives are being tested, and some water-soluble *N*-glycosides also show high activity.

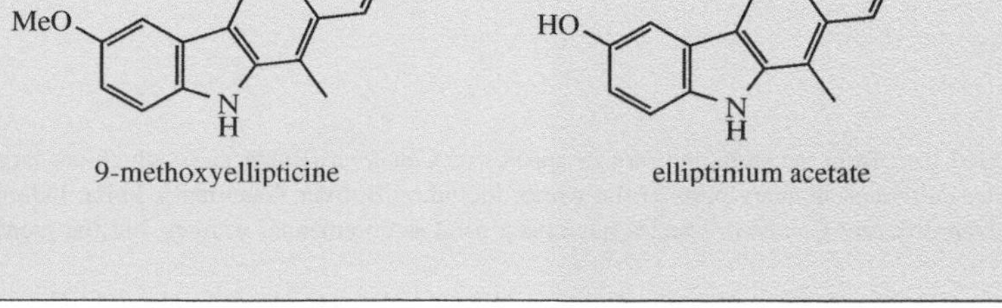

9-methoxyellipticine   elliptinium acetate

Figure 6.91

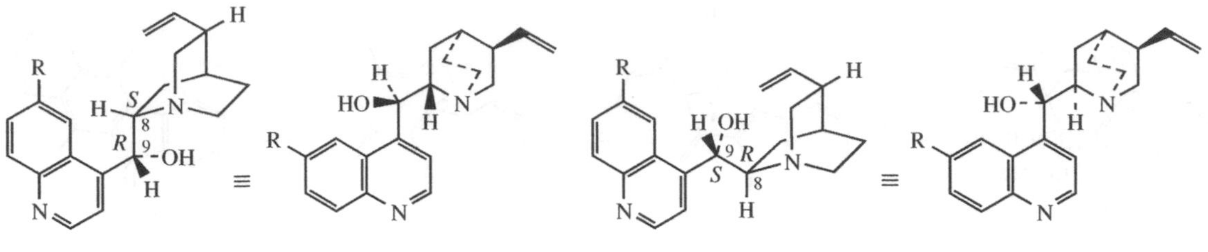

R = OMe, (−)-quinine
R = H,　(−)-cinchonidine

R = OMe, (+)-quinidine
R = H,　(+)-cinchonine

Figure 6.92

Quinoline Alkaloids

Some of the most remarkable examples of terpenoid indole alkaloid modifications are to be found in the genus *Cinchona* (Rubiaceae), in the alkaloids **quinine, quinidine, cinchonidine**, and **cinchonine** (Figure 6.92), long prized for their antimalarial properties [Box 6.16]. These structures are remarkable in that the indole nucleus is no longer present; it has been rearranged into a quinoline system (Figure 6.93). The relationship was suspected quite early on, however, since the indole derivative **cinchonamine** (Figure 6.94) was known to co-occur with these quinoline alkaloids.

An outline of the pathway from the *Corynanthe*-type indole alkaloids to cinchonidine is shown in Figure 6.94. The conversion is dependent on the reversible processes by which amines plus aldehydes or ketones, imines, and amine reduction products of imines are related in nature (see page 19).

Suitable modification of strictosidine leads to an aldehyde (compare the early reactions in the ajmalicine

pathway; Figure 6.80). Hydrolysis/decarboxylation would initially remove one carbon from the iridoid portion and produce **corynantheal**; the demonstrated involvement of this compound shows that the methoxycarbonyl group is lost at an early stage. **Cinchonaminal**, an oxidized version of cinchonamine, would result if the tryptamine side-chain were cleaved adjacent to the nitrogen, and if this nitrogen was then bonded to the acetaldehyde function. Ring opening in the indole heterocyclic ring could generate new amine and keto functions. The new quinoline heterocycle would then be formed by combining this amine with the aldehyde produced in the tryptamine side-chain cleavage, giving **cinchonidinone**. The two epimers **cinchonidinone** and **cinchoninone** equilibrate readily in the plant through keto–enol tautomerism. Finally, reduction of the ketone group gives **cinchonidine** or **cinchonine**. Hydroxylation and methylation at some stage allows biosynthesis of **quinine** and **quinidine**.

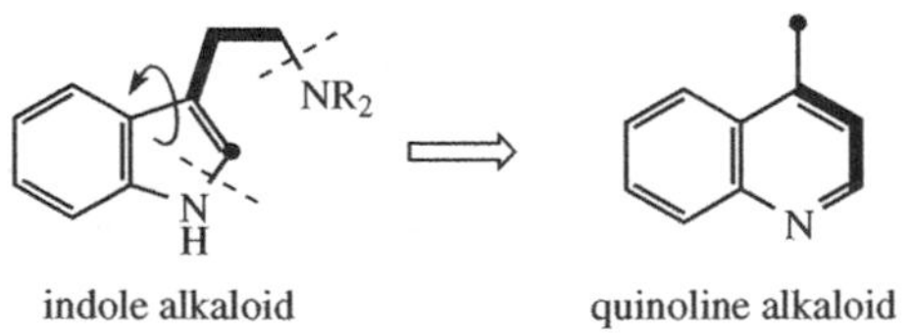

Quinine and quinidine, or cinchonidine and cinchonine, are pairs of diastereoisomers, and have opposite chiralities at two centres, C-8 and C-9 (Figure 6.92). The stereochemistry at C-8 is easily reversed by tautomerism in cinchonidinone and cinchoninone as described above. The stereochemistry adjacent to the quinoline ring (C-9) is controlled by the reduction step. An enzyme catalysing the reduction of cinchoninone produces an unequal mixture of cinchonine and cinchonidine, showing that the stereochemistry of reduction may somehow depend upon the substrate (Figure 6.94).

indole alkaloid　　　quinoline alkaloid

Figure 6.93

Box 6.16

Cinchona

Cinchona bark is the dried bark from the stem and root of species of *Cinchona* (Rubiaceae), which are large trees indigenous to South America. Trees are cultivated in many parts of the world, including Bolivia, Guatemala, India, Indonesia, Zaire, Tanzania, and Kenya. About a dozen different *Cinchona* species have been used as commercial sources, but the great variation in alkaloid

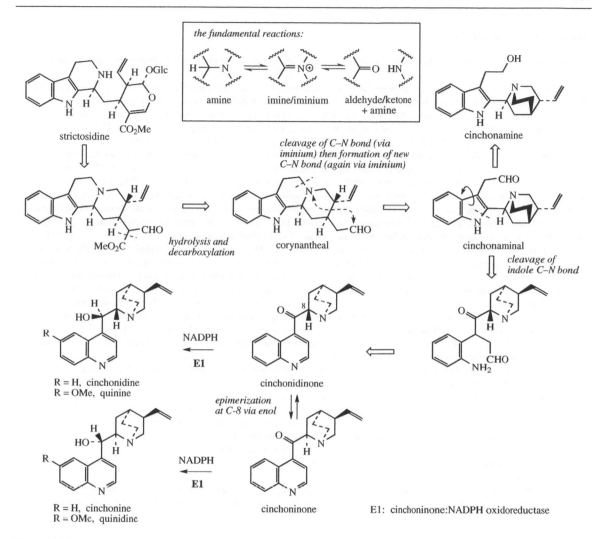

Figure 6.94

Box 6.16   (continued)

content and the range of alkaloids present has favoured cultivation of three main species, together with varieties, hybrids, and grafts. *Cinchona succirubra* provides what is called 'red' bark (alkaloid content 5–7%), *Cinchona ledgeriana* gives 'brown' bark (alkaloid content 5–14%), and *Cinchona calisaya* 'yellow' bark with an alkaloid content of 4–7%. Selected hybrids can yield up to 17% total alkaloids. Bark is stripped from trees which are about 8–12 years old, the trees being totally uprooted by tractor for the process.

A considerable number of alkaloids have been characterized in cinchona bark, four of which account for some 30–60% of the alkaloid content. These are quinine, quinidine, cinchonidine, and cinchonine, quinoline-containing structures representing two pairs of diastereoisomers (Figure 6.92). Quinine and quinidine have opposite configurations at two centres. Cinchonidine and cinchonine are demethoxy analogues of quinine and quinidine respectively; unfortunately, use of the *-id-* syllable in the nomenclature does not reflect a particular stereochemistry. Quinine is usually the major component (half to two-thirds total alkaloid content), but the proportions of the four alkaloids vary according to species or hybrid. The alkaloids are often present in the bark in salt combination with quinic acid (see page 138) or a tannin material called cinchotannic acid. Cinchotannic acid

Box 6.16 (continued)

decomposes due to enzymic oxidation during processing of the bark to yield a red pigment, which is particularly prominent in the 'red' bark.

Cinchona and its alkaloids, particularly **quinine**, have been used for many years in the treatment of malaria, a disease caused by protozoa, of which the most troublesome is *Plasmodium falciparum*. The beneficial effects of cinchona bark were first discovered in South America in the 1630s, and the bark was then brought to Europe by Jesuit missionaries. Religious intolerance initially restricted its universal acceptance, despite the widespread occurrence of malaria in Europe and elsewhere. The name cinchona is a misspelling derived from Chinchon. In an often quoted tale, now historically disproved, the Spanish Countess of Chinchon, wife of the viceroy of Peru, was reputedly cured of malaria by the bark. For many years, the bark was obtained from South America, but cultivation was eventually established by the English in India, and by the Dutch in Java, until just before the Second World War, when almost all the world's supply came from Java. When this source was cut off by Japan in the Second World War, a range of synthetic antimalarial drugs was hastily produced as alternatives to quinine. Many of these compounds were based on the quinine structure. Of the wide range of compounds produced, **chloroquine, primaquine**, and **mefloquine** (Figure 6.95) are important antimalarials. Primaquine is exceptional in having an 8-aminoquinoline structure, whereas chloroquine and mefloquine retain the 4-substituted quinoline as in quinine. The acridine derivative **mepacrine** (Figure 6.95), though not now used for malaria treatment, is of value in other protozoal infections. At one time, synthetic antimalarials had almost entirely superseded natural quinine, but the emergence of *Plasmodium falciparum* strains resistant to the synthetic drugs, especially the widely used inexpensive prophylactic chloroquine, has resulted in reintroduction of quinine. Mefloquine is currently active against chloroquine-resistant strains, but whilst 10 times as active as quinine, it does produce gastrointestinal upsets and dizziness, and it can trigger psychological problems such as depression, panic, or psychosis in some patients. The ability of *Plasmodium falciparum* to develop resistance to modern drugs means malaria still remains a huge health problem, and is probably the major single cause of deaths in the modern world. It is estimated that 200–500 million people are affected by malaria, and some 2–3 million die each year from this disease. **Chloroquine** and its derivative **hydroxychloroquine** (Figure 6.95), although antimalarials, are also used to suppress the disease process in rheumatoid arthritis.

Quinine (Figure 6.92), administered as free base or salts, continues to be used for treatment of multidrug-resistant malaria, though it is not suitable for prophylaxis. The specific mechanism of action is still not thoroughly understood, though it is believed to prevent polymerization of toxic haemoglobin breakdown products formed by the parasite (see artemisinin, page 219). Vastly larger amounts of the alkaloid are consumed in beverages, including vermouth and tonic water. It is amusing to realize that gin was originally added to quinine to make the bitter antimalarial more palatable. Typically, the quinine dosage was up to 600 mg three times a day. Quinine in tonic water is now the mixer added to gin, though the amounts of quinine used (about 80 mg l^{-1}) are well below that providing antimalarial protection. Quinine also has a skeletal muscle relaxant effect with a mild curare-like action. It thus finds use in the prevention and treatment of nocturnal leg cramps, a painful condition affecting many individuals, especially the elderly.

Until recently, **quinidine** (Figure 6.92) was used to treat cardiac arrhymias. It inhibits fibrillation, the uncoordinated contraction of muscle fibres in the heart. However, it is rapidly absorbed by the gastrointestinal tract and overdose can be hazardous, leading to diastolic arrest. This has effectively curtailed its use.

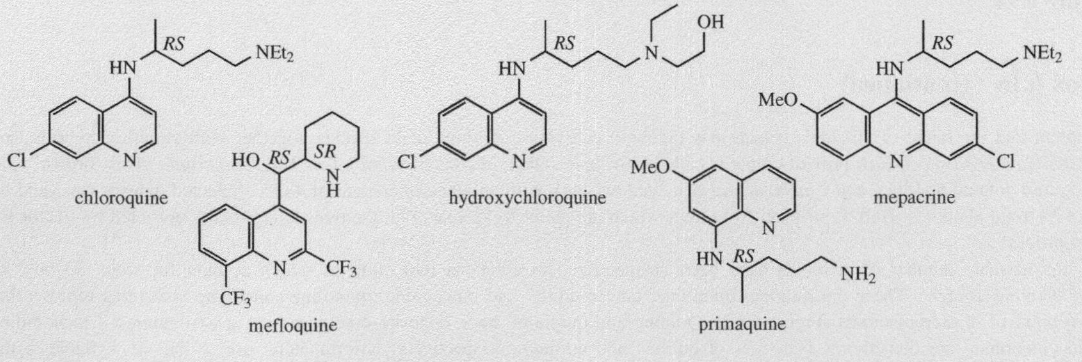

Figure 6.95

Box 6.16 (continued)

Quinidine, cinchonine, and cinchonidine also have antimalarial properties, but these alkaloids are not as effective as quinine. The cardiac effect makes quinidine unsuitable as an antimalarial. However, mixtures of total *Cinchona* alkaloids, even though low in quinine content, are acceptable antimalarial agents. This mixture, termed totaquine, has served as a substitute for quinine during shortages. Quinine-related alkaloids, especially quinidine, are also found in the bark of *Remija pendunculata* (Rubiaceae).

In organic chemistry, quinine and related alkaloids are widely used as chiral ligands for asymmetric synthesis, as resolving agents for acids, and in the elaboration of chiral phases for chromatography.

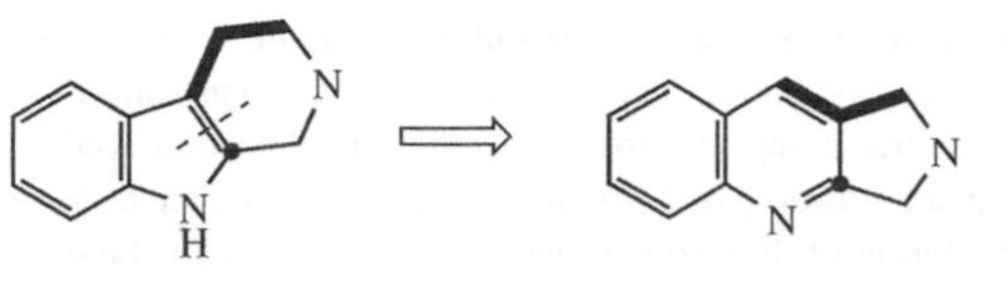

β-carboline alkaloid
(pyrido*indole*)

pyrrolo*quinoline* alkaloid

Figure 6.96

Camptothecin (see Figure 6.97) from *Camptotheca acuminata* (Nyssaceae) is a further example of a quinoline-containing structure that is derived in nature by skeletal modification of an indole system. The main rearrangement process is that an original β-carboline

6–5–6 ring system becomes a 6–6–5 pyrroloquinoline by ring expansion of the indole heterocycle (Figure 6.96). In camptothecin, the C_{10} iridoid portion as seen in strictosidine is effectively still intact, and the original ester function is utilized in forming an amide (lactam) linkage to the secondary amine. This occurs relatively early, in that **strictosamide** is an intermediate. An effectively identical reaction is seen in the formation of demethylalangiside in the ipecac alkaloids (see page 362). **Pumiloside** and **deoxypumiloside**, both found in *Ophiorrhiza pumila* (Rubiaceae), another plant producing camptothecin, are potential intermediates. Steps beyond are not yet defined, but involve relatively straightforward oxidation and reduction processes (Figure 6.97). Camptothecin derivatives have provided some useful anticancer drugs [Box 6.17].

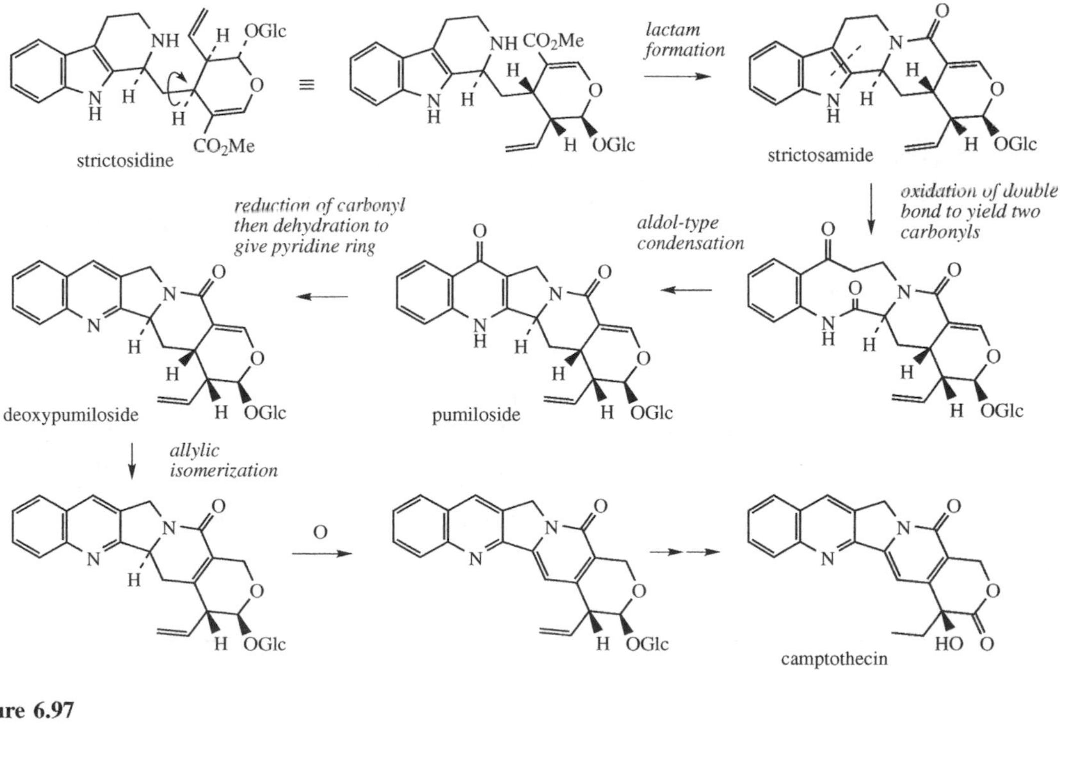

Figure 6.97

Box 6.17

Camptothecin

Camptothecin (Figure 6.97) and derivatives are obtained from the Chinese tree *Camptotheca acuminata* (Nyssaceae). Seeds yield about 0.3% camptothecin, bark about 0.2%, and leaves up to 0.4%. *Camptotheca acuminata* is found only in Tibet and west China, but other sources of camptothecin, such as *Nothapodytes foetida* (formerly *Mappia foetida*) (Icacinaceae), *Merilliodendron megacarpum* (Icacinaceae), *Pyrenacantha klaineana* (Icacinaceae), *Ophiorrhiza pumila* (Rubiaceae), and *Ervatmia heyneana* (Apocynaceae), have been discovered. In limited clinical trials, camptothecin showed broad-spectrum anticancer activity, but toxicity and poor solubility were problems. The natural 10-hydroxycamptothecin (about 0.05% in the bark of *Camptotheca acuminata*) is more active than camptothecin, and is used in China against cancers of the neck and head. Semi-synthetic analogues 9-aminocamptothecin (Figure 6.98) and the water-soluble derivatives **topotecan** and **irinotecan** (Figure 6.98) showed good responses in a number of cancers; topotecan and irinotecan are now available for the treatment of ovarian cancer and colorectal cancer respectively. These drugs are currently made from natural camptothecin, extracted from bark and seeds of *Camptotheca acuminata* and *Nothapodytes foetida*. Irinotecan is a carbamate pro-drug of 10-hydroxy-7-ethylcamptothecin, and is converted into the active drug by liver enzymes. These agents act by inhibition of the enzyme topoisomerase I, which is involved in DNA replication and reassembly, by binding to and stabilizing a covalent DNA–topoisomerase complex (see page 155). Camptothecin has also been shown to have potentially useful activity against pathogenic protozoa such as *Trypanosoma brucei* and *Leishmania donovani*, which cause sleeping sickness and leishmaniasis respectively. Again, this is due to topoisomerase I inhibition.

The camptothecin group of compounds are currently of considerable interest as anticancer agents. Many analogues and pro-drugs have been synthesized; several agents are in clinical trials, and **belotecan** (Figure 6.98) has recently been introduced in Korea. The silicon-containing variant **karenitecin** has shown reduced sensitivity to common tumour-mediated drug resistance mechanisms. The *S*-configuration at C-20 appears essential for activity, but this centre is part of an α-hydroxy-δ-lactone ring that is rapidly hydrolysed under physiological conditions to an open-chain carboxylate form; this is biologically almost inactive. Camptothecin analogues (homocamptothecins) with a seven-membered β-hydroxy-ε-lactone ring retain activity and display enhanced stability towards hydrolysis; **diflomotecan** (Figure 6.94) is a promising drug in this class undergoing clinical trials.

Low concentrations of camptothecin have also been detected in cultures of an unidentified fungus isolated from the inner bark of *Nothapodytes foetida*.

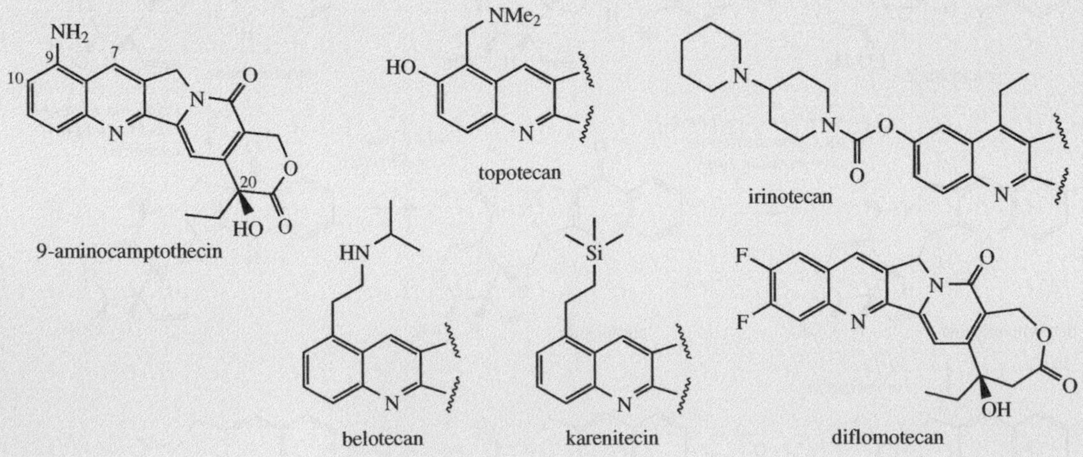

Figure 6.98

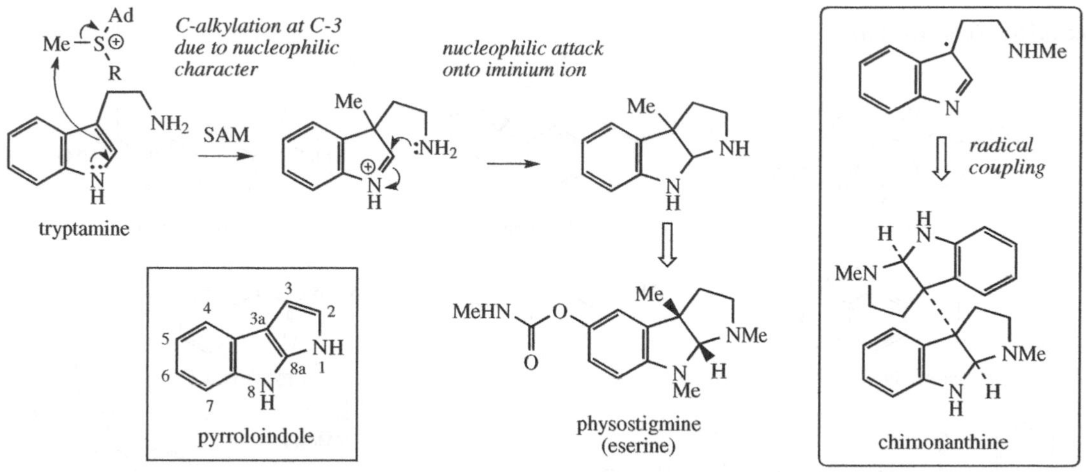

Figure 6.99

Pyrroloindole Alkaloids

Both C-2 and C-3 of the indole ring can be regarded as nucleophilic, but reactions involving C-2 appear to be the most common in alkaloid biosynthesis. There are some examples where the nucleophilic character of C-3 is exploited, however, and the pyrroloindole skeleton typified by **physostigmine** (**eserine**; Figure 6.99) is a likely case. A suggested pathway to physostigmine is by C-3 methylation of tryptamine, followed by ring formation involving attack of the primary amine function onto the iminium ion (Figure 6.95). Further substitution is then necessary. Dimers with this ring system are also

known, e.g. **chimonanthine** (Figure 6.99) from *Chimonanthus fragrans* (Calycanthaceae), the point of coupling being C-3 of the indole, and an analogous radical reaction may be proposed. Pyrroloindole alkaloids are quite rare, though examples have been detected in plants, amphibians, and marine algae. Physostigmine is found in seeds of *Physostigma venenosum* (Leguminosae/Fabaceae) and has played an important role in pharmacology because of its anticholinesterase activity [Box 6.18]. The inherent activity is, in fact, derived from the carbamate side-chain rather than the heterocyclic ring system, and this has led to a range of synthetic materials being developed.

Box 6.18

Physostigma

Physostigma venenosum (Leguminosae/Fabaceae) is a perennial woody climbing plant found on the banks of streams in West Africa. The seeds are known as Calabar beans (from Calabar, now part of Nigeria) and have an interesting history in the native culture as an ordeal poison. The accused was forced to swallow a potion of the ground seeds, and if the mixture was subsequently vomited, they were judged innocent and set free. If the poison took effect, the prisoner suffered progressive paralysis and died from cardiac and respiratory failure. It is said that slow consumption allows the poison to take effect, whilst emesis is induced by a rapid ingestion of the dose.

The seeds contain several alkaloids (alkaloid content about 1.5%), the major one (up to 0.3%) being physostigmine (eserine; Figure 6.99). The unusual pyrroloindole ring system is also present in some of the minor alkaloids, e.g. eseramine (Figure 6.100), whilst physovenine (Figure 6.100) contains an undoubtedly related furanoindole system. Another alkaloid, geneserine (Figure 6.101), is an artefact produced by oxidation of physostigmine, probably by formation of an *N*-oxide; in salt form, it retains the pyrroloindole ring system, but under basic conditions it takes up an oxazinoindole form, incorporating the hydroxyl oxygen into the ring system. Interconversion may occur via the indolium cation shown. Solutions of physostigmine are not particularly stable in the presence of air and light, especially under alkaline conditions, oxidizing to a red quinone, rubeserine (Figure 6.101).

Box 6.18   (continued)

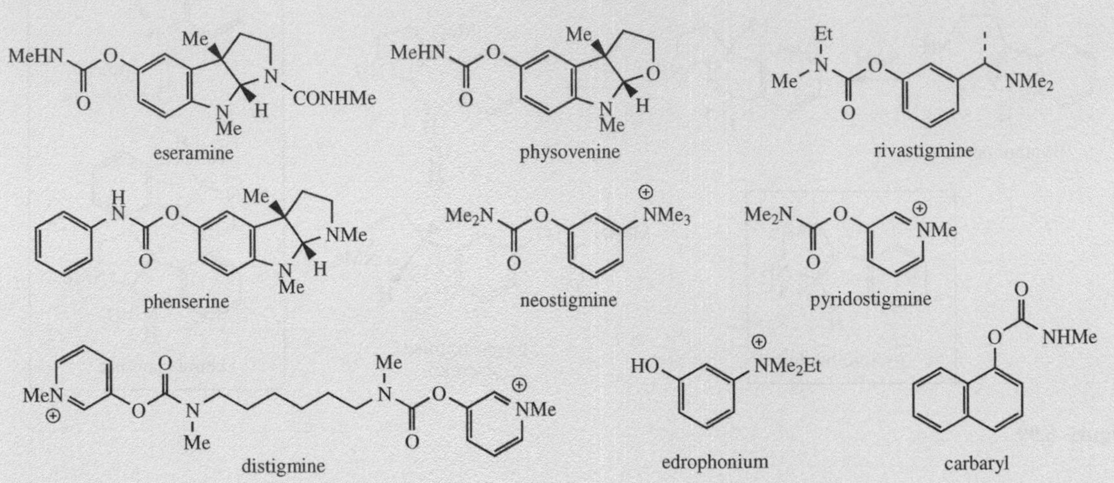

eseramine   physovenine   rivastigmine

phenserine   neostigmine   pyridostigmine

distigmine   edrophonium   carbaryl

Figure 6.100

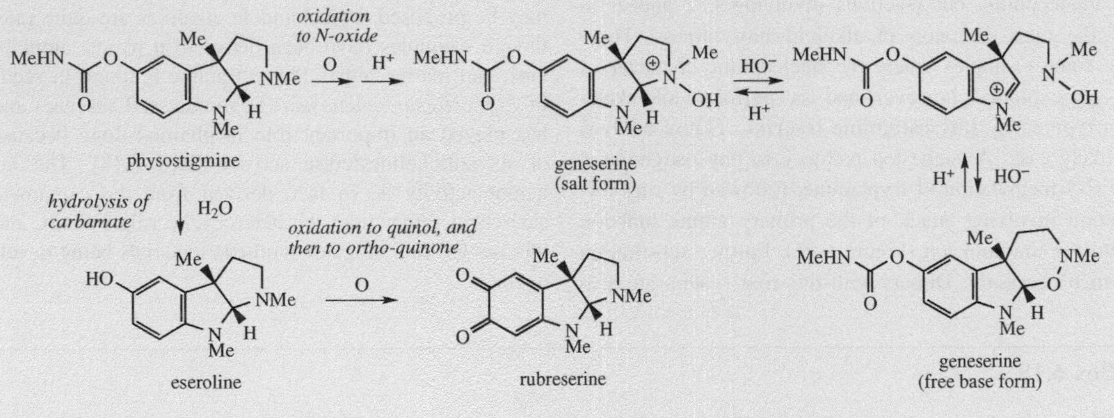

physostigmine

oxidation to N-oxide

geneserine (salt form)

hydrolysis of carbamate

eseroline

oxidation to quinol, and then to ortho-quinone

rubreserine

geneserine (free base form)

Figure 6.101

Physostigmine (eserine) is a reversible inhibitor of acetylcholinesterase, preventing normal destruction of acetylcholine and, thus, enhancing cholinergic activity. Though it is now rarely used as a drug, it has played a pivotal role in investigating the function of acetylcholine as a neurotransmitter. Its major use has been as a miotic, to contract the pupil of the eye, often to combat the effect of mydriatics such as atropine (see page 318). It also reduces intraocular pressure in the eye by increasing outflow of the aqueous humour, and provided a valuable treatment for glaucoma, often in combination with pilocarpine (see page 399). Because it prolongs the effect of endogenous acetylcholine, physostigmine can be used as an antidote to anticholinergic poisons such as hyoscyamine/atropine (see page 318), and it also reverses the effects of competitive muscle relaxants such as curare, tubocurarine, atracurium, etc. (see page 344). Acetylcholinesterase-inhibiting drugs are also of value in the treatment of Alzheimer's disease, which is characterized by a dramatic decrease in functionality of the central cholinergic system. Use of acetylcholinesterase inhibitors can result in significant memory enhancement in patients, and analogues of physostigmine are presently in use (e.g. **rivastigmine**) or have been tested in clinical trials (e.g. **phenserine**; Figure 6.100). These analogues have a longer duration of action, less toxicity, and better bioavailability than physostigmine. Rivastigmine is also used to treat mild to moderate dementia associated with Parkinson's disease.

Box 6.18 (continued)

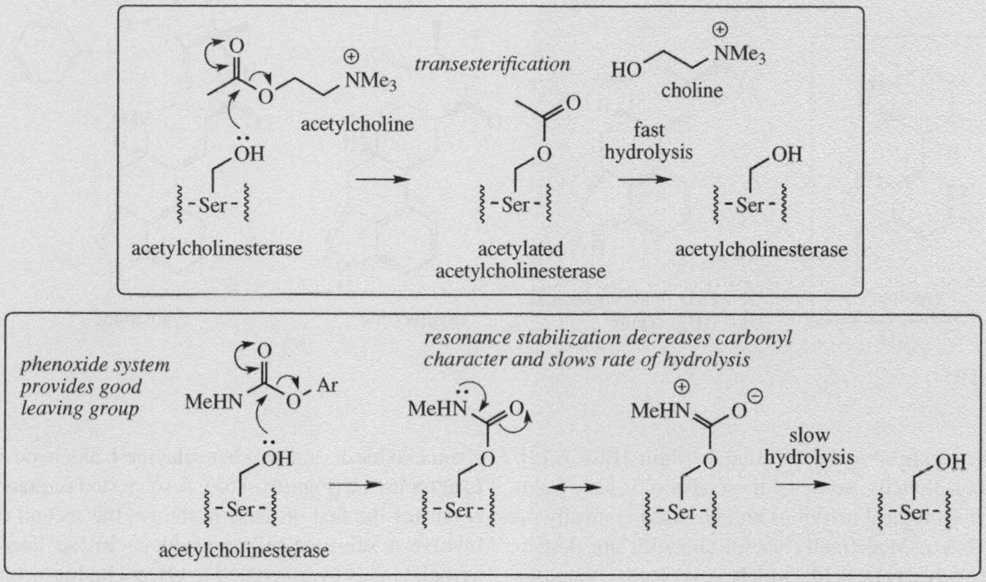

Figure 6.102

The biological activity of physostigmine resides primarily in the carbamate portion, which is transferred to the hydroxyl group of an active site serine in acetylcholinesterase (Figure 6.102). The resultant carbamoyl–enzyme intermediate then hydrolyses only very slowly (minutes rather than microseconds), effectively blocking the active site for most of the time. The slower rate of hydrolysis of the serine carbamate ester is a consequence of decreased carbonyl character resulting from resonance stabilization. Synthetic analogues of physostigmine which have been developed retain the carbamate residue, and an aromatic ring to achieve binding and to provide a good leaving group, whilst ensuring water solubility through possession of a quaternary ammonium system. **Neostigmine, pyridostigmine**, and **distigmine** (Figure 6.100) are examples of synthetic anticholinesterase drugs used primarily for enhancing neuromuscular transmission in the rare autoimmune condition myasthenia gravis, in which muscle weakness is caused by faulty transmission of nerve impulses. **Edrophonium**, though not a carbamate, is a competitive blocker of the acetylcholinesterase active site; it binds to the anionic site of the enzyme, and its action is only very brief. This compound is used mainly for the diagnosis of myasthenia gravis. Neostigmine is also routinely used to reverse the effects of non-depolarizing muscle relaxant drugs, e.g. atracurium and pancuronium (see page 345), after surgery.

A number of carbamate insecticides, e.g. carbaryl (Figure 6.100), also depend on inhibition of acetylcholinesterase for their action, insect acetylcholinesterase being more susceptible to such agents than the mammalian enzyme. Physostigmine displays little insecticidal action because of its poor lipid solubility.

Ergot Alkaloids

Ergot is a fungal disease commonly found on many wild and cultivated grasses that is caused by species of *Claviceps* [Box 6.19]. The disease is eventually characterized by the formation of hard, seed-like 'ergots' instead of normal seeds; these structures, called sclerotia, are the resting stage of the fungus. The poisonous properties of ergots in grain, especially rye, for human or animal consumption have long been recognized, and the causative agents are known to be a group of indole alkaloids, referred to collectively as the ergot alkaloids or ergolines (Figure 6.103). Under natural conditions the alkaloids are elaborated by a combination of fungal and plant metabolisms, but they can be synthesized in cultures of suitable *Claviceps* species. Ergoline alkaloids have also been found in fungi belonging to genera *Aspergillus, Rhizopus*, and *Penicillium*, as well as *Claviceps*, and simple examples are also found in some plants of the Convolvulaceae such as *Ipomoea* and *Rivea* (morning glories),

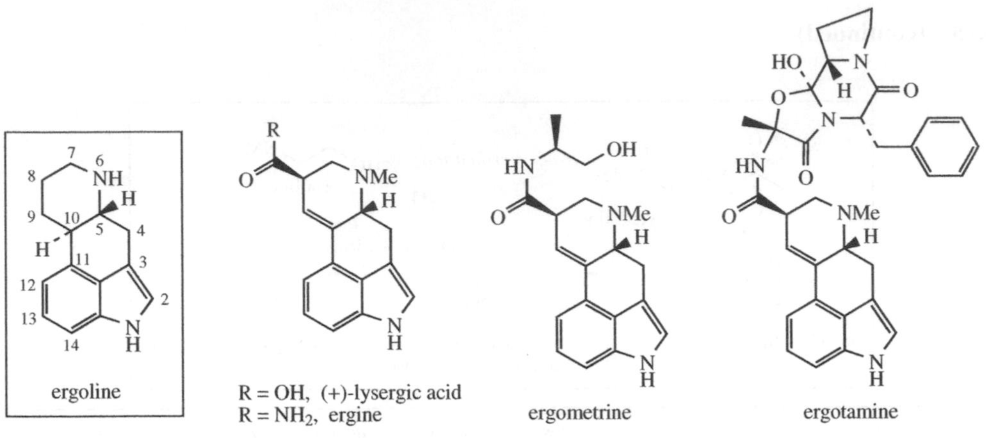

ergoline

R = OH, (+)-lysergic acid
R = NH$_2$, ergine

ergometrine

ergotamine

Figure 6.103

though they appear to be of fungal origin [Box 6.19]. Despite their toxicity, some of these alkaloids have valuable pharmacological activities and are used clinically on a routine basis. Medicinally useful alkaloids are derivatives of (+)-**lysergic acid**, which is typically bound as an amide with an amino alcohol as in **ergometrine**, or with a small polypeptide structure as in **ergotamine** (see Figure 6.103).

The building blocks for lysergic acid are tryptophan (less the carboxyl group) and an isoprene unit (Figure 6.104). Alkylation of tryptophan with dimethylallyl diphosphate gives 4-dimethylallyl-L-tryptophan, which then undergoes *N*-methylation (Figure 6.105). Formation of the tetracyclic ring system of lysergic acid is known to proceed through **chanoclavine-I** and **agroclavine**, though the mechanistic details are far from clear, and the enzymes are only partially characterized. Labelling studies have established that the double bond in the dimethylallyl substituent must become a single bond on two separate occasions, allowing rotation to occur as new rings are established. This gives the appearance of *cis–trans* isomerizations as 4-dimethylallyl-L-tryptophan is transformed into

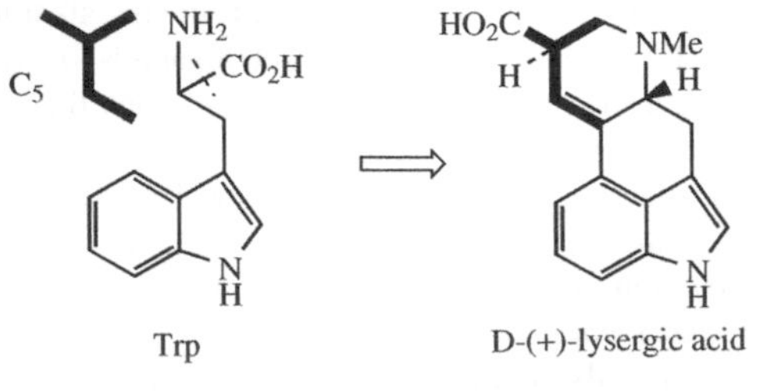

chanoclavine-I, and as chanoclavine-I aldehyde cyclizes to agroclavine (Figure 6.105). A suggested sequence to account for the first of these is shown; the second one may involve a substrate–enzyme adduct. In the later stages, agroclavine is hydroxylated to **elymoclavine**, with further oxidation of the primary alcohol to **paspalic acid**, both reactions catalysed by cytochrome P-450 systems. **Lysergic acid** then results from a spontaneous allylic isomerization.

Simple derivatives of lysergic acid require the formation of amides; for example, **ergine** (Figure 6.103) in *Rivea* and *Ipomoea* species is **lysergic acid amide**, whilst **ergometrine** from *Claviceps purpurea* is the amide with 2-aminopropanol. The more complex structures containing peptide fragments, e.g. **ergotamine** (Figure 6.103), are formed by sequentially adding amino acid residues to thioester-bound lysergic acid, giving a linear lysergyl–tripeptide covalently attached to the enzyme complex (Figure 6.106). Peptide formation involves the same processes seen in the non-ribosomal biosynthesis of peptides (see page 438), and is catalysed by a typical non-ribosomal peptide synthase. The enzyme activates lysergic acid and the three amino acids through an ATP-mediated mechanism prior to attachment to the enzyme complex through thioester linkages. A phosphopantetheine arm is used to enable the growing chain to reach the various active sites (compare fatty acid biosynthesis, page 42). The enzyme is known to consist of four modules, each housing domains for adenylation, thiolation, and condensation reactions, and is comprised of two subunits. The smaller subunit is responsible for lysergic acid activation. Cyclization in the tripeptide residue is readily rationalized by the formation of a lactam (amide) which releases the product from the enzyme, followed by hydroxylation of

Trp

D-(+)-lysergic acid

Figure 6.104

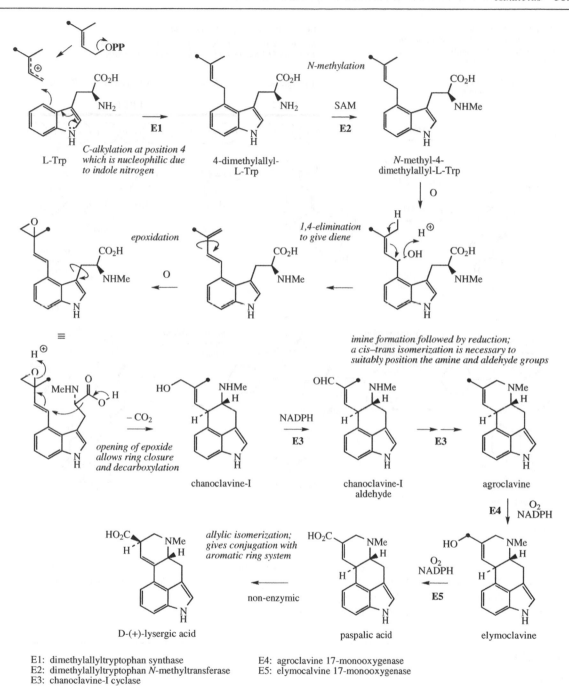

L-Trp — *C-alkylation at position 4 which is nucleophilic due to indole nitrogen* — **E1** → 4-dimethylallyl-L-Trp — *N-methylation* SAM **E2** → *N*-methyl-4-dimethylallyl-L-Trp

epoxidation O ← *1,4-elimination to give diene*

imine formation followed by reduction; a cis–trans isomerization is necessary to suitably position the amine and aldehyde groups

opening of epoxide allows ring closure and decarboxylation − CO$_2$ → chanoclavine-I — NADPH **E3** → chanoclavine-I aldehyde — **E3** → agroclavine

E4 O$_2$ NADPH

D-(+)-lysergic acid ← *non-enzymic* — *allylic isomerization; gives conjugation with aromatic ring system* ← paspalic acid ← O$_2$ NADPH **E5** ← elymoclavine

E1: dimethylallyltryptophan synthase
E2: dimethylallyltryptophan *N*-methyltransferase
E3: chanoclavine-I cyclase

E4: agroclavine 17-monooxygenase
E5: elymocalvine 17-monooxygenase

Figure 6.105

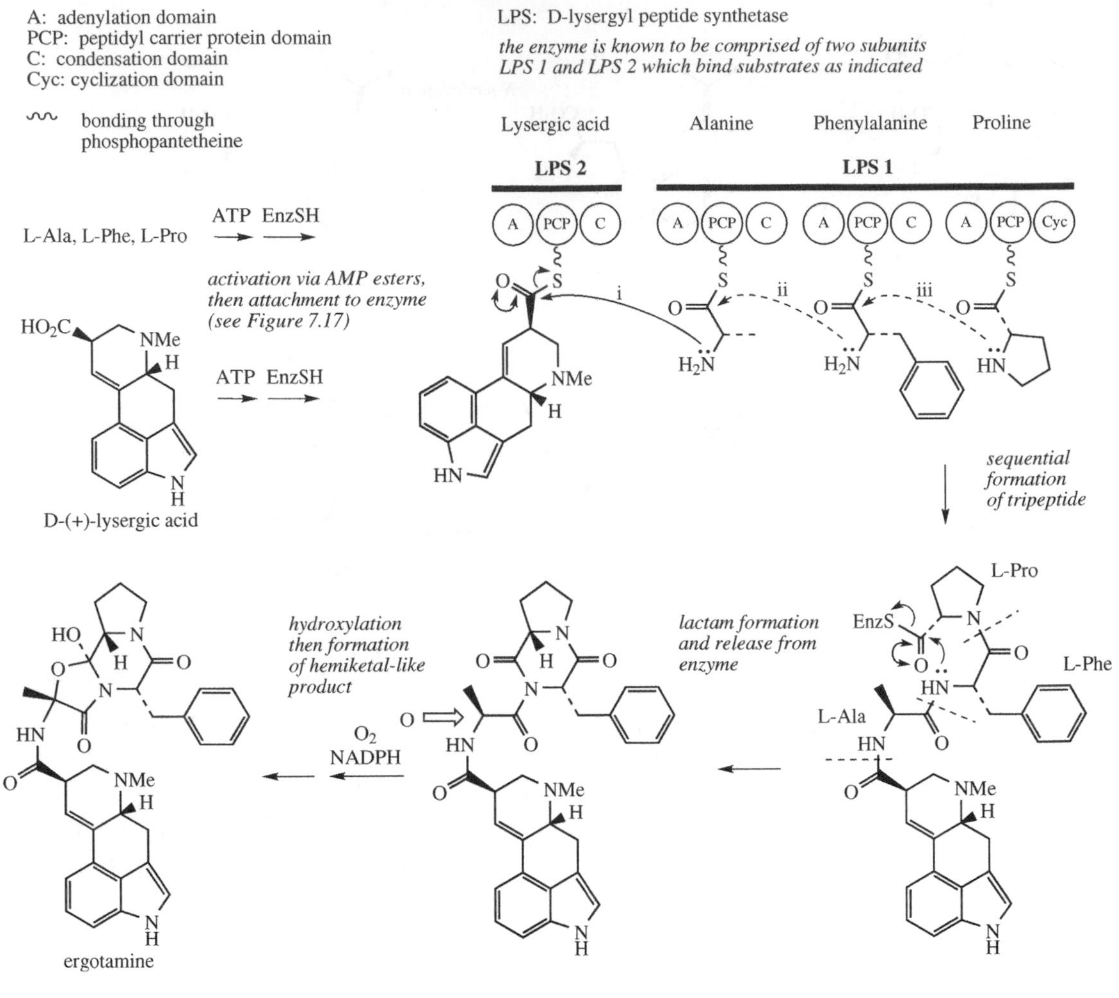

A: adenylation domain
PCP: peptidyl carrier protein domain
C: condensation domain
Cyc: cyclization domain

~~   bonding through
     phosphopantetheine

LPS: D-lysergyl peptide synthetase

*the enzyme is known to be comprised of two subunits
LPS 1 and LPS 2 which bind substrates as indicated*

Figure 6.106

the first amino acid of the chain, and generation of a hemiketal-like linkage as shown (Figure 6.106). Precursor studies have shown that the simple amide ergometrine is formed from lysergyl-alanine; this suggests a close similarity with peptide alkaloid biosynthesis, perhaps via a non-ribosomal peptide synthase that adds just a single amino acid.

Box 6.19

Ergot

Medicinal ergot is the dried sclerotium of the fungus *Claviceps purpurea* (Clavicipitaceae) developed on the ovary of rye, *Secale cereale* (Graminae/Poaceae). Ergot is a fungal disease of wild and cultivated grasses, and initially affects the flowers. In due course, a dark sclerotium, the resting stage of the fungus, is developed instead of the normal seed. This protrudes from the seed head, the name ergot deriving from the French word argot (a spur). The sclerotia fall to the ground, germinating in the spring and reinfecting grasses or grain crops by means of spores. Two types of spore are recognized: ascospores, which are formed in

Box 6.19 (continued)

the early stages and are dispersed by the wind, and later on conidiospores are produced, which are insect distributed. The flowers are only susceptible to infection before pollination. Ergots may subsequently be harvested with the grain and contaminate flour or animal feed. The consumption of ergot-infected rye has resulted in the disease ergotism, which has a long, well-documented history.

There are three broad clinical features of ergot poisoning which are due to the alkaloids present and the relative proportions of each component:

- Alimentary upsets, e.g. diarrhoea, abdominal pains, and vomiting.
- Circulatory changes, e.g. coldness of hands and feet due to a vasoconstrictor effect, a decrease in the diameter of blood vessels, especially those supplying the extremeties.
- Neurological symptoms, e.g. headache, vertigo, convulsions, psychotic disturbances, and hallucinations.

These effects usually disappear on removal of the source of poisoning, but much more serious problems develop with continued ingestion, or with doses of heavily contaminated food. The vasoconstrictor effect leads to restricted blood flow in small terminal arteries, death of the tissue, the development of gangrene, and even the shedding of hands, feet, or limbs. Gangrenous ergotism was known as St Anthony's Fire, the Order of St Anthony traditionally caring for sufferers in the Middle Ages. The neurological effects were usually manifested by severe and painful convulsions. Outbreaks of the disease in both humans and animals were relatively frequent in Europe in the Middle Ages, but once the cause had been established, it became relatively simple to avoid contamination. Separation of the ergots from grain, or the use of fungicides during cultivation of the crop, has removed most of the risks, though infection of crops is still common.

The ergot sclerotia contain from 0.15–0.5% alkaloids, and more than 50 have been characterized. The medicinally useful compounds are derivatives of (+)-lysergic acid (Figure 6.107) and can be separated into two groups: the water-soluble amino alcohol derivatives (up to about 20% of the total alkaloids) and water-insoluble peptide derivatives (up to 80% of total alkaloids).

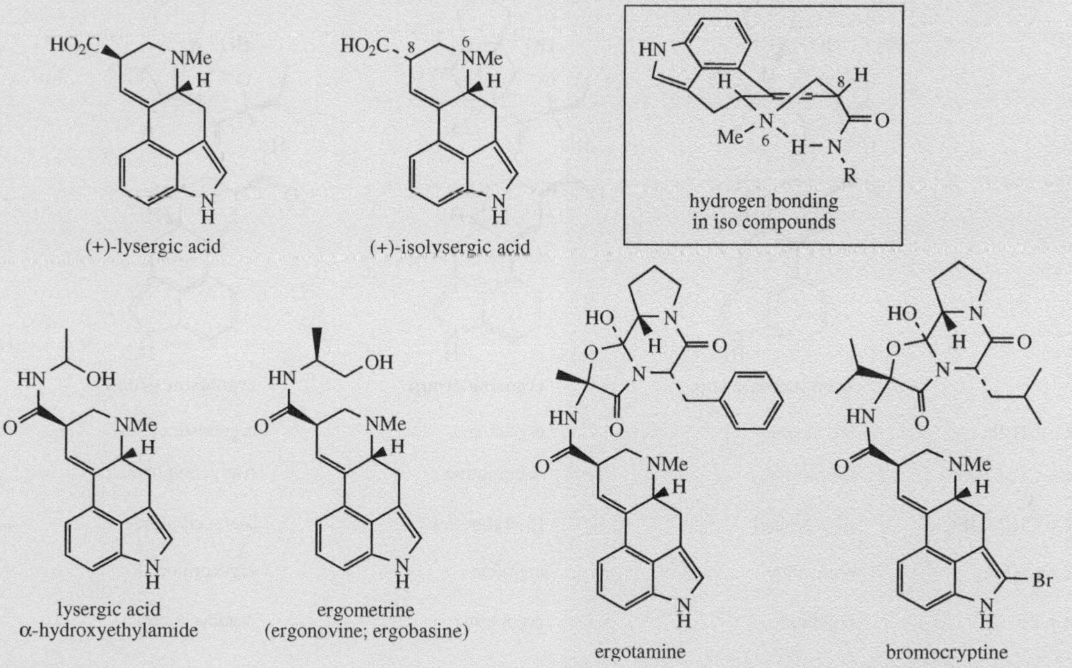

(+)-lysergic acid

(+)-isolysergic acid

hydrogen bonding
in iso compounds

lysergic acid
α-hydroxyethylamide

ergometrine
(ergonovine; ergobasine)

ergotamine

bromocryptine

Figure 6.107

Box 6.19　(continued)

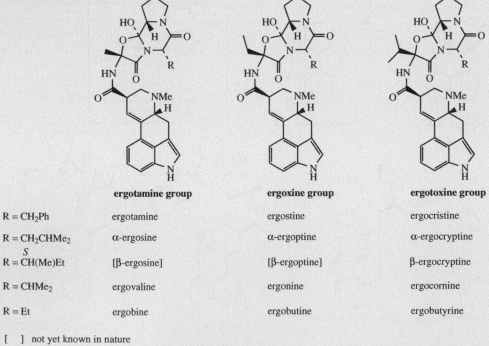

cabergoline　lisuride (lysuride)　pergolide　methysergide　lysergic acid diethylamide (lysergide; LSD)

Figure 6.107 (*continued*)

Ergometrine (Figure 6.107), also known as ergonovine in the USA and ergobasine in Switzerland, is an amide of lysergic acid and 2-aminopropanol, and is the only significant member of the first group.

The peptide derivatives contain a cyclized tripeptide fragment bonded to lysergic acid via an amide linkage. Based on the nature of the three amino acids, these structures can be subdivided into three groups: the ergotamine group, the ergoxine group, and the ergotoxine group (Figure 6.108). The amino acids involved are alanine, valine, leucine, isoleucine, phenylalanine, proline,

Peptide alkaloids in Ergot:

	ergotamine group	**ergoxine group**	**ergotoxine group**
R = CH₂Ph	ergotamine	ergostine	ergocristine
R = CH₂CHMe₂	α-ergosine	α-ergoptine	α-ergocryptine
R = CH(Me)Et *(S)*	[β-ergosine]	[β-ergoptine]	β-ergocryptine
R = CHMe₂	ergovaline	ergonine	ergocornine
R = Et	ergobine	ergobutine	ergobutyrine

[　]　not yet known in nature

Figure 6.108

Box 6.19 (continued)

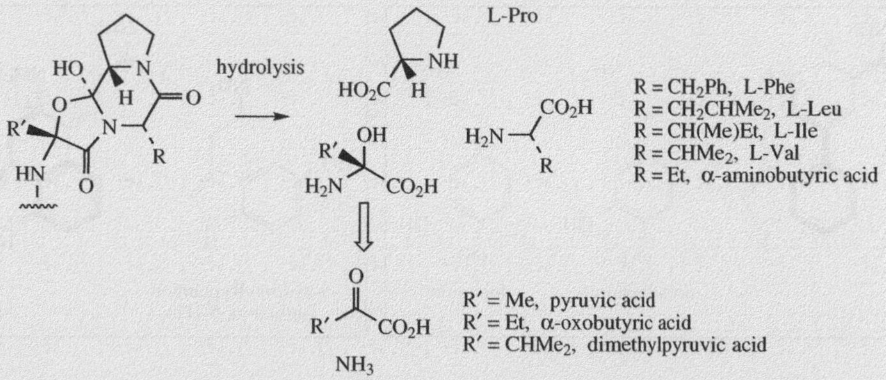

Figure 6.109

and α-aminobutyric acid, in various combinations (see Figure 6.109). All contain proline in the tripeptide, and one of the amino acids is effectively incorporated into the final structure in the form of an α-hydroxy-α-amino acid. Thus, ergotamine incorporates alanine, phenylalanine, and proline residues in its peptide portion. Hydrolysis gives (+)-lysergic acid, proline, and phenylalanine, together with pyruvic acid and ammonia, the latter hydrolysis products a consequence of the additional hydroxylation involving alanine (Figure 6.109). Hydrolysis of the ergotoxine group of alkaloids results in the proximal valine unit being liberated as dimethylpyruvic acid (not systematic nomenclature) and ammonia, and the ergoxine group similarly yields α-oxobutyric acid from the α-aminobutyric acid fragment. The alkaloid 'ergotoxine' was originally thought to be a single compound, but was subsequently shown to be a mixture of alkaloids. The proposed structures β-ergosine and β-ergoptine which complete the combinations shown in Figure 6.108 have not yet been isolated as natural products.

Medicinal ergot is cultivated in the Czech Republic, Germany, Hungary, Switzerland, Austria, and Poland. Fields of rye are infected artificially with spore cultures of *Claviceps purpurea*, either by spraying or by a mechanical process which uses needles dipped in a spore suspension. The ergots are harvested by hand, by machine, or by separation from the ripe grain by flotation in a brine solution. By varying the strain of the fungal cultures, it is possible to maximize alkaloid production (0.4–1.2%), or give alkaloid mixtures in which particular components predominate. Ergots containing principally ergotamine in concentrations of about 0.35% can be cultivated. More recently, ergot of wheat (*Triticum aestivum*) and the wheat–rye hybrid triticale (*Triticosecale*) have been produced commercially; the latter is now the preferred crop for field cultivation.

Alternatively, the ergot alkaloids can be produced by culturing the fungus. Initially, cultures of the rye parasite *Claviceps purpurea* in fermentors did not give the typical alkaloids associated with the sclerotia, e.g. ergometrine and ergotamine. These medicinally useful compounds appear to be produced only in the later stages of development of the fungus. Instead, the cultures produced alkaloids which were not based on lysergic acid, and are now recognized as intermediates in the biosynthesis of lysergic acid, e.g. chanoclavine-I, agroclavine, and elymoclavine (Figure 6.105). Ergot alkaloids which do not yield lysergic acid on hydrolysis have been termed clavine alkaloids. Useful derivatives based on lysergic acid can be obtained by fermentation growth of another fungal species, namely *Claviceps paspali*. Although some strains are available which produce peptide alkaloids in culture, other strains produce high yields of simple lysergic acid derivatives. These include lysergic acid α-hydroxyethylamide (Figure 6.107), lysergic acid amide (ergine; Figure 6.103) which is also an acid-catalysed decomposition product from lysergic acid α-hydroxyethylamide, and the $\Delta^{8,9}$-isomer of lysergic acid, paspalic acid (Figure 6.105). Lysergic acid is obtained from the first two by hydrolysis, or from paspalic acid by allylic isomerization. Other alkaloids, e.g. ergometrine and ergotamine, can then be produced semi-synthetically. High-yielding fermentation methods have also been developed for direct production of ergotamine and the ergotoxine group of peptide alkaloids. About 60% of ergot alkaloid production is now via fermentation.

The pharmacologically active ergot alkaloids are based on (+)-lysergic acid (Figure 6.107), but since one of the chiral centres in this compound (and its amide derivatives) is adjacent to a carbonyl, the configuration at this centre can be changed as a result of enolization brought about by heat or base (compare tropane alkaloids, page 318; again note that enolization is favoured by conjugation with the aromatic ring). The new diastereomeric form of (+)-lysergic acid is (+)-isolysergic acid (Figure 6.107), and alkaloids based on this compound are effectively pharmacologically inactive. They are frequently found along with the (+)-lysergic acid derivatives, in significant amounts if unsuitable isolation techniques are employed or if old ergot samples are

Box 6.19   (continued)

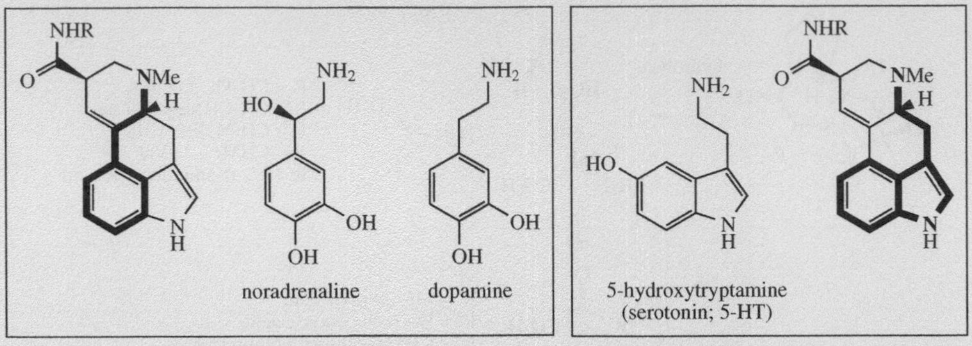

Figure 6.110

processed. In the biologically active lysergic acid derivatives, the amide group occupies an 8-equatorial position, whilst this group is axial in the inactive iso-forms. However, since the tetrahydropyridine ring adopts a half-chair conformation, hydrogen bonding from the amide N–H to the heterocyclic nitrogen at position 6 can occur, and this considerably stabilizes the otherwise unfavourable axial form (see Figure 6.107). Derivatives of (+)-isolysergic acid are named by adding the syllable *–in–* to the corresponding (+)-lysergic acid compound, e.g. ergometri*nin*e, ergotami*nin*e.

The ergot alkaloids owe their pharmacological activity to their ability to act at α-adrenergic, dopaminergic, and serotonergic receptors. The relationship of the general alkaloid structure to those of noradrenaline, dopamine, and 5-HT (serotonin) is shown in Figure 6.110. The pharmacological response may be complex. It depends on the preferred receptor to which the compound binds, though all may be at least partially involved, and on whether the alkaloid is an agonist or antagonist.

Despite the unpleasant effects of ergot as manifested by St Anthony's Fire, whole ergot preparations have been used since the 16th century to induce uterine contractions during childbirth and to reduce haemorrhage following the birth. 'Mothercorn' was a common name for ergot, and is still part of the German language (mutterkorn = ergot). This oxytocic effect (oxytocin is the pituitary hormone that stimulates uterine muscle, see page 430) is still medicinally valuable, but is now achieved through use of the isolated alkaloid ergometrine. The deliberate use of ergot to achieve abortions is dangerous and has led to fatalities.

Ergometrine (ergonovine; Figure 6.107) is used as an oxytocic and is injected during the final stages of labour and immediately following childbirth, especially if haemorrhage occurs. Bleeding is reduced because of its vasoconstrictor effects, and it is valuable after Caesarian operations. It is sometimes administered in combination with oxytocin itself (see page 430). Ergometrine is also orally active. It produces faster stimulation of uterine muscle than do the other ergot alkaloids, and probably exerts its effect by acting on α-adrenergic receptors, though it may also stimulate 5-HT receptors.

Ergotamine (Figure 6.107) is a partial agonist of α-adrenoceptors and 5-HT receptors. It is not suitable for obstetric use because it also produces a pronounced peripheral vasoconstrictor action. This property is exploited in the treatment of acute attacks of migraine, where it reverses the dilatation of cranial blood vessels. Ergotamine is effective orally, or by inhalation in aerosol form, and may be combined with caffeine, which is believed to enhance its action.

A number of semi-synthetic lysergic acid derivatives act by stimulation of dopamine receptors in the brain, and are of value in the treatment of neurological disorders such as Parkinson's disease. **Bromocriptine** (2-bromo-α-ergocryptine), **cabergoline, lisuride (lysuride)**, and **pergolide** (Figure 6.107) are all used in this way. Bromocriptine and cabergoline find wider use, in that they also inhibit release of prolactin by the pituitary and can thus suppress lactation and be used in the treatment of breast tumours. **Methysergide** (Figure 6.107) is a semi-synthetic analogue of ergometrine having a modified amino alcohol side-chain and an *N*-methyl group on the indole ring. It is a potent 5-HT antagonist and, as such, is employed in the prophylaxis of severe migraine headaches, though its administration has to be very closely supervised.

Prolonged treatment with any of the ergot alkaloids is undesirable and it is vital that the clinical features associated with ergot poisoning are recognized. Treatment must be withdrawn immediately if any numbness or tingling develops in the fingers or toes. Side-effects will disappear on withdrawal of the drug, but there have been many cases where misdiagnosis has unfortunately led to foot or toe rot, and the necessity for amputation of the dead tissue.

Undoubtedly the most notorious of the lysergic acid derivatives is lysergide (lysergic acid diethylamide or LSD; Figure 6.107). This widely abused hallucinogen, known as 'acid', is probably the most active and specific psychotomimetic known, and is a mixed agonist–anatagonist at 5-HT receptors, interfering with the normal processes. An effective oral dose is from 30 to 50 μg.

Box 6.19 (continued)

It was synthesized from lysergic acid, and even the trace amounts absorbed during its handling were sufficient to give its creator quite dramatic hallucinations. LSD intensifies perceptions and distorts them. How the mind is affected depends on how the user is feeling at the time, and no two 'trips' are alike. Experiences can vary from beautiful visions to living nightmares, sometimes lasting for days. Although the drug is not addictive, it can lead to schizophrenia and there is danger of serious physical accidents occurring whilst the user is under the influence of the drug.

Morning Glories

Lysergic acid derivatives have also been characterized in the seeds of morning glory (*Ipomoea violacea*), *Turbinia corymbosa* (syn. *Rivea corymbosa*), and other members of the Convolvulaceae. Such seeds formed the ancient hallucinogenic drug Ololiuqui still used by the Mexican Indians in religious and other ceremonies. The active constituent has been identified to be principally ergine (lysergic acid amide; Figure 6.103), and this has an activity about one-twentieth that of LSD, but is more narcotic than hallucinogenic. The alkaloid content of the seeds is usually low, at about 0.05%, but higher levels (0.5–1.3%) have been recorded. Minor ergot-related constituents include ergometrine (Figure 6.107), lysergic acid α-hydroxyethylamide (Figure 6.107), the inactive isolysergic acid amide (erginine), and some clavine alkaloids, e.g. agroclavine and elymoclavine. However, it has now been shown that these alkaloids are not synthesized by the plant itself, but are the product of a plant-associated fungus, transmitted via the seeds of the plant. Plants grown in the presence of systemic fungicides no longer produce ergot alkaloids. *Ipomoea asarifolia*, *Ipomoea violacea*, and *Turbinia corymbosa* all accumulate ergot alkaloids via associated fungi.

Since morning glories are widely cultivated ornamentals and seeds are readily available, deliberate ingestion by thrill-seekers has been considerable. Although the biological activity is well below that of LSD, the practice is potentially dangerous.

ALKALOIDS DERIVED FROM ANTHRANILIC ACID

Anthranilic acid (Figure 6.111) is a key intermediate in the biosynthesis of L-tryptophan (see page 145) and so contributes to the elaboration of indole alkaloids. During this conversion, the anthranilic acid residue is decarboxylated, so that only the C_6N skeleton is utilized. However, there are also many examples of where anthranilic acid itself functions as an alkaloid precursor, using processes which retain the full skeleton and exploit the carboxyl group (Figure 6.111). It should also be appreciated that, in mammals, L-tryptophan can be degraded back to anthranilic acid (see page 332), but this is not a route of importance in plants.

Quinazoline Alkaloids

Peganine (Figure 6.112) is a quinazoline alkaloid found in *Peganum harmala* (Zygophyllaceae) where it co-occurs with the β-carboline alkaloid harmine (see page 369). It

is also responsible for the bronchodilator activity of *Justicia adhatoda* (*Adhatoda vasica*) (Acanthaceae), a plant used in the treatment of respiratory ailments. As a result, the alternative name **vasicine** is also sometimes used for peganine. Studies in *Peganum harmala* have clearly demonstrated peganine to be derived from anthranilic acid, the remaining part of the structure being a pyrrolidine ring supplied by ornithine (see page 313). The peganine skeleton is readily rationalized as a result of nucleophilic attack from the anthranilate nitrogen onto the pyrrolinium cation, followed by amide (lactam) formation (Figure 6.112). Remarkably, this pathway is not operative in *Justicia adhatoda*, and a much less predictable sequence from *N*-acetylanthranilic acid and aspartic acid is observed (Figure 6.112). **Bromhexine** (Figure 6.112) is an expectorant used in veterinary practice developed from the structure of peganine.

Febrifugine (Figure 6.113) is also a quinazoline alkaloid, though no details about its biosynthetic origins are

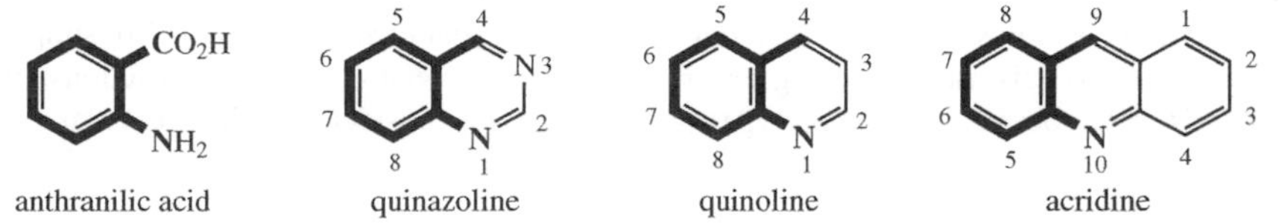

anthranilic acid   quinazoline   quinoline   acridine

Figure 6.111

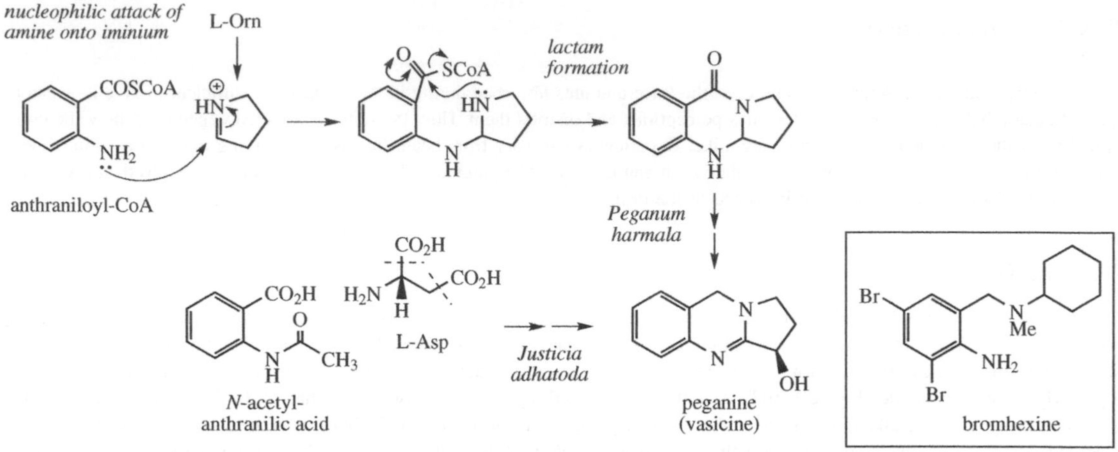

Figure 6.112

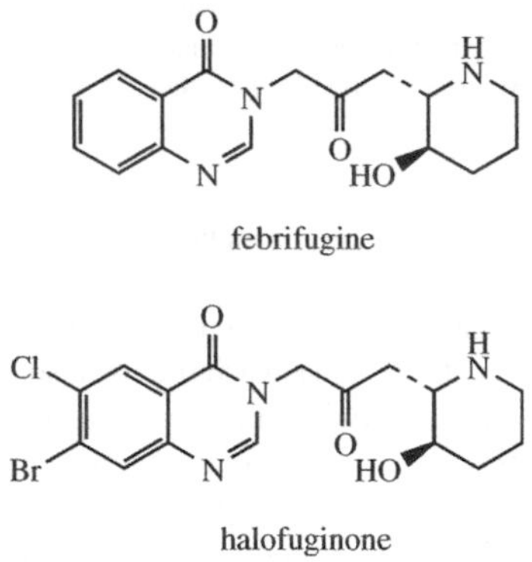

febrifugine

halofuginone

Figure 6.113

available. This alkaloid was isolated from the roots of a Chinese plant *Dichroa febrifuga* (Saxifragaceae), traditionally employed in the treatment of malaria fevers. It is also found in the unrelated *Hydrangea umbellata* (Hydrangeaceae). Febrifugine has powerful antimalarial activity, some 100–200 times greater than that of quinine (see page 382), but unacceptable side-effects, including liver toxicity and strong emetic properties, have precluded its use as a drug. It is currently a template for development of safer synthetic analogues. The halogenated derivative **halofuginone** (Figure 6.113) shows particular promise; it is already widely used as a coccidiostat for poultry, though it has potential for control of diseases involving excessive collagen synthesis, scleroderma, and certain types of cancer.

Quinoline and Acridine Alkaloids

Alkaloids derived from anthranilic acid undoubtedly occur in greatest abundance in plants from the Rutaceae family. Particularly well represented are alkaloids based on quinoline and acridine skeletons (Figure 6.111). Some quinoline alkaloids, such as quinine and camptothecin, have been established to arise by fundamental rearrangement of indole systems and have their origins in tryptophan (see pages 380, 383). A more direct route to the quinoline ring system is by the combination of anthranilic acid and acetate/malonate, and an extension of this process also accounts for the origins of the acridine ring system (see Figure 3.91, page 117). Thus, anthraniloyl-CoA (Figure 6.114) can act as a starter unit for chain extension via one molecule of malonyl-CoA, and amide (lactam) formation generates the heterocyclic system, which will adopt the more stable 4-hydroxy-2-quinolone tautomeric form (Figure 6.114). Position 3 is highly nucleophilic and susceptible to alkylation, especially via dimethylallyl diphosphate in the case of these alkaloids. This allows formation of additional six- and five-membered oxygen heterocyclic rings, as seen with other systems, e.g. coumarins and isoflavonoids (see pages 164 and 176). By an analogous series of reactions, the dimethylallyl derivative can act as a precursor of furoquinoline alkaloids, such as **dictamnine** and **skimmianine** (Figure 6.114). These alkaloids are found in both *Dictamnus albus* and *Skimmia japonica* (Rutaceae). To simplify the mechanistic interpretation of these reactions, it is more convenient to consider the di-enol form of the quinolone system.

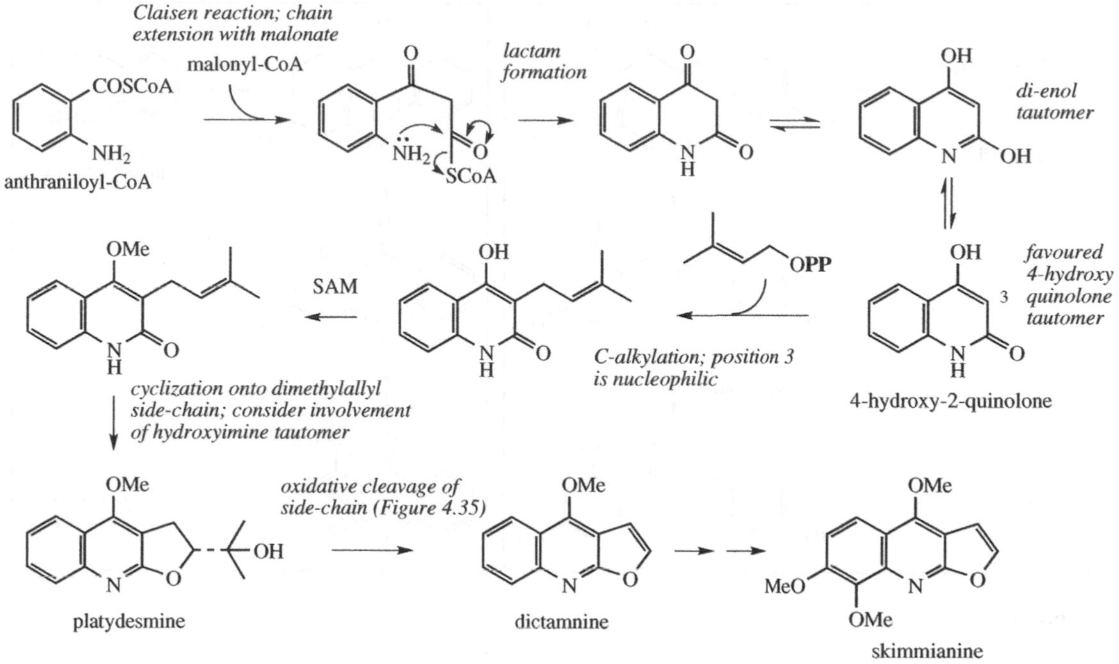

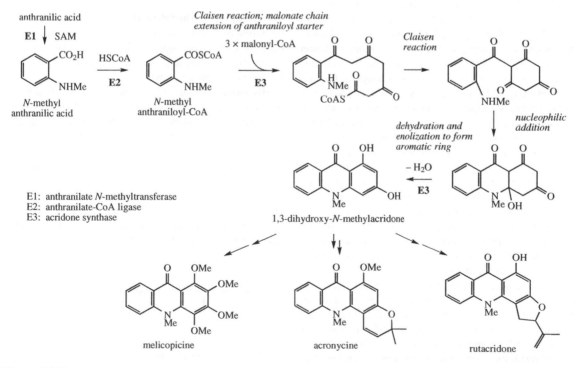

Figure 6.114

E1: anthranilate *N*-methyltransferase
E2: anthranilate-CoA ligase
E3: acridone synthase

Figure 6.115

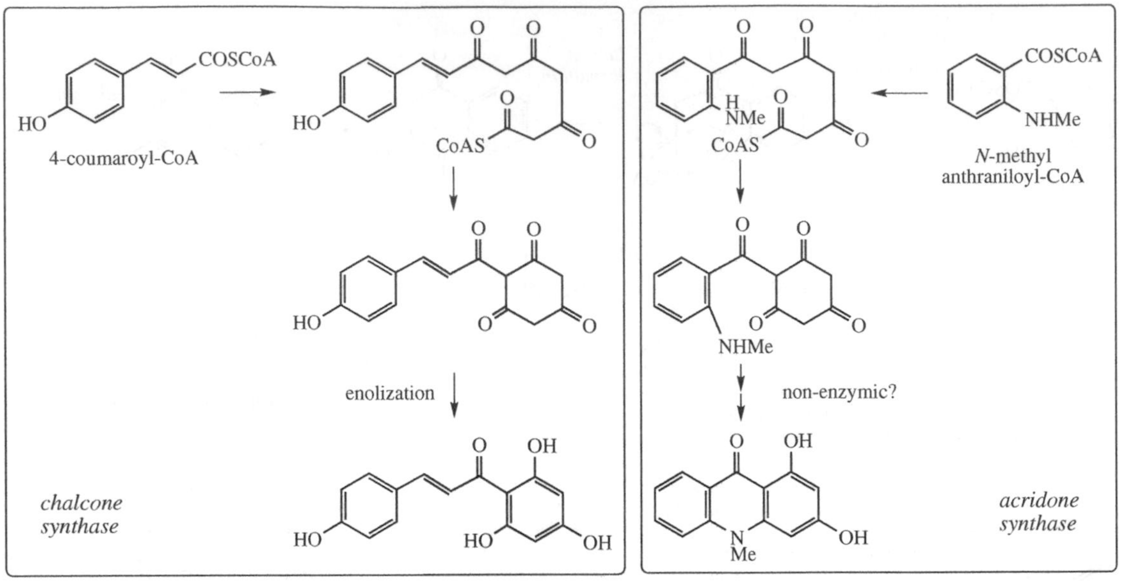

Figure 6.116

Should chain extension of anthraniloyl-CoA (as the *N*-methyl derivative) incorporate three acetate/malonate units, a polyketide would result (Figure 6.115). The acridine skeleton is then produced by sequential Claisen reaction and C–N linkage by an addition reaction, dehydration, and enolization, leading to the stable aromatic tautomer 1,3-dihydroxy-*N*-methylacridone. The enzyme acridone synthase catalyses all these steps and belongs to the family of plant type III polyketide synthases (PKSs). It is closely related structurally to chalcone synthase, the enzyme catalysing chalcone formation from cinnamoyl-CoA and three malonyl-CoA units and involved in flavonoid formation (see page 169) (Figure 6.116). Indeed, acridone synthase also displays a modest chalcone synthase activity, and replacement of amino acids in three critical positions was sufficient to change its activity completely to chalcone synthase, so that it no longer accepted *N*-methylanthraniloyl-CoA as substrate. This may also mean the latter ring formation steps are non-enzymic. In contrast to chalcone synthases, acridone synthase appears to be confined to the Rutaceae family.

Again, the acetate-derived ring, with its alternate oxygenation, is susceptible to electrophilic attack, and this can lead to alkylation (with dimethylallyl diphosphate) or further hydroxylation. Alkaloids **melicopicine** from *Melicope fareana*, **acronycine** from *Acronychia baueri*, and **rutacridone** from *Ruta graveolens* (Rutaceae) typify some of the structural variety which may then ensue (Figure 6.115).

ALKALOIDS DERIVED FROM HISTIDINE

Imidazole Alkaloids

The amino acid L-**histidine** (Figure 6.117) contains an imidazole ring, and is thus the likely presursor of alkaloids containing this ring system. There are relatively few examples, however, and definite evidence linking them to histidine is often lacking.

Histamine (Figure 6.117) is the decarboxylation product from histidine and is often involved in human allergic responses, e.g. to insect bites or pollens. Stress stimulates the action of the enzyme histidine decarboxylase and histamine is released from mast cells. It then produces its typical response by interaction with specific histamine receptors, of which there are several types. H_1 receptors are associated with inflammatory and allergic reactions, and H_2 receptors are found in acid-secreting cells in the stomach. The term antihistamine usually relates to H_1 receptor antagonists. Topical antihistamine creams are valuable for

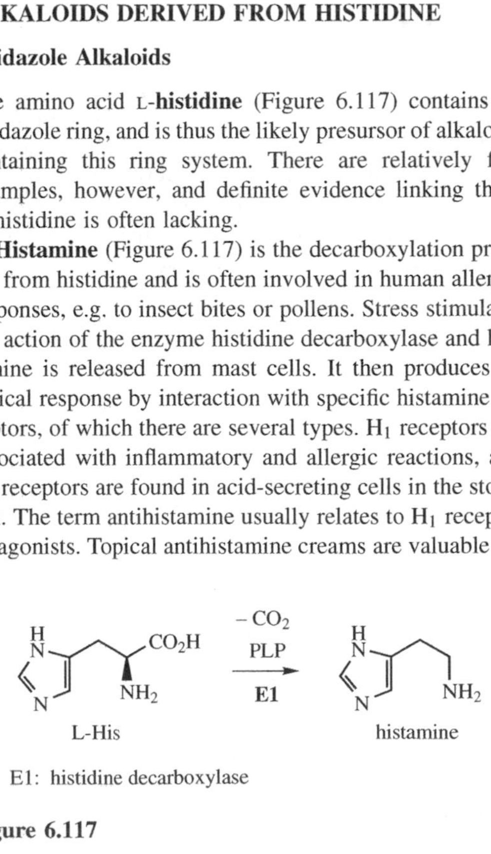

E1: histidine decarboxylase

Figure 6.117

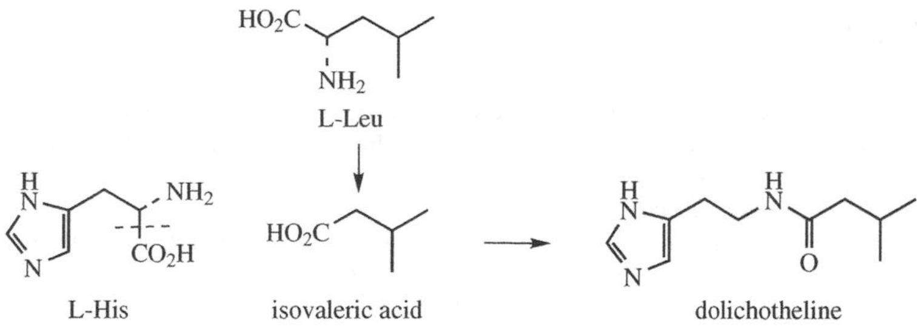

Figure 6.118

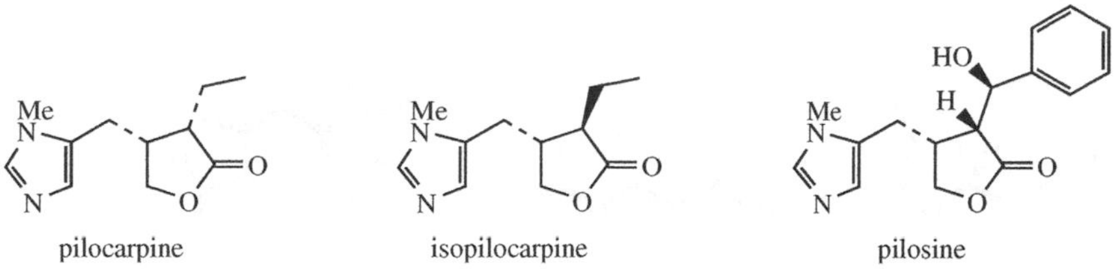

Figure 6.119

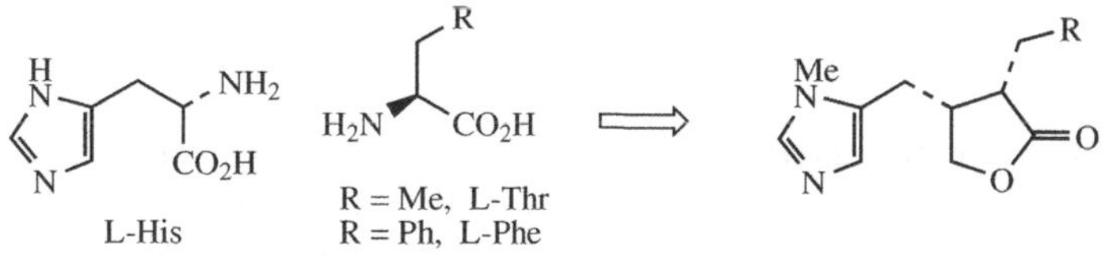

Figure 6.120

pain relief, and oral antihistamines are widely prescribed for nasal allergies such as hay-fever. Major effects of histamine include dilation of blood vessels, inflammation and swelling of tissues, and narrowing of airways. In serious cases, life-threatening anaphylactic shock may occur, caused by a dramatic fall in in blood pressure.

Histidine is a proven precursor of **dolichotheline** (Figure 6.118) in *Dolichothele sphaerica* (Cactaceae), the remaining carbon atoms originating from leucine via isovaleric acid (see page 56). The imidazole alkaloids found in Jaborandi leaves (*Pilocarpus microphyllus* and *Pilocarpus jaborandi*; Rutaceae) are also probably derived from histidine, but experimental data are lacking. Jaborandi leaves contain primarily **pilocarpine** and **pilosine** (Figure 6.119). Pilocarpine is valuable in ophthalmic work as a miotic and as a treatment for glaucoma [Box 6.20]. Additional carbon atoms may originate from acetate or perhaps the amino acid threonine in the case of pilocarpine, whilst pilosine may incorporate a phenylpropane C_6C_3 unit (Figure 6.120).

Box 6.20

Pilocarpus

Pilocarpus or jaborandi consists of the dried leaflets of *Pilocarpus jaborandi, Pilocarpus microphyllus*, or *Pilocarpus pennatifolius* (Rutaceae), small shrubs from Brazil and Paraguay. *Pilocarpus microphyllus* is currently the main source. The alkaloid content (0.5–1.0%) consists principally of the imidazole alkaloid pilocarpine (Figure 6.119), together with small amounts of pilosine

Box 6.20 (continued)

(Figure 6.119) and related structures. Isomers such as isopilocarpine (Figure 6.119) and isopilosine are readily formed if base or heat is applied during extraction of the alkaloids. This is a result of enolization in the lactone ring, followed by adoption of the more favourable *trans* configuration rather than the natural *cis*. However, the iso alkaloids lack biological activity. The alkaloid content of the leaf rapidly deteriorates on storage.

Pilocarpine salts are valuable in ophthalmic practice and are used in eyedrops as miotics and for the treatment of glaucoma. Pilocarpine is a cholinergic agent and stimulates the muscarinic receptors in the eye, causing constriction of the pupil and enhancement of outflow of aqueous humour. The structural resemblance to muscarine and acetylcholine is shown in Figure 6.121. Pilocarpine gives relief for both narrow-angle and wide-angle glaucoma. However, the ocular bioavailability of pilocarpine is low and it is rapidly eliminated, thus resulting in a rather short duration of action. Pilocarpine is antagonistic to atropine (see page 318). Pilocarpine gives relief for dryness of the mouth that is very common in patients undergoing radiotherapy for mouth and throat cancers, and is now prescribed for this purpose.

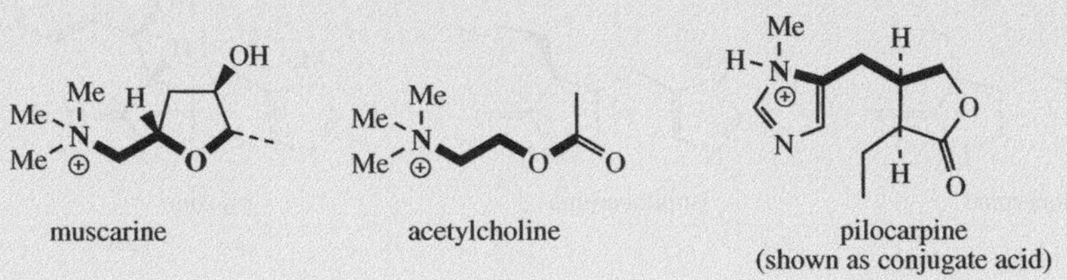
muscarine
acetylcholine
pilocarpine
(shown as conjugate acid)

Figure 6.121

ALKALOIDS DERIVED BY AMINATION REACTIONS

The majority of alkaloids are derived from amino acid precursors by processes which incorporate into the final structure the nitrogen atom together with the amino acid carbon skeleton or a large proportion of it. Many alkaloids do not conform to this description, however, and are synthesized primarily from non-amino acid precursors, with the nitrogen atom being inserted into the structure at a relatively late stage. The term 'pseudoalkaloid' is sometimes used to distinguish this group. Such structures are frequently based on terpenoid or steroidal skeletons, though some relatively simple alkaloids also appear to be derived by similar late amination processes. In most of the examples studied, the nitrogen atom is donated from

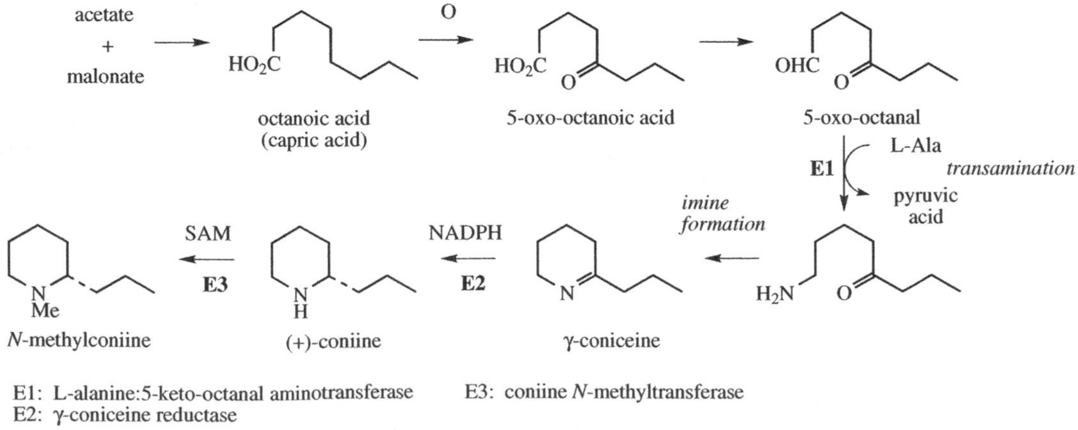
acetate
+
malonate
octanoic acid
(capric acid)
5-oxo-octanoic acid
5-oxo-octanal
L-Ala
E1
transamination
pyruvic
acid
imine
formation
NADPH
E2
SAM
E3
N-methylconiine
(+)-coniine
γ-coniceine
E1: L-alanine:5-keto-octanal aminotransferase
E2: γ-coniceine reductase
E3: coniine N-methyltransferase

Figure 6.122

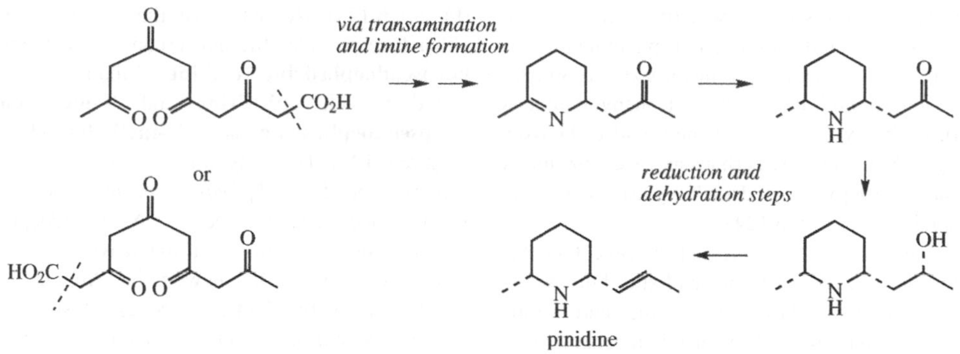

Figure 6.123

an amino acid source through a transamination reaction with a suitable aldehyde or ketone (see page 20).

Acetate-derived Alkaloids

The poison hemlock (*Conium maculatum*; Umbelliferae/Apiaceae) accumulates a range of simple piperidine alkaloids, e.g. **coniine** and **γ-coniceine** (Figure 6.122) [Box 6.21]. These alkaloids would appear to be related to simple lysine-derived compounds such as pelletierine (see page 327); surprisingly, however, a study of their biosynthetic origins excluded lysine as a precursor, and demonstrated instead the sequence shown in Figure 6.122.

A fatty acid precursor, capric (octanoic) acid, is utilized, and this is transformed into 5-oxo-octanal by successive oxidation and reduction steps. This ketoaldehyde is then the substrate for a transamination reaction, the amino group originating from L-alanine. Subsequent transformations are imine formation, giving the heterocyclic ring of γ-coniceine, reduction to **coniine**, then methylation to *N*-methylconiine. **Pinidine** (Figure 6.123) from *Pinus* species is found to have a rather similar origin in acetate, and most likely a poly-β-keto acid. During the sequence outlined in Figure 6.123, the carboxyl group is lost. Note that an alternative folding of the poly-β-keto acid and loss of carboxyl might be formulated.

Box 6.21

Conium maculatum

Conium maculatum (Umbelliferae/Apiaceae) or poison hemlock is a large biennial herb indigenous to Europe and naturalized in North and South America. As a common poisonous plant, recognition is important, and this plant can be differentiated from most other members of the Umbelliferae/Apiaceae by its tall, smooth, purple-spotted stems. The dried unripe fruits were formerly used as a pain reliever and sedative, but have no medicinal use now. The ancient Greeks are said to have executed condemned prisoners, including Socrates, using poison hemlock. The poison causes gradual muscular paralysis followed by convulsions and death from respiratory paralysis. All parts of the plant are poisonous due to the alkaloid content, though the highest concentration of alkaloids is found in the green fruit (up to 1.6%). The major alkaloid (about 90%) is the volatile liquid coniine (Figure 6.122), with smaller amounts of structurally related piperidine alkaloids, including *N*-methylconiine and γ-coniceine (Figure 6.122).

In North America, the name hemlock refers to species of *Tsuga* (Pinaceae), a group of coniferous trees, which should not be confused with the poison hemlock.

Phenylalanine-derived Alkaloids

Whilst the aromatic amino acid L-tyrosine is a common and extremely important precursor of alkaloids (see page 336), L-**phenylalanine** is less frequently utilized; usually it contributes only carbon atoms, e.g. C_6C_3, C_6C_2 or C_6C_1

units, without providing a nitrogen atom from its amino group (see colchicine, page 360, lobeline, page 327, etc.). **Ephedrine** (Figure 6.124), the main alkaloid in species of *Ephedra* (Ephedraceae) and a valuable nasal decongestant and bronchial dilator, is a prime example [Box 6.22].

Whilst ephedrine contains the same carbon and nitrogen skeleton as seen in phenylalanine, and L-phenylalanine is a precursor, only seven carbon atoms, a C_6C_1 fragment, are actually incorporated. It is found that phenylalanine is metabolized, probably through cinnamic acid to benzoic acid (see page 157), and this, perhaps as its coenzyme A ester, is acylated with pyruvate, decarboxylation occurring during the addition (Figure 6.124).

The use of pyruvate as a nucleophilic reagent in this way is unusual in secondary metabolism, but occurs in primary metabolism during isoleucine and valine biosynthesis. A thiamine PP-mediated mechanism is suggested (Figure 6.125), i.e. decarboxylation precedes the nucleophilic attack (compare decarboxylation of pyruvate, page 23, and formation of deoxyxylulose phosphate, page 191). This process yields the diketone, and a transamination reaction would then give **cathinone**

(Figure 6.124). Reduction of the carbonyl group from either face provides the diastereomeric **norephedrine** or **norpseudoephedrine** (**cathine**). Finally, N-methylation would provide **ephedrine** and **N-methylephedrine**, or **pseudoephedrine** and **N-methylpseudoephedrine** (Figure 6.124). Typically, all of the latter six compounds can be found in *Ephedra* species, the proportions varying according to species. Norpseudoephedrine is also a major constituent of the leaves of khat (*Catha edulis*; Celastraceae), chewed in African and Arab countries as a stimulant [Box 6.22]. Most of the central nervous system stimulant action comes from the more active cathinone, the corresponding carbonyl derivative. These natural compounds are structurally similar to the synthetic amfetamine/dexamfetamine (amphetamine/dexamphetamine) (see Figure 6.127), and have similar properties.

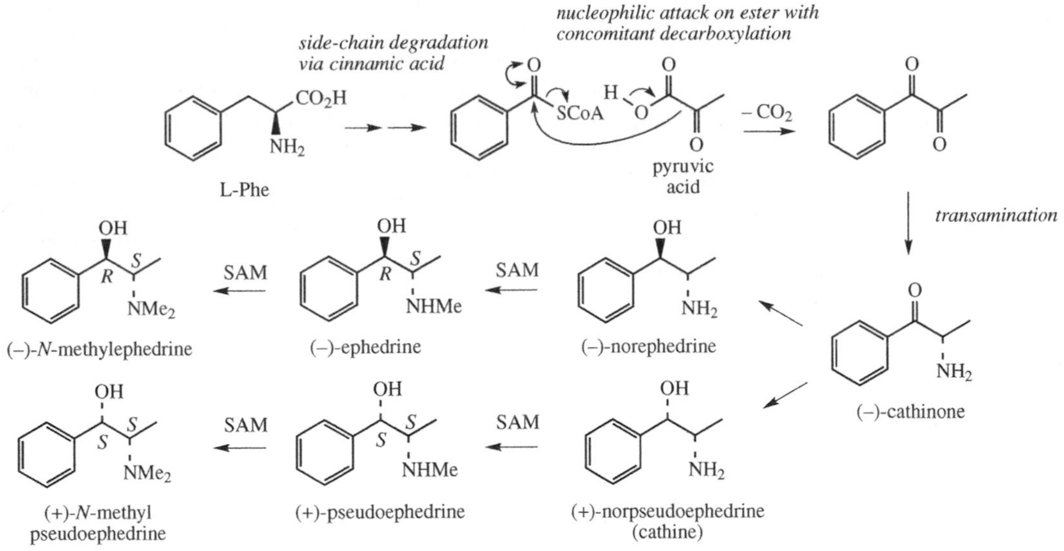

Figure 6.124

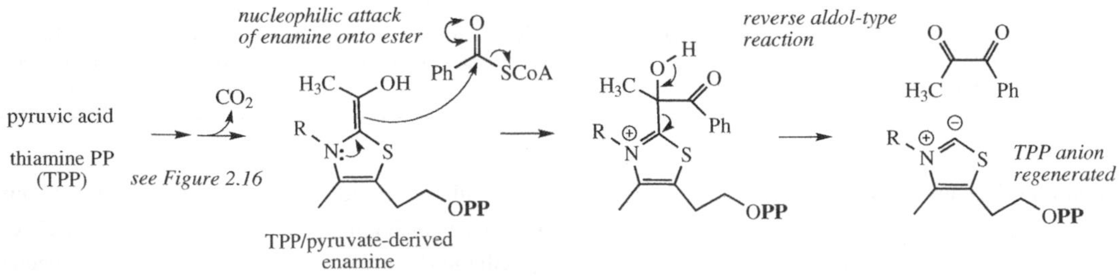

Figure 6.125

Box 6.22

Ephedra

Ephedra or Ma Huang is one of the oldest known drugs, having being used by the Chinese for at least 5000 years. It consists of the entire plant or tops of various *Ephedra* species (Ephedraceae), including *Ephedra sinica* and *Ephedra equisetina* from China, and *Ephedra geriardiana, Ephedra intermedia* and *Ephedra major* from India and Pakistan. The plants are small bushes with slender aerial stems and minute leaves, giving the appearance of being effectively leafless. The plants typically contain 0.5–2.0% of alkaloids, according to species. There are three pairs of optically active diastereomeric alkaloids: (−)-ephedrine and (+)-pseudoephedrine, (−)-methylephedrine and (+)-methylpseudoephedrine, and (−)-norephedrine and (+)-norpseudoephedrine (Figure 6.124). Typically, from 30 to 90% of the total alkaloids is (−)-ephedrine. In *Ephedra intermedia*, the proportion of pseudoephedrine exceeds that of ephedrine.

Ephedrine is an indirectly acting sympathomimetic amine active at both α- and β-adrenergic receptors with effects similar to noradrenaline (see page 336). Lacking the phenolic groups of the catecholamines, it has only weak action on adrenoreceptors, but it is able to displace noradrenaline from storage vesicles in the nerve terminals, which can then act on receptors. It is orally active and has a longer duration of action than noradrenaline. It also has bronchodilator activity, giving relief in asthma, plus a vasoconstrictor action on mucous membranes, making it an effective nasal decongestant. **Pseudoephedrine** is also widely used in compound cough and cold preparations and as a decongestant. The ephedrine and pseudoephedrine used medicinally are usually synthetic. One commercial synthesis of ephedrine involves a fermentation reaction on benzaldehyde using brewer's yeast (*Saccharomyces* sp.), giving initially an alcohol, reductive condensation with methylamine then yields (−)-ephedrine with very high enantioselectivity (Figure 6.126). The fermentation reaction is similar to that shown in Figure 6.125, in that an activated acetaldehyde bound to TPP is produced by the yeast through decarboxylation of pyruvate, and this unit is added stereospecifically to benzaldehyde in an aldol-like reaction.

The herbal drug ephedra/Ma Huang is currently being traded as 'herbal ecstasy'. Consumption gives central nervous system stimulation, but in high amounts it can lead to hallucinations, paranoia, and psychosis. Dietary supplements containing Ma Huang are sold for weight loss and endurance enhancement, but, because of misuse and abuse, these have been regulated or even banned in some countries.

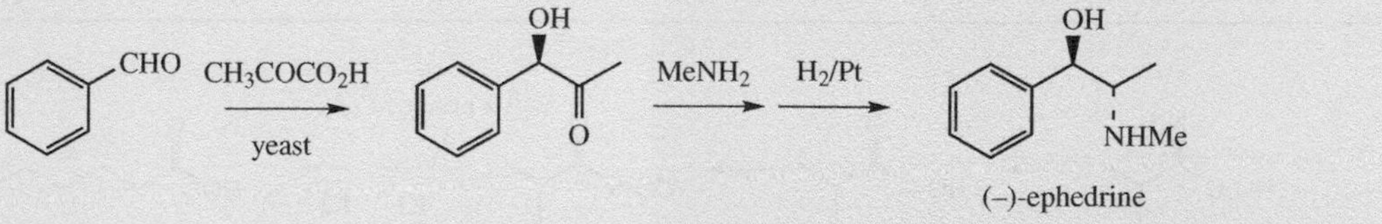

Figure 6.126

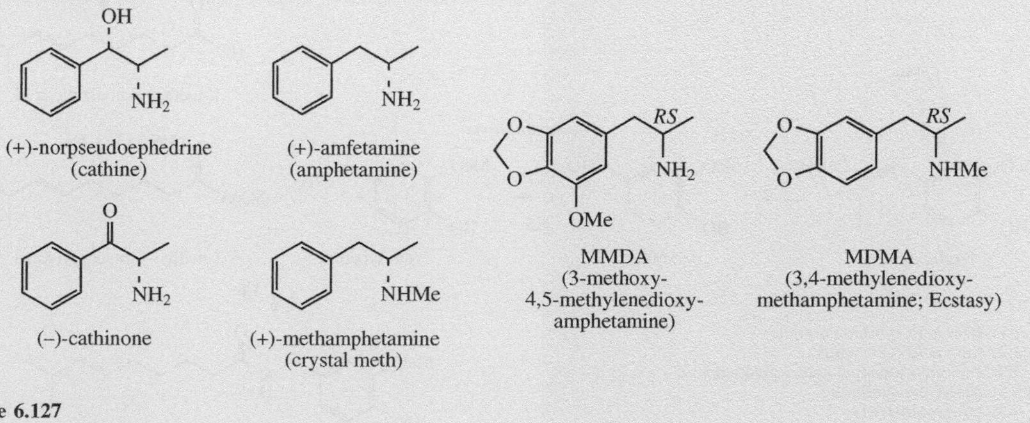

Figure 6.127

Box 6.22   (continued)

Khat

Khat, or Abyssinian tea, consists of the fresh leaves of *Catha edulis* (Celastraceae), a small tree cultivated in Ethiopia, East and South Africa, and in the Yemen. The leaves are widely employed in African and Arabian countries, where they are chewed for a stimulant effect. This traditional use alleviates hunger and fatigue, but also gives a sensation of general well-being (compare coca, page 322). Users become cheerful and talkative, and khat has become a social drug. Prolonged usage can lead to hypertension, insomnia, or even mania. Khat consumption may lead to pyschological dependence, but not normally physical dependence. There is presently little usage outside of Africa and Arabia, although this is increasing due to immigration from these areas. However, for maximum effects, the leaves must be fresh, and this somewhat restricts international trade. Young fresh leaves contain 0.1–0.3% (−)-cathinone (Figure 6.127) as the principal central nervous system stimulant. Cathinone is relatively unstable, decomposing to (+)-norpseudoephedrine (cathine; Figure 6.127) and norephedrine (Figure 6.124) after harvesting or as the leaves are dried. Cathinone has similar pharmacological properties as the synthetic central nervous system stimulant (+)-amfetamine/dexamfetamine (amphetamine/dexamphetamine; Figure 6.127), with a similar potency. Both compounds act by inducing release of catecholamines.

Medicinal use of amfetamine has declined markedly as drug dependence and the severe depression generated on withdrawal have been appreciated. Nevertheless, amfetamine abuse is significant. Amfetamines are taken orally, sniffed, or injected to give a long period of central nervous system stimulation (from hours to days). Users often then take a depressant drug (alcohol, barbiturates, or opioids) to terminate the effects; users rapidly become dependent and develop tolerance. The consumption of khat is not yet restricted in the UK, even though both cathine and cathinone are now controlled drugs. It remains to be seen whether khat will be reclassified and its use restricted in any way. Other amfetamine-like derivatives of note are methamphetamine, methoxymethylenedioxyamphetamine (MMDA) and methylenedioxymethamphetamine (MDMA) (Figure 6.127). Methamphetamine is commonly known as crystal meth, speed, or ice, and is a very addictive and potent psychostimulant. It is usually made illicitly from ephedrine or pseudoephedrine. MMDA is thought to be formed in the body after ingestion of nutmeg (*Myristica fragrans*; Myristicaceae), by an amination process on myristicin (see page 156), and it may be the agent responsible for the euphoric and hallucinogenic effects of nutmeg. MDMA is the illicit drug Ecstasy, a synthetic amfetamine-like stimulant popular among young people. MDMA enhances release of the amine neurotransmitters serotonin, noradrenaline, and dopamine in the brain, and generates multiple short- and long-term neuropsychiatric effects. The use of Ecstasy has resulted in a number of deaths, brought about by subsequent heatstroke and dehydration.

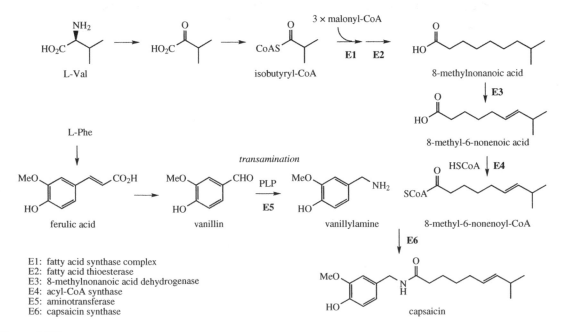

E1: fatty acid synthase complex
E2: fatty acid thioesterase
E3: 8-methylnonanoic acid dehydrogenase
E4: acyl-CoA synthase
E5: aminotransferase
E6: capsaicin synthase

Figure 6.128

The amide **capsaicin** (Figure 6.128) constitutes the powerfully pungent principal in chilli peppers (*Capsicum annuum*; Solanaceae). Apart from its culinary importance, it is also used medicinally in creams to counter neuralgia caused by herpes infections and in other topical pain-relieving preparations [Box 6.23]. The initial burning effect of capsaicin is found to affect the pain receptors, making them less sensitive. The aromatic portion of capsaicin is derived from phenylalanine through ferulic acid and vanillin (Figure 6.128, compare page 160), this aldehyde being the substrate for transamination to give vanillylamine. The acid portion of the amide structure is of polyketide origin, with a branched-chain fatty acid being produced by chain extension of isobutyryl-CoA. This starter unit is valine derived (see page 56). Some, but not all, of the fatty acid synthase component genes have been characterized.

Terpenoid Alkaloids

A variety of alkaloids based on mono-, sesqui-, di-, and tri-terpenoid skeletons have been characterized, but information about their formation in nature is still somewhat sparse. Monoterpene alkaloids are, in the main, structurally related to iridoid materials (see page 206), the oxygen heterocycle being replaced by a nitrogen-containing ring. β-**Skytanthine** from *Skytanthus acutus* (Apocynaceae) and **actinidine** from *Actinidia polygama* (Actinidiaceae) serve as examples (Figure 6.130). The iridoid loganin, so important in the biosynthesis of terpenoid indole alkaloids (see page 370) and the ipecac alkaloids (see page 363), is not a precursor of these structures, and a modified series of reactions starting from geraniol is proposed (Figure 6.130). The formation of the dialdehyde follows closely elaboration of its stereoisomer in loganin biosynthesis (see page 207). This could then act as a substrate for amination via an amino acid, followed by ring formation as seen with coniine (see page 400). Reduction and methylation would yield β-**skytanthine**, whereas further oxidation could provide the pyridine ring of **actinidine**.

Gentianine (Figure 6.130) is probably the most common of the monoterpene alkaloids, but it is frequently formed as an artefact when a plant extract containing suitable iridoid structures is treated with acid and then ammonia, the procedure commonly used during isolation of alkaloids. Thus, the secoiridoid **gentiopicroside** from *Gentiana lutea* (Gentianaceae) is hydrolysed and then reacts with ammonia to give a heterocyclic system that readily dehydrates to a pyridine, generating gentianine

(Figure 6.131). Other iridoid structures are known to react with ammonia to produce alkaloid artefacts. In some plants, however, gentianine can be found when no ammonia treatment has been involved, and one may speculate that it is indeed a natural alkaloid.

Perhaps the most dramatic examples of terpenoid alkaloids from a structural and pharmacological point of view are those found in aconite (*Aconitum* species; Ranunculaceae), commonly known as monkshood or wolfsbane, and species of *Delphinium* (Ranunculaceae). Whilst *Aconitum napellus* has had some medicinal use, mainly for external relief of pain, plants of both genera owe their highly toxic nature to diterpenoid alkaloids. Aconite in particular is regarded as extremely toxic, due to the presence of **aconitine** (Figure 6.132) and related C_{19} norditerpenoid alkaloids. Species of *Delphinium* accumulate diterpenoid alkaloids such as **atisine** (Figure 6.132), which tend to be less toxic than aconitine. These alkaloids appear to behave as neurotoxins by acting on sodium channels. Little experimental evidence about their origins is available, though their structural relationship to diterpenes, e.g. *ent*-**kaurene** (see page 229), can be appreciated. This is most apparent in the veatchine-type diterpenoid alkaloids; the other skeletons involve substantial rearrangement reactions.

Steroidal Alkaloids

Many plants in the Solanaceae accumulate steroidal alkaloids based on a C_{27} cholestane skeleton, e.g. **solasodine** and **tomatidine** (Figure 6.133). These are essentially nitrogen analogues of steroidal saponins (see page 263) and have already been briefly considered along with these

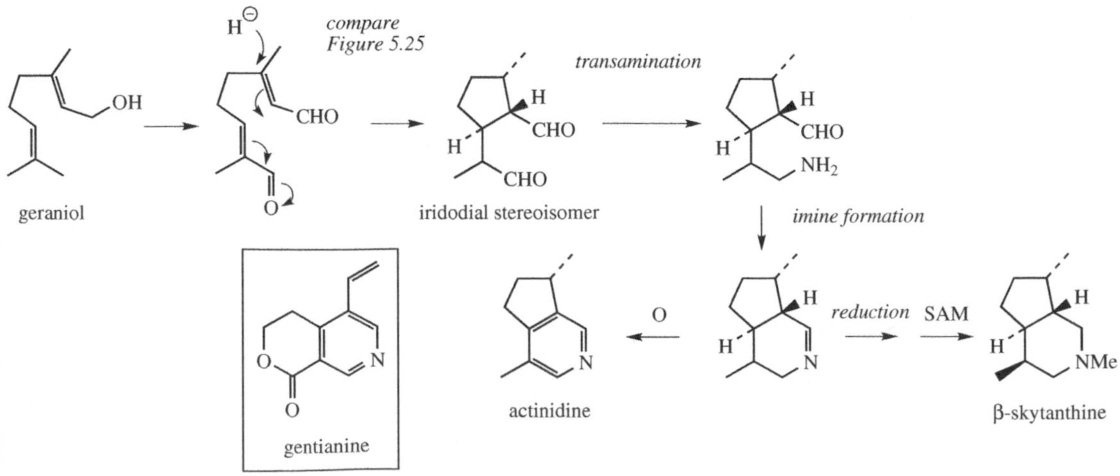

Figure 6.130

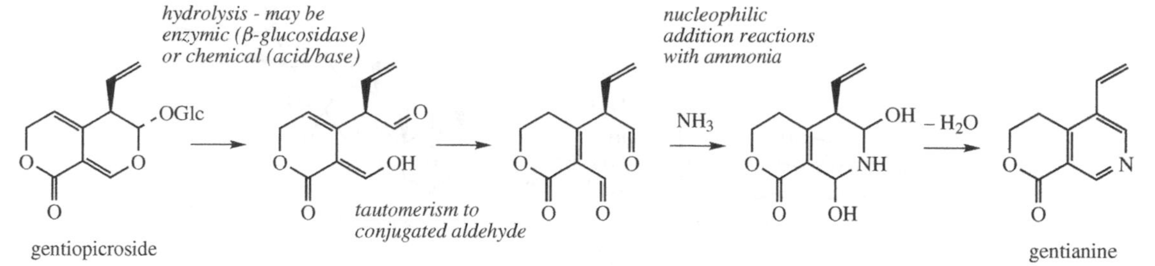

Figure 6.131

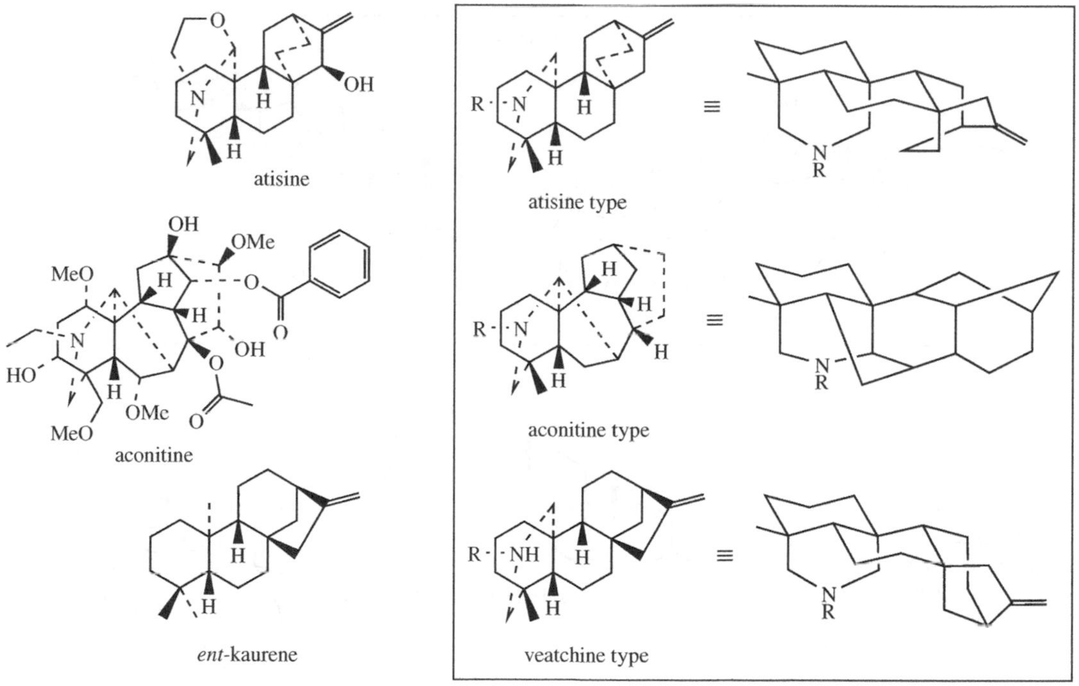

Figure 6.132

compounds. In contrast to the oxygen analogues, these compounds all have the same stereochemistry at C-25 (methyl always equatorial), but C-22 isomers do exist, as solasodine and tomatidine exemplify. They are usually present as glycosides, which have surface activity and haemolytic properties, as do the saponins, but these compounds are also toxic if ingested. α-**Solasonine** from *Solanum* species and α-**tomatine** (Figure 6.133) from tomato (*Lycopersicon esculente*) are typical examples of such glycosides.

As with the sapogenins, this group of steroidal alkaloids is derived from cholesterol, with appropriate side-chain modifications during the sequence

(Figure 6.134). Amination appears to employ L-arginine as the nitrogen source, probably via a substitution process on 26-hydroxycholesterol. A second substitution allows 26-amino-22-hydroxycholesterol to cyclize, generating a piperidine ring. After 16β-hydroxylation, the secondary amine is oxidized to an imine, and the spiro-system can be envisaged as the result of a nucleophilic addition of the 16β-hydroxyl onto the imine (or iminium via protonation). Whether the $22R$ (as in **solasodine**) or $22S$ (as in **tomatidine**) configuration is established may depend on this reaction.

A variant on the way the cholesterol side-chain is cyclized can be found in **solanidine** (Figure 6.133),

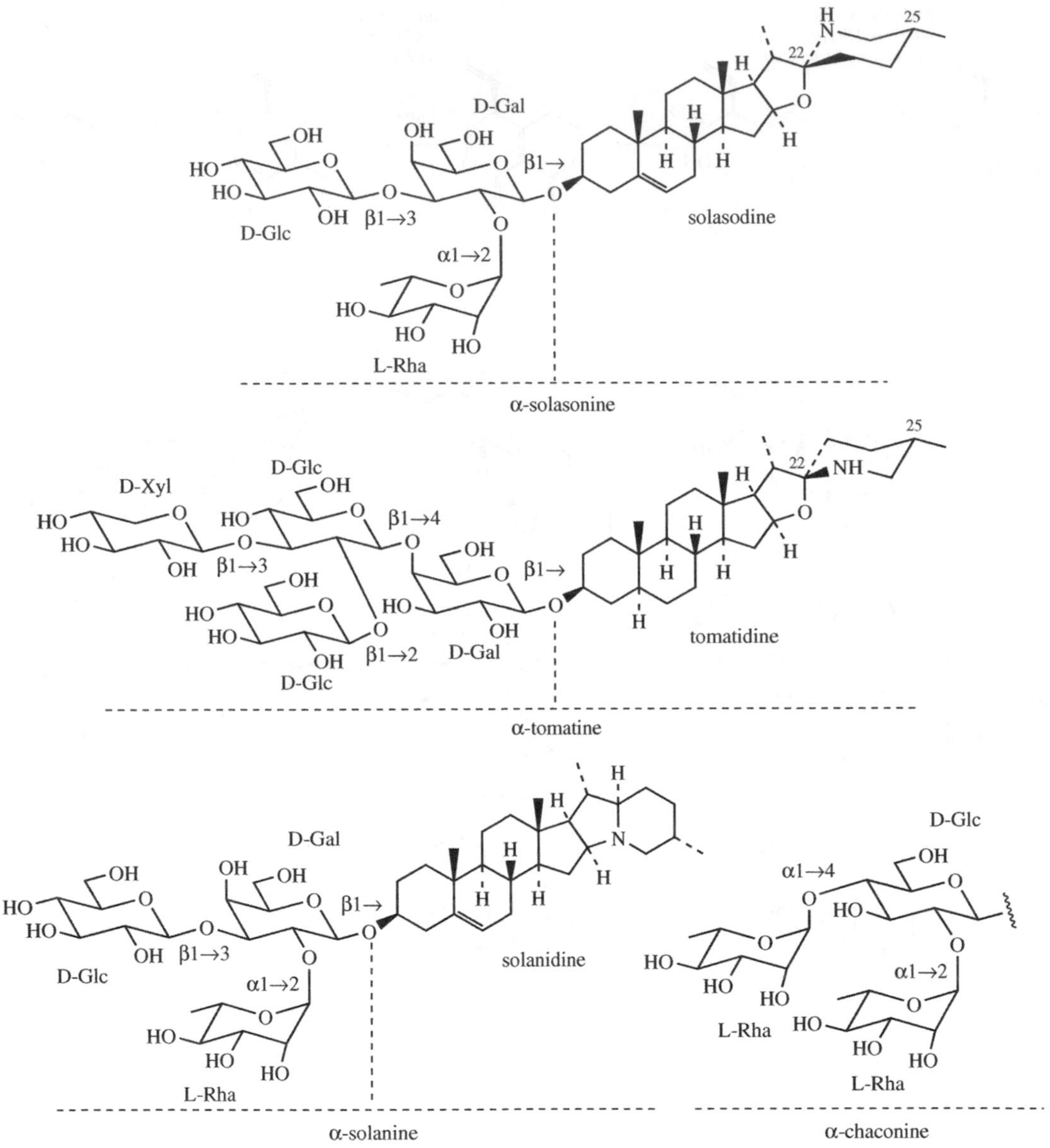

Figure 6.133

which contains a condensed ring system with nitrogen at the bridgehead. Solanidine is found in potatoes (*Solanum tuberosum*) typically as the glycosides **α-solanine** and **α-chaconine** (Figure 6.133). This condensed ring system appears to be produced by a branch from the main pathway to the solaso-dine/tomatidine structures. Thus, a substitution process will allow generation of the new ring system (Figure 6.135).

Enzymic data relating to the formation of the steroidal alkaloid aglycones are not available, but both enzymic and genetic studies have clarified sequences for elaboration of various glycoside side-chains. Thus, in potato, solanidine is converted into α-solanine by way of γ- and β-solanine. Alternatively, a different sequence of glycosylation reactions leads to α-chaconine (Figure 6.136). Toxicity appears to increase as the glycoside chain is extended.

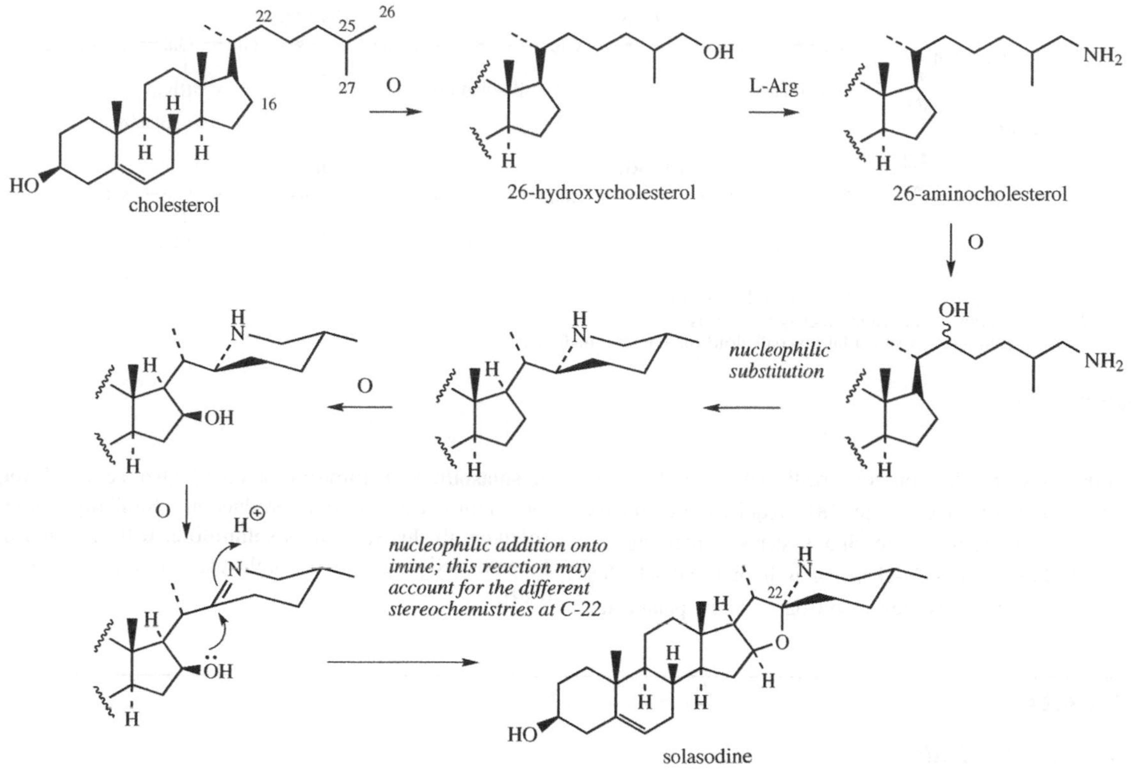

Figure 6.134

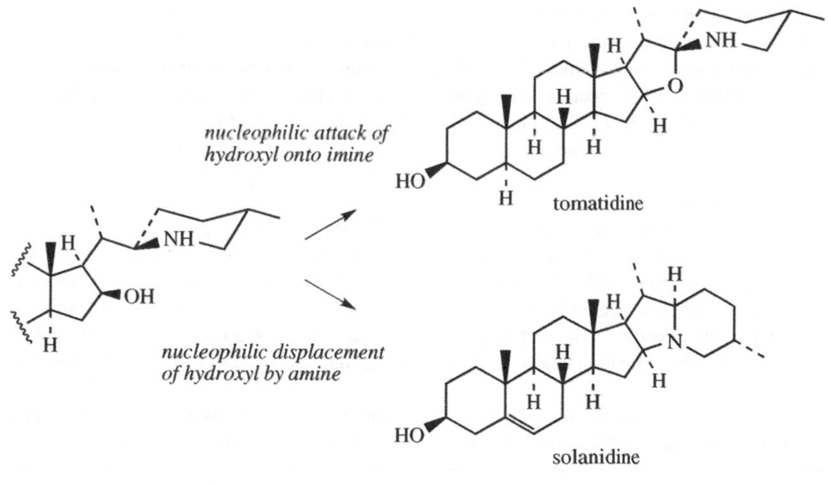

Figure 6.135

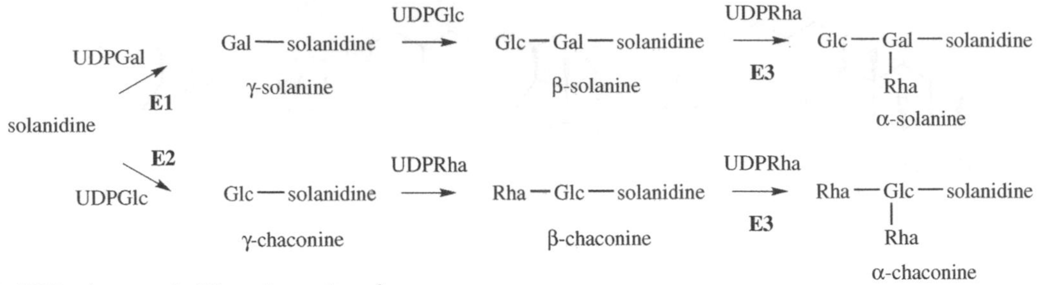

E1: UDP-galactose:solanidine galactosyltransferase
E2: UDP-glucose:solanidine glucosyltransferase
E3: UDP-rhamnose:β-steroidal glycoalkaloid rhamnosyltransferase

Figure 6.136

Since the production of medicinal steroids from steroidal saponins (see page 281) requires preliminary degradation to remove the ring systems containing the original cholesterol side-chain, it is immaterial whether these rings contain oxygen or nitrogen. Thus, plants rich in **solasodine** or **tomatidine** could also be employed for commercial steroid production. Similarly, other *Solanum* alkaloids, such as **solanidine**, with nitrogen in a condensed ring system might also be exploited [Box 6.24].

Box 6.24

Solanum Alkaloids

The toxicity of steroidal alkaloid glycosides is of some concern, in that several plant species that can accumulate them are major food crops. Tomatoes, peppers, and especially potatoes fall in this group; the production of glycoalkaloids in potato (*Solanum tuberosum*) has been studied extensively. α-Solanine and α-chaconine (Figure 6.133) account for up to 95% of the glycoalkaloids in potato, and α-chaconine is regarded as the more toxic. The alkaloids inhibit acetylcholinesterase and butyrylcholinesterase, disrupt cell membranes, and may be teratogenic. Mild clinical symptoms of glycoalkaloid poisoning include abdominal pain, vomiting, and diarrhoea. At higher levels, more severe symptoms may occur, including fever, rapid pulse, low blood pressure, rapid respiration, and neurological disorders. The highest glycoalkaloid level in potato plants is found in flowers and sprouts, followed by the leaves, and the lowest amounts are detected in stems and tubers. However, the amount of glycoalkaloids in tubers increases upon wounding and light exposure; green tubers have a significantly higher alkaloid content and are considered unsuitable for human consumption.

The major alkaloidal component in many *Solanum* species is solasodine (Figure 6.133). It is present as glycosides in the leaves, and especially in the unripe fruits. Solasodine may be converted into progesterone by means of the Marker degradation shown in Figure 5.118 (see page 281). Trial cultivations of a number of *Solanum* species, including *Solanum laciniatum* and *Solanum aviculare* (indigenous to New Zealand), *Solanum khasianum* (from India), and *Solanum marginatum* (from Ecuador), have been conducted in various countries. Alkaloid levels of 1–2% have been obtained. These plants are especially suitable for long-term cultivation if the fruits provide suitable quantities, being significantly easier to cultivate than disogenin-producing *Dioscorea* species.

Cultivation of tomato fruits (*Lycopersicon esculentum*) is carried out on a huge scale as a food crop. The aerial parts, currently waste plant material after harvesting, contain about 0.1% tomatidine (Figure 6.133) which may also be processed.

Several plants in the Liliaceae, notably the genus *Veratrum* (Liliaceae/Melanthiaceae), contain a remarkable group of steroidal alkaloids in which a fundamental change to the basic steroid nucleus has taken place. This change expands ring D by one carbon at the expense of ring C, which consequently becomes five-membered. The resulting skeleton is termed a C-nor-D-homosteroid in keeping with these alterations in ring size.

Cholesterol is a precursor of this group of alkaloids, and a possible mechanism accounting for the ring

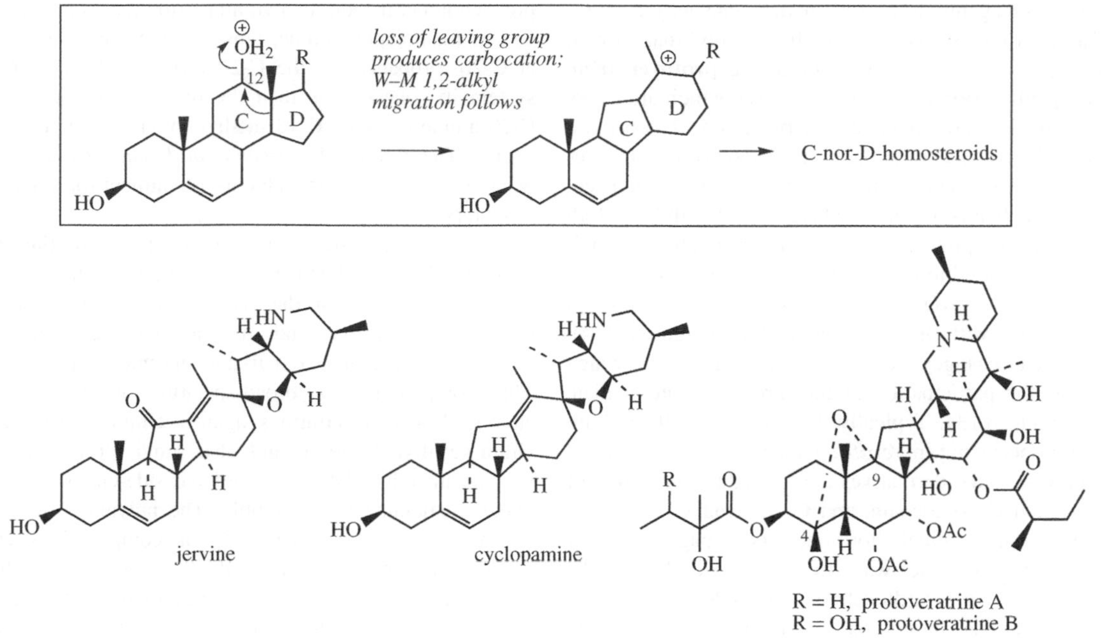

Figure 6.137

modifications is shown in Figure 6.137, where the changes are initiated by loss of a suitable leaving group from C-12. Typical representatives of C-nor-D-homosteroids are **jervine** and **cyclopamine** (Figure 6.137) from *Veratrum californicum*, toxic components in this plant which are responsible for severe teratogenic effects. Animals grazing on *Veratrum californicum* and some other species

of *Veratrum* frequently give birth to offspring with cyclopia, a malformation characterized by a single eye in the centre of the forehead. The teratogenic effects of jervine, cyclopamine, and cyclopamine glucoside (cycloposine) on the developing fetus are now well established. Other *Veratrum* alkaloids, especially those found in *Veratrum album* and *Veratrum viride*, have been employed medicinally as

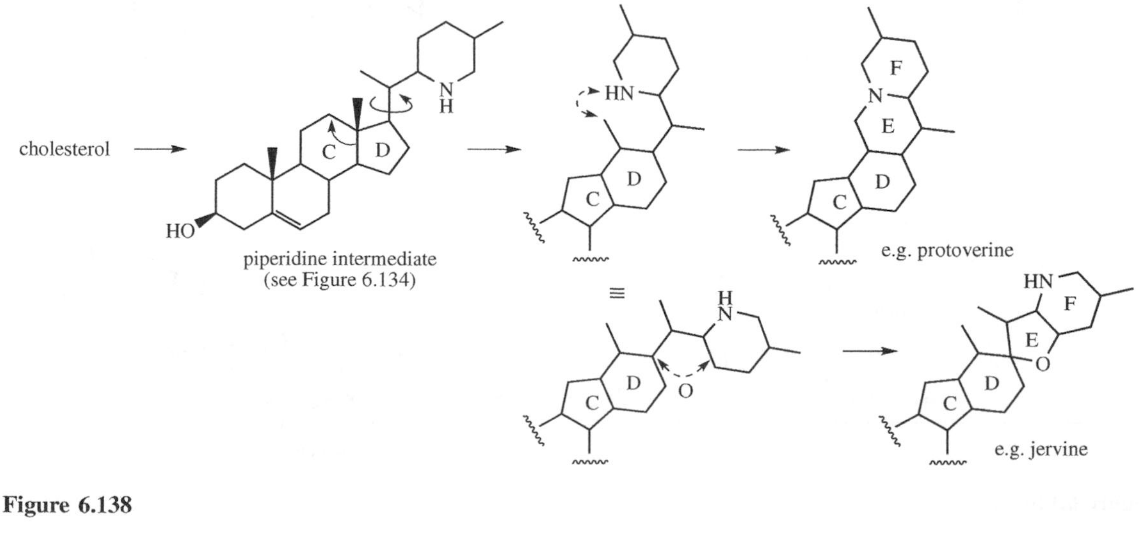

Figure 6.138

hypotensive agents, and used in the same way as *Rauwolfia* alkaloids (see page 372), often in combination with *Rauwolfia*. These medicinal alkaloids, e.g. **protoveratrine A** and **protoveratrine B** (Figure 6.137), which are esters of protoverine, are characterized by fusion of two more six-membered rings onto the C-nor-D-homosteroid skeleton. This hexacyclic system is extensively oxygenated, and a novel hemiketal linkage bridges C-9 with C-4. Both the jervine and protoverine skeletons are readily rationalized through additional cyclization reactions involving a piperidine ring, probably formed by processes analogous to those seen with the *Solanum* alkaloids (Figure 6.134). The skeletal changes are outlined in Figure 6.138, which suggests the participation of the piperidine intermediate from Figure 6.134. Typically, both types of alkaloid are found co-occurring in *Veratrum* species.

Many steroidal derivatives are formed by truncation of the original C_8 side-chain, and C_{21} pregnane derivatives are important animal hormones (see page 287) or intermediates on the way to other natural steroidal derivatives, e.g. cardioactive glycosides (see page 265). Alkaloids based on a pregnane skeleton are found in plants, particularly in the Apocynaceae and Buxaceae, and **pregnenolone** (Figure 6.139) is usually involved in their production. **Holaphyllamine** from *Holarrhena floribunda* (Apocynaceae) is obtained from pregnenolone by replacement of the 3-hydroxyl with an amino group (Figure 6.139). **Conessine** (Figure 6.139) from *Holarrhena antidysenterica* is also derived from

pregnenolone; it requires two amination reactions, one at C-3 as for holaphyllamine, plus a further one, originally at C-20, probably via the C-20 alcohol. The new ring system in conessine is then the result of attack of the C-20 amine onto the C-18 methyl, suitably activated, of course. The bark of *Holarrhena antidysenterica* has long been used, especially in India, as a treatment for amoebic dysentery.

The novel steroidal polyamine **squalamine** (Figure 6.140) has been isolated in very small amounts (about 0.001%) from the liver of the dogfish shark (*Squalus acanthias*), and has attracted considerable attention because of its remarkable antimicrobial activity. This compound is a broad-spectrum agent effective at very low concentrations against Gram-positive and Gram-negative bacteria, and also fungi, protozoa, and viruses, including HIV. The sulfated side-chain helps to make squalamine water soluble. The polyamine portion is spermidine (see page 313), a compound widely distributed in both animals and plants. Squalamine has since been found to possess a range of other biological activities, including antiangiogenic properties; synthetic squalamine is in clinical trials as an anticancer agent against solid tumours. Related aminosterol derivatives with similar high antimicrobial activity have also been isolated from the shark liver extracts. These compounds exhibit structural variation in the C_8 side-chain, with spermidine (or in one example, spermine) attached at C-3.

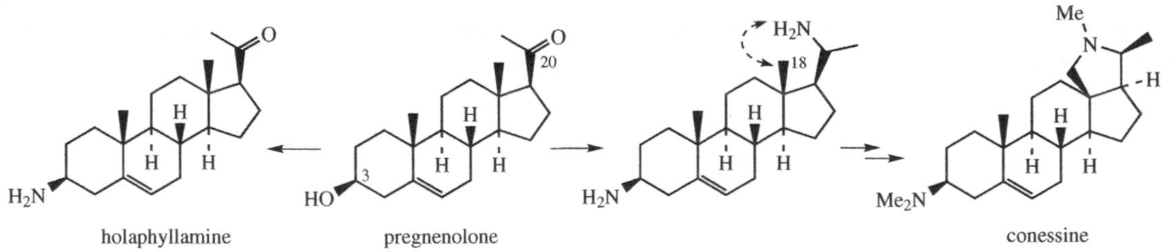

Figure 6.139

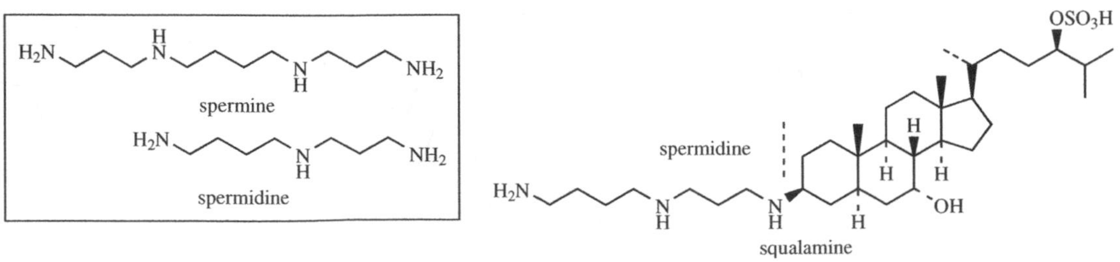

Figure 6.140

PURINE ALKALOIDS

Caffeine

The purine derivatives **caffeine, theobromine**, and **theophylline** (Figure 6.141) are usually referred to as purine alkaloids. They have a rather limited distribution, and their origins are very closely linked with those of the purine bases adenine and guanine, fundamental components of nucleosides, nucleotides, and the nucleic acids. Caffeine, in the form of beverages such as tea, coffee, and cola, is one of the most widely consumed and socially accepted natural stimulants. It is also used medicinally, but theophylline is more important as a drug compound because of its muscle relaxant properties, utilized in the relief of bronchial asthma. Theobromine is a major constituent of cocoa and related chocolate products [Box 6.25].

The purine ring is gradually elaborated by piecing together small components from primary metabolism. The largest component incorporated is glycine, which provides a C_2N unit, whilst the remaining carbon atoms come from formate (by way of N^{10}-formyl-tetrahydrofolate; see page 144) and bicarbonate. Two of

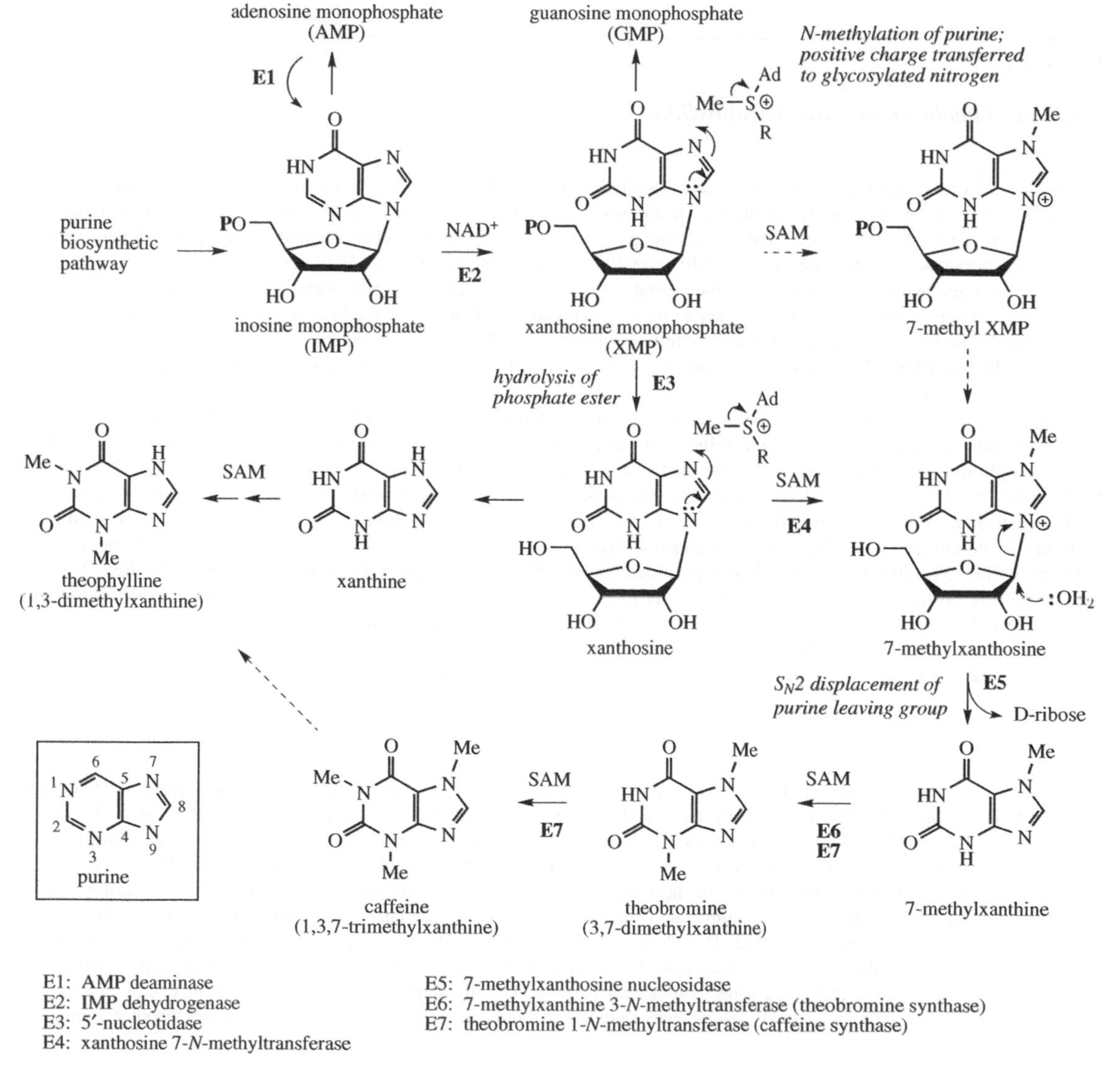

adenosine monophosphate (AMP)
guanosine monophosphate (GMP)
N-methylation of purine; positive charge transferred to glycosylated nitrogen
E1
purine biosynthetic pathway
NAD+
E2
SAM
inosine monophosphate (IMP)
xanthosine monophosphate (XMP)
7-methyl XMP
hydrolysis of phosphate ester
E3
E4
theophylline (1,3-dimethylxanthine)
xanthine
xanthosine
7-methylxanthosine
SN2 displacement of purine leaving group
E5
D-ribose
purine
caffeine (1,3,7-trimethylxanthine)
theobromine (3,7-dimethylxanthine)
7-methylxanthine
E6
E7
E1: AMP deaminase
E2: IMP dehydrogenase
E3: 5′-nucleotidase
E4: xanthosine 7-N-methyltransferase
E5: 7-methylxanthosine nucleosidase
E6: 7-methylxanthine 3-N-methyltransferase (theobromine synthase)
E7: theobromine 1-N-methyltransferase (caffeine synthase)

Figure 6.141

the four nitrogen atoms are supplied by glutamine and a third by aspartic acid. Synthesis of the nucleotides adenosine 5′-monophosphate (AMP) and guanosine 5′-monophosphate (GMP) is by way of inosine 5′-monophosphate (IMP) and xanthosine 5′-monophosphate (XMP) (Figure 6.141), and the purine alkaloids then branch away through XMP. AMP, if available, can also serve as a source of IMP. Methylation and then loss of phosphate generates the nucleoside **7-methylxanthosine**, which is then released from the sugar. Successive methylations on the nitrogen atoms give **caffeine** by way of **theobromine**, whilst a different methylation sequence can account for the formation of **theophylline**. Theophylline can also be produced by demethylation of caffeine as part of a degradative pathway. Some of the *N*-methyltransferases display rather broad substrate specificity, and this allows minor pathways to operate in certain plants, e.g. the alternative sequence to 7-methylxanthosine via 7-methyl XMP shown in Figure 6.141. In addition, the enzyme caffeine synthase in coffee (*Coffea arabica*; Rubiaceae) has dual functionality, and methylates both theobromine and 7-methylxanthine; a tea (*Camellia sinensis*; Theaceae) enzyme is specific for theobromine.

Box 6.25

Caffeine, Theobromine, and Theophylline

The purine alkaloids caffeine, theobromine, and theophylline (Figure 6.141) are all methyl derivatives of xanthine and they commonly co-occur in a particular plant. The major sources of these compounds are the beverage materials such as tea, coffee, cocoa, and cola, which owe their stimulant properties to these water-soluble alkaloids. They competitively inhibit the phosphodiesterase that degrades cyclic AMP (cAMP). The resultant increase in cAMP levels thus mimics the action of catecholamines and leads to a stimulation of the central nervous system, a relaxation of bronchial smooth muscle, and induction of diuresis, as major effects. These effects vary in the three compounds. **Caffeine** is the best central nervous system stimulant, and has weak diuretic action. **Theobromine** has little stimulant action, but has more diuretic activity and also muscle relaxant properties. **Theophylline** also has low stimulant action and is an effective diuretic, but it relaxes smooth muscle better than caffeine or theobromine.

Caffeine is used medicinally as a central nervous system stimulant, usually combined with another therapeutic agent, as in compound analgesic preparations. **Theobromine** is of value as a diuretic and smooth muscle relaxant, but is not now routinely used. **Theophylline** is an important smooth muscle relaxant for relief of bronchospasm; it is frequently dispensed in slow-release formulations to reduce side-effects. It is also available as **aminophylline**, a more soluble preparation containing theophylline with ethylenediamine in a 2:1 ratio. The alkaloids may be isolated from natural sources, or obtained by total or partial synthesis.

It has been estimated that beverage consumption may provide the following amounts of caffeine per cup or average measure: coffee, 30–150 mg (average 60–80 mg); instant coffee, 20–100 mg (average 40–60 mg); decaffeinated coffee, 2–4 mg; tea, 10–100 mg (average 40 mg); cocoa, 2–50 mg (average 5 mg); cola drink, 25–60 mg. The maximal daily intake should not exceed about 1 g to avoid unpleasant side-effects, e.g. headaches and restlessness. An acute lethal dose is about 5–10 g. The biological effects produced from the caffeine ingested via the different drinks can vary, since its bioavailability is known to be modified by the other constituents present, especially the amount and nature of polyphenolic tannins.

Coffee

Coffee consists of the dried ripe seed of *Coffea arabica*, *Coffea canephora*, *Coffea liberica*, or other *Coffea* species (Rubiaceae). The plants are small evergreen trees, widely cultivated in various parts of the world, e.g. Brazil and other South American countries and Kenya. The fruit is deprived of its seed coat, then dried and roasted to develop its characteristic colour, odour, and taste. Coffee seeds contain 1–2% of caffeine and traces of theophylline and theobromine. These are mainly combined in the green seed with chlorogenic acid (see page 150) (5–7%); roasting releases them and also causes some decomposition of chlorogenic acid to quinic acid and caffeic acid. Trigonelline (*N*-methylnicotinic acid) is present in green seeds to the extent of about 0.25–1%; during roasting, this is extensively converted into nicotinic acid (vitamin B$_3$, see page 31). Volatile oils and tannins provide odour and flavour. A proportion of the caffeine may sublime off during the roasting process, providing some commercial caffeine. Decaffeinated coffee, containing up to 0.08% caffeine, is obtained by removing caffeine, usually by aqueous percolation prior to roasting. This process provides another source of caffeine.

Box 6.25 (continued)

Tea

Tea is the prepared leaves and leaf buds of *Camellia sinensis* (*Thea sinensis*) (Theaceae), an evergreen shrub cultivated in China, India, Japan, and Sri Lanka. For black tea, the leaves are allowed to ferment, allowing enzymic oxidation of the polyphenols, whilst green tea is produced by steaming and drying the leaves to prevent oxidation. During oxidation, colourless catechins (up to 40% in dried leaf; see page 171) are converted into intensely coloured theaflavins and thearubigins. Oolong tea is semi-fermented. Tea contains 1–4% caffeine and small amounts (up to 0.05%) of both theophylline and theobromine. Astringency and flavour come from tannins and volatile oils, the latter containing monoterpene alcohols (geraniol, linalool) and aromatic alcohols (benzyl alcohol, 2-phenylethanol). Theaflavins (see page 172) are believed to act as radical scavengers/antioxidants, and to provide beneficial effects against cardiovascular disease, cancers, and the ageing process generally. Green tea, in particular, contains significant amounts of epigallocatechin gallate (see page 172), a very effective antioxidant regarded as one of the more desirable dietary components. Tea leaf dust and waste is a major source of caffeine.

Cola

Cola, or kola, is the dried cotyledon from seeds of various species of *Cola* (Sterculiaceae), e.g. *Cola nitida* and *Cola acuminata*, trees cultivated principally in West Africa and the West Indies. Seeds are prepared by splitting them open and drying. Cola seeds contain up to 3% caffeine and about 0.1% theobromine, partly bound to tannin materials. Drying allows some oxidation of polyphenols, formation of a red pigment, and liberation of free caffeine. Fresh cola seeds are chewed in tropical countries as a stimulant, and vast quantities of dried seeds are processed for the preparation of cola drinks, e.g. Coca-Cola® and Pepsi-Cola®.

Cocoa

Although cocoa as a drink is now rather unfashionable, it provides the raw material for the manufacture of chocolate and is commercially very important. Cocoa (or cacao) is derived from the roasted seeds of *Theobroma cacao* (Sterculiaceae), a tree widely cultivated in South America and West Africa. The fruits develop on the trunk of the tree; the seeds from them are separated, allowed to ferment, and are then roasted to develop the characteristic chocolate flavour. The kernels are then separated from the husks, ground up, and processed in various ways to give chocolate, cocoa, and cocoa butter.

Cocoa seeds contain 35–50% of oil (cocoa butter or theobroma oil), 1–4% theobromine and 0.2–0.5% caffeine, plus tannins and volatile oils. During fermentation and roasting, most of the theobromine from the kernel passes into the husk, which thus provides a convenient source of the alkaloid. Theobroma oil or cocoa butter is obtained by hot expression from the ground seeds as a whitish solid with a mild chocolate taste. It is a valuable formulation aid in pharmacy, where it is used as a suppository base (see page 48).

Maté Tea

Maté tea is consumed in South America as a stimulant drink. Maté or Paraguay tea consists of the leaves of *Ilex paraguensis* (Aquifoliaceae), South American shrubs of the holly genus. The dried leaf contains 0.8–1.7% caffeine and smaller amounts of theobromine (0.3–0.9%) with little or no theophylline. Considerable amounts (10–16%) of chlorogenic acid (see page 150) are also present.

Guarana

The seeds of the Brazilian plant *Paullinia cupana* (Sapindaceae) are used to make a stimulant drink. Crushed seeds are mixed with water to a paste, which is then sun-dried. Portions of this are then boiled with hot water to provide a refreshing drink. The principal constituent, previously called guaranine, has been shown to be identical to caffeine, and the seeds may contain 3–5%. Small amounts of theophylline (0–0.25%) and theobromine (0.02–0.06%) are also present. Guarana is widely available as tablets and capsules, or as extracts, in health food shops, where it is promoted to relieve mental and physical fatigue. Labels on such products frequently show the active constituent to be guaranine, but may not indicate that this is actually caffeine.

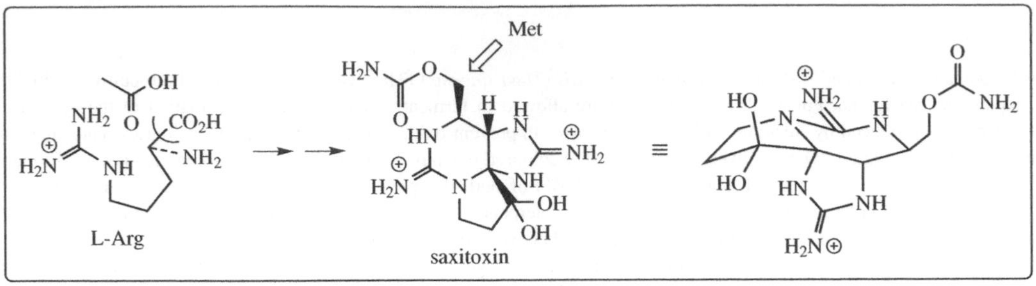

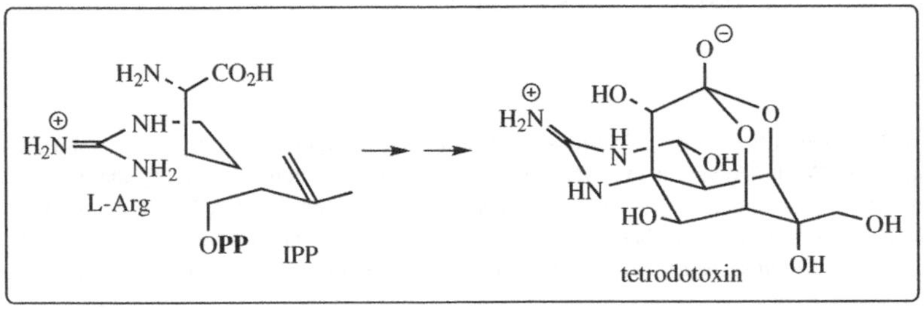

Figure 6.142

Saxitoxin and Tetrodotoxin

The structure of **saxitoxin** (Figure 6.142) contains a reduced purine ring system, but it is not biosynthetically related to the purine alkaloids described above. Not all features of its biosynthetic origin have been established, but the amino acid supplying most of the ring system is known to be L-arginine (Figure 6.142). Acetate and a C_1 unit from methionine are also utilized. Saxitoxin contains two highly polar guanidino functions, one of which is provided by arginine, and is a fast-acting neurotoxin inhibiting nerve conduction by blocking sodium channels [Box 6.26]. It is one of a group of marine toxins referred to as paralytic shellfish poisons, found in

a range of shellfish, but ultimately derived from toxic strains of dinoflagellates consumed by the shellfish. Arginine is also a precursor for **tetrodotoxin** (Figure 6.142), another marine neurotoxin containing a polar guanidino group. Tetrodotoxin exists as a zwitterion involving the guanidino group and a hemilactal function. It has been established that the remainder of the carbon skeleton in tetrodotoxin is a C_5 isoprene unit, probably supplied as isopentenyl diphosphate (Figure 6.142). Tetrodotoxin is well known as the toxic principle in the puffer fish (*Tetraodon* species), regarded as delicacy in Japanese cuisine [Box 6.26]. As potent sodium channel blockers, both saxitoxin and tetrodotoxin are valuable pharmacological tools.

Box 6.26

Saxitoxin

Saxitoxin (Figure 6.142) was first isolated from the Alaskan butter clam (*Saxidomus giganteus*) and has since been found in many species of shellfish, especially bivalves such as mussels, scallops, and oysters. These filter feeders consume dinoflagellates (plankton) and can accumulate toxins synthesized by these organisms, particularly during outbreaks known as red tides, when conditions favour formation of huge blooms of the dinoflagellates (see also brevetoxin A, page 93). Species of the dinoflagellate *Gonyaulax* in marine locations or of the cyanobacterium *Aphanizomenon* in freshwater have been identified among the causative organisms, and the problem is encountered widely in temperate and tropical areas (including Europe, North America, and Japan). Commercial production of shellfish is routinely monitored for toxicity, which will slowly diminish as conditions change and the causative organism disappears from the water. About a dozen natural saxitoxin-related structures have been characterized,

Box 6.26   (continued)

and mixtures in various proportions are typically synthesized by a producer, with the possibility that the shellfish may also structurally modify the toxins further. Acute and often fatal poisonings caused by the consumption of contaminated shellfish are termed paralytic shellfish poisoning, which involves paralysis of the neuromuscular system, with death resulting from respiratory failure. Saxitoxin is a cationic molecule which binds to sodium channels to block the influx of sodium ions through excitable nerve membranes, and it is a valuable pharmacological tool for the study of this process. Saxitoxin and tetrodotoxin (below) are some of the most potent non-protein neurotoxins known, and are active at very low concentrations ($\mu g\ kg^{-1}$).

Tetrodotoxin

Tetrodotoxin (Figure 6.142) is traditionally associated with the puffer fish, known in Japan as fugu, a highly prized but risky culinary delicacy; tetrodotoxin is named after the puffer family Tetraodontidae. Preparation of fugu is a skilled operation in which organs containing the highest levels of toxin, e.g. liver, ovaries, testes, are carefully separated from the flesh. Even so, deaths from fugu poisoning are not uncommon, and the element of risk presumably heightens culinary appreciation of the fish. As with saxitoxin, tetrodotoxin appears to be produced by microorganisms, and symbiotic marine bacteria, e.g. *Vibrio* species, have been implicated as the synthesizers. In addition to fugu, several other species of fish, octopus, newts, and frogs have been found to accumulate tetrodotoxin or related structures. The mode of action of tetrodotoxin is exactly the same as that of saxitoxin above, though there are some subtle differences in the mechanism of binding. **Tetrodotoxin** is currently being investigated for use in treatment of cancer pain and management of opiate withdrawal symptoms. It is also being tested as a local and topical anaesthetic in procedures where general anaesthesia is not appropriate.

FURTHER READING

Biosynthesis, General

Facchini PJ (2001) Alkaloid biosynthesis in plants: biochemistry, cell biology, molecular regulation, and metabolic engineering applications. *Annu Rev Plant Physiol Plant Mol Biol* **52**, 29–66.

Facchini PJ and St-Pierre B (2005) Synthesis and trafficking of alkaloid biosynthetic enzymes. *Curr Opin Plant Biol* **8**, 657–666.

Goossens A and Rischer H (2007) Implementation of functional genomics for gene discovery in alkaloid producing plants. *Phytochem Rev* **6**, 35–49.

Hashimoto T and Yamada Y (2003) New genes in alkaloid metabolism and transport. *Curr Opin Biotechnol* **14**, 163–168.

Herbert RB (2003) The biosynthesis of plant alkaloids and nitrogenous microbial metabolites. *Nat Prod Rep* **20**, 494–508.

Zenk MH and Juenger M (2007) Evolution and current status of the phytochemistry of nitrogenous compounds. *Phytochemistry* **68**, 2757–2772.

Polyamines

Casero RA and Woster PM (2001) Terminally alkylated polyamine analogues as chemotherapeutic agents. *J Med Chem* **44**, 1–26.

Wallace HM, Fraser AV and Hughes A (2003) A perspective of polyamine metabolism. *Biochem J* **376**, 1–14.

Walters DR (2003) Polyamines and plant disease. *Phytochemistry* **64**, 97–107.

Tropane Alkaloids

Bekkouchea K, Daali Y, Cherkaoui S, Veuthey J-L and Christen P (2001) Calystegine distribution in some solanaceous species. *Phytochemistry* **58**, 455–462.

Dräger B (2004) Chemistry and biology of calystegines. *Nat Prod Rep* **21**, 211–223.

Dräger B (2006) Tropinone reductases, enzymes at the branch point of tropane alkaloid metabolism. *Phytochemistry* **67**, 327–337.

Hanuš LO, Řezanka T, Spížek J and Dembitsky VM (2005) Substances isolated from *Mandragora* species. *Phytochemistry* **66**, 2408–2417.

Humphrey AJ and O'Hagan D (2001) Tropane alkaloid biosynthesis. A century old problem unresolved. *Nat Prod Rep* **18**, 494–502.

Schimming T, Jenett-Siems K, Mann P, Tofern-Reblin B, Milson J, Johnson RW, Deroin T, Austin DF and Eich E (2005) Calystegines as chemotaxonomic markers in the Convolvulaceae. *Phytochemistry* **66**, 469–480.

Cocaine

Bieri S, Brachet A, Veuthey J-L and Christen P (2006) Cocaine distribution in wild *Erythroxylum* species. *J Ethnopharmacol* **103**, 439–447.

Carrera MRA, Meijler MM and Janda KD (2004) Cocaine pharmacology and current pharmacotherapies for its abuse. *Bioorg Med Chem* **12**, 5019–5030.

Zheng F and Zhan C-G (2008) Structure-and-mechanism-based design and discovery of therapeutics for cocaine overdose and addiction. *Org Biomol Chem* **6**, 836–843.

Pyrrolizidine Alkaloids

Hartmann T (2004) Plant-derived secondary metabolites as defensive chemicals in herbivorous insects: a case study in chemical ecology. *Planta* **219**, 1–4.

Piperidine Alkaloids

Felpin F-X and Lebreton J (2004) History, chemistry and biology of alkaloids from *Lobelia inflata*. *Tetrahedron* **60**, 10127–10153.

Scott IM, Jensen HR, Philogène BJR and Arnason JT (2008) A review of *Piper* spp. (Piperaceae) phytochemistry, insecticidal activity and mode of action. *Phytochem Rev* **7**, 65–75.

Quinolizidine and Indolizidine Alkaloids

Stead D and O'Brien P (2007) Total synthesis of the lupin alkaloid cytisine: comparison of synthetic strategies and routes. *Tetrahedron* **63**, 1885–1897.

Watson AA, Fleet GWJ, Asano N, Molyneux RJ and Nash RJ (2001) Polyhydroxylated alkaloids – natural occurrence and therapeutic applications. *Phytochemistry* **56**, 265–295.

Tobacco

Häkkinen ST, Tilleman S, Światek A, De Sutter V, Rischer H, Vanhoutte I, Van Onckelen H, Hilson P, Inzé D, Oksman-Caldentey K-M and Goossens A (2007) Functional characterisation of genes involved in pyridine alkaloid biosynthesis in tobacco. *Phytochemistry* **68**, 2773–2785.

Hecht SS (2006) Smoking and lung cancer – a new role for an old toxicant? *Proc Natl Acad Sci USA* **103**, 15725–15726.

Nugroho LH and Verpoorte R (2002) Secondary metabolism in tobacco. *Plant Cell Tiss Org Cult* **68**, 105–125.

Wagner FF and Comins DL (2007) Recent advances in the synthesis of nicotine and its derivatives. *Tetrahedron* **63**, 8065–8082.

Catecholamines

Szopa J and Kulma A (2007) Catecholamines are active compounds in plants. *Plant Sci* **172**, 433–440.

Isoquinoline Alkaloids

Grycová L, Dostál J and Marek R (2007) Quaternary protoberberine alkaloids. *Phytochemistry* **68**, 150–175.

Liscombe DK, MacLeod BP, Loukanina N, Nandi OI and Facchini PJ (2005) Evidence for the monophyletic evolution of benzylisoquinoline alkaloid biosynthesis in angiosperms. *Phytochemistry* **66**, 1374–1393. Errata: p. 2500.

Minami H, Kim J-S, Ikezawa N, Takemura T, Katayama T, Kumagai H and Sato F (2008) Microbial production of plant benzylisoquinoline alkaloids. *Proc Natl Acad Sci USA* **105**, 7393–7398.

Zhang A, Zhang Y, Branfman AR, Baldessarini RJ and Neumeyer JL (2007) Advances in development of dopaminergic aporphinoids. *J Med Chem* **50**, 171–181.

Opium Alkaloids

Blakemore PR and White JD (2002) Morphine, the Proteus of organic molecules. *Chem Commun* 1159–1168.

Carroll FI (2003) Monoamine transporters and opioid receptors. Targets for addiction therapy. *J Med Chem* **46**, 1775–1794.

Coop A and MacKerell AD (2002) The future of opioid analgesics. *Am J Pharm Educ* **66**, 153–156.

Facchini PJ, Hagel JM, Liscombe DK, Loukanina N, MacLeod BP, Samanani N and Zulak KG (2007) Opium poppy: blueprint for an alkaloid factory. *Phytochem Rev* **6**, 97–124.

Schiff PL (2002) Opium and its alkaloids. *Am J Pharm Educ* **66**, 186–194.

Wolfson W (2005) Janus-faced drugs: the double-edged synthetic opiate trade. *Chem Biol* **12**, 1055–1056.

Colchicine

Schmalz H-G and Graening T (2004) Total syntheses of colchicine in comparison: a journey through 50 years of synthetic organic chemistry. *Angew Chem Int Ed* **43**, 3230–3256.

Amaryllidaceae Alkaloids

Heinrich M and Teoh HL (2004) Galanthamine from snowdrop – the development of a modern drug against Alzheimer's disease from local Caucasian knowledge. *J Ethnopharmacol* **92**, 147–162.

Houghton PJ, Ren Y and Howes MJ (2006) Acetylcholinesterase inhibitors from plants and fungi. *Nat Prod Rep* **23**, 181–199.

Kornienko A and Evidente A (2008) Chemistry, biology, and medicinal potential of narciclasine and its congeners. *Chem Rev* **108**, 1982–2014.

Marco-Contelles J, do Carmo Carreiras M, Rodríguez C, Villarroya M and García AG (2006) Synthesis and pharmacology of galantamine. *Chem Rev* **106**, 116–133.

Unver N (2007) New skeletons and new concepts in Amaryllidaceae alkaloids. *Phytochem Rev* **6**, 125–135.

Serotonin

Glennon RA (2003) Higher-end serotonin receptors: 5-HT$_5$, 5-HT$_6$, and 5-HT$_7$. *J Med Chem* **46**, 2795–2812.

Langlois M and Fischmeister R (2003) 5-HT$_4$ receptor ligands: applications and new prospects. *J Med Chem* **46**, 319–344.

Nichols DE and Nichols CD (2008) Serotonin receptors. *Chem Rev* **108**, 1614–1641.

Zheng W and Cole PA (2002) Serotonin *N*-acetyltransferase: mechanism and inhibition. *Curr Med Chem* **9**, 1187–1199.

β-Carboline Alkaloids

Cao R, Peng W, Wang Z and Xu A (2007) β-Carboline alkaloids: biochemical and pharmacological functions. *Curr Med Chem* **14**, 479–500.

Terpenoid Indole Alkaloids

O'Connor SE and Maresh JJ (2006) Chemistry and biology of monoterpene indole alkaloid biosynthesis. *Nat Prod Rep* **23**, 532–547.

Stöckigt J and Panjikar S (2007) Structural biology in plant natural product biosynthesis – architecture of enzymes from monoterpenoid indole and tropane alkaloid biosynthesis. *Nat Prod Rep* **24**, 1382–1400.

Rauwolfia

Chen F-E and Huang J (2005) Reserpine: a challenge for total synthesis of natural products. *Chem Rev* **105**, 4671–4706.

Stöckigt J, Panjikar S, Ruppert M, Barleben L, Ma X, Loris E and Hill M (2007) The molecular architecture of major enzymes from ajmaline biosynthetic pathway. *Phytochem Rev* **6**, 15–34.

Yohimbine

Zanolari B, Ndjoko K, Ioset J-R, Marston A and Hostettmann K (2003) Qualitative and quantitative determination of yohimbine in authentic yohimbe bark and in commercial aphrodisiacs by HPLC-UV-API/MS methods. *Phytochem Anal* **14**, 193–201.

Catharanthus Alkaloids

El-Sayed M and Verpoorte R (2007) *Catharanthus* terpenoid indole alkaloids: biosynthesis and regulation. *Phytochem Rev* **6**, 277–305.

Hisiger S and Jolicoeur M (2007) Analysis of *Catharanthus roseus* alkaloids by HPLC. *Phytochem Rev* **6**, 207–234.

Loyola-Vargas VM, Galaz-Ávalos RM and Kú-Cauich R (2007) *Catharanthus* biosynthetic enzymes: the road ahead. *Phytochem Rev* **6**, 307–339.

Sottomayor M, Lopes Cardoso I, Pereira LG and Ros Barceló A (2004) Peroxidase and the biosynthesis of terpenoid indole alkaloids in the medicinal plant *Catharanthus roseus*. *Phytochem Rev* **3**, 159–171.

Van der Heijden R, Jacobs DI, Snoeijer W, Hallard D and Verpoorte R (2004) The *Catharanthus* alkaloids: pharmacognosy and biotechnology. *Curr Med Chem* **11**, 607–628.

Verpoorte R, Contin A and Memelink J (2002) Biotechnology for the production of plant secondary metabolites. *Phytochem Rev* **1**, 13–25.

Zárate R and Verpoorte R (2007) Strategies for the genetic modification of the medicinal plant *Catharanthus roseus*. *Phytochem Rev* **6**, 475–491.

Zhao J and Verpoorte R (2007) Manipulating indole alkaloid production by *Catharanthus roseus* cell cultures in bioreactors: from biochemical processing to metabolic engineering. *Phytochem Rev* **6**, 435–457.

Ibogaine

Alper KR, Lotsof HS and Kaplan CD (2008) The ibogaine medical subculture. *J Ethnopharmacol* **115**, 9–24.

Strychnine

Bonjoch J and Solé D (2000) Synthesis of strychnine. *Chem Rev* **100**, 3455–3482.

Philippe G and Angenot L (2005) Recent developments in the field of arrow and dart poisons. *J Ethnopharmacol* **100**, 85–91.

Cinchona

Egan T (2001) Structure–function relationships in chloroquine and related 4-aminoquinoline antimalarials. *Mini Rev Med Chem* **1**, 113–123.

Kaufman TS and Rúveda EA (2005) The quest for quinine: those who won the battles and those who won the war. *Angew Chem Int Ed* **44**, 854–885.

Marcelli T, van Maarseveen JH and Hiemstra H (2006) Cupreines and cupreidines: an emerging class of bifunctional cinchona organocatalysts. *Angew Chem Int Ed* **45**, 7496–7504.

Wiesner J, Ortmann R, Jomaa H and Schlitzer M (2003) New antimalarial drugs. *Angew Chem Int Ed* **42**, 5274–5293.

Camptothecin

Beretta GL, Perego P and Zunino F (2006) Mechanisms of cellular resistance to camptothecins. *Curr Med Chem* **13**, 3291–3305.

Du W (2003) Towards new anticancer drugs: a decade of advances in synthesis of camptothecins and related alkaloids. *Tetrahedron* **59**, 8649–8687.

Li Q-Y, Zu Y-G, Shi R-Z and Yao L-P (2006) Review camptothecin: current perspectives. *Curr Med Chem* **13**, 2021–2039.

Lorence A and Nessler CL (2004) Molecules of interest. Camptothecin, over four decades of surprising findings. *Phytochemistry* **65**, 2735–2749.

Thomas CJ, Rahier NJ and Hecht SM (2004) Camptothecin: current perspectives. *Bioorg Med Chem* **12**, 1585–1604.

Physostigmine

Houghton PJ, Ren Y and Howes MJ (2006) Acetylcholinesterase inhibitors from plants and fungi. *Nat Prod Rep* **23**, 181–199.

Steven A and Overman LE (2007) Total synthesis of complex cyclotryptamine alkaloids: stereocontrolled construction of quaternary carbon stereocenters. *Angew Chem Int Ed* **46**, 5488–5508.

Ergot

Ahimsa-Müller MA, Markert A, Hellwig S, Knoop W, Steiner U, Drewke C and Leistner E (2007) Clavicipitaceous fungi associated with ergoline alkaloid-containing Convolvulaceae. *J Nat Prod* **70**, 1955–1960.

Komarova EL and Tolkachev ON (2001) The chemistry of peptide ergot alkaloids. Part 1. Classification and chemistry of ergot peptides. *Pharm Chem J* **35**, 504–513.

Schiff PL (2006) Ergot and its alkaloids. *Am J Pharm Educ* **70**, 1–10.

Tudzynski P, Correia T and Keller TU (2001) Biotechnology and genetics of ergot alkaloids *Appl Microbiol Biotechnol* **57**, 593–605.

Quinazoline Alkaloids

Claeson UP, Malmfors T, Wikman G and Bruhn JG (2000) *Adhatoda vasica*: a critical review of ethnopharmacological and toxicological data. *J Ethnopharmacol* **72**, 1–20.

Mhaske, SB and Argade NP (2006) The chemistry of recently isolated naturally occurring quinazolinone alkaloids. *Tetrahedron* **62**, 9787–9826.

Quinoline and Acridine Alkaloids

Tillequin F (2007) Rutaceous alkaloids as models for the design of novel antitumor drugs. *Phytochem Rev* **6**, 65–79.

Ephedra, Khat

Abourashed EA, El-Alfy AT, Khan IA and Walker L (2003) Ephedra in perspective – a current review. *Phytother Res* **17**, 703–712.

Al-Motarreb A, Baker K and Broadley KJ (2002) Khat: pharmacological and medical aspects and its social use in Yemen. *Phytother Res* **16**, 403–413.

Capsaicin

Appendino G, Minassi A and Daddario N (2005) Hot cuisine as a source of anti-inflammatory drugs. *Phytochem Rev* **4**, 3–10.

Díaz J, Pomar F, Bernal A and Merino F (2004) Peroxidases and the metabolism of capsaicin in *Capsicum annuum*. *Phytochem Rev* **3**, 141–157.

Szallasi A and Appendino G (2004) Vanilloid receptor TRPV1 antagonists as the next generation of painkillers. Are we putting the cart before the horse? *J Med Chem* **47**, 2717–2723.

Terpenoid and Steroidal Alkaloids

Friedman M (2002) Tomato glycoalkaloids: role in the plant and in the diet. *J Agric Food Chem* **50**, 5751–5780.

Friedman M (2006) Potato glycoalkaloids and metabolites: roles in the plant and in the diet. *J Agric Food Chem* **54**, 8655–8681.

Kalinowska M, Zimowski J, Płczkowski C and Wojciechowski ZA (2005) The formation of sugar chains in triterpenoid saponins and glycoalkaloids. *Phytochem Rev* **4**, 237–257.

Korpan YI, Nazarenko IV, Skryshevskaya IV, Martelet C, Jaffrezic-Renault N and El'skaya AV (2004) Potato glycoalkaloids: true safety or false sense of security? *Trends Biotechnol* **22**, 147–151.

Li H-J, Jiang Y and Li P (2006) Chemistry, bioactivity and geographical diversity of steroidal alkaloids from the Liliaceae family. *Nat Prod Rep* **23**, 735–752.

Reina M and González-Coloma A (2007) Structural diversity and defensive properties of diterpenoid alkaloids. *Phytochem Rev* **6**, 81–95.

Reynolds T (2005) Hemlock alkaloids from Socrates to poison aloes. *Phytochemistry* **66**, 1399–1406.

Purine Alkaloids

Ashihara H, Sano H and Crozier A (2008) Caffeine and related purine alkaloids: biosynthesis, catabolism, function and genetic engineering. *Phytochemistry* **69**, 841–856.

Zrenner R, Stitt M and Sonnewald U (2006) Pyrimidine and purine biosynthesis and degradation in plants. *Annu Rev Plant Biol* **57**, 805–836.

Saxitoxin, Tetrodotoxin

Anger T, Madge DT, Mulla M and Riddall D (2001) Medicinal chemistry of neuronal voltage-gated sodium channel blockers. *J Med Chem* **44**, 115–137.

Daly JW (2004) Marine toxins and nonmarine toxins: convergence or symbiotic organisms? *J Nat Prod* **67**, 1211–1215.

Llewellyn LE (2006) Saxitoxin, a toxic marine natural product that targets a multitude of receptors. *Nat Prod Rep* **23**, 200–222.

7

PEPTIDES, PROTEINS, AND OTHER AMINO ACID DERIVATIVES

Although the participation of amino acids in the biosynthesis of some shikimate metabolites and particularly in the pathways leading to alkaloids has already been explored in Chapters 4 and 6, amino acids are also the building blocks for other important classes of natural products. The elaboration of shikimate metabolites and alkaloids utilized only a limited range of amino acid precursors. Peptides, proteins, and the other compounds considered in this chapter are synthesized from a very much wider range of amino acids. Peptides and proteins represent another grey area between primary metabolism and secondary metabolism, in that some materials are widely distributed in nature and found, with subtle variations, in many different organisms, whilst others are of very restricted occurrence.

PEPTIDES AND PROTEINS

Peptides and proteins are both polyamides composed of α-amino acids linked through their carboxyl and α-amino functions (Figure 7.1). In biochemistry, the amide linkage is traditionally referred to as a peptide bond. Whether the resultant polymer is classified as a peptide or a protein is not clearly defined; generally, a chain length of more than 40 residues confers protein status, whilst the term polypeptide can be used to cover all chain lengths. Although superficially similar, peptides and proteins display a wide variety of biological functions and many have marked physiological properties. For example, they function as structural molecules in tissues, as enzymes, as antibodies, as neurotransmitters, and when acting as hormones can control many physiological processes, ranging from gastric acid secretion and carbohydrate metabolism to growth itself. The toxic components of snake and spider venoms are usually peptide in nature, as are some plant toxins. These different activities arise as a consequence of the sequence of amino acids in the peptide or protein (the primary structure), the three-dimensional structure which the molecule then adopts as a result of this sequence (the secondary and tertiary structures), and the specific nature of individual side-chains in the molecule. Many structures have additional modifications to the basic polyamide system shown in Figure 7.1, and these features may also contribute significantly to their biological activity.

The tripeptide formed from L-alanine, L-phenylalanine, and L-serine in Figure 7.2 by two condensation reactions is alanyl–phenylalanyl–serine, commonly represented as Ala–Phe–Ser, using the standard three-letter abbreviations for amino acids as shown in Table 7.1, which gives the structures of the 20 L-amino acids which are encoded by DNA. By convention, the left-hand amino acid in this sequence is the one with a free amino group, the N-terminus, whilst the right-hand amino acid has the free carboxyl, the C-terminus. Sometimes, the termini are emphasized by showing H– and –OH (Figure 7.2); H– represents the amino group and –OH the carboxyl group. In cyclic peptides, this convention can have no significance, so arrows are incorporated into the sequence to indicate peptide bonds in the direction CO→NH. As sequences become longer, one letter abbreviations for amino acids are commonly used instead of the three-letter abbreviations, thus Ala–Phe–Ser becomes AFS. The amino acid components of peptides and proteins predominantly have the L-configuration, but many peptides contain one or more D-amino acids in their structures. Abbreviations thus assume the L-configuration applies, and D-amino acids must be specifically noted, e.g. Ala–D-Phe–Ser. Some amino acids which are are frequently encountered in

Medicinal Natural Products: A Biosynthetic Approach. 3rd Edition Paul Dewick
© 2009 John Wiley & Sons, Ltd

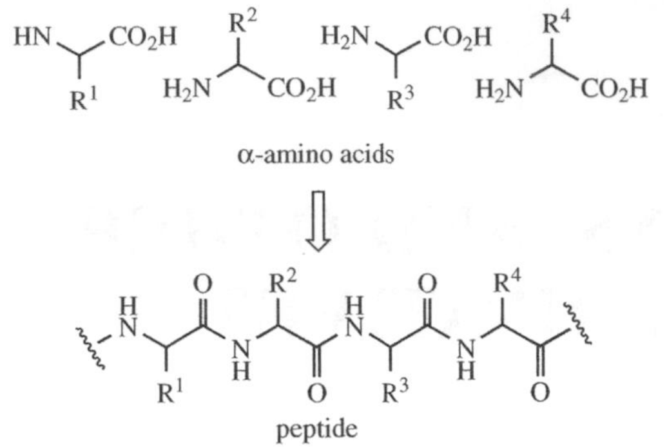

Figure 7.1

peptides, but not encoded by DNA, are also shown in Table 7.1, along with their appropriate abbreviations. Modified amino acids may be represented by a variation of the normal abbreviation, e.g. *N*-methyltyrosine as Tyr(Me). A frequently encountered modification is the conversion of the C-terminal carboxyl into an amide, and this is represented as Phe–NH$_2$ for example; this must not be interpreted as an indication of the N-terminus.

RIBOSOMAL PEPTIDE BIOSYNTHESIS

Protein biosynthesis takes place on the ribosomes, and a simplified representation of the process as characterized in the bacterium *Escherichia coli* is shown in Figure 7.3. The messenger RNA (mRNA) contains a transcription of one of the genes of DNA and carries the information

necessary to direct the biosynthesis of a specific protein. The message is stored as a series of three-base sequences (codons) in its nucleotides, and is read (translated) in the 5′ to 3′ direction along the mRNA molecule. The mRNA is bound to the smaller 30S subunit of the bacterial ribosome. Initially, the amino acid is activated by an ATP-dependent process and it then binds via an ester linkage to an amino acid-specific transfer RNA (tRNA) molecule through a terminal adenosine group, giving an aminoacyl-tRNA (Figure 7.4). This is mechanistically equivalent to the formation of amino acid thioesters for non-ribosomal peptide biosynthesis (see page 440). The aminoacyl-tRNA contains in its nucleotide sequence a combination of three bases (the anticodon) which allows binding via hydrogen bonding to the appropriate codon on mRNA. In prokaryotes, the first amino acid encoded in the sequence is *N*-formylmethionine, and the corresponding aminoacyl-tRNA is thus bound and positioned at the P (for peptidyl) site on the ribosome (Figure 7.3). The next aminoacyl-tRNA (Figure 7.3 shows one specific for alanine) is also bound via a codon–anticodon interaction and is positioned at an adjacent A (for aminoacyl) site on the ribosome. This allows peptide bond formation to occur, the amino group of the amino acid in the A site attacking the activated ester in the P site. The peptide chain is thus initiated and has become attached to the tRNA located in the A site. The tRNA at the P site is no longer required and is released from the ribosome. Then the peptidyl-tRNA at the A site is translocated to the P site by the ribosome moving along the mRNA a codon at a time, exposing the A site for a new aminoacyl-tRNA appropriate to the

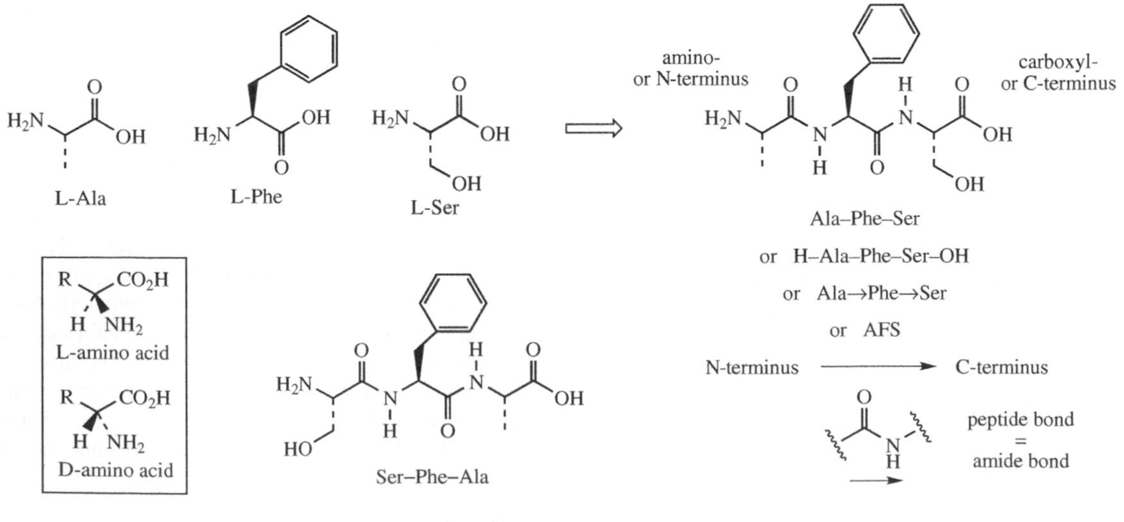

Figure 7.2

Table 7.1 Amino acids: structures and standard abbreviations

Amino acids encoded by DNA

Amino acid		Abbr.		Amino acid		Abbr.	
Alanine		Ala	A	Leucine		Leu	L
Arginine		Arg	R	Lysine		Lys	K
Asparagine		Asn	N	Methionine		Met	M
Aspartic acid		Asp	D	Phenylalanine		Phe	F
Cysteine		Cys	C	Proline		Pro	P
Glutamic acid		Glu	E	Serine		Ser	S
Glutamine		Gln	Q	Threonine		Thr	T
Glycine		Gly	G	Tryptophan		Trp	W
Histidine		His	H	Tyrosine		Tyr	Y
Isoleucine		Ile	I	Valine		Val	V

Some common amino acids not encoded by DNA

Amino acid		Abbr.	Amino acid		Abbr.
Pyroglutamic acid (5-oxoproline)		Glp oxoPro <Glu	Ornithine		Orn
Hydroxyproline		HPro	Sarcosine (N-methylglycine)		Sar

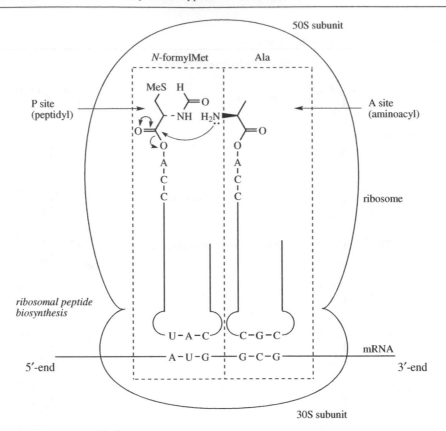

Figure 7.3

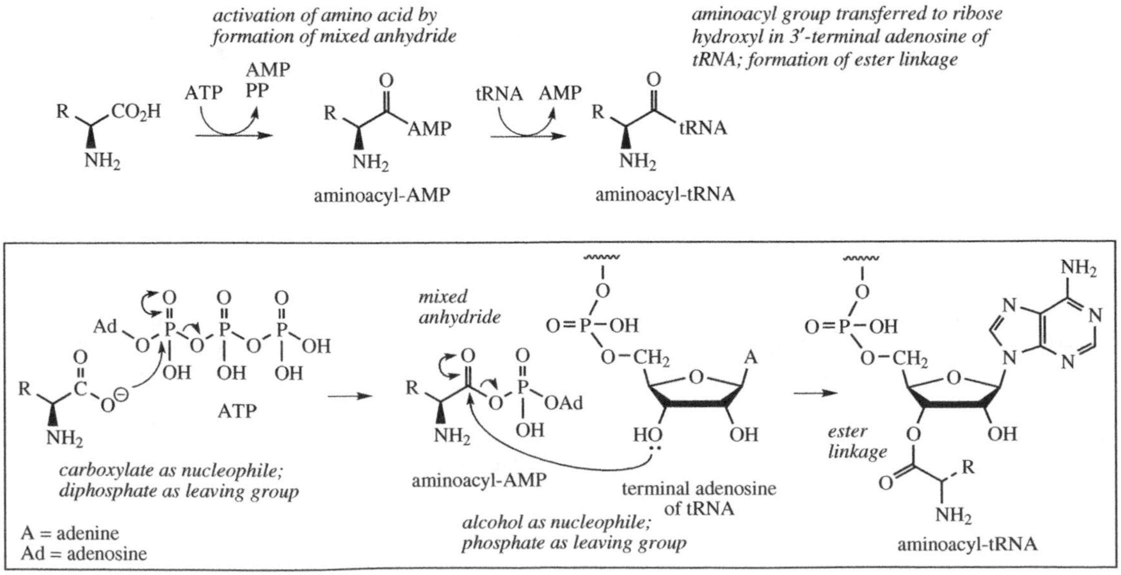

aminoacyl group transferred to ribose
hydroxyl in 3'-terminal adenosine of
tRNA; formation of ester linkage

aminoacyl-AMP

aminoacyl-tRNA

carboxylate as nucleophile;
diphosphate as leaving group

mixed
anhydride

aminoacyl-AMP

alcohol as nucleophile;
phosphate as leaving group

terminal adenosine
of tRNA

ester
linkage

aminoacyl-tRNA

A = adenine
Ad = adenosine

Figure 7.4

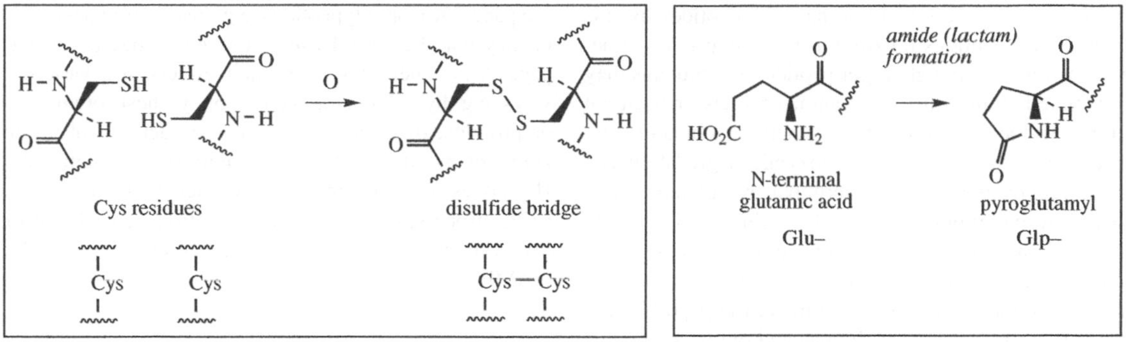

Figure 7.5

particular codon, and a repeat of the elongation process occurs. The cycles of elongation and translocation continue until a termination codon is reached, and the peptide or protein is then hydrolysed and released from the ribosome. The peptide is synthesized from the N-terminus towards the C-terminus. The individual steps of protein biosynthesis all seem susceptible to disruption by specific agents. Many of the antibiotics used clinically are active by their ability to inhibit protein biosynthesis in bacteria. They may interfere with transcription (e.g. rifampicin, see page 89), the binding of the aminoacyl-tRNA to the A site (e.g. tetracyclines, see page 128), the formation of the peptide bond (e.g. chloramphenicol, see page 148), or the translocation step (e.g. erythromycin, see page 77).

Many of the peptides and proteins synthesized on the ribosome then undergo enzymic post-translational modifications. In prokaryotes, the *N*-formylmethionine initiator may be hydrolysed to methionine, and in both prokaryotes and eukaryotes a number of amino acids from the N-terminus may be removed altogether, thus shortening the original chain. This type of chain shortening is a means by which a protein or peptide can be stored in an inactive form and then transformed into an active form when required, e.g. the production of insulin from proinsulin (see page 432). **Glycoproteins** are produced by adding sugar residues via *O*-glycoside linkages to the hydroxyl groups of serine and threonine residues or via *N*-glycoside linkages to the amino of asparagine. **Phosphoproteins** have the hydroxyl groups of serine or threonine phosphorylated. Importantly, post-translational modifications allow those amino acids which are not encoded by DNA

yet are found in peptides and proteins to be formed by the transformation of encoded ones. These include the hydroxylation of proline to hydroxyproline and of lysine to hydroxylysine, *N*-methylation of histidine, and the oxidation of the thiol groups of two cysteine residues to form a **disulfide bridge**, allowing cross-linking of polypeptide chains. This latter process (Figure 7.5) will form loops in a single polypeptide chain, or may join separate chains together as in insulin (see page 433). Many peptides contain a **pyroglutamic acid** residue (Glp) at the N-terminus, a consequence of intramolecular cyclization between the γ-carboxylic acid and the α-amino of an N-terminal glutamic acid (Figure 7.5). The C-terminal carboxylic acid may also frequently be converted into an amide. Both modifications are exemplified in the structure of gastrin (Figure 7.6), a peptide hormone which stimulates secretion of HCl in the stomach. Such terminal modifications comprise a means of protecting a peptide from degradation by exopeptidases, which remove amino acids from the ends of peptides and, thus, prolong its period of action.

With the rapid advances made in genetic engineering, it is now feasible to produce relatively large amounts of ribosome-constructed polypeptides, especially enzymes, by isolating or constructing DNA sequences encoding the particular product, and inserting these into suitable organisms, commonly *Escherichia coli*. Of course, such procedures will not duplicate any post-translational modifications, and these will have to be carried out on the initial polypeptide by chemical means or by use of suitable enzymes if these are available. Not all of the transformations can be carried out both efficiently and selectively,

Glp–Gly–Pro–Trp–Leu–Glu–Glu–Glu–Glu–Glu–Ala–Tyr–Gly–Trp–Gly–Trp–Met–Asp–**Phe–NH$_2$**

human gastrin

Figure 7.6

thus restricting access to important polypeptides by this means. Small peptides for drug use are generally synthesized chemically, though larger peptides and proteins may be extracted from human and animal tissues or bacterial cultures. In the design of semi-synthetic or synthetic analogues for potential drug use, enzymic degradation can be reduced by the use of N-terminal pyroglutamic acid and C-terminal amide residues, the inclusion of D-amino acids, and the removal of specific residues. These ploys to change recognition by specific degradative peptidases can increase the activity and lifetime of the peptide. Very few peptide drugs may be administered orally, since they are rapidly inactivated or degraded by gastrointestinal enzymes and they must, therefore, be given by injection.

PEPTIDE HORMONES

Hormones are mammalian metabolites released into the blood stream to elicit specific responses on a target tissue or organ. They are chemical messengers, and may be simple amino acid derivatives, e.g. adrenaline (epinephrine;

see page 336), or polypeptides, e.g. insulin (see page 433), or they may be steroidal in nature, e.g. progesterone (see page 287). They may exert their effects in a variety of ways, e.g. by influencing the rate of synthesis of enzymes or proteins, by affecting the catalytic activity of an enzyme, or by altering the permeability of cell membranes. Hormones are not enzymes; they act by regulating existing processes. Frequently, this action depends on the involvement of a second messenger, such as cyclic AMP (cAMP).

Thyroid Hormones

Thyroid hormones are necessary for the development and function of cells throughout the body. The thyroid hormones **thyroxine** and **tri-iodothyronine** (Figure 7.7) are not peptides, but are actually simple derivatives of tyrosine [Box 7.1]. However, they are believed to be derived by degradation of a larger protein molecule. One hypothesis for their formation invokes tyrosine residues in the protein thyroglobulin which are iodinated to di-iodotyrosine

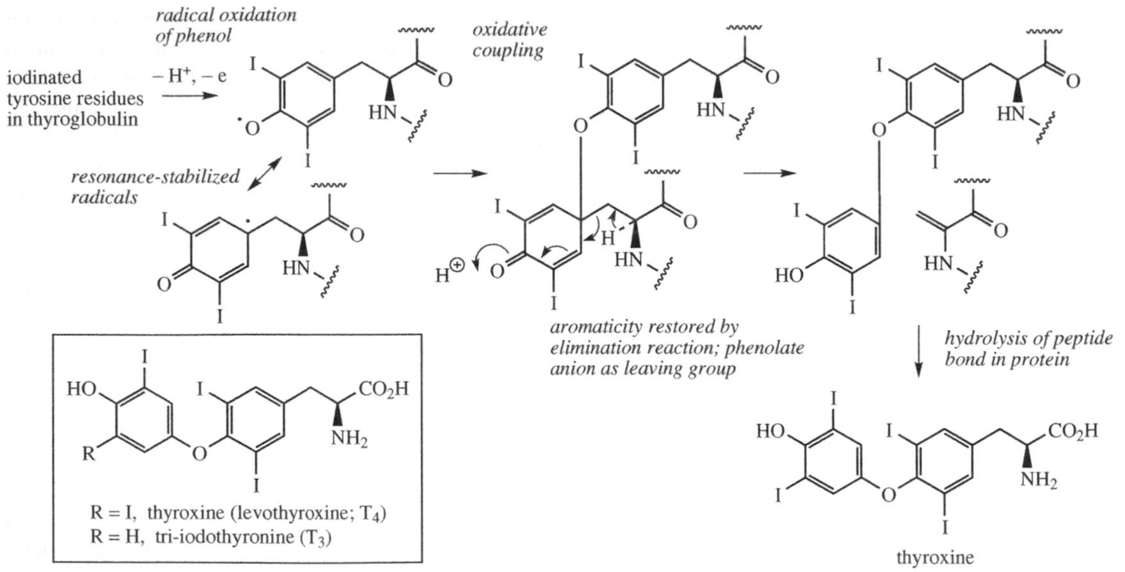

Figure 7.7

Cys–**Gly**–Asn–Leu–Ser–Thr–Cys–**Met**–Leu–Gly–
Thr–**Tyr**–**Thr**–Gln–**Asp**–**Phe**–**Asn**–Lys–**Phe**–**His**–
Thr–**Phe**–Pro–**Gln**–Thr–**Ala**–**Ile**–Gly–**Tyr**–Gly–
Ala–Pro–NH$_2$

human calcitonin

Cys–**Ser**–Asn–Leu–Ser–Thr–Cys–**Val**–Leu–Gly–
Lys–**Leu**–**Ser**–Gln–**Glu**–**Leu**–**His**–Lys–**Leu**–**Gln**–
Thr–**Tyr**–Pro–**Arg**–Thr–**Asn**–**Thr**–Gly–**Ser**–Gly–
Thr–Pro–NH$_2$

calcitonin (salmon), salcatonin

Figure 7.8

with suitably placed residues reacting together by a phenolic oxidative coupling process catalysed by thyroid peroxidase (Figure 7.7). Re-aromatization is achieved by an elimination reaction in the side-chain of one residue, and thyroxine (or triiodothyronine) is released from the protein by proteolytic cleavage. Alternative mechanisms have been proposed, however. **Calcitonin** (Figure 7.8) is also produced in the thyroid gland and is involved in the regulation of bone turnover and maintenance of calcium balance [Box 7.1]. This is a relatively simple peptide structure containing a disulfide bridge and a C-terminal amide function.

Box 7.1

Thyroxine

The thyroid hormones thyroxine (T_4) and tri-iodothyronine (T_3) (Figure 7.7) are derivatives of tyrosine and are necessary for development and function of cells throughout the body. They increase protein synthesis in almost all types of body tissue and increase oxygen consumption dependent upon Na^+/K^+ ATPase (the Na pump). Excess thyroxine causes hyperthyroidism, with increased heart rate, blood pressure, overactivity, muscular weakness, and loss of weight. Too little thyroxine may lead to cretinism in children, with poor growth and mental deficiency, or myxoedema in adults, resulting in a slowing down of all body processes. The main hormone thyroxine is converted into its active form tri-iodothyronine by enzymic mono-deiodination in tissues outside the thyroid. Tri-iodothyroxine is actually three- to five-fold more active than thyroxine. **Levothyroxine (thyroxine)** and **liothyronine (tri-iodothyronine)** are both used to supplement thyroid hormone levels and may be administered orally. Levothyroxine is the treatment of choice for maintenance therapy, whilst liothyronine is used when a rapid response is required. These materials are readily produced by chemical synthesis.

Calcitonin

Also produced in the thyroid gland, calcitonin is involved along with parathyroid hormone and $1\alpha,25$-dihydroxyvitamin D_3 (see page 258) in the regulation of bone turnover and maintenance of calcium balance. Under the influence of high blood calcium levels, calcitonin is released to lower these levels by inhibiting uptake of calcium from the gastrointestinal tract, and by promoting its storage in bone. It also suppresses loss of calcium from bone when levels are low. **Calcitonin** is used to treat weakening of bone tissue and hypercalcaemia. Human calcitonin (Figure 7.8) is a 32-residue peptide with a disulfide bridge, and synthetic material may be used. However, synthetic or recombinant salmon calcitonin – **calcitonin (salmon)**; **salcatonin** (Figure 7.8) – is found to have greater potency and longer duration than human calcitonin. Calcitonins from different sources have quite significant differences in their amino acid sequences; the salmon peptide shows 16 changes from the human peptide.

Hypothalamic Hormones

Hypothalamic hormones (Figure 7.9) can modulate a wide variety of actions throughout the body, via the regulation of anterior pituitary hormone secretion [Box 7.2]. **Thyrotrophin-releasing hormone (TRH)** is a tripeptide with an N-terminal pyroglutamyl residue and a C-terminal prolineamide, whilst **luteinizing hormone-releasing hormone (LH-RH)** is a straight-chain decapeptide which also has both N- and C-termini blocked, through pyroglutamyl and glycineamide respectively. **Somatostatin (growth hormone-release inhibiting factor, GHRIH)** merely displays a disulfide bridge in its 14-amino-acid chain. One of the largest of the hypothalamic hormones is **corticotropin-releasing hormone (CRH)**, which contains 41 amino acids, with only the C-terminal blocked as an amide. This hormone controls release from the anterior pituitary of **corticotropin (ACTH**; see page 429), which in turn is responsible for production of corticosteroids.

Box 7.2

Thyrotrophin-releasing Hormone

Thyrotrophin-releasing hormone (TRH; Figure 7.9) acts directly on the anterior pituitary to stimulate release of thyroid-stimulating hormone (TSH), prolactin, and growth hormone (GH). Many other hormones may be released by direct or indirect effects. Synthetic material (known as **protirelin**) is used to assess thyroid function and TSH reserves.

Box 7.2 (continued)

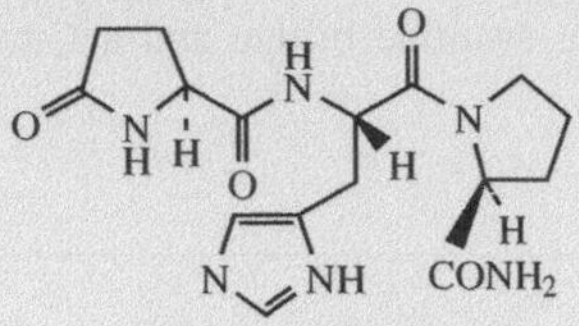

Glp–His–Pro–NH$_2$

thyrotrophin-releasing hormone (TRH)
(protirelin)

Tyr–Ala–Asp–Ala–Ile–Phe–Thr–Asn–Ser–Tyr–
Arg–Lys–Val–Leu–Gly–Gln–Leu–Ser–Ala–Arg–
Lys–Leu–Leu–Gln–Asp–Ile–Met–Ser–Arg–NH$_2$

sermorelin

6 10
Glp–His–Trp–Ser–Tyr–**Gly**–Leu–Arg–Pro–**Gly**–NH$_2$

luteinizing hormone-releasing factor (LH-RH)
(gonadorelin)

Glp–His–Trp–Ser–Tyr–**D-Trp**–Leu–Arg–Pro–Gly–NH$_2$

triptorelin

Glp–His–Trp–Ser–Tyr–**D-Ser(^{t}Bu)**–Leu–Arg–Pro–NHEt

buserelin

Glp–His–Trp–Ser–Tyr–**D-Ser(^{t}Bu)**–Leu–Arg–Pro–NH–NH–CONH$_2$

goserelin

Glp–His–Trp–Ser–Tyr–**D-Leu**–Leu–Arg–Pro–NHEt

leuprorelin

Glp–His–Trp–Ser–Tyr–**3-(2-naphthyl)-D-Ala**–Leu–Arg–Pro–Gly–NH$_2$

nafarelin

Ala–Gly–Cys–Lys–Asn–Phe–**Phe**–**Trp**
| ↓
Cys–Ser–Thr–Phe–**Thr**–**Lys**

growth hormone-release inhibiting factor (GHRIH)
(somatostatin)

D-Phe–Cys–**Phe**–**D-Trp**
| ↓
Thr(ol)–Cys–**Thr**–**Lys**

octreotide

3-(2-naphthyl)-D-Ala– Cys–**Tyr**–**D-Trp**
| ↓
H$_2$N–Thr–Cys–**Val**–**Lys**

lanreotide

Ser–Gln–Glu–Pro–Pro–Ile–Ser–Leu–Asp–Leu–
Thr–Phe–His–Leu–Leu–Arg–Glu–Val–Leu–Glu–
Met–Thr–Lys–Ala–Asp–Gln–Leu–Ala–Gln–Gln–
Ala–His–Ser–Asn–Arg–Lys–Leu–Leu–Asp–Ile–
Ala–NH$_2$

corticotropin-releasing hormone (CRH)

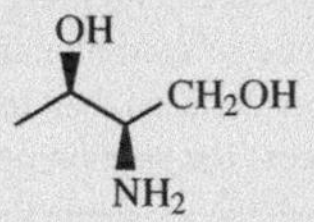

Thr(ol)

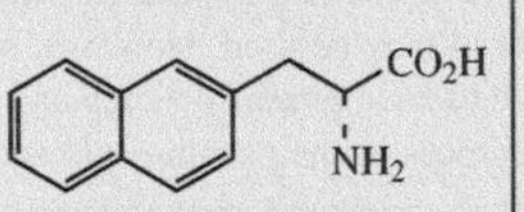

3-(2-naphthyl)-D-Ala

Figure 7.9

Luteinizing Hormone-releasing Hormone

Luteinizing hormone-releasing hormone (LH-RH) (also gonadotrophin-releasing hormone, GnRH; Figure 7.9) is the mediator of gonadotrophin secretion from the anterior pituitary, stimulating both luteinizing hormone (LH) and follicle-stimulating hormone (FSH) release (see page 430). These are both involved in controlling male and female reproduction, inducing the production of oestrogens and progestogens in the female and of androgens in the male. LH is essential for causing ovulation and for the development and maintenance of the corpus luteum in the ovary, whilst FSH is required for maturation of both ovarian follicles in women and of the testes in men. **Gonadorelin** is synthetic LH-RH and is used for the assessment of pituitary function, and also for the treatment of infertility, particularly in women. Analogues of LH-RH, e.g. **triptorelin, buserelin, goserelin, leuprorelin,** and **nafarelin** (Figure 7.9) have been developed and find use to inhibit ovarian steroid secretion and to deprive cancers such as prostate and breast cancers of essential steroid hormones. All of these analogues include a D-amino acid residue at position 6, and modifications to the terminal residue 10, including omission of this residue in some cases. Such changes increase activity and half-life compared with gonadorelin, e.g. leuprorelin is 50 times more potent and half-life is increased from 4 min to 3–4 h.

Box 7.2 (continued)

Growth Hormone-releasing Hormone/Factor

Growth hormone-releasing hormone/factor (GHRH/GHRF) contains 40–44 amino acid residues and stimulates secretion of growth hormone (GH; see page 430) from the anterior pituitary. Synthetic material containing the first 29 amino acid sequence has the full activity and potency of the natural material. This peptide (**sermorelin**; Figure 7.9) is used as a diagnostic aid to test the secretion of GH.

Somatostatin

Somatostatin (growth hormone-release inhibiting factor; GHRIH; Figure 7.9) is a 14 amino acid peptide containing a disulfide bridge, but is derived from a larger precursor protein. It is found in the pancreas and gastrointestinal tract, as well as in the hypothalamus. It inhibits the release of growth hormone (see page 430) and thyrotrophin from the anterior pituitary, and also secretions of hormones from other endocrine glands, e.g. insulin and glucagon. Receptors for somatostatin are also found in most carcinoid tumours. Somatostatin has a relatively short duration of action (half-life 2–3 min), whereas the synthetic analogue **octreotide** (Figure 7.9) is much longer-acting (half-life 60–90 min) and is currently in drug use for the treatment of various endocrine and malignant disorders, especially treatment of neuroendocrine tumours and the pituitary-related growth condition acromegaly. Octreotide contains only eight residues, but retains a crucial Phe–Trp–Lys–Thr sequence, even though the Trp residue now has the D-configuration. This prevents proteolysis between between Trp and Lys, a major mechanism in somatostatin degradation. The C-terminus is no longer an amino acid, but an amino alcohol related to threonine. Radiolabelled octreotide has considerable potential for the visualization of neuroendocrine tumours. **Lanreotide** (Figure 7.9) is a recently introduced somatostatin analogue used in a similar way to octreotide. Lanreotide retains only a few of the molecular characteristics of octreotide, with the Phe–D-Trp–Lys–Thr sequence now modified to Tyr–D-Trp–Lys–Val and the N-terminal amino acid is the synthetic analogue 3-(2-napthyl)-D-alanine, also seen in the LH-RH analogue nafarelin.

Anterior Pituitary Hormones

The main hormones of the anterior pituitary are **corticotropin** and **growth hormone (GH),** which each consist of a single, long polypeptide chain, and the **gonadotrophins**, which are glycoproteins containing two polypeptide chains. These hormones regulate release of glucocorticoids, human growth, and sexual development respectively [Box 7.3]. A further hormone **prolactin**, which controls milk production in females, is structurally related to GH.

Box 7.3

Corticotropin

Corticotropin (corticotrophin; adrenocorticotrophin; ACTH) is a straight-chain polypeptide with 39 amino acid residues; its function is to control the activity of the adrenal cortex, particularly the production of corticosteroids. Secretion of the hormone is controlled by corticotropin-releasing hormone (CRH) from the hypothalamus. ACTH was formerly used as an alternative to corticosteroid therapy in rheumatoid arthritis, but its value was limited by variable therapeutic response. ACTH may be used to test adrenocortical function. It has mainly been replaced for this purpose by the synthetic analogue **tetracosactide (tetracosactrin**; Figure 7.10), which contains the first 24 amino acid residues of ACTH and is preferred because of its shorter duration of action and lower allergenicity.

Ser–Tyr–Ser–Met–Glu–His–Phe–Arg–Trp–Gly–Lys–Pro–Val–Gly–Lys–Lys–Arg–Arg–Pro–Val–Lys–Val–Tyr–Pro

tetracosactide (tetracosactrin)

Figure 7.10

Box 7.3   (continued)

Growth Hormone

Growth hormone (GH, human growth hormone, HGH, or **somatotrophin)** is necessary for normal growth characteristics, especially the lengthening of bones during development. A lack of HGH in children results in dwarfism, whilst continued release can lead to gigantism, or acromegaly, in which only the bones of the hands, feet, and face continue to grow. GHs from animal sources are very species-specific, so it has not been possible to use animal hormones for drug use. HGH contains 191 amino acids with two disulfide bridges, one of which creates a large loop (bridging residues 53 and 165) and the other a very small loop (bridging residues 182 and 189) near the C-terminus. It is synthesized via a prohormone containing 26 extra amino acids. Production of material with the natural amino acid sequence has become possible as a result of recombinant DNA technology; this is termed **somatropin**. This drug is used to improve linear growth in patients whose short stature is known to be caused by a lack of pituitary growth hormone. There is also some abuse of the drug by athletes wishing to enhance performance. Although HGH increases skeletal mass and strength, its use can result in some abnormal bone growth patterns.

Prolactin

Prolactin has structural similarities to GH, in that it is a single-chain polypeptide (198 amino acid residues) with three loops created by disulfide bonds. It is synthesized via a prohormone containing 29 extra amino acids. GH itself can bind to the prolactin receptor, but this is not significant under normal physiological conditions. Prolactin release is controlled by dopamine produced by the hypothalamus, and its main function is to control milk production. Prolactin has a synergistic action with oestrogen to promote mammary tissue proliferation during pregnancy, then at parturition, when oestrogen levels fall, prolactin levels rise and lactation is initiated. New nursing mothers have high levels of prolactin, and this also inhibits gonadotrophin release and/or the response of the ovaries to these hormones. As a result, ovulation does not usually occur during breast feeding, preventing further conception. No prolactin derivatives are currently used in medicine, but the ergot alkaloid derivatives bromocriptine and cabergoline (see page 394) are dopamine agonists employed to inhibit prolactin release by pituitary tumours. They are not now recommended for routine suppression of lactation.

Gonadotrophins

Follicle-stimulating hormone (FSH) and **luteinizing hormone (LH)** are involved in controlling both male and female reproduction (see LH-RH, page 428). These are glycoproteins both composed of two polypeptide chains of 89 and 115 amino acid residues. The shorter chains are essentially identical, and differences in activity are caused by differences in the longer chain. Each chain has asparagine-linked oligosaccharide residues: the short chains each have one, the long chains two in the case of FSH or one for LH. **Lutropin alfa** is a recombinant LH for infertility treatment. FSH and LH together (**human menopausal gonadotrophins; menotrophin**) are purified from the urine of post-menopausal women and used in the treatment of female infertility. FSH alone is also available for this purpose, either natural material from urine termed **urofollitropin (urofollitrophin)**, or the recombinant proteins **follitropin alfa** and **follitropin beta**. Chorionic gonadotrophin (**human chorionic gonadotrophin; HCG**) is a gonad-stimulating glycoprotein hormone which is obtained for drug use from the urine of pregnant women. It is used in infertility treatment and also to stimulate testosterone production in males with delayed puberty.

Posterior Pituitary Hormones

The two main hormones of the posterior pituitary are **oxytocin**, which contracts the smooth muscle of the uterus, and **vasopressin**, the antidiuretic hormone (ADH) [Box 7.4]. These are nonapeptides containing a disulfide bridge and are structurally very similar, differing in only two amino acid residues (Figure 7.11). Structurally related peptides are classified as belonging to the vasopressin family when the amino acid residue at position 8 is basic, e.g. Arg or Lys, or to the oxytocin family when this amino acid is neutral.

Box 7.4

Oxytocin

Oxytocin (Figure 7.11) stimulates the pregnant uterus, causing contractions, and also brings about ejection of milk from the breasts. It thus plays a major role in the normal onset of labour at the end of pregnancy. Oxytocin for drug use is produced by synthesis and is employed to induce or augment labour, as well as to minimize subsequent blood loss.

Box 7.4   (continued)

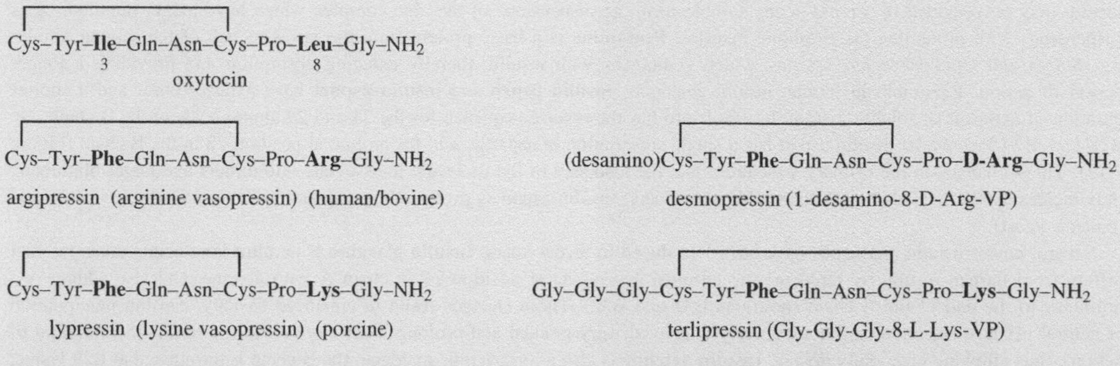

Cys–Tyr–Ile–Gln–Asn–Cys–Pro–Leu–Gly–NH$_2$
    3   8
                oxytocin

Cys–Tyr–Phe–Gln–Asn–Cys–Pro–Arg–Gly–NH$_2$

argipressin (arginine vasopressin) (human/bovine)

(desamino)Cys–Tyr–Phe–Gln–Asn–Cys–Pro–D-Arg–Gly–NH$_2$

desmopressin (1-desamino-8-D-Arg-VP)

Cys–Tyr–Phe–Gln–Asn–Cys–Pro–Lys–Gly–NH$_2$

lypressin (lysine vasopressin) (porcine)

Gly–Gly–Gly–Cys–Tyr–Phe–Gln–Asn–Cys–Pro–Lys–Gly–NH$_2$

terlipressin (Gly-Gly-Gly-8-L-Lys-VP)

Figure 7.11

Vasopressin

Vasopressin (antidiuretic hormone, ADH) is a hormone which has an antidiuretic action on the kidney, regulating the reabsorption of water. A deficiency in this hormone leads to diabetes insipidus, where the patient suffers increased urine output and intense thirst, typically consuming enormous quantities of fluid. Vasopressin is used to treat this condition. At high dosage, vasopressin promotes contraction of arterioles and capillaries, and brings about an increase in blood pressure. The structure of human and bovine vasopressin (arginine vasopressin; **argipressin**; Figure 7.11) differs from that of oxytocin only in two amino acid residues. Lysine vasopressin (**lypressin**; Figure 7.11) from pigs differs from arginine vasopressin in the second amino acid from the C-terminus. Both bovine and porcine peptides have been used medicinally, but these have been replaced by synthetic materials. The 1-desamino-8-D-Arg-vasopressin analogue **desmopressin** (Figure 7.11) has a longer duration of action than vasopressin and may also be administered orally. In contrast to the natural hormone, desmopressin has no vasoconstrictor effect. **Terlipressin** (Figure 7.11) is a lypressin pro-drug in which the polypeptide chain has been extended by three glycine residues. Enzymic hydrolysis liberates lypressin. It is mainly used for control of oesophageal bleeding.

Box 7.5

Insulin

Insulin is a hormone produced by the pancreas that plays a key role in the regulation of carbohydrate, fat, and protein metabolism. In particular, it has a hypoglycaemic effect, lowering the levels of glucose in the blood. If a malfunctioning pancreas results in a deficiency in insulin synthesis or secretion, the condition known as diabetes mellitus ensues. This results in increased amounts of glucose in the blood and urine, diuresis, depletion of carbohydrate stores, and subsequent breakdown of fat and protein. Incomplete breakdown of fat leads to the accumulation of ketones in the blood, severe acidosis, coma, and death. Where the pancreas is still functioning, albeit less efficiently, the condition is known as type 2 diabetes (non-insulin-dependent diabetes) and can be controlled by a regulated diet or oral antidiabetic drugs. In type 1 diabetes (insulin-dependent diabetes), pancreatic cells no longer function, and injections of insulin are necessary, one to four times daily, depending on the severity of the condition. These need to be combined with a controlled diet and regular monitoring of glucose levels, but do not cure the disease, so treatment is lifelong.

Mammalian insulins from different sources are very similar and may be used to treat diabetes. **Porcine insulin** and **bovine insulin** (Figure 7.13) are extracted from the pancreas of pigs and cattle respectively. **Human insulin** (Figure 7.13) is produced by the use of recombinant DNA technology in *Escherichia coli* to obtain the two polypeptide chains, and linking these chemically to form the disulfide bridges (such material is coded 'crb'), or by modification of proinsulin produced in genetically modified *Escherichia coli* (coded 'prb'). Human insulin may also be obtained from porcine insulin by semi-synthesis, replacing the terminal alanine in chain B with threonine by enzymic methods (coded 'emp'). Human insulin does not appear to be less immunogenic than animal insulin, but genetic engineering offers significant advantages over animal sources for obtaining highly purified material.

Box 7.5   (continued)

Insulin may be provided in a rapid-acting soluble form, as suspensions of the zinc complex which have longer duration, or as suspensions with protamine (as **isophane insulin**). **Protamine** is a basic protein from the testes of fish of the salmon family, e.g. *Salmo* and *Onchorhynchus* species, which complexes with insulin, thereby reducing absorption and providing a longer period of action. Recombinant human insulin analogues **insulin lispro** and **insulin aspart** have a faster onset and a shorter duration of action than soluble insulin. Insulin lispro has the reverse sequence for the 28 and 29 amino acids in the B chain, i.e. B28-Lys–B29-Pro, whilst insulin aspart has a single substitution of aspartic acid for proline at position 28 in the B chain (Figure 7.13). These changes in the primary structure affect the tendency of the molecule to associate into dimers and larger oligomers, thus increasing the availability of absorbable monomers. Insulin aspart is produced by expression in *Saccharomyces cerevisiae* (baker's yeast).

Several newer insulin analogues have been introduced in recent years. **Insulin glargine** is an ultra-long-acting analogue that differs from human insulin by replacing the terminal asparagine at position 21 in chain A with glycine, and also adding two arginines to the end of the B chain (positions B31 and B32). These changes result in enhanced basicity, causing precipitation at neutral pH post-injection, and consequently a delayed, very gradual and prolonged activity profile (up to 24–48 h duration of action), thus allowing once-daily dosing. **Insulin detemir** is also a long-acting analogue; the B chain is terminated at B29 lysine, to which the fatty acid myristic acid is added via an amide linkage. The fatty acid allows binding to albumin, from which it is slowly released. **Insulin glulisine** is a rapid-acting version with two amino acid changes in the B chain, B3 asparagine to lysine and B29 lysine to glutamic acid.

Glucagon

Glucagon (Figure 7.13) is a straight-chain polypeptide hormone containing 29 amino acids. It is secreted by the pancreas when blood sugar levels are low, thus stimulating breakdown of glycogen in the liver, and is the counter-regulatory hormone to insulin. It may be isolated from animal pancreas or is produced by recombinant DNA processes using *Saccharomyces cerevisiae*. It may be administered for the emergency treatment of diabetes patients suffering from hypoglycaemia as a result of building up dangerously high insulin levels. Normally, a patient would counter this by eating some glucose or sucrose, but hypoglycaemia can rapidly cause unconsciousness, requiring very prompt action.

Pancreatic Hormones

The hormone **insulin** plays a key role in the regulation of carbohydrate, fat, and protein metabolism [Box 7.5]. In particular, it has a hypoglycaemic effect, lowering the levels of glucose in the blood. A deficiency in insulin synthesis leads to the condition diabetes, treatment of which requires daily injections of insulin. Insulin is composed of two straight-chain polypeptides joined by disulfide bridges. This structure is known to arise from a straight-chain polypetide preproinsulin containing 100 amino acid residues. This loses a 16-residue portion of its chain and forms proinsulin with disulfide bridges connecting the terminal portions of the chain in a loop (Figure 7.12). A central portion of the loop (the C chain) is

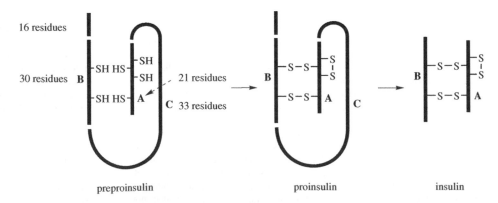

preproinsulin   proinsulin   insulin

Figure 7.12

A chain

Gly–Ile–Val–Glu–Gln–Cys–Cys–**Thr–Ser–Ile**–Cys–Ser–Leu–Tyr–Gln–Leu–Glu–Asn–Tyr–Cys–Asn
8   9   10   21

B chain

Phe–Val–Asn–Gln–His–Leu–Cys–Gly–Ser–His–Leu–Val–Glu–Ala–Leu–Tyr–Leu–Val–Cys–Gly–Glu–Arg–Gly–Phe–Phe–Tyr–**Pro–Lys–Thr**
28   29   30

human insulin

	A			B		
	8	9	10	28	29	30
insulin (human)	Thr	Ser	Ile	Pro	Lys	Thr
insulin (porcine)	Thr	Ser	Ile	Pro	Lys	**Ala**
insulin (bovine)	Ala	Ser	**Val**	Pro	Lys	**Ala**
insulin lispro	Thr	Ser	Ile	**Lys**	**Pro**	Thr
insulin aspart	Thr	Ser	Ile	**Asp**	Lys	Thr

His–Ser–Gln–Gly–Thr–Phe–Thr–Ser–Asp–Tyr–
Ser–Lys–Tyr–Leu–Asp–Ser–Arg–Arg–Ala–Gln–
Asp–Phe–Val–Gln–Trp–Leu–Met–Asn–Thr

glucagon

Figure 7.13

then cleaved out, leaving the A chain (21 residues) bonded to the B chain (30 residues) by two disulfide bridges. This is the resultant insulin. Mammalian insulins (Figure 7.13) from different sources are very similar, showing variations in the sequence of amino acid residues 8–10 in chain A, and at amino acid 30 in chain B. **Glucagon** (Figure 7.13) is a straight-chain polypeptide hormone containing 29 amino acids that is secreted by the pancreas when blood sugar levels are low, thus stimulating breakdown of glycogen in the liver [Box 7.5]. It is a counter-regulatory

hormone to insulin. Unlike insulin, its structure is identical in all animals.

Interferons

The **interferons** are a family of proteins secreted by animal cells in response to viral and parasitic infections, and are part of the host's defence mechanism [Box 7.6]. They display multiple activities, affecting the functioning of the immune system, cell

---

Box 7.6

Interferons

Interferons were originally discovered as proteins that interfered with virus replication. When mice were injected with antibodies to interferons, they became markedly susceptible to virus-mediated disease, including virus-related tumour induction. Interferons can be detected at low levels in most human tissues, but amounts increase upon infection with viruses, bacteria, protozoa, and exposure to certain growth factors. Interferons were initially classified according to the cellular source, but recent nomenclature is based primarily on sequencing data. Thus, leukocyte interferon (a mixture of proteins) is now known as interferon alfa, fibroblast interferon as interferon beta, and immune interferon as interferon gamma. These typically range in size from 165 to 172 amino acid residues. The interferon system can often impair several steps in viral replication, and can modulate the immune system, and affect cell growth and differentiation. They have potential in treating many human diseases, including leukaemias and solid tumours, and viral conditions such as chronic infection with hepatitis B and C. Much of their current drug use is still research based.

Interferon alfa is a protein containing 166 amino acid residues with two disulfide bridges and is produced by recombinant DNA techniques using *Escherichia coli*, or by stimulation of specific human cell lines. Variants with minor differences in sequence may be obtained according to the gene used, and are designated as alfa-2a, alfa-2b, etc. Interferon alfa is employed as an antitumour drug against certain lymphomas and solid tumours, and in the management of chronic hepatitis. **Interferon beta** is of value in the treatment of multiple sclerosis patients, though not all patients respond. Variants are designated beta-1a (a glycoprotein with 165 amino acid residues and one disulfide bridge) or beta-1b (a non-glycosylated protein with 164 amino acid residues and one disulfide bridge). **Interferon gamma** produced by recombinant DNA methods contains an unbridged polypeptide chain, and is designated gamma-1a (146 residues) or gamma-1b (140 residues) according to sequence. The immune interferon interferon gamma-1b is used to reduce the incidence of serious infection in patients with chronic granulomatous disease.

Tyr–Gly–Gly–Phe–Met–Thr–Ser–Glu–Lys–Ser–
Gln–Thr–Pro–Leu–Val–Thr–Leu–Phe–Lys–Asn–
Ala–Ile–Val–Lys–Asn–Ala–His–Lys–Lys–Gly–
Gln

β-endorphin

Tyr–Gly–Gly–Phe–Met

Met-enkephalin

Tyr–Gly–Gly–Phe–Leu

Leu-enkephalin

Tyr–Gly–Gly–Phe–Leu–Arg–Arg–Ile–Arg–Pro–Lys–Leu–Lys–Trp–Asp–Asn–Gln

dynorphin A

Tyr–Pro–Trp–Phe–NH$_2$

endomorphin-1

Tyr–Pro–Phe–Phe–NH$_2$

endomorphin-2

Figure 7.14

proliferation, and cell differentiation, primarily by inducing the synthesis of other proteins. Accordingly, they have potential as antiviral, antiprotozoal, immunomodulatory, and cell growth regulatory agents.

Opioid Peptides

Although the pain-killing properties of morphine and related compounds have been known for a considerable time (see page 349), the existence of endogenous peptide ligands for the receptors to which these compounds bind is a more recent discovery. It is now appreciated that the body produces a family of endogenous opioid peptides which bind to a series of receptors in different locations. These peptides include **enkephalins, endorphins**, and **dynorphins**, and are produced primarily, but not exclusively, in the pituitary gland. The pentapeptides **Met-enkephalin** and **Leu-enkephalin** (Figure 7.14) were isolated from mammalian brain tissue and were the first to be characterized. The largest peptide is **β-endorphin** ('*end*ogenous m*orphine*'; Figure 7.14), which is several times more potent than morphine in relieving pain. Although β-endorphin contains the sequence for Met-enkephalin, the latter peptide and Leu-enkephalin are derived from the larger peptide proenkephalin A, whilst β-endorphin itself is formed by cleavage of pro-opiomelanocortin. The proenkephalin A structure contains four Met-enkephalin sequences and one of Leu-enkephalin. The dynorphins, e.g. **dynorphin A** (Figure 7.14), are also produced by cleavage of a larger precursor, namely proenkephalin B (prodynorphin), and all contain the Leu-enkephalin sequence. Some 20 opioid ligands have now been characterized.

When released, these endogenous opioids act upon opioid receptors, inducing analgesia, and depressing respiratory function and several other processes. Three main receptor types are recognized: μ, δ, and κ. The individual peptides bind to all three types, but show a degree of specificity towards different receptors. Though β-endorphin binds to μ and δ receptors with similar affinity, the enkephalins are considered endogenous ligands for δ receptors, and dynorphins for κ receptors. Two tetrapeptides that bind specifically to μ receptors have been termed **endomorphins**, e.g. endomorphin-1 and endomorphin-2 (Figure 7.14); morphine itself acts primarily at the μ receptor. The opioid peptides are implicated in analgesia brought about by acupuncture, since opiate antagonists can reverse the effects. The hope of exploiting similar peptides as ideal, non-addictive analgesics has yet to be attained. Though many potent and selective opioid agonists have been developed, poor penetration of the blood–brain barrier leads to poor bioavailability.

RIBOSOMAL PEPTIDE TOXINS

Chemicals that are toxic to man and animals are found throughout nature, and long-term survival of a species is dependent upon the recognition and avoidance of toxin-containing materials, be they plants, fungi, or other toxin-containing life-forms, such as venomous snakes or spiders. Many of the plant and fungal toxins are small molecules, and where relevant, these have been discussed in earlier chapters. However, some of the most dangerous natural toxins are peptide structures, and formed via typical ribosomal peptide biosynthesis sequences. Some of those that are especially hazardous to man, or have potential value in medicine, are described in [Box 7.7].

Box 7.7

Death Cap (Amanita phalloides)

The death cap, *Amanita phalloides*, is a highly poisonous European fungus with a mushroom-like fruiting body. The death cap has a whitish-green cap and white gills; it has a superficial similarity to the common mushroom, *Agaricus campestris*, and may

Box 7.7 (continued)

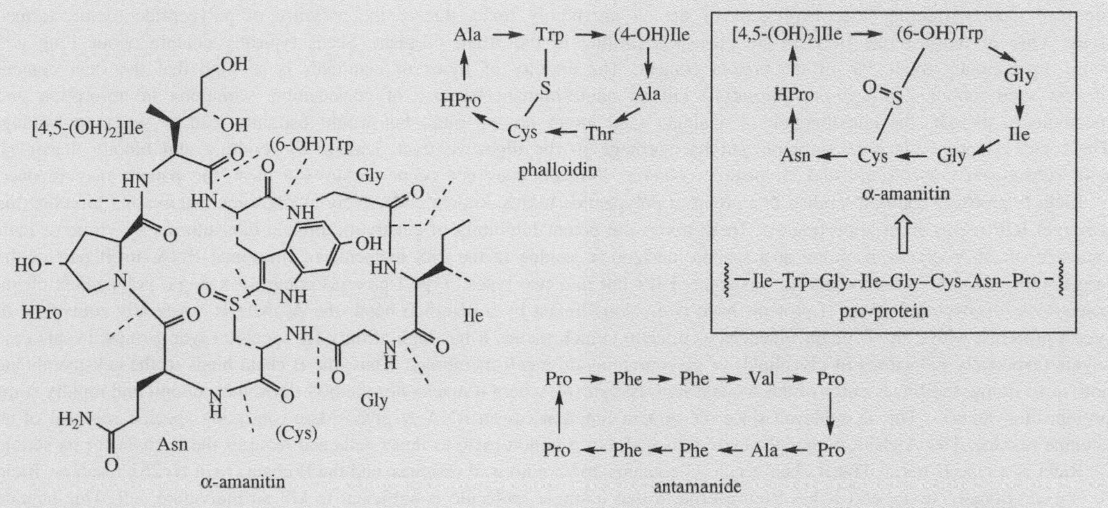

Figure 7.15

sometimes be collected in error. Some 90% of human fatalities due to mushroom poisoning are attributed to the death cap. Identification of the death cap as a member of the genus *Amanita*, which includes other less poisonous species, is easily achieved by the presence of a cup-like membranous structure (volva) at the base of the stem. The volva is the remains of the universal veil in which the immature fruiting body was enclosed. Ingestion of the death cap produces vomiting and diarrhoea during the first 24 h, followed after 3–5 days by coma and death. Some recoveries do occur, but the fatality rate is probably from 30 to 60%. There is no guaranteed treatment for death cap poisoning, though removal of material from the gastrointestinal tract, replacement of lost fluids, blood dialysis, and blood transfusions may be undertaken. The antihepatotoxic agent silybin (see page 174) has been used successfully.

The toxic principles are cyclic polypeptides which bring about major degeneration of the liver and kidneys. At least 10 toxins have been identified, which may be subdivided into two groups: the phallotoxins and the amatoxins. The most extensively studied compounds are phalloidin, a phallotoxin, and α-amanitin, an amatoxin (Figure 7.15). The phallotoxins are much less toxic than the amatoxins since, after ingestion, they are not well absorbed into the bloodstream. When injected, they can cause severe damage to the membranes of liver cells, releasing potassium ions and enzymes. The amatoxins are extremely toxic when ingested, with a lethal dose of 5–7 mg for an adult human; an average specimen of the death cap contains about 7 mg. The amatoxins cause lesions to the stomach and intestine, and then irreversible loss of liver and kidney function. α-Amanitin has been shown to be a powerful inhibitor of RNA polymerase, blocking the elongation step. Structurally, the two groups of toxins appear similar. They both contain a 4-hydroxylated amino acid essential for toxic action, and have a sulfur bridge between a cysteine residue and the 2-position of a tryptophan or 6-hydroxytryptophan residue (Figure 7.15). In the case of the phallotoxins, this is a simple sulfide bridge, but in the amatoxins it is in the form of a sulfoxide. Interestingly, a rather simpler cyclic peptide antamanide (anti-amanita peptide) has also been isolated from *Amanita phalloides*. When pre-administered to laboratory animals, antamanide (Figure 7.15) provided prophylactic protection from the lethal affects of the phallotoxins.

Contrary to expectations, the *Amanita* toxins are synthesized on ribosomes; most cyclic peptides are produced by non-ribosomal peptide synthetases (see page 438). α-Amanitin is synthesized via a pro-protein of 35 amino acids, from which an eight-residue peptide is released by a protease enzyme. In all cases, cleavage is adjacent to a proline residue; all the *Amanita* toxins contain a proline (as hydroxyproline) residue, and this derives from the terminal amino acid of the excised peptide. The pro-protein also has a proline adjacent to the other end of the heptapeptide. Hydroxylations and construction of the sulfur cross-bridge must be late modifications.

Ricin

The distinctive mottled-brown seeds of the castor oil plant *Ricinus communis* (Euphorbiaceae) are crushed to produce castor oil, which is predominantly composed of glycerides of ricinoleic acid (see page 46 and Table 3.1). The seed itself and

Box 7.7   (continued)

the seed cake remaining after expression of the oil are highly toxic, due to the presence of polypeptide toxins, termed ricins. One or more forms of ricin are present according to the strain of plant. Seeds typically contain about $1\,\mathrm{mg\ g^{-1}}$ ricin, representing about 5% of the protein content. The toxicity of ricins to mammals is so high that the ricin content of one seed (about $250\,\mu\mathrm{g}$) is suffcient to kill an adult human; because of considerable variations in absorption and metabolism, though, the consumption of a single seed might not be fatal, but would certainly lead to severe poisoning. The toxic symptoms include irritation and haemorrhage in the digestive tract, leading to vomiting and bloody diarrhoea, with subsequent convulsions and circulatory collapse. If death does not occur within 3–5 days, the patient may recover.

Ricin is probably the best-studied of a group of polypeptide toxins, known collectively as ribosome-inactivating proteins (the acronym RIP seems most appropriate!). These toxins are potent inhibitors of eukaryotic protein biosynthesis by virtue of their cleavage of an N-glycosidic bond at a specific nucleotide residue in the 28S molecule of ribosomal RNA, itself part of the larger (60S) subunit of the eukaryotic ribosome. RIPs fall into two types. Type I proteins comprise a single polypeptide chain, sometimes glycosylated. Type II proteins have two chains linked by a disulfide bond: the A chain is essentially equivalent to type I proteins, whilst the B chain functions as a lectin, which means it has high affinity for specific sugar groups, in this case sugars (especially galactose) in glycolipids or glycoproteins on a cell membrane. Thus, the B chain binds to the cell membrane and, in so doing, facilitates entry of the A chain into the cytosol, where it inactivates the 60S ribosomal subunit and rapidly stops protein biosynthesis. This is achieved since the protein functions as an RNA N-glycosidase catalysing specific removal of an adenine residue. The A chain is thus the toxic principal, but it is non-toxic to intact cells and requires the B chain for its action.

Ricin is a type II toxin. The A chain (ricin A) contains 267 amino acid residues, and the B chain (ricin B) 262 residues. Ricin A is exceptionally toxic, and it has been estimated that a single molecule is sufficient to kill an individual cell. This peptide can be prepared by genetic engineering using *Escherichia coli*. The potent action of this material on eukaryotic cells has been investigated in anticancer therapy. Ricin A has been coupled to monoclonal antibodies and successfully delivered specifically to the tumour cells. However, *in vitro* toxicity of ricin A-based immunotoxins is enhanced significantly if ricin B is also present.

Similar toxic RIPs are found in other plants. Trichosanthin is a type I toxin from the root tubers of *Trichosanthes kirilowii* (Cucurbitaceae). Type II toxins are exemplified by abrin from the small brightly coloured red and black jequirity seeds (*Abrus precatorius*; Leguminosae/Fabaceae), viscumin from the leaf and stems of mistletoe (*Viscum album*, Loranthaceae), and cinnamomin from seeds of the camphor tree (*Cinnamomum camphora*; Lauraceae). Bacterial and fungal RIPs include diphtheria toxin, pseudomonas exotoxin, and Shiga toxin.

Botulinum Toxin

The Gram-positive bacterium *Clostridium botulinum* produces one of the most toxic materials known to man, botulinum toxin. Poisoning by the neurotoxins from this source, known as botulism, is not uncommon and is a life-threatening form of food poisoning. It has been estimated that as many as 50 million people could be killed by 1 g of the toxin. *Clostridium botulinum* is an anaerobic organism that is significantly heat resistant, though botulinum toxin is easily destroyed by heat. Food poisoning is almost always associated with foods such as canned meats and fish that have been incompletely sterilized, allowing growth of the bacterium, after which the food is then consumed without further cooking. Botulinum toxin is an extremely potent neurotoxin that acts by blocking calcium-dependent acetylcholine release at the peripheral neuromuscular junctions. Poisoning leads to paralysis and death from respiratory failure. Damage to the nervous system is usually preceded by vomiting, diarrhoea, and severe abdominal pains.

Seven different neurotoxins, types A–G, have been characterized, though only four of these, types A, B, E, and F, are clearly associated with human poisoning. A particular strain of the bacterium usually produces only one type of toxin; type A is the most toxic. Each of these proteins is produced as a single-chain polypeptide (mass about 150 kDa) that is cleaved by a protease into two subunits, a 'light' subunit with a mass of about 50 kDa and a 'heavy' subunit with a mass of about 100 kDa, though these subunits are still linked by at least one disulfide bridge. The heavy subunit is responsible for toxin binding, whilst the light subunit possesses zinc metalloprotease activity, cleaving one of the proteins involved in the docking and release of synaptic vesicles. There is considerable structural similarity between botulinum toxins and tetanus toxin. **Botulism antitoxin** is available for the treatment of botulism food poisoning. This is a mixture of globulins raised against types A, B, and E toxins, those responsible for most human poisonings.

Botulinum toxin A and **botulinum toxin B** are currently employed medicinally to counter involuntary facial muscle spasms, e.g. around the eye, spasm of neck muscles (cervical dystonia) and others. The type B toxin is more stable and may be stored at room temperature. Very small (nanogram) amounts are injected locally and result in the destruction of the acetylcholine release mechanism at the neuromuscular junction. Since new nerve junctions will gradually be formed over 2 months or so, the result is not permanent and the treatment will need to be repeated. It has also been found useful in easing muscle spasticity in children

Box 7.7 (continued)

with cerebral palsy. In recent years, Botox® (type A toxin) has enjoyed considerable popularity in cosmetic procedures to relax muscles and remove facial wrinkles and lines; again, this can only be a temporary effect. This is not without risk; botulinum toxin is potentially fatal, and even the antitoxin can cause immunological side-effects.

Tetanus toxin is closely related in structure to botulinum toxin; it is produced by the anaerobic bacterium *Clostridium tetani* found in soil and animal manure. Tetanus is characterized by development of muscle spasms in the jaws, hence the common name lockjaw, and if left untreated can prove fatal. The usual source of infection is through open wounds or animal bites, and tetanus is considered a significant health threat to humans. However, tetanus vaccines are routinely used for individuals most at risk.

Conotoxins

The genus *Conus* comprises some 700 species of predatory cone snails, probably the largest single genus of marine invertebrates. About 10% of species prey exclusively on fish. These snails use a complex mixture of peptide toxins, conotoxins, to capture their prey, defend against predators, and compete with other animals in their environment. The active components of the venoms are mainly small peptides, 10–30 amino acids in length, most of which are extensively cross-linked with disulfide bridges. The disulfide bridges add rigidity and stability to the peptide and constrain its conformation for optimal interaction with specific receptors. The peptides appear to be formed by proteolytic cleavage of longer (80–100 amino acid) precursors and may then involve some post-translational changes. Each *Conus* species may produce 50–200 different peptides, and these are proving to be a rich source of compounds with potential pharmacological activity. The major known targets of *Conus* venom peptides are ion channels, primarily in neurons. *Conus* peptides are becoming widely used tools in neuroscience and have promising therapeutic applications, with several examples in clinical trials; one such compound is now established as a commercial drug. This is **ziconotide** (Figure 7.16), which is synthetic ω-conotoxin MVIIA, the natural peptide produced by the magician's cone snail *Conus magus*. Ziconotide is a powerful analgesic for chronic severe pain that acts on voltage-gated Ca^{2+} channels. Several other *Conus* peptides in development are also potential analgesics for chronic pain, working through five distinct mechanisms, none of which is opioid-like. Indeed, ziconotide may be administered at the same time as an opioid.

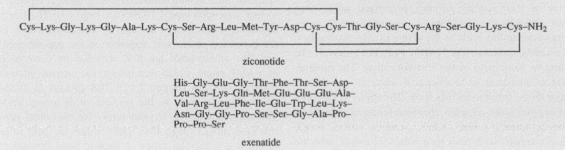

Cys–Lys–Gly–Lys–Gly–Ala–Lys–Cys–Ser–Arg–Leu–Met–Tyr–Asp–Cys–Cys–Thr–Gly–Ser–Cys–Arg–Ser–Gly–Lys–Cys–NH₂

ziconotide

His–Gly–Glu–Gly–Thr–Phe–Thr–Ser–Asp–
Leu–Ser–Lys–Gln–Met–Glu–Glu–Glu–Ala–
Val–Arg–Leu–Phe–Ile–Glu–Trp–Leu–Lys–
Asn–Gly–Gly–Pro–Ser–Ser–Gly–Ala–Pro–
Pro–Pro–Ser

exenatide

Figure 7.16

Snake Venoms

It is estimated that some 1300 of the 3200 known species of snake are venomous. Snake venoms are used to immobilize prey and to facilitate its digestion. Most of the material is polypeptide in nature, and can include enzymes and polypeptide toxins. A number of enzymes have been identified in all venoms; these include hyaluronidase (see Table 7.2), which facilitates the distribution of the other venom components through the tissues. Peptidases, phosphodiesterases, phospholipases, ribonuclease, and deoxyribonuclease are all hydrolytic enzymes designed to digest the tissue of the prey. Some enzymes induce direct toxic effects; for example, L-amino acid oxidase liberates hydrogen peroxide, a powerful oxidizing agent. In some venoms, the enzyme acetylcholinesterase disturbs the normal physiological response of the prey by hydrolysing acetylcholine. Major groups of polypeptide toxins found in snake venoms may be classified as neurotoxins, cardiotoxins, dendrotoxins, proteinase inhibitors, or acetylcholinesterase inhibitors. α-Neurotoxins (curaremimetic neurotoxins) found in many mambas (*Dendroaspis*) and cobras (*Naja*) are capable of interacting with nicotinic acetylcholine receptors in the post-synaptic membranes of skeletal muscles, leading to paralysis, an action similar to that of curare (see page 344). κ-Neurotoxins, on the other hand, are selective for neuronal nicotinic acetylcholine receptors.

Box 7.7   (continued)

Typically, the neurotoxins contain 60–74 amino acid residues with four or five disulfide bridges. Cardiotoxins are present in cobra (*Naja*) and ringhal (*Hemachatus*) venoms and produce a rapid effect on the heart and circulation, though the mode of action is not well established. Most have 60–62 amino acid residues. Dendrotoxins from mambas are characterized by their ability to facilitate the release of acetylcholine from nerve endings, and also act as highly potent and selective blockers of potassium channels. These contain about 60 amino acid residues, with three disulfide bridges. Anticholinesterase toxins have also been found in mamba venoms, typically with about 60 amino acids and four disulfide bridges.

Gila Monster Venom

The Gila monster (*Heloderma suspectum*) is a large poisonous lizard found in the southwestern United States and northern Mexico. Its bite is painful, but not usually fatal to adult humans, and is defensive rather than predatory. Its venom, present in the saliva, contains a variety of peptides and proteins. One of these peptides, exendin-4, was found to have a structure similar to glucagon-like peptide-1 (GLP-1), a hormone in the human digestive tract that increases the production of insulin when blood sugar levels are high; exendin-4 has similar properties. Synthetic exendin-4 (**exenatide**), a 39-amino acid peptide (Figure 7.16), is now available for the treatment of type 2, non-insulin-dependent, diabetes (see page 431). It primarily stimulates the pancreas to produce insulin and reduces the production of glucagon. Like GLP-1, exenatide also reduces the rate at which food passes from the stomach into the intestine, and acts on the brain to cause a feeling of fullness. It is injected using a pen device as an adjunct to oral antidiabetic drugs.

ENZYMES

Enzymes are proteins that act as biological catalysts. They facilitate chemical modification of substrate molecules by virtue of their specific binding properties, which arise from particular combinations of functional groups in the constituent amino acids at the active site. In many cases, an essential cofactor, e.g. NAD^+, PLP, TPP, may also be bound to participate in the transformation. The involvement of enzymes in biochemical reactions has been a major theme throughout this book. The ability of enzymes to carry out quite complex chemical reactions, rapidly, at room temperature, and under essentially neutral conditions is viewed with envy by synthetic chemists, who are making rapid progress in harnessing this ability for their own uses. Several enzymes are currently of importance commercially, or for medical use, and these are described in Table 7.2. Enzymes are typically larger than most of the polypeptides discussed above and are usually extracted from natural sources. Recombinant DNA procedures are likely to make a very significant contribution in the future.

NON-RIBOSOMAL PEPTIDE BIOSYNTHESIS

In marked contrast to the ribosomal biosynthesis of peptides and proteins, where a biological production line interprets the genetic code, many natural peptides are known to be synthesized by a more individualistic sequence of enzyme-controlled processes, in which each amino acid is added as a result of the specificity of the enzyme involved. The many stages of the whole process are carried out by a multifunctional enzyme (non-ribosomal peptide synthase, NRPS) with a modular arrangement comparable to that seen with type I PKSs (see page 67). The linear sequence of modules in the enzyme then corresponds to the generated amino acid sequence in the peptide product. The amino acids are first activated by conversion into AMP-esters, which then bind to the enzyme through thioester linkages (Figure 7.17). This process is mechanistically analogous to the conversion of amino acids into aminoacyl-tRNA oxygen esters for ribosomal peptide biosynthesis (page 424). The residues are held so as to allow a sequential series of peptide bond formations (Figure 7.18 gives a simplified representation), until the peptide is finally released from the enzyme.

A typical elongation module consists of an adenylation (A) domain, a peptidyl carrier protein (PCP) domain, and a condensation (C) domain, typically arranged in the order C–A–PCP. The A domain activates a specific amino acid as an aminoacyl adenylate, which is then transferred to the PCP domain, forming an aminoacyl thioester. Pantothenic acid (vitamin B_5, see page 32), bound to the enzyme as pantetheine, is used to carry the growing peptide chain through its thiol group (Figure 7.18). The significance of this is that the long 'pantetheinyl arm' allows different active sites on the enzyme to be reached in the chain assembly process (compare biosynthesis of fatty acids, page 42, and polyketides, page 70). Nucleophilic attack by the amino group of the neighbouring aminoacyl thioester is catalysed by the C domain and results in

Table 7.3 Pharmaceutically important enzymes

Enzyme	Action	Source	Use
Hydrolytic enzymes			
Chymotrypsin	hydrolysis of proteins	bovine pancreas	zonal lysis in cataract removal
Hyaluronidase	hydrolysis of mucopolysaccharides	mammalian testes	renders tissues more permeable for subcutaneous or intramuscular injections
Pancreatin	hydrolysis of starch (amylase), fat (lipase), and protein (protease)	porcine pancreas	digestive aid
Papain	hydrolysis of proteins	papaya fruit (*Carica papaya*; Caricaceae)	meat tenderizer; cleaning of contact lenses
Pepsin	hydrolysis of proteins	porcine stomach	digestive aid
Trypsin	hydrolysis of proteins	bovine pancreas	wound and ulcer cleansing
Fibrinolytic enzymes			
Alteplase (recombinant tissue-type plasminogen activator; rt-PA)	a protease which binds to fibrin converting it to a potent plasminogen activator; only active at the surface of the blood clot	recombinant genetic engineering: human gene expressed in Chinese hamster ovary cells	treatment of acute myocardial infarction
Reteplase	a fibrinolytic protease; a genetically engineered human tissue-type plasminogen activator differing from alteplase at four amino acid residues	recombinant genetic engineering	treatment of acute myocardial infarction
Streptokinase	no enzymic activity, until it complexes with and activates plasminogen in blood plasma to produce the proteolytic enzyme plasmin, which hydrolyses fibrin clots	*Streptococcus haemolyticus*	treatment of venous thrombosis and pulmonary embolism
Tenecteplase	a fibrinolytic protease; a genetically engineered human tissue-type plasminogen activator differing from alteplase at six amino acid residues	recombinant genetic engineering	treatment of acute myocardial infarction
Urokinase	a protease which activates plasminogen in blood plasma to form plasmin, which hydrolyses fibrin clots	human urine or human kidney tissue cultures	treatment of venous thrombosis and pulmonary embolism; thrombolysis in the eye; removal of fibrin clots in catheters and cannulas
Others			
Crisantaspase (asparaginase)	degradation of L-asparagine	*Erwinia chrysanthemi*	treatment of acute lymphoblastic leukaemia; results in death of those tumour cells which require increased levels of exogenous L-asparagine; side-effects nausea and vomiting, allergic reactions and anaphylaxis

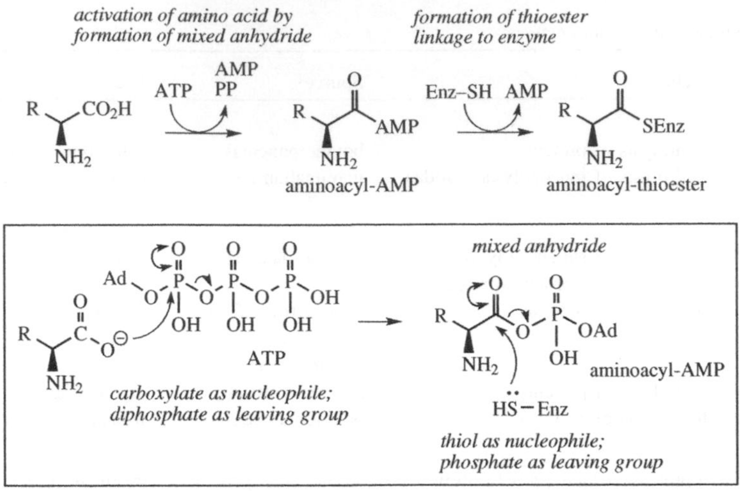

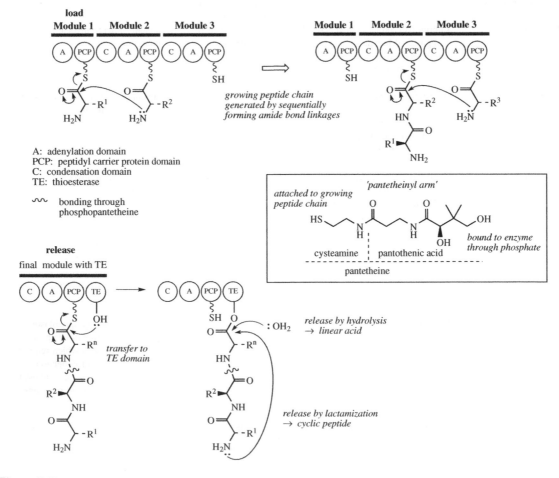

Figure 7.17

A: adenylation domain
PCP: peptidyl carrier protein domain
C: condensation domain
TE: thioesterase

〰 bonding through
    phosphopantetheine

Figure 7.18

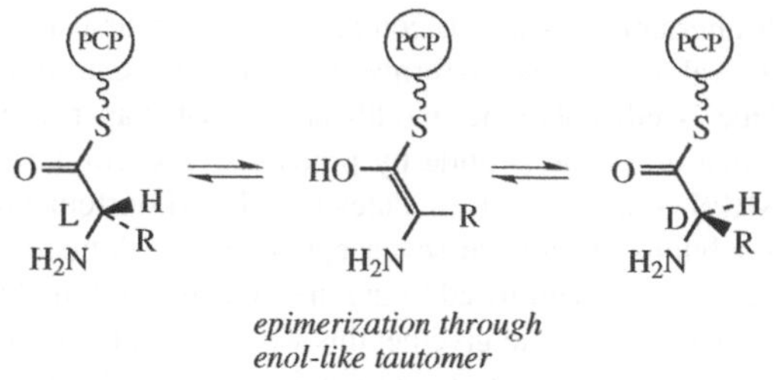

Figure 7.19

amide (peptide) bond formation. The first module is a dedicated loading module, and usually carries just A and T domains; in some instances, a three-domain module is encountered. Chain termination is achieved by having a thioesterase (TE) domain in the final module. This can liberate a typical linear peptide chain, but cyclic peptides may result if release involves the terminal amino group (Figure 7.18). Cyclization may alternatively be the result of ester or amide linkages, utilizing side-chain functionalities in the constituent amino acids. Enzyme-controlled biosynthesis in this manner is a feature of many microbial peptides, especially those containing unusual amino acids not encoded by DNA and where post-translational modification is unlikely, and also for the cyclic structures which are frequently encountered. As well as activating the amino acids and catalysing formation of the peptide linkages, NRPSs may possess other domains that are responsible for significant changes to the amino acid residues. Almost all non-ribosomally synthesized peptides contain one or more D-amino acids. Epimerization domains convert the PCP-tethered aminoacyl substrate from L- to D-forms, probably through enol-like tautomers (Figure 7.19). Epimerization is the most common approach to D-amino acid residues, and only occasionally does an A domain exclusively incorporate a D-amino acid supplied by an external racemase enzyme. Methylation domains are also found in some NRPSs; such domains catalyse transfer of a methyl group from SAM, most commonly to the amino acid nitrogen. The presence of D- and *N*-methyl-amino acids helps to protect the peptide from proteolytic hydrolysis. Other domains are known that are responsible for oxidation, reduction, and heterocyclic ring formation processes. Peptides derived by non-ribosomal

1 2 3 4 5 6 7 8 9 10 11 12 13 14 15 16

OHC–**X**–Gly–Ala–D-Leu–Ala–D-Val–Val–D-Val–Trp–D-Leu–**Y**–D-Leu–Trp–D-Leu–Trp–$NH(CH_2)_2OH$

	X	Y
gramicidin A		
Val-gramicidin A	Val	Trp
Ile-gramicidin A	Ile	Trp
gramicidin B		
Val-gramicidin B	Val	Phe
Ile-gramicidin B	Ile	Phe
gramicidin C		
Val-gramicidin C	Val	Tyr
Ile-gramicidin C	Ile	Tyr

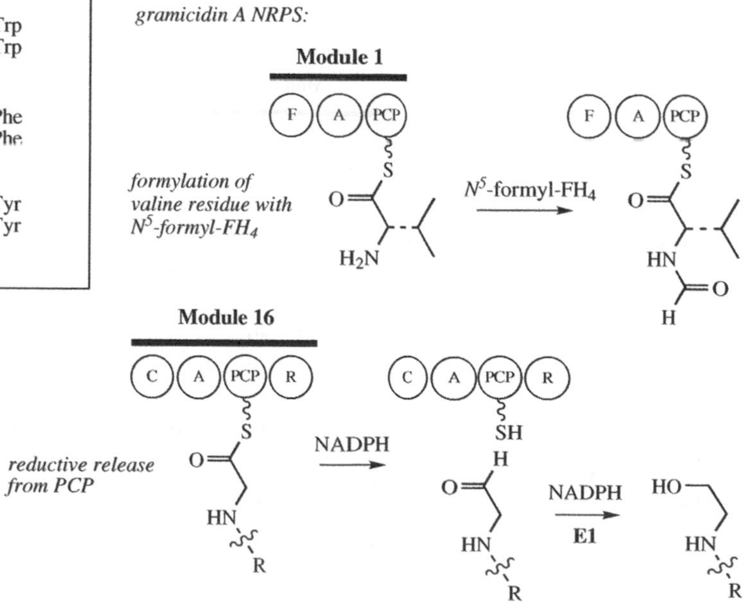

F: formylation domain
R: reductase domain

gramicidin A NRPS:
modules 1–2: lgrA
modules 3–6: lgrB
modules 7–12: lgrC
modules 13–16: lgrD
E1: LgrE (aldoreductase)

Figure 7.20

processes are usually much smaller than those produced by the ribosomal sequence.

As with PKSs, genetic manipulation of non-ribosomal peptide synthases allows production of peptide derivatives in which rational modifications may be programmed according to the genes encoded. There is also considerable scope for combining PKS and NRPS modules to generate hybrid molecules. Some of the PKS systems already discussed show that this occurs naturally. Thus, enzymes involved in the biosynthesis of ascomycin and tacrolimus (see page 83), rapamycin (see page 84), epothilones (see page 85), and rifamycins (see page 88), though predominantly PKS in nature, also contain NRPS-like domains that incorporate amino acid-related units. More examples follow, e.g. bleomycin (see page 449) and streptogramin A (see page 450), where predominantly NRPS enzymes also carry PKS-like domains and incorporate acetate-like units.

Though most examples to be considered are cyclic peptides, **gramicidin** from *Bacillus brevis* is a mixture of linear peptides (Figure 7.20). Linear gramicidins (gramicidin S is cyclic, see below) have 15 amino acid residues with alternating L- and D-configuration. The first amino acid (Val or Ile) is *N*-formylated, and the C-terminal residue is ethanolamine, modifications that may prevent degradation of the peptide by proteases. The gramicidin A NRPS consists of 16 modules distributed on four protein subunits. There are seven epimerization domains, a formylation domain fused to the first module, and module 16 appears to activate glycine; this would initially provide the C-terminal residue of the peptide chain, prior to reduction to ethanolamine. It has been shown that there is no thioesterase domain, but instead an adjacent reductase domain releases the peptide chain from its PCP by reduction of the thioester to an aldehyde. A separate reductase enzyme is then responsible for producing the alcohol (Figure 7.20). The formylating agent is N^5-formyl-tetrahydrofolate (see page 144).

Bacillus brevis also produces a series of cyclic decapeptides called **tyrocidines** (Figure 7.21). The tyrocidine A NRPS has 10 modules on three protein subunits. There are two epimerization domains, the first of which is in module 1; hence, the chain starts as a D-Phe residue. Module 10 contains a thioesterase domain, and this is responsible

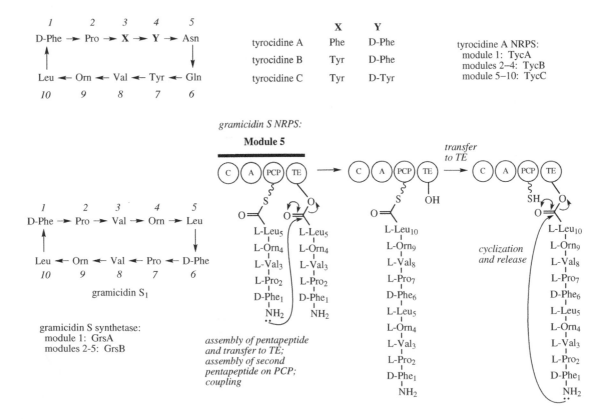

Figure 7.21

for terminating chain assembly and producing the cyclic product. Another cyclic decapeptide, **gramicidin S$_1$** (Figure 7.21), is the principal component from another strain of *Bacillus brevis*. However, there is a subtle structural difference between gramicidin S$_1$ and the tyrocidines, in that there is an element of symmetry in the amino acid sequence. Remarkably, this arises because gramicidin S synthetase assembles two separate pentapeptides and then couples these together. The TE domain is responsible for the coupling of the two pentapeptides and the subsequent cyclization/release (Figure 7.21). The adenylation domain of module 4 is specific for L-ornithine; the origins of this amino acid are discussed in Chapter 6 (see page 311).

The **bacitracins** from *Bacillus licheniformis* are also cyclic peptides, but cyclization does not encompass all the component amino acids; the side-chain amino function of a lysine residue is utilized to make an amide linkage with the carboxyl terminus (Figure 7.22). Twelve amino acid residues are incorporated, and bacitracin A NRPS consists of 12 modules on three protein subunits. Four epimerization domains are present; one of these converts L-ornithine into D-ornithine. The thioesterase in module 12 uses the ε-amino group of Lys$_6$ to link with Asn$_{12}$.

Also present in the structure of bacitracin A is a thiazoline (dihydrothiazole) ring, comparable to the thiazole ring seen in epothilones (see page 86). The thiazole ring in epothilones was constructed from cysteine and malonate using a combination of NRPS and PKS modules. In bacitracins, the ring system derives from a combination of cysteine and isoleucine, and is the result of amide formation, followed by nucleophilic attack of sulfur on the carbonyl, then dehydration (Figure 7.22). A cyclization domain (Cy) is present in module 2 to catalyse these reactions; cyclization domains are structurally very similar to condensation domains and are likely to initiate reaction by using a basic function to generate an appropriate anionic nucleophile. Though the two-stage reaction in Figure 7.22 is shown as amide formation followed by attack of sulfur, the order of these two processes might be reversed.

Capreomycin (Figure 7.23) from *Streptomyces capreolus* is also characterized by the presence of some unusual amino acids in its structure. These are elaborated from standard amino acids before incorporation into the polypeptide structure. Dap (2,3-diaminopropionic acid) is formed from serine via a PLP-dependent elimination/addition mechanism involving dehydroalanine. The

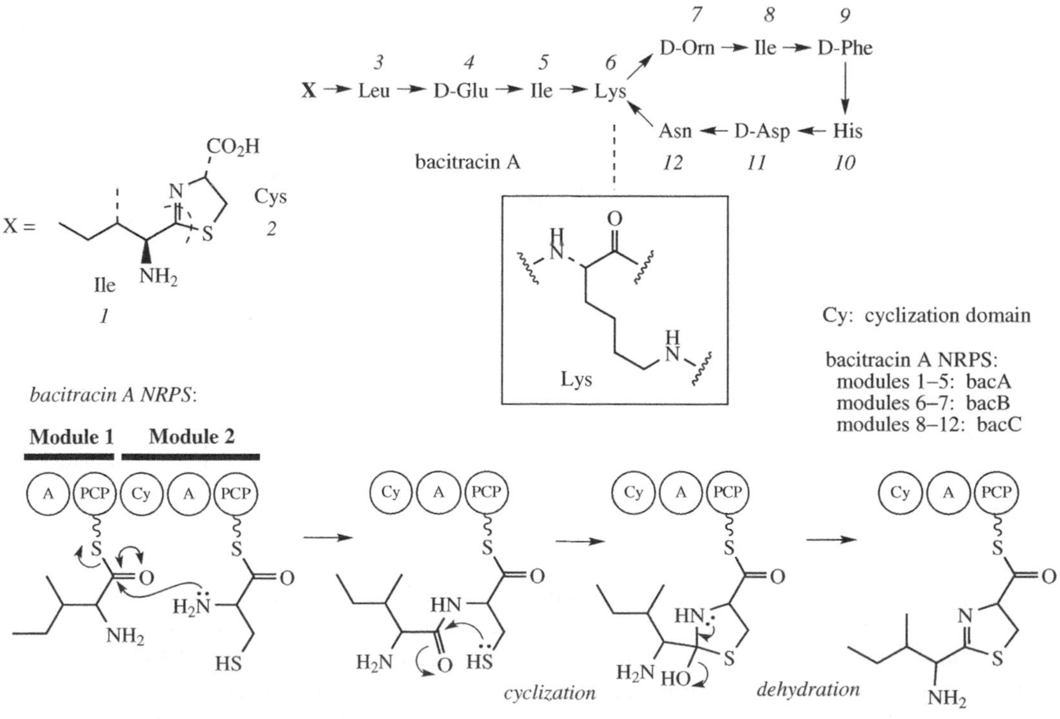

Figure 7.22

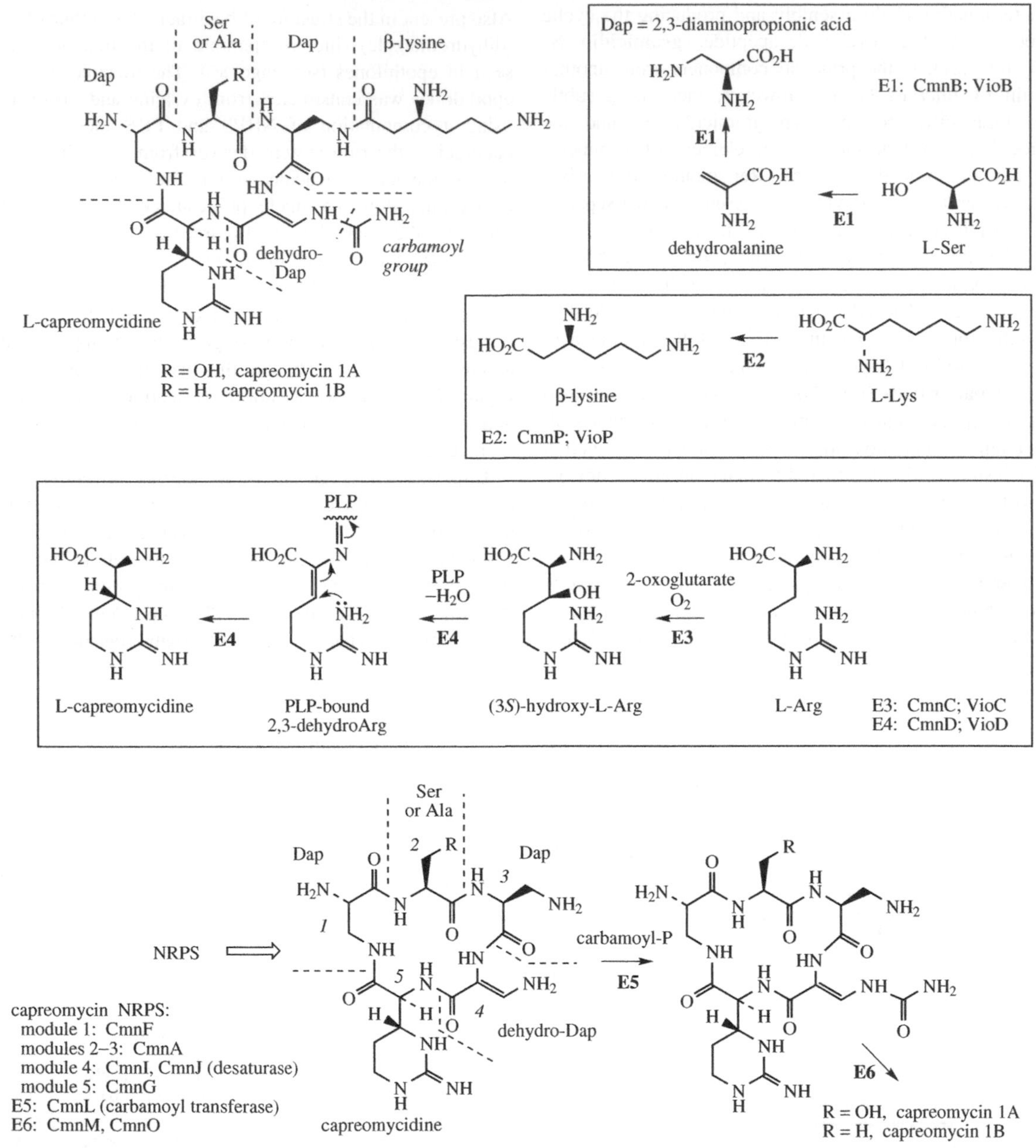

Figure 7.23

heterocycle capreomycidine is derived from arginine, and also requires PLP in the sequence (Figure 7.23). A third unusual amino acid is β-lysine, formed from lysine by the action of a 2,3-aminomutase, that repositions the amino group. A phenylalanine aminomutase converting phenylalanine into β-phenylalanine features in taxol

biosynthesis (see page 225). However, this residue is not incorporated via the NRPS, but added later on in the sequence. The product from the NRPS is a cyclic pentapeptide that combines capreomycidine, serine or alanine, and three molecules of Dap. One of these Dap residues is modified during assembly by a desaturase component

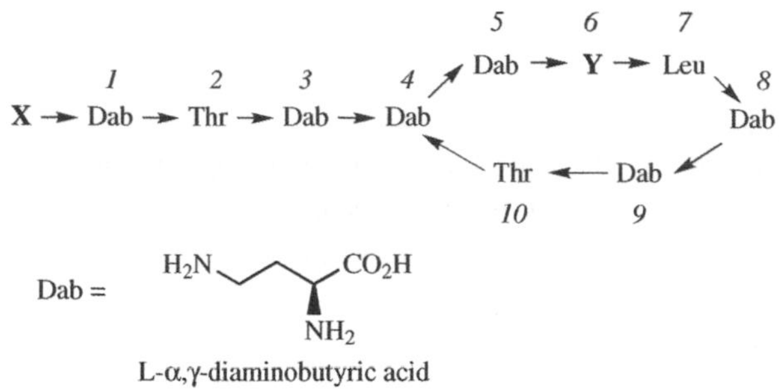
X → Dab → Thr → Dab → Dab → Dab → Y → Leu → Dab → Dab → Thr
Dab =
L-α,γ-diaminobutyric acid

	X	Y
polymyxin B_1	6-methyloctanoic acid	D-Phe
polymyxin B_2	6-methylheptanoic acid	D-Phe
polymyxin E_1 (colistin A)	6-methyloctanoic acid	D-Leu
polymyxin E_2 (colistin B)	6-methylheptanoic acid	D-Leu

Figure 7.24

of the enzyme, and is thus found as dehydro-Dap. This residue is subsequently carbamoylated; as the last step, the β-lysine group is added by the combined action of two proteins to give the capreomycins.

The **polymyxins** are a group of cyclic polypeptide antibiotics produced by species of *Bacillus*; structures of some of these are shown in Figure 7.24. These molecules each contain 10 amino acids, six of which are L-α, γ-diaminobutyric acid (L-Dab), with a fatty acid (6-methyloctanoic acid or 6-methylheptanoic acid) bonded to the N-terminus. The cyclic peptide portion is constructed via an amide bond between the carboxyl terminus and the γ-amino of one of the Dab residues, comparable to the bonding seen in bacitracins. The major novel feature here is the fatty acyl component. This would appear to be bonded via its CoA ester to the first

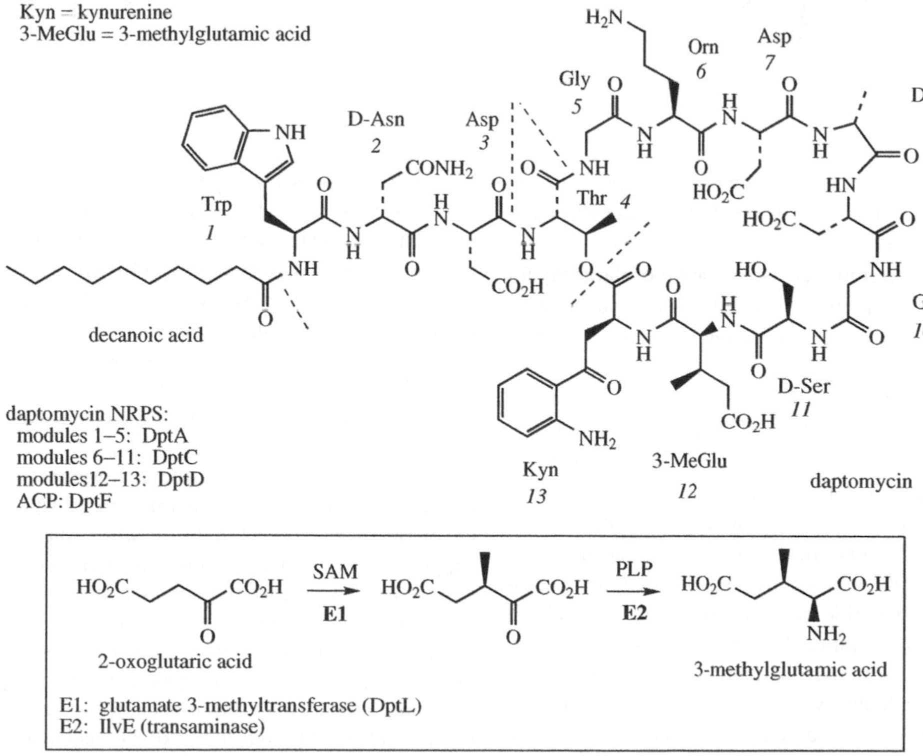
Kyn = kynurenine
3-MeGlu = 3-methylglutamic acid
decanoic acid
Trp 1
D-Asn 2
Asp 3
Thr 4
Gly 5
Orn 6
Asp 7
D-Ala 8
Asp 9
Gly 10
D-Ser 11
3-MeGlu 12
Kyn 13
daptomycin
daptomycin NRPS:
modules 1–5: DptA
modules 6–11: DptC
modules12–13: DptD
ACP: DptF
SAM
E1
PLP
E2
2-oxoglutaric acid
3-methylglutamic acid
E1: glutamate 3-methyltransferase (DptL)
E2: IlvE (transaminase)

Figure 7.25

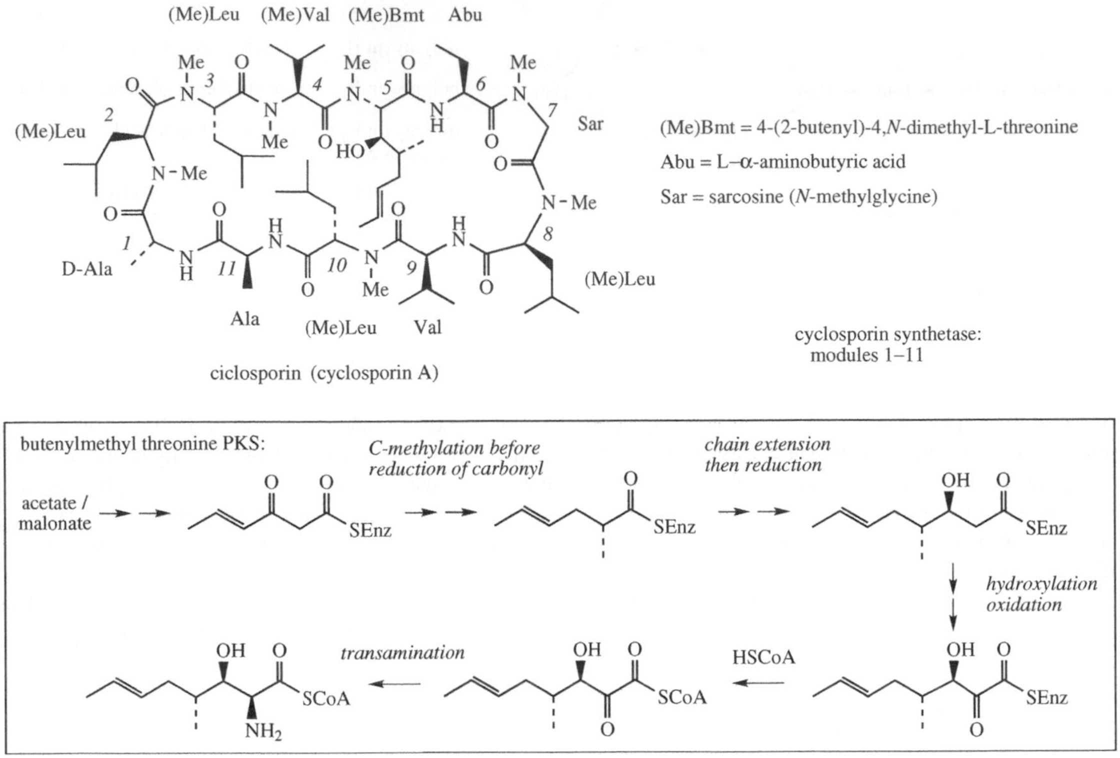

(Me)Bmt = 4-(2-butenyl)-4,*N*-dimethyl-L-threonine

Abu = L–α-aminobutyric acid

Sar = sarcosine (*N*-methylglycine)

cyclosporin synthetase: modules 1–11

ciclosporin (cyclosporin A)

Figure 7.26

amino acid in the sequence, namely Dab, prior to the NRPS catalysing chain extension. Little is known about the origins of Dab.

A fatty acyl starter unit is also a feature of **daptomycin** and some related structures collectively termed lipopeptides. The producing organism, *Streptomyces roseosporus*, normally synthesizes a mixture of antibiotics with differing fatty acyl chains, but the structure shown in Figure 7.25 is the predominant product if decanoic acid is provided during fermentation. This compound shows activity against life-threatening pathogens that are resistant to all current treatments, including vancomycin (see below), and represents the first new class of natural antibiotic to reach the clinic in many years. Structurally, daptomycin is a depsipeptide (essentially a peptide cyclized via a lactone) containing 13 amino acids, with a decapeptide lactone ring derived from cyclization of a threonine side-chain hydroxyl onto the C-terminal carboxyl group. In addition to the decanoyl fatty acid attached to the N-terminus, the structure has three D-amino acids as well as uncommon residues, including L-kynurenine (Kyn), L-ornithine (Orn), and L-3-methylglutamic acid.

The NRPS carries three epimerization domains to provide residues with the D-configuration, and the three uncommon residues are all preformed before incorporation. Kynurenine is an amino acid derived from tryptophan and part of the sequence to anthranilic acid (see page 332), whilst 3-methylglutamic acid is produced via methylation of 2-oxoglutaric acid as shown in Figure 7.25. A stand-alone ACP is used to activate the fatty acyl-CoA and facilitate amino acid chain extension via the NRPS.

Some 25 naturally occurring cyclosporins have been characterized, the most studied being **cyclosporin A** (ciclosporin), a cyclic peptide possessing valuable immunosuppressive properties. Cyclosporin A (Figure 7.26) contains several *N*-methylated amino acid residues, together with the less common L-α-aminobutyric acid and an *N*-methylated butenylmethyl-threonine. Assembly of the polypeptide chain by the enzyme cyclosporin synthetase is known to start from the D-alanine residue. The D-Ala component is not produced by NRPS-mediated epimerization, but is formed externally by way of alanine racemase. Modules 2, 3, 4, 5, 7, 8, and 10 all contain additional *N*-methyltransferase domains; thus, the additional methyl groups are introduced as part of the extension cycle. The butenylmethyl-threonine residue (module 5) has been shown to originate via the acetate pathway, and it effectively comprises a C_8 polyketide chain plus a methyl group from SAM. The sequence shown in Figure 7.26 is consistent with experimental data and is analogous to other acetate-derived compounds (see Chapter 3). The polyketide appears to be released from the PKS as its CoA ester prior to the transamination step.

The glycopeptide vancomycin from cultures of *Amycolatopsis orientalis* (formerly *Streptomyces orientalis*) is a particularly important antibiotic. It is frequently the last-resort agent in the control of methicillin-resistant *Staphylococcus aureus* (MRSA), since many strains have become resistant to all other antibiotics. A novel feature of **vancomycin**, and several other related antibiotics, is the tricyclic structure generated by three phenolic oxidative coupling reactions (Figure 7.27). The vancomycin NRPS produces a linear peptide, probably as shown in Figure 7.27, and utilizes L-leucine, L-tyrosine, L-asparagine, and two unusual amino acids L-4-hydroxyphenylglycine (Hpg) and L-3,5-dihydroxyphenylglycine (Dpg). Despite their apparent similarity, the different hydroxylation patterns in these two compounds are indicative of their quite different origins. 3,5-Dihydroxyphenylglycine is actually PKS derived with a malonate starter, thus allowing ready formation of the aromatic ring (Figure 7.27), followed by appropriate side-chain modifications. On the other hand, 4-hydroxyphenylglycine derives from the shikimate pathway, via 4-hydroxyphenylpyruvic acid; transamination allows tyrosine to supply this keto acid (see page 146). A dioxygenase catalyses oxidative decarboxylation and converts the keto acid into 4-hydroxymandelic acid. This enzyme is similar to the dioxygenase that converts 4-hydroxyphenylpyruvic acid into homogentisic acid (see page 178), but hydroxylates the side-chain rather than the aromatic ring.

Two β-hydroxytyrosine residues in vancomycin originate from L-tyrosine, though hydroxylation probably occurs during NRPS processing. Modules 2, 4, and 5 contain epimerization domains, in keeping with the observed D-amino acids residues in the vancomycin. However, no epimerization domain is associated with module 1, and how the D-leucine residue is formed has yet to be ascertained. In addition, the precise stage at which chlorination of the β-hydroxytyrosine residues occurs needs to be clarified.

Three cytochrome P-450-dependent oxygenase proteins are involved in coupling the aromatic rings, forming two ether linkages and one C–C bond. These are typical phenolic oxidative coupling reactions. *N*-Methylation of the D-leucine residue is probably a late step, perhaps

just prior to the final stages which are the glycosylation reactions.

Bleomycin is a mixture of glycopeptide antibiotics isolated from cultures of *Streptomyces verticillus*, and used for its anticancer activity. The major components of the mixture are bleomycin A_2 and bleomycin B_2, which differ only in the terminal amine functions (Figure 7.28). Bleomycin synthetase is a hybrid NRPS–PKS system containing nine NRPS modules and one PKS module in a protein comprised of eight subunits. The basic structure also shows several quite novel features that are elaborated by the NRPS during the assembly process. The first two amino acids are not linked by an amide bond, but by conjugate addition of the α-amino group of asparagine (residue 2) onto dehydroserine, produced by dehydration from serine (residue 1) (Figure 7.28). The residual thioester bond at the terminus is then converted into an amide. Asn_2 then contributes to formation of a pyrimidine ring by reaction with Asn_3. The initially formed ring is modified further by oxidation and transamination and, somewhat later, by methylation, all involving enzymes outside of the modular NRPS system. Further chain extension adds His_4, Ala_5, and a C_2 fragment from malonate, though the presence of methyltransferase and ketoreductase domains on the PKS module means this fragment then carries methyl and hydroxyl functions. The integration of NRPS and PKS modules is illustrated in Figure 7.28. The remaining residues added are Thr_7, β-Ala_8, Cys_9, and Cys_{10}, with the final three amino acids contributing to a pair of thiazole ring systems. The mechanism for formation of these rings has already been noted in the biosynthesis of epothilones (see page 86) and bacitracins (see page 443). Oxidase domains are responsible for production of thiazole rings (as in epothilones), rather than dihydrothiazole rings (as in bacitracins). In due course, sugar units and an amine side chain are added; the terminal amide function is thus formed in a similar manner to that at the beginning of the chain. The introduction of a hydroxyl group onto the histidine residue, allowing glycosylation, is achieved quite late on in the sequence, after the polypeptide chain is complete. The amide-forming enzyme has relatively broad substrate specificity, and several different amine substrates can be accepted, allowing production of a range of bleomycin analogues just by adding different precursor amines to the culture medium.

The names **streptogramin, virginiamycin, pristinamycin**, and others have been applied to antibiotic mixtures isolated from strains of *Streptomyces virginiae*, and individual components have acquired multiple synonyms; as a family, these antibiotics have now been termed streptogramin antibiotics. The compounds fall into two distinct groups: group A, containing a

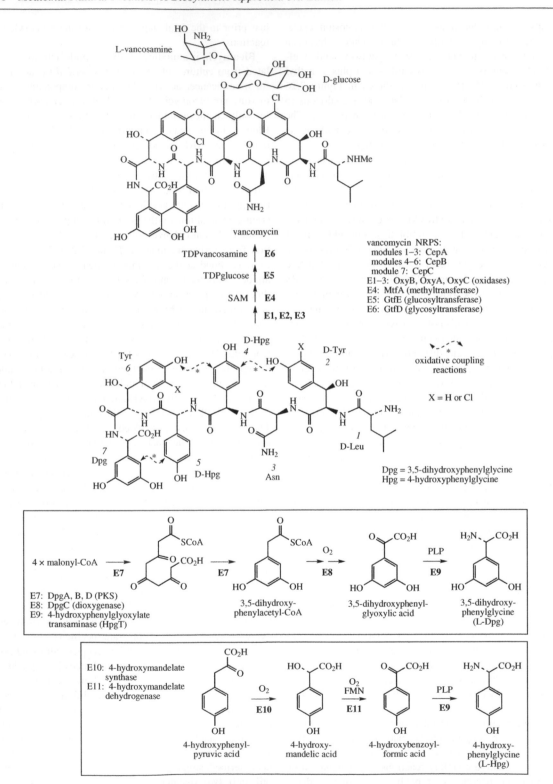

Figure 7.27

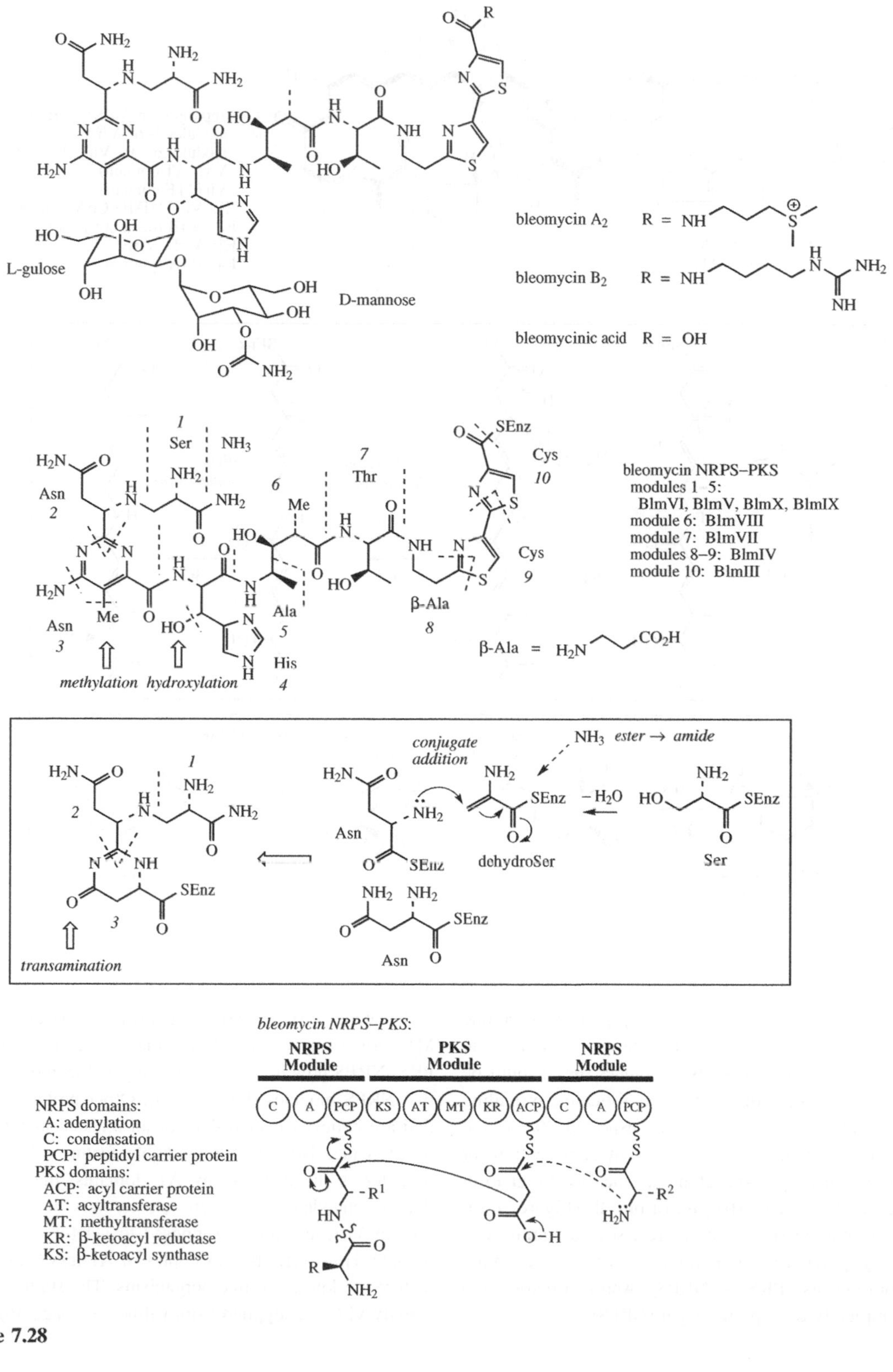
L-gulose
D-mannose
bleomycin A2 R = NH
bleomycin B2 R = NH
bleomycinic acid R = OH
Ser
Asn
Thr
Cys
Me
Ala
His
β-Ala
methylation hydroxylation
bleomycin NRPS–PKS
modules 1–5:
BlmVI, BlmV, BlmX, BlmIX
module 6: BlmVIII
module 7: BlmVII
modules 8–9: BlmIV
module 10: BlmIII
β-Ala = H2N CO2H
conjugate addition
NH3 ester → amide
dehydroSer
– H2O
SEnz
transamination
bleomycin NRPS–PKS:
NRPS Module
PKS Module
NRPS Module
C
A
PCP
KS
AT
MT
KR
ACP
NRPS domains:
A: adenylation
C: condensation
PCP: peptidyl carrier protein
PKS domains:
ACP: acyl carrier protein
AT: acyltransferase
MT: methyltransferase
KR: β-ketoacyl reductase
KS: β-ketoacyl synthase

Figure 7.28

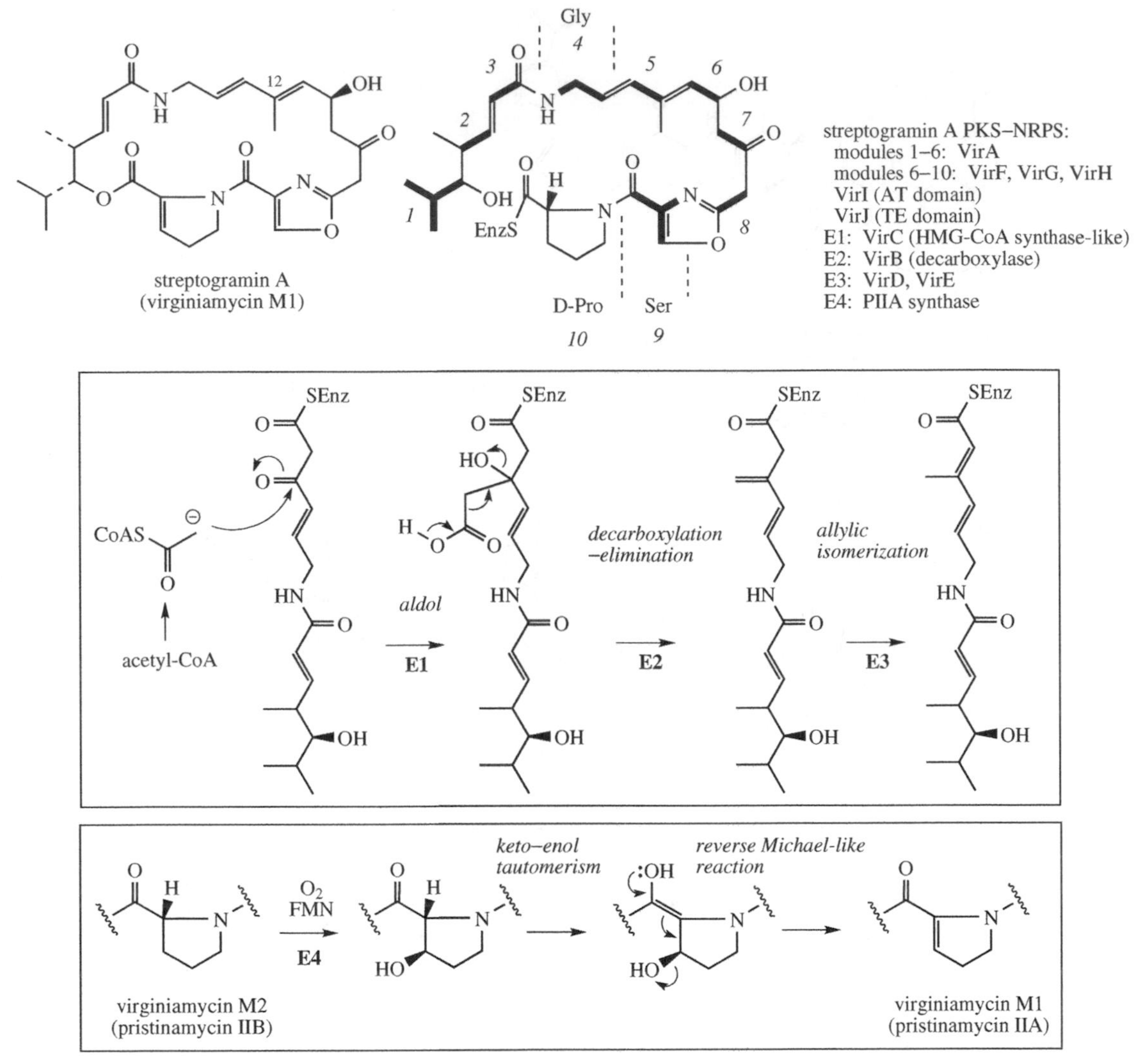

Figure 7.29

23-membered unsaturated ring with peptide and lactone bonds, and group B, which are depsipeptides. Despite their structural differences, as antibiotics they should be considered together: they act synergistically, providing greater antibiotic activity than expected from the separate components. Further, they provide an option for combating vancomycin-resistant bacteria. **Streptogramin A** (Figure 7.29) and **streptogramin B** (Figure 7.30) are representative of the structures; streptogramin A is essentially polyketide in nature, but with some amino acid components (PKS + NRPS), whilst streptogramin B is completely of peptide origin (NRPS).

The gene cluster for **streptogramin A** (=virginiamycin M1; pristinamycin IIA) biosynthesis encodes a hybrid PKS–NRPS protein, consisting of eight PKS modules and two NRPS modules (Figure 7.29). Other proteins identified are believed responsible for an unusual methylation reaction (see below). A remarkable feature of the PKS modules is the lack of any AT domains; the only AT domain encoded by the gene cluster was in a separate protein VirI, and this is supposed to act for every PKS module. However, other examples of AT-less PKS architecture are known in microorganisms. The starter unit is isobutyryl-CoA, supplied from valine (see page 56), then

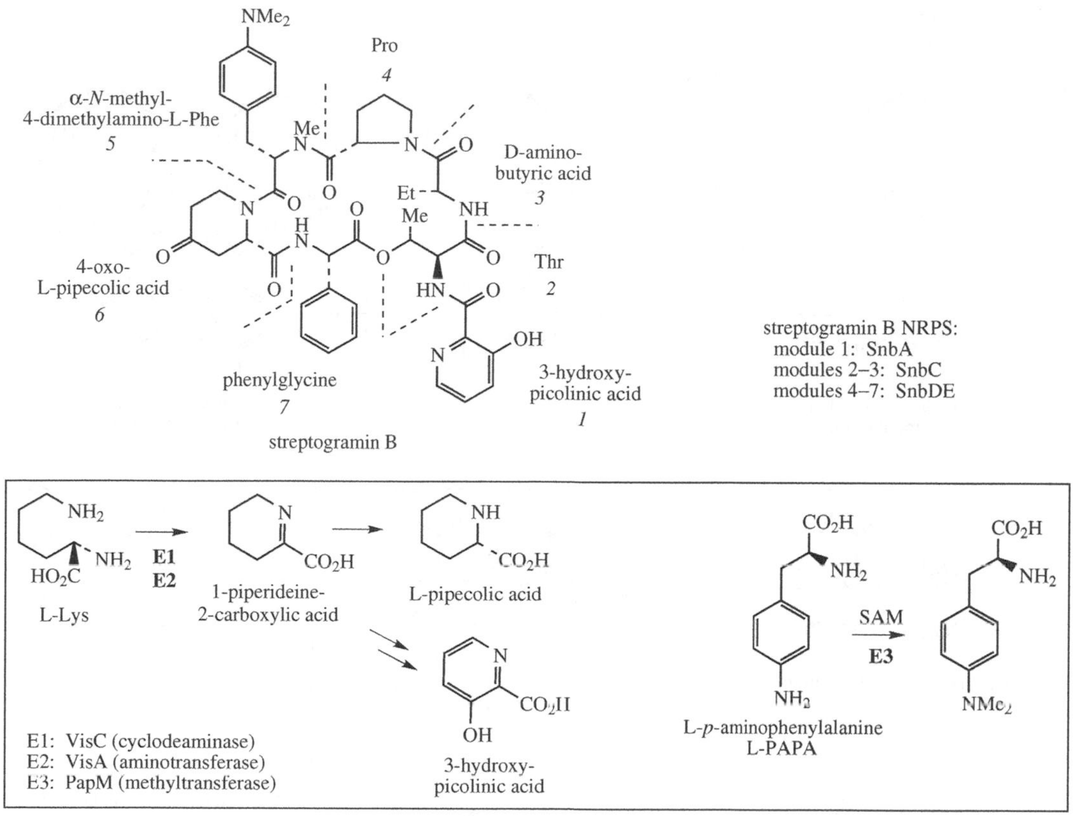

Figure 7.30

two rounds of chain extension with malonate occur. An NRPS module then incorporates glycine, followed by two further malonate extensions. At this stage, other enzyme activities incorporate a methyl group at position 12 (streptogramin numbering). This originates from the methyl of an acetate precursor. Since the protein VirC bears considerable structural similarity to HMG-CoA synthase, the enzyme involved in mevalonate biosynthesis (see page 190), a similar aldol addition mechanism is proposed in this reaction (Figure 7.29). Decarboxylation–elimination (again compare mevalonate biosynthesis) catalysed by VirB is then followed by an allylic elimination. This sequence has also been proposed to occur during mupirocin biosynthesis (see page 91). Again, this 'methylation' looks much more complicated than an SAM-mediated reaction, but is a rather neat way of inserting a methyl group onto a carbonyl, which is not possible with SAM. The oxazole ring is simply analogous to the thiazole ring seen in epothilone (see page 86) and bleomycin (see above), and is formed from malonate and the adjacent serine extender; this is catalysed by Cy (cyclization) domains on the Ser_9 NRPS

module. The last residue to be introduced is D-proline, and this ring is hydroxylated and dehydrated to a dehydroproline. To accommodate the apparent *cis* elimination, the dehydration step is perhaps a reverse Michael-type reaction.

Streptogramin B biosynthesis is rather more straightforward, and performed by a purely NRPS system. However, it is characterized by the utilization of several unusual amino acid precursors that are synthesized prior to peptide assembly (Figure 7.30). The starter amino acid is the pyridine derivative 3-hydroxypicolinic acid, which is derived from lysine via 1-piperideine-2-carboxylic acid (see page 330). 1-Piperideine-2-carboxylic acid is also on the pathway to pipecolic acid, which provides the module 6 residue; 4-oxygenation of this residue is thought to occur post-cyclization. Separate proteins VisC (a cyclodeaminase) and VisA (a PLP-dependent aminotransferase) are involved in producing 1-piperideine-2-carboxylic acid for pipecolic acid and 3-hydroxypicolinic acid biosynthesis. Of the other amino acids, dimethylaminophenylalanine is produced

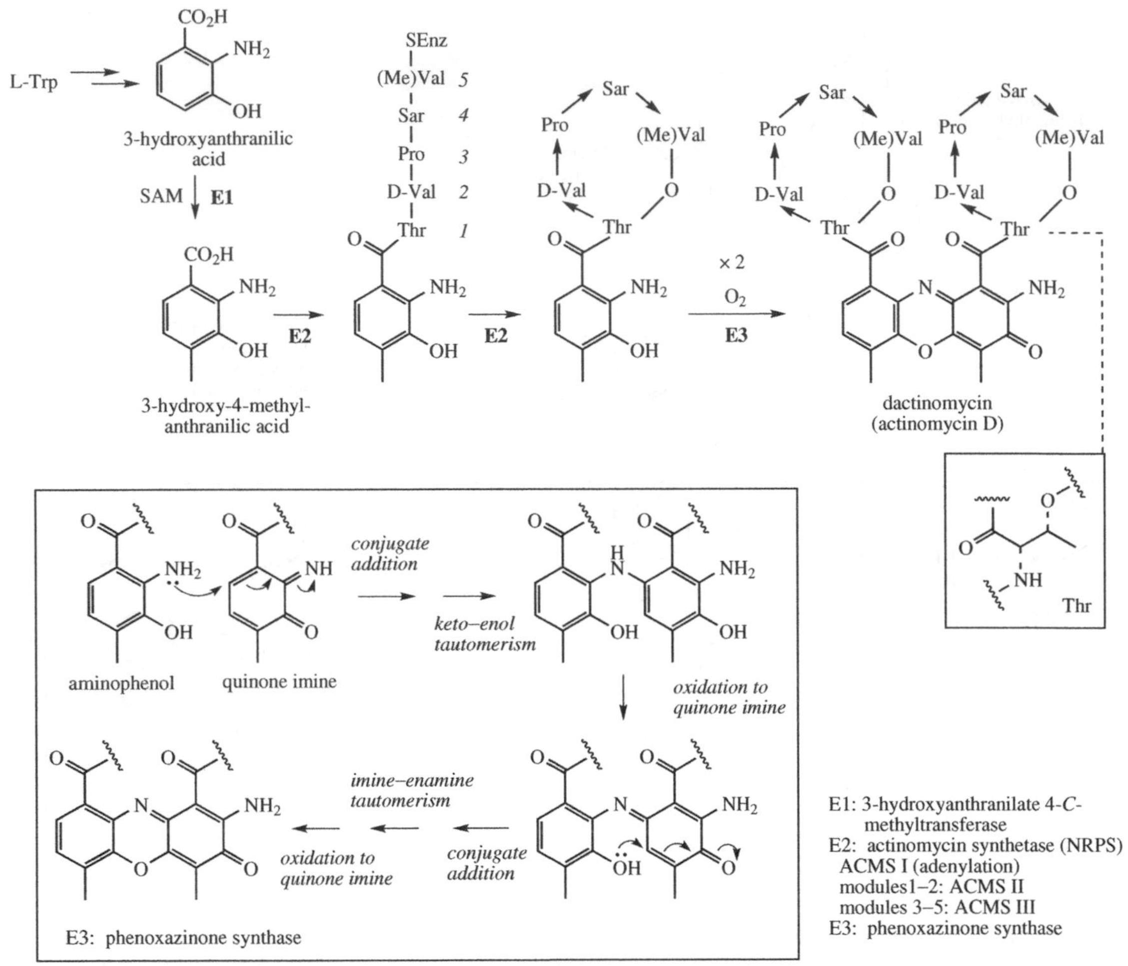

Figure 7.31

by methylation of *p*-aminophenylalanine, already met as a precursor of chloramphenicol (see page 147). *N*-Methylation of the peptide bond between this amino acid and Pro4 is carried out by an *N*-methyltransferase domain of module 5. Phenylglycine is phenylalanine derived, and is probably formed by the series of modifications as seen for 4-hydroxyphenylglycine in vancomycin biosynthesis. The D-aminobutyric acid residue is formed from the L-amino acid by epimerization during assembly. The linear peptide precursor is finally cyclized by nucleophilic attack of the threonine hydroxyl onto the C-terminal carboxyl group of L-phenylglycine, thus releasing it from the synthetase as a lactone.

The structure of **dactinomycin (actinomycin D)** from *Streptomyces parvullus* (formerly *Streptomyces antibioticus*) contains two identical cyclic pentapeptides linked by a planar phenoxazinone dicarboxylic acid system (Figure 7.31). Several related natural actinomycins are known, in which the two peptides are not necessarily identical. Only dactinomycin is used medicinally, and though it has antibacterial and antifungal activity, its high toxicity limits its use to anticancer therapy. The peptide portions are relatively uncomplicated: *N*-methylated-valine and glycine (sarcosine) residues arise via NRPS methyl transferase domains and an epimerization domain accounts for the D-valine isomer. Cyclization is by a lactone linkage utilizing the hydroxyl group of threonine. The phenoxazinone ring system is known to be formed by fusing together two 3-hydroxy-4-methylanthranilic acid-containing substrates (Figure 7.31). This acid arises by *C*-methylation of 3-hydroxyanthranilic acid, a metabolite of tryptophan by the kynurenine pathway (see page 332), and becomes the starter for the NRPS system. The two aromatic rings are subsequently coupled to the phenoxazinone by an oxidizing enzyme, phenoxazinone synthase. A mechanism proposed involves one molecule being oxidized to a quinone imine and acting as substrate for

nucleophilic attack by a second molecule of the amino phenol; both nitrogen and oxygen act as nucleophiles, and the process requires intermediate oxidation steps to regenerate quinone-like electrophiles. An alternative mechanism coupling two molecules of the quinone imine has also been suggested.

Further information concerning medicinally useful peptide antibiotics and related compounds formed by NRPS enzymes are described in Box 7.8.

Box 7.8

Non-ribosomal Peptide Antibiotics

Tyrothricin and Gramicidins

Tyrothricin is a mixture of polypeptide antibiotics produced by cultures of *Bacillus brevis*. The mixture contains about 20–30% linear polypeptides called gramicidins (Figure 7.20), and 70–80% of cyclic structures called tyrocidines (Figure 7.21). Tyrothricin is active against many Gram-positive bacteria, with the linear gramicidins being more active than the cyclic tyrocidines. The two groups are readily separated by solvent fractionation; the **gramicidin** fraction, sometimes termed gramicidin D, a mixture of at least eight closely related compounds, is used principally in ophthalmic preparations. The gramicidins are neutral polypeptides having the N-terminal amino group formylated and the carboxyl group linked to ethanolamine. Most of the gramicidin mixture is composed of valine-gramicidin A (about 80%; Figure 7.20). Apart from the glycine residue, these compounds have a sequence of alternating D- and L-amino acids. Gramicidins adopt a β-helix secondary structure with all amino acid side-chains extending outwards and hydrogen-bonded carbonyl groups lining the core. This arrangement provides an ion channel that traverses bacterial membranes, allowing univalent ions to diffuse out. The tyrocidines are too toxic for therapeutic use on their own, but the tyrothricin mixture is incorporated into lozenges for relief of throat infections.

Gramicidin S is a mixture of cyclic peptides obtained from another strain of *Bacillus brevis*. Its main component is gramicidin S_1 (Figure 7.21), a symmetrical decapeptide known to be formed by the joining together of two separate chains. Gramicidin S is fairly toxic, so its use is restricted to topical preparations, usually in combination with other antibacterials. Gramicidin S also acts on bacterial membranes, increasing permeability and loss of barrier function.

Bacitracins

Bacitracin is a mixture of at least nine peptides produced by cultures of *Bacillus subtilis* and *Bacillus licheniformis*, with the principal component being bacitracin A (Figure 7.22). Bacitracin is active against a wide range of Gram-positive bacteria and appears to affect biosynthesis of the bacterial cell wall by binding to and sequestering a polyprenyl diphosphate carrier of intermediates; this binding also requires a divalent metal ion, with zinc being especially active. It is rarely used systemically because some bacitracin components are nephrotoxic, but as zinc bacitracin, it is a component of ointment formulations for topical application. The vast majority of bacitracin manufactured is used at sub-therapeutic doses as an animal feed additive, to increase feed efficiency, and at therapeutic dosage to control a variety of disorders in poultry and animals.

Capreomycin

Capreomycin is a mixture of cyclic polypeptides obtained from cultures of *Streptomyces capreolus*. It contains about 90% of capreomycins I, principally capreomycin IB (Figure 7.23). This antibiotic is given intramuscularly to treat tuberculosis patients who do not respond to first-line drugs, e.g. rifampicin, or where patients are sensitive to streptomycin. It can cause irreversible hearing loss and impair kidney function. Capreomycin inhibits protein biosynthesis at the translocation step in sensitive bacteria.

Polymyxins

The polymyxins are a group of cyclic polypeptide antibiotics produced by species of *Bacillus*. Polymyxins A–E were isolated from *Bacillus polymyxa*, though polymyxin B and polymyxin E were both subsequently shown to be mixtures of two components.

Box 7.8   (continued)

A polypeptide mixture called colistin isolated from *Bacillus colistinus* was then found to be identical to polymyxin E. **Polymyxin B** and **colistin (polymyxin E)** are both used clinically. These antibiotic mixtures respectively contain principally polymyxin B_1 with small amounts of polymyxin B_2, or predominantly polymyxin E_1 ($\equiv$ colistin A) with small amounts of polymyxin E_2 ($\equiv$ colistin B) (Figure 7.24). The γ-amino groups of the α,γ-diaminobutyric acid (Dab) residues confer a strongly basic character to the antibiotics. This results in detergent-like properties and allows them to bind to and damage bacterial membranes. These peptides have been used for the treatment of infections with Gram-negative bacteria such as *Pseudomonas aeruginosa*, but are seldom used now because of neurotoxic and nephrotoxic effects. However, they are included in some topical preparations, such as ointments, eye drops, and ear drops, frequently in combination with other antibiotics.

Daptomycin

Daptomycin (Figure 7.25) is a cyclic lipopeptide antibiotic produced by *Streptomyces roseosporus*. This organism normally synthesizes a mixture of antibiotics with differing fatty acyl chains, but the structure shown in Figure 7.25 is the predominant product if decanoic acid is provided as a component of the culture medium. Daptomycin has potent antibacterial activity *in vitro* against Gram-positive pathogens, including vancomycin-resistant *Staphylococcus aureus*, MRSA, penicillin-resistant *Streptococcus pneumoniae*, vancomycin-resistant enterococci, and other antibiotic-resistant strains. It is currently used to treat complicated skin and soft-tissue infections caused by Gram-positive bacteria, including MRSA. It needs to be combined with other agents for mixed infections involving Gram-negative bacteria.

Daptomycin and related lipopeptides contain several amino acid residues with acidic side-chains. These acidic groups coordinate calcium ions and are necessary for bioactivity. Calcium binding facilitates aggregation and penetration into the cytoplasmic membranes of Gram-positive bacteria. This is postulated to generate pores in the membrane, which results in the loss of potassium ions, membrane depolarization, and ultimately cell death. This group of antibiotics has become a major focus of attention, since daptomycin represents the first new structural class of natural antimicrobial agents to be approved for clinical use in over 30 years. However, it is expected that resistance to daptomycin will develop, and that further new antibiotics will be needed in the future.

Cyclosporins

The cyclosporins are a group of cyclic peptides produced by fungi such as *Cylindrocarpon lucidum* and *Tolypocladium inflatum*. These agents showed a narrow range of antifungal activity, but high levels of immunosuppressive and anti-inflammatory activities. The main component from the culture extracts is cyclosporin A (**ciclosporin**, **cyclosporin**; Figure 7.26), but some 25 naturally occurring cyclosporins have been characterized. Many of the other natural cyclosporin structures differ only with respect to a single amino acid (the α-aminobutyric acid residue) or the amount of *N*-methylation. Of all the natural analogues, and many synthetic ones produced, cyclosporin A is the most valuable for drug use. It is now widely exploited in organ and tissue transplant surgery to prevent rejection following bone-marrow, kidney, liver, pancreas, lung, and heart transplants. It has revolutionized organ transplant surgery, substantially increasing survival rates in transplant patients. It may be administered orally or by intravenous injection, and the primary side-effect is nephrotoxicity, necessitating careful monitoring of kidney function. Cyclosporin A has the same mode of action as the macrolide FK-506 (tacrolimus; see page 85) and is believed to inhibit T-cell activation in the immunosuppressive mechanism by first binding to a receptor protein, giving a complex that then inhibits a phosphatase enzyme

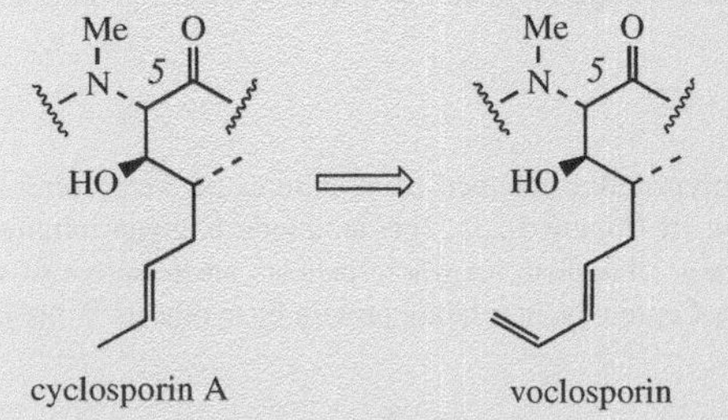

Figure 7.32

Box 7.8 (continued)

called calcineurin. The resultant aberrant phosphorylation reactions prevent appropriate gene transcription and subsequent T-cell activation. Cyclosporin A also finds use in the specialist treatment of severe resistant psoriasis and severe eczema. It is also being used in cases of severe rheumatoid arthritis when other therapies are ineffective.

A semi-synthetic derivative of cyclosporin A, **voclosporin** (Figure 7.32), is in advanced clinical trials. This has a modified module 5 component and is more potent than cyclosporin A, as well as being less toxic. This drug is also successful in treatment of the eye condition uveitis, an inflammatory disease of the iris and focusing muscle.

Vancomycin and Teicoplanin

Vancomycin (Figure 7.27) is a glycopeptide antibiotic produced in cultures of *Amycolatopsis orientalis* (formerly *Streptomyces orientalis*) and has activity against Gram-positive bacteria, especially resistant strains of staphylococci, streptococci, and enterococci. It is an important agent reserved for the control of MRSA, with some strains now being sensitive only to vancomycin or teicoplanin (below). Vancomycin is not absorbed orally, so must be administered by intravenous injection. However, it can be given orally in the treatment of pseudomembranous colitis caused by *Clostridium difficile*, which may occur after administration of other antibiotics. The antibiotic may cause toxicity to ear and kidneys. Vancomycin acts by its ability to form a complex with terminal –D-Ala–D-Ala residues of growing peptidoglycan chains (see Figure 7.42 and page 496). This prevents cross-linking to adjacent strands and thus inhibits bacterial cell wall biosynthesis. The –D-Ala–D-Ala residues are accommodated in a 'carboxylate-binding pocket' in the vancomycin structure. By preventing peptidoglycan polymerization and cross-linking, it weakens the bacterial cell wall and ultimately causes cell lysis.

The teicoplanins (Figure 7.33) possess the same basic structure as vancomycin, but the N-terminal (4-hydroxyphenylglycine) and third (3,5-dihydroxyphenylglycine) amino acids are also aromatic, this allowing further phenolic oxidative coupling and generation of yet another ring system. **Teicoplanin** for drug use is a mixture of five teicoplanins produced by cultures of *Actinoplanes teichomyceticus* which differ only in the nature and length of the fatty acid chain attached to the sugar residue. Teicoplanin has similar antibacterial activity to vancomycin, but has a longer duration of action and may be administered by intramuscular as well as by intravenous injection. It is also used against Gram-positive pathogens resistant to established antibiotics.

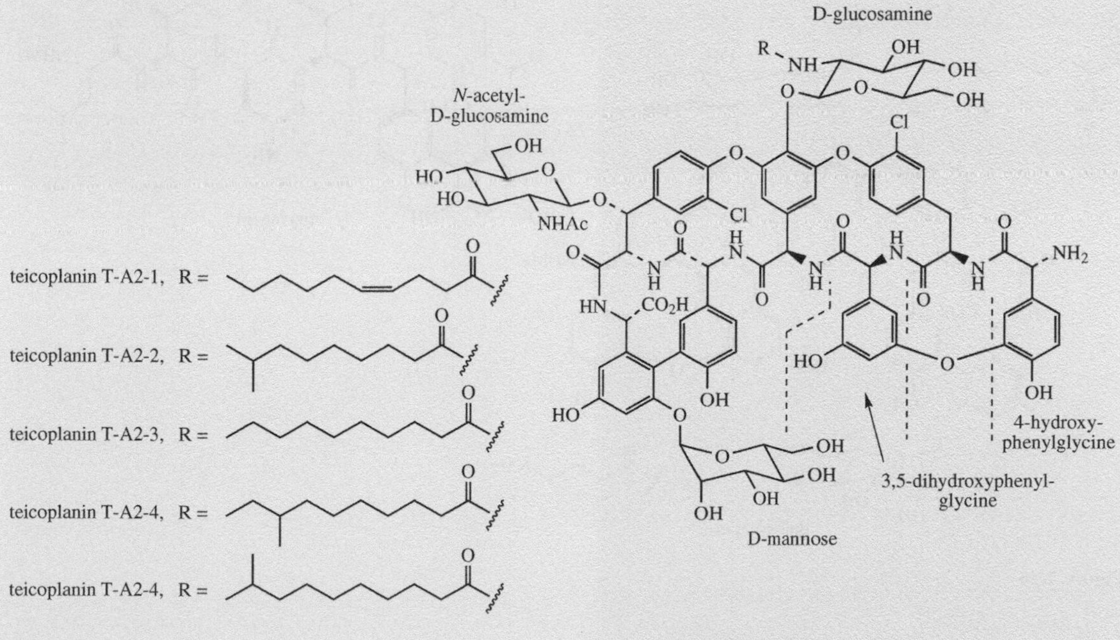

Figure 7.33

Box 7.8   (continued)

   Vancomycin, teicoplanin, and structurally related glycopeptides are often referred to as dalbaheptides (from D-*al*anyl-D-alanine-*b*inding *hepta*pept*ide*), reflecting their mechanism of action and their chemical nature. Unfortunately, with increasing use of vancomycin and teicoplanin, there have even been reports of these agents becoming ineffective because resistant bacterial strains have emerged, particularly in enterococci. In resistant strains, the terminal amide-linked –D-Ala–D-Ala residues, to which the antibiotic normally binds, have become replaced by ester-linked –D-Ala–D-lactate. The –D-Ala–D-lactate sequence results in loss of crucial hydrogen-bonding interactions and a thousand-fold reduction in binding efficiency for vancomycin; however, this variant still be used by the bacteria in peptidoglycan cross-linking.

   The need for new antibiotics becomes progressively urgent as bacteria steadily become resistant to current drugs. Oritavancin and dalbavancin (Figure 7.34) are two promising new agents from the vancomycin/teicoplanin group. **Oritavancin** is a semi-synthetic *N*-aryl derivative of the natural product chloroeremomycin from *Amycolatopsis orientalis* that differs from vancomycin in having two 3-epivancosamine sugars. Oritavancin shows strong bactericidal properties under conditions where vancomycin is bacteriostatic. **Dalbavancin** is a semi-synthetic derivative of a teicoplanin family member termed A40926 from a *Nonomuraea* species. Like teicoplanin, it has a long fatty acyl moiety, in this case a C_{12} terminally branched chain, in amide linkage to the glucosamine, but the *N*-acetylglucosamine sugar is missing. The synthetic modification is amidation of the terminal carboxylic acid. Dalbavancin has enhanced potency over vancomycin and teicoplanin against susceptible enterococci and MRSA. Other variants are being explored, including a covalently bonded vancomycin–cephalosporin combination.

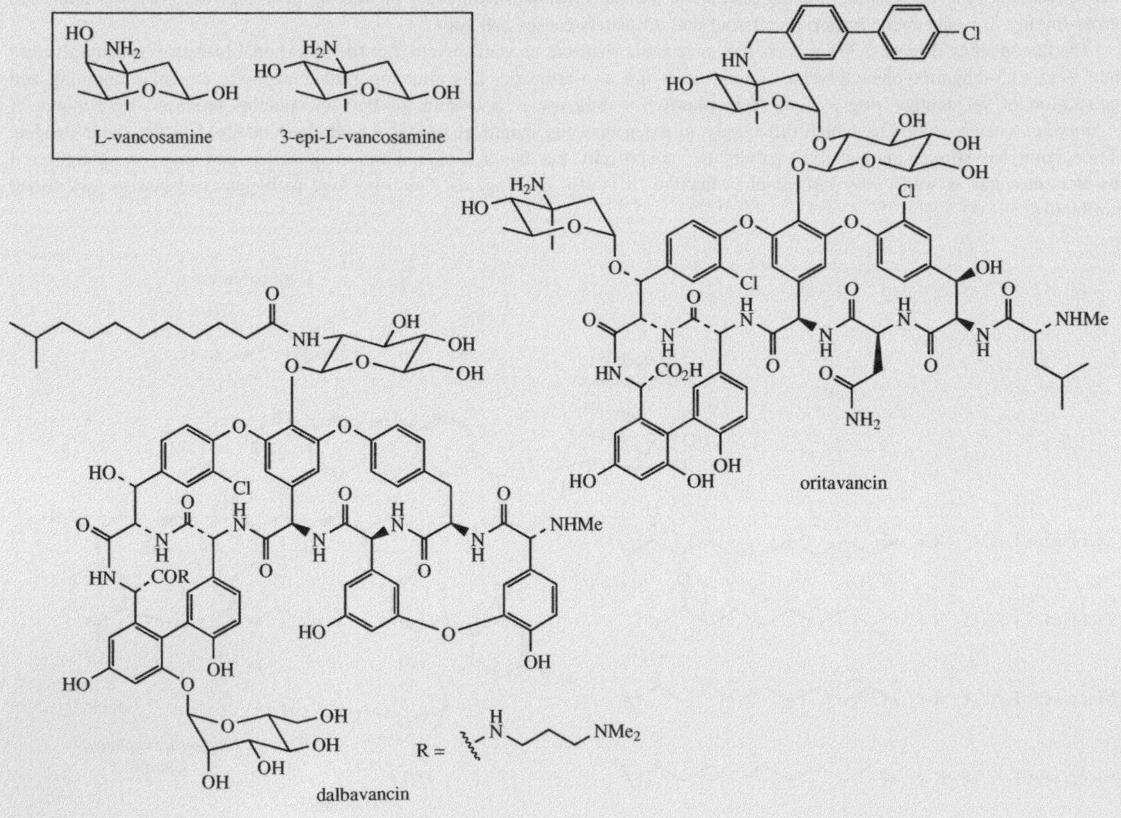

Figure 7.34

Box 7.8 (continued)

Bleomycin

Bleomycin is a mixture of glycopeptide antibiotics isolated from cultures of *Streptomyces verticillus*, used for its anticancer activity. The major component (55–70%) of the mixture is bleomycin A_2 (Figure 7.28), with bleomycin B_2 constituting about 30%. The various bleomycins differ only in their terminal amine functions, the parent compound bleomycinic acid (Figure 7.28) being inactive. Bleomycin is a DNA-cleaving drug, causing single and double strand breaks in DNA. The bithiazole system is involved in binding to DNA, probably by intercalation, whilst other parts of the molecule near the N-terminus are involved in chelating a metal ion, usually Fe^{2+}, and oxygen, which are necessary for the DNA degradation reaction. More recently, bleomycin A_2 has been shown to cleave RNA as well as DNA.

Bleomycin is used alone, or in combination with other anticancer drugs, to treat squamous cell carcinomas of various organs, lymphomas, and some solid tumours. It is unusual amongst antitumour antibiotics in producing very little bone-marrow suppression, making it particularly useful in combination therapies with other drugs which do cause this response. However, there is some lung toxicity associated with bleomycin treatment. Various analogues have been made by adding different precursor amines to the culture medium, or by semi-synthesis from bleomycinic acid, though none has yet proved clinically superior to bleomycin.

Streptogramins

The names streptogramin, virginiamycin, pristinamycin, and others have been applied to antibiotic mixtures isolated from strains of *Streptomyces virginiae*, and individual components have thus acquired multiple synonyms; as a family, these antibiotics have now been termed streptogramin antibiotics. These compounds fall into two distinct groups: group A, containing a 23-membered unsaturated ring with peptide and lactone bonds, and group B, which are depsipeptides (essentially peptides cyclized via a lactone). Until recently, most commercial production of these antibiotics was directed towards animal feed additives, but the growing emergence of antibiotic-resistant bacterial strains has led to the drug use of some streptogramin antibiotics. Thus, **dalfopristin** and **quinupristin** (Figure 7.35) are water-soluble drugs that may be used in combination for treating infections caused by Gram-positive bacteria that have failed to respond to other antibiotics, including MRSA, vancomycin-resistant enterococci and staphylococci, and drug-resistant *Streptococcus pneumoniae*. They may also need to be combined with other agents where mixed infections involve Gram-negative organisms.

Dalfopristin is a semi-synthetic sulfonyl derivative of streptogramin A (also termed virginiamycin M1, mikamycin A, pristinamycin IIA, and other names; Figure 7.29) and quinupristin is a modified form of streptogramin B (also mikamycin B, pristinamycin IA, and other names; Figure 7.30). Members of the A group tend to be less powerful antibiotics than those of the B group, but together they act synergistically, providing greater activity than the combined activity expected from the separate

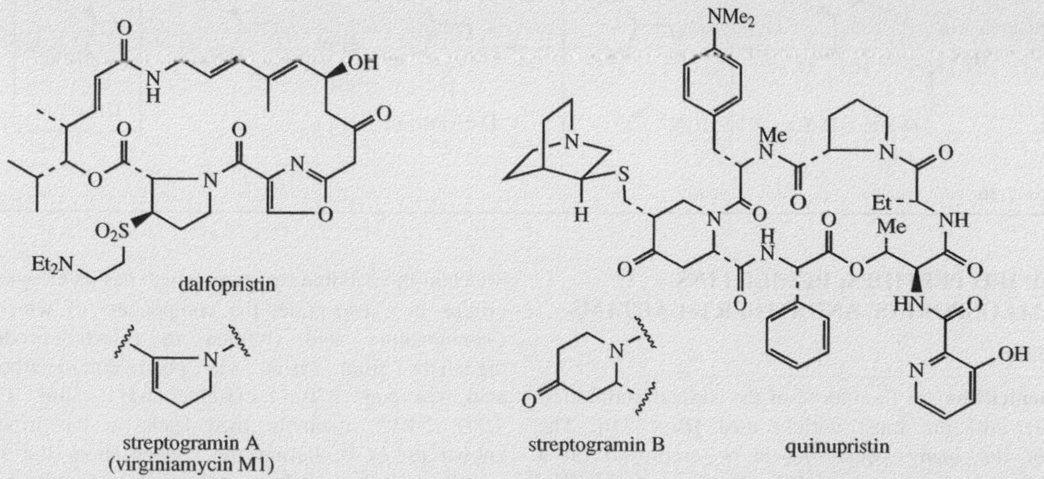

dalfopristin

streptogramin A
(virginiamycin M1)

streptogramin B

quinupristin

Figure 7.35

Box 7.8   (continued)

components. The dalfopristin and quinupristin combination is supplied in a 70:30 ratio, which provides maximum synergy (a 100-fold increase in activity compared with the single agents), and also corresponds to the natural proportion of group A to group B antibiotics in the producer organism. The streptogramins bind to the peptidyl transferase domain of the 50S ribosomal subunit; the remarkable synergism arises because initial binding of the group A derivative causes a conformational change to the ribosome, increasing affinity for the group B derivative and formation of an extremely stable ternary complex. This makes the streptogramin combination bactericidal, whereas the single agents provide only bacteriostatic activity. Other streptogramin-related combinations are being tested clinically.

Dactinomycin

Dactinomycin (**actinomycin D**; Figure 7.31) is an antibiotic produced by *Streptomyces parvullus* (formerly *Streptomyces antibioticus*), which has antibacterial and antifungal activity, but whose high toxicity limits its use to anticancer therapy. Several related natural actinomycins are known, but only dactinomycin is used medicinally. Dactinomycin has a planar phenoxazinone dicarboxylic acid system in its structure, to which are attached two identical cyclic pentapeptides. In other actinomycins, the two peptides are not necessarily identical. This planar phenoxazinone ring intercalates with double-stranded DNA, inhibiting DNA-dependent RNA polymerases, but can also cause single-strand breaks in DNA. It is principally used to treat paediatric cancers, including Wilms' tumour of the kidney, but may produce several serious and painful side-effects. However, as a selective inhibitor of DNA-dependent RNA synthesis (transcription), it has become an important research tool in molecular biology.

Cycloserine

D-**Cycloserine** (Figure 7.36) is probably the simplest substance with useful antibiotic activity, and though not a peptide, is amino acid related. It is produced by cultures of *Streptomyces garyphalus* and *Streptomyces lavendulae*, though drug material is prepared synthetically. Cycloserine is water soluble and has a broad spectrum of antibacterial activity, but it is only employed for its activity against *Mycobacterium tuberculosis*. It behaves as a structural analogue of D-alanine, inhibiting the incorporation of D-alanine into bacterial cell walls by inhibition of D-Ala–D-Ala ligase. It is also an inhibitor of alanine racemase, preventing the formation of D-alanine for the cross-linking reactions (see page 464). Since it can produce neurotoxicity in patients, it is only used occasionally and is reserved for infections resistant to first-line drugs.

The heterocyclic nitrogen of D-cycloserine is known to originate from hydroxyurea, and *O*-ureido-D-serine is the immediate precursor (Figure 7.36).

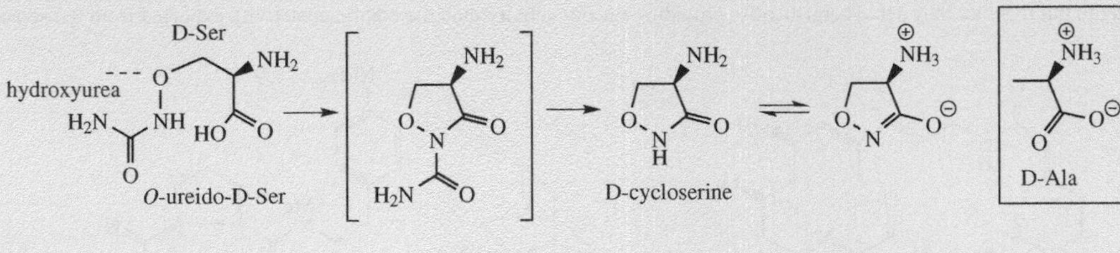

Figure 7.36

MODIFIED PEPTIDES: PENICILLINS, CEPHALOSPORINS, AND OTHER β-LACTAMS

Penicillins

The **penicillins** are the oldest of the clinical antibiotics, but are still the most widely used [Box 7.9]. The first of the many penicillins to be employed on a significant scale was **penicillin G** (**benzylpenicillin**; Figure 7.37), obtained from the fungus *Penicillium chrysogenum* by fermentation in a medium containing corn-steep liquor. Penicillins contain a fused β-lactam-thiazolidine structure, which has its biosynthetic origins in a tripeptide, the components of which are L-aminoadipic acid (formed in β-lactam-producing organisms from lysine via piperideine-6-carboxylic acid; see page 330), L-cysteine and L-valine (Figure 7.37). NRPS assembly then leads to the tripeptide known as ACV; during the condensation, the valine residue is epimerized to the D-form. [*Caution*: ACV is an acronym; it does not refer to the systematic abbreviation described on page 423; ACV refers to δ-(L-α-aminoadipyl)–L-cysteinyl–D-valine.]

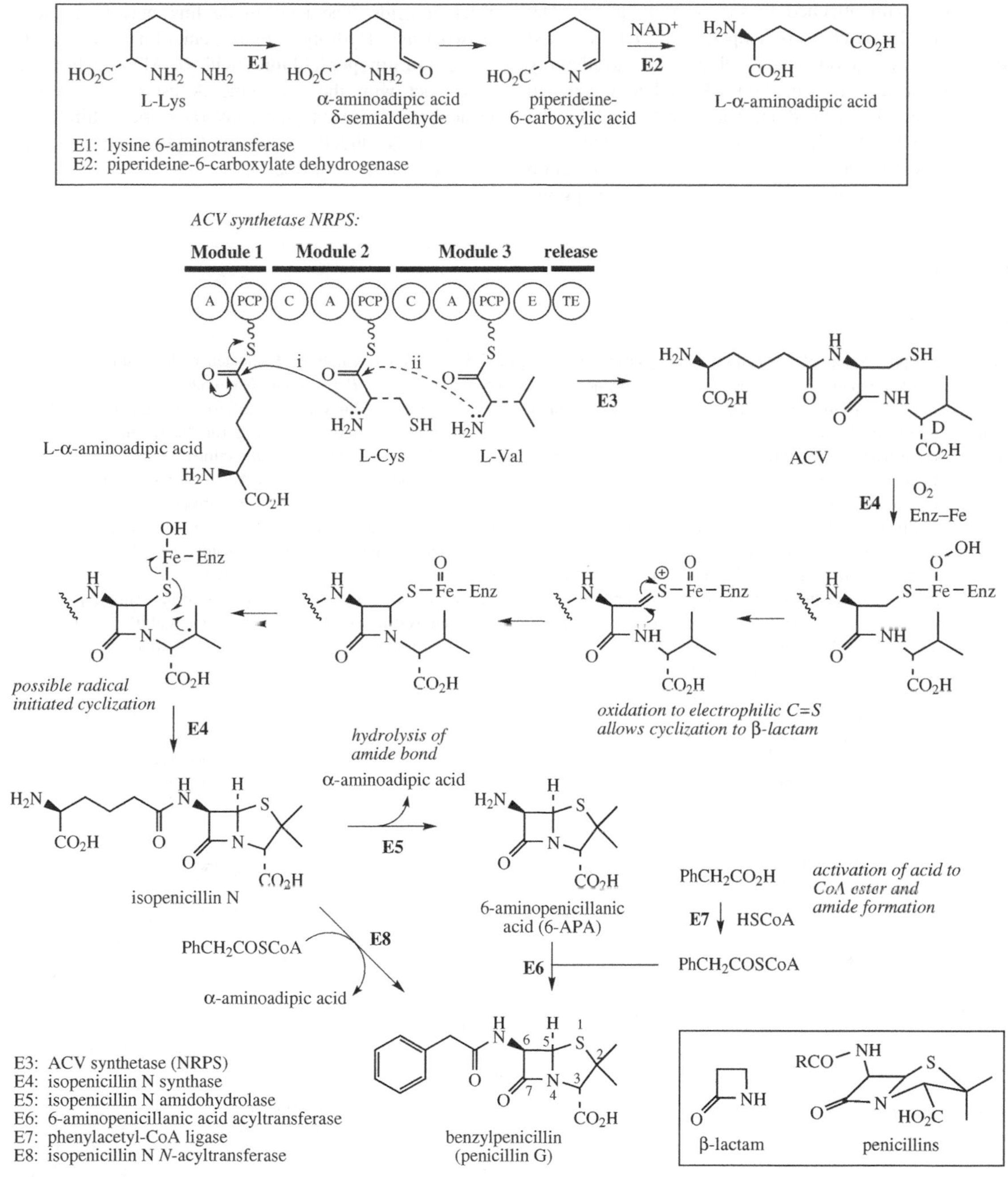

Figure 7.37

ACV is then cyclized to **isopenicillin N**, with a single enzyme isopenicillin N synthase catalysing formation of the bicyclic ring system characteristic of the penicillins. The reaction is oxidative, requires molecular oxygen, and

there is evidence that the four-membered β-lactam ring is formed first. The mechanism shown in Figure 7.37 is a simplistic outline for what is a quite complex reaction. **Penicillin G** differs from isopenicillin N by the nature of the side-chain attached to the 6-amino group. The α-aminoadipyl side-chain of isopenicillin N is removed and replaced by another according to its availability from the fermentation medium. Phenylethylamine in the corn-steep liquor medium was transformed by the fungus into phenylacetic acid, which then reacted as its coenzyme A ester to produce the new amide penicillin G. Several other penicillins are accessible by supplying different acids. The new amide link may be achieved in two ways. Hydrolysis of isopenicillin N releases the amine **6-aminopenicillanic acid** (6-APA), which can then react with the coenzyme A ester. Alternatively, an acyltransferase enzyme converts isopenicillin N into penicillin G directly, without 6-APA actually being released from the enzyme.

Box 7.9

Penicillins

Commercial production of **benzylpenicillin (penicillin G**; Figure 7.37) is by fermentation of selected high-yielding strains of *Penicillium chrysogenum* in the presence of phenylacetic acid. Though wild-type *Penicillium chrysogenum* cultures produced about 70 mg l^{-1}, commercial strains can yield more than 50 g l^{-1}. Benzylpenicillin was the earliest commercially available member of the penicillin group of antibiotics, and it still remains an important and useful drug for the treatment of many Gram-positive bacteria, including streptococcal, pneumococcal, gonococcal, and meningococcal infections.

Benzylpenicillin is destroyed by gastric acid; thus, it is not suitable for oral administration and is best given as intramuscular or intravenous injection of the water-soluble sodium salt. Decomposition under acidic conditions leads to formation of penicillic acid and/or penicillenic acid, depending on pH (Figure 7.38). The β-lactam ring is opened by a mechanism in which the side-chain carbonyl participates, resulting in formation of an oxazolidine ring. Penicillic acid arises as the result of nucleophilic attack of the thiazolidine nitrogen onto the iminium function, followed by expulsion of the carboxylate leaving group. Alternatively, elimination of thiol accounts for formation of penicillenic acid. At higher pH values, benzylpenicillin suffers simple β-lactam ring opening and gives penicilloic acid (Figure 7.38). The strained β-lactam (cyclic amide) ring is more susceptible to hydrolysis than the unstrained side-chain amide function, since the normal stabilizing effect of the lone pair from the adjacent nitrogen is not possible due to the geometric restrictions (Figure 7.39).

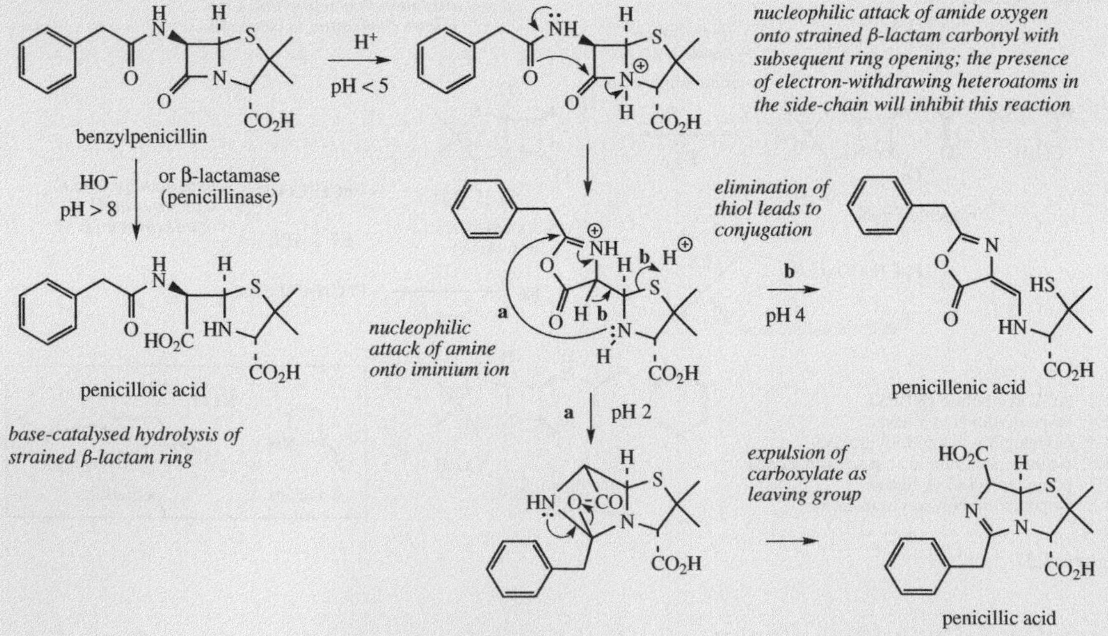

Figure 7.38

Box 7.9 (continued)

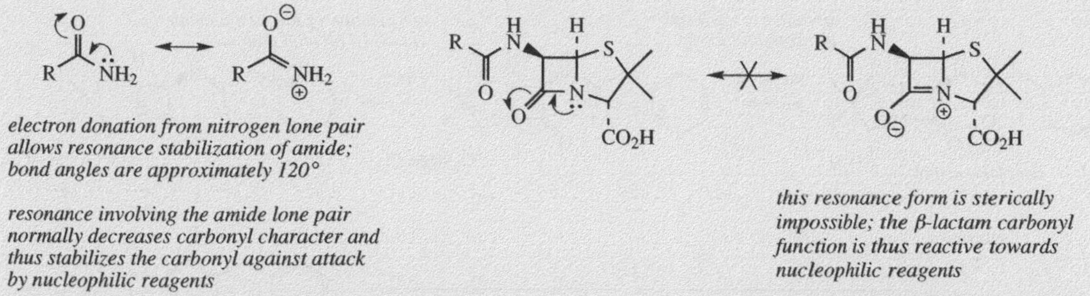

*electron donation from nitrogen lone pair
allows resonance stabilization of amide;
bond angles are approximately 120°*

*resonance involving the amide lone pair
normally decreases carbonyl character and
thus stabilizes the carbonyl against attack
by nucleophilic reagents*

*this resonance form is sterically
impossible; the β-lactam carbonyl
function is thus reactive towards
nucleophilic reagents*

Figure 7.39

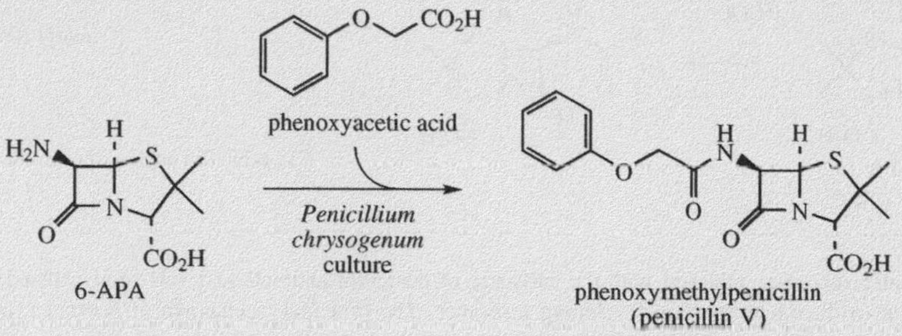

phenoxyacetic acid

*Penicillium
chrysogenum*
culture

6-APA

phenoxymethylpenicillin
(penicillin V)

Figure 7.40

Supplementation of the fermentation medium with acids other than phenylacetic acid was used to provide structurally modified penicillins, though the scope was limited to a series of monosubstituted acetic acids by the specificity of the fungal enzymes involved in activation of the acids to their coenzyme A esters. The most important new penicillin produced was **phenoxymethylpenicillin (penicillin V)**, a result of adding phenoxyacetic acid to the culture (Figure 7.40). This new penicillin had the great advantage of being acid resistant, since the introduction of an electron-withdrawing heteroatom into the side-chain inhibits participation of the side-chain carbonyl in the reaction shown in Figure 7.38. Thus, penicillin V is suitable for oral administration, and still has particular value for respiratory tract infections and tonsillitis.

A much wider range of penicillins, many of which have become clinically useful, may be produced by semi-synthesis from 6-APA. A multistage, but high yielding, procedure has been developed to hydrolyse chemically a primary fermentation product like benzylpenicillin to 6-APA (Figure 7.41). This exploits the ability of the side-chain amide to adopt a resonance form, thus allowing conversion into an imidyl chloride and then an imidyl ether, which is readily hydrolyzed. A new side-chain can then be added by simple esterification (Figure 7.41). Hydrolysis of penicillin G or penicillin V may also be accomplished enzymically in very high yield by using bacterial enzyme preparations from *Escherichia coli*, or species of *Fusarium* or *Erwinia*. Certain strains of *Penicillium chrysogenum* accumulate 6-APA, so that this compound may be produced by fermentation, though this is commercially less economic than the hydrolysis approach.

Clinically useful penicillins produced by semi-synthesis or total synthesis are listed in Table 7.3. Penicillins with side-chains containing a basic amino group, e.g. **ampicillin** and **amoxicillin (amoxycillin)**, are also acid resistant, since this nitrogen becomes preferentially protonated. In addition, these agents were found to have a broader spectrum of activity than previous materials, particularly activity against some Gram-negative bacteria which were not affected by penicillins G and V. The polar side-chain improves water solubility and cell penetration into these microorganisms. Amoxicillin shows better oral absorption properties than ampicillin. Broad-spectrum activity is also found in penicillins containing a carboxyl group in the side-chain, e.g. **ticarcillin**; although this compound is still acid sensitive and, thus, orally inactive, it demonstrates activity against pseudomonads, especially *Pseudomonas aeruginosa*. The acylureido penicillins, e.g. **piperacillin**, are much more active against *Pseudomonas aeruginosa*, and they are also active against other Gram-negative bacteria, such as *Klebsiella pneumoniae* and *Haemophilus influenzae*. **Pivmecillinam** is an acyloxymethyl ester pro-drug and is hydrolysed to mecillinam after oral ingestion. It is unusual in not having an acyl side-chain; instead it is an amidino derivative. It has significant activity towards many Gram-negative bacteria.

Box 7.9   (continued)

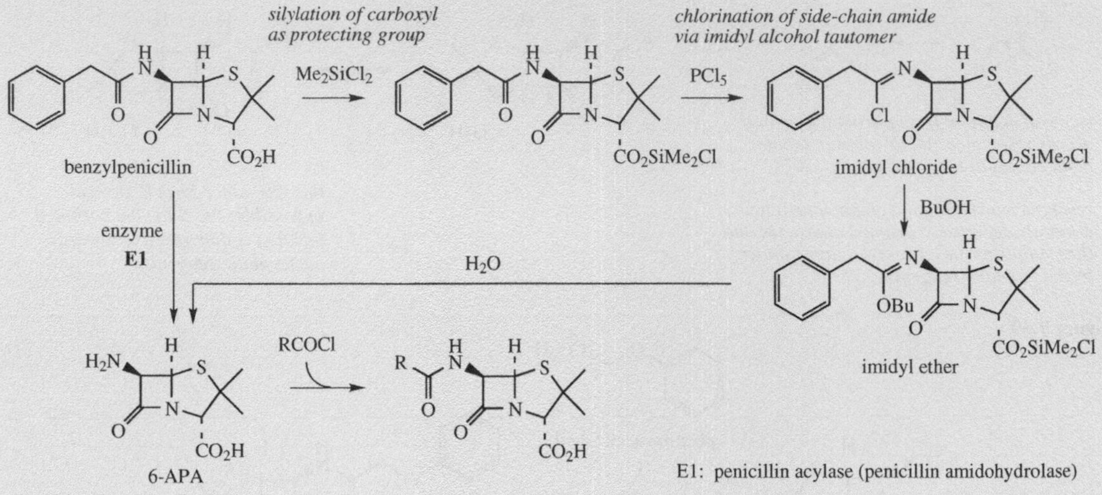

Figure 7.41

Despite the dramatic successes achieved with the early use of penicillin antibiotics, it was soon realized that many bacteria previously susceptible to these agents were developing resistance. The principal mechanism of resistance lies in the ability of organisms to produce β-lactamase (penicillinase) enzymes capable of hydrolysing the β-lactam ring in the same manner as shown for the base-catalysed hydrolysis in Figure 7.38. Several distinct classes of bacterial β-lactamases are recognized, the main division being into serine enzymes and zinc enzymes. The former have an active site serine residue which attacks the β-lactam carbonyl, forming an acyl–enzyme intermediate. On the basis of characteristic amino acid sequences, they are then subdivided into three classes: A, C and D. The zinc metallo-enzymes form class B and appear to involve only non-covalently bound intermediates. Class A β-lactamases are the most common amongst pathogenic bacteria. Most staphylococci are now resistant to benzylpenicillin. The discovery of new penicillins that were not hydrolysed by bacterial β-lactamases was thus a major breakthrough. **Methicillin** (Table 7.3), though no longer used, was the first commercial β-lactamase-resistant penicillin, and the steric bulk of the side-chain appears to contribute to this valuable property, hindering the approach of β-lactamase enzymes. Methicillin is acid sensitive, since it lacks an electron-withdrawing side-chain, but other penicillins were developed which combined bulk and electron-withdrawing properties, and could thus be used orally. These include a group of isoxazole derivatives termed the oxacillins, of which cloxacillin and, in due course, its fluoro derivative **flucloxacillin** became first-choice agents against β-lactamase-producing *Staphylococcus aureus*. **Temocillin** also has excellent resistance to β-lactamases as well as high activity towards Gram-negative organisms. It differs from all the other penicillins described in possessing a 6α-methoxyl group (compare the cephamycins, page 465). Another way of overcoming the penicillin-degrading effects of β-lactamase is to combine a β-lactamase-sensitive agent, e.g. amoxicillin or ticarcillin, with clavulanic acid (see page 470) which is a specific inhibitor of β-lactamase. Other mechanisms of resistance which have been encountered include modification of the binding sites on penicillin-binding proteins (see below) thus reducing their affinity for the penicillin, and decreased cell permeability leading to reduced uptake of the antibiotic. Strains of *Staphylococcus aureus* resistant to both methicillin and isoxazolylpenicillins, e.g. cloxacillin and flucloxacillin, are known to have modified and insensitive penicillin-binding proteins; such strains are termed methicillin-resistant *Staphylococcus aureus* (MRSA). Some 60% of clinically isolated strains of *Staphylococcus aureus* are methicillin resistant, and this has become a major cause of concern in today's hospitals (see vancomycin, page 456).

Penicillins and other β-lactam drugs exert their antibacterial effects by binding to proteins (penicillin-binding proteins), peptidase enzymes that are involved in the late stages of the biosynthesis of the bacterial cell wall. Cross-linking of the peptidoglycan chains which constitute the bacterial cell wall (see page 496) involves a terminal –D-Ala–D-Ala intermediate which, in its transition state conformation, closely resembles the penicillin molecule (Figure 7.42). As a result, the penicillin occupies the active site of the enzyme and becomes bound via an active-site serine residue, this binding causing irreversible enzyme inhibition and cessation of cell wall biosynthesis. Growing cells are then killed due to rupture of the cell membrane

Box 7.9 (continued)

Table 7.3 Semi-synthetic and synthetic penicillins

R (or full structure)	**Name**	**Notes**
Ph⌒	benzylpenicillin (penicillin G)	acid-sensitive, narrow spectrum
Ph–O⌒	phenoxymethylpenicillin (penicillin V)	acid-resistant, narrow spectrum
Ph (NH₂)	ampicillin	acid-resistant, broad spectrum
HO–C₆H₄ (NH₂)	amoxicillin (amoxycillin)	acid-resistant, broad spectrum; better absorption than ampicillin; may be used in combination with β-lactamase inhibitor clavulanic acid
thiophene (CO₂H, RS)	ticarcillin	acid-sensitive, broad spectrum; used in combination with β-lactamase inhibitor clavulanic acid
piperazinedione structure	piperacillin	broad spectrum, more active than ticarcillin against *P. aeruginosa*; used by injection in combination with β-lactamase inhibitor tazobactam for serious infections
azepine structure	pivmecillinam	orally active ester pro-drug; β-lactamase sensitive, active against Gram-negative organisms (except *P. aeruginosa*)
(OMe)₂C₆H₃	methicillin	β-lactamase-resistant, acid-sensitive superseded
fluoro-chloro isoxazole	flucloxacillin	β-lactamase- and acid-resistant; may be used in combination with ampicillin
thiophene full structure	temocillin	β-lactamase-resistant; active against Gram-negative (except *P. aeruginosa*) but not Gram-positive organisms

Box 7.9   (continued)

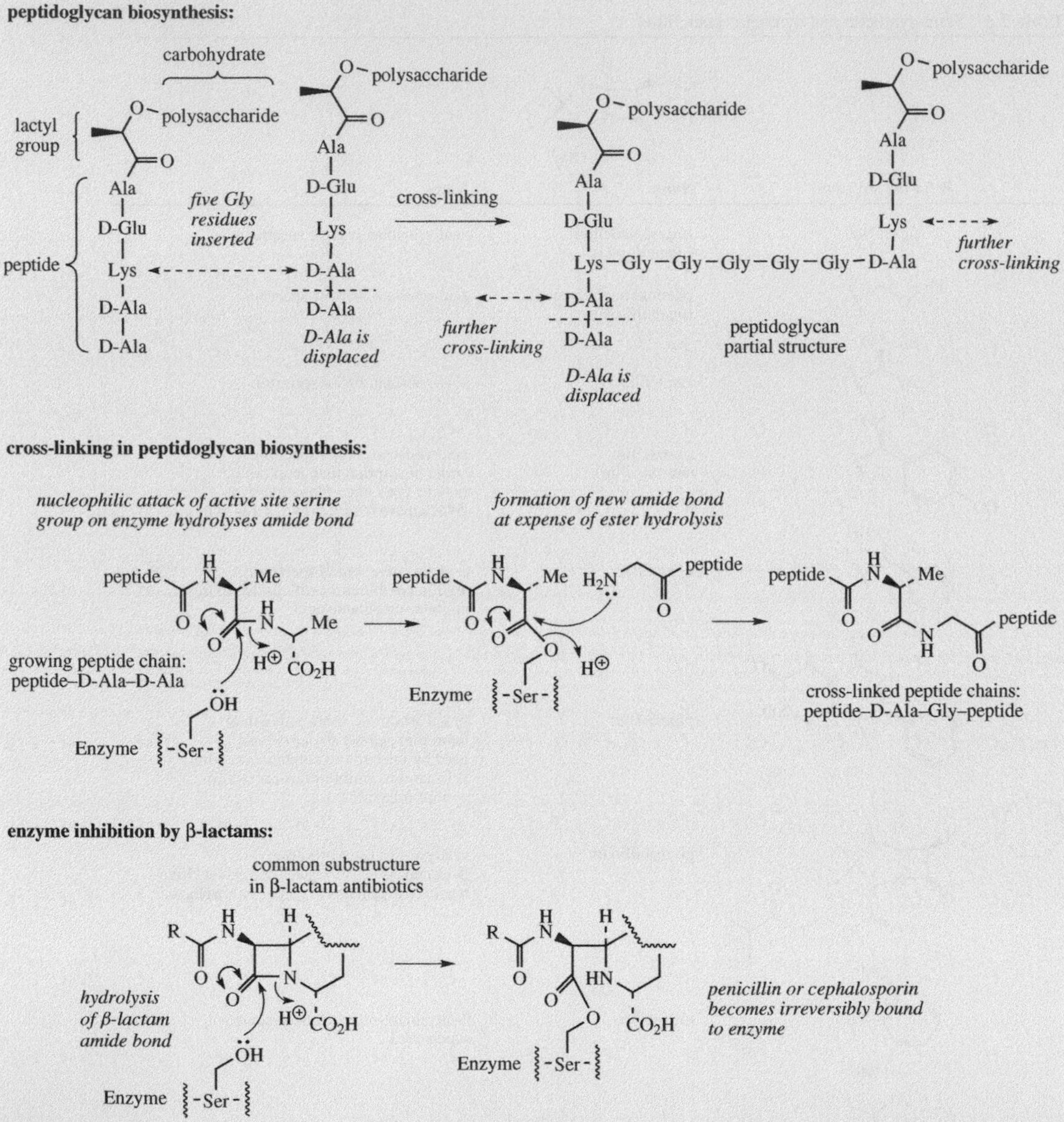

peptidoglycan biosynthesis:

cross-linking in peptidoglycan biosynthesis:

nucleophilic attack of active site serine group on enzyme hydrolyses amide bond

formation of new amide bond at expense of ester hydrolysis

growing peptide chain:
peptide–D-Ala–D-Ala

Enzyme –Ser–

cross-linked peptide chains:
peptide–D-Ala–Gly–peptide

enzyme inhibition by β-lactams:

common substructure in β-lactam antibiotics

hydrolysis of β-lactam amide bond

penicillin or cephalosporin becomes irreversibly bound to enzyme

Enzyme –Ser–

Figure 7.42

and loss of cellular contents. β-Lactamases are thought to have evolved from penicillin-binding proteins. The binding reaction is chemically analogous, but in the case of β-lactamases, penicilloic acid (see Figure 7.38) is subsequently released and the enzyme can continue to function. The penicillins are very safe antibiotics for most individuals. The bacterial cell wall has no counterpart in mammalian cells, and the action is thus very specific. However, a significant proportion of patients can experience allergic responses ranging from a mild rash to fatal anaphylactic shock. Cleavage of the β-lactam ring through nucleophilic attack of an amino group in a protein is believed to lead to the formation of antigenic substances causing the allergic response.

Cephalosporins

The **cephalosporins**, e.g. **cephalosporin C** (Figure 7.43), are a penicillin-related group of antibiotics having a fused β-lactam-dihydrothiazine ring system and are produced by species of *Cephalosporium* [Box 7.10]. The six-membered dihydrothiazine ring is produced from the five-membered thiazolidine ring of the penicillin system by an oxidative process of ring expansion, incorporating one of the methyl groups. The pathway (Figure 7.43) diverges from that to penicillins at **isopenicillin N**, which is first epimerized in the α-aminoadipyl side-chain to give **penicillin N**. The fungal epimerization process is different from that in bacterial β-lactam producers: there appears to be direct epimerization in bacteria, but fungi employ a longer sequence via conversion to the CoA ester, epimerization,

and then hydrolysis. Ring expansion then occurs, incorporating one of the methyl groups into the heterocyclic ring, though the mechanism for this is not clearly defined. A radical mechanism is suggested in Figure 7.43 to rationalize the transformation. Hydroxylation of the remaining methyl gives deacetylcephalosporin C. In *Acremonium chrysogenum* (formerly *Cephalosporium acremonium*), a single bifunctional protein catalyses both ring expansion and the hydroxylation. **Cephalosporin C** is the subsequent acetyl ester, whilst a further group of antibiotics termed the **cephamycins** are characterized by a 7α-methoxy group, and are produced by hydroxylation/methylation, and in the case of **cephamycin C** by introduction of a carbamate group from carbamoyl phosphate onto the hydroxymethyl function.

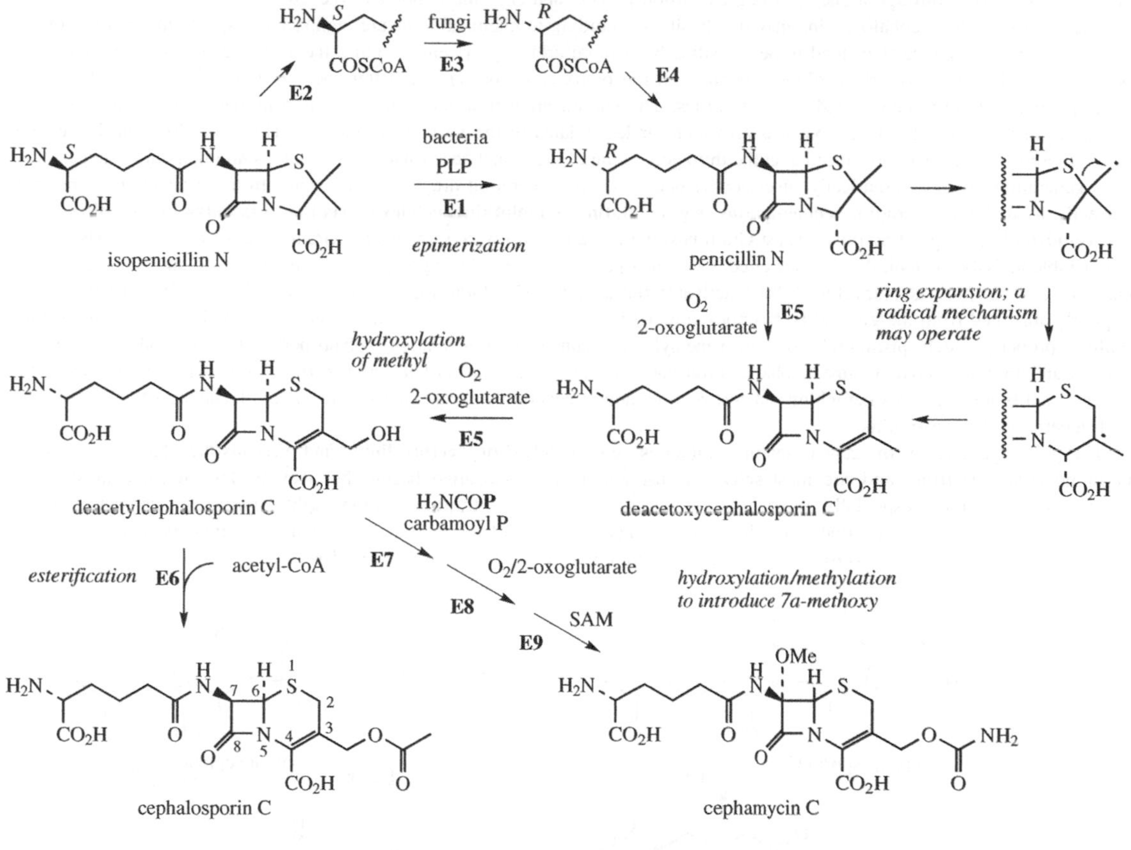

E1: isopenicillin N epimerase
E2: isopenicillin N-CoA synthetase
E3: isopenicillin N-CoA epimerase
E4: thioesterase
E5: deacetoxycephalosporin C synthetase-deacetylcephalosporin C synthetase (bifunctional)

E6: deacetylcephalosporin C acetyltransferase
E7: deacetylcephalosporin C *O*-carbamoyltransferase
E8: CmcJ (cephalosporin 7α-hydroxylase)
E9: CmcI (methyltransferase)

Figure 7.43

Box 7.10

Cephalosporins

Cephalosporin C (Figure 7.43) is produced commercially by fermentation using cultures of a high-yielding strain of *Acremonium chrysogenum* (formerly *Cephalosporium acremonium*). Initial studies of the antibiotic compounds synthesized by *Cephalosporium acremonium* identified penicillin N (originally called cephalosporin N) as the major component, with small amounts of cephalosporin C. In contrast to the penicillins, cephalosporin C was stable under acidic conditions and also was not attacked by penicillinase (β-lactamase). Antibacterial activity was rather low, however, and the antibiotic was poorly absorbed after oral administration. However, the structure offered considerable scope for side-chain modifications, more so than with the penicillins, since it has two side-chains, and this has led to a wide variety of cephalosporin drugs, many of which are currently in clinical use. As with the penicillins, removal of the amide side-chain by the hydrolysis of cephalosporin C to 7-aminocephalosporanic acid (7-ACA; Figure 7.44) was the key to semi-synthetic modifications, and this may be achieved chemically by the procedure used for the penicillins (compare Figure 7.41). Removal of this side-chain by suitable microorganisms or enzymes has proved more challenging, though a two-enzyme process has been developed using a D-amino acid oxidase and a glutarylacylase (Figure 7.44). The ester side-chain at C-3 may be hydrolysed enzymically by fermentation with a yeast; alternatively, the acetoxy group is easily displaced by nucleophilic reagents. It is also possible to convert readily available benzylpenicillin into the deacetoxy derivative of 7-ACA through a chemical ring expansion process and enzymic removal of the side-chain.

The semi-synthetic cephalosporins may be classified according to chemical structure, antibacterial spectrum, or β-lactamase resistance, but in practice they tend to be classified by a more arbitrary system, dividing them into 'generations' (Table 7.4). Note that all the cephalosporin antibiotics begin with the prefix *ceph-* or *cef-*, the latter spelling now being preferred, though both spellings are still encountered for some drugs. The classification into generations is based primarily on the antibacterial spectrum displayed by the drugs, but it is also more or less related to the year of introduction. However, drugs in the second generation may have been introduced after the third generation of drugs had been established. There is no intention to suggest that third-generation drugs automatically supersede second- and first-generation drugs; indeed, agents from all generations are still currently used. First-generation cephalosporins, e.g. **cefalotin (cephalothin)**, **cefalexin (cephalexin)**, **cefradine (cephradine)**, and **cefadroxil**, have good activity against Gram-positive bacteria but low activity against Gram-negative organisms. They have comparable activity to ampicillin and are effective against penicillinase-producing *Staphylococcus*. However, another β-lactamase enzyme (cephalosporinase) developed that inactivated these agents. Cefalotin, the first modified cephalosporin to be marketed, is poorly absorbed from the gut and is thus not orally active. However, cefalexin, cefradine, and cefadroxil may be administered orally, a property which appears related to the 3-methyl side-chain. Second-generation cephalosporins show a broader spectrum of activity and are more active against aerobic Gram-negative bacteria, like *Haemophilus influenzae* and *Neisseria gonorrhoeae*. This group of antibiotics includes **cefachlor** and **cefuroxime**, and in general displays better resistance to β-lactamases that inactivated first-generation cephalosporins.

The third generation of cephalosporin antibiotics, e.g. **cefotaxime, ceftazidime**, and **ceftriaxone**, has an extended Gram-negative spectrum, and are most active against enteric Gram-negative bacilli, but may be less active against some Gram-positive bacteria, especially *Staphylococcus aureus*. Many of the third-generation cephalosporins are characterized by an aminothiazole ring on the amide side-chain, which appears to impart the high activity against Gram-negative bacteria. The *O*-substituted oxime group also improves potency and confers resistance to β-lactamases. The oximes with *syn* stereochemistry,

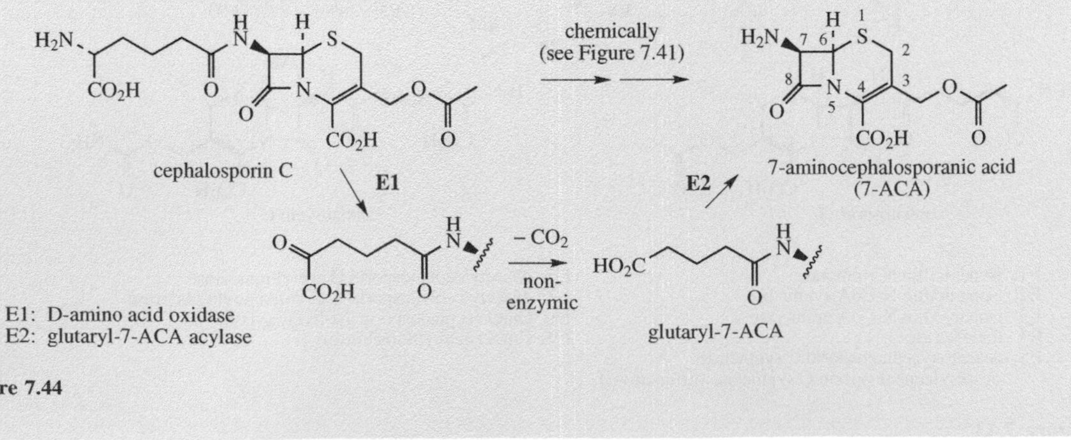

E1: D-amino acid oxidase
E2: glutaryl-7-ACA acylase

Figure 7.44

Box 7.10 (continued)

Table 7.4 Cephalosporin antibiotics

R^1	R^2	Name	Notes
First generation			
	$-CH_2OCOMe$	cefalotin (cephalothin)	R^2 group unstable to mammalian esterases; generally superseded
	$-CH_3$	cefalexin (cephalexin)	orally active
	$-CH_3$	cefradine (cephradine)	orally active
	$-CH_3$	cefadroxil	orally active
Second generation			
	$-Cl$	cefaclor	orally active
	$-CH_2OCONH_2$	cefuroxime	high resistance to β-lactamases; resistant to mammalian esterases
Third generation			
	$-CH_2OCOMe$	cefotaxime	unstable to mammalian esterases, but deacetyl metabolite still has considerable antimicrobial activity
		ceftazidime	broad-spectrum Gram-negative activity; good activity towards *Pseudomonas*

(continued overleaf)

Box 7.10 (continued)

Table 7.4 *continued*

R¹	R²	Name	Notes
		ceftriaxone	broad-spectrum Gram-negative activity; longer half-life than other cephalosporins
	$-CH=CH_2$	cefixime	2nd/3rd generation, orally active; long duration of action

Pro-drugs

		cefuroxime-axetil	2nd generation, orally active; hydrolysed by esterases to liberate cefuroxime
		cefpodoxime-proxetil	2nd/3rd generation, orally active; hydrolysed by esterases to liberate cefpodoxime

as shown in Table 7.4, are considerably more active than the *anti* isomers. A disadvantage of many of the current cephalosporin drugs is that they are not efficiently absorbed when administered orally. This is to some extent governed by the nature of the side-chain on C-3. The orally active **cefixime** has a spectrum of activity between the second and third generations. Orally active pro-drugs, such as **cefuroxime-axetil** and **cefpodoxime-proxetil**, have been developed with an additional ester function on the C-4 carboxyl. These compounds are hydrolysed to the active agents by esterases.

Cephalosporin antibiotics are especially useful for treating infections in patients who are allergic to penicillins. Hypersensitivity to cephalosporins is much less common, and only about 5–10% of penicillin-sensitive patients will also be allergic to cephalosporins. Several new structures are in clinical development.

Cephamycins

The antibiotic cephamycin C (Figure 7.43) was isolated from *Streptomyces clavuligerus* and shown to have a 7α-methoxy group on the basic cephalosporin ring system. Although cephamycin C and other natural cephamycins have only weak antibacterial activity, they are resistant to β-lactamase hydrolysis, a property conferred by the increased steric crowding from the additional

Box 7.10 (continued)

methoxy group. Semi-synthetic analogues have been obtained either by modification of the side-chains of natural cephamycins or by chemical introduction of the 7α-methoxyl. Currently, the only cephamycin in general use is **cefoxitin** (Figure 7.45), which is active against bowel flora, including *Bacterioides fragilis*, and is used for treatment of peritonitis.

Carbacephems

Although cephalosporin analogues where the sulfur heteroatom has been replaced with carbon are not known naturally (contrast carbapenems, natural penicillin analogues, page 471), synthetically produced carbacephems have shown good antibacterial activity with considerably improved chemical stabilty over cephalosporins. The first of these to be produced for drug use is **loracarbef** (Figure 7.45), which has similar antibacterial activity to cefaclor, but considerably greater stability, a longer half-life, and better oral bioavailability.

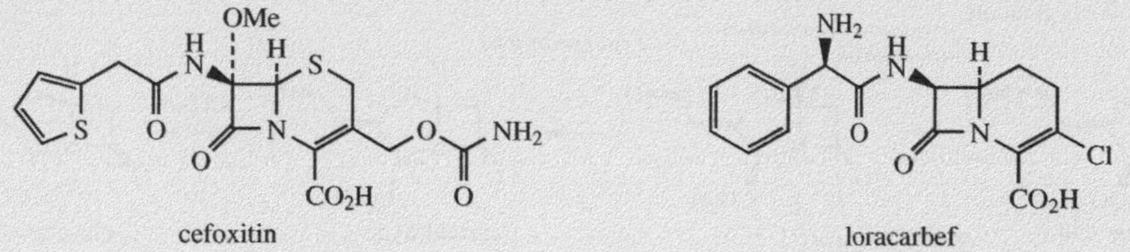

cefoxitin   loracarbef

Figure 7.45

Other β-Lactams

The fused β-lactam skeletons found in penicillins and cephalosporins are termed **penam** and **cephem** respectively (Figure 7.46). Other variants containing the basic β-lactam ring system are also found in nature. Although these are derived from amino acid precursors, in contrast to the penicillins and cephalosporins, they are not usually the products of NRPS enzymes. Of particular importance is the **clavam** (Figure 7.46) or oxapenam fused-ring system typified by **clavulanic acid** (Figure 7.47) from *Streptomyces clavuligerus*. The weak antibacterial activity of

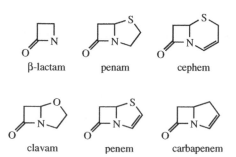

Figure 7.46

clavulanic acid is unimportant, for this compound is valuable as an efficient inhibitor of β-lactamases from both Gram-positive and Gram-negative bacteria [Box 7.11]. Despite the obvious structural similarity between clavulanic acid and the penicillins, they are not derived from common precursors, and there are some novel aspects associated with clavulanic acid biosynthesis. All carbon atoms are provided by two precursors, arginine and glyceraldehyde 3-P, which are coupled in a TPP-dependent reaction. Glyceraldehyde 3-P supplies the β-lactam carbon atoms, whilst the α-amino group of arginine provides the β-lactam nitrogen. In contrast to the penicillins, the β-lactam ring is formed via an acyl-AMP-activated intermediate. The sequence of reactions shown in Figure 7.47 leads to the monocyclic β-lactam **proclavaminic acid**, which is the substrate for oxidative cyclization to provide the oxazolidine ring and then dehydrogenation to **clavaminic acid**. The three 2-oxoglutarate-dependent oxidations in this sequence are all catalysed by a single enzyme. The mechanistic details for oxazolidine ring formation are still to be clarified. The final transformation into **clavulanic acid** requires oxidative deamination of the terminal amine to an aldehyde and then reduction to an alcohol, but, more intriguingly, the two chiral centres C-3 and C-5 have to be epimerized. These reactions are

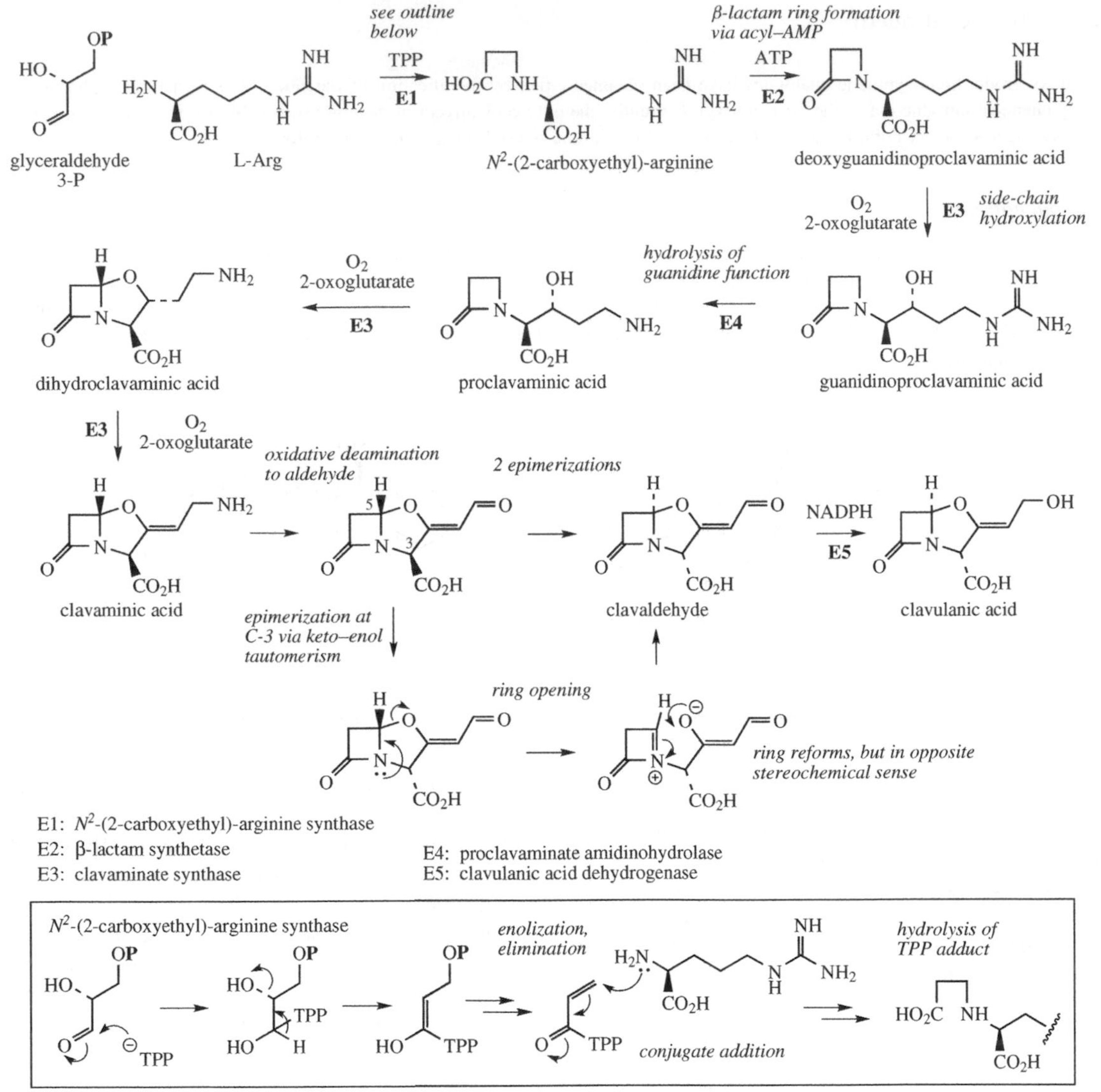

Figure 7.47

E1: N^2-(2-carboxyethyl)-arginine synthase
E2: β-lactam synthetase
E3: clavaminate synthase

E4: proclavaminate amidinohydrolase
E5: clavulanic acid dehydrogenase

also not fully established. The unsaturated aldehyde is easily susceptible to inversion of configuration at C-3 via keto–enol tautomerism, but a change of chirality at C-5 (which actually retains the C-5 hydrogen) must invoke opening of the oxazolidine ring (Figure 7.47).

Box 7.11

Clavulanic Acid

Clavulanic acid (Figure 7.47) is produced by cultures of *Streptomyces clavuligerus*, the same actinomycete that produces cephamycin C (see page 468). Although it has only weak antibacterial activity, it is capable of reacting with a wide variety of β-lactamase enzymes, opening the β-lactam ring, in a process initially analogous to that seen with penicillins (Figure 7.48).

Box 7.11 (continued)

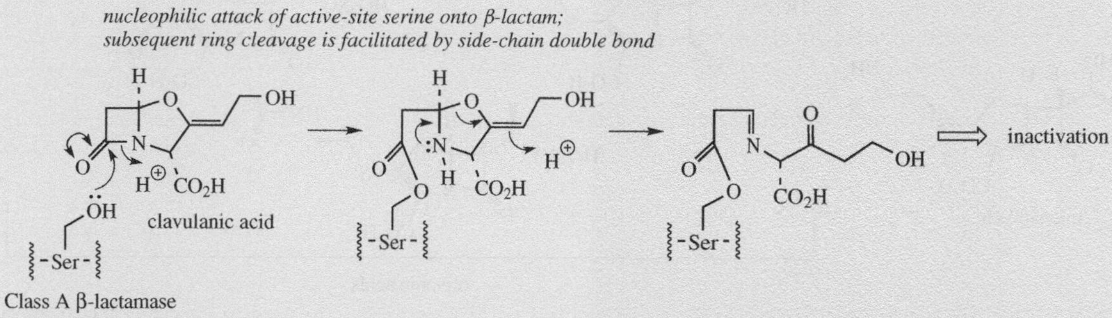

nucleophilic attack of active-site serine onto β-lactam;
subsequent ring cleavage is facilitated by side-chain double bond

clavulanic acid

Class A β-lactamase

inactivation

Figure 7.48

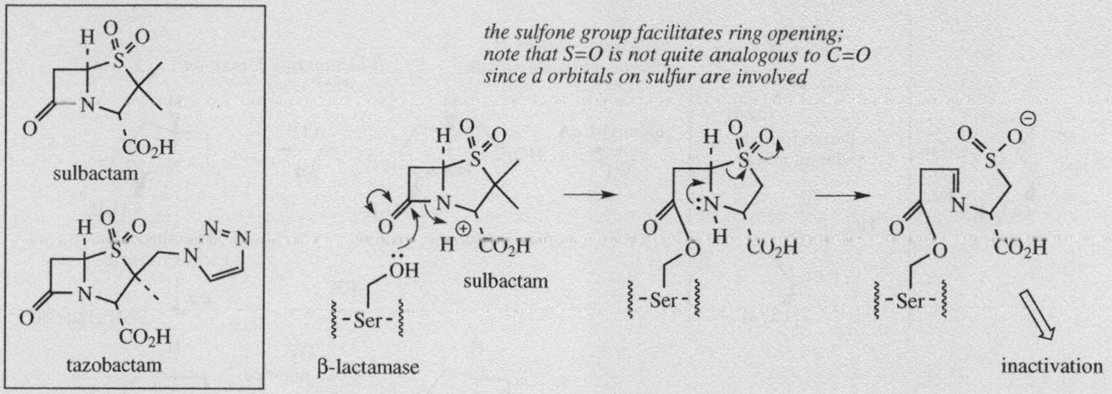

sulbactam

tazobactam

the sulfone group facilitates ring opening;
note that S=O is not quite analogous to C=O
since d orbitals on sulfur are involved

sulbactam

β-lactamase

inactivation

Figure 7.49

However, binding to the enzyme becomes irreversible and the β-lactamase is inactivated. It seems likely that the side-chain with its double bond may contribute to this effect, causing further ring opening and subsequent changes to give intermediates which react at the active site of the enzyme (Figure 7.48). Thus, dihydroclavulanic acid is no longer an enzyme inhibitor.

Clavulanic acid is usually combined with a standard penicillin, e.g. with amoxicillin (as **co-amoxyclav**) or with ticarcillin, to act as a suicide substrate and provide these agents with protection against class A β-lactamases, thus extending their effectiveness against a wider range of organisms. Similar success has been achieved using the semi-synthetic penicillanic acid sulfone, **sulbactam** (Figure 7.49), which is also a potent inhibitor of β-lactamase initiated by a similar double ring-opening mechanism (Figure 7.49). Sulbactam was used in combination with ampicillin, but has been superseded by the related sulfone **tazobactam** (Figure 7.49), which is a β-lactamase inhibitor used in combination with piperacillin.

Another variant on the penicillin penam ring system is found in a group of compounds termed **carbapenems** (Figure 7.46), where the sulfur heteroatom has been replaced by carbon [Box 7.12]. This is exemplified by **thienamycin**, an antibiotic isolated from cultures of *Streptomyces cattleya*, and the **olivanic acids** from *Streptomyces olivaceus* (Figure 7.50). The sulfur-containing side-chain is a feature of many of the natural examples, and this may feature sulfide or sulfone moieties. The olivanic acids

are potent β-lactamase inhibitors, especially towards the cephalosporinases, which are poorly inhibited by clavulanic acid.

The pathway to the simplest example of carbapenems, namely **carbapen-2-em-3-carboxylic acid**, has now been established in the bacterium *Erwinia carotovora* (Figure 7.51). The pyrrolidine ring is essentially proline derived. Oxidation of proline provides glutamate semialdehyde, which condenses with malonate to give

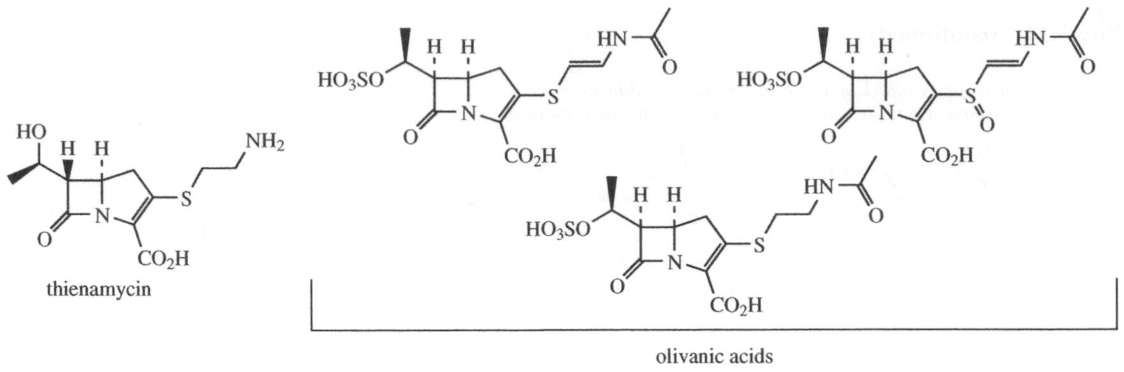

olivanic acids

Figure 7.50

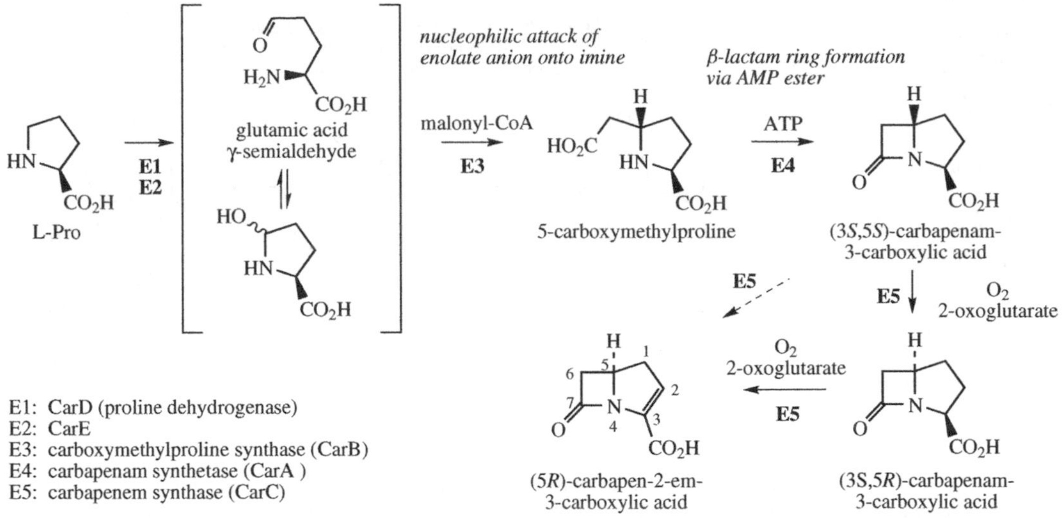

E1: CarD (proline dehydrogenase)
E2: CarE
E3: carboxymethylproline synthase (CarB)
E4: carbapenam synthetase (CarA)
E5: carbapenem synthase (CarC)

Figure 7.51

5-carboxymethylproline. The β-lactam ring is then formed via an acyl-AMP ester, the same mechanism as seen with clavulanic acid biosynthesis (see page 470). Further modifications are then controlled by a 2-oxoglutarate-dependent oxygenase enzyme carbapenem synthase, which has much similarity to clavaminic acid synthase. It carries out dehydrogenation in the pyrollidine ring (penem), but remarkably also epimerizes the fused-ring system at position 5. The saturated but epimerized metabolite (penam) is also produced in the enzymic reaction, and is also a substrate for the enzyme. It is not yet clear whether a two-step or one-step conversion is normally involved. Epimerization is suggested to proceed by removal of a hydrogen atom from C-5 via a radical intermediate.

The **thienamycin** structure is constructed in a similar manner, but two additional modifications are necessary. There is a two-carbon side-chain at C-2 known to be derived from methionine, the result of a double methylation sequence reminiscent of the processes used for alkylating the side-chains of sterols (see page 252). One methyl is introduced quite early on in the pathway, alkylating the malonyl-derived residue prior to β-lactam formation (Figure 7.52). The second is introduced after construction of the fused ring, and will necessitate oxidation of the methyl group introduced earlier. The C-2 cysteaminyl side-chain is supplied by the amino acid cysteine, and probably requires participation of an intermediate penem substrate.

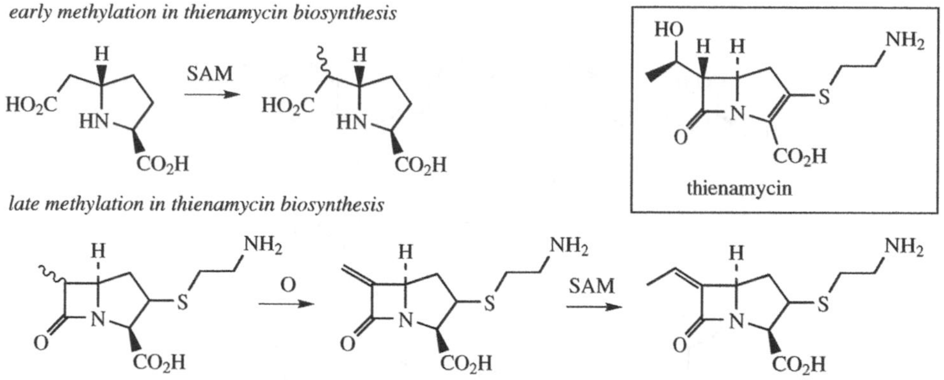

early methylation in thienamycin biosynthesis

late methylation in thienamycin biosynthesis

Figure 7.52

Box 7.12

Carbapenems

Thienamycin (Figure 7.50) is produced by cultures of *Streptomyces cattleya*, but in insufficient amounts for commercial use. This compound is thus obtained by total synthesis. In addition, thienamycin is relatively unstable, its side-chain primary amino group reacting as a nucleophile with other species, including the β-lactam group in other molecules. For drug use, therefore, thienamycin is converted into its more stable *N*-formimidoyl derivative **imipenem** (Figure 7.53). Imipenem has a broad spectrum of activity, which includes activity towards many aerobic and anaerobic Gram-positive and Gram-negative bacteria. It is resistant to hydrolysis by most classes of β-lactamases, and it also possesses β-lactamase inhibitory activity. However, it is partially inactivated in the kidney by human renal dehydropeptidase, and is thus administered in combination with cilastatin (Figure 7.53), a specific inhibitor of this enzyme. **Meropenem** (Figure 7.53) is stable to renal dehydropeptidase by virtue of the extra 1β-methyl group and can thus be administered as a single agent. It has good activity against all clinically significant aerobes and anaerobes, except MRSA and *Enterococcus faecium*, and stability to serine-based β-lactamases. The newer agent **ertapenem** (Figure 7.53) is also a 1β-methyl derivative. It is a broad-spectrum carbapenem antibiotic that has proved effective against the growing number of cephalosporin-resistant bacteria. It also has an improved pharmacokinetic profile in comparison with other carbapenems, allowing single-agent therapy and once-daily dosing. Other penem derivatives are undergoing clinical trials; **doripenem** (Figure 7.53) has recently been approved for complicated urinary tract and intra-abdominal infections.

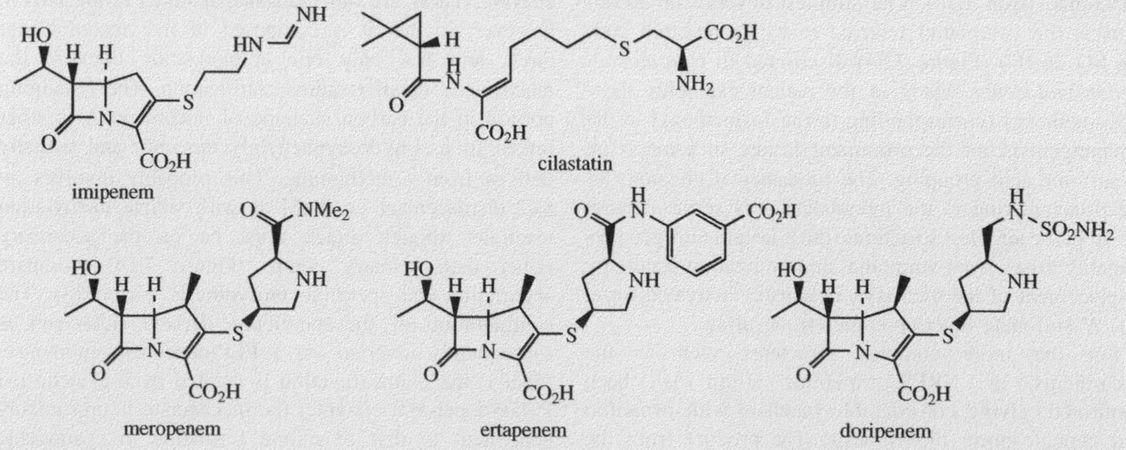

Figure 7.53

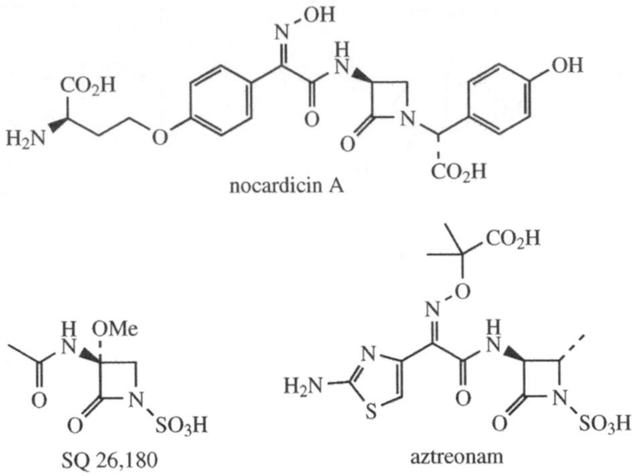

nocardicin A

SQ 26,180

aztreonam

Figure 7.54

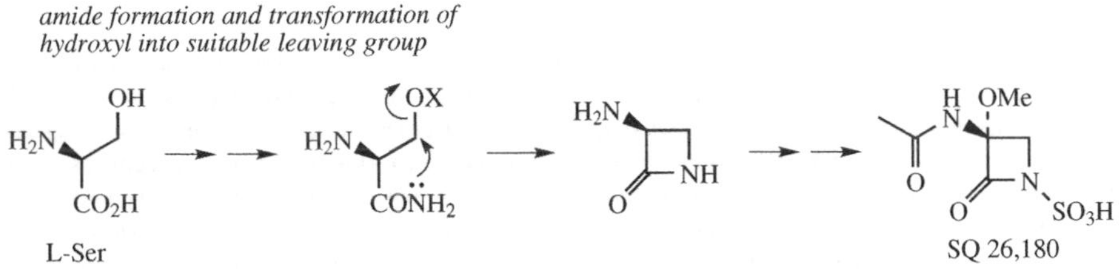

amide formation and transformation of hydroxyl into suitable leaving group

L-Ser

SQ 26,180

Figure 7.55

The simple non-fused β-lactam ring is encountered in a number of natural structures, such as the **nocardicins**, e.g. nocardicin A (Figure 7.54), from *Nocardia uniformis*; these structures are termed **monobactams** (monocylic β-lactams) [Box 7.13]. The simplest of these monobactams is the compound referred to by its research coding SQ 26,180 (Figure 7.54) discovered in *Chromobacterium violaceum*. Many of the natural examples show a 3α-methoxyl (corresponding to the 7α-methoxyl in the cephamycins), but the prominent feature in some is an *N*-sulfonic acid grouping. The fundamental precursor of the β-lactam ring in the monobactams is serine (Figure 7.55). In the simplest structures, the β-lactam nitrogen presumably arises from ammonia, and cyclization occurs by displacement of the hydroxyl, in suitably activated form. The *N*-sulfonate function comes from sulfate.

For the more complex structures such as the nocardicins, an NRPS tripeptide origin has been confirmed, giving considerable parallels with penicillin and cephalosporin biosynthesis. The product from the nocardicin NRPS is nocardicin G (Figure 7.56). The amino acids accepted are L-serine and two molecules of L-4-hydroxyphenylglycine, the latter having its origins in L-tyrosine (compare vancomycin, page 448). Thus, the postulated D,L,D-tripeptide is also cyclized by the enzyme. There are other unusual features in the NRPS, however, in that it is composed of five modules, not three, and has only one epimerization domain; the mechanistic detail requires clarification. The remaining portion in the carbon skeleton of nocardicin A is ether linked to a 4-hydroxyphenylglycine unit, and actually derives from L-methionine. This probably involves an S_N2 displacement on SAM as with simple methylation reactions, though attack must be on the secondary rather than primary centre (Figure 7.56) (compare spermidine and spermine biosynthesis, page 313). The configuration of the methionine-derived side-chain is subsequently inverted via a PLP-dependent epimerase. Finally, the oxime function is created by a cytochrome P-450-dependent enzyme; the mechanism is presumably equivalent to that of oxime formation in cyanogenic glycosides (see page 477).

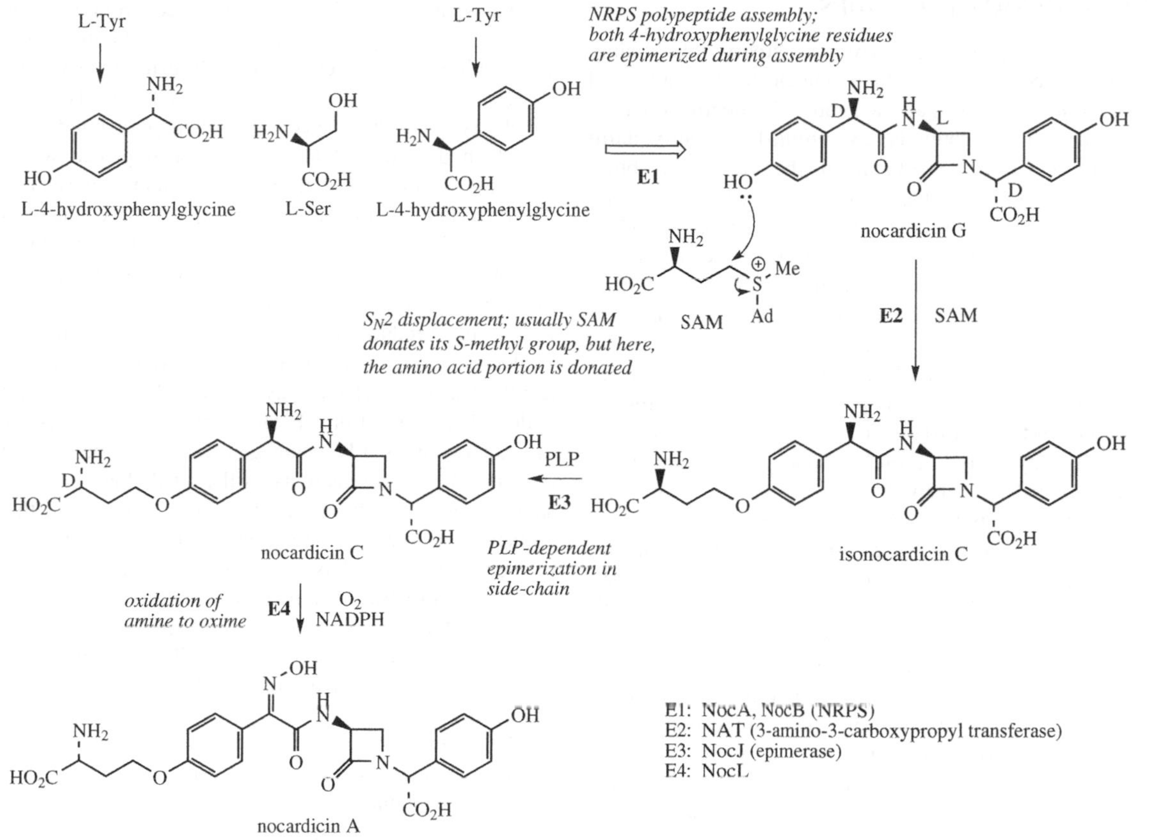

Figure 7.56

Box 7.13

Monobactams

The naturally occurring monobactams show relatively poor antibacterial activity, but alteration of the side-chain, as with penicillins and cephalosporins, has produced many potent new compounds. Unlike those structures, the non-fused β-lactam ring is readily accessible by synthesis, so all analogues are produced synthetically. The first of these to be used clinically is **aztreonam** (Figure 7.54), which combines the side-chain of the cephalosporin ceftazidime (Table 7.4) with the monobactam nucleus. Aztreonam is very active against Gram-negative bacteria, including *Pseudomonas aeruginosa*, *Haemophilus influenzae*, and *Neisseria meningitidis*, but has little activity against Gram-positive organisms. It also displays a high degree of resistance to enzymatic hydrolysis by most of the common β-lactamases. Oral absorption is poor, and this drug is administered by injection. A lysine salt formulation for inhalation is in clinical trials.

CYANOGENIC GLYCOSIDES

Cyanogenic glycosides are a group of mainly plant-derived materials which liberate hydrocyanic acid (HCN) on hydrolysis, and are thus of concern as natural toxicants. The group is exemplified by **amygdalin** (Figure 7.57), a constituent in the kernels of bitter almonds (*Prunus amygdalus* var. *amara*; Rosaceae) and other *Prunus* species, such as apricots, peaches, cherries, and plums. When plant tissue containing a cyanogenic glycoside is crushed, glycosidase enzymes also in the plant, but usually located in different cells, are brought into contact with the glycoside and begin to hydrolyse it. Thus, amygdalin is hydrolysed sequentially by β-glucosidase-type enzymes to **prunasin** and then **mandelonitrile**, the latter compound actually being the cyanohydrin of benzaldehyde (Figure 7.57). Mandelonitrile is then hydrolysed to its component parts, benzaldehyde and toxic HCN, by the action of a further enzyme. The kernels of bitter almonds, which contain amygdalin and the hydrolytic enzymes, are thus potentially toxic if ingested, whilst kernels of sweet almonds (*Prunus amygdalus* var. *dulcis*) are not toxic, containing the enzymes but no cyanogenic glycoside. Amygdalin itself is not especially toxic to animals; toxicity depends on the co-ingestion of the hydrolytic enzymes. Although formed by the hydrolysis of amygdalin, prunasin is also a natural cyanogenic glycoside and may be found in seeds of black cherry (*Prunus serotina*) and in the seeds and leaves of cherry laurel (*Prunus laurocerasus*). The food plant cassava (or tapioca) (*Manihot esculenta*; Euphorbiacae) also produces the cyanogenic glycosides **linamarin** and **lotaustralin** (see Figure 7.59), and preparation of the starchy tuberous roots involves prolonged hydrolysis and boiling to release and drive off the HCN before they are suitable for consumption.

Cyanogenic glycosides are produced from a range of amino acids by a common pathway (Figure 7.58). The amino acid precursor of **dhurrin** in sorghum (*Sorghum bicolor*; Graminae/Poaceae) is tyrosine, which is *N*-hydroxylated and then converted into the aldehyde oxime (aldoxime) by a sequence which involves further *N*-hydroxylation and subsequent decarboxylation–elimination, with all of these reactions catalysed by a single cytochrome P-450-dependent enzyme. The nitrile is formed by dehydration of the oxime, but this reaction actually proceeds on the *Z*-aldoxime produced by isomerization of the first-formed *E*-aldoxime. (*S*)-4-Hydroxymandelonitrile is then the result of a stereoselective cytochrome P-450-dependent hydroxylation reaction. Finally, glycosylation occurs, the sugar unit usually being glucose, as in the case of dhurrin. The stereoselectivity of the nitrile hydroxylation step varies depending on the plant system, so that epimeric cyanohydrins are found in nature, though not in the same plant. Thus, the (*R*)-enantiomer of dhurrin (called **taxiphyllin**) is found in arrowgrass (*Triglochin maritima*; Juncaginaceae) and several species of bamboo. The (*S*)-enantiomer of prunasin (called **sambunigrin**) is found in the leaves of elder (*Sambucus nigra*; Caprifoliaceae).

The main amino acids utilized in the biosynthesis of cyanogenic glycosides are phenylalanine (e.g. prunasin, sambunigrin, and amygdalin), tyrosine (e.g. dhurrin and taxiphyllin), valine (e.g. linamarin from flax (*Linum usitatissimum*; Linaceae)), isoleucine (e.g. lotaustralin, also from flax), and leucine (e.g. **heterodendrin** from *Acacia* species (Leguminosae/Fabaceae)) (Figure 7.59). Although cyanogenic glycosides are widespread, they are particularly found in the families Rosaceae, Leguminosae/Fabaceae, Graminae/Poaceae, Araceae, Compositae/Asteraceae, Euphorbiaceae, and

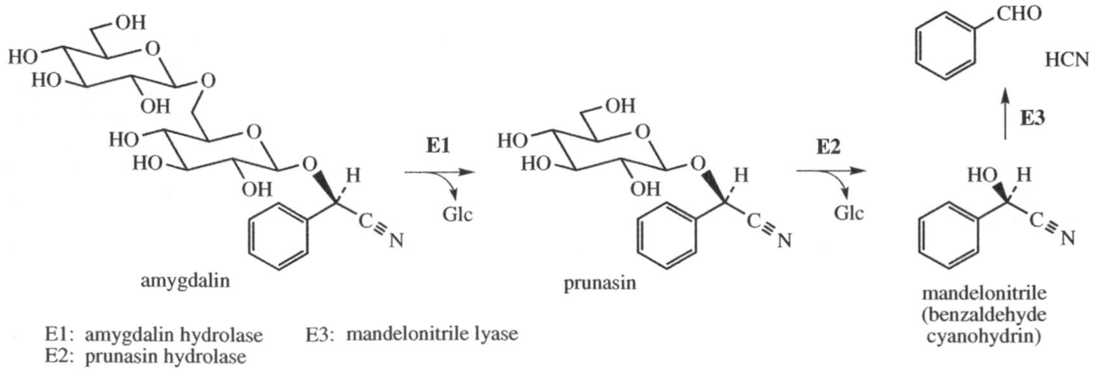

E1: amygdalin hydrolase E3: mandelonitrile lyase
E2: prunasin hydrolase

Figure 7.57

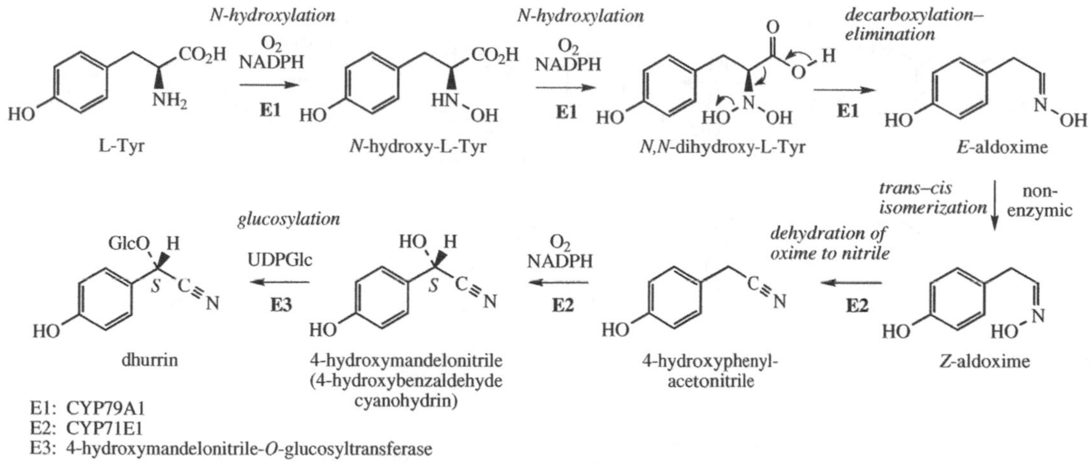

E1: CYP79A1
E2: CYP71E1
E3: 4-hydroxymandelonitrile-*O*-glucosyltransferase

Figure 7.58

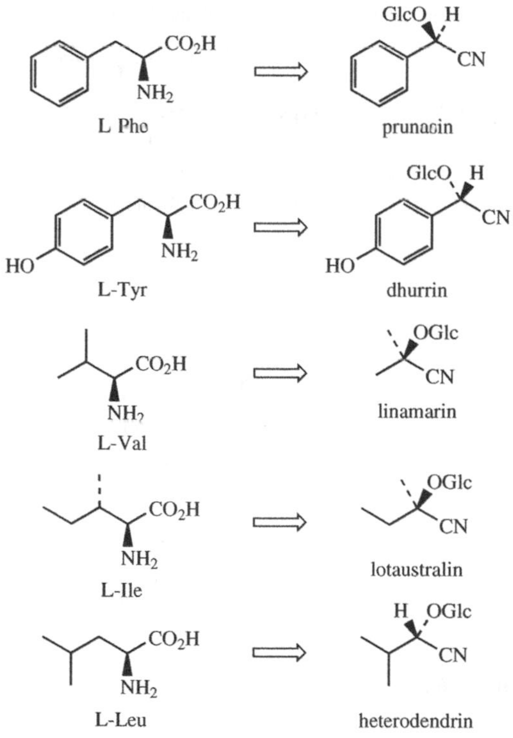

Figure 7.59

Passifloraceae. It is highly likely that plants synthesize these compounds as protecting agents against herbivores. Some insects also accumulate cyanogenic glycosides in their bodies, again as a protective device. Whilst many insects obtain these compounds by feeding on suitable plant sources, it is remarkable that others are known to synthesize cyanogenic glycosides themselves from amino acid precursors. There is hope that cyanogenesis may provide a means of destroying cancer cells. By targeting cancer cells with linamarase via a retrovirus and then supplying linamarin, it has been possible to selectively generate toxic HCN in cancer cells. Hydroxynitrile lyases, the hydrolytic enzymes that normally release HCN from cyanohydrins, are proving useful in synthetic chemistry. They are able to catalyse the reverse reaction and can, therefore, be used to produce cyanohydrins in good yield and with defined stereochemistry.

GLUCOSINOLATES

Glucosinolates have several features in common with the cyanogenic glycosides. They, too, are glycosides, in this case *S*-glycosides, which are enzymically hydrolysed in damaged plant tissues, giving rise to potentially toxic materials, and they share the early stages of the cyanogenic glycoside biosynthetic pathway for their formation in plants. A typical structure is **sinalbin** (Figure 7.60), found in seeds of white mustard (*Sinapis alba*; Cruciferae/Brassicaceae). Addition of water to the crushed or powdered seeds results in hydrolysis of the *S*-glucoside bond via the enzyme myrosinase (a thioglucosidase) to give a thiohydroximate sulfonate (Figure 7.60). This compound usually yields the isothiocyanate **acrinylisothiocyanate** by a Lössen-type rearrangement as shown, prompted by loss of the sulfate leaving group. Under certain conditions, dependent on pH, or the presence of metal

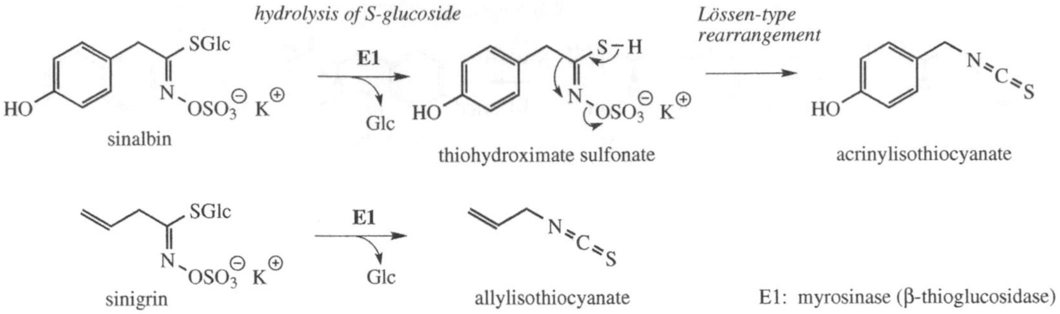

Figure 7.60

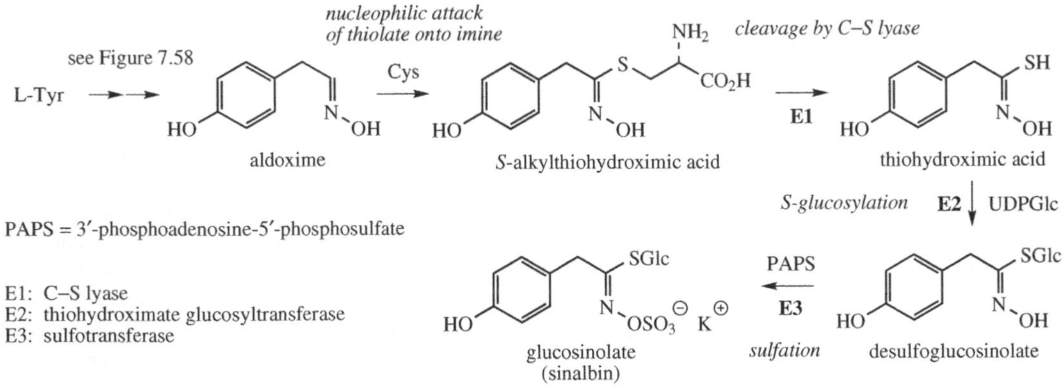

Figure 7.61

ions or other enzymes, related compounds such as thiocyanates (RSCN) or nitriles (RCN) may be formed from glucosinolates. Acrinylisothiocyanate is a pungent-tasting material (mustard oil) typical of many plants in the Cruciferae/Brassicaceae that are used as vegetables (e.g. cabbage, radish) and condiments (e.g. mustard, horseradish). Black mustard (*Brassica nigra*) contains **sinigrin** (Figure 7.60) in its seeds, which by a similar sequence is hydrolysed to **allylisothiocyanate**. Allylisothiocyanate is considerable more volatile than acrinylisothiocyanate, so that condiment mustard prepared from black mustard has a pungent aroma as well as taste.

The biosynthesis of **sinalbin** from tyrosine is indicated in Figure 7.61. The aldoxime is produced from the amino acid by the early part of the cyanogenic glycoside pathway shown in Figure 7.58. This aldoxime incorporates sulfur from cysteine to give the thiohydroximic acid, most likely by attack of the thiolate ion onto the imine system. This reaction is the only part of the pathway yet to be characterized, but resembles a glutathione *S*-transferase process. The

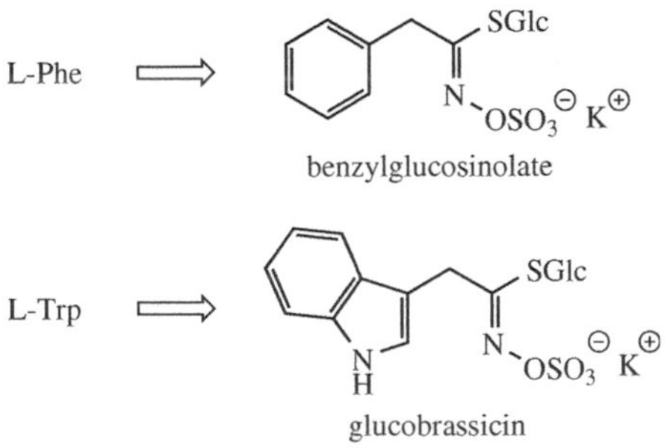

Figure 7.62

S-alkylthiohydroximate undergoes bond cleavage via a C–S lyase and the thiol group is then *S*-glucosylated using UDPglucose. Sulfation features as the last step in the pathway; in nature, sulfate groups are provided by PAPS (3'-phosphoadenosine-5'-phosphosulfate). Similarly, phenylalanine is the precursor of **benzyl-glucosinolate** (Figure 7.62) in nasturtium (*Tropaeolum*

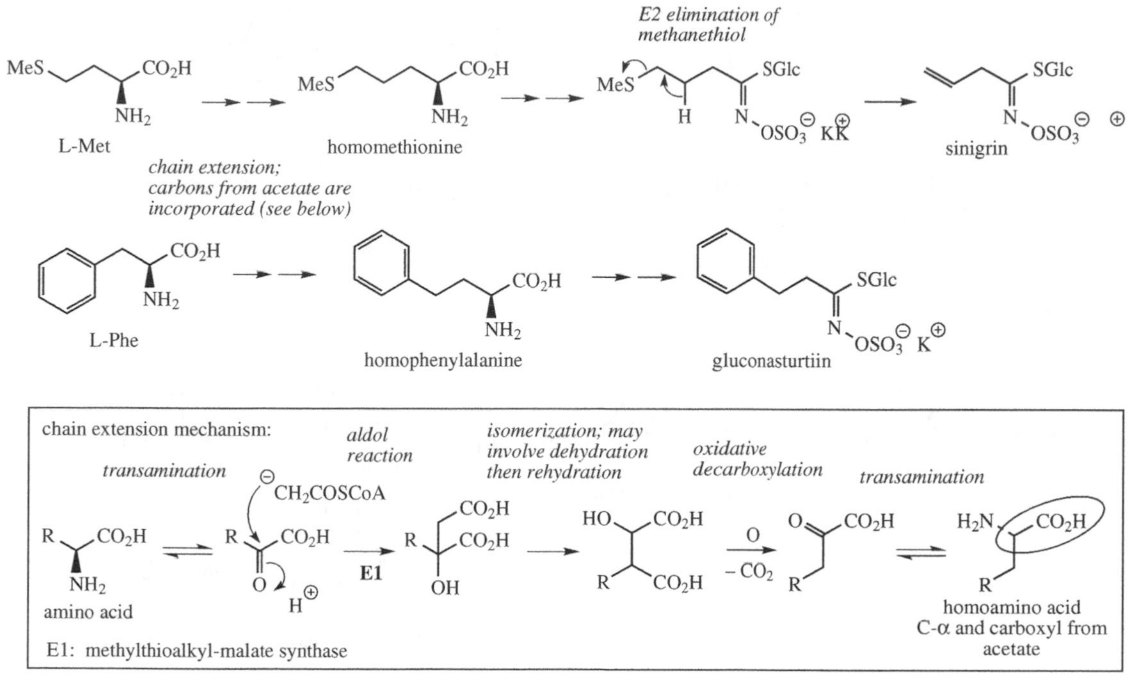

Figure 7.63

majus; Tropaeolaceae), and tryptophan yields **gluco-brassicin** in horseradish (*Armoracia rusticana*; Cruciferae/Brassicaceae).

Interestingly, chain extension of methionine to **homomethionine** and of phenylalanine to **homophenylalanine** is involved in the formation of **sinigrin** in *Brassica nigra* and **gluconasturtiin** in rapeseed (*Brassica napus*) respectively (Figure 7.63). Two carbon atoms derived from acetate are incorporated into the side-chain in each case and the original carboxyl is lost, allowing biosynthesis of the appropriate glucosinolates. These reactions are encountered elsewhere; they exactly parallel the formation of the amino acid leucine from valine in primary metabolism. The processes may also be repeated to incorporate further carbon atoms into the chain (see dihomomethionine, below). For elaboration of the allyl side-chain in the biosynthesis of sinigrin, loss of methanethiol occurs as a late step (Figure 7.63).

Glucosinolates are found in many plants of the Cruciferae/Brassicaceae, Capparidaceae, Euphorbiacae, Phytolaccaceae, Resedaceae, and Tropaeolaceae, contributing to the pungent properties of their crushed tissues. They are often at their highest concentrations in seeds rather than leaf tissue. These compounds and their degradation products presumably deter some predators, but may actually attract others, e.g. caterpillars

on cabbages and similar crops. There is evidence that consumption of the hydrolysis products from glucosinolates in food crops may induce goitre, an enlargement of the thyroid gland. Thus, **progoitrin** in oil seed rape (*Brassica napus*; Cruciferae/Brassicaceae) on hydrolysis yields the oxazolidine-2-thione **goitrin** (Figure 7.64) which is a potent goitrogen, inhibiting iodine incorporation and thyroxine formation (see page 426). The goitrogenic effects of glucosinolates cannot be alleviated merely by the administration of iodine. This severely limits economic utilization for animal foodstuffs of the rapeseed meal remaining after oil expression unless strains with very low levels of the glucosinolate are employed. On the other hand, **sulforaphane** (Figure 7.64), formed from the glucosinolate **glucoraphanin** in broccoli (*Brassica oleracea italica*; Cruciferae/Brassicaceae), has been shown to have beneficial medicinal properties, in that it induces carcinogen-detoxifying enzyme systems and accelerates the removal of xenobiotics. Though young sprouted seedlings contain some 10–100 times as much glucoraphanin as the mature plant, broccoli may be regarded as a valuable dietary vegetable. Glucoraphanin is derived from methionine via dihomomethionine, in a process involving two chain elongation cycles.

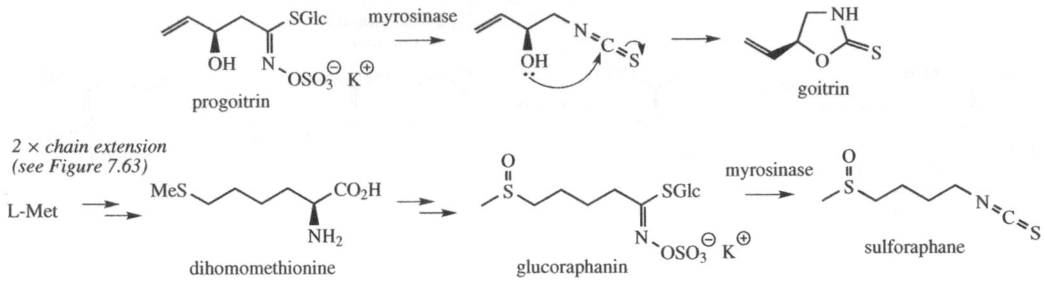

Figure 7.64

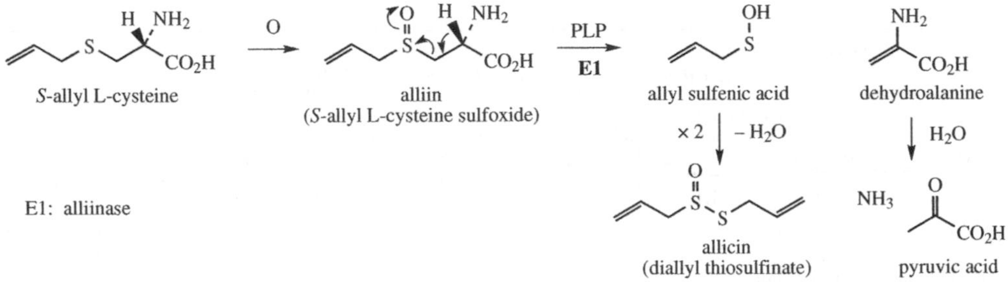

Figure 7.65

CYSTEINE SULFOXIDES

The major flavour component of garlic (*Allium sativum*; Liliaceae/Alliaceae) is a thiosulfinate called **allicin** (Figure 7.65). This compound is formed when garlic tissue is damaged as a hydrolysis product of *S*-allyl cysteine sulfoxide (**alliin**) brought about by the pyridoxal phosphate-dependent C–S lyase enzyme alliinase (Figure 7.65). Under these conditions, alliin is cleaved by an elimination reaction and two molecules of the sulfenic acid combine to form allicin. Pyruvic acid and ammonia are the other hydrolysis products. Allicin has considerable antibacterial and antifungal properties. There is widespread use of garlic (fresh, dried, or as garlic oil) as a beneficial agent to lower cholesterol levels and reduce the risk of heart attacks [Box 7.14]. The origins of the alliin precursor *S*-allyl cysteine are still to be established completely. *S*-Alkyl cysteine sulfoxides are characteristic components of the onion (*Allium*) genus. All *Allium* species contain *S*-methyl cysteine sulfoxide (**methiin**), though the *S*-propyl analogue **propiin** (Figure 7.66) predominates in chives (*Allium schoenoprasum*), the *S*-1-propenyl derivative **isoalliin** in onions (*Allium cepa*), and the *S*-allyl compound alliin in garlic. Propenyl sulfenic acid derived by hydrolysis of *S*-1-propenyl cysteine sulfoxide provides a common kitchen hazard: it rearranges to form the lachrymatory

Figure 7.66

Figure 7.67

factor of onions (Figure 7.67). The cysteine sulfoxides themselves are odourless and non-volatile, with smell and taste developing only upon hydrolytic and other reactions.

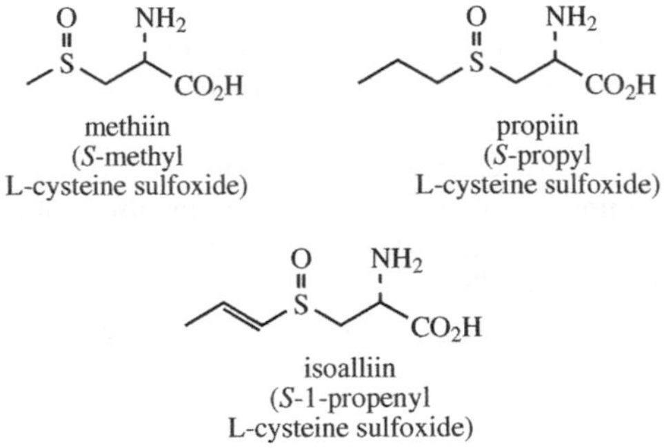

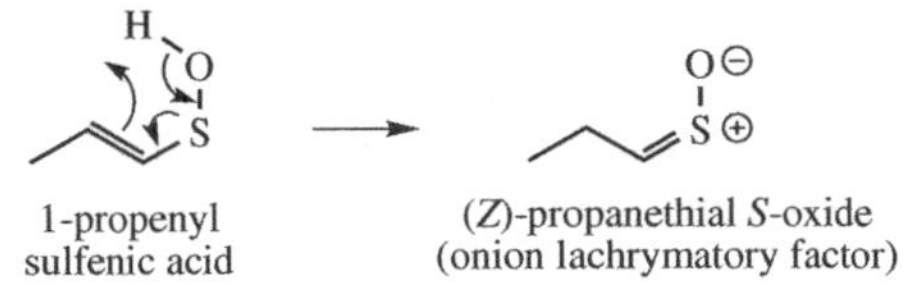

Box 7.14

Garlic

Garlic (*Allium sativum*; Liliaceae/Alliaceae) has a long history of culinary and medicinal use. The compound bulb is composed of several smaller sections termed cloves; this term should not be confused with the spice cloves (see page 158). Allicin (Figure 7.65) is considered to be the most important of the biologically active components in the crushed bulb. It is not present in fresh garlic, but is rapidly produced when the precursor alliin is cleaved by the action of the enzyme alliinase upon crushing the tissue. Both alliin and alliinase are stable when dry, and dried garlic still has the potential for releasing allicin when subsequently moistened. However, allicin itself is very unstable to heat or organic solvents, degrading to many other compounds, including diallyl sulfides (mono-, di-, and oligo-sulfides), vinyldithiins, and ajoenes (Figure 7.68). Processed garlic preparations typically contain a range of different sulfur compounds. Garlic preparations used medicinally include steam-distilled oils, garlic macerated in vegetable oils (e.g. soybean oil), dried garlic powder, and gel-suspensions of garlic powder. Analyses indicate wide variations in the nature and amounts of constituents in the various preparations. Thus, freshly crushed garlic cloves typically contain allicin (about 0.4%) and other thiosulfinates (about 0.1%, chiefly allyl methyl thiosulfinate). Garlic powder usually produces less of these materials, though high-quality samples may be similar to the fresh samples. Oil-macerated powders appear to lose up to 80% of their sulfur compounds, and vinyldithiins and ajoenes predominate. Steam-distilled garlic oil has dialk(en)yl sulfides (i.e. diallyl sulfide, allyl methyl sulfide, etc.) as the major sulfur components (0.1–0.5%). Bad breath and perspiration odours which often follow the ingestion of garlic, either medicinally or culinarily, are due to sulfides: allyl methyl sulfide and disulfide, diallyl sulfide and disulfide, and 2-propenethiol.

Garlic is used for a variety of reasons, and some of the attributes associated with it, e.g. for cancer prevention or to reduce heart attacks, may not be substantiated. Other properties, such as antimicrobial activity, effects on lipid metabolism, and platelet aggregation inhibitory action, have been demonstrated. Stabilized formulations of allicin are being employed in antibacterial creams and show high activity towards multi-drug-resistant strains of MRSA. Ajoene has been shown to be a potent antithrombotic agent through inhibition of platelet aggregation.

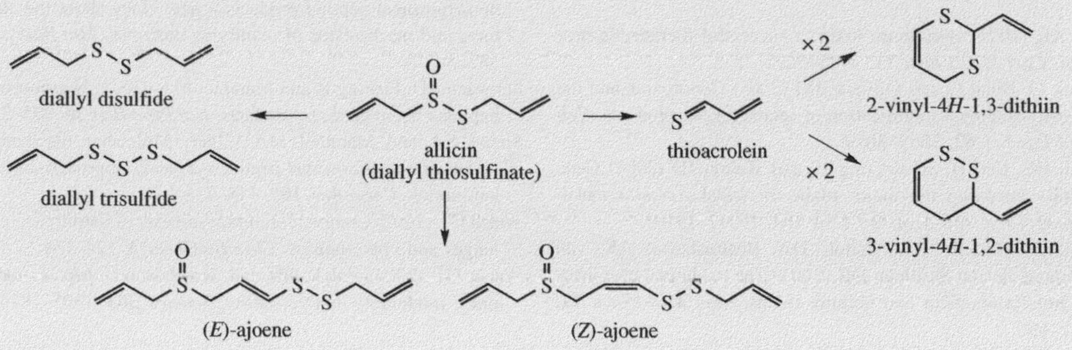

Figure 7.68

FURTHER READING

Peptides, General

Freidinger RM (2003) Design and synthesis of novel bioactive peptides and peptidomimetics. *J Med Chem* **46**, 5553–5566.

Lien S and Lowman HB (2003) Therapeutic peptides. *Trends Biotechnol* **21**, 556–562.

Rademann J (2004) Organic protein chemistry: drug discovery through the chemical modification of proteins. *Angew Chem Int Ed* **43**, 4554–4556.

Ribosomal Peptide Biosynthesis

Davidson VL (2007) Protein-derived cofactors. Expanding the scope of post-translational modifications. *Biochemistry* **46**, 5283–5292.

Spirin AS (2002) Ribosome as a molecular machine. *FEBS Lett* **514**, 2–10.

Wang L and Schultz PG (2005) Expanding the genetic code. *Angew Chem Int Ed* **44**, 34–66.

Wilson DN and Nierhaus KH (2003) The ribosome through the looking glass. *Angew Chem Int Ed* **42**, 3464–3486.

Insulin

McCarthy AA (2004) New approaches to diabetes disease control, insulin delivery, and monitoring. *Chem Biol* **11**, 1597–1598.

Mehanna AS (2005) Insulin and oral antidiabetic agents. *Am J Pharm Educ* **69**, 1–11.

Notkins AL (2002) Immunologic and genetic factors in type 1 diabetes. *J Biol Chem* **277**, 43545–43548.

Ross SA, Gulve EA and Wang M (2004) Chemistry and biochemistry of type 2 diabetes. *Chem Rev* **104**, 1255–1282.

Skyler JS (2004) Diabetes mellitus: pathogenesis and treatment strategies. *J Med Chem* **47**, 4113–4117.

Interferons

Maher SG, Romero-Weaver AL, Scarzello AJ and Gamero AM (2007) Interferon: cellular executioner or white knight? *Curr Med Chem* **14**, 1279–1289.

Pestka S (2007) The interferons: 50 years after their discovery, there is much more to learn. *J Biol Chem* **282**, 20047–20051.

Opioid Peptides

Janecka A, Staniszewska R and Fichna J (2007) Endomorphin analogs. *Curr Med Chem* **14**, 3201–3208.

Ribosomal Peptide Toxins

Aoki KR (2004) Botulinum toxin: a successful therapeutic protein. *Curr Med Chem* **11**, 3085–3092.

Buczek O, Bulaj G and Olivera BM (2005) Conotoxins and the post-translational modification of secreted gene products. *Cell Mol Life Sci* **62**, 3067–3079.

Hallen HE, Luo H, Scott-Craig JS and Walton JD (2007) Gene family encoding the major toxins of lethal *Amanita* mushrooms. *Proc Natl Acad Sci USA* **104**, 19097–19101.

Hicks RP, Hartell MG, Nichols DA, Bhattacharjee AK, van Hamont JE and Skillman DR (2005) The medicinal chemistry of botulinum, ricin and anthrax toxins. *Curr Med Chem* **12**, 667–690.

Hogg RC (2006) Novel approaches to pain relief using venom-derived peptides. *Curr Med Chem* **13**, 3191–3201.

Kini RM (2006) Anticoagulant proteins from snake venoms: structure, function and mechanism. *Biochem J* **397**, 377–387.

Livett BG, Gayler KR and Khalil Z (2004) Drugs from the sea: conopeptides as potential therapeutics. *Curr Med Chem* **11**, 1715–1723.

Miljanich GP (2004) Ziconotide: neuronal calcium channel blocker for treating severe chronic pain. *Curr Med Chem* **11**, 3029–3040.

Narayanan S, Surendranath K, Bora N, Surolia A and Karande AA (2005) Ribosome inactivating proteins and apoptosis. *FEBS Lett* **579**, 1324–1331.

Olivera BM (2006) *Conus* peptides: biodiversity-based discovery and exogenomics. *J Biol Chem* **281**, 31173–31177.

Park S-W, Vepachedu R, Sharma N and Vivanco JM (2004) Ribosome-inactivating proteins in plant biology. *Planta* **219**, 1093–1096.

Schiavo G, Matteoli M and Montecucco C (2000) Neurotoxins affecting neuroexocytosis. *Physiol Rev* **80**, 717–766.

Terlau H and Olivera BM (2004) *Conus* venoms: a rich source of novel ion channel-targeted peptides. *Physiol Rev* **84**, 41–68.

Tsetlina VI and Hucho F (2004) Snake and snail toxins acting on nicotinic acetylcholine receptors: fundamental aspects and medical applications. *FEBS Lett* **557**, 9–13.

Non-ribosomal Peptide Biosynthesis

Doekel S and Marahiel MA (2001) Biosynthesis of natural products on modular peptide synthetases. *Metab Eng* **3**, 64–77.

Du L, Sánchez C and Shen B (2001) Hybrid peptide–polyketide natural products: biosynthesis and prospects toward engineering novel molecules. *Metab Eng* **3**, 78–95.

Finking R and Marahiel MA (2004) Biosynthesis of nonribosomal peptides. *Annu Rev Microbiol* **58**, 453–488.

Fischbach MA and Walsh CT (2006) Assembly-line enzymology for polyketide and nonribosomal peptide antibiotics: logic, machinery, and mechanisms. *Chem Rev* **106**, 3468–3496.

Kohli RM and Walsh CT (2003) Enzymology of acyl chain macrocyclization in natural product biosynthesis. *Chem Commun* 297–307.

Kopp F and Marahiel MA (2007) Macrocyclization strategies in polyketide and nonribosomal peptide biosynthesis. *Nat Prod Rep* **24**, 735–749.

Samel SA, Marahiel MA and Essen L-A (2008) How to tailor non-ribosomal peptide products – new clues about the structures and mechanisms of modifying enzymes. *Mol BioSyst* **4**, 387–393.

Schwarzer D, Finking R and Marahiel MA (2003) Nonribosomal peptides: from genes to products. *Nat Prod Rep* **20**, 275–287.

Sieber SA and Marahiel MA (2005) Molecular mechanisms underlying nonribosomal peptide synthesis: approaches to new antibiotics. *Chem Rev* **105**, 715–738.

Walsh CT (2002) Combinatorial biosynthesis of antibiotics: challenges and opportunities. *ChemBioChem* **3**, 124–134.

Walsh CT (2004) Polyketide and nonribosomal peptide antibiotics: modularity and versatility. *Science* **303**, 1805–1810.

Non-ribosomal Peptide Antibiotics

Bacitracin

Wagner B, Schumann U, Koert U and Marahiel MA (2006) Rational design of bacitracin A derivatives by incorporating natural product derived heterocycles. *J Am Chem Soc* **128**, 10513–10520.

Bleomycin

Galm U, Hager MH, Van Lanen SG, Ju J, Thorson JS and Shen B (2005) Antitumor antibiotics: bleomycin, enediynes, and mitomycin. *Chem Rev* **105**, 739–758.

Shen B, Du L, Sanchez C, Edwards DJ, Chen M and Murrell JM (2002) Cloning and characterization of the bleomycin biosynthetic gene cluster from *Streptomyces verticillus* ATCC150031. *J Nat Prod* **65**, 422–431.

Capreomycin

Felnagle EA, Rondon MR, Berti AD, Crosby HA and Thomas MG (2007) Identification of the biosynthetic gene cluster and an additional gene for resistance to the antituberculosis drug capreomycin. *Appl Environ Microbiol* **73**, 4162–4170.

Cyclosporin

Mann J (2001) Natural products as immunosuppressive agents. *Nat Prod Rep* **18**, 417–430.

Dactinomycin

Schauwecker F, Pfennig F, Grammel N and Keller U (2000) Construction and in vitro analysis of a new bi-modular polypeptide synthetase for synthesis of *N*-methylated acyl peptides. *Chem Biol* **7**, 287–297.

Daptomycin

Baltz RH, Miao V and Wrigley SK (2005) Natural products to drugs: daptomycin and related lipopeptide antibiotics. *Nat Prod Rep* **22**, 717–741.

Miao V, Coöffet-LeGal M-F, Brian P, Brost R, Penn J, Whiting A, Martin S, Ford R, Parr I, Bouchard M, Silva CJ, Wrigley SK and Baltz RH (2005) Daptomycin biosynthesis in *Streptomyces roseosporus*: cloning and analysis of the gene cluster and revision of peptide stereochemistry. *Microbiology* **151**, 1507–1523.

Gramicidin

Hoyer KM, Mahlert C and Marahiel MA (2007) The iterative gramicidin S thioesterase catalyzes peptide ligation and cyclization. *Chem Biol* **14**, 13–22.

Schoenafinger G, Schracke N, Linne U and Marahiel MA (2006) Formylation domain: an essential modifying enzyme for the nonribosomal biosynthesis of linear gramicidin. *J Am Chem Soc* **128**, 7406–7407.

Streptogramin

Mukhtar TA and Wright GD (2005) Streptogramins, oxazolidinones, and other inhibitors of bacterial protein synthesis. *Chem Rev* **105**, 529–542.

Tyrothricin

Schwarzer D, Mootz D and Marahiel MA (2001) Exploring the impact of different thioesterase domains for the design of hybrid peptide synthetases. *Chem Biol* **8**, 997–1010.

Vancomycin, Teicoplanin

Gao Y (2002) Glycopeptide antibiotics and development of inhibitors to overcome vancomycin resistance. *Nat Prod Rep* **19**, 100–107.

Hubbard BK and Walsh CT (2003) Vancomycin assembly: nature's way. *Angew Chem Int Ed* **42**, 730–765.

Kahne D, Leimkuhler C, Lu W and Walsh C (2005) Glycopeptide and lipoglycopeptide antibiotics. *Chem Rev* **105**, 425–448.

Malabarba A and Ciabatti R (2001) Glycopeptide derivatives. *Curr Med Chem* **8**, 1759–1773.

Welzel P (2005) Syntheses around the transglycosylation step in peptidoglycan biosynthesis. *Chem Rev* **105**, 4610–4660.

Penicillins and Cephalosporins

Brakhage AE, Al-Abdallah Q, Tüncher A and Spröte P (2005) Evolution of β-lactam biosynthesis genes and recruitment of *trans*-acting factors. *Phytochemistry* **66**, 1200–1210.

Krebs C, Fujimori DG, Walsh CT and Bollinger JM (2007) Non-heme Fe(IV)–oxo intermediates. *Acc Chem Res* **40**, 484–492.

Ullán, RV, Campoy S, Casqueiro J, Fernández FJ and Martín JF (2007) Deacetylcephalosporin C production in *Penicillium chrysogenum* by expression of the isopenicillin N epimerization, ring expansion, and acetylation genes. *Chem Biol* **14**, 329–339.

β-Lactamases

Crowder MW, Spencer J and Vila AJ (2006) Metallo-β-lactamases: novel weaponry for antibiotic resistance in bacteria. *Acc Chem Res* **39**, 721–728.

Fisher JF, Meroueh SO and Mobashery S (2005) Bacterial resistance to β-lactam antibiotics: compelling opportunism, compelling opportunity. *Chem Rev* **105**, 395–424.

Sandanayaka VP and Prashad AS (2002) Resistance to β-lactam antibiotics: structure and mechanism based design of β-lactamase inhibitors. *Curr Med Chem* **9**, 1145–1165.

Clavams, Carbapenems, Monobactams

Gunsior M, Breazeale SD, Lind AJ, Ravel J, Janc JW and Townsend CA (2004) The biosynthetic gene cluster for a monocyclic β lactam antibiotic, nocardicin A. *Chem Biol* **11**, 927–938.

Kershaw NJ, Caines MEC, Sleeman MC and Schofield CJ (2005) The enzymology of clavam and carbapenem biosynthesis. *Chem Commun* 4251–4263.

Núñez LE, Méndez C, Braña AF, Blanco G and Salas JA (2003) The biosynthetic gene cluster for the β-lactam carbapenem thienamycin in *Streptomyces cattleya*. *Chem Biol* **10**, 301–311.

Townsend CA (2002) New reactions in clavulanic acid biosynthesis. *Curr Opinion Chem Biol* **6**, 583–589.

Cyanogenic Glycosides

Bak S, Paquette SM, Morant M, Morant AV, Saito S, Bjarnholt N, Zagrobelny M, Jørgensen K, Osmani S, Simonsen HT, Pérez RS, van Heeswijck TB, Jørgensen B and Møller BL (2006) Cyanogenic glycosides: a case study for evolution and application of cytochromes P450. *Phytochem Rev* **5**, 309–329.

Bjarnholt N and Møller BL (2008) Molecules of interest. Hydroxynitrile glucosides. *Phytochemistry* **69**, 1947–1961.

Sánchez-Pérez R, Jørgensen K, Olsen CE, Dicenta F and Møller BL (2008) Bitterness in almonds. *Plant Physiol* **146**, 1040–1052.

Zagrobelny M, Bak S and Møller BL (2008) Cyanogenesis in plants and arthropods. *Phytochemistry* **69**, 1457–1468.

Zagrobelny M, Bak S, Rasmussen AV, Jørgensen B, Naumann CM and Møller BL (2004) Cyanogenic glucosides and plant–insect interactions. *Phytochemistry* **65**, 293–306.

Glucosinolates

Bones AM and Rossiter JT (2006) The enzymic and chemically induced decomposition of glucosinolates. *Phytochemistry* **67**, 1053–1067.

Cartea ME and Velasco P (2008) Glucosinolates in *Brassica* foods: bioavailability in food and significance for human health. *Phytochem Rev* **7**, 213–229.

Fahey JW, Zalcmann AT and Talalay P (2001) The chemical diversity and distribution of glucosinolates and isothiocyanates among plants. *Phytochemistry* **56**, 5–51; corrigendum (2002) **59**, 237.

Halkier BA and Gershenzon J (2006) Biology and biochemistry of glucosinolates. *Annu Rev Plant Biol* **57**, 303–333.

Holst B and Williamson G (2004) A critical review of the bioavailability of glucosinolates and related compounds. *Nat Prod Rep* **21**, 425–447.

Johnson IT (2002) Glucosinolates in the human diet. Bioavailability and implications for health. *Phytochem Rev* **1**, 183–188.

Cysteine Sulfoxides, Garlic

Banerjee SK, Mukherjee PK and Maulik SK (2003) Garlic as an antioxidant: the good, the bad and the ugly. *Phytother Res* **17**, 97–106.

Fritsch RM and Keusgen M (2006) Occurrence and taxonomic significance of cysteine sulphoxides in the genus *Allium* L. (Alliaceae). *Phytochemistry* **67**, 1127–1135.

Griffiths G, Trueman L, Crowther T, Thomas B and Smith B (2002) Onions – a global benefit to health. *Phytother Res* **16**, 603–615.

Rose P, Whiteman M, Moore PK and Zhu YZ (2005) Bioactive *S*-alk(en)yl cysteine sulfoxide metabolites in the genus *Allium*: the chemistry of potential therapeutic agents. *Nat Prod Rep* **22**, 351–368.

8

CARBOHYDRATES

Carbohydrates are among the most abundant constituents of plants, animals, and microorganisms. Polymeric carbohydrates function as important food reserves and as structural components in cell walls. Animals and most microorganisms are dependent upon the carbohydrates produced by plants for their very existence. Carbohydrates are the first products formed in photosynthesis, and are the products from which plants synthesize their own food reserves and other chemical constituents. These materials then become the foodstuffs of other organisms. The main pathways of carbohydrate biosynthesis and degradation comprise an important component of primary metabolism that is essential for all organisms. Secondary metabolites are also ultimately derived from carbohydrate metabolism, and the relationships of the acetate, shikimate, mevalonate, and methylerythritol phosphate pathways to primary metabolism have already been indicated. Many of the medicinally important secondary metabolites described in the earlier chapters have been seen to contain clearly recognizable carbohydrate portions in their structures; for example, note the frequent occurrence of glycosides. In this chapter, some of the important natural materials which can be grouped together because they are composed entirely or predominantly of carbohydrate units are discussed. Because of their widespread use in medicinal preparations, some materials with no inherent biological activity, and which are clearly of primary metabolic status (e.g. sucrose, starch, alginic acid) are also included.

The name **carbohydrate** was introduced because many of the compounds had the general formula $C_x(H_2O)_y$, and thus appeared to be hydrates of carbon. The terminology is now commonly used in a much broader sense to denote polyhydroxy aldehydes and ketones, and their derivatives. **Sugars** or saccharides are other terms used in a rather broad sense to cover carbohydrate materials. Though these words link directly to compounds with sweetening properties, application of the terms extends considerably beyond this. A **monosaccharide** is a carbohydrate usually in the range C_3–C_9, whilst **oligosaccharide** covers small polymers comprised of 2–10 monosaccharide units. The term **polysaccharide** is used for larger polymers.

MONOSACCHARIDES

Six-carbon sugars (**hexoses**) and five-carbon sugars (**pentoses**) are the most frequently encountered monosaccharide carbohydrate units in nature. Photosynthesis produces initially the three-carbon sugar 3-phosphoglyceraldehyde, two molecules of which are used to synthesize the hexose glucose 6-phosphate by a sequence which effectively achieves the reverse of the glycolytic reactions (Figure 8.1). Alternatively, by the complex reactions of the Calvin cycle, 3-phosphoglyceraldehyde may be used in the construction of the pentoses ribose 5-phosphate, ribulose 5-phosphate, and xylulose 5-phosphate. Although these sequences need not be considered, they do incorporate a range of fundamental transformations which are used in the biochemical manipulation of monosaccharide structures, and which will be encountered here and elsewhere:

- Intramolecular transfer of a group (mutation), e.g. the isomerization of glucose 6-phosphate and glucose 1-phosphate (Figure 8.2), which is actually achieved via an intermediate diphosphate.
- Epimerization to change the stereochemistry at one of the chiral centres, e.g. the interconversion of ribulose 5-phosphate and xylulose 5-phosphate (Figure 8.3). This reaction involves epimerization adjacent to a carbonyl group and probably proceeds through a common enol tautomer, but some other epimerizations are known to

Medicinal Natural Products: A Biosynthetic Approach. 3rd Edition   Paul Dewick
© 2009 John Wiley & Sons, Ltd

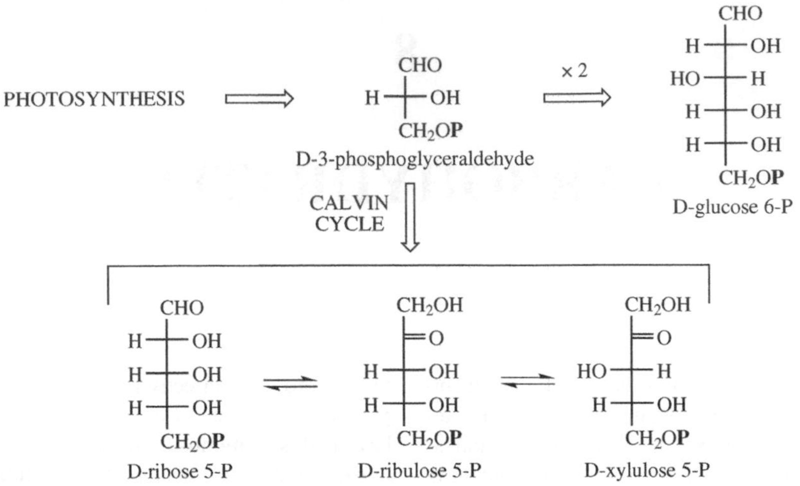

Figure 8.1

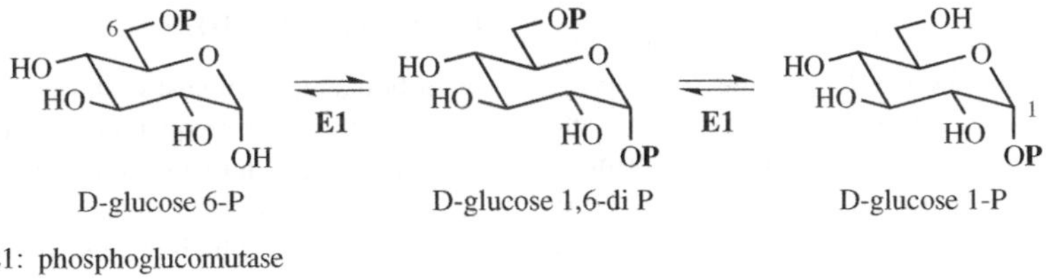

E1: phosphoglucomutase

Figure 8.2

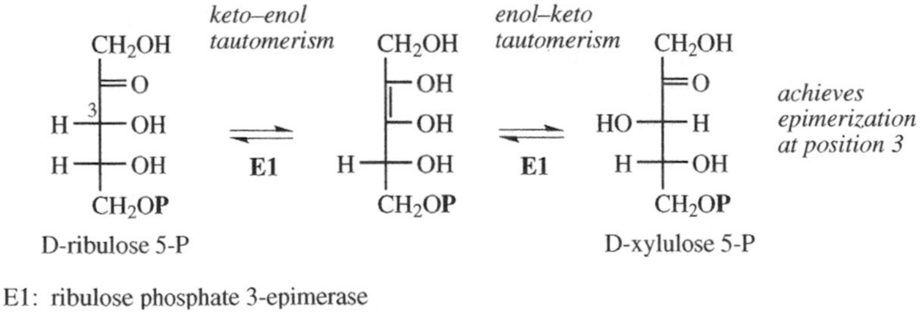

E1: ribulose phosphate 3-epimerase

Figure 8.3

proceed through oxidation to an intermediate carbonyl, followed by reduction to give the opposite configuration. The substrate for epimerization is often the UDPsugar rather than the monosaccharide phosphate.

- Aldose–ketose interconversions, e.g. glucose 6-phosphate to fructose 6-phosphate (Figure 8.4), also proceed through a common enol intermediate.
- Transfer of C_2 and C_3 units in reactions catalysed by transketolase and transaldolase respectively modify

the chain length of the sugar. Transketolase removes a two-carbon fragment from ketols such as fructose 6-phosphate (alternatively xylulose 5-phosphate or sedoheptulose 7-phosphate) through the participation of thiamine diphosphate. Nucleophilic attack of the thiamine diphosphate anion onto the carbonyl results in an addition product which then fragments by a reverse aldol reaction, generating the chain-shortened aldose erythrose 4-phosphate, and the two carbon carbanion

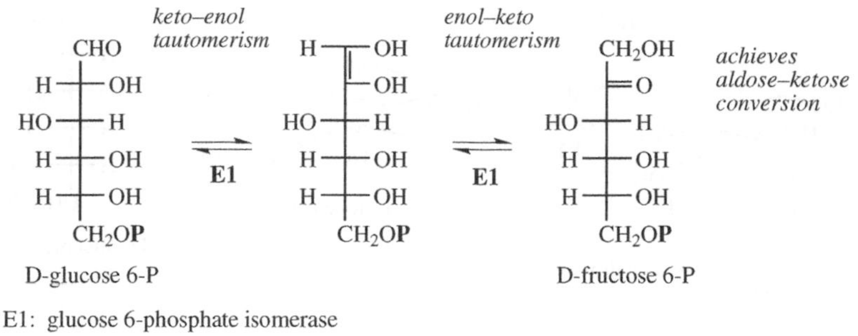

Figure 8.4

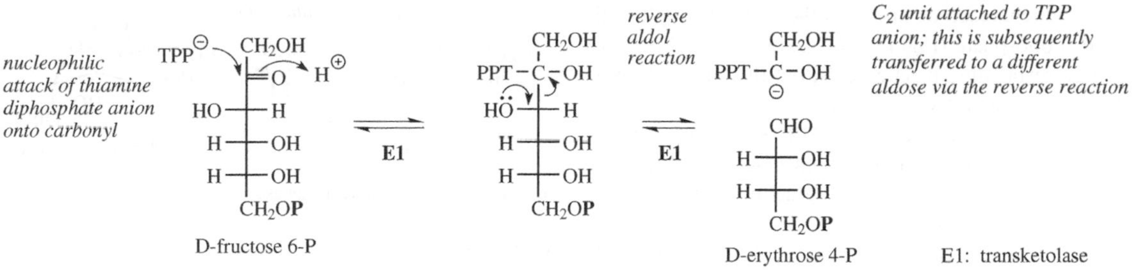

Figure 8.5

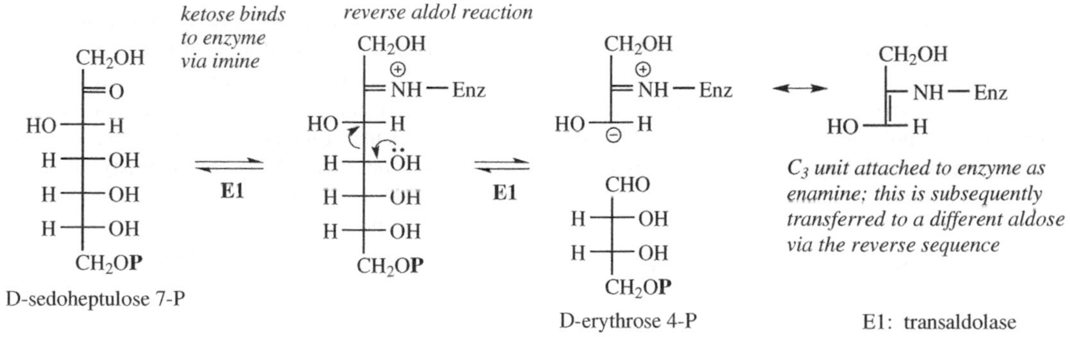

Figure 8.6

unit attached to TPP (Figure 8.5) (compare the role of TPP in the decarboxylation of α-keto acids, page 23). Then, in what is formally the reverse of this reaction, this carbanion can attack another aldose such as ribose 5-phosphate (alternatively erythrose 4-phosphate or glyceraldehyde 3-phosphate), thus extending its chain length by two carbon atoms. Transaldolase removes a three-carbon fragment from a ketose, such as sedoheptulose 7-phosphate (alternatively fructose 6-phosphate), in a reverse aldol reaction, though

this requires formation of an imine between the carbonyl group and an active site lysine of the enzyme (Figure 8.6). Again, the reaction is completed by a reversal of this process, but transferring the C_3 carbanion to another aldose, such as glyceraldehyde 3-phosphate (alternatively erythrose 4-phosphate or ribose 5-phosphate), and thus increasing its chain length.

• Oxidation and reduction reactions, typically employing the NAD/NADP nucleotides, alter the oxidation state of the substrate. Oxidation at C-1 converts an

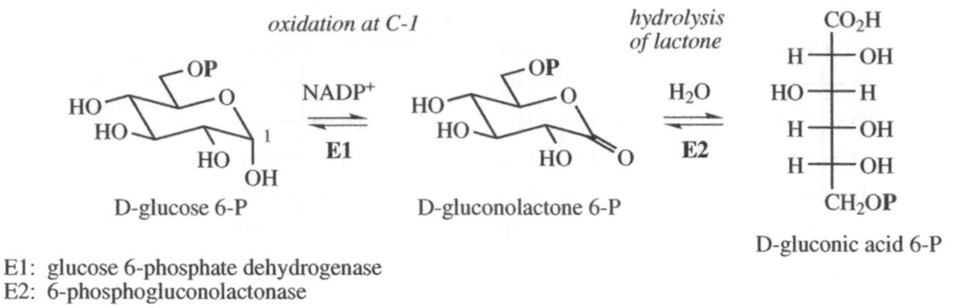

E1: glucose 6-phosphate dehydrogenase
E2: 6-phosphogluconolactonase

Figure 8.7

aldose into an aldonic acid, e.g. glucose 6-phosphate gives gluconolactone and then the open-chain gluconic acid 6-phosphate (Figure 8.7). Oxidation at C-6 yields the corresponding uronic acids, but this takes place on UDPsugar derivatives (see page 30), e.g. UDPglucose to UDPglucuronic acid (Figure 8.8). Reduction is exemplified by the conversion of both glucose and fructose into the sugar alcohol sorbitol (glucitol) and of fructose into mannitol (Figure 8.9).

• Transamination reactions (see page 20) on appropriate keto sugars allow the introduction of amino groups. In some systems, nucleoside diphosphate (UDP or TDP) sugars are involved, as shown for the amino sugar glucosamine (Figure 8.10). The major route to glucosamine derivatives, however, is via conversion of fructose 6-phosphate (Figure 8.4) into glucosamine 6-phosphate by a transaminase–isomerase enzyme. Amino sugars, as their *N*-acetyl derivatives, are part of the structures of several natural polysaccharides (see page 495), and other uncommon amino sugars are components of the aminoglycoside antibiotics (see page 499).

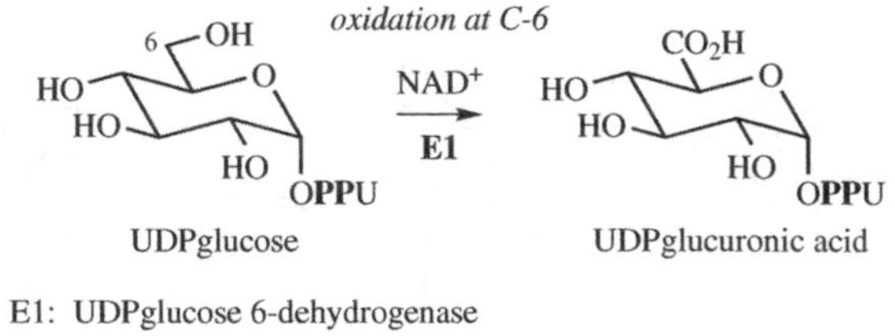

Monosaccharide structures may be depicted in open-chain forms showing their carbonyl character, or in

E1: UDPglucose 6-dehydrogenase

Figure 8.8

cyclic hemiacetal or hemiketal forms. The compounds exist predominantly in the cyclic forms, which result from nucleophilic attack of an appropriate hydroxyl onto the carbonyl (Figure 8.11). Both six-membered pyranose and five-membered furanose structures are encountered, a particular ring size usually being characteristic for any one sugar. Since the carbonyl group may be attacked from either side, two epimeric structures (anomers) are possible in each case, and in solution, the two forms are frequently in equilibrium. In natural product structures, sugar units are most likely (but not always)

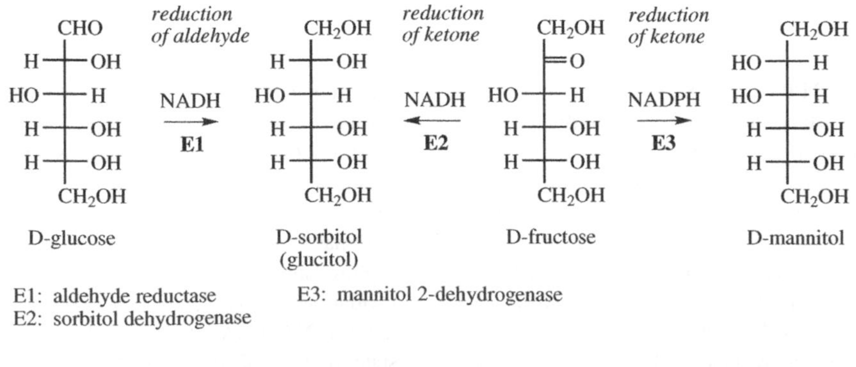

E1: aldehyde reductase
E2: sorbitol dehydrogenase
E3: mannitol 2-dehydrogenase

Figure 8.9

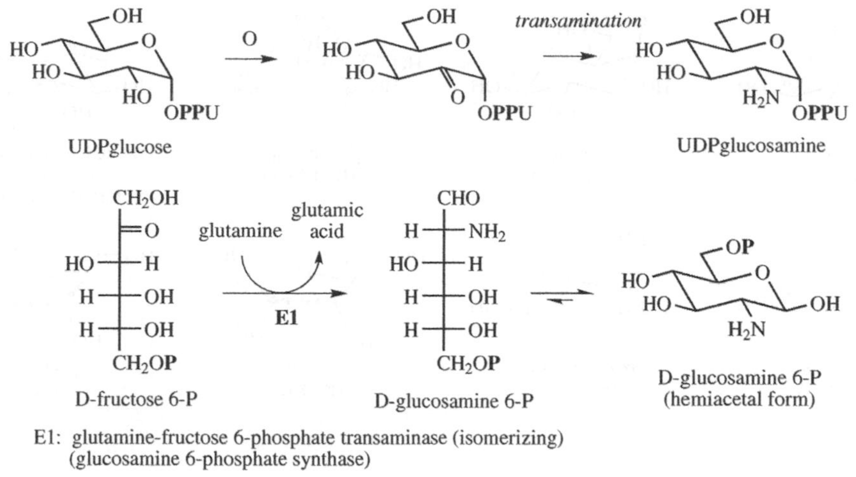

Figure 8.10

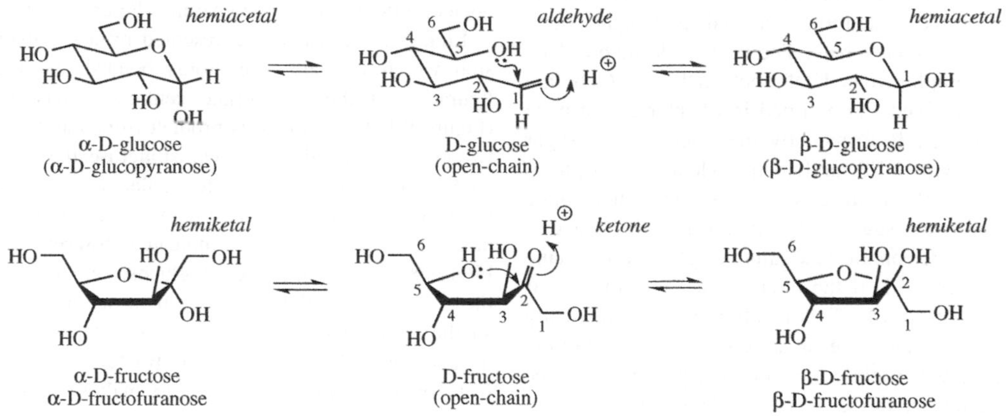

Figure 8.11

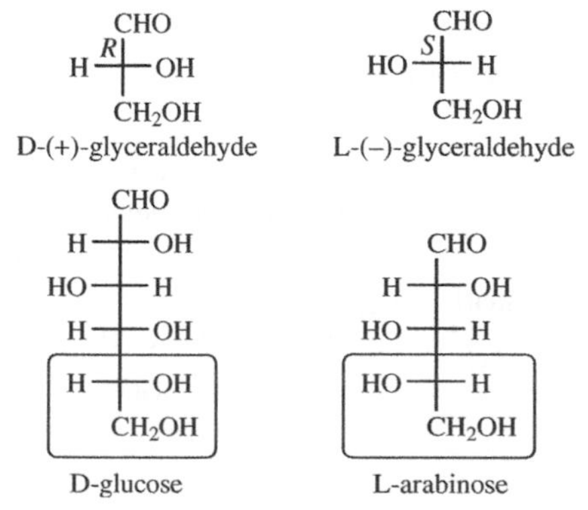

Figure 8.12

to be encountered in just one of the epimeric forms. The two forms are designated α or β on the basis of the chiralities at the anomeric centre and at the highest numbered chiral centre. If these are the same (using the *RS* convention), then the anomer is termed β, or α if they are different. In practice, this translates to the anomeric hydroxyl being 'up' in the case of β-D-sugars and α-L-sugars. Note that D- and L- prefixes are assigned on the basis of the chirality (as depicted in Fischer projection) at the highest numbered chiral centre and its relationship to D-(*R*)-(+)-glyceraldehyde or L-(*S*)-(-)-glyceraldehyde (Figure 8.12). The most commonly encountered monosaccharides and their usual anomers are shown in Figure 8.13.

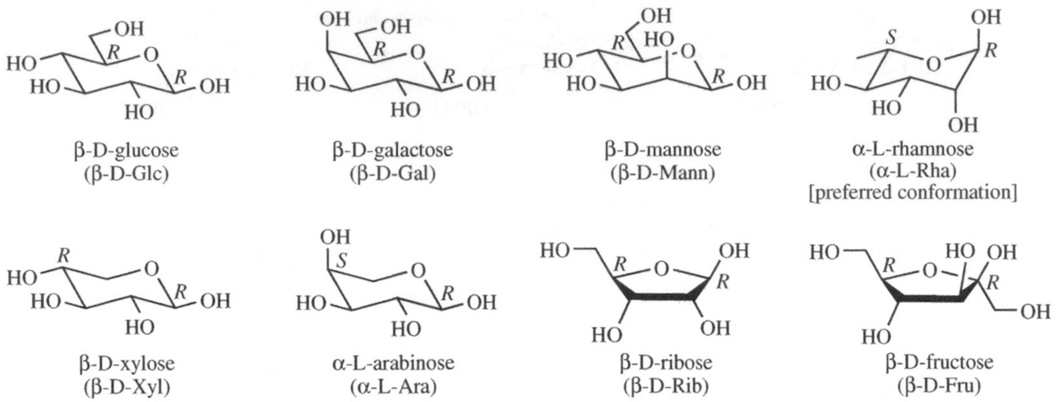

β-D-glucose
(β-D-Glc)

β-D-galactose
(β-D-Gal)

β-D-mannose
(β-D-Mann)

α-L-rhamnose
(α-L-Rha)
[preferred conformation]

β-D-xylose
(β-D-Xyl)

α-L-arabinose
(α-L-Ara)

β-D-ribose
(β-D-Rib)

β-D-fructose
(β-D-Fru)

Figure 8.13

OLIGOSACCHARIDES

The formation of oligosaccharides and polysaccharides is dependent on the generation of an activated sugar bound to a nucleoside diphosphate. The nucleoside diphosphate most often employed is UDP, but ADP and GDP are sometimes involved. As outlined in Chapter 2 (see page 30), a UDPsugar is formed by the reaction of a sugar 1-phosphate with UTP, and then nucleophilic displacement of the UDP leaving group by a suitable nucleophile generates the new sugar derivative. This will be a glycoside if the nucleophile is a suitable aglycone molecule or an oligosaccharide if the nucleophile is another sugar molecule (Figure 8.14). This reaction, if mechanistically of S_N2 type, should give an inversion of configuration at C-1 in the electrophile, generating a product with the β-configuration in the case of UDPglucose as shown. Many of the linkages formed between glucose monomers

actually have the α-configuration, and it is believed that a double S_N2 mechanism operates, via participation of a nucleophilic group on the enzyme (Figure 8.14).

Linkages are usually represented by a shorthand version which indicates the atoms bonded and the configuration at the appropriate centre(s). Thus, **maltose** (Figure 8.15), a hydrolysis product from starch, contains two glucoses linked α1→4, whilst **lactose**, the main sugar component of cow's milk, has galactose linked β1→4 to glucose. In the systematic names, the ring size (pyranose or furanose) is also indicated. **Sucrose** ('sugar'; Figure 8.15) is composed of glucose and fructose, but these are both linked through their anomeric centres, so the shorthand representation becomes α1→β2. This means that both the hemiacetal structures are prevented from opening, and in contrast to maltose and lactose, there can be no open-chain form in equilibrium with the

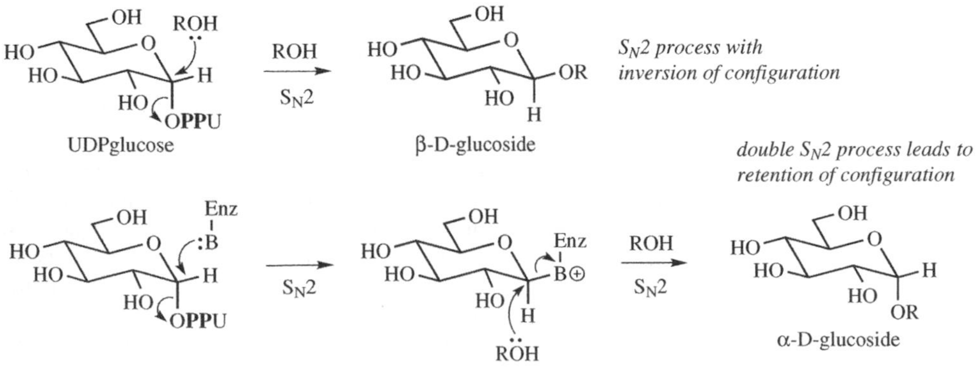

UDPglucose

β-D-glucoside

S_N2 process with inversion of configuration

double S_N2 process leads to retention of configuration

α-D-glucoside

Figure 8.14

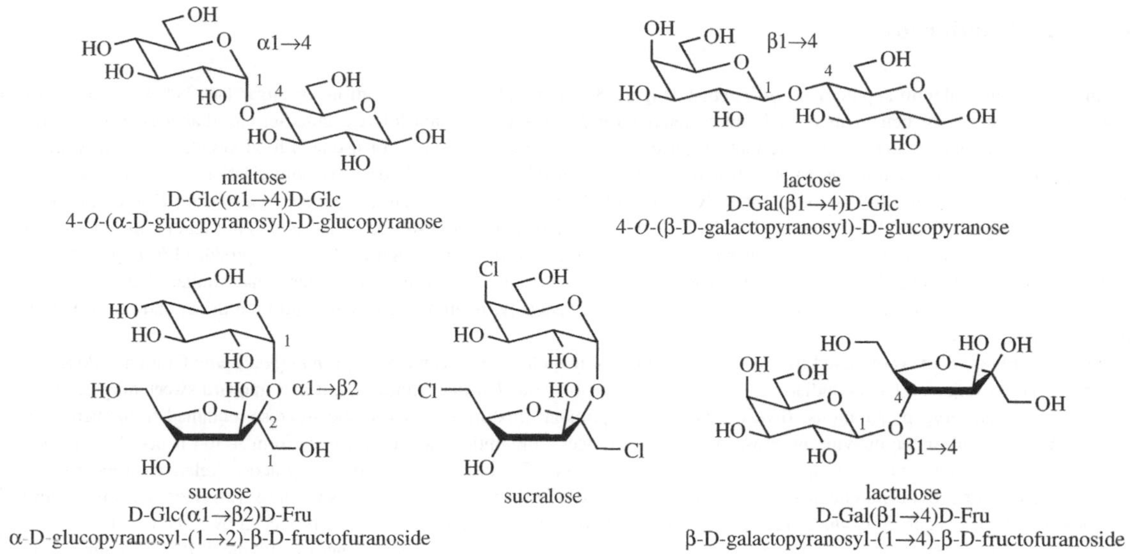

Figure 8.15

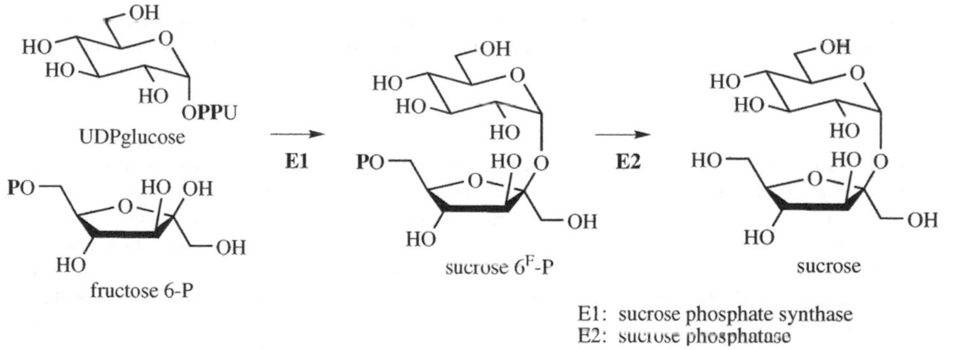

E1: sucrose phosphate synthase
E2: sucrose phosphatase

Figure 8.16

cyclic form. Therefore, sucrose does not display any of the properties usually associated with the masked carbonyl group, e.g. it is not a reducing sugar. Sucrose is known to be formed predominantly by a slightly modified form of the sequence shown in Figure 8.16, in that UDPglucose is attacked by fructose 6-phosphate and that the first-formed product is sucrose 6^F-phosphate (F indicating the numbering refers to the fructose ring). Hydrolysis of the phosphate then generates sucrose (Figure 8.16). Mono- and di-saccharides of general or medicinal importance are described in Box 8.1.

Box 8.1

Monosaccharides and Disaccharides

D-**Glucose** (**dextrose**; Figure 8.9) occurs naturally in grapes and other fruits. It is usually obtained by enzymic hydrolysis of starch, and is used as a nutrient, particularly in the form of an intravenous infusion. Chemical oxidation of glucose converts the aldehyde function to a carboxylic acid and produces gluconic acid. The soluble calcium salt **calcium gluconate** is used

Box 8.1   (continued)

as an intravenous calcium supplement. D-**Fructose** (Figure 8.9) is usually obtained from invert sugar (see below) separating it from glucose, and is of benefit as a food and sweetener for patients who cannot tolerate glucose, e.g. diabetics. Fructose has the sweetness of sucrose and about twice that of glucose. High fructose corn syrup for use as a food sweetener is a mixture of fructose and glucose containing up to 90% fructose and is produced by enzymic hydrolysis/isomerization of starch. The sugar alcohol D-**sorbitol** (Figure 8.9) is found naturally in the ripe berries of the mountain ash (*Sorbus aucuparia*; Rosaceae), but is prepared semi-synthetically from glucose. It is half as sweet as sucrose, is not absorbed orally, and is not readily metabolized in the body. It finds particular use as a sweetener for diabetic products. D-**Mannitol** (Figure 8.9) is also produced from glucose, but occurs naturally in manna, the exudate of the manna ash *Fraxinus ornus* (Oleaceae). This material has similar characteristics to sorbitol, but is used principally as a diuretic. It is injected intravenously, is eliminated rapidly into the urine, and removes fluid by an osmotic effect.

Sucrose (Figure 8.15) is obtained from a variety of sources, including sugar cane (*Saccharum officinarum*; Graminae/Poaceae), sugar beet (*Beta vulgaris*; Chenopodiaceae), and sugar maple (*Acer saccharum*; Aceraceae). It is a standard sweetening agent for foods, syrups, and drug preparations. **Invert sugar** is an equimolar mixture of glucose and fructose, obtained from sucrose by hydrolysis with acid or the enzyme invertase. During this process, the optical activity changes from + to– hence the reference to inversion. The high sweetness of fructose (relative sweetness 1.7) combined with that of glucose (relative sweetness 0.7) means invert sugar provides a cheaper, less-calorific food sweetener than sucrose (relative sweetness 1.0). Honey is also mainly composed of invert sugar. **Sucralose** (Figure 8.15) is a semi-synthetic derivative of sucrose, produced by selective chlorination. Sucralose is some 600 times sweeter than sucrose and is used as a sweetening agent. **Lactose** (Figure 8.15) can comprise up to 8% of mammalian milk; it is extracted from cow's milk, often as a by-product from cheese manufacture. It is only faintly sweet, and its principal use is as a diluent in tablet formulations. Lactose intolerance is a condition in certain adults who are unable to tolerate milk products in the diet. This is a consequence of very low levels of the hydrolytic enzyme lactase, and ingestion of lactose can lead to adverse reactions such as gastric upsets. **Lactulose** (Figure 8.15) is a semi-synthetic disaccharide prepared from lactose and composed of galactose linked $\beta1{\to}4$ to fructose. It is not absorbed from the gastrointestinal tract and is predominantly excreted unchanged. It helps to retain fluid in the bowel by osmosis, and is thus used as a laxative.

Vitamin C

Vitamin C (ascorbic acid; Figure 8.17) can be synthesized by most animals except humans, other primates, guinea pigs, bats, and some birds; for these it is obtained via the diet. Citrus fruits, peppers, guavas, rose hips, and blackcurrants are especially rich sources, but it is present in most fresh fruit and vegetables. Raw citrus fruits provide a good daily source. It is a water-soluble acidic compound (hence ascorbic acid), readily losing a proton from the 3-hydroxyl and leading to a resonance-stabilized anion. It is rapidly degraded during cooking in the presence of air. Vitamin C deficiency leads to scurvy, characterized by muscular pain, skin lesions, fragile blood vessels, bleeding gums, and tooth loss. The vitamin is essential for the formation of collagen, the principal structural protein in skin, bone, tendons, and ligaments, being a cofactor in the hydroxylation of proline to 4-hydroxyproline and of lysine to 5-hydroxylysine (see page 425), which account for up to 25% of the collagen structure. These reactions are catalysed by 2-oxoglutarate dioxygenases (see page 27), and the ascorbic acid requirement is to reduce an enzyme-bound iron–oxygen complex. Skin lesions characteristic of scurvy are a direct result of low levels of hydroxylation in the collagen structure synthesized in the absence of ascorbic acid. Ascorbic acid is also associated with the hydroxylation of tyrosine in the pathway to catecholamines (see page 335) and in the biosynthesis of homogentisic acid, the precursor of tocopherols and plastoquinones (see page 178). Ascorbic acid is usually prepared synthetically and is used to treat or prevent deficiency. Natural ascorbic acid is extracted from rose hips, persimmons, and citrus fruits. Large doses have been given after surgery or burns to promote healing by increasing collagen synthesis. The benefits of consuming large doses of vitamin C to alleviate the common cold and other viral infections are not proven. Some sufferers believe it to be beneficial in the prevention and therapy of cancer. Vitamin C does have valuable antioxidant properties, and these are exploited commercially in the food industries. Its main function as an antioxidant is to provide a regenerating system for tocopherol (vitamin E, see page 181).

In animals, ascorbic acid is synthesized in the liver from glucose, by a pathway which initially involves oxidation to glucuronic acid via UDP derivatives. This is followed by reduction of the carbonyl function, lactone formation, oxidation of the secondary alcohol to a carbonyl, and subsequent enolization (Figure 8.17). Man and other primates appear to lack the enzyme gulonolactone oxidase, and are thus dependent on a dietary source of vitamin C. Several different pathways have been observed in plants, though the predominant one (Figure 8.18) proceeds from L-galactose and does not involve an early uronic acid derivative. Yeast and some fungi synthesize a five-carbon analogue of ascorbic acid, D-erythroascorbic acid (Figure 8.18), from L-arabinose. This compound has the same biological properties as ascorbic acid.

Box 8.1 (continued)

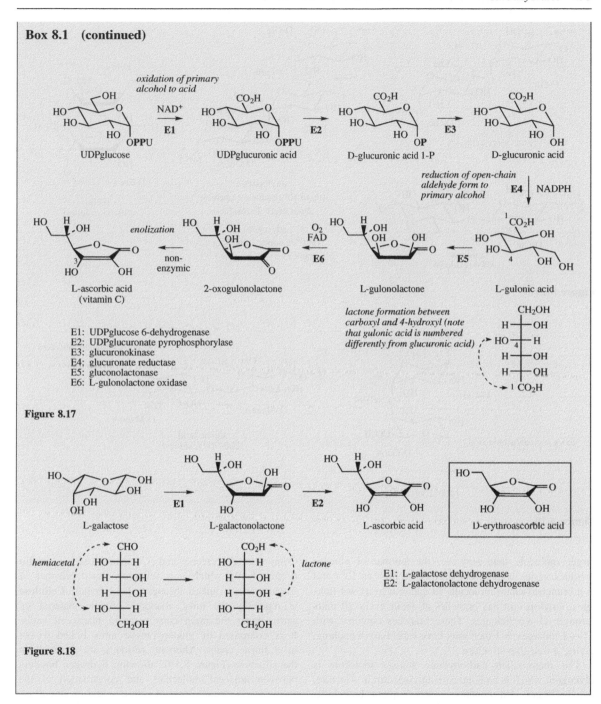

oxidation of primary alcohol to acid
UDPglucose
NAD+
E1
UDPglucuronic acid
E2
D-glucuronic acid 1-P
E3
D-glucuronic acid
reduction of open-chain aldehyde form to primary alcohol
E4
NADPH
L-gulonic acid
E5
L-gulonolactone
O2
FAD
E6
2-oxogulonolactone
enolization
non-enzymic
L-ascorbic acid
(vitamin C)
lactone formation between carboxyl and 4-hydroxyl (note that gulonic acid is numbered differently from glucuronic acid)
E1: UDPglucose 6-dehydrogenase
E2: UDPglucuronate pyrophosphorylase
E3: glucuronokinase
E4: glucuronate reductase
E5: gluconolactonase
E6: L-gulonolactone oxidase

Figure 8.17

L-galactose
E1
L-galactonolactone
E2
L-ascorbic acid
D-erythroascorbic acid
hemiacetal
lactone
E1: L-galactose dehydrogenase
E2: L-galactonolactone dehydrogenase

Figure 8.18

POLYSACCHARIDES

Polysaccharides fulfil two main functions in living organisms: as food reserves and as structural elements. Plants accumulate starch as their main food reserve, a material that is composed entirely of glucopyranose units, but in two types of molecule. **Amylose** (Figure 8.19) is a linear polymer containing some 1000–2000 glucopyranose units linked $\alpha 1 \rightarrow 4$. **Amylopectin** (Figure 8.19) is a much

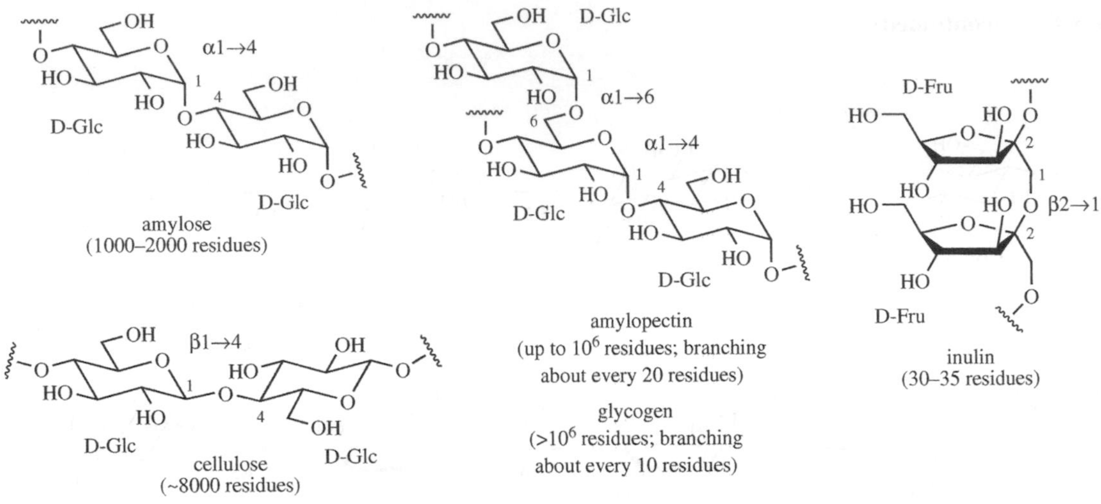

amylose
(1000–2000 residues)

cellulose
(~8000 residues)

amylopectin
(up to 10^6 residues; branching
about every 20 residues)

glycogen
($>10^6$ residues; branching
about every 10 residues)

inulin
(30–35 residues)

Figure 8.19

GalA = galacturonic acid

pectin
(400–1000 residues)

MannA = mannuronic acid

alginic acid
(200–900 residues)

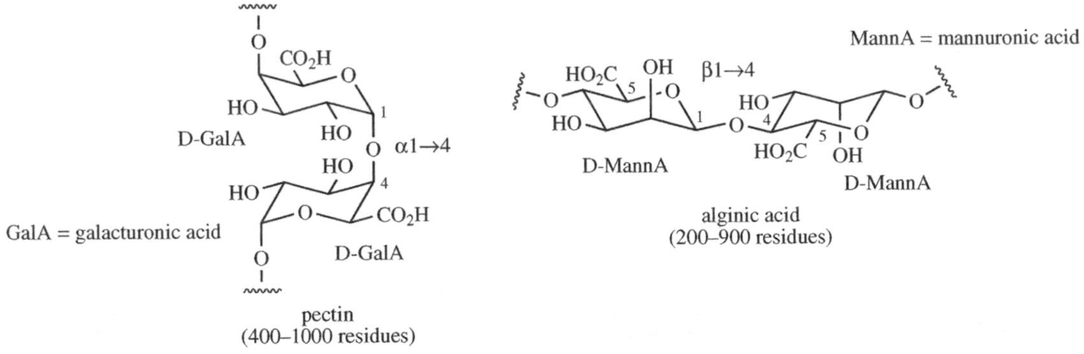

Figure 8.20

larger molecule than amylose (the number of glucose residues varies widely, but may be as high as 10^6) and is a branched-chain molecule. In addition to $\alpha1{\rightarrow}4$ linkages, amylopectin has branches at about every 20 units through $\alpha1{\rightarrow}6$ linkages. These branches continue with $\alpha1{\rightarrow}4$ linkages, but then may have subsidiary branching, giving a tree-like structure.

The mammalian carbohydrate storage molecule is **glycogen**, which is analogous to amylopectin in structure, but is larger and contains more frequent branching, about every 10 residues. The branching in amylopectin and glycogen is achieved by the enzymic removal of a portion of the $\alpha1{\rightarrow}4$ linked straight chain consisting of several glucose residues, then transferring this short chain to a suitable 6-hydroxyl group. A less common storage polysaccharide found in certain plants of the

Compositae/Asteraceae and Campanulaceae is **inulin** (Figure 8.19), which is a relatively small polymer of fructofuranose, linked through $\beta2{\rightarrow}1$ bonds. **Cellulose** is reputedly the most abundant organic material on earth, being the main constituent in plant cell walls. It is composed of glucopyranose units linked $\beta1{\rightarrow}4$ in a linear chain. Alternate residues are 'rotated' in the structure (Figure 8.19), allowing hydrogen bonding between adjacent molecules and construction of the strong fibres characteristic of cellulose, e.g. as in cotton.

Polymers of uronic acids are encountered in **pectins**, which are essentially chains of galacturonic acid residues linked $\alpha1{\rightarrow}4$ (Figure 8.20), though some of the carboxyl groups are present as methyl esters. These materials are present in the cell walls of fruit, and the property of aqueous solutions under acid conditions forming gels

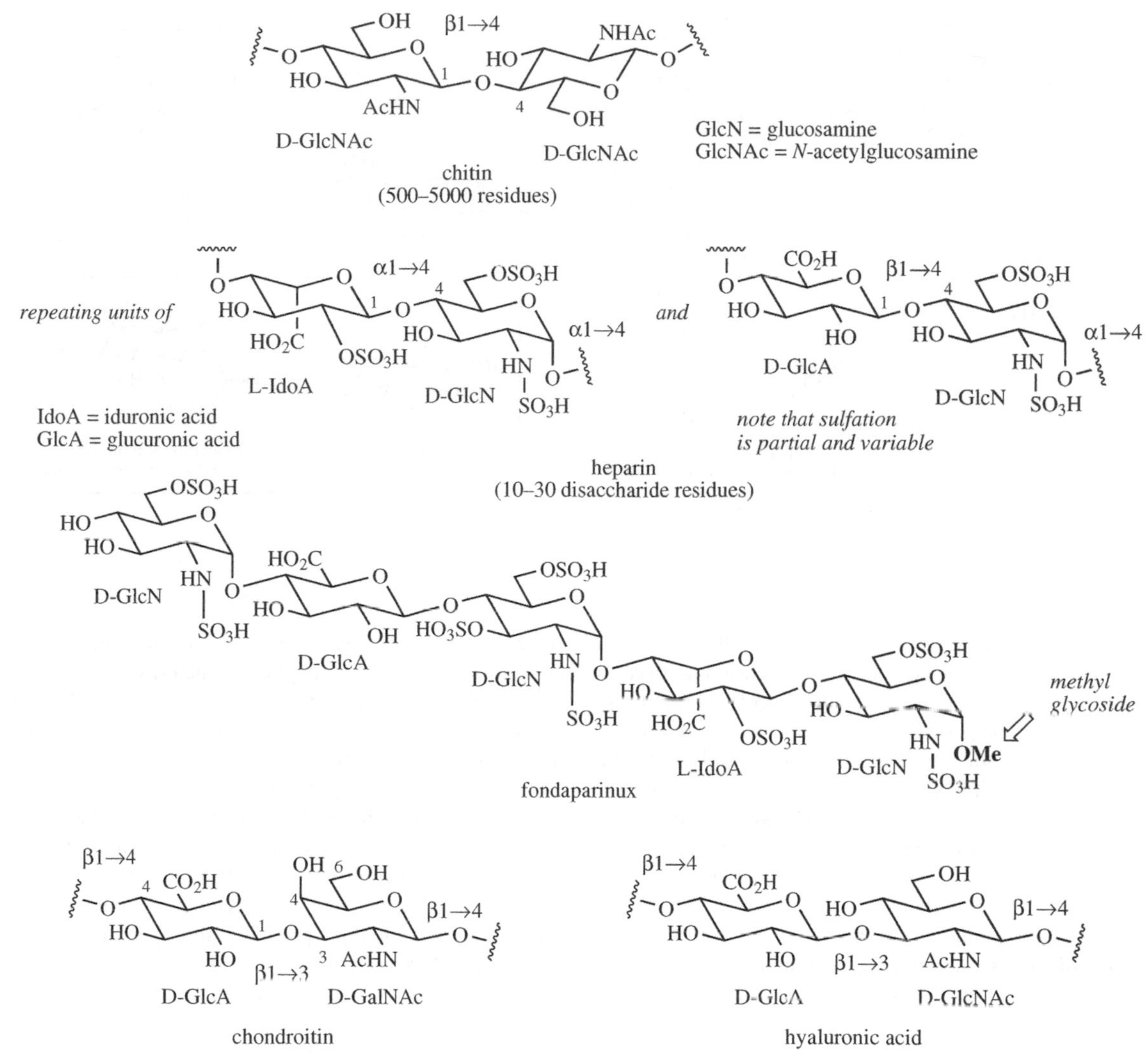

Figure 8.21

is the basis of jam making. **Alginic acid** (Figure 8.20) is a linear polysaccharide formed principally by $\beta1\rightarrow4$ linkage of D-mannuronic acid residues, though sometimes residues of the C-5 epimer L-guluronic acid are also part of the structure. Alginic acid is the main cell wall constituent of brown algae (seaweeds). Some bacteria also produce alginates; the human pathogen *Pseudomonas aeruginosa* secretes alginates as a viscous coating when it infects lung tissue in cystic fibrosis. Salts of algal alginic acid are valuable thickening agents in the food industry, and the insoluble calcium salt is the basis of absorbable alginate surgical dressings.

The structure of **chitin** (Figure 8.21) is rather similar to cellulose, though it is composed of amino sugar residues, *N*-acetylglucosamine linked $\beta1\rightarrow4$. Chitin is a major constituent in the shells of crustaceans, e.g. crabs and lobsters, and insect skeletons, and, as with cellulose, its strength again depends on hydrogen bonding between adjacent molecules, producing rigid sheets. Chemical deacetylation of chitin provides **chitosan**, a valuable industrial material used for water purification because of its chelating properties and in wound-healing preparations. The mammalian blood anticoagulant **heparin** (Figure 8.21) is also a carbohydrate polymer containing glucosamine derivatives, but these alternate

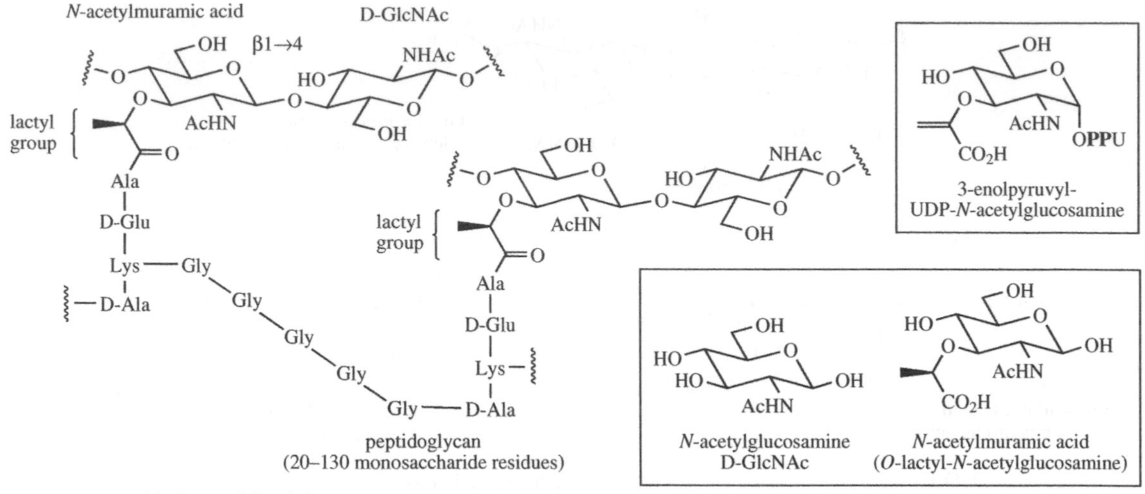

Figure 8.22

with uronic acid residues. Polymers of this kind are known as mucopolysaccharides or glycosaminoglycans. Heparin consists of two repeating disaccharide units, in which the amino functions and some of the hydroxyls are sulfated, producing a heterogeneous polymer. The carboxyls and sulfates together make heparin a strongly acidic water-soluble material. **Heparan sulfate** is structurally related to heparin, but the chains are longer and more heterogeneous. Heparin is secreted only by mast cells, whereas heparan sulfate is more widely distributed and found in different cell types and tissues. **Chondroitin** is another glycosaminoglycan widely distributed in animal tissues, particularly connective tissue, and is a major component of cartilage, lubricating and cushioning joints. It is based on a repeating disaccharide unit of glucuronic acid linked $\beta1{\rightarrow}3$ to N-acetylgalactosamine, with the disaccharide units then connected $\beta1{\rightarrow}4$. It is typically found as chondroitin sulfate, the polymer having varying sulfation patterns, though mainly monosulfation at positions 4 and 6 of the GalNAc residues (Figure 8.21). **Hyaluronic acid** (Figure 8.21) is a similar polymer based on glucuronic acid and N-acetylglucosamine, though not modified by sulfation; it is a major component of the extracellular matrix and plays an important role in tissue regeneration and wound healing.

Bacterial cell walls contain **peptidoglycan** structures in which carbohydrate chains are composed of alternating $\beta1{\rightarrow}4$ linked N-acetylglucosamine and O-lactyl-N-acetylglucosamine (also called

N-acetylmuramic acid) residues. These chains are cross-linked via short peptide structures of variable composition and complexity according to species. Part of the peptidoglycan of *Staphylococcus aureus* is illustrated in Figure 8.22, showing the involvement of the lactyl group of the N-acetylmuramic acid in linking the peptide with the carbohydrate via an amide/peptide bond. The lactyl function derives from phosphoenolpyruvate; reduction of 3-enolpyruvyl-UDP-N-acetylglucosamine yields 3-O-lactyl-UDP-N-acetylglucosamine. During the cross-linking process, the peptide chains from the N-acetylmuramic acid residues have a terminal –Lys–D-Ala–D-Ala sequence, and the lysine from one chain is bonded to the penultimate D-alanine of another chain through five glycine residues, at the same time displacing the terminal D-alanine (see Figure 7.42, page 464). The biological activities of the β-lactam antibiotics, e.g. penicillins and cephalosporins (see page 464) and of the last-resort antibiotic vancomycin (see page 455), stem from an inhibition of the cross-linking mechanism during the biosynthesis of the bacterial cell wall, and relate to this terminal –D-Ala–D-Ala sequence during biosynthesis. The subdivision of bacteria into Gram-positive or Gram-negative reflects the ability of the peptidoglycan cell wall to take up Gram's dye stain. In Gram-negative organisms, an additional lipopolysaccharide cell membrane surrounding the peptidoglycan prevents attack of the dye.

Polysaccharides of medicinal importance are discussed in detail in Box 8.2.

Box 8.2

Polysaccharides

Starch is widely used in the food industry and finds considerable applications in medicine. For medicinal and pharmaceutical use, starch may be obtained from a variety of plant sources, including maize (*Zea mays*; Gramineae), wheat (*Triticum aestivum*; Gramineae), potato (*Solanum tuberosum*; Solanaceae), rice (*Oryza sativa*; Gramineae/Poaceae), and arrowroot (*Maranta arudinacea*; Marantaceae). Most contain about 25% amylose and 75% amylopectin (Figure 8.19), but these proportions can vary according to the plant tissue. Its absorbent properties make it ideal for dusting powders, and its ability to swell in water makes it a valuable formulation aid, being the basis for tablet disintegrants. **Soluble starch** is obtained by partial acid hydrolysis and is completely soluble in hot water.

Cellulose (Figure 8.19) may be extracted from wood pulp and is usually partially hydrolysed with acid to give micro-crystalline cellulose. These materials are used as tablet diluents. Semi-synthetic derivatives of cellulose, e.g. **methylcellulose, hydroxymethylcellulose**, and **carboxymethylcellulose**, are used as emulsifying and suspending agents. **Cellulose acetate phthalate** is cellulose with about half the hydroxyl groups acetylated and the remainder esterified with phthalic acid. It is used as an acid-resistant enteric coating for tablets and capsules.

Alginic acid (Figure 8.20) is obtained by alkaline (Na_2CO_3) extraction of a range of brown seaweeds, chiefly species of *Laminaria* (Laminariaceae) and *Ascophyllum* (Phaeophyceae) in Europe, and species of *Macrocystis* (Lessoniaceae) on the Pacific coast of the USA. The carbohydrate material constitutes 20–40% of the dry weight of the algae. The acid is usually converted into its soluble sodium salt or insoluble calcium salt. Sodium alginate finds many applications as a stabilizing and thickening agent in a variety of industries, particularly food manufacture, and also the pharmaceutical industry, where it is of value in the formulation of creams, ointments, and tablets. Calcium alginate is the basis of many absorbable haemostatic surgical dressings. Alginic acid or alginates are incorporated into many aluminium- and magnesium-containing antacid preparations to protect against gastro-oesophageal reflux. Alginic acid released by the action of gastric acid helps to form a barrier over the gastric contents.

Agar is a carbohydrate extracted using hot dilute acid from various species of red algae (seaweeds), including *Gelidium* (Gelidiaceae) and *Gracilaria* (Gracilariaceae) from Japan, Spain, Australasia, and the USA. Agar is a heterogeneous polymer which may be fractionated into two main components, agarose and agaropectin. Agarose yields D- and L-galactose on hydrolysis, and contains alternating $\beta1\rightarrow3$ linked D-galactose and $\alpha1\rightarrow4$ linked L-galactose, with the L-sugar in a 3,6-anhydro form. Agaropectin has a similar structure but some of the residues are methylated, sulfated, or in the form of a cyclic ketal with pyruvic acid. Agar's main application is in bacterial culture media, where its gelling properties are exploited. It is also used to some extent as a suspending agent and a bulk laxative. Agarose is important as a support in affinity chromatography.

Carrageenan is a carbohydrate polymer extracted from the red alga *Chondrus crispus* (Gigartinaceae) (chondrus, or Irish moss) collected from Irish and other Atlantic coasts in Europe, also North America. Related species of algae, e.g. *Gigartina*, may also be used. Several types of linear polymer based on D-galactose, distinguished by their different repeating units, are present. Three important ones are κ- (kappa), ι- (iota), and λ- (lambda) carrageenans, based primarily on D-galactose and 3,6-anhydro-D-galactose as mono- and di-sulfates. κ-Carrageenans form strong rigid gels, ι-carrageenans soft gels, and λ-carrageenans form gels when mixed with proteins, e.g. dairy products. They are widely used in the food and pharmaceutical industries as thickening agents. Laboratory studies suggest that carrageenans can function as topical microbicides, blocking sexually transmitted viruses such as human papillomavirus and herpes, though not HIV.

Tragacanth is a dried gummy exudate obtained from *Astragalus gummifer* (Leguminosae/Fabaceae) and other *Astragalus* species, small shrubs found in Iran, Syria, Greece, and Turkey. It is usually obtained by deliberate incision of the stems. This material swells in water to give a stiff mucilage with an extremely high viscosity, and provides a useful suspending and binding agent. It is chemically a complex material, and yields D-galacturonic acid, D-galactose, L-fucose, L-arabinose, and D-xylose on hydrolysis. Some of the uronic acid carboxyls are methylated.

Acacia (gum arabic) is a dried gum from the stems and branches of the tree *Acacia senegal* (Leguminosae/Fabaceae), abundant in the Sudan and Central and West Africa. Trees are tapped by removing a portion of the bark. The gum is used as a suspending agent and as an adhesive and binder for tablets. The carbohydrate is a complex branched-chain material which yields L-arabinose, D-galactose, D-glucuronic acid, and L-rhamnose on hydrolysis. Occluded enzymes (oxidases, peroxidases, and pectinases) can cause problems in some formulations, unless inactivated by heat.

Karaya or **sterculia gum** is a dried gum obtained from the trunks of the tree *Sterculia urens* (Sterculiaceae) or other *Sterculia* species found in India. It exudes naturally or may be obtained by incising through the bark. It contains a branched polysaccharide comprising L-rhamnose, D-galactose, and with a high proportion of D-galacturonic acid and D-glucuronic acid residues. The molecule is partially acetylated, and the gum typically has an odour of acetic acid. It is used as a bulk laxative and as a suspending agent. It has proved particularly effective as an adhesive for stomal appliances, rings of the purified gum being used to provide a non-irritant seal between the stomal bag and the patient's skin.

Box 8.2   (continued)

Heparin is usually extracted from the intestinal mucosa of pigs or cattle, where it is present in the mast cells. It is a blood anticoagulant and is used clinically to prevent or treat deep-vein thrombosis. It is administered by injection or intravenous infusion and provides rapid action. It is also active *in vitro*, and is used to prevent the clotting of blood in research preparations. Heparin acts by complexing with enzymes in the blood which are involved in the clotting process. Although not strictly an enzyme inhibitor, its presence enhances the natural inhibition process between thrombin and antithrombin III by forming a ternary complex. A specific pentasaccharide sequence containing a 3-*O*-sulfated D-glucosamine residue is essential for functional binding of antithrombin. Natural heparin is a mixture of glycosaminoglycans (Figure 8.21), with only a fraction of the molecules having the required binding sequence, and it has a relatively short duration of action (half-life about 1 h). Partial hydrolysis of natural heparin by chemical or enzymic means has resulted in a range of low molecular weight heparins having similar activity but with a longer duration of action. **Bemiparin**, **dalteparin**, **enoxaparin**, and **tinzaparin** are examples of these currently being used clinically. More significantly, the natural pentasaccharide sequence has been produced synthetically, stabilized as its methyl glycoside. This compound, **fondaparinux sodium** (Figure 8.21), is now used as an antithrombotic drug for treating venous thromboembolisms following surgery, deep-vein thrombosis, and pulmonary embolisms. Its half-life is about 17 h, allowing once-daily administration. **Protamine**, a basic protein from the testes of fish of the salmon family, e.g. *Salmo* and *Onchorhynchus* species (see insulin, page 432), is a heparin antagonist, which may be used to counteract haemorrhage caused by overdosage of heparin.

Chondroitin sulfate (Figure 8.21) is usually derived from pig or cow cartilage, though shark and fish cartilage is also used. It may possibly play a role as a protective agent in joints, and is thus used as a dietary supplement to minimize arthritis and cartilage problems. It is often combined with **glucosamine**, considered a potential precursor of beneficial cartilage glycosaminoglycans. Glucosamine is usually obtained by hydrolysis of the shells from various shellfish, being liberated from the chitin component. **Hyaluronic acid** (hyaluronan; Figure 8.21) is extracted from rooster comb or *Streptococcus* bacterial cultures. It is injected to reduce pain in osteoarthritis, and is also used as a component of some specialized wound dressings.

AMINOSUGARS AND AMINOGLYCOSIDES

Amino sugars are readily produced from keto sugars by transamination processes (see page 488); in many cases UDPsugars or TDPsugars are involved in the conversions. Glucosamine or *N*-acetylglucosamine has featured in several of the polysaccharide structures described (see page 495), though the major route to glucosamine for these compounds tends to be via a more complicated process involving conversion of fructose 6-phosphate into glucosamine 6-phosphate by a transaminase–isomerase enzyme (see page 489). Other less common aminosugars will be encountered in the aminoglycoside antibiotics. In addition, there are some further structures where the newly introduced amino group subsequently becomes part of a heterocyclic ring system, generating compounds termed variously iminocyclitols, azasugars, or iminosugars. These arise by using the amino group as a nucleophile to generate an aminohemiacetal linkage. This, of course, is simply the initial addition step in the formation of an imine (see page 19). Should the anomeric hydroxyl then be removed in subsequent modifications, such as imine formation, the product will be a polyhydroxy-piperidine or -pyrrolidine. Any confusion with ornithine/lysine-derived alkaloids (see pages 311, 326) should be dispelled by the characteristic sugar-like polyhydroxy substitution. The piperidine structures **deoxynojirimycin** and **deoxymannojirimycin** (Figure 8.23) from *Streptomyces subrutilis* are good examples.

The pathways to deoxynojirimycin and deoxymannojirimycin (Figure 8.23) in microorganisms start from the keto sugar fructose, which is aminated and then oxidized to **mannojirimycin**. This can then form a cyclic aminohemiacetal. Dehydration to the imine can follow, and reduction yields deoxymannojirimycin. **Nojirimycin** is an epimer of mannojirimycin, and analogous modifications then give deoxynojirimycin. Deoxynojirimycin is found in various strains of *Streptomyces* and *Bacillus*, as well as some plants, e.g. *Morus* spp. (Moraceae) and *Commelina communis* (Commelinaceae). However, the pathway to deoxynojirimycin is found to be different in *Commelina*, in that a shorter route that does not require an epimerization step is functioning (Figure 8.23). It does mean, however, that different atoms are cyclized in microorganisms or plants. Compounds in this group are inhibitors of glycosidase enzymes (compare indolizidine alkaloids such as castanospermine, page 330) and have considerable potential as agents against several diseases, including diabetes, cancer, and bacterial and viral infections, especially against HIV. They are suggested to bind to glycosidases by mimicking the shape and charge of the postulated oxonium intermediate for the glycosidic bond cleavage reaction (see page 331). By altering the constitution of glycoproteins on the surface of the virus, such compounds interfere with the binding of the HIV particle to components of the immune system. Two deoxynojirimycin derivatives are currently being

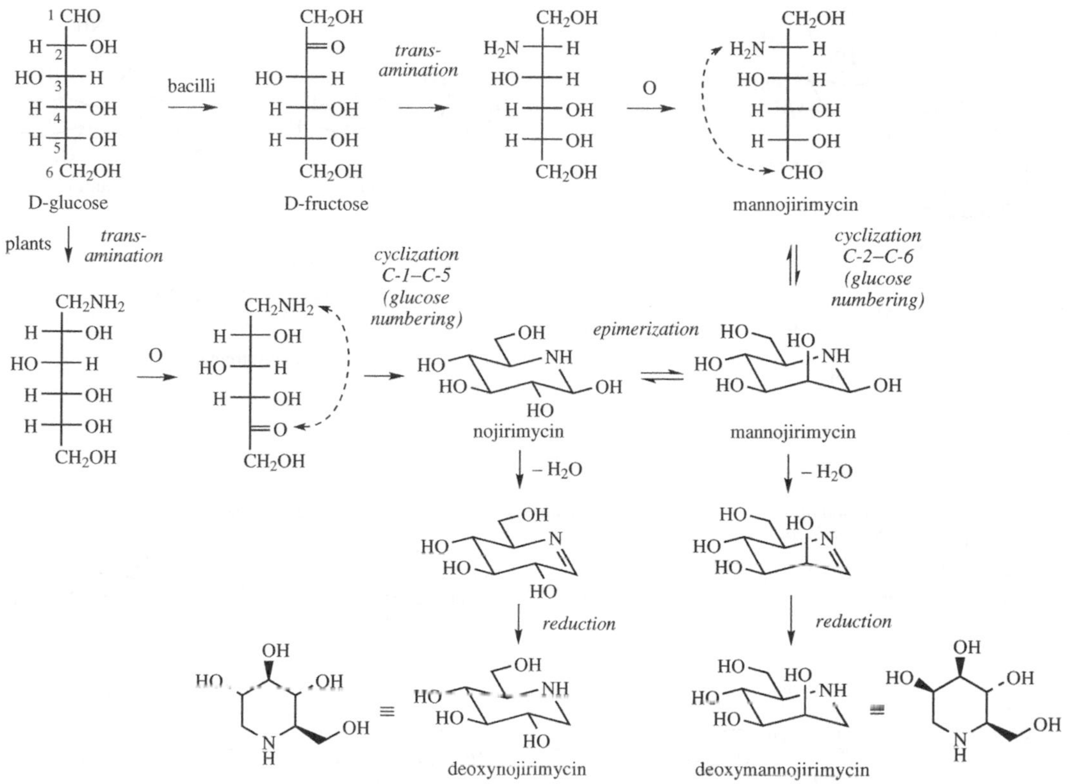

Figure 8.23

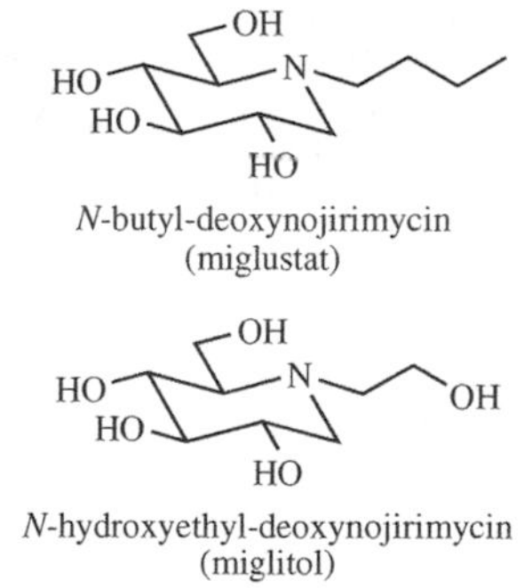

N-butyl-deoxynojirimycin
(miglustat)

N-hydroxyethyl-deoxynojirimycin
(miglitol)

Figure 8.24

used as drugs. *N*-Butyl-deoxynojirimycin (**miglustat**; Figure 8.24) is an inhibitor of glucosylceramide synthase, an enzyme essential for glycosphingolipid biosynthesis, and is used to treat Gaucher's disease, a condition characterized by a build-up of fatty deposits in various organs. **Miglitol** (*N*-hydroxyethyl-deoxynojirimycin; Figure 8.24) is an oral antidiabetic drug; it is an inhibitor of α-glucosidase.

The **aminoglycosides** form an important group of antibiotic agents [Box 8.3] and are immediately recognizable as modified carbohydrate molecules. Typically, they have two or three uncommon sugars, mainly aminosugars, attached through glycoside linkages to an aminocyclitol, i.e. an amino-substituted cyclohexane system, which also has carbohydrate origins. The first of these agents to be discovered was **streptomycin** (see Figure 8.25) from *Streptomyces griseus*, whose structure contains the aminocyclitol **streptamine**, though both amino groups in streptamine are bound as guanidino substituents, making **streptidine**. Other medicinally useful aminoglycoside antibiotics are based on the aminocyclitol **2-deoxystreptamine** (Figure 8.25), e.g. gentamicin C_1 (see Figure 8.29).

Streptamine and 2-deoxystreptamine are both derived from glucose 6-phosphate. The route to the **streptamine** system can be formulated to involve oxidation (in the acyclic form) of the 5-hydroxyl, allowing removal of a proton from C-6 and generation of an enolate anion (Figure 8.26). The cyclohexane ring is then formed by attack of this enolate anion onto the C-1 carbonyl. Reduction and hydrolysis of the phosphate produces

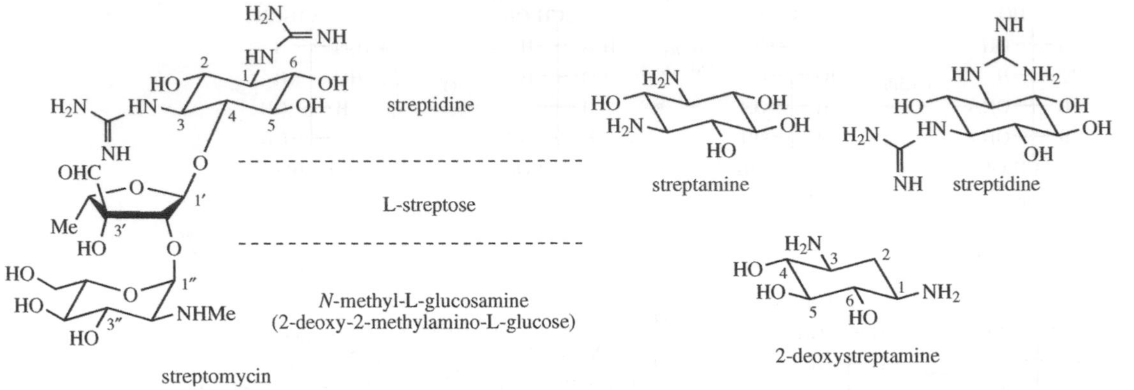

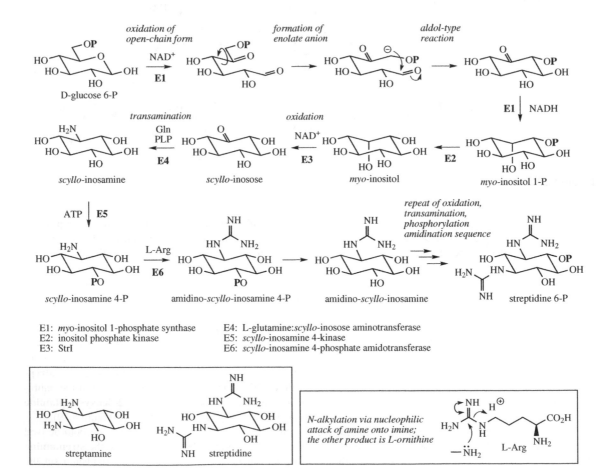

streptidine
L-streptose
N-methyl-L-glucosamine
(2-deoxy-2-methylamino-L-glucose)
streptomycin
streptamine
streptidine
2-deoxystreptamine

Figure 8.25

oxidation of open-chain form
formation of enolate anion
aldol-type reaction
D-glucose 6-P
transamination
oxidation
scyllo-inosamine
scyllo-inosose
myo-inositol
myo-inositol 1-P
repeat of oxidation, transamination, phosphorylation amidination sequence
scyllo-inosamine 4-P
amidino-scyllo-inosamine 4-P
amidino-scyllo-inosamine
streptidine 6-P
E1: myo-inositol 1-phosphate synthase
E2: inositol phosphate kinase
E3: StrI
E4: L-glutamine:scyllo-inosose aminotransferase
E5: scyllo-inosamine 4-kinase
E6: scyllo-inosamine 4-phosphate amidotransferase
streptamine
streptidine
N-alkylation via nucleophilic attack of amine onto imine; the other product is L-ornithine
L-Arg

Figure 8.26

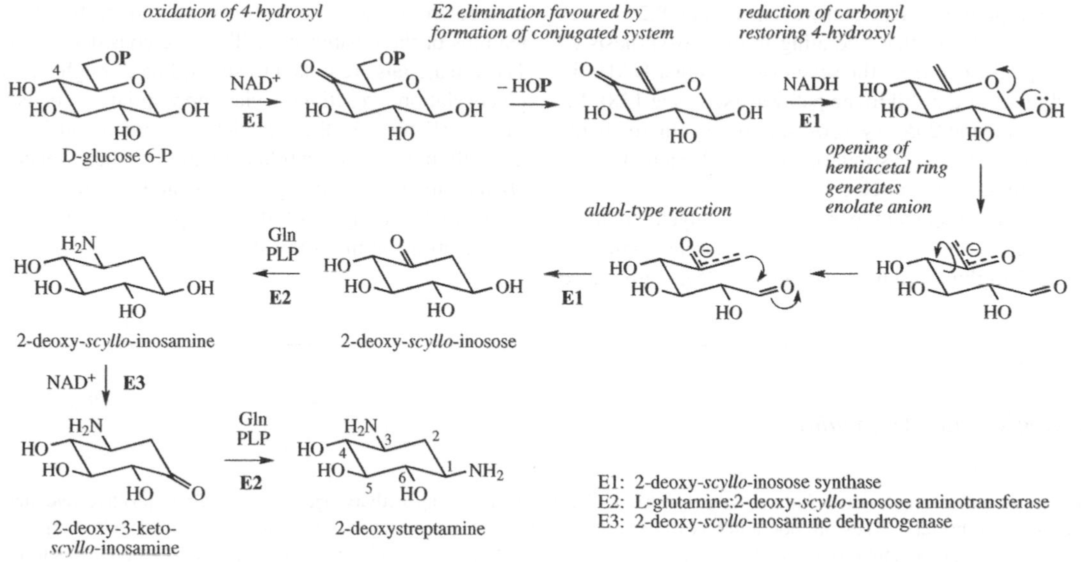

Figure 8.27

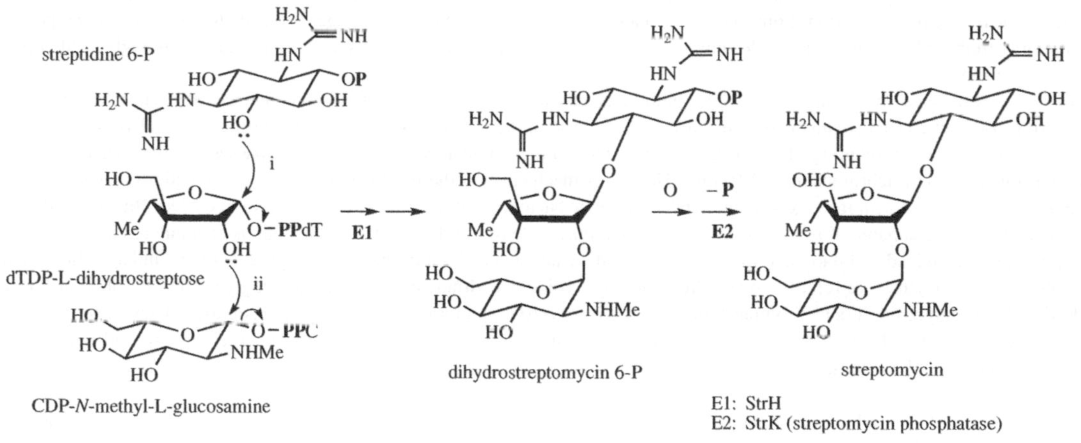

Figure 8.28

myo-**inositol**. The amino groups, as in streptamine, are then introduced by oxidation/transamination reactions. **Streptidine** incorporates amidino groups from arginine, by nucleophilic attack of the aminocyclitol amino group onto the imino function of arginine (Figure 8.26). However, streptamine itself is not a precursor, and the first guanidino side-chain is built up before the second amino group is introduced. The *N*-alkylation steps also involve aminocyclitol phosphate substrates, and streptidine 6-phosphate is subsequently the precursor of streptomycin.

The biosynthesis of **2-deoxystreptamine** shares similar features, but the sequence involves loss of the oxygen function from C-6 of glucose 6-phosphate in an elimination reaction (Figure 8.27). The elimination is facilitated by oxidation of the 4-hydroxyl, which thus allows a conjugated enone to develop in the elimination step, and the original hydroxyl is reformed by reduction after the elimination. The cyclohexane ring is then formed by attack of an enolate anion onto the C-1 carbonyl giving a tetrahydroxycyclohexanone 2-deoxy-*scyllo*-inosose; oxidation and transamination reactions allow formation of

2-deoxystreptamine. The pathway in Figure 8.27 is remarkably similar to that operating in the biosynthesis of dehydroquinic acid from the seven-carbon sugar DAHP in the early part of the shikimate pathway (see page 138). Indeed, the enzyme 2-deoxy-*scyllo*-inosose synthase shares considerable sequence similarity to dehydroquinate synthases from several sources.

The other component parts of streptomycin, namely L-streptose and 2-deoxy-2-methylamino-L-glucose (*N*-methyl-L-glucosamine) (Figure 8.25), are also derived from D-glucose 6-phosphate, though the detailed features of these pathways will not be considered further. These materials are linked to streptidine through stepwise glycosylation reactions via appropriate nucleoside sugars (Figure 8.28); the less common nucleosides deoxythymidine and cytidine feature in these reactions. Oxidation of the primary alcohol at C-3″ to a formyl group and hydrolysis of the phosphate at C-6 are late modifications in the pathway.

Box 8.3

Aminoglycoside Antibiotics

The aminoglycoside antibiotics have a wide spectrum of activity, including activity against some Gram-positive and many Gram-negative bacteria. They are not absorbed from the gut, so for systemic infections they must be administered by injection. However, they can be administered orally to control intestinal flora. The widespread use of aminoglycoside antibiotics is limited by their nephrotoxicity, which results in impaired kidney function, and by their ototoxicity, which is a serious side-effect and can lead to irreversible loss of hearing. They are thus reserved for treatment of serious infections where less toxic antibiotics have proved ineffective. The aminoglycoside antibiotics interfere with protein biosynthesis by acting on the smaller 30S subunit of the bacterial ribosome. Streptomycin is known to interfere with the initiation complex, but most agents block the translocation step as the major mechanism of action. Some antibiotics can also induce a misreading of the genetic code to yield membrane proteins with an incorrect amino acid sequence, leading to altered membrane permeability. This actually increases aminoglycoside uptake and leads to rapid cell death.

Bacterial resistance to the aminoglycoside antibiotics has proved to be a problem, and this has also contributed to their decreasing use. Several mechanisms of resistance have been identified. These include changes in the bacterial ribosome so that the affinity for the antibiotic is significantly decreased, reduction in the rate at which the antibiotic passes into the bacterial cell, and plasmid transfer of extrachromosomal R-factors. This latter mechanism is the most common and causes major clinical problems. Bacteria are capable of acquiring genetic material from other bacteria, and in the case of the aminoglycosides, this has led to the organisms becoming capable of producing enzymes which inactivate the antibiotic. The modifications encountered are acetylation, phosphorylation, and adenylylation. (Note: adenylic acid = adenosine 5′-phosphate; adenylylation infers linkage through the phosphate.) The enzymes are referred to as AAC (aminoglycoside acetyltransferase), APH (aminoglycoside phosphotransferase), and ANT (aminoglycoside nucleotidyltransferase) (sometimes AAD, aminoglycoside adenylyltransferase). They differ in respect of the reaction catalysed, the position of derivatization (see numbering scheme in gentamicin, Figure 8.29), and the range of substrates attacked. Thus, some clinically significant inactivating enzymes are:

- AAC(3) and AAC(6′), which acetylate the 3- and 6′-amino functions respectively in gentamicin, tobramycin, kanamycin, neomycin, amikacin, and netilmicin.
- ANT(2″), which adenylylates the 2″-hydroxy group in gentamicin, tobramycin, and kanamycin.
- APH(3′), which phosphorylates the 3′-hydroxyl in neomycin and kanamycin.
- APH(3″), which phosphorylates the 3″-hydroxyl of streptomycin.

Other changes which may be imparted include acetylation of groups at position 2′, adenylylation of position 4′ substituents, and phosphorylation of the position 2″ substituent. Position 6 in the streptamine portion of streptomycin is also susceptible to adenylylation and phosphorylation.

Aminoglycoside antibiotics are produced in culture by strains of *Streptomyces* and *Micromonospora*. Compounds obtained from *Streptomyces* have been given names ending in -*mycin*, whilst those from *Micromonospora* have names ending in -*micin*.

Streptamine-containing Antibiotics

Streptomycin (Figure 8.25) is produced by cultures of a strain of *Streptomyces griseus*, and is mainly active against Gram-negative organisms. Because of its toxic properties it is rarely used in modern medicine except against resistant strains of *Mycobacterium tuberculosis* in the treatment of tuberculosis.

Box 8.3 (continued)

2-Deoxystreptamine-containing Antibiotics

Gentamicin is a mixture of antibiotics obtained from *Micromonospora purpurea*. Fermentation yields a mixture of gentamicins A, B, and C, from which gentamicin C is separated for medicinal use. This is also a mixture, the main component being gentamicin C_1 (50–60%; Figure 8.29), with smaller amounts of gentamicin C_{1a} and gentamicin C_2. These three components differ in respect to the side-chain in the purpurosamine sugar. Gentamicin is clinically the most important of the aminoglycoside antibiotics, and is widely used for the treatment of serious infections, often in combination with a penicillin when the infectious organism is unknown. It has a broad spectrum of activity, but is inactive against anaerobes. It is active against pathogenic enterobacteria such as *Enterobacter, Escherichia* and *Klebsiella*, and also against *Pseudomonas aeruginosa*. Compared with other compounds in this group, its component structures contain fewer functionalities that may be attacked by inactivating enzymes, which means that gentamicin may be more effective than some other agents.

Sisomicin (Figure 8.29) is a dehydro analogue of gentamicin C_{1a} and is produced by cultures of *Micromonospora inyoensis*. It is used medicinally in the form of the semi-synthetic *N*-ethyl derivative **netilmicin** (Figure 8.29), which has a similar activity to gentamicin, but causes less ototoxicity.

The **kanamycins** (Figure 8.30) are a mixture of aminoglycosides produced by *Streptomyces kanamyceticus*, but have been superseded by other drugs. **Amikacin** (Figure 8.29) is a semi-synthetic acyl derivative of kanamycin A, the introduction of the 4-amino-2-hydroxybutyryl group helping to protect the antibiotic against enzymic deactivation at several positions, whilst still maintaining the activity of the parent molecule. It is stable to many of the aminoglycoside inactivating enzymes and is valuable for the treatment of serious infections caused by Gram-negative bacteria that are resistant to gentamicin. **Tobramycin** (Figure 8.30; also called nebramycin factor 6) is an analogue of kanamycin B isolated from *Streptomyces tenebrarius* and is also less prone to deactivation, in that it lacks the susceptible 3′-hydroxyl group. It is slightly more active towards *Pseudomonas aeruginosa* than gentamicin, but shows less activity against other Gram-negative bacteria.

Neomycin is a mixture of neomycin B (**framycetin**; Figure 8.31) and its epimer neomycin C, the latter component accounting for some 5–15% of the mixture. It is produced by cultures of *Streptomyces fradiae* and, in contrast to the other clinically useful aminoglycosides described, contains three sugar residues linked to 2-deoxystreptamine. One of these is the common sugar D-ribose. Neomycin has good activity against Gram-positive and Gram-negative bacteria, but is very ototoxic. Its use is thus restricted to oral treatment of intestinal infections (it is poorly absorbed from the digestive tract) and topical applications in eyedrops, eardrops, and ointments. **Paromomycin** (Figure 8.31) is a new broad-spectrum aminoglycoside antibiotic currently being used as an oral formulation to treat intestinal parasites, including visceral leishmaniasis. It is produced by *Streptomyces rimosus* var. *paromomycinus* and is structurally related to neomycin B, with glucosamine replacing the neosamine fragment.

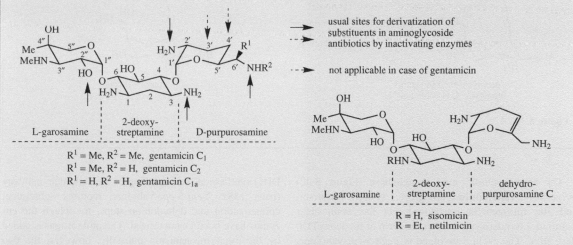

Figure 8.29

Box 8.3   (continued)

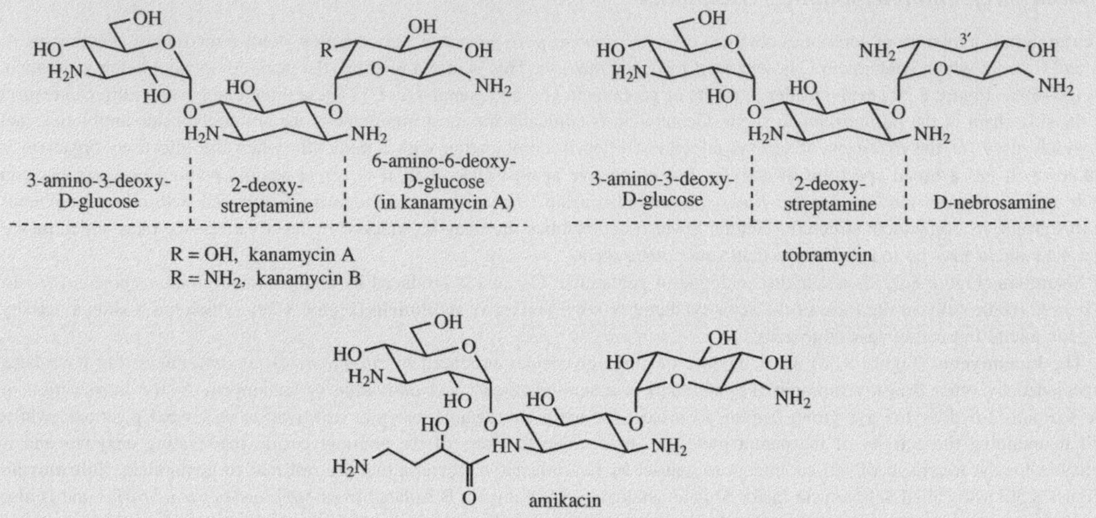

R = OH, kanamycin A
R = NH₂, kanamycin B

tobramycin

Figure 8.30

amikacin

neomycin B (framycetin)
[neomycin C is epimer at *]

paromomycin

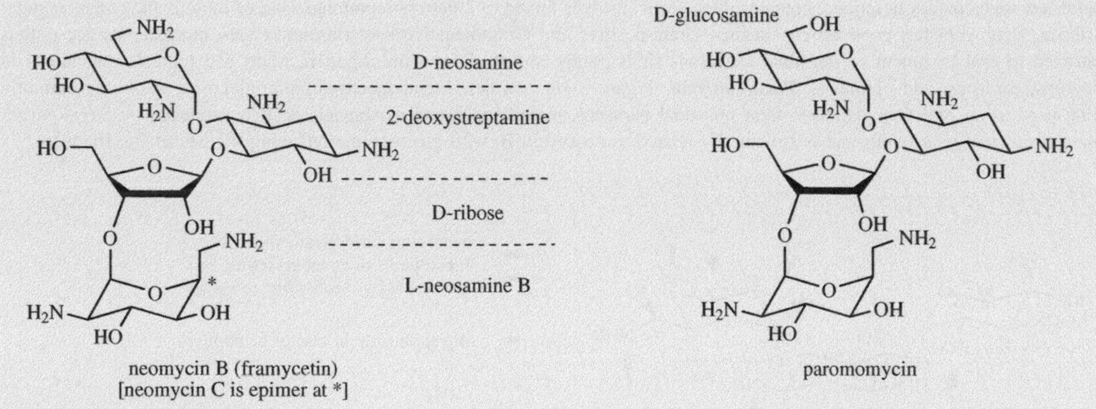

Figure 8.31

The aminocyclitol found in **acarbose** (Figure 8.32) is based on **valienamine**, though this is not a precursor, and the nitrogen is introduced via the aminosugar 4-amino-4,6-dideoxyglucose in the form of its deoxyTDP derivative. The cyclitol involved is 2-*epi*-5-*epi*-valiolone, and this appears to be produced from the seven-carbon sugar derivative sedoheptulose 7-phosphate. The reaction sequence is exactly analogous to that seen in the transformation of DAHP into 3-dehydroquinic acid by DHQ synthase at the beginning of the shikimate pathway (page 138). 2-*epi*-5-*epi*-Valiolone requires subsequent epimerization and dehydration steps, for which the enzymes have been characterized. The aminosugar is added by the usual mechanism; unusually, however, the two further glucose units are not added sequentially, but via the preformed dimer maltose (see page 491). Acarbose is produced by strains of an *Actinoplanes* sp. and is of clinical importance in the treatment of diabetes [Box 8.4].

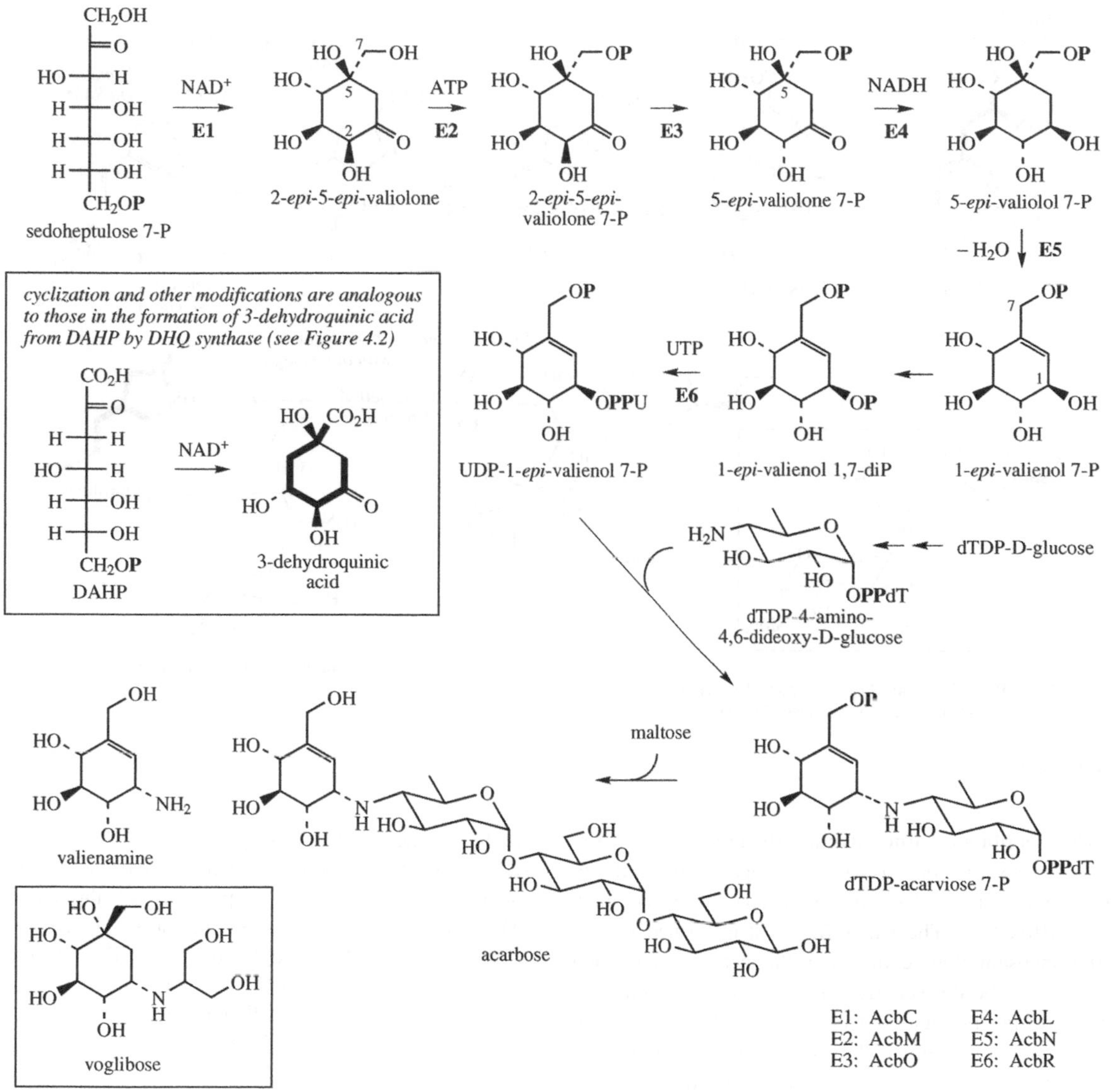

Figure 8.32

Box 8.4

Acarbose

Acarbose is obtained commercially from fermentation cultures of selected strains of an undefined species of *Actinoplanes*. It is an inhibitor of α-glucosidase, the enzyme that hydrolyses starch and sucrose. It is employed in the treatment of diabetic patients, allowing better utilization of starch- or sucrose-containing diets, by delaying the digestion of such foods and thus slowing down the intestinal release of α-D-glucose. It has a small but significant effect in lowering blood glucose, and is used either on its own or alongside oral hypoglycaemic agents in cases where dietary control with or without drugs has proved inadequate. Flatulence is a common side-effect.

A more recent agent **voglibose** (Figure 8.32) is structurally related to the valienamine fragment of acarbose; it is said to be considerably more effective than acarbose as an enzyme inhibitor.

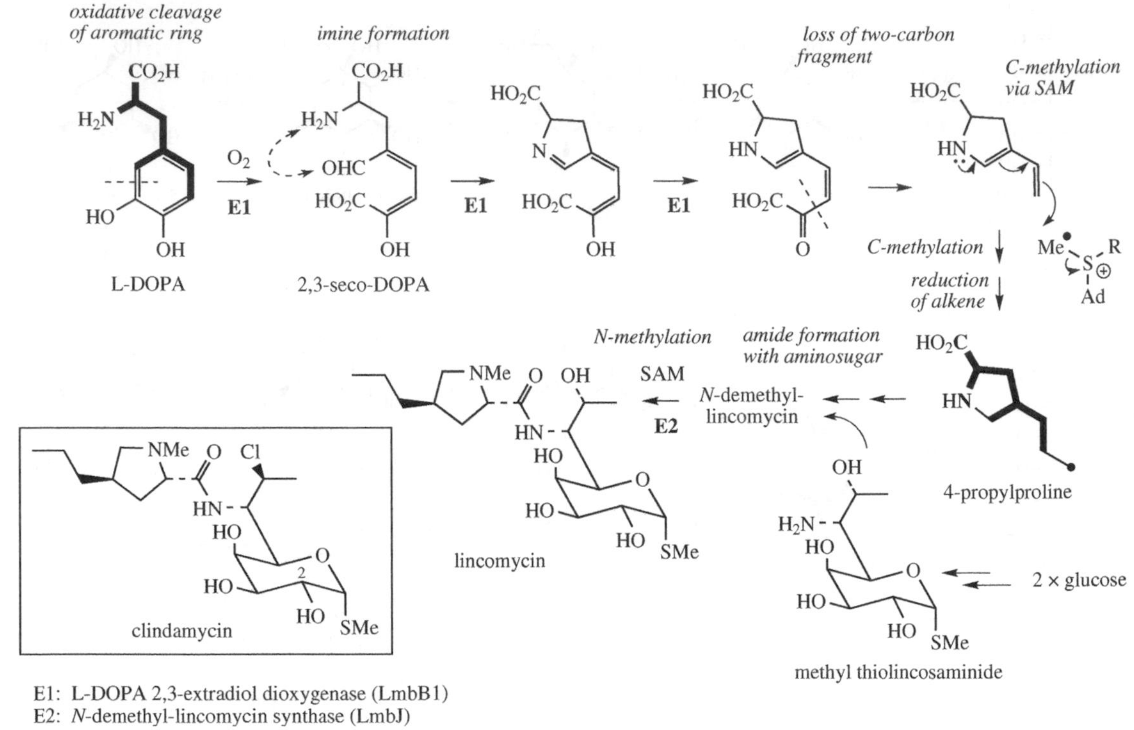

Figure 8.33

E1: L-DOPA 2,3-extradiol dioxygenase (LmbB1)
E2: *N*-demethyl-lincomycin synthase (LmbJ)

The antibiotic **lincomycin** (Figure 8.33) from *Streptomyces lincolnensis* bears a superficial similarity to the aminoglycosides, but has a rather more complex origin [Box 8.5]. The sugar fragment is termed methyl α-thiolincosaminide, contains a thiomethyl group, and is known to be derived from two molecules of glucose, one of which provides a five-carbon unit and the other a three-carbon unit. The 4-propyl-*N*-methylproline fragment does not originate from proline, but is actually a metabolite from the aromatic amino acid L-DOPA (Figure 8.33). Oxidative cleavage of the aromatic ring (see page 28) provides all the carbon atoms for the pyrrolidine ring, the carboxyl, and two carbon atoms of the propyl side-chain. The terminal carbon of the propyl is supplied by L-methionine, as are the *N*-methyl, and the *S*-methyl in the sugar fragment. Two carbon atoms from DOPA are lost during the biosynthesis.

Box 8.5

Lincomycin and Clindamycin

Lincomycin (Figure 8.33) is obtained from cultures of *Streptomyces lincolnensis* var. *lincolnensis*. The semi-synthetic derivative **clindamycin** (Figure 8.33), obtained by chlorination of the lincomycin with resultant inversion of stereochemistry, is more active and better absorbed from the gut, and has effectively replaced lincomycin. Both antibiotics are active against most Gram-positive bacteria, including penicillin-resistant staphylococci. Their use is restricted by side-effects. These include diarrhoea and occasionally serious pseudomembraneous colitis, caused by overgrowth of resistant strains of *Clostridium difficile*, which can cause fatalities in elderly patients. However, this may be controlled by the additional administration of vancomycin (see page 455). Clindamycin finds particular application in the treatment of staphylococcal joint and bone infections, such as osteomyelitis,

Box 8.5 (continued)

since it readily penetrates into bone. It is sometimes used to treat malarial infections in combination with an antimalarial drug, e.g. quinine or chloroquine. **Clindamycin 2-phosphate** is also of value, especially in the topical treatment of acne vulgaris and vaginal infections. Lincomycin and clindamycin inhibit protein biosynthesis by blocking the peptidyltransferase site on the 50S subunit of the bacterial ribosome. Microbial resistance may develop slowly, and in some cases has been traced to adenylylation of the antibiotic (see page 502).

FURTHER READING

Vitamin C

Smirnoff N, Conklin PL and Loewus FA (2001) Biosynthesis of ascorbic acid in plants: a renaissance. *Annu Rev Plant Physiol Plant Mol Biol* **52**, 437–467.

Wolucka BA and Van Montagu M (2007) The VTC2 cycle and the *de novo* biosynthesis pathways for vitamin C in plants: an opinion. *Phytochemistry* **68**, 2602–2613.

Polysaccharides

Damonte EB, Matulewicz MC and Cerezo AS (2004) Sulfated seaweed polysaccharides as antiviral agents. *Curr Med Chem* **11**, 2399–2419.

Joshi CP and Mansfield SD (2007) The cellulose paradox – simple molecule, complex biosynthesis. *Curr Opin Plant Biol* **10**, 220–226.

Kumar RMNV, Muzzarelli RAA, Sashiwa H and Domb AJ (2004) Chitosan chemistry and pharmaceutical perspectives. *Chem Rev* **104**, 6017–6084.

Lerouxe O, Cavalier DM and Liepman AH (2006) Biosynthesis of plant cell wall polysaccharides – a complex process. *Curr Opin Plant Biol* **9**, 621–630.

Paulsen BS (2002) Biologically active polysaccharides as possible lead compounds. *Phytochem Rev* **1**, 379–387.

Ridley BL, O'Neill MA and Mohnen D (2001) Pectins: structure, biosynthesis, and oligogalacturonide-related signaling. *Phytochemistry* **57**, 929–967.

Stern R and Jedrzejas MJ (2006) Hyaluronidases: their genomics, structures, and mechanisms of action. *Chem Rev* **106**, 818–839.

Sugahara K, Mikami T, Uyama T, Mizuguchi S, Nomura K and Kitagawa H (2003) Recent advances in the structural biology of chondroitin sulfate and dermatan sulfate. *Curr Opin Struct Biol* **13**, 612–620.

Volpi N (2006) Therapeutic applications of glycosaminoglycans. *Curr Med Chem* **13**, 1799–1810.

Weigel PH and DeAngelis PL (2007) Hyaluronan synthases: a decade-plus of novel glycosyltransferases. *J Biol Chem* **282**, 36777–36781.

Yip VLY and Withers SG (2004) Nature's many mechanisms for the degradation of oligosaccharides. *Org Biomol Chem* **2**, 2707–2713.

Yu H and Chen X (2007) Carbohydrate post-glycosylational modifications. *Org Biomol Chem* **5**, 865–872.

Heparin

Linhardt RJ (2003) Heparin: structure and activity. *J Med Chem* **46**, 2551–2564.

Noti C and Seeberger PH (2005) Chemical approaches to define the structure–activity relationship of heparin-like glycosaminoglycans. *Chem Biol* **12**, 731–756.

Petitou M and van Boeckel CAA (2004) A synthetic antithrombin III binding pentasaccharide is now a drug! What comes next? *Angew Chem Int Ed* **43**, 3118–3133.

Rabenstein DL (2002) Heparin and heparan sulfate: structure and function. *Nat Prod Rep* **19**, 312–331.

Sasisekharan R and Venkataraman G (2000) Heparin and heparan sulfate: biosynthesis, structure and function. *Curr Opin Chem Biol* **4**, 626–631.

Whitelock JM and Iozzo RV (2005) Heparan sulfate: a complex polymer charged with biological activity. *Chem Rev* **105**, 2745–2764.

Aminoglycoside Antibiotics

Busscher GF, Rutjes FPJT and van Delft FL (2005) 2-Deoxystreptamine: central scaffold of aminoglycoside antibiotics. *Chem Rev* **105**, 775–791.

Flatt PK and Mahmud T (2007) Biosynthesis of aminocyclitol-aminoglycoside antibiotics and related compounds. *Nat Prod Rep* **24**, 358–392.

Hainrichson M, Nudelman I and Baasov T (2008) Designer aminoglycosides: the race to develop improved antibiotics and compounds for the treatment of human genetic diseases. *Org Biomol Chem* **6**, 227–239.

Huang F, Haydock SF, Mironenko T, Spiteller D, Li Y and Spencer JB (2005) The neomycin biosynthetic gene cluster of *Streptomyces fradiae* NCIMB 8233: characterisation of an aminotransferase involved in the formation of 2-deoxystreptamine. *Org Biomol Chem* **3**, 1410–1419.

Kharel MK, Subba B, Basnet DB, Woo JS, Lee HC, Liou K and Sohng JK (2004) A gene cluster for biosynthesis of kanamycin from *Streptomyces kanamyceticus*: comparison with gentamicin biosynthetic gene cluster. *Arch Biochem Biophys* **429**, 204–214.

Kudo F, Yamamoto Y, Yokoyama K, Eguchi T and Kakinuma K (2005) Biosynthesis of 2-deoxystreptamine by three crucial enzymes in *Streptomyces fradiae* NBRC 12773. *J Antibiot* **58**, 766–774.

Llewellyn NM and Spencer JB (2006) Biosynthesis of 2-deoxystreptamine-containing aminoglycoside antibiotics. *Nat Prod Rep* **23**, 864–874.

Magnet S and Blanchard JS (2005) Molecular insights into aminoglycoside action and resistance. *Chem Rev* **105**, 477–497.

Silva JG and Carvalho I (2007) New insights into aminoglycoside antibiotics and derivatives. *Curr Med Chem* **14**, 1101–1119.

Unwin J, Standage S, Alexander D, Hosted T, Horan AC and Wellington EM (2004) Gene cluster in *Micromonospora echinospora* ATCC15835 for the biosynthesis of the gentamicin C complex. *J Antibiot* **57**, 436–445.

Acarbose

Bai L, Li L, Xu H, Minagawa K, Yu Y, Zhang Y, Zhou X, Floss HG, Mahmud T and Deng Z (2006) Functional analysis of the validamycin biosynthetic gene cluster and engineered production of validoxylamine A. *Chem Biol* **13**, 387–397.

Chen X, Zheng Y and Shen Y (2006) Voglibose (Basen®, AO-128), one of the most important α-glucosidase inhibitors. *Curr Med Chem* **13**, 109–116.

Mahmud T (2003) The C_7N aminocyclitol family of natural products. *Nat Prod Rep* **20**, 137–166.

Mahmud T, Flatt PM and Wu X (2007) Biosynthesis of unusual aminocyclitol-containing natural products. *J Nat Prod* **70**, 1384–1391.

Lincomycin

Spížek J and Řezanka T (2004) Lincomycin, cultivation of producing strains and biosynthesis. *Appl Microbiol Biotechnol* **63**, 510–519.

INDEX

Bold type has been used for major sections and monograph material; asterisks indicate structural formulae are presented.

Printed and bound by CPI Group (UK) Ltd, Croydon, CR0 4YY

27/10/2024

14580377-0005